The Nucleolus

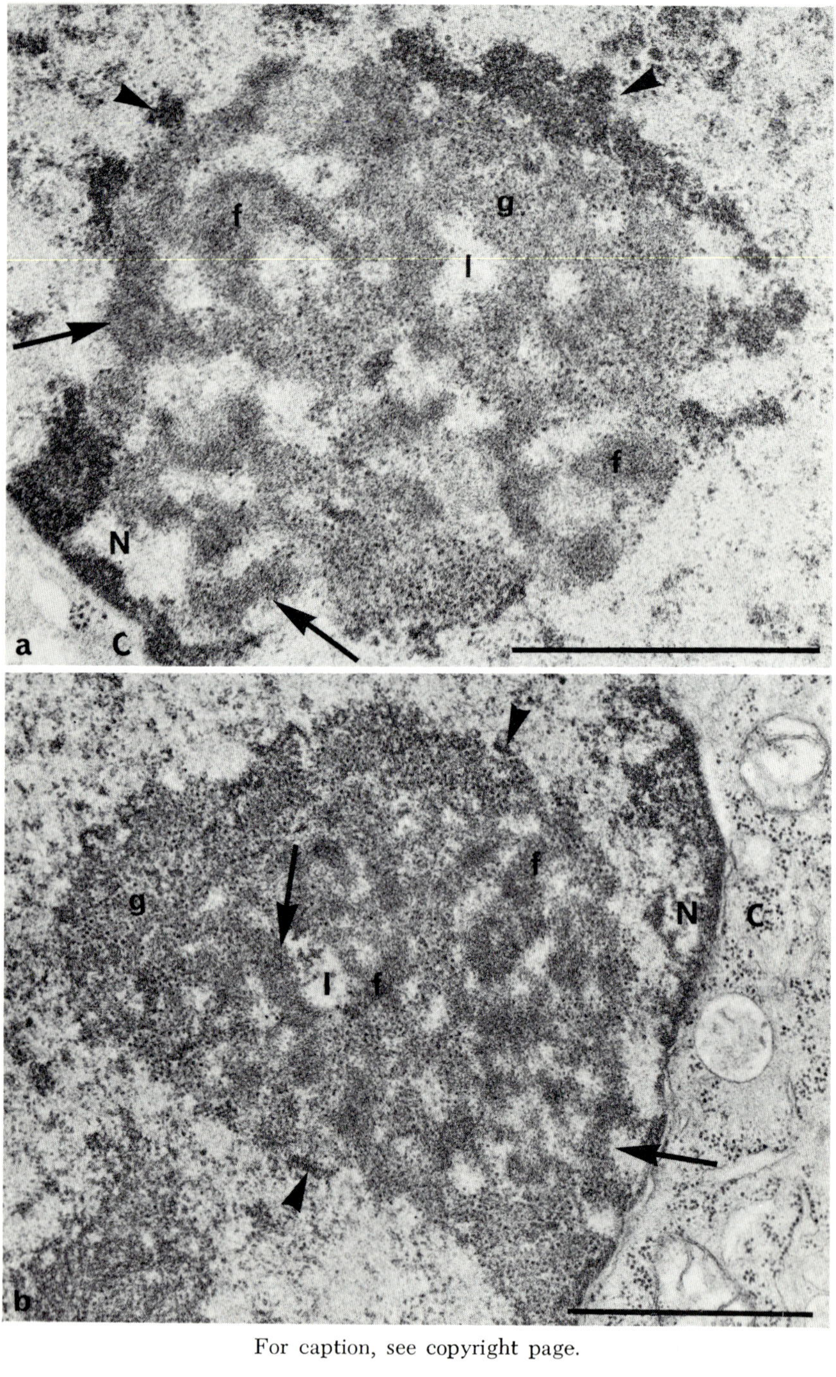

For caption, see copyright page.

The Nucleolus

HARRIS BUSCH

DEPARTMENT OF PHARMACOLOGY
BAYLOR COLLEGE OF MEDICINE
HOUSTON, TEXAS

AND

KAREL SMETANA

LABORATORY FOR THE ULTRASTRUCTURE RESEARCH OF CELLS AND TISSUES
CZECHOSLOVAK ACADEMY OF SCIENCES
PRAHA, CZECHOSLOVAKIA

ACADEMIC PRESS New York and London 1970

Frontispiece

a. Nucleolus of a rat liver cell with nucleolonemas (arrows) and prominent nucleolus associated chromatin (points). From T. Unuma, H. P. Morris, and H. Busch, *Cancer Res.* **27**, 2221 (1967). ×41,000.

b. Nucleolus of a Walker tumor cell with densely packed nucleolonemas (arrows). ×32,000. N—nucleus; C—cytoplasm; f—fibrillar components; g—granular components; l—light areas; points—chromatin structures.

ACADEMIC PRESS, INC.
111 Fifth Avenue, New York, New York 10003

United Kingdom Edition published by
ACADEMIC PRESS, INC. (LONDON) LTD.
Berkeley Square House, London W1X 6BA

LIBRARY OF CONGRESS CATALOG CARD NUMBER: 74-84238

PRINTED IN THE UNITED STATES OF AMERICA

ERRATUM

THE NUCLEOLUS

by

Harris Busch and Karel Smetana

The illustration that now appears on page 534 is Fig. 11.8 and should appear on page 536. The illustration appearing on page 536 is Fig. 11.7A and should appear on page 534. The illustration captions are correct as printed.

Contents

II. The Cytochemistry and Ultrastructure of the Nucleolus

III. Nucleolar Organizers, Nucleolar Size, Shape, and Number

IV. Nucleolar DNA

V. Nucleolar RNA of High Molecular Weight

Preface

The nucleolus is an intranuclear organelle which has the important function of producing the vast bulk (80–90%) of cellular RNA and probably serves as a site of organization of ribonucleoproteins which are subsequentially transported to the remainder of the nucleus and to the cytoplasm. Convincing evidence has been presented from morphological, biochemical, genetic, and other studies that the nucleolus produces most, if not all, of the ribosomal RNA of the cell and provides basic patterns of ribosomal organization.

In the past decade both the organization and the function of the nucleolus have been considerably clarified although it is clear that much remains to be learned. The factors controlling the number of nucleoli in any given cell, the nucleolar organization, and the rates of nucleolar activity present interesting vistas of research for the future. It is clear that each cell contains a definitive number of NOR's, or "nucleolus organizing regions," which may function simultaneously or independently depending on the gene controls in any given cell. The interaction of these controlling mechanisms with the NOR's will be a subject of importance, particularly now that good techniques are available for the use of drugs and other stimuli and repressors of nucleolar function.

Studies on the nucleoli have represented a junctional point for analyses of normal and pathological cell function. Because nucleoli have been shown to differ remarkably in morphology in cancer cells, the role of the nucleolus in cancer and in other diseases has been a topic of considerable interest in our laboratories as well as in several others. The possibility that nucleoli of cancer cells may produce abnormal products that may affect the responsiveness of cancer cells both to internal and external control systems makes it important that not only the ultrastructural morphology of these nucleoli be defined with the greatest possible precision, but also that in the future each of the components of the nucleolus be isolated, purified, analyzed for primary, secondary, and tertiary structure, and, finally, reconstructed into functional mosiacs of the nucleolus. Such studies would include analysis of nucleolar enzyme function to which increasing importance is being assigned. Recently, some very important enzymes, including NAD synthetase and an RNA polymerase, have been found to be localized in the nucleolus.

The objective of this book is to provide a basic summary of the status of the field as it existed in early 1970. It should be pointed out that so much work has been done on nucleoli of many different cells that it is almost impossible to totally cover the literature on the nucleolus and as a result, many references have not been adequately reviewed, but it is hoped that their content is reflected in the overall descriptions in the chapters and in the many books and reviews that have been cited.

Acknowledgments

We wish to express our sincere appreciation for critical reading of the manuscripts and most helpful suggestions to Dr. D. D. Brown, Professor of Biology, University of Maryland; Dr. F. Gyorkey, Professor of Pathology, Baylor College of Medicine and the Veteran's Administration Hospital; Dr. T. C. Hsu, Professor of Cell Biology, M. D. Anderson Hospital and Tumor Institute; Dr. Robert Liebelt, Professor of Anatomy, Baylor College of Medicine; Dr. Barbara McClintock, Professor of Biology, Carnegie Institution of Washington; Professor J. Wolf, Czechoslovak Academy of Sciences; Dr. Max Birnstiel, Senior Lecturer, Department of Anatomy, University of Edinburgh; Dr. B. R. Brinkley, Assistant Biologist, Department of Biology, Section of Cell Biology, M. D. Anderson Hospital and Tumor Institute; Dr. Archie Prestayko, Instructor, Department of Pharmacology, Baylor College of Medicine; Dr. Tae Suk Ro-Choi, Instructor, Department of Pharmacology, Baylor College of Medicine; Dr. Egon Hidvegi, Visiting Associate Professor, Department of Pharmacology, Baylor College of Medicine; and Dr. G. Quagliarotti, Postdoctoral Fellow, Department of Pharmacology, Baylor College of Medicine.

We wish to acknowledge the aid of the U. S. Public Health Service Cancer Research Center Grant (CA 10893) and the American Cancer Society which made possible the original research presented in a number of sections of this book.

We wish to express our appreciation to the following for permission to reproduce figures and tables: Dr. N. J. Lane for Figure 1.1; Dr. B. R. Brinkley for Figure 1.3; Dr. Rollin Reynolds and Dr. P. O'B. Montgomery for Figures 1.7, 10.1, 10.2, 10.4, and 10.5; Dr. Sophie Guttes and Dr. Edmund Guttes for Figures 1.8, 1.9, and 1.20; Dr. Daniel Szollosi for Figure 1.10; Dr. Hans Stich for Figures 1.11–1.13; Dr. K. Yamamoto for Figures 1.14 and 1.15; Dr. John A. Terzakis for Figures 1.16, 1.17, and 1.18; Dr. Herbert A. Kordan for Figure 1.19; Dr. Robert Love for Figure 1.21; Dr. S.

Liao for Figure 1.22; Professor Y. Miura for Figure 1.23; Dr. B. B. Hyde for Figures 2.10d and 2.16a; Professor G. Attardi for Figures 3.2 and 3.3; Dr. J. G. Gall and Dr. H. G. Callan for Figures 3.4 and 4.13; Dr. J. Jacob and Dr. J. L. Sirlin for Figure 3.5; Dr. J. L. Sirlin and Dr. G. R. Knight for Figure 3.9; Dr. O. L. Miller for Figure 4.12 and Dr. O. L. Miller and Dr. B. A. Beatty for Figures 4.12, 4.14, and 5.3; Dr. S. Fogel and Dr. James Madison for Figure 6.1; Dr. T. S. Ro-Choi for Figures 6.2 and 6.12; Dr. A. W. Prestayko for Figure 6.3; Dr. Y. Moriyama for Figure 6.14; Dr. S. Fogel and Dr. P. S. Sypherd for Figure 7.3; Dr. W. Caley for Figure 9.26; Professor E. Pannese for Figure 9.27; Dr. V. Smejkal for Figure 9.28; Dr. G. Seman for Figure 9.30; Dr. John Lazlo for Figure 9.35; Dr. D. Svoboda, Dr. A. Racela, and Dr. J. Higginson for Figures 10.6, 10.7, and 10.9; Dr. K. Tokuyasu for Figure 10.17; Dr. T. C. Hsu for Figure 3.1; Dr. H. J. Barr for Figures 4.1 and 4.2; Dr. J. E. Dworak for Figure 7.4; and Dr. F. Györkey for Figure 9.31.

The Nucleolus

CHAPTER I

Introduction

I. The Relationships of the Nucleolus to the Overall Functions of the Nucleus

To students of modern biology, particularly molecular biology, it hardly seems possible that little more than a decade ago the cell nucleus was regarded as an amorphous, rather uninteresting cellular organelle to which had been assigned the vague function of "governor of the cell." Recently, knowledge of its function has been advanced with the recognition of the central role of the polysome in protein synthesis, the improvements in biological methods, and the chemical methods that led to Nobel-Prize-winning studies on nucleic acids. The present direction in such research relates to the elucidation of the operation of the genes, particularly the regulation of the gene readouts which act to meet the demands of environment with the functional capacities built into cells over the long history of evolution. Since the nucleolus represents one of the most active sites of RNA synthesis in the cell and contains the genes (rDNA) coding for ribosomal RNA, it has been a subject of special interest to geneticists and molecular biologists (see General Reviews at the end of this chapter).

At present, the best understood basic functional elements of the nucleus are the DNA templates which function for ultimate protein synthesis through RNA intermediates. The enzymes that "read" the DNA and the "regulatory factors," possibly histones and other nuclear proteins (*14a, 92*) are only now becoming a subject of refined biochemical analysis. As a result of the actions of these multiple elements, the cell nucleus produces messenger and ribosomal RNA of the polysomes as well as transfer and other RNA's. The polysome consists of messenger RNA and ribosomes; the ribosomes are composed of a number of proteins and specific types of RNA which include 18 S rRNA, 28 S rRNA, and one or more species of 5 S and possibly 8 S RNA (*62*). At the appropriate moments in protein

TABLE 1.1

NUCLEAR RNA PRODUCTS

Extranucleolar Nuclear RNA Products
I. Messenger RNA
Precursors 100 S RNA, 85 S RNA, 60 S RNA, 45 S RNA
Monomers: 10 S RNA – 45 S RNA
II. Transfer RNA (tRNA or amino acid acceptor RNA)
III. Low molecular weight RNA's localized to the nucleus
Uridylic acid-rich RNA's
RNA's with approximately equal concentrations of the four common nucleotides.
Nucleolar RNA Products (Chapters V, VI)
I. Ribosomal RNA precursors
Precursors: 55 S RNA, 45 S RNA, 35 S RNA
Nucleolar monomer 28 S RNA (→ 28 S rRNA)
Nucleolar monomer 23 S RNA (→ 18 S rRNA)
II. Low molecular weight ribosomal RNA: 5 S RNA, 7 S RNA; uridylic acid-rich nucleolar RNA
III.? Messenger RNA: short- and long-lived templates for ribosomal protein synthesis and/or for nucleolar enzyme synthesis

synthesis, amino acid-"charged" tRNA's or transfer RNA's are hydrogen-bonded to the messenger RNA's. The bulk of these RNA's are formed in the nucleus and can be divided into nucleolar and extranucleolar products as shown in Table 1.1.

A. Role of the Nucleolus in Synthesis of Ribosomal RNA

The role of the nucleolus that has been definitely established is the synthesis of the high molecular weight ribosomal RNA (Table 1.2). On the other hand, some nuclear functions clearly do not involve the nucleolus; i.e., synthesis of tRNA and synthesis of most messenger RNA's (*113a, 137a*). As will be noted in Chapter VI, there are low molecular weight RNA's concentrated in the nucleolus but the role of the nucleolus in their synthesis is not yet clear.

Although at one time it was believed that the nucleolus served as some sort of collecting station, the weight of evidence indicates that the nucleolus is the site of synthesis of much of the cellular RNA (70–90% in many cells). The vast bulk of this RNA is transported to the cytoplasm as ribosomal components.

B. Does the Nucleolus Make Whole Ribosomes?

Although the nucleolus synthesizes ribosomal RNA, the sites and mechanisms for assembly of the whole ribosomes are just now being

TABLE 1.2

EVIDENCE THAT THE NUCLEOLUS IS THE SITE OF SYNTHESIS OF RIBOSOMAL RNA

1. The nucleolus contains rapidly synthesized 28 S RNA which is very similar in composition to ribosomal 28 S RNA (*87, 88*).
2. If the nucleolar RNA is labeled with a pulse of radioactive precursor, the label is initially found in 45 S RNA and then is transferred from 45 S to 35 S to 28 S RNA. *In vivo* treatment of cells with actinomycin D results in a rapid loss of 45 S RNA initially, 35 S later, and 28 S finally (*97–101*). The general pathway is 45 S RNA → 35 S RNA → 28 S RNA. Since in some nucleolar preparations 85 S and 60 S RNA are labeled more rapidly than is 45 S RNA, these may be oligomers of nucleolar 45 S RNA and/or 28 S RNA (*87, 88, 97–102*).
3. Isolated nucleoli contain a DNA-dependent RNA polymerase capable of biosynthesis of RNA (*115, 116, 121*).
4. Inhibition of biosynthesis of ribosomal RNA results from UV-microbeam irradiation of nucleoli (*101–104*).
5. Synthesis of ribosomes or ribosomal RNA does not occur in mutants that do not contain nucleolus organizers (*13*).
6. There is selective hybridization of 28 S ribosomal RNA with nucleolus organizers and/or nucleolar DNA (*71e, 108a, 122a*).
7. Hybridization of ribosomal RNA occurs with DNA of *Drosophila melanogaster* in direct proportion to the number of nucleolus organizers present (*113, 114*).
8. Hybridization of nucleolar 28 S RNA with nucleolar DNA is inhibited 85% by ribosomal 28 S RNA and not by ribosomal 18 S RNA (*108a*). Ribosomal 18 S and 28 S RNA inhibit hybridization of nucleolar 45 S RNA by 15% and 40%, respectively (*108a, 122a*).

analyzed. Electron microscopic studies have revealed the presence in nucleoli of a number of types of particulate bodies (*4a, 122b*) which are believed by some authors (*52, 113*) to be ribosomes. At high magnifications of the electron microscope, some of these "particles" appear to be parts of long chains of ribonucleoproteins that are twisted and coiled into small ribonucleoprotein elements that may then be "pinched off" the mass of ribonucleoproteins. This possibility has been an intriguing one for those students of the nucleolus who have likened this activity of the nucleoli to that of a silkworm which also produces a very long repeating thread.

The presence of ribosomes within nucleoli has been reported (*122*). However, these structures are relatively few in number and may be functional in nucleolar protein synthesis rather than being products for export. Interestingly, in the cytoplasmic polysomes, there is only one RNA fiber between the bodies of the ribosomes, but frequently there are two or more fibers connecting the RNP masses of the "nucleolar ribosomes."

Although most workers agree that the vast bulk of the ribosomal RNA is a nucleolar product (*16, 93–104, 108a*), one unanswered question is whether the RNA in the nucleolus is "mature" ribosomal RNA or only ribosomal precursor RNA (*69a*). Much evidence (Chapter V) supports the concept that it is precursor and not completed product. The nucleolar 28 S RNA which apparently is one final nucleolar product contains the gene readouts for both the 28 S rRNA and a small RNA fragment (*108a*). Assuming 23 S nucleolar RNA is a final nucleolar product, at least one more "cleavage" or maturation step would be necessary to produce ribosomal 18 S rRNA from 23 S nucleolar RNA. Since 18 S RNA is not present in nucleoli, this step presumably occurs outside the nucleolus, possibly in the nuclear ribonucleoprotein network (Chapter VIII). In oocytes, ribosomes are apparently completed somewhere between the

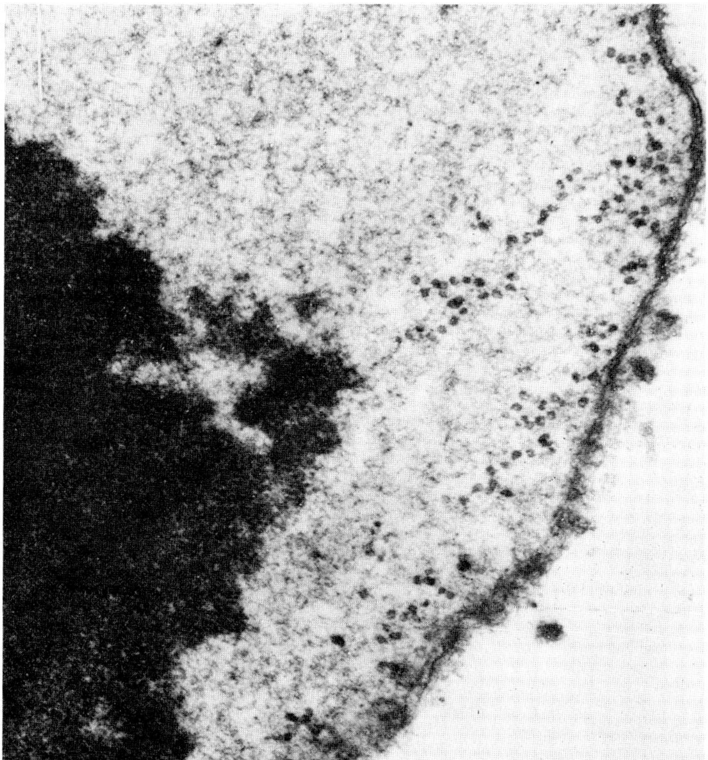

Fig. 1.1. Ribosomelike particles apparently migrating from the nucleolus toward the nuclear envelope in oocytes of the axolotl *A. mexicanum* (courtesy of Nancy J. Lane, Yale University).

nucleolus and the region of the nuclear membrane (*61*); in some pictures, they can be seen "marching single file" toward the cytoplasm (Fig. 1.1).

C. Does the Nucleolus Produce RNA Other than Ribosomal RNA?

Evidence presently available does not clearly establish a role for the nucleolus in the synthesis of either messenger RNA (*25*) or low molecular weight nucleolar or nuclear RNA. The finding of low molecular weight RNA rich in uridylic acid in nucleoli (*50, 85, 90;* see also Chapter VI) serves to show that there is nonribosomal nucleolar RNA since this RNA is not found in ribosomes. However, the origin of this RNA has not been determined either by kinetic or blocking studies.

When nucleoli of tumors are treated with actinomycin D, another type of 45 S or high molecular weight RNA is found in the nucleolus which is neither ribosomal nor extranucleolar nuclear RNA as indicated by its nucleotide composition (*25*). However, direct tests of this RNA for template activity have not been reported.

In summary, the nucleolus definitely synthesizes ribosomal precursor RNA, may make at least one type of low molecular weight RNA, and in addition may produce some types of messenger RNA (Table 1.2).

D. Protein Synthesis in the Nucleolus

Maggio and others (*14a, 70*) have presented evidence for the synthesis of proteins in the nucleolus. If this is true, the nucleolus may be the site of ribosomal protein synthesis. Other studies which favor the view that nuclear and presumably ribosomal proteins are synthesized in the cytoplasm imply that the nucleolus has a special role only in ribosome assembly (Chapter VII). In fact, the ratio of protein to RNA in the nucleolus is far greater than in ribosomes; accordingly, either the nucleolus contains large amounts of its own special structural proteins or enzymes or is a reservoir of ribosomal precursor proteins.

Several types of studies have shown that after incubation of cells with labeled amino acids, nucleolar proteins become labeled but the results on kinetics of labeling of nucleolar proteins do not permit a definitive conclusion on whether they are synthesized within the nucleolus proper or are carried into the nucleolus from elsewhere in the cell. The finding that *in vitro* incubation with labeled amino acids results in little labeling of the nucleolar proteins and the low specific activities of the nucleolar proteins after *in vivo* incubations with labeled amino acids supports the idea that a pool of protein migrates into the nucleolus from the extranucleolar portions of the nucleus or the cytoplasm.

On the other hand, the high concentrations of some enzymes in nucleoli suggest that they are synthesized or concentrated within the nucleolus

(*121*). RNA polymerase and some ribonucleases are in this category (Chapter VII). Theoretically, it should be possible to prove whether their localization results from their "concentration" by the nucleolus or synthesis in the nucleolus.

II. Historical

The extraordinarily rapid advances in science, particularly in cellular and molecular biology, are well illustrated by the recent development of knowledge of the structure, function, composition, and pathology of the nucleolus (Table 1.3). About 200 years ago in 1774 the nucleolus was first recognized by Fontana (*83*); in keeping with the pace of science of the time little more information developed until the late 19th century. Although it seems certain that the very first microscopists peering into the problems of cell structure and function must have seen the nucleoli of cells, the variety of cell types and their cytoplasmic components excited more interest than the nuclei and their inclusions.

A. Visualization of Nucleoli

With the development of dye chemistry in the 19th century and particularly with the development of the basic dyes, various structures of the cell, including the nucleoli, became readily visible (Fig. 1.2).

TABLE 1.3

Historical Events in Research of the Nucleolus

Year	Event	Reference
1774	Recognition of the nucleolus by light microscopy	*83*
1898	Compendium on nucleolar morphology in many species	*83*
1934	Demonstration of structural irregularities of nucleoli of cancer cells	*67–69*
1935	Demonstration of nucleolus organizers, secondary chromosomal constrictions, and "nucleolar mutants"	*49a, 71c, 71d*
1938	Correlation of functional activity of cells in protein synthesis and nucleolar size	*7–12, 20–23*
1951	Demonstration of nucleolonemas	*33a*
1963	Demonstration of the granular and fibrillar elements of nucleoli by electron microscopy	*4a, 71b, 122b, 123*
1963	Methods for isolation of nucleoli	*19, 70, 71, 89*
1963	Demonstration of 45 S RNA precursor of ribosomal RNA and specific inhibition of its formation	*93–104, 117*
1966	Isolation of nucleolar 45 S RNA, the GC-rich precursor of ribosomal RNA	*86–88*
1967	Isolation of the satellite DNA of the *Xenopus*	*5, 6*

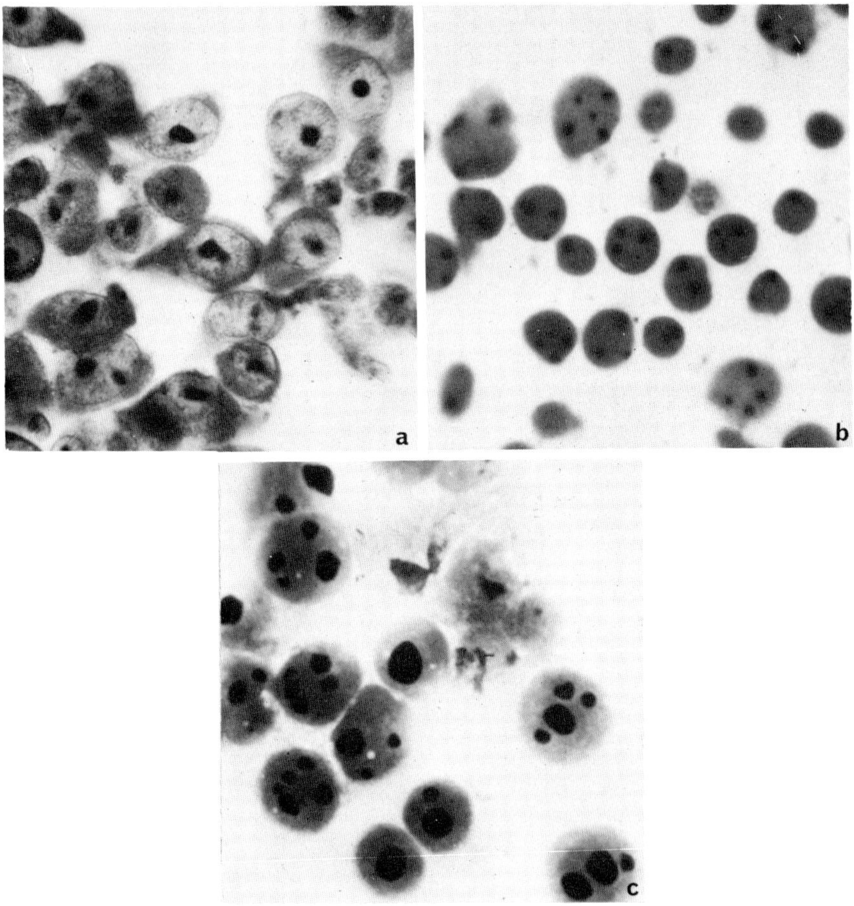

Fig. 1.2. Nucleoli in nuclear preparations. (a) Nuclear preparation of Walker 256 carcinosarcoma stained with azure C. (b) Normal liver nuclei stained with toluidine blue. (c) Nuclei of thioacetamide (50 mg/kg)-treated rat livers stained with toluidine blue. From (*19*).

Fascinating, unproved hypotheses of function were developed for the various organelles.

The availability of various stains led Montgomery (*83*) to a thoroughgoing analysis of the number and appearance of nucleoli in many species. Present-day procedures employing living or fixed tissues have extended the idea that normal cells have striking variations in the number and size of their nucleoli. In addition, remarkable variations in nucleolar size and shape are noted frequently in tumor cells and other growing cells (*16, 47, 56*).

This variability in number and size of the nucleoli has led to many attempts to define the constituents of the nucleoli of various cell types both in chemical and ultrastructural terms and has heightened interest in the nucleolar organizers and their associated "nucleolar chromosomes" (see Chapter III).

B. Phase Microscopy

The general uncertainties that emerged about staining during the late 19th century raised the question of whether the nucleoli were real or artifactual structures. Fortunately, phase microscopy unequivocally established their existence in living cells. With phase microscopy, nucleoli are visible in smears of cells in plasma or tissue juices or when the cells or tissues are placed in saline solutions or other artificial media. Nucleoli may also be seen very clearly in isolated nuclei (see Chapter XI) particularly when isolated in media containing calcium ion or other divalent ions. With ordinary phase microscopy, nucleoli appear to have a yellowish cast. Since these structures turn either the expected blue or orange color in phase when appropriate dyes are added (see Chapter XII), "nucleoli" seen by phase and light microscopy are identical. On close observation with the phase microscope, it is possible to observe some aspects of nucleolar substructure including stippled elements of some of the nucleoli; i.e., "nucleolini" (*66*) and vacuoles (*16*). As will be noted in subsequent chapters neither the vacuoles nor the "nucleolini" have been isolated and it is not known whether they contain specific nucleolar components.

With the aid of phase microscopy, it is also possible to carry out microdissection of the nuclei to obtain a nucleolar product and in addition, to reimplant nucleoli into other cells (*51, 57, 58*). Correlated studies have been made on the overall appearance of the nucleoli and their numbers in living (vital) tissues and cells and on fixed and stained materials.

C. Microspectrophotometry and Biochemical Analysis

Quantitative analysis by microspectrophotometry of both unstained and stained cells and nucleoli has provided information on nucleic acids and proteins which has been extended by later biochemical studies (*7–12, 20–23*). Studies on individual nucleoli by light microscopy and electron microscopy are extremely valuable and cannot readily be supplanted by biochemical techniques which require masses of materials for adequacy of analysis. Such biochemical analyses are by necessity averages derived from millions to billions of nucleoli without regard to the individual nucleoli.

Perhaps the earliest correlations developed with respect to the nucleolus and cell functions arose in connection with the observed difference in nucleolar size in certain types of cells. For example, in the central nervous system, particularly in the cells of the cerebellum, there are very large nucleoli. In the cells of skeletal muscle and cardiac muscle as well as in mature lymphocytes, the nucleoli are small. On the basis of these specific differences in size and localization in tissues and organs of large nucleoli, the suggestion was made that the nucleolus was related in function to certain of the activities of the cells and their nuclei.

III. Nucleolar Chromosomes and Mutants

A. Nucleolar Chromosomes

All of the fundamental morphological methods were available by the end of the nineteenth century, but relatively few workers were interested in nucleolar function or structure. Inasmuch as other problems were more pressing, research on the nucleolus remained in the shadow of the great developments that occurred in the area of chromosome structure and function. The studies on the chromosomes resulted in visualization of the nucleolus and among the by-products of such studies were the following: (1) the nucleolus contains little (or no) chromatin as visualized by the Feulgen stain or other stains for DNA; (2) nucleoli remain attached to specific chromosomes in prophase but they disappear during metaphase when the chromosomes separate and the cells are readied for division; and (3) the nucleoli are reconstituted in early telophase. These findings served as springboards for some interesting ideas, one of which relates to the question of competition of chromosomes for the nucleolus and its products. One idea (*16, 58*) is that chromosomes vie with one another for dominant roles in the cell. In cells which have only a few nucleoli or one nucleolus, the chromosomes were thought to compete for the product of the nucleolus or for "control" of the nucleolus. The chromosomes which ultimately control the function of the cell were thought to be those which have the ability to capture or monopolize nucleolar function.

This concept is neither proved nor disproved at present. According to Harris and others (*43–46, 118*) many genes or perhaps most are read by RNA polymerases, but the messenger RNA and/or other nuclear RNA products are rapidly hydrolyzed unless the proper types of bonding are effected with polysomal precursors. Thus, the stabilization of messenger RNA's may require their bonding to newly synthesized nucleolar ribosomal precursors (see p. 46 on hormones).

B. Secondary Chromosomal Constrictions

Because the nucleoli persist during prophase and during this period the chromosomes are undergoing contraction and are assuming definitive shapes, it has been possible to identify the "nucleolar chromosomes." Heitz (*49a*) reported that only specific regions of the nucleolar chromosomes which apparently "lacked DNA," and which he referred to as SAT (sine acido thymonucleico) regions were linked to the nucleoli. These regions do not really lack DNA as shown in Fig. 1.3, but have undefined regions that appear to be "secondary chromosomal constrictions" around which there is a layer of undefined substances.

C. Nucleolar Mutants

One of the most satisfying discoveries of the 1930's was made by McClintock (*71a*) in studies on mutant strains of maize (corn). She found that "secondary chromosomal constrictions" were related to both the number and the type of nucleoli present. In some of her mutant strains, these regions were deleted. If the corn was homozygous with respect to the presence of the constriction, large nucleoli were present. If the corn was heterozygous or lacked the constriction, there was a decreased number or a failure of formation of the nucleoli. McClintock called these "secondary chromosomal constrictions," "nucleolus-organizer," regions, or NOR's. Her pioneering studies on nucleolar mutants have been extended many ways (Chapters III, IV) by such notable studies as those of Longwell and Svihla (*65*), Brown and Gurdon (*13*), Birnstiel and his associates (*6*), and more recently by Ritossa and Spiegelman (*114*) all of whom employed mutants containing greater or lesser numbers of nucleoli depending upon their chromosomal constitution.

IV. Autoradiography

With the aid of autoradiography alone, Goldstein and his associates (*34–37*) were able to establish that the nucleus has a rapid rate of labeling of RNA with appropriate nucleotides and purine and pyrimidine precursors. The very rapid biosynthetic reactions which they reported are now known to be catalyzed by RNA polymerase, discovered by Weiss

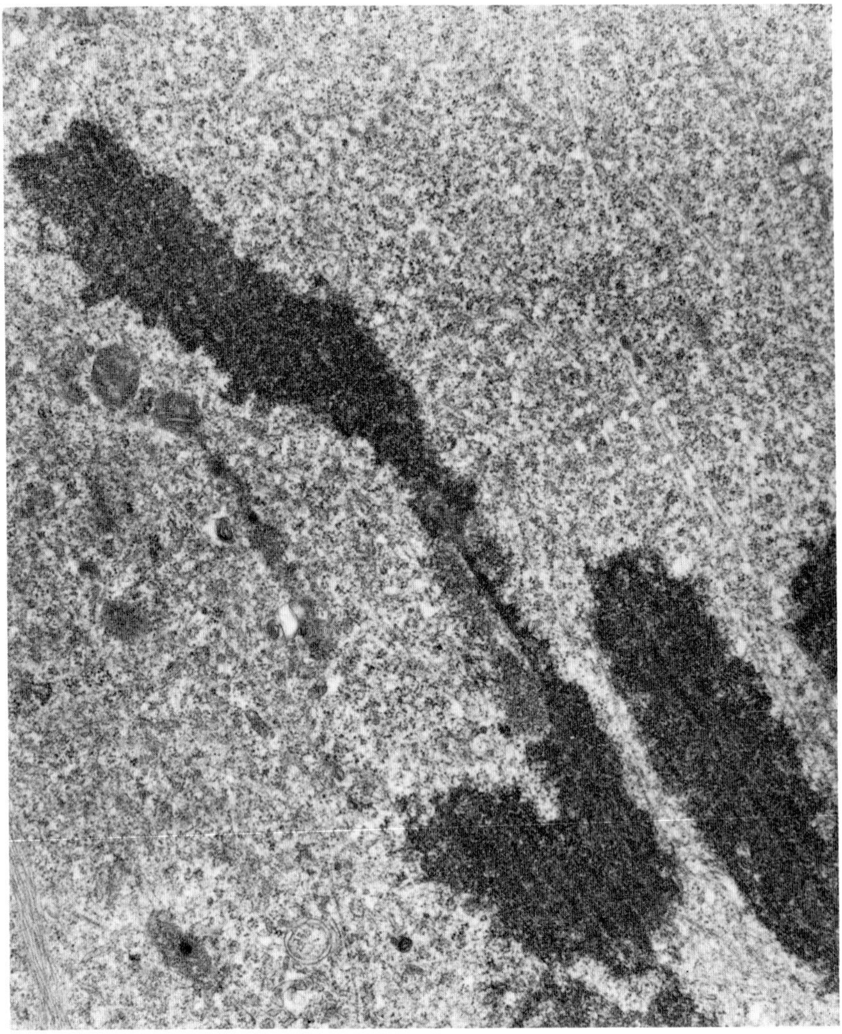

Fig. 1.3. The X-chromosome of the kangaroo rat (*Potorous tridactylus*) showing the nucleolus organizer region (secondary constriction). The continuity of the dense deoxyribonucleoprotein structure in the region of the secondary constriction is apparent. There is another element of lighter density around the constriction. The preparation was fixed in glutaraldehyde, postfixed in osmium, and embedded in Epon. ×25,700 (courtesy of Dr. B. R. Brinkley, University of Texas M. D. Anderson Hospital and Tumor Institute).

and his associates (*136, 137*) which has been found to be highly concentrated in nucleoli (*115, 116, 121*). The use of autoradiography and its quantitative counterpart of grain counting (Fig. 1.4) enabled a host of workers to establish that the nucleus and in many cells the nucleolus was labeled most rapidly with RNA precursors, that the highest specific activity of the RNA (grains per unit area) was generally in the nucleolus and that the flow of newly formed RNA was unidirectional, from the nucleus to the cytoplasm (Fig. 1.5). Finally, the nucleus was shown to contain two major types of RNA with respect to metabolic activity one of which was rapidly labeled and the other very slowly labeled.

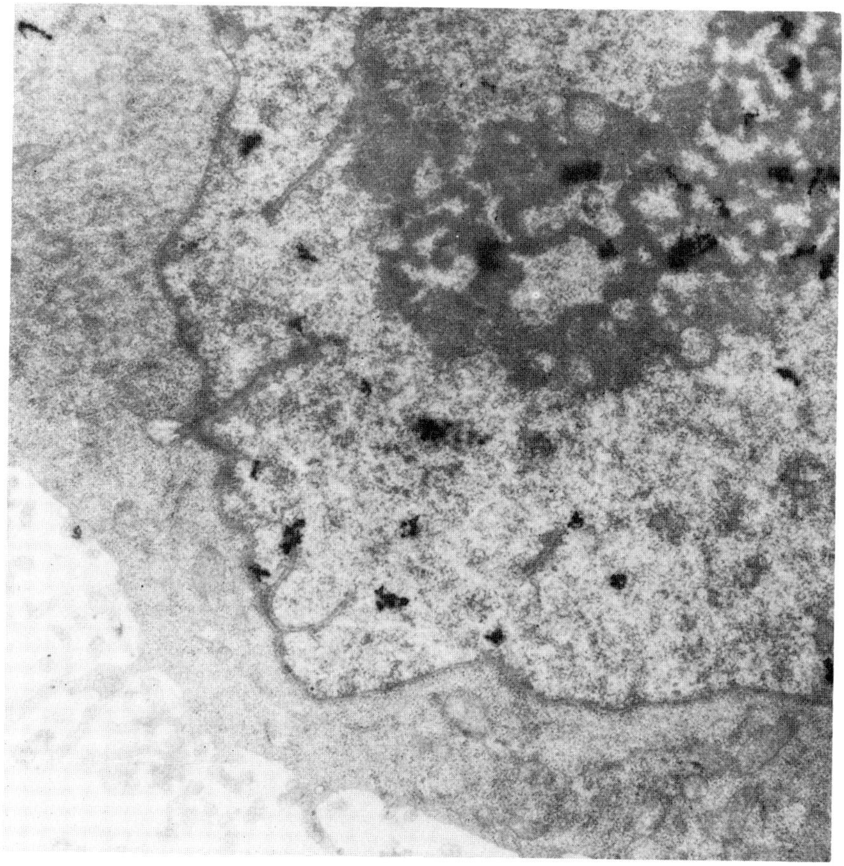

Fig. 1.4. Electron microscopic autoradiograph of a Novikoff hepatoma ascites cell showing the nuclear uptake of tritiated uridine (15-minute pulse). There is a large concentration of isotope within the nucleolus. The specimen was fixed with glutaraldehyde and osmium, embedded with Epon-Araldite, and stained with uranyl acetate and lead citrate. ×15,000. From (*126a*).

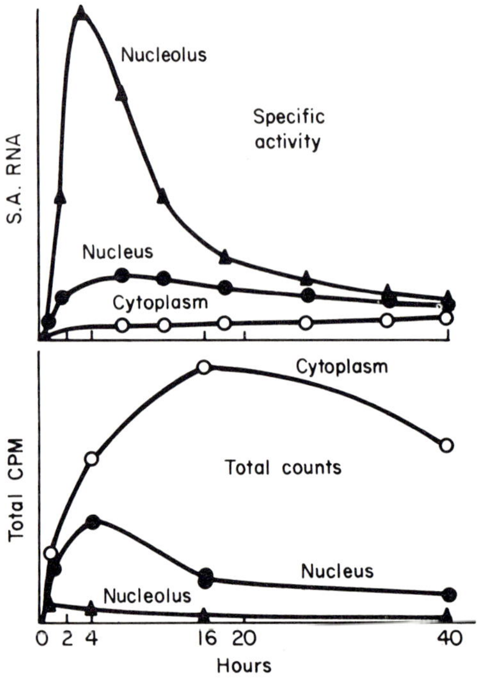

Fig. 1.5. Schematic representation of data obtained by autoradiographic analysis after incubation of tissues with RNA precursors. The kinetics of labeling are shown for both specific activity and ^{14}C incorporated into RNA of the nucleus, nucleolus, and cytoplasm. From (*16*).

Autoradiography and Selective Radiation of Nucleoli

One of the most interesting procedures devised for studies on the biosynthetic activities of nucleoli was that of specific or selective radiation evolved by the group working with Montgomery (*73–82*) in Dallas and that of Perry *et al.* (*97–104, 121*) working initially in Brussels and later at Oak Ridge. The technique of microirradiation initially worked out by Zirkle (*140*) was also utilized for studies on chromosomal structures. The microbeam produced by the system is only 0.2 μ in diameter and has the effect of inhibition of the function or biosynthetic activity of the nuclear region irradiated. In combination with autoradiography, it has been used to analyze the fate of nucleolar RNA as well as its rate of biosynthesis.

Interestingly, "enucleolation" by ultraviolet microbeam irradiation blocks the synthesis of only 30% of the total nuclear RNA but inhibits the synthesis of 70% of the total cellular RNA (*102–104*) suggesting that 70% of cellular RNA is synthesized in the nucleolus.

V. Isolation of Nucleoli

Among the methods for analysis of nucleolar function noted earlier in this chapter are: (a) Microscopic observations on unfixed material; (b) microscopic observations on stained nucleolar material; (c) studies on mutants and quantitative histochemical and cytochemical observations on nucleoli. A newer technique was the isolation of nucleoli which only recently has become routine (*18, 19, 86, 89*). There were many disputed reports on the isolation of the nucleoli from both mammalian and plant species because most of the pioneers in this field worked against obstacles that were not appreciated at the time. Logically, they concentrated on the morphological rather than the biochemical problems involved. As a result of these problems, few of the early techniques have survived in their original form but many of their aspects have been incorporated into more modern technology. In view of the apparent shortcomings of present-day procedures which will be discussed later, it is more than likely that some even better procedures for studies on nucleoli will be developed.

Among the first isolations of nucleoli was that of Vincent (*129–132*) who isolated nucleoli from starfish eggs. Both mechanical means and differential centrifugation were employed. The nucleoli of starfish eggs are large and readily visible and as a result the technique led to many dividends from the point of view of isolation. What was particularly important in these studies was that not only was it possible to provide some characterization of nucleolar protein but also it was possible to demonstrate the presence of both RNA and DNA in the nucleolar preparations.

Improved procedures for microdissection were developed by Kopac and Mateyko (*57, 58*). From their studies, it was possible to appreciate the remarkable pleomorphism of isolated nucleoli of neoplastic cells. These painstaking micromanipulative studies are demanding of skill and patience. From the standpoint of biochemical analysis of nucleolar function, this was not a practical procedure. Later, micromanipulation was incorporated into studies of Edström who added "string" microphoresis to separate nucleotides and determined base ratios (*32, 33*).

In the first reported large scale isolation of nucleoli, Dounce and his associates (*28, 84*) reported that they had isolated nucleoli from rat liver cells. A number of serious problems harassed these studies, including the lack of information on nuclear and nucleolar ribonucleases, the presence of large particulate structures in the cytoplasm, and the difficulties in the preparation of the nuclei of cells.

The nucleoli were separated from nuclei of liver cells by differential sedimentation through gum acacia (*84*) at room temperature for 24 hours. Since cold precautions, RNase inhibitors, and phenol were not used, the RNA in their nucleolar preparations was almost completely lost. When they reported that DNA was the main nucleic acid in nucleoli, histochemists carefully rechecked their methods for analysis of RNA and their early results on nucleolar RNA (*7–12, 23*). After enzymatic analysis with RNase, DNase, and proteinases, the histochemists concluded that the nucleolus contained RNA as the primary nucleic acid and further that the procedure of Dounce and his associates was unsatisfactory.

The nearest approximation of presently used methods for isolation of nucleoli was the procedure of Maggio *et al.* (*71*) which employed the principle of sonic oscillation with nuclei obtained from guinea pig liver. Minor modifications were made in the procedure of Chauveau *et al.* (*24*) for isolation of nuclei from guinea pig liver and then the nuclei were subjected to sonic oscillation for periods up to 20 minutes. Differential centrifugation used to sediment selectively the nucleoli was based on the principle of Baltus (*2–4*) who showed that only the nucleoli obtained from starfish eggs had sufficient density to penetrate through 2.2 *M* sucrose under specified conditions of centrifugation (*26, 27*).

Many of these procedures have been incorporated into the methods now used as standard procedures (*18, 19, 89*) (see Chapter XI) which produce highly purified nucleoli (Fig. 1.6). Although subsequent chapters will show that current methods are of great value for studies on nucleolar components, completely adequate methods have not been developed. Advocates of the use of nonaqueous systems (*120*) noted that employment of aqueous techniques for isolation of any intracellular components is fraught with the danger of absorption of substances that were not normally present in the particle. At the same time, it is clear that losses of soluble substances can occur. If one adds to these normal hazards additional problems of treatment of nuclei with detergents, such as Triton-X and sodium dodecyl sulfate, as well as hypotonic solutions, such as have been used in some procedures (*93–96, 105–107*), it is clear that the product may bear only a superficial resemblance to the intracellular structure and many chemically important elements may be lost.

VI. Electron Microscopic Studies on Effects of Drugs and Other Pharmacological Agents

A. Electron Microscopy of the Nucleolus

With the advent of electron microscopy it became possible to define nucleolar components in a much more satisfactory way than had been

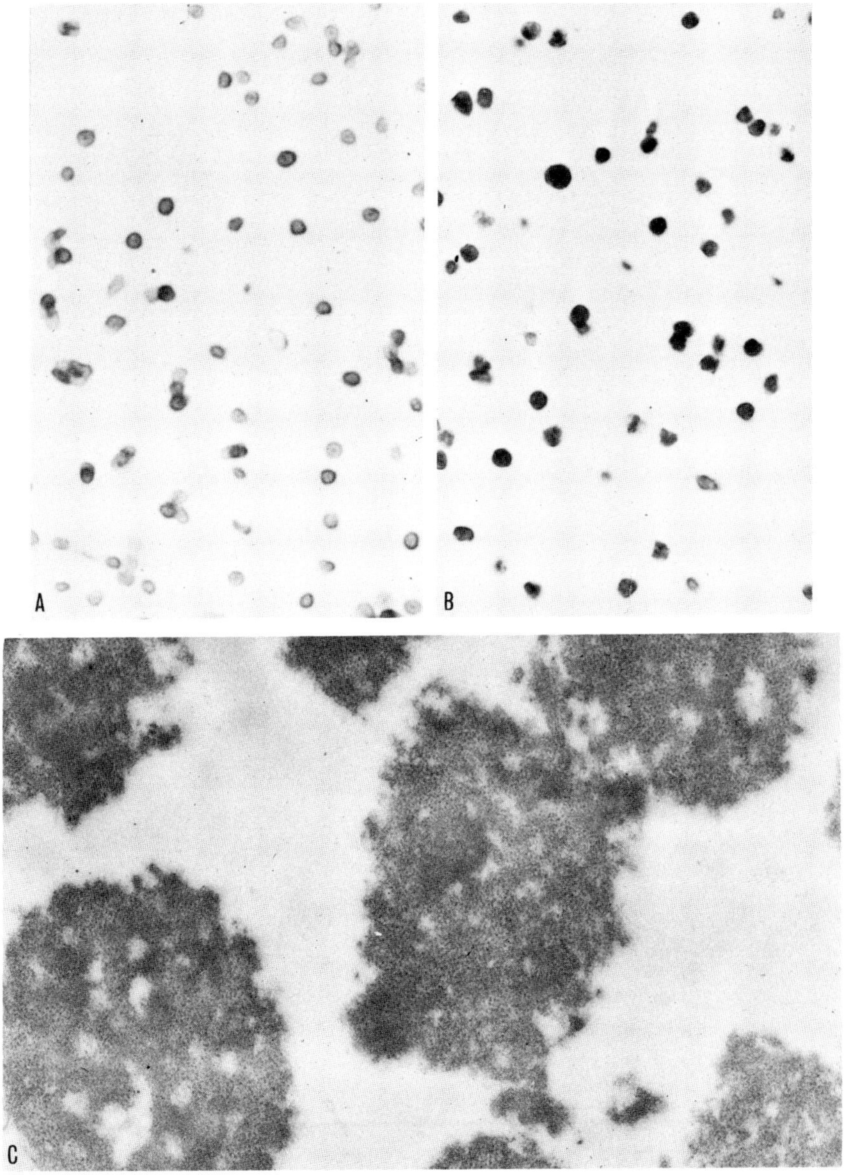

Fig. 1.6. Isolated nucleoli. The nucleoli were prepared from (A) Morris hepatoma 7794 and (B) Morris hepatoma 9121 by the technique discussed in Chapter XI. They were stained with azure C. The electron micrograph (C) shows the preservation of granular and fibrillar elements in isolated nucleoli of normal liver cells. $\times$23,000.

possible with light microscopy (Chapter II). Studies on ultrastructure showed that the nucleolus lacks a definite membrane but contains dense granular elements and fibrillar elements which have a density that differs from the remainder of the nucleolus (Frontispiece). The granular elements are frequently found in skeins referred to as "nucleonemas" (*4a, 33a*), which are separated from one another by "light spaces." Frequently the nucleonemas contain fibrillar elements interspersed with granular elements. The organization of the granular and fibrillar elements and light spaces varies widely from cell to cell (*4a, 16, 122b*) and gives definitive character to many types of nucleoli (Chaps. II, IX). When combined with such techniques as autoradiography and quantitative analyses of planimetry, important findings have been made by electron microscopy about the distribution of DNA and RNA in nucleoli (*126a*).

B. Nucleolar Inhibitors

Actinomycin D, one of the main tools in the studies on nucleolar RNA was initially studied as an antibiotic. It was initially isolated from *Streptomyces griseus* by Waksman and his collaborators (*133, 134*) and later was shown by Reich and his colleagues (*109–111*) to be a specific inhibitor of DNA-primed RNA synthesis by coupling with the deoxyguanosine moieties of DNA. Perry (*97–100*) reported that this drug and azaguanine markedly and specifically inhibited the biosynthesis of nucleolar RNA.

These and other studies (*25*), which showed that the specificity of inhibition of nucleolar RNA synthesis occurred only with the use of low doses of the drug, led to incisive studies on the metabolism of nucleolar RNA. Much of the synthesis of nucleolar RNA results from a complex process in which very rapidly sedimenting and presumably very large RNA molecules are formed first and then are either broken down (*97–101, 117*) or structurally reorganized (*122*) to yield the smaller ribosomal precursors. However, even now, little is known about the nucleotide sequences or tertiary structure of nucleolar 45 S RNA (see Chapter V).

It is now appreciated that a number of substances exert inhibitory effects on the nucleolus (Chapter X); they include aflatoxins, nitroquinolines, proflavin, a number of antibiotics of the type that combine with DNA, such as actinomycin D, and other substances such as viral products. These substances do not all have common properties either with respect to their effects on DNA or their effects on the nucleolus. Fascinating studies have been made on their effects on nucleolar morphology. Among the most interesting effects to emerge were those reported by Journey and Goldstein (*53*) and Reynolds *et al.* (*112*) who studied the

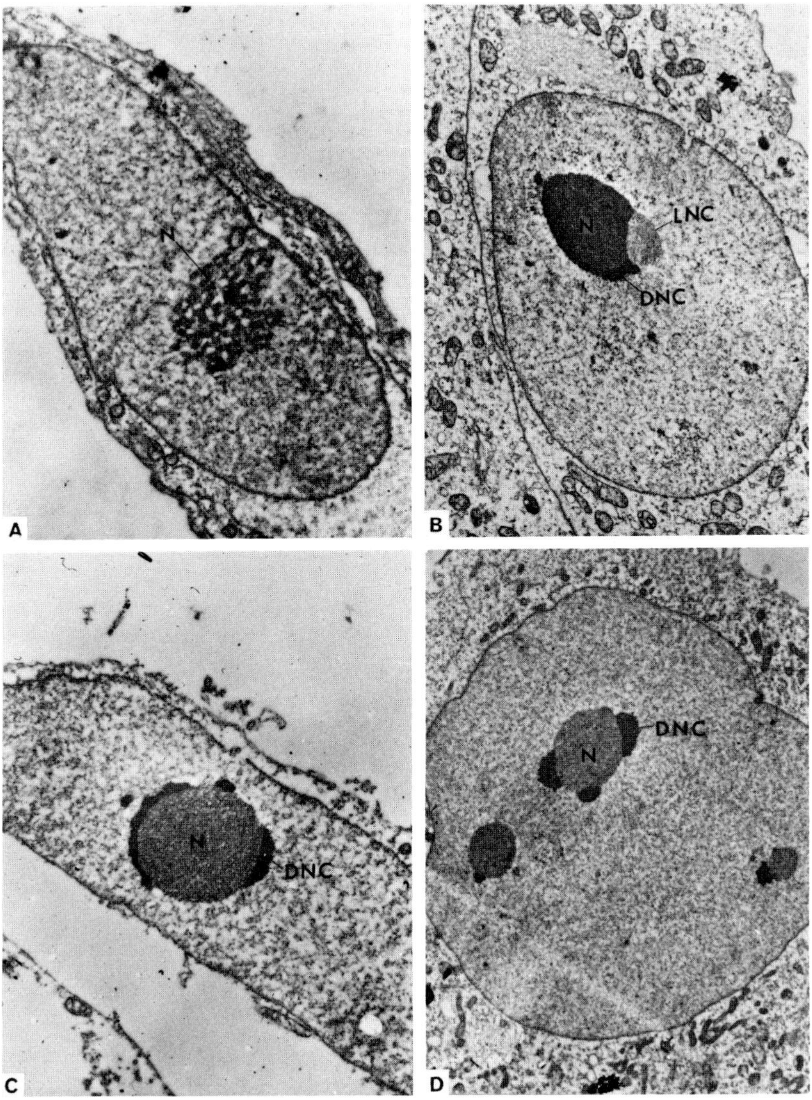

Fig. 1.7. Variation in types of segregation of nucleolar granular and fibrillar elements after treatment of liver cells *in vitro* with actinomycin D. The upper left-hand corner contains a normal nucleolus showing nucleonemas and light spaces. DNC = dark nucleolar cap; LNC = light nucleolar cap; N = nucleolus (courtesy of Dr. Rollin Reynolds and Dr. P. O'B. Montgomery, Southwestern Medical School, Dallas, Texas).

morphological changes introduced in nucleoli of Chang liver cells in cell cultures. They noted the phenomenon of formation of "nucleolar caps" or blebs and described both dark and light caps (Fig. 1.7).

The evidence that there are a number of types of nucleolar caps has been interesting from the point of view that these caps (Fig. 1.7) contain different types of nucleolar components including: (1) granular ribonucleoprotein elements (referred to by some as "ribosomes"); (2) fibrillar ribonucleoprotein elements (of unknown function but containing 4–7 S RNA); (3) elements of the "pars amorpha;" and (4) deoxyribonucleoproteins. Normally, in the nucleolus all of these elements are somewhat mixed together in what usually looks to be quite a hodgepodge of components (Frontispiece). Their separation as entities can be readily detected in electron micrographs. After the administration of actinomycin D, aflatoxin, or some other agent, the nucleolus seems to separate into its constituent parts; the fibrillar regions move to one section of the nucleolus, the granular to another, the deoxyribnucoleoproteins to a third. This "segregation" results in appearance of the "nucleolar caps" described by Reynolds *et al.* (*112*) and led to the conclusion that the nucleolus contains a number of types of subelements which can be differentiated from one another morphologically if not biochemically. Moreover, as time passes after administration of these agents the granular elements decrease and the fibrillar elements predominate.

C. Stimulation of the Nucleolus

A drug that produces remarkable stimulatory effects on the nucleolus and its biosynthesis of RNA is the very simple compound thioacetamide, CH_3CSNH_2. This very small molecule exerts destructive effects on the cytoplasm of the cell as pointed out by Laird (*60*) with the end result that the amounts of ribosomes and cytoplasmic ribonucleoprotein decrease to the point where they constitute as little as 25–50% of the original amount present. Along with this very marked reduction in the size of the cytoplasmic ribosomal protein and RNA pool, there occurs a gigantic increment in the size of the nucleolus primarily due to increases in the granular elements. In laboratory jargon, these have been referred to as "grapefruit," "watermelon," or "Texas-sized" nucleoli as they enlarge to fill half and three-quarters of the total nuclear volume. Although studies with this compound permit further analysis of nucleolar components, the mechanism of action of thioacetamide is not yet defined fully. The suggestion has been made that the system involved in the control of nucleolar biosynthetic reactions is a "feedback system" in which either the ribosomes themselves or a ribosomal product may act as a gene suppressor (see Chapter IX).

D. Drug Combinations

One very interesting experiment carried out in the author's laboratory was that in which actinomycin D was injected into rats that were pretreated with thioacetamide (*91*). The enormous increase in the size of nucleoli produced by thioacetamide is due to the increased number of granular elements. The fibrillar component constitutes a very small percentage of the total nucleolar mass or area. Actinomycin D produces a "balling-up" of the fibrillar elements of thioacetamide-treated nucleoli. The "microspherules" produced were initially present throughout the nucleolus. As the interval after actinomycin treatment increased, the microspherules migrated out of the nucleolar mass, disappearing initially from the center, leaving a nucleolar residue that appeared to be composed virtually completely of granular elements (see Chapter II).

E. Nucleolar Polarity

There is apparently some "polarity" of the nucleolus inasmuch as the direction of flow of the elements seems to be from the center of the periphery. In low doses, actinomycin D alone produces "central light areas" in liver nucleoli ("bulls-eye" nucleoli). Since the center of the nucleolus becomes less dense initially, it is reasonable to assume that the cessation of biosynthetic events is followed by a migration of nucleolar components out of the nucleolus toward their goal in either the nucleus or the cytoplasm. The analogy has been drawn that the nucleolus has a maturing center from which products are removed by some unwinding process as from the center of a ball of string.

VII. Nucleolar Origin and Fate

The nucleolus is not a permanent cellular structure in the sense of DNA, mitochondria, ribosomes, and other components that survive during mitosis and separation of daughter cells. The nucleoli disappear as organized bodies during prophase; they re-form during telophase, and reach a large size in early interphase when they have a roughened and irregular appearance suggesting that they are extremely active at this point.

Apparently, the chromosomes forming the nucleoli are the last to cease functioning during mitosis and among the first to initiate synthetic reactions in telophase. Nucleoli persist during early prophase when they may be visible as still functioning masses. Late in prophase, they are seen in association with one or more "nucleolar" chromosomes presumably linked to them in the specific sites of the "nucleolus-organizing regions"

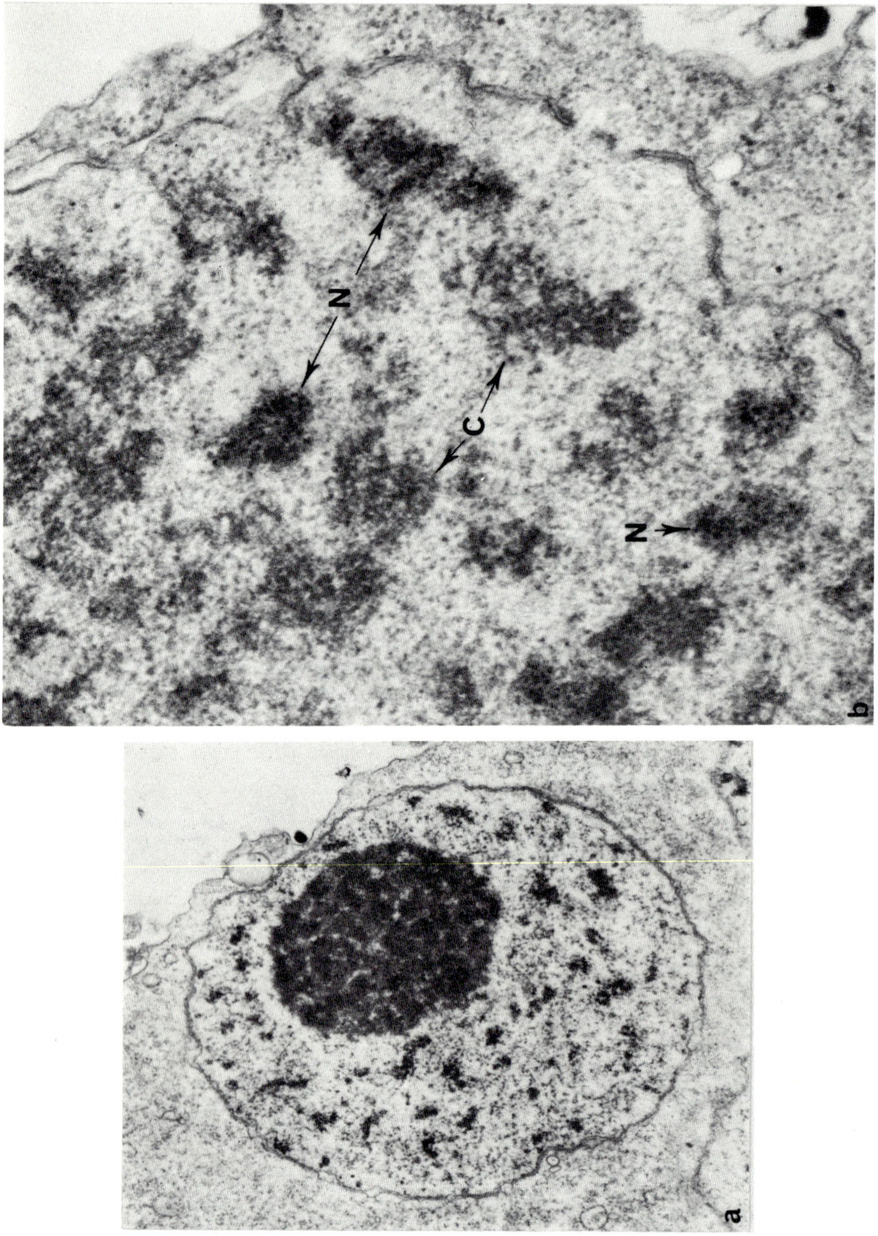

Fig. 1.8. (a) Nucleus of *Physarum polycephalum* at late interphase. Nucleolus beginning to assume eccentric position. ×8600. (b) Prophase (8 minutes before metaphase). Nucleolar fragments (N) mingled with chromosomes (C). The nucleolar fragments (N) are identifiable by their electron-dense granules. ×34,000.

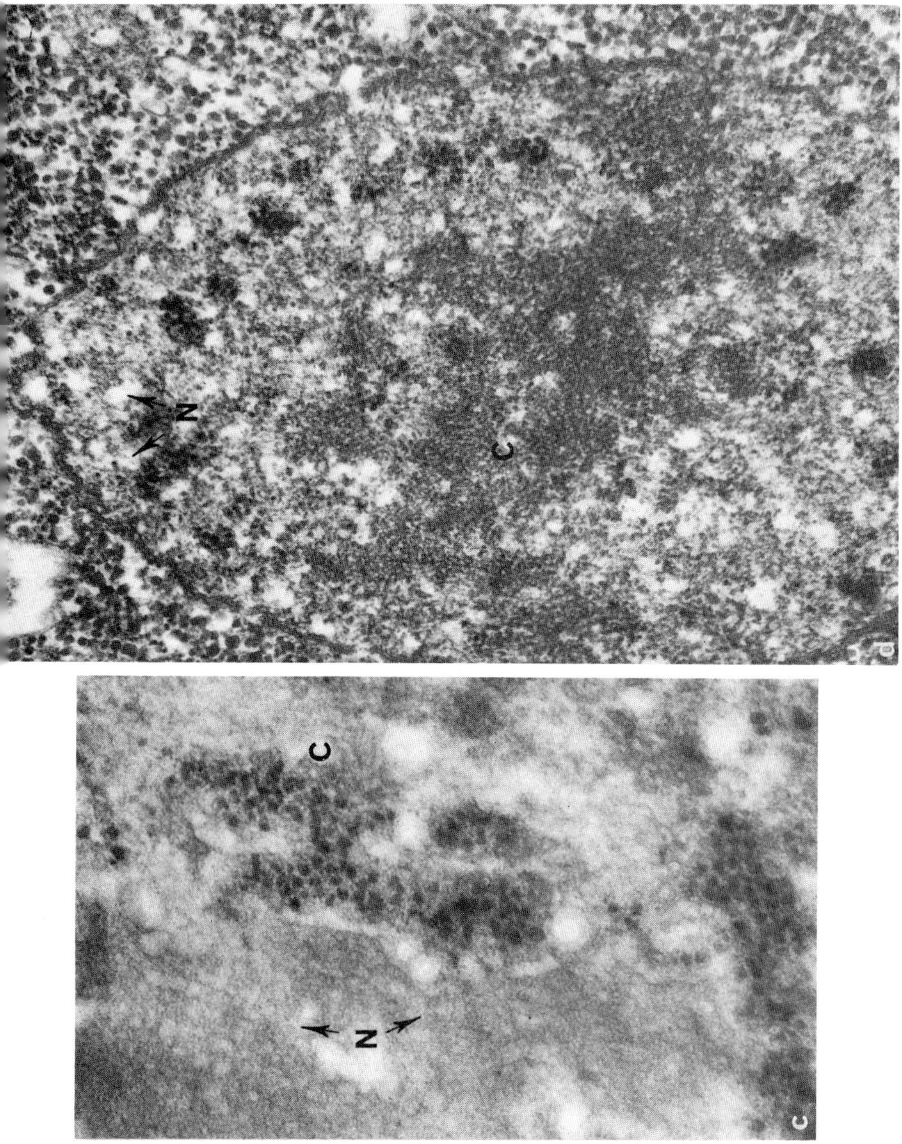

Fig. 1.8. (c) Late prophase. Nucleolar remnants (N) closely adjacent to the chromosomes (C). ×82,500. (d) Metaphase, part only. Chromosomal plate (C) and nucleolar remnants (N). The latter are characterized by electron-opaque granules. This preparation contained numerous glycogen granules in the cytoplasm. ×28,800 (courtesy of Dr. Sophie Guttes and Dr. Edmund Guttes, Loyola University).

or NOR's. These regions are identifiable as "secondary constrictions" which are specific loci of the genes that code for synthesis of the nucleolus and rRNA.

Although the nucleoli seem to disintegrate, the phenomenon is not completely understood. Some electron microscopic studies have been carried out on nucleolar disintegration in prophase in *Physarum* by Guttes *et al.* (*41*). They found cells in which a nucleolar condensation was occurring at a time when there was spindle formation in the region of the nucleolus. At 8 minutes before metaphase, the nucleolus appears only as a remnant (Fig. 1.8a–c). During metaphase, there are nucleolar remnants in the nucleus. Although not clearly organized as nucleoli, there are masses of opaque elements in the nucleolar regions. In these preparations, the nucleolar elements are distinguished as bodies containing dense masses (Fig. 1.8b–d).

The basic mechanisms involved in the resynthesis of the nucleolus are as interesting and unknown as the shut-off mechanisms that are involved in the cessation of its synthesis. It is certain that the nucleolus is not randomly re-formed but is rather the readout of a specific series of genes, some of which apparently set up a lattice work of fibrillar elements of the nucleolonemas. Within these morphological sequences of events there must be a host of undefined biochemical reactions of great interest, from both an enzymatic point of view and the point of view of biosynthesis of nucleic acids and nuclear proteins.

The rate of reconstitution of the nucleolus has been studied. In the *Physarum*, there is nucleolar reconstitution within 10 minutes after metaphase, or at least by this time nucleolar masses seem to coalesce (Fig. 1.9a). Defined nucleoli appear 20 minutes after metaphase (Fig. 1.9b). In some nuclei, horseshoe nucleoli are seen about 30 minutes after metaphase (Fig. 1.9c) and by 40 minutes, normal nucleoli are seen (Fig. 1.9d).

In reviewing the question of whether the nucleolus that forms in late telophase and metaphase constitutes the formation of a totally new structure *de novo* or simply reflects the reconstruction of a nucleolar mass from a pre-existing matrix such as that of Figs. 1.8 (c,d) and 1.9 (a,b), Guttes *et al.* (*41*) pointed out that the results in *Physarum* apparently favor the latter view. However, the persistence of nucleolar material might simply reflect an interaction between late-functioning nucleolar organizer and extrachromosomal sites of RNA synthesis (*123;* also, see Chapter III).

The bulk of the nucleolar mass formed at the NOR in most cells is probably the product of new synthetic activity of that locus. Chouinard (*25b*) indicated that the residual portions of the nucleoli are not functional in the formation of new nucleoli in *Allium cepa*. There is no coalescence of nucleolar material and the interchromosomal material disappears long before the nucleolus reaches its final size. Moreover, there is no connection between the nucleolus and patches of interchromosomal material as contrasted to *Physarum polycephalum* where the nucleolus is apparently formed by coalescence of small nucleolar bodies (*40, 41*). Thus, it seems that there may be differences of the mechanisms that lead to the reformation of the nucleolus in various species.

NUCLEOLAR ACTIVITY IN INTERPHASE, G_1, S, AND G_2 PHASES

Recently, Gonzalez and Nardone (*38*) and others (*47, 48, 56*) have investigated the activity and morphology of the nucleoli of L cells during the G_1, S, and G_2 phases. Between G_1 and mitosis, there was a decrease in the number of nucleoli from 5.7 to 4 which apparently resulted from nucleolar fusion. During S phase, the nucleoli were oval shaped and centrally located and did not exhibit movement. On the other hand, there was an irregular morphology of the nucleoli and association with the nuclear membrane during G_1 and G_2 phases. In their cells, the nucleoli were diffuse and distended. Such findings are not easy to compare with the states of the nucleoli in many cell types. Somewhat similar results have been reported by Hay (*47*) and Hay and Revel (*48, 49*). In the G_1 phase, Maisel and King (*71a*) found that nucleolar labeling with tritiated uridine reached its peak coincidentally with the peak of non-nucleolar nuclear labeling.

VIII. Chemical Contents of Nucleoli of Cells Active in Protein Synthesis

A. NUCLEOLAR RNA

It is largely due to the observations of Brachet (*7–12*) and Caspersson (*20–23*) that a clearer picture emerged of the interrelationships of the nucleolus and the activity of cells in protein synthesis. In essence, Brachet and his associates made use of the Unna stain for DNA (methyl green) and RNA (pyronin) of the nucleus to show that the nucleolus was rich in RNA. More importantly, cells active in the synthesis of proteins had larger numbers and/or sizes of nucleoli than cells that were less active or inactive in protein synthesis. It was fortunate that about the same time, Caspersson and his associates (*20–23*) developed cytospectrophotometric methods for analysis of RNA; they found that in the nucleus, RNA was largely localized to the nucleolus. These studies permitted the important

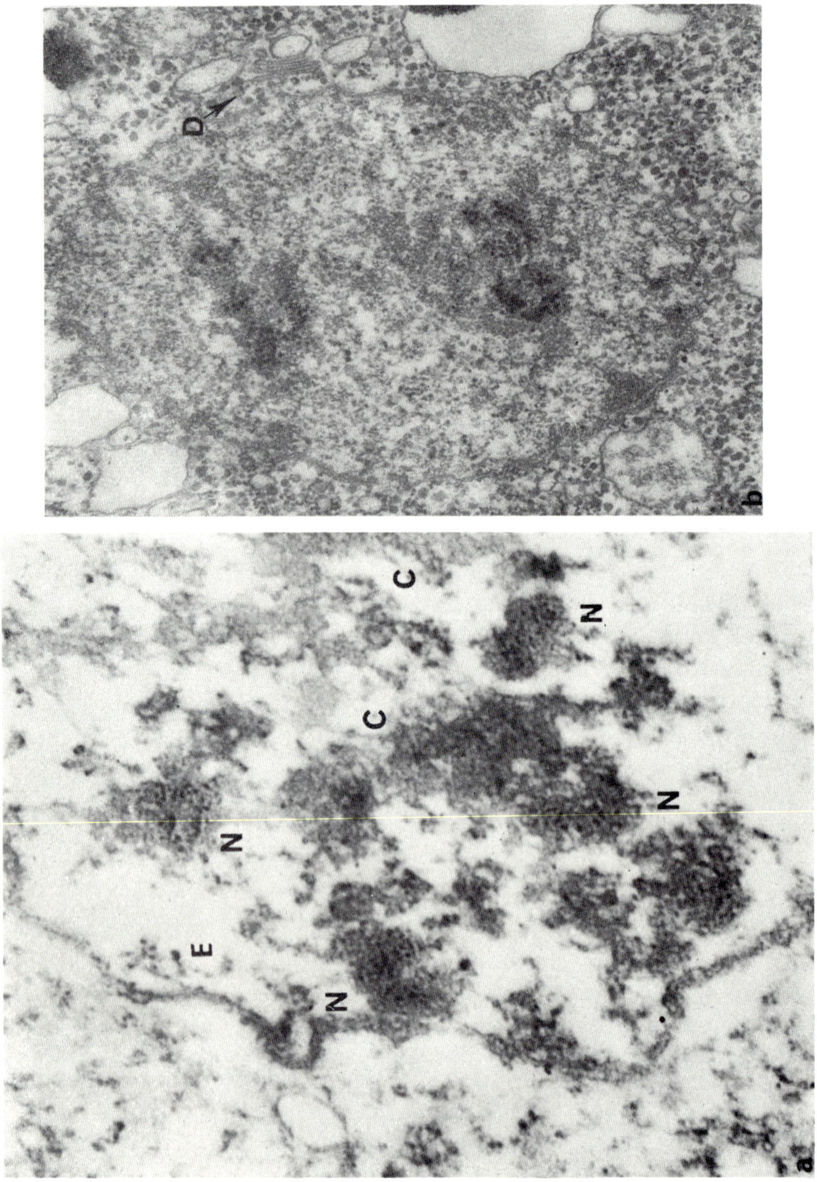

Fig. 1.9. (a) Nuclear reconstitution in *Physarum polycephalum* (10 minutes after metaphase). Part of a nucleus. Underprinted to show prenucleolar material (N) containing electron-opaque granules. C, Chromosomes; E, nuclear envelope. ×47,000. (b) Nuclear reconstitution (20 minutes after metaphase). At this stage the nucleus contains several small nucleoli. D, dictyosome. ×11,400.

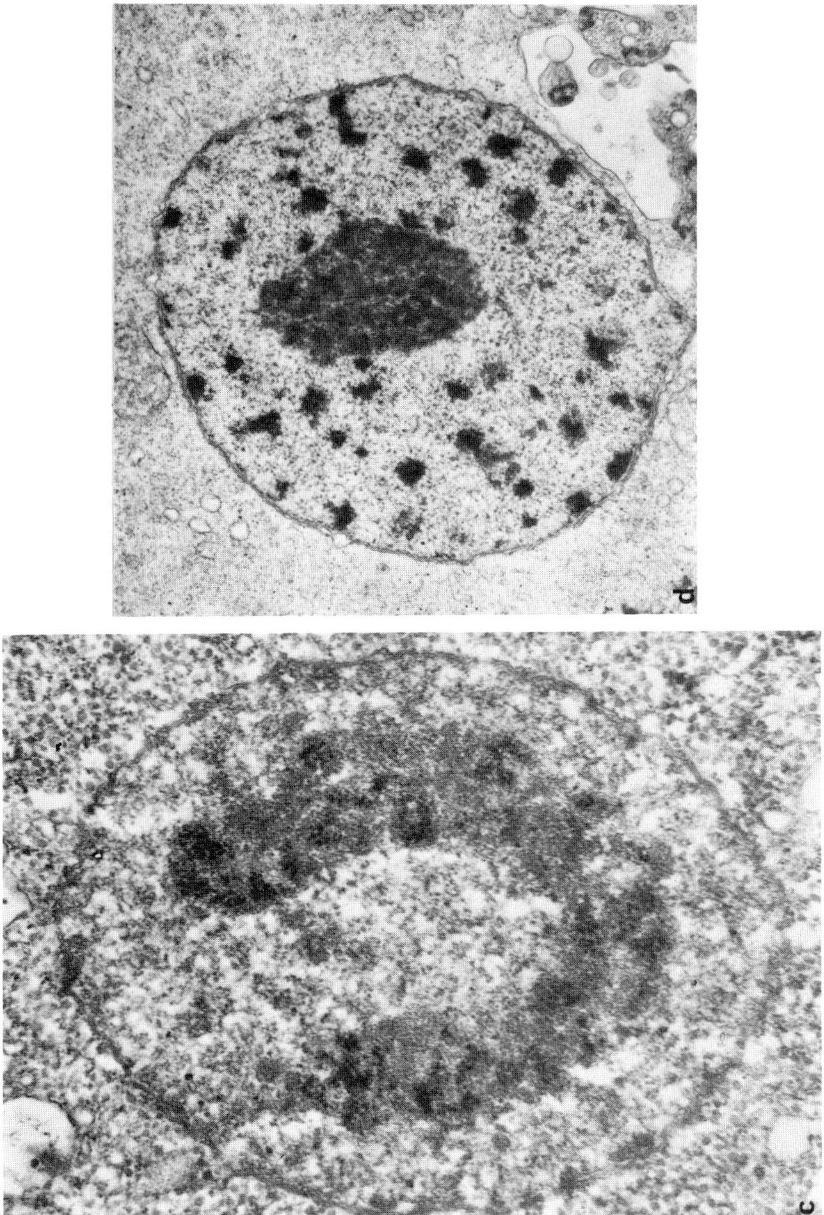

Fig. 1.9. (c) Reconstitution of daughter nuclei (30 minutes after metaphase). Coalescence almost complete. Nuclei with horseshoe-shaped nucleoli are frequently seen at this stage. ×14,100. (d) Early interphase (40 minutes after metaphase): one central, spherical nucleolus. ×14,400 (courtesy of Dr. Sophie Guttes and Dr. Edmund Guttes, Loyola University).

conclusion that the RNA content of the cells, and particularly that of growing or dividing cells, was associated with a high rate of protein synthesis, but there was little clarity regarding the relationship between RNA synthesis and protein synthesis at that time.

The development of molecular biology and the methods for isolation of nucleoli provided biochemical extensions of the morphological observations on nucleoli with regard to nucleolar function (*119*). Among the methods of molecular biology that have been of value in the study of the nucleolar activity are: (1) isolation and fractionation of the nucleolar RNA into discrete sedimentation classes; (2) analysis of the nucleolar RNA with respect to UV and ^{32}P nucleotide compositions; and (3) methods for structural analysis of nucleolar RNA (Chapters V, VI).

Scherrer *et al.* (*117*) found that the RNA formed initially in the nuclei of HeLa cells (a human tumor cell) was a very rapidly sedimenting species of RNA (45 S or greater). This RNA had a high content of guanylic and cytidylic acids and presumably represented a precursor of ribosomal RNA. Perry and his associates (*97–104*) then found that the biosynthesis of the newly synthesized, rapidly labeled RNA of HeLa cells in tissue culture could be selectively blocked by actinomycin D because of the specific effect of these compounds on the nucleolus. On the basis of data obtained with their elegant techniques, it was generalized that the rapidly labeled and rapidly sedimenting RNA was synthesized first and was the probable precursor of the 18 and 28 S ribosomal RNA (Tables 1.1 and 1.2 and Chapter V). Although their evidence was very indirect and did not satisfactorily distinguish the rapidly sedimenting RNA of the nucleolus and the nucleus as a whole, their conclusions are still valid; i.e., the ribosomal RNA is largely synthesized in the nucleolus (see General Reviews, also *29–31*).

In addition to the rapidly sedimenting 28 S, 35 S and 45 S RNA, the nucleolus contains a newly discovered series of 4–7 S RNA's (*90, 108*). Three major RNA components have been found, one of which, U3, is of particular interest because of its high content of uridylic acid. Although their roles are not yet clear, some of the U-rich RNA's are apparently localized to the nucleoli; others are primarily found in the chromatin and the nuclear sap (Chapter VI). Current studies in this field are directed toward the isolation and purification of these RNA species.

B. Nucleolar DNA

Although the history of the nucleolus and its products clearly begins with DNA, "nucleolar DNA," or rDNA containing the templates for synthesis of ribosomal RNA was not isolated in satisfactory form until the recent studies of Birnstiel *et al.* on the DNA of *Xenopus laevis* (*6*). None of the nucleoli isolated from mammalian species are a satisfactory source of the nucleolar ribosomal template DNA inasmuch as they also contain DNA in the perinucleolar chromatin which has not been chemically separated (see Chapter IV).

Birnstiel *et al.* (*6*) found a satellite DNA in the wild type of *Xenopus laevis* that codes for the synthesis of ribosomal RNA. This rDNA (Chapter IV) has a different rate of sedimentation in cesium chloride or in sucrose gradients than the large mass of DNA. This DNA (rDNA) is present in reduced amount in the mutant which contains only one nucleolus and is absent from the mutant that contains no nucleoli; i.e., the anucleolate mutant of the *Xenopus* (see Chapter IV).

C. Nucleolar Proteins

Nucleoli contain histones, but they are very similar to those of the whole nucleus (Chapter VII). The only protein fraction that has been isolated from the nucleoli that has been defined is the saline-soluble "nucleolar acidic proteins." Some of these recently analyzed proteins (*14a*) are of particular interest because of their large number of acidic groups (Chapter VII), i.e., the total number of acidic amino acid residues in some of the proteins is in the range of 35–40% of the total number of residues.

The nucleoli also contain a series of acidic proteins insoluble in 0.15 *M* NaCl, 2 *M* NaCl, or 0.25 *N* HCl. In amino acid composition, they strongly resemble ribosomal proteins (*14a*) but proof of their identity with ribosomal proteins is not yet available. The possibility exists that these proteins are bound either in hydrophobic or covalent bonds to the lipids or other insoluble components of the nucleolus. Their insolubility is a matter of some interest inasmuch as there is now evidence that some of the important biosynthetic reactions of the nucleolus occur in "solid state" matrices (*25a*).

D. Nucleolar Enzymes

In addition to such important biosynthetic enzymes as RNA polymerase (*115, 116, 135*) and NAD pyrophosphorylase (Chapter VII,) the nucleoli contain some enzymes destructive to RNA; i.e., RNases (*121, 127, 128*). The presence of these enzymes in the nucleolus is definite but the role of such enzymes in nucleolar function is a mystery. In some states in which there is an increased biosynthesis of nucleolar RNA, the amount of this enzyme in the nucleolus increases remarkably; i.e., in regenerating liver, fetal liver, and in thioacetamide-treated liver. Some of these enzymes may function in the nucleolus as cleavage enzymes for high molecular weight RNA that is being converted to ribosomal precursors.

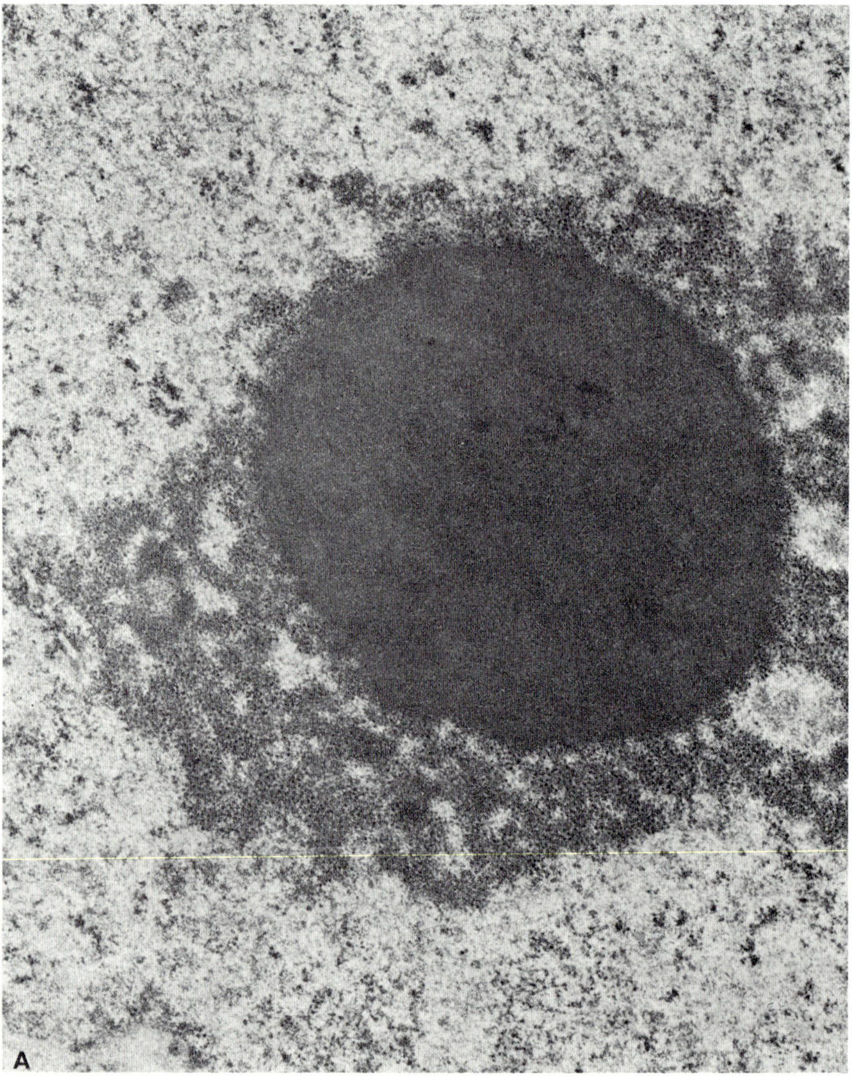

Fig. 1.10. Nucleoli of rat embryos. (A) 8-cell stage. ×31,000. (B) 16-cell stage. ×23,000 (courtesy of Dr. Daniel Szollosi, University of Washington).

Other important nucleolar enzymes and methylases apparently add methyl groups to ribose and RNA bases specifically (see Chapter VII).

IX. The Nucleolus of Various Cell Types

The nucleolus is in some respects a relatively late evolutionary development. Although ribosomes are present in virtually every type of cell, many lower species lack nucleoli including bacteria and yeast. Even

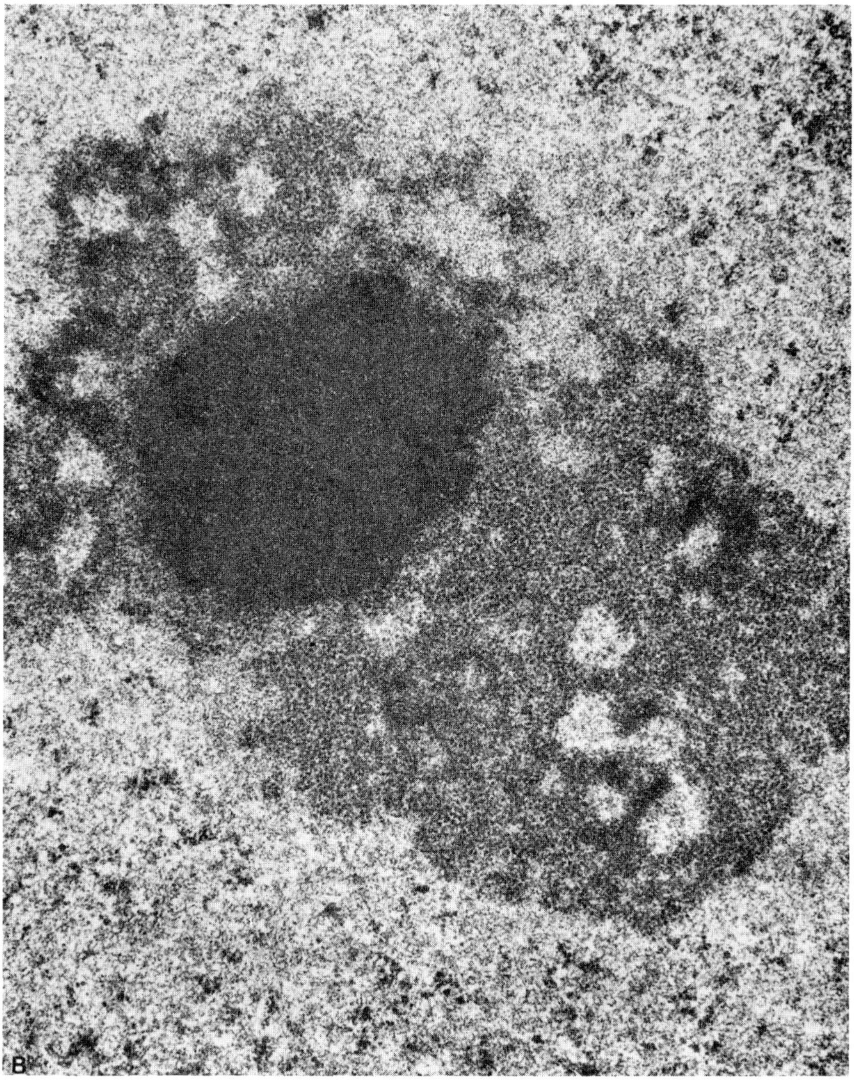

Fig. 1.10. (*Continued*).

some important mammalian cells lack nucleoli. These include red blood cells, reticulocytes, and some specialized reproductive cells, such as spermatozoa, which can generally be considered end-product cells.

A. Embryonic Nucleoli

Relatively few studies have been made on either the ultrastructural appearance or the differentiation of the nucleoli during embryogenesis.

Szollosi (*124*) has made some interesting observations (Fig. 1.10A,B) on the ultrastructure of rat embryonic nucleoli. Although normal nucleoli of the adult rat liver cell and other cells contain intermixed fibrillar and granular elements, in the 8-cell rat embryo, the nucleolus apparently consists of a large central fibrillar mass surrounded by a differentiating layer of nucleolonemas. The fibrillar portion is relatively homogeneous and the perifibrillar zone contains fibrillar and granular elements as well as pars amorpha.

In the later 16-cell stage, the fibrillar portion is considerably smaller and the granular and fibrillar region increases so much in size that the fibrillar mass appears as a kind of central residue. Inasmuch as the cells have undergone division at this stage, the fibrillar mass is not likely to be a product left from an early functional mass but rather seems to be a differentiated structure of undefined function. This fibrillar structure is of interest because it may represent a cellular attempt at differentiation

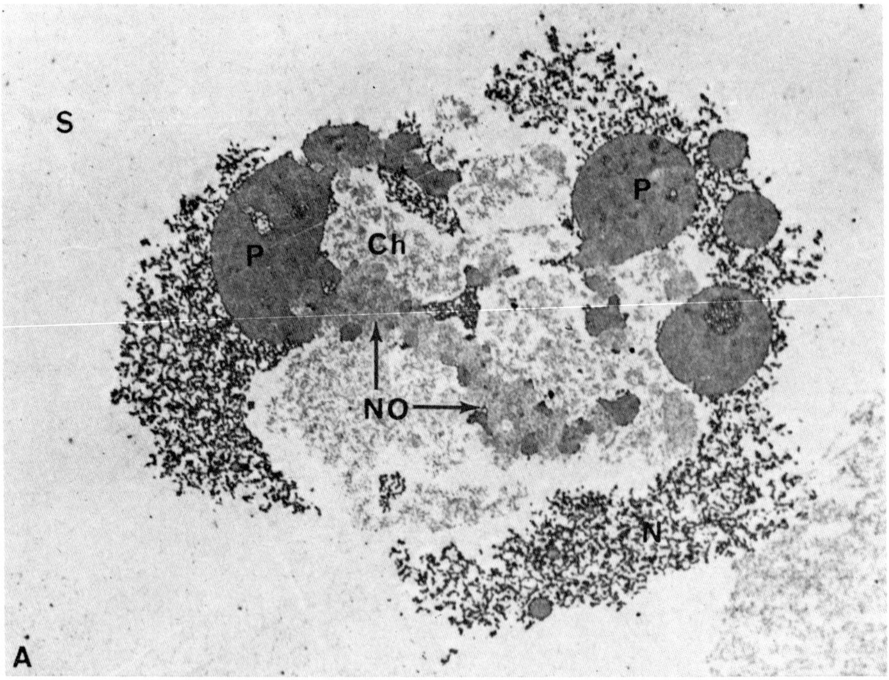

Fig. 1.11. (A) Nucleolus surrounding the chromosome (Ch) in form of a ring and the nucleolar organizer (NO) of the fourth chromosome of *C. plumosus*. Nucleolar material consisting of strands of nucleolonema (N) and pars amorpha (P). Section perpendicular to the longitudinal axis of the chromosome. Nuclear sap (S). ×3800.

and because of its similarity to the types of structures produced in nonmammalian species.

B. Nucleolar Ultrastructure in Midges

Although superficially nucleoli are quite similar in their appearance in the light or phase microscope, there are remarkable variations in the ultrastructure of nucleoli of different species under the influence of different drugs and in their relationships in various cells. One demonstration of these variations is that reported by Kalnins *et al.* (*54, 55*).

In *Chironomus* and *Glypotendipes* the nucleolus frequently has a fibrillar center that looks like the center of embryonic nucleoli (Figs. 1.11–1.13). Peripheral to the fibrillar center there is a fibrous or weblike layer of nucleolonemal strands containing granular elements presumably

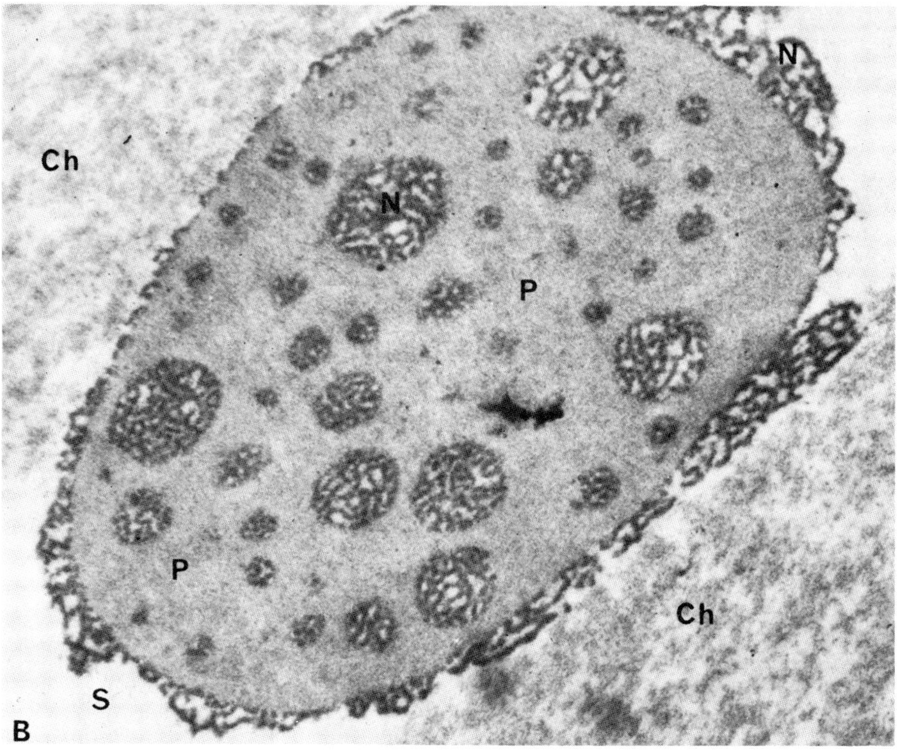

Fig. 1.11. (B) Nucleolus and parts of chromosome (Ch) of *G. lobiferus* showing strands of nucleolonemas (N) forming a peripheral network. Islands of nucleolonemal strands (N) are embedded in pars amorpha (P). Nuclear sap (S). ×12,000 (courtesy of Dr. Hans Stich, McMaster University).

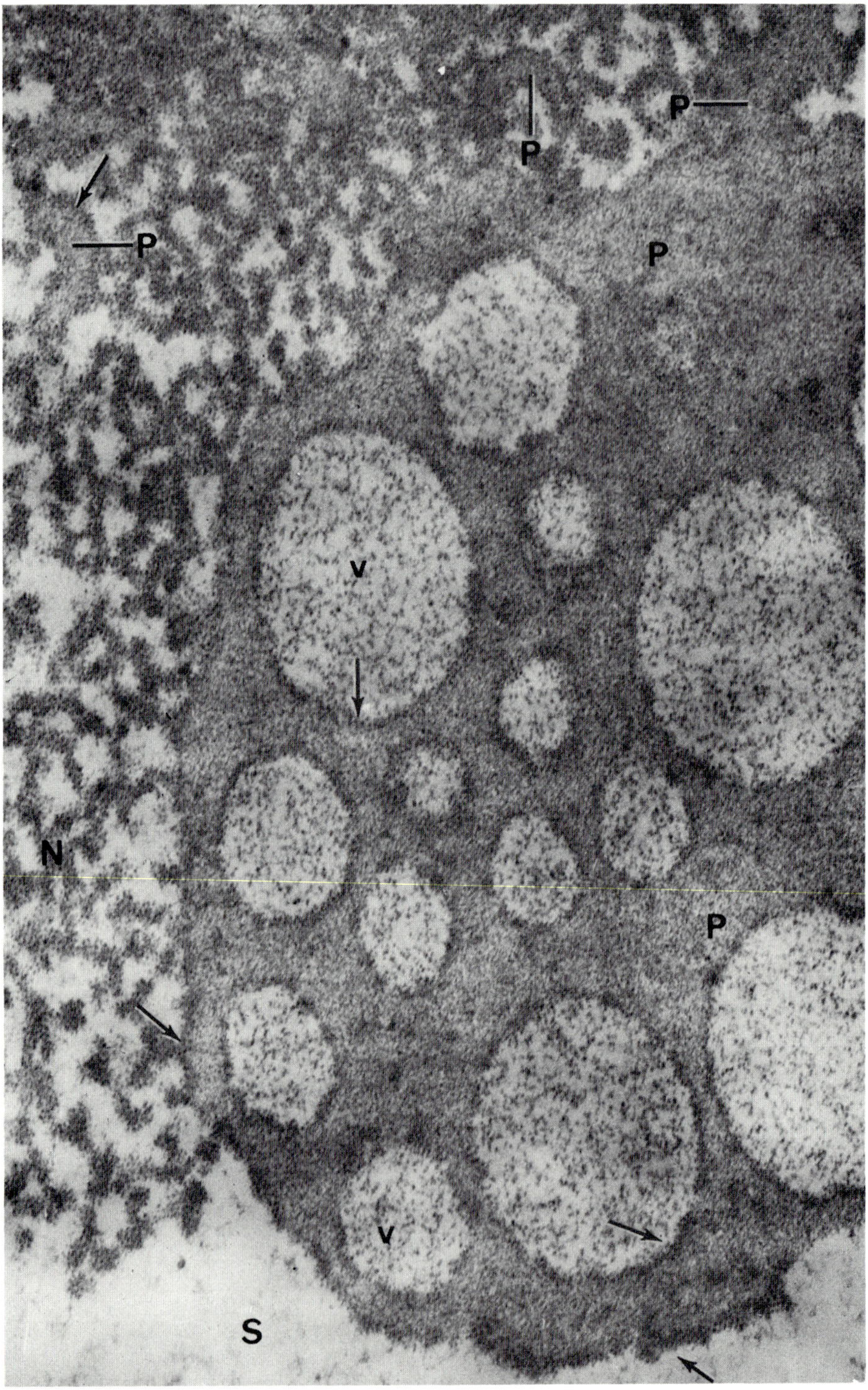
P
P
P
P
v
N
P
v
S

maturing to ribosomal precursors. Although these nucleoli are quite different in ultrastructure from those of mammalian cells, the differences may reflect maturation more than composition.

There are very interesting loops of nucleolonemas in the nucleoli of the flies (Fig. 1.11) which are not seen in mammalian nucleoli. They produce a moth-eaten appearance of the fibrillar mass of the nucleolus. The nucleolonemal mass on the surface of the nucleolus does not seem to communicate directly with the mass in the center of the fibrillar regions although it may do so on the ends of the nucleoli (Fig. 1.12). When the *Chironomus* is treated with actinomycin D, there is a loss of both intranucleolar and perinucleolar granular elements so that the nucleolus is largely fibrillar.

Unlike nucleoli, the Balbiani rings contain dense granular masses which are quite unlike any of the nucleolar products (Chapter III); they do not produce ribosomal RNA. Whether their dense granular elements (Fig. 1.12) are similar to the dense granules of mammalian cells is unclear at present.

X. Nucleolar Specialization

An important aspect of nucleolar physiology is the specialization of the nucleolus in specific species and specific tissues. Among the best examples of such differentiation are those of the amphinucleolus of the slug (*Limax flavus*) and the interesting uterine nucleolar channel system.

A. Nucleoli of the Egg Cells of the Slug (*Limax flavus*)

As shown in Fig. 1.14 the nucleus of the oocyte of *Limax flavus* (*138*, *139*) initially contains two nucleoli which are very similar in size and shape. Both contain a small polar vesicle but only one undergoes enlargement. With time a vesicle appears in its center and the dense nucleolar material forms a distinct surface mass like a polar cap (Fig. 1.14). The vesicle- or nucleolus-associated corpuscle or cap (NAC) is initially two spheres, one within the other, but the internal sphere diminishes in size while the external sphere increases in size.

Later, the nucleolar corpuscle moves to the nuclear membrane and persists there during a "contact" stage. After a time, some small and

Fig. 1.12. Part of a nucleolus of *G. barbipes* showing strands of nucleolonema (N) forming a network, pars amorpha (P), and nucleolar vacuoles (V) which contain fibers and ribosomelike granules. The pars amorpha is bordered by thin strands of nucleolonema (arrows) and extends into the peripheral nucleolonemas (upper part of the micrograph). Nuclear sap (S). ×30,000 (courtesy of Dr. Hans Stich, McMaster University).

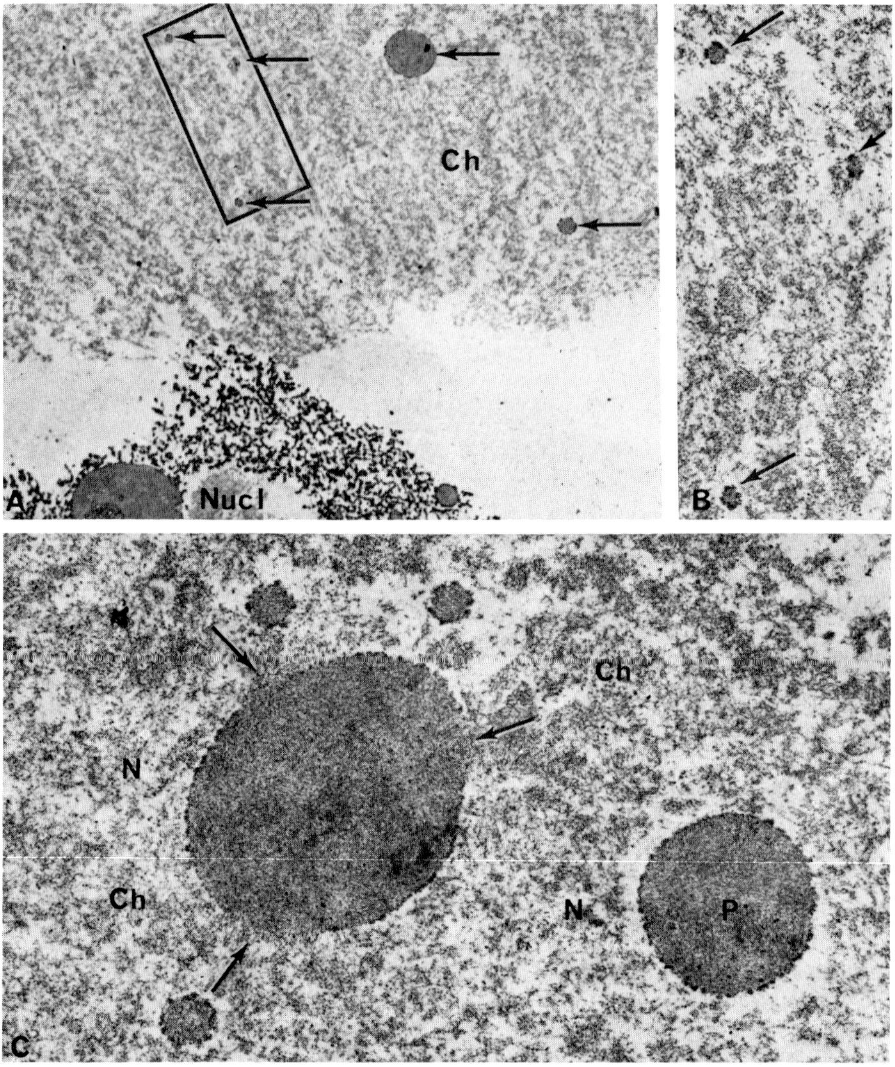

Fig. 1.13. (A) Micronucleoli (arrows) inside one of the long chromosomes (Ch). The nucleolar material (Nucl) is that of the nucleolus of the fourth chromosome of *C. plumosus*. ×3800. (B) An enlargement of the enclosed area of Fig. 1.13A, showing some of the smaller micronucleoli, about 3000 Å in diameter (arrows). ×10,000. (C) Five micronucleoli in a chromosome different from that shown in Fig. 13A but in the same nucleus. Strands of nucleolonemas (N) are absent in regions (arrows) where chromosome material (Ch) seems to be associated with pars amorpha (P). *C. plumosus*. ×10,000 (courtesy of Dr. Hans Stich, McMaster University).

faint granules appear in the mass of the corpuscle. The functions of these bodies are not understood, but they may be related to nucleolar RNA synthesis inasmuch as labeling with tritiated uridine appears later most densely in a granular region (Fig. 1.15). At a later time, the RNA formed appears to move out of the central formative body and into the mass of the nucleolus-associated corpuscle.

These two nucleoli differ in staining characteristics. Both stain with pyronine and methylene blue. The basophilic nucleolus which does not form the corpuscle does not stain with Sakaguchi reagent (for arginine) and Feulgen stain for DNA. The PAS reaction is marked with the corpuscular nucleolus, as is the stain with the Sudan black B. Further details of this fascinating example of nucleolar specialization are under study (*138, 139*).

B. Nucleoli of Endometrial Cells

Terzakis (*126*) showed that the nucleoli of endometrial cells contained a series of tubular channels during the early to the mid-secretory part of the menstrual cycle. He referred to this specialized system as "the nucleolar channel system." Although others had reported that the channel system existed in the nucleolus, a clear relationship between this structure, the remainder of the nucleus, and possibly also of the cytoplasm was documented by Terzakis (Figs. 1.16, 1.17, 1.18). It seems that an arm of the cytoplasm invaginates into the nucleus as a semidense mass and this mass associates with the dense nucleolar matrix. Figures 1.16 and 1.17 show that the channels are present as a regularly arranged honeycombed structure. These channels are part of an elongated loop that seems to coil in wormlike fashion through the nucleolus.

The function of this channel is unknown but Terzakis (*126*) raised the possibility that subunits of ribosomal RNA may be moved to the cytoplasm through this mechanism or that special substances may be synthesized within the nucleolus and the nucleus by the aid of this specialized structure. This nucleolar specialization is clearly a morphological manifestation of highly specialized gene readouts.

Unique structures have been reported to be present in nucleoli of various types. Among such examples are optically anisotropic nucleolar material found in lemon fruit nucleoli which is birefringent when viewed between crossed polars (Fig. 1.19a,b). The birefringent nucleolar granules were orange-yellow under white light (*59*). They are possibly part of a filamentous structure. In any event, they are not believed to be artifacts in view of their ready demonstration under a variety of conditions.

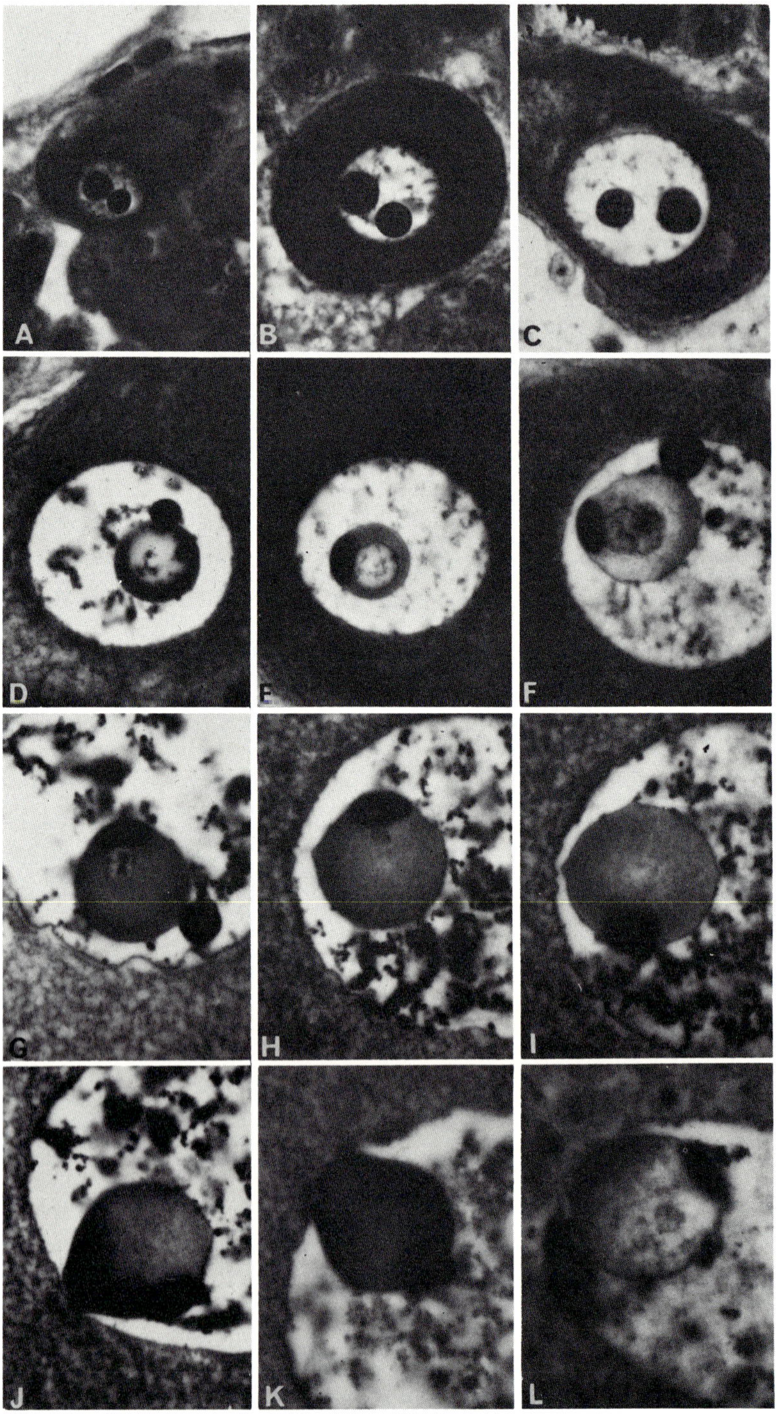
A
B
C
D
F
G
H
I
J
K
L

Similarly, Guttes and Guttes (*40*) have demonstrated fibrillar elements in nucleoli (Fig. 1.20) which may be similar in character to the elements found in lemon fruit nucleoli or they may be like the nucleolonemas or nucleolini (Fig. 1.21) found by Love (*66*). They are said to be formed in anaphase and to aggregate within the nucleoli during nucleolar reconstitution. In prophase, they apparently disintegrate.

Other types of bodies within the nucleolus have been found by Love (*66*). Using special fixation methods dense granular elements which have been referred to as "nucleolini" can be seen by light microscopy. The possibility exists that these elements represent condensed nucleonemas or nucleolar vacuoles (Fig. 1.21). Love has indicated that the sizes and shapes of the nucleolini are regular in normal cells (isonucleolinosis) and abnormal in tumor cells (anisonucleolinosis).

C. Nucleoli of Cancer Cells

One development of the 1930's that captured the attention of oncologists and later of molecular biologists was the report first from the Mayo clinic (*67–69*) and later from Sweden (*22*) that the morphology of nucleoli was aberrant in cancer cells. Large numbers of studies have been made on the morphology of neoplastic cells ever since the students of Virchow suggested that the "only constant feature of neoplasia was

---

Fig. 1.14. Explanations of morphological findings of the plate figures. A. At the early indifferent stage of nucleoli, an oocyte shows two nucleoli indistinguishable in shape and size. B. A little later stage of the nucleoli than in Fig. 14A. C. Nucleoli containing vesicles at the later indifferent stage of nucleoli. D. At the stage of formation of the nucleolus-associated corpuscle (NAC), a nucleolar vesicle is growing in one nucleolus (plasmosome). The other nucleolus remains unchanged. E. A deeply staining lens-shaped body (MN) is distinct at one pole of the nucleolus (NAC). F. At the middle period of the formation of the NAC, there are two phases in the nucleolus: one is the plasmosome and the other the NAC in which a roughly surfaced vesicle still remains. The other nucleolus, homogeneously and deeply staining sphere body, is present. G. A little later stage of the nucleoli than in Fig. 1.14F. The size of the vesicle decreases, and the intravesicular structure is distinct forming granular bodies. H. At the following stage of nucleoli of Fig. 1.14G, the vesicle is reduced and the NAC is nearly completed. I. The complex of MN and NAC (amphinucleolus) is completed. The other nucleolus (uniform nucleolus, UN) is in contact with the plasmosomal part (MN) of the complex (amphinucleolus). Chromatin threads are attached to the NAC. The membranelike envelope surrounding the NAC is clearly visible. J. The oocyte nucleoli in the early contact stage. The NAC protrudes to the nuclear membrane and fused with it. The part of the NAC in contact is deeply and homogeneously stained. K. At the middle period of the contact stage, the NAC is homogeneously stained without prominent structures. L. At the later period of the contact stage, a granular structure appears within the NAC, and the envelope is clearly visible. The other part of the NAC is slightly stained. Magnification A–L ×1000 (courtesy of Dr. Keizo Yamamoto, Nara Medical College, Nara, Japan).

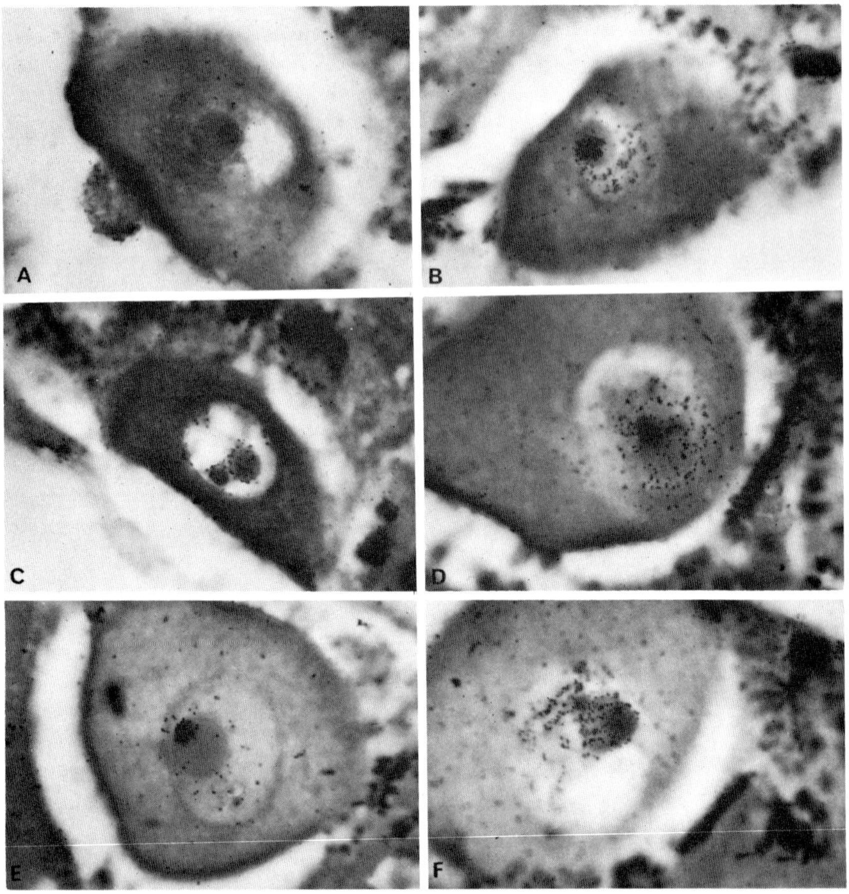

Fig. 1.15. Explanations of radioautographic findings of the plate figures. Magnification Figs. A–F ×500. A. Immediately after the injection of ^{3}H-uridine, several grains are strictly limited in MN. B. In the nucleoli in an indifferent stage, the grains are prominent on the larger nucleolus but not prominent on the chromatin and cytoplasm 4 minutes after the injection of ^{3}H-uridine. C. In the complex of MN and NAC, the MN is more heavily labeled than the chromatin and NAC 15 minutes after the injection of ^{3}H-uridine. D. At the same stage as Fig. 15C, the MN and chromatin are more heavily labeled than the cytoplasm 30 minutes after the injection of ^{3}H-uridine. E. The grains are distributed in the MN, NAC, chromatin of the nucleus, and perinuclear part of the cytoplasm 60 minutes after the injection of ^{3}H-uridine. The number of grains on the MN decreases less than that of Fig. 15D, while the grain number of the NAC and cytoplasm increases. F. The NAC is heavily labeled and the number of grains increases much more than that of Fig. 15E 120 minutes after the injection of ^{3}H-uridine (courtesy of Dr. Keizo Yamamoto, Nara Medical College, Nara, Japan).

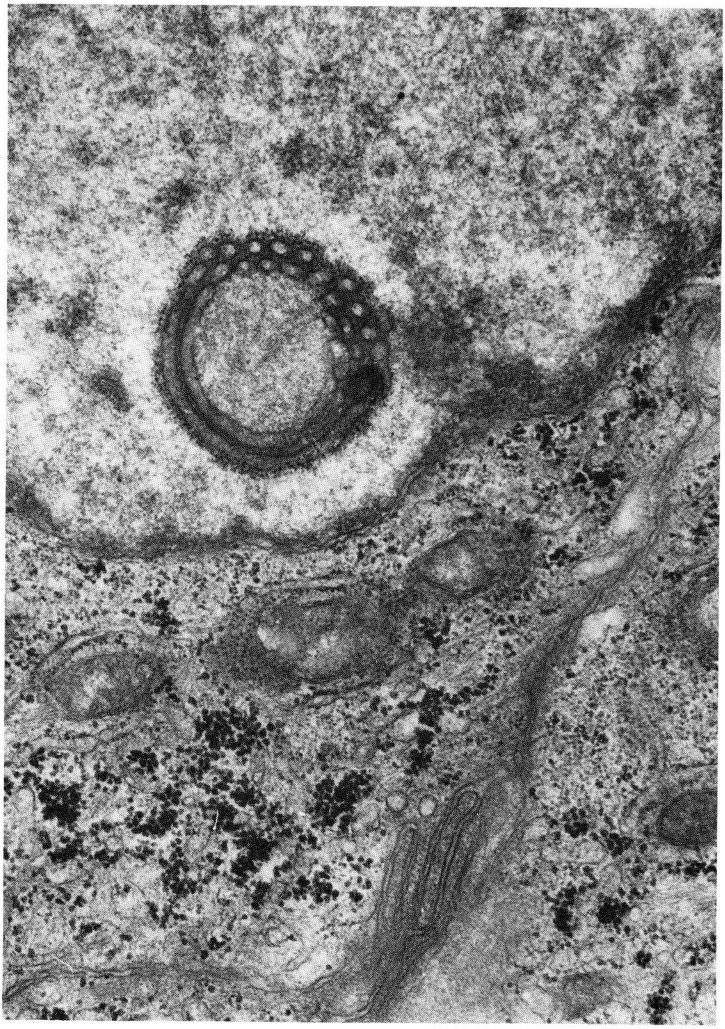

Fig. 1.16. A portion of an endometrial secretory cell nucleus and cytoplasm. The nucleolar channel system is in the form of a hollow sphere in this case which shows what are apparently small vesicles connected to a hemispherical channel system. On the right-hand side there is an apparent connection of the nucleolus to the nuclear membrane by an amorphous darkened mass. ×33,000 (courtesy of Dr. John A. Terzakis, Columbia University).

nucleolar aberration" in morphology and in nucleolar function (*16*). Large, irregularly shaped nucleoli few in number but with increased density are found in tumors. Both of the main features of nucleolar morphology are abnormal: (a) the nucleolar shape is pleomorphic;

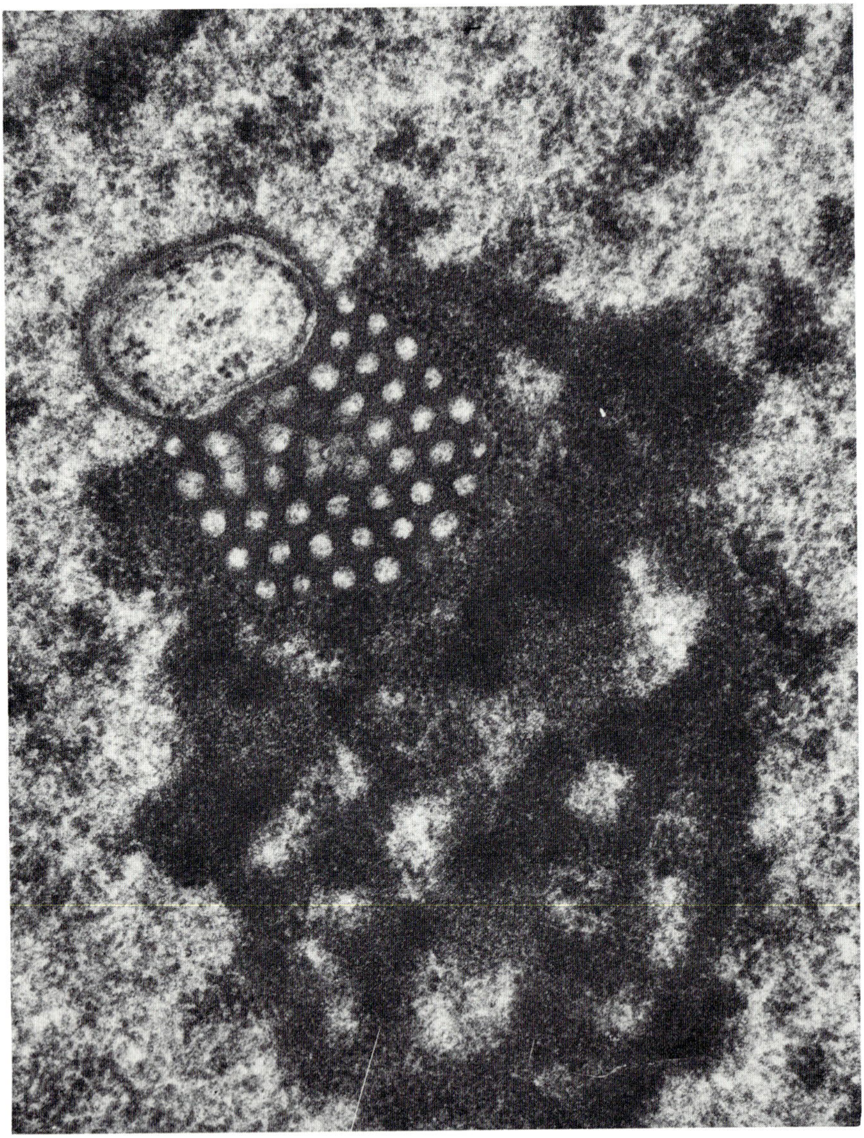

Fig. 1.17. A portion of an endometrial secretory cell nucleolus showing the nucleolar channel system at the periphery of the nucleolus. The nucleolar channel system appears in this case to have a honeycomblike structure in a circular profile about 500–600 Å in diameter. Within the channels are amorphous material of low density and small granules. The channels are embedded in a rather dense, amorphous matrix material. Just peripheral to the matrix is a row of densely staining 150-Å granules that are somewhat denser than the nucleolar granular elements. Note that an apparent invagination of cytoplasm is present in a region just above and to the left of the nucleolar channel system. ×53,000 (courtesy of Dr. John A. Terzakis, Columbia University).

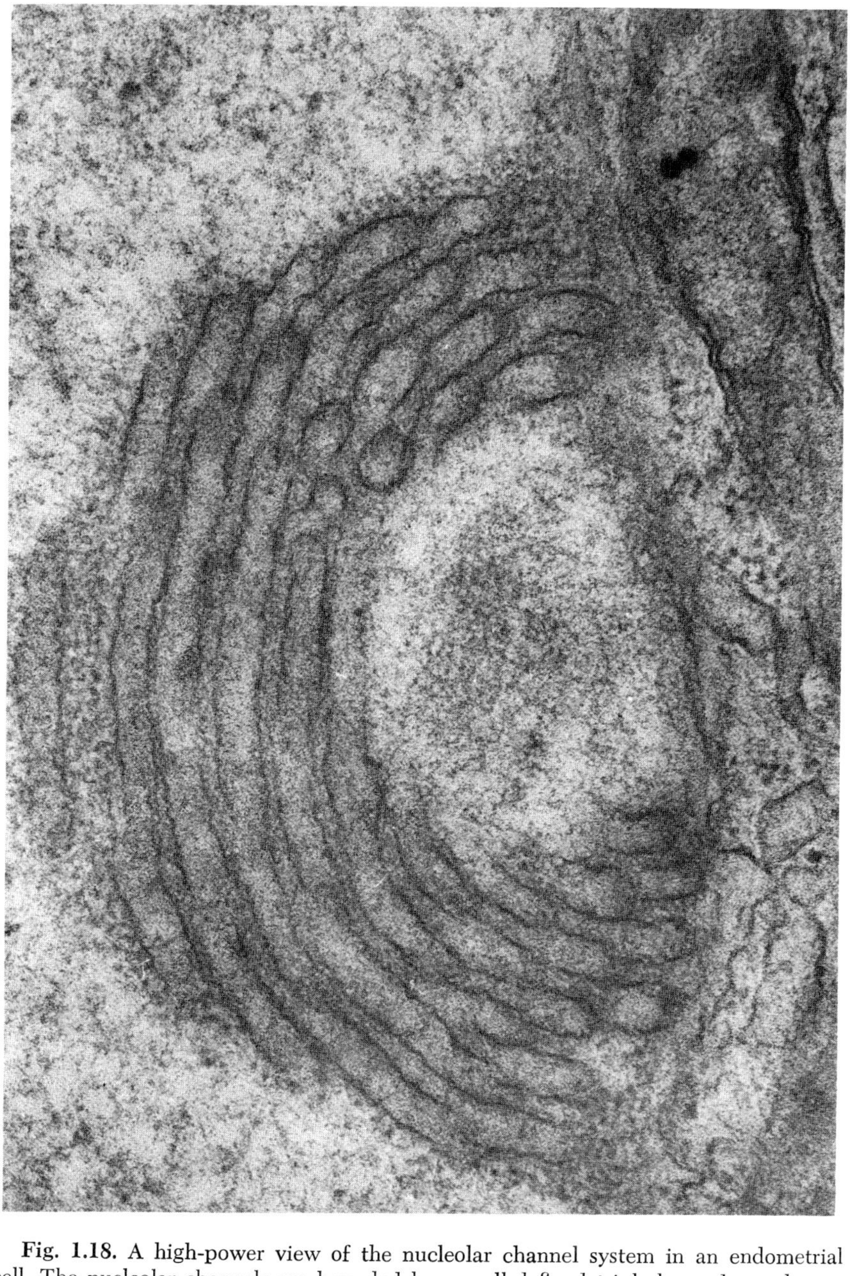

Fig. 1.18. A high-power view of the nucleolar channel system in an endometrial cell. The nucleolar channels are bounded by a well-defined triple-layered membrane. Within a channel, long fibrils are present. ×100,000 (courtesy of Dr. John A. Terzakis, Columbia University).

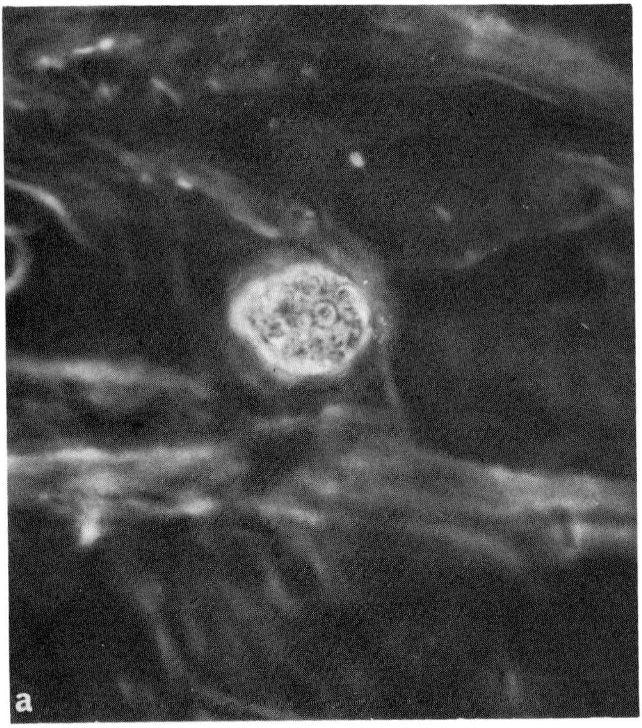
a

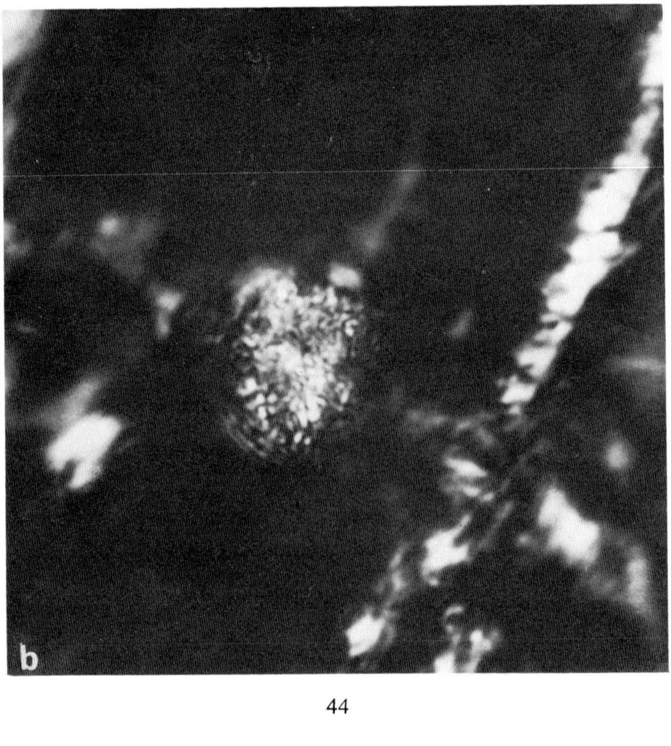
b

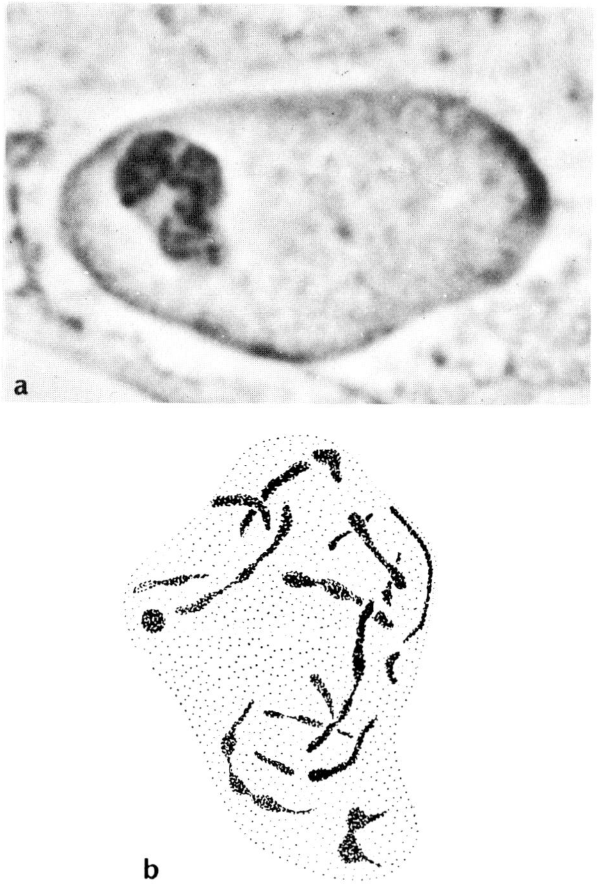

Fig. 1.20. (a) Nucleolonemas in a flattened nucleolus of a chick embryo fibroblast. (b) Drawing of nucleolus to show the position and the number of the filaments (courtesy of Drs. Edmund and Sophie Guttes, Loyola University).

while characteristic of rapidly growing tumors, this pleomorphism has been found to a limited extent in other rapidly growing tissues; (b) the size is large, which is also found in a number of nontumor tissues as well as tumors. These morphological distortions are of interest, but it is

---

Fig. 1.19. (a) Nucleus and nucleolus of unfixed lemon vesicle stalk after 48 hours *in vitro* showing the prominent nucleolar inclusions. Anoptral phase-contrast microscopy. ×1900. (b) Nucleolus of "a" viewed between crossed polars showing the birefringent nature of the inclusions. The radial symmetry of many of the granules is obscured as a result of the partial juxtaposition of one granule with another. ×1600 (courtesy of Dr. Herbert A. Kordan, Gustavus Adolphus College, St. Peter, Minnesota).

apparent that the cancer problem must be approached with more incisive systems than with the tools of gross morphology. These early studies pointed a very important direction for cancer research (*14–18*) which only recently has been carried out at a molecular level.

Additional studies on the labeling of nucleolar RNA of tumor cells have shown that the ^{32}P nucleotide compositions and oligonucleotide frequencies differ markedly from those of nucleoli of normal liver and regenerating liver which are the optimal sources of nucleoli of nontumor tissues (Chapter V). At present, it seems possible that some of the differences between cancer cells and other cells may reside in the nucleolar gene readouts.

XI. Cytoplasmic Aspects of Nucleolar Function

The varied functional states of the nucleoli in various types of cells in their ranges of physiological activity reflect the readouts of the genomes of these cells as influenced by a variety of substances including cytoplasmic factors. In the cells of any individual (including in a restricted sense, spermatozoa, which lack nucleoli) there is a common genome which is under a variety of controls. In the cells of various species so carefully studied by Montgomery (*83*), the reactions of the genome in production of nucleoli are conditioned in part by the stimuli to the nucleolus organizer genes and in part by the cytoplasmic functions of the cells (*125a*).

Recently, Gurdon and Woodland (*39*) have focused attention on the cytoplasmic regulation of gene activity, particularly the sets involved in the synthesis of DNA and RNA. It appears that a number of soluble "gene inducers" for DNA and RNA synthesis are present in the cytoplasm.

Specific evidence for the presence of nucleolar gene inhibitors has been obtained by Yamana and Shiokawa (*139a*). They found that a specific soluble cytoplasmic substance inhibits nucleolar activity in amphibian embryos (see Chapter IV). Although this substance has not been characterized, its presence poses the possibility that such molecules may play a general regulatory role in nucleolar function.

Effects of Hormones on Nucleolar RNA Synthesis

Hormones, including estrogens, androgens, growth hormone, thyroid hormone, and hydrocortisone, produce large increments in the synthesis of ribosomal RNA and other types of RNA (*125*). The synthesis of rRNA seems to be most sensitive to the stimulatory effect of hormones (*125, 137b*). Cell regeneration and chemical carcinogenesis also increase rRNA

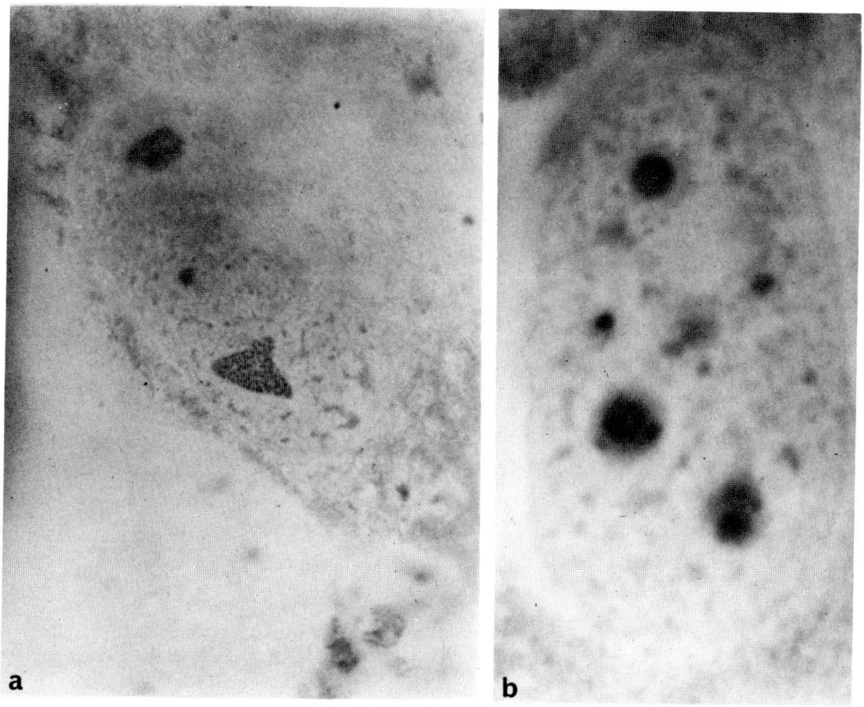

Fig. 1.21. (a) Nucleolini or dense nucleolar granules fixed within the nucleus of a normal cell (WI-38). All of the nucleolini are of approximately equal size. (b) Nucleolini of a tumor cell showing anisonucleolinosis; i.e., differences in the sizes of nucleolini (courtesy of Dr. Robert Love, Jefferson Medical College).

synthesis. The mechanism of increase in synthesis of nucleolar RNA is not yet clear although presumably the synthesis of the new ribosomal precursors represents an obligatory step in the synthesis of new endoplasmic reticulum and polysomes. The possibility has been suggested that an intermediate such as cyclic 3′,5′-AMP is produced by hormones and in turn increases nucleolar synthesis (*125*). Some hormones may increase the permeability of the cells and change the intracellular fluid and electrolyte concentrations; the hormones may then serve as stimuli to nucleolar synthetic reactions via secondary effects on fluid and electrolyte.

Liao *et al.* (*1, 63, 64*) showed that after the incubation of prostatic tissue with androgens, the type of RNA produced was largely the very GC-rich rRNA type characteristic of nucleoli. Nearest-neighbor frequencies showed that the GC-rich dinucleotides increased in isotope con-

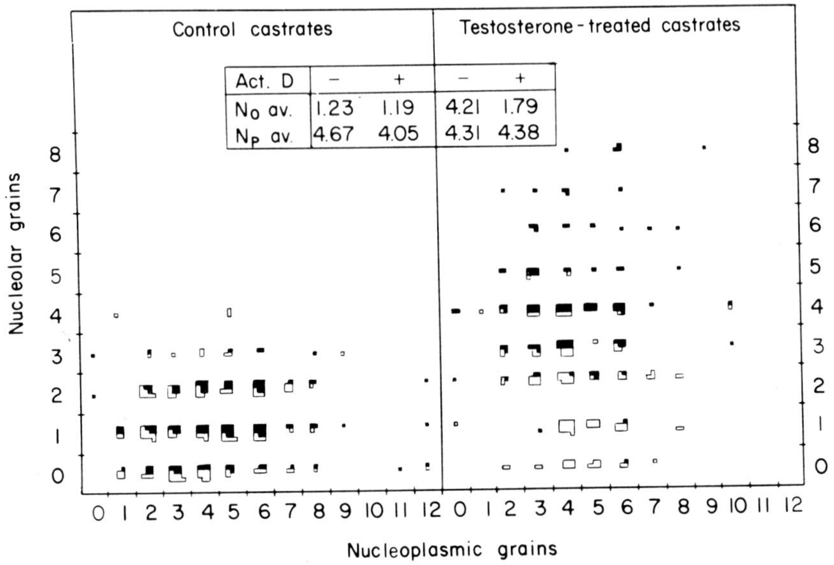

Fig. 1.22. Increased nucleolar grains in the prostatic cells of testosterone-treated castrate rats and the inhibitory effect of actinomycin on the accumulation of these grains. The shaded portion represents controls and the unshaded areas represent the actinomycin D-treated nucleoli. The areas also indicate the number of samples studied (courtesy of Dr. Shutsung Liao, University of Chicago).

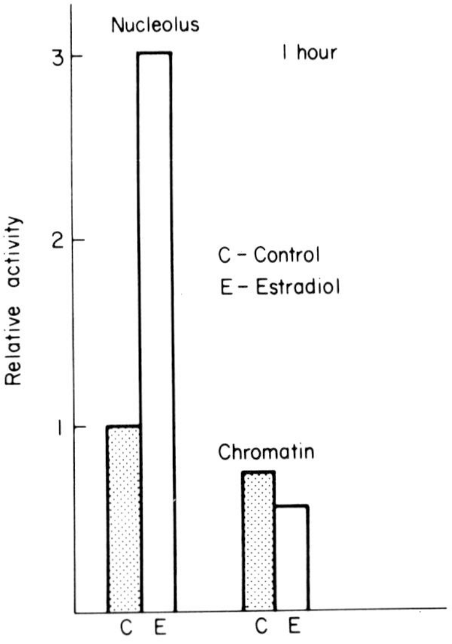

Fig. 1.23. Increased nucleolar RNA polymerase induced by estradiol in estrogen-sensitive tumor cells (courtesy of Prof. Y. Miura, Chiba University School of Medicine, Japan).

tent. Subsequent studies by autoradiographic analysis of the nucleoli of prostatic cells indicated that after treatment of the prostate with testosterone there was a remarkable increase in the labeling of the nucleoli (Fig. 1.22).

Significantly, the only tissue in which this effect of androgens was noted was the prostate which is stimulated by androgens. This effect of the androgens is somewhat similar to that observed for estrogens by Hamilton *et al.* (*42*).

In studies of the effects of estradiol, Miura and his associates (*72, 72a*) found a very large increase in the nucleolar RNA polymerase activity in estrogen-sensitive tumors after administration of estradiol (Fig. 1.23). In other tissues, there was little or no change in the nucleolar RNA polymerase. Presumably as a result of the increased activity of the nucleolar RNA synthetic system, there was a large increase in the amount of rRNA in these cells (*72*). Yamanaka *et al.* (*139b*) suggested that there is a selective synthesis of ribosomal RNA in kidney nuclei at one stage following testosterone administration.

XII. Summary

The history of studies on the nucleoli has shown that each advance in technique for investigation of nucleolar function brought about an increased understanding of structure and functions of this organelle. Important findings made with (1) light and electromicroscopic studies; (2) genetic methods; (3) techniques of molecular biology; (4) isolation techniques; and (5) drugs and enzyme inhibitors.

The morphology of nucleoli varies markedly from cell type to cell type and from species to species. In addition, in any given cell line under defined conditions of growth and functional activity, remarkable changes in morphology and functional activity are produced by drugs and physical factors that have permitted improved analyses of nucleolar composition. The recent advent of methods for isolation of nucleoli has provided a basis for analyses of nucleolar constituents, including RNA, DNA, proteins, and special enzymes, which are important to the role of this organelle. Future investigations will be directed to the factors that initiate, control, and terminate nucleolar functions in various functional stages of cells and cell division.

REFERENCES

1. Anderson, K. M., and Liao, S., Selective retention of dihydrotestosterone by prostatic nuclei. *Nature* **219**, 277–279 (1968).

2. Baltus, E., Observations sur le rôle biochimique du nucléole. *Arch. Intern. Physiol.* **62,** 281–282 (1954).
3. Baltus, E., Observations sur le rôle biochimique du nucléole. *Biochim. Biophys. Acta* **15,** 263–267 (1954).
4. Baltus, E., Etude de la ribonucléase de nucléolus isolés à partie des oocytes d'astéries. *Biochim. Biophys. Acta* **55,** 82–87 (1962).
4a. Bernhard, W., and Granboulan, N., The fine structure of the cancer cell nucleus. *Exptl. Cell Res. Suppl.* **9,** 19–53 (1958).
5. Birnstiel, M. L., Speirs, J., Purdom, I., Jones, K., and Loening, U. E., Properties and composition of the isolated ribosomal DNA satellite of *Xenopus laevis. Nature* **219,** 454–463 (1968).
6. Birnstiel, M. L., Wallace, H., Sirlin, J. L., and Fischberg, M., Localization of the ribosomal DNA complements in the nucleolar organizer region of *Xenopus laevis. Natl. Cancer Inst. Monograph* **23,** 431–448 (1966).
7. Brachet, J., La localisation des acides pentosenucléiques dans les tissus animaux et les oeufs d'amphibiens en voie de developpement. *Arch. Biol.* **53,** 207–257 (1942).
8. Brachet, J., Biochemical and physiological interrelations between nucleus and cytoplasm during early development. *Growth* **11,** 309–324 (1947).
9. Brachet, J., The role of the cell nucleus in oxidations and phosphorylations. *Biochim. Biophys. Acta* **9,** 221–222 (1952).
10. Brachet, J., "The Biological Role of Ribonucleic Acids." Elsevier, Amsterdam, 1960.
11. Brachet, J., Ribonucleic acids and the synthesis of cellular proteins. *Nature* **186,** 194–199 (1960).
12. Brachet, J., Nucleocytoplasmic interactions in unicellular organisms. *In* "The Cell" (J. Brachet and A. E. Mirsky, eds.), Vol. 2, pp. 696–745. Academic Press, New York, 1961.
13. Brown, D. D., and Gurdon, J. B., Absence of ribosomal RNA synthesis in the anucleolate mutant of *Xenopus laevis. Proc. Natl. Acad. Sci. U. S.* **51,** 139–146 (1964).
14. Busch, H., "An Introduction to the Biochemistry of the Cancer Cell." Academic Press, New York, 1962.
14a. Busch, H., "Histones and Other Nuclear Proteins." Academic Press, New York, 1965.
15. Busch, H., Isolation of nucleoli. *In* "Methods in Enzymology," (L. Grossman and K. Moldave, eds.), Vol. XIIa, pp. 448–464. Academic Press, New York, 1967.
16. Busch, H., Byvoet, P., and Smetana, K., The nucleolus of the cancer cell: A review. *Cancer Res.* **23,** 313–339 (1963).
17. Busch, H., and Desjardins, R., A continuous tissue homogenizer. *Exptl. Cell Res.* **40,** 353–359 (1965).
18. Busch, H., Lane, M., Adams, H. R., DeBakey, M. E., and Muramatsu, M., Isolation of nucleoli from human tumors. *Cancer Res.* **25,** 225–233 (1965).
19. Busch, H., Muramatsu, M., Adams, H. R., Smetana, K., Steele, W. J., and Liau, M. C., Isolation of nucleoli. *Exptl. Cell Res.* Suppl. 9, 150–163 (1963).
20. Caspersson, T. O., "Cell Growth and Cell Function. A Cytochemical Study." Norton, New York, 1950.
21. Caspersson, T. O., Farber, S., Foley, G. E., and Killander, D., Cytochemical observations on the nucleolus-ribosome system. *Exptl. Cell Res.* **32,** 529–552 (1963).

22. Caspersson, T. O., and Santesson, L., Studies on protein metabolism in the cells of epithelial tumors. *Acta Radiol.* Suppl. 46, 1–105 (1942).
23. Caspersson, T. O., and Schultz, J., Ribonucleic acids in both nucleus and cytoplasm and the function of the nucleolus. *Proc. Natl. Acad. Sci. U. S.* **26**, 507–515 (1940).
24. Chauveau, J., Moule, Y., and Rouiller, C., Isolation of pure and unaltered liver nuclei, morphology and biochemical composition. *Exptl. Cell Res.* **11**, 317–321 (1956).
25. Choi, Y. C., and Busch, H., Effects of actinomycin D on the oligonucleotide composition of nucleolar 45 S RNA of Novikoff hepatoma ascites cells. *Biochim. Biophys. Acta* **174**, 766–769 (1969).
25a. Choi, Y. C., Smetana, K., and Busch, H., Studies on RNA isolated from nucleolar fractions of Novikoff hepatoma ascites cells by sequential extractions with dilute and concentrated salt solutions. *Exptl. Cell Res.* **53**, 582–602 (1969).
25b. Chouinard, L. A., Nucleolar architecture in root meristematic cells of *Allium cepa*. *Natl. Cancer Inst. Monograph* **23**, 125–143 (1966).
26. Desjardins, R., Smetana, K., and Busch, H., A large scale procedure for isolation of nucleoli from rat liver. *Exptl. Cell Res.* **40**, 127–137 (1965).
27. Desjardins, R., Smetana, K., Grogan, D., Higashi, K., and Busch, H., A large scale procedure for isolation of nucleoli from Walker tumor. *Cancer Res.* **26**, 97–104 (1966).
28. Dounce, A. L., The isolation and composition of cell nuclei and nucleoli. *In* "The Nucleic Acids" (E. Chargaff and J. N. Davidson, eds.), Vol. 2, pp. 93–154. Academic Press, New York, 1955.
29. Edström, J.-E., The composition of ribonucleic acid from various parts of spider oocytes. *J. Biophys. Biochem. Cytol.* **8**, 47–51 (1960).
30. Edström, J.-E., Chromosomal RNA and other nuclear RNA fractions. *In* "The Role of Chromosomes in Development" (M. Locke, ed.), pp. 137–152. Academic Press, New York, 1965.
31. Edström, J.-E., The base composition of RNA from nucleoli formed at two different organizers in *Chironomus tentans*. *Biochem. Biophys. Res. Commun.* **18**, 341–344 (1965).
32. Edström, J.-E., and Gall J. G., The base composition of ribonucleic acid in lampbrush chromosomes, nucleoli, nuclear sap and cytoplasm of *Triturus* oocytes. *J. Cell Biol.* **19**, 279–284 (1963).
33. Edström, J.-E., Grampp, W., and Schor, N., The intracellular distribution and heterogeneity of ribonucleic acid in starfish oocytes. *J. Biophys. Biochem. Cytol.* **11**, 549–557 (1961).
33a. Estable, C., and Sotelo, J. R., Una nueva estructura celulare: El nucleolonema. *Publ. Invest. Sci. Biol.* **1**, 105–126 (1951).
34. Goldstein, L., Cailleau, R., and Crocker, T. T., Nuclear-cytoplasmic relationships in human cells in tissue culture. II. The microscopic behavior of enucleate human cell fragments. *Exptl. Cell Res.* **19**, 332–342 (1960).
35. Goldstein, L., and Micou, J., Nuclear-cytoplasmic relationships in human cells in tissue culture. III. Autoradiographic study of interrelation of nuclear and cytoplasmic ribonucleic acid. *J. Biophys. Biochem. Cytol.* **6**, 1–6 (1959).
36. Goldstein, L., and Micou, J., On the primary site of nuclear RNA synthesis. *J. Biophys. Biochem. Cytol.* **6**, 301–304 (1959).
37. Goldstein, L., Micou, J., and Crocker, T. T., Nuclearcytoplasmic relationships in human cells in tissue culture. IV. A study of some aspects of nucleic acid

and protein metabolism in enucleate cells. *Biochim. Biophys. Acta* **45,** 82–86 (1960).

38. Gonzalez, P., and Nardone, R. M., Cyclic nucleolar changes during the cell cycle. I. Variations in number, size, morphology and position. *Exptl. Cell Res.* **50,** 599–615 (1968).
39. Gurdon, J. B., and Woodland, H. R., The cytoplasmic control of nuclear activity in animal development. *Biol. Rev.* **43,** 233–267 (1968).
40. Guttes, E., and Guttes, S., On the filamentous elements in the nucleoli of chick embryo fibroblasts. *Quart. J. Microscop. Sci.* **103,** 23–26 (1961).
41. Guttes, S., Guttes, E., and Ellis, R. A., Electron microscope study of mitosis in *Physarum polycephalum. J. Ultrastruct. Res.* **22,** 508–509 (1968).
42. Hamilton, T. H., Widnell, C. C., and Tata, J. R., Synthesis of ribonucleic acid during early estrogen action. *J. Biol. Chem.* **243,** 408–417 (1968).
43. Harris, H., Turnover of nuclear and cytoplasmic ribonucleic acid in two types of animal cell, with some further observations on the nucleolus. *Biochem. J.* **73,** 362–369 (1959).
44. Harris, H., The relationship between the synthesis of protein and the synthesis of ribonucleic acid in the connective tissue cell. *Biochem. J.* **74,** 276–279 (1960).
45. Harris, H., Nuclear ribonucleic acids. *Progr. Nucleic Acid Res.* **2,** 19–59 (1963).
46. Harris, H., The ribonucleic acids in the nucleus and cytoplasm of animal cells. *Endeavour* **24,** 50–56 (1965).
47. Hay, E. D., Structure and function of the nucleolus in developing cells. *In* "The Nucleus" (A. J. Dalton and F. Haguenau, eds.), pp. 1–79. Academic Press, New York, 1968.
48. Hay, E. D., and Revel, J. P., The DNA component of the nucleolus, studied in autoradiographs viewed with the electron microscope. *Proc. 5th Intern. Congr. Electron Microscopy, Philadelphia, 1962* Vol. II, pp. 1–7. Academic Press, New York, 1962.
49. Hay, E. D., and Revel, J. P., The fine structure of the DNP component of the nucleus. *J. Cell Biol.* **16,** 29–51 (1963).

49a. Heitz, E., Uber totale und partielle somatische heteropyknose, sowie strukturelle geschlectschromosemen bie Drosophila funebris. *Z. Zellforsch. Mikroskop. Anat.* **19,** 720–742 (1933).

50. Hodnett, J. L., and Busch, H., Isolation and characterization of the uridylic acid-rich RNA of rat liver nuclei. *J. Biol. Chem.* **243,** 6336–6344 (1968).
51. Hoskins, G. C., and Montgomery, P. O'B., Nucleolar phagocytosis by HeLa and Chang liver cells in tissue culture as studied by time lapse phase contrast cinematography. *Exptl. Cell Res.* **26,** 534–540 (1962).
52. Hyde, B. B., Sankaranarayanan, K., and Birnstiel, M. L., Observation on fine structure in pea nucleoli *in situ* and isolated. *J. Ultrastruct. Res.* **12,** 652–667 (1965).
53. Journey, L. J., and Goldstein, M. N., Electron microscope studies on HeLa cell lines sensitive and resistant to actinomycin D. *Cancer Res.* **21,** 929–932 (1961).
54. Kalnins, V. I., Stich, H. F., and Bencosme, S. A., Fine structure of nucleoli and RNA-containing chromosome regions of salivary gland chromosomes of chironomids and their interrelationship. *Can. J. Zool.* **42,** 1147–1155 (1964).
55. Kalnins, V. I., Stich, H. F., and Bencosme, S. A., Fine structure of the nucleolar organizer of salivary gland chromosome of *Chironomids. J. Ultrastruct. Res.* **11,** 282–291 (1964).

56. Karasaki, S., The ultrastructure and RNA metabolism of nucleoli in early sea urchin embryos. *Exptl. Cell Res.* **52**, 13–26 (1968).
57. Kopac, M. J., and Mateyko, G. M., Malignant nucleoli: Cytological studies and perspectives. *Ann. N. Y. Acad. Sci.* **73**, 237–282 (1958).
58. Kopac, M. J., and Mateyko, G. M., Nucleolar chromosomes: Structures, interactions and perspectives. *Advan. Cancer Res.* **8**, 121–190 (1964).
59. Kordan, H. A., and Preston, R. D., Birefringence in unfixed lemon fruit nucleoli. *Nature* **216**, 1105–1106 (1967).
60. Laird, A. K., Nuclear changes induced in rat liver cells by thioacetamide. *Arch. Biochem. Biophys.* **46**, 119–127 (1953).
61. Lane, N. J., Spheroidal and ring nucleoli in amphibian oocytes. *J. Cell Biol.* **35**, 421–434 (1967).
62. Larsen, C.-J., Galibert, F., Lelong, J.-C., and Boiron, M., Caracterisation dans la fraction nucleaire de cellules KB d'un ARN metaboliquement stable de faible poids moleculaire. *Compt. Rend.* **264**, 1523–1526 (1967).
63. Liao, S., Barton, R. W., and Lin, A. H., Differential synthesis of ribonucleic acid in prostatic nuclei: Evidence for selective gene transcription induced by androgens. *Proc. Natl. Acad. Sci. U. S.* **55**, 1593–1600 (1966).
64. Liao, S., and Lin, A. H., Prostatic nuclear chromatin: An effect on testosterone on the synthesis of ribonucleic acid rich in cytidylyl (3′,5′)guanosine. *Proc. Natl. Acad. Sci. U. S.* **57**, 379–386 (1967).
65. Longwell, A. C., and Svihla, G., Specific chromosomal control of the nucleolus and of the cytoplasm in wheat. *Exptl. Cell Res.* **20**, 294–312 (1960).
66. Love, R., Anisonucleolinosis in mammalian cell cultures. *Natl. Cancer Inst. Monograph.* **23**, 167–180 (1966).
67. MacCarty, W. C., The value of the macronucleolus in the cancer problem. *Am. J. Cancer* **26**, 529–532 (1936).
68. MacCarty, W. C., Further observations on the macronucleolus of cancer. *Am. J. Cancer* **31**, 104–106 (1937).
69. MacCarty, W. C., and Haumeder, E., Has the cancer cell any differential characteristics? *Am. J. Cancer* **20**, 403–407 (1934).
69a. Maden, B. E. H., Ribosome formation in animal cells. *Nature* **219**, 685–689 (1968).
70. Maggio, R., Progress report on the characterization of nucleoli from guinea pig liver. *Natl. Cancer Inst. Monograph* **23**, 213–222 (1966).
71. Maggio, R., Siekevitz, P., and Palade, G. E., Studies on isolated nuclei. II. Isolation and chemical characterization of nucleolar and nucleoplasmic subfractions. *J. Cell Biol.* **18**, 293–313 (1963).
71a. Maisel, J. C., and King, D. W., Nucleolar nucleic acid synthesis in G-1 and early S. *Lab. Invest.* **20**, 292–297 (1969).
71b. Marinozzi, V., and Bernhard, W., Presence dans le nucléole des deux types de ribonucléoprotéines morphologiquément distinctes. *Exptl. Cell Res.* **32**, 595–598 (1963).
71c. McClintock, B., The relation of a particular chromosomal element to the development of the nucleoli in Zea mays. *Z. Zellforsch. Mikroskop. Anat.* **21**, 294–328 (1934).
71d. McClintock, B., The relation of a particular chromosomal element to the development of the nucleoli in Zea mays. *Am. Naturalist* **95**, 265–328 (1961).
71e. McConkey, E. H., and Hopkins, J. W., The relationship of the nucleolus to the synthesis of ribosomal RNA in HeLa cells. *Proc. Natl. Acad. Sci. U. S.* **51**, 1197–1204 (1964).

72. Miura, Y., Akamatsu, N., Moriyama, Y., Maeda, H., and Fukui, N., Effect of hormones on nucleolar RNA-polymerase in malignant cells. *Gunma Symp. Endocrinol.* **5**, 23–30 (1968).

72a. Miura, Y., Hormonal effect on nucleolar RNA polymerase in normal liver and hepatoma cells. *In* "Advances in Enzyme Regulation" (G. Weber, ed.), Vol. 2, pp. 439–451. Pergamon Press, Oxford and New York, 1968.

73. Montgomery, P. O'B., and Bonner, W. A., A new technique for ultraviolet microbeam irradiation of living cells. *A.M.A. Arch. Pathol.* **66**, 418–421 (1958).

74. Montgomery, P. O'B., and Bonner, W. A., A "flying-spot" microscope. *Sci. Am.* **198**, 2–7 (1958).

75. Montgomery, P. O'B., and Bonner, W. A., The ultraviolet flying-spot television microscope. *IRE, Trans. Med. Electron.* **6**, 186–189 (1959).

76. Montgomery, P. O'B., Bonner, W. A., and Roberts, F. F., Ultra-violet damage to living HeLa cells as recorded by time-lapse motion picture studies. *Proc. Soc. Exptl. Biol. Med.* **95**, 589–591 (1957).

77. Montgomery, P. O'B., Bonner, W. A., and Roberts, F. F., Ultra-violet flying spot time lapse motion picture observations of living cells. *Texas Rept. Biol. Med.* **15**, 386–395 (1957).

78. Montgomery, P. O'B., Bonner, W. A., Hundley, L. L., and Ashworth, C. T., Biological effects of U.V. irradiation in Chaos Chaos. *J. Roy. Microscop. Soc.* [3] **80**, 19–24 (1961).

79. Montgomery, P. O'B., and Hundley, L. L., The use of television and scanning techniques for ultraviolet irradiation studies of living cells. *IRE, Trans. Med. Electron.* **7**, 135–138 (1960).

80. Montgomery, P. O'B., and Hundley, L. L., Effects of selective ultra-violet irradiation of cytoplasm of living cells. *Proc. Soc. Exptl. Biol. Med.* **105**, 117–120 (1960).

81. Montgomery, P. O'B., and Hundley, L. L., Ultra-violet microbeam irradiation of the nucleoli of living cells. *Exptl. Cell Res.* **24**, 1–5 (1961).

82. Montgomery, P. O'B., Roberts, F., and Bonner, W., The flying-spot mono chromatic ultra-violet television microscope. *Nature* **177**, 1172 (1956).

83. Montgomery, T. H., Comparative cytological studies with special regard to the nucleolus. *J. Morphol.* **15**, 265–564 (1898).

84. Monty, K. J., Litt, M., Kay, E. R. M., and Dounce, A. L., Isolation and properties of liver cell nuclei. *J. Biophys. Biochem. Cytol.* **2**, 127–145 (1956)

85. Moriyama, Y., Hodnett, J. L., Prestayko, A. W., and Busch, H., Studies on the nuclear 4–7 S RNA of the Novikoff hepatoma. *J. Mol. Biol.* **39**, 335–349 (1969).

86. Muramatsu, M., and Busch, H., Isolation, composition and function of nucleoli of tumors and other tissues. *Methods Cancer Res.* **2**, 303–359 (1967).

87. Muramatsu, M., Hodnett, J. L., and Busch, H., Base composition of fractions of nuclear and nucleolar ribonucleic acid obtained by sedimentation and chromatography. *J. Biol. Chem.* **241**, 1544–1550 (1966).

88. Muramatsu, M., Hodnett, L., Steele, W. J., and Busch, H., Molecular properties of RNA of isolated nucleoli of rat liver. I. Synthesis of 28 S RNA in the nucleolus. *Biochim. Biophys. Acta* **123**, 116–125 (1966).

89. Muramatsu, M., Smetana, K., and Busch, H., Quantitative aspects of isolation of nucleoli of the Walker carcinosarcoma and liver of the rat. *Cancer Res.* **23**, 510–518 (1963).

90. Nakamura, T., Prestayko, A., and Busch, H., Studies on nucleolar 4–6 S

ribonucleic acid of the Novikoff hepatoma. *J. Biol. Chem.* **243**, 1368–1375 (1968).

91. Narayan, K. S., Steele, W. J., and Busch, H., Evidence that the granular and fibrillar elements of nucleoli contain 28 S and 6 S RNA, respectively. *Exptl. Cell Res.* **43**, 483–492 (1966).
92. Ohshima, Y., Tomizawa, J., and Horiuchi, T., Isolation of temperature-sensitive Lac repressors. *J. Mol. Biol.* **34**, 195–198 (1968).
93. Penman, S., RNA metabolism in the HeLa nucleus. *J. Mol. Biol.* **17**, 117–130 (1966).
94. Penman, S., Scherrer, K., Becker, Y., and Darnell, J. E., Polysomes in normal polio virus-infected HeLa cells and their relationship to messenger-RNA. *Proc. Natl. Acad. Sci. U. S.* **49**, 654–661 (1963).
95. Penman, S., Smith, I., and Holtzman, E., Ribosomal RNA synthesis and processing in a particulate site in the HeLa cell nucleus. *Science* **154**, 786–789 (1966).
96. Penman, S., Smith, I., Holtzman, E., and Greenberg, H., RNA metabolism in the HeLa cell nucleus and nucleolus. *Natl. Cancer Inst. Monograph.* **23**, 489–512 (1966).
97. Perry, R. P., The cellular sites of synthesis of ribosomal and 4 S RNA. *Proc. Natl. Acad. Sci. U. S.* **48**, 2179–2186 (1962).
98. Perry, R. P., Selective effects of actinomycin D on the intracellular distribution of RNA synthesis in tissue culture cells. *Exptl. Cell Res.* **29**, 400–406 (1963).
99. Perry, R. P., Role of the nucleolus in ribonucleic acid metabolism and other cellular processes. *Natl. Cancer Inst. Monograph* **14**, 73–89 (1964).
100. Perry, R. P., The nucleus and the synthesis of ribosomes. *Natl. Cancer Inst. Monograph* **18**, 325–340 (1965).
101. Perry, R. P., On ribosome biogenesis. *Natl. Cancer Inst. Monograph* **23**, 527–546 (1966).
102. Perry, R. P., and Errera, M., The influence of nucleolar ribonucleic acid metabolism in that of the nucleus and cytoplasm. *In* "The Cell Nucleus" (J. S. Mitchell, ed.), pp. 24–29. Butterworth, London and Washington, D. C., 1960.
103. Perry, R. P., Errera, M., Hell, A., and Durwald, H., Kinetics of nucleoside incorporation into nuclear and cytoplasmic RNA. *J. Biophys. Biochem. Cytol.* **11**, 1–13 (1961).
104. Perry, R. P., Hell, A., and Errera, M., The role of the nucleolus in ribonucleic acid and protein synthesis. I. Incorporation of cytidine into normal and nucleolar inactivated HeLa cells. *Biochim. Biophys. Acta* **49**, 47–57 (1961).
105. Poort, C., A new homogenizer for the isolation of nuclei in concentrated glycerine. *Biochim. Biophys. Acta* **25**, 32–34 (1957).
106. Poort, C., Electrophoretic comparison of nuclear and nucleolar proteins. I. Beef pancreas. *Biochim. Biophys. Acta* **46**, 373–380 (1961).
107. Poort, C., Electrophoretic comparison of nuclear and nucleolar proteins. II. Rat pancreas. *Biochim. Biophys. Acta* **51**, 236–240 (1961).
108. Prestayko, A., and Busch, H., Low molecular weight RNA of the chromatin fraction from Novikoff hepatoma and rat liver nuclei. *Biochim. Biophys. Acta* **169**, 338–349 (1968).

108a. Quagliarotti, G., Hidvegi, E., Wikman, J., and Busch, H., Structural analysis of nucleolar precursors of ribosomal RNA. II. Comparative hybridization of nucleolar and ribosomal RNA with nucleolar DNA. *J. Biol. Chem.* **245**, April (1970).

109. Reich, E., Actinomycin: Correlation of structure and function of its complexes with purines and DNA. *Science* **143,** 684–689 (1964).
110. Reich, E., Franklin, R. M., Shatkin, A. J., and Tatum, E. L., Action of actinomycin D on animal cells and viruses. *Proc. Natl. Acad. Sci. U. S.* **48,** 1238–1245 (1962).
111. Reich, E., Goldberg, I. H., and Rabinowitz, M., Structure-activity correlations of actinomycins and their derivatives. *Nature* **196,** 743–748 (1962).
112. Reynolds, R. C., Montgomery, P. O'B., and Karney, D. H., Nucleolar "caps"—a morphologic entity produced by the carcinogen 4-nitroquinoline N-oxide. *Cancer Res.* **23,** 535–538 (1963).
113. Rho, J. H., and Bonner, J., The site of ribonucleic acid synthesis in the isolated nucleus. *Proc. Natl. Acad. Sci. U. S.* **47,** 1611–1619 (1961).
113a. Ritossa, F. M., Atwood, K., Lindsley, D., and Spiegelman, S., On the chromosomal distribution of DNA complementary to ribosomal and soluble RNA. *Natl. Cancer Inst. Monograph* **23,** 449–472 (1966).
114. Ritossa, F. M., and Spiegelman, S., Localization of DNA complementary to ribosomal RNA in the nucleolus organizer region of *Drosophila melanogaster*. *Proc. Natl. Acad. Sci. U. S.* **53,** 737–745 (1965).
115. Ro, T. S., and Busch, H., *In vitro* labeling of RNA in isolated nucleoli of the Walker tumor and liver. *Cancer Res.* **24,** 1630–1633 (1964).
116. Ro, T. S., Muramatsu, M., and Busch, H., Labeling of RNA of isolated nucleoli with UTP ^{14}C. *Biochem. Biophys. Res. Commun.* **14,** 149–155 (1964).
117. Scherrer, K., Latham, H., and Darnell, J. E., Demonstration of an unstable RNA and of a precursor to ribosomal RNA in HeLa cells. *Proc. Natl. Acad. Sci. U. S.* **49,** 240–248 (1963).
118. Shearer, R. W., and McCarthy, B. J., Evidence for ribonucleic acid molecules restricted to the cell nucleus. *Biochemistry* **6,** 283–289 (1967).
119. Sibatani, A., DeKloet, S. R., Allfrey, V. G., and Mirsky, A. E., Isolation of a nuclear RNA fraction resembling DNA in its base composition. *Proc. Natl. Acad. Sci. U. S.* **48,** 471–477 (1962).
120. Siebert, G., and Humphrey, G. B., Enzymology of the nucleus. *Advan. Enzymol.* **27,** 239 (1965).
121. Siebert, G., Villalobos, J., Jr., Ro, T. S., Steele, W. J., Lindenmayer, G., Adams, H. R., and Busch, H., Enzymatic studies on isolated nucleoli of rat liver. *J. Biol. Chem.* **241,** 71–78 (1966).
122. Smetana, K., Unuma, T., and Busch, H., Ultrastructural studies on nucleic acids of nucleolar granular components in Novikoff hepatoma cells. *Exptl. Cell Res.* **51,** 105–122 (1968).
122a. Steele, W. J., Localization of deoxyribonucleic acid complementary to ribosomal ribonucleic acid and preribosomal ribonucleic acid in the nucleolus of rat liver. *J. Biol. Chem.* **243,** 3333–3341 (1968).
122b. Swift, H., Cytochemical studies on nuclear fine structure. *Exptl. Cell Res.* Suppl. 9, 54–67 (1958).
123. Swift, H., and Stevens, B. J., Nucleolar-chromosomal interaction in microspores of maize. *Natl. Cancer Inst. Monograph* **23,** 145–166 (1966).
124. Szollosi, D., Extrusion of nucleoli from pronuclei of the rat. *J. Cell Biol.* **25,** 545–562 (1965).
125. Tata, J. R., Hormonal regulation of growth and protein synthesis. *Nature* **219,** 331–337 (1968).
125a. de Terra, N., Cytoplasmic control over the nuclear events of cell reproduc-

tion. *In* "International Review of Cytology" (G. H. Bourne and J. F. Danielli, eds.), Vol. 25, pp. 1–29. Academic Press, New York, 1969.

126. Terzakis, J. A., The nucleolar channel system of human endometrium. *J. Cell Biol.* **27**, 293–304 (1965).

126a. Unuma, T., Arendell, J. P., and Busch, H., High resolution autoradiographic studies of the uptake of ^{3}H-5-uridine into condensed and dispersed chromatin of nuclei and granular and fibrillar components of nucleoli of Novikoff hepatoma ascites cells. *Exptl. Cell Res.* **52**, 429–438 (1968).

127. Villalobos, J., Jr., Steele, W. J., and Busch, H., Increased ribonuclease activity of nuclei of liver cells of rats treated with thioacetamide. *Biochem. Biophys. Res. Commun.* **17**, 723–729 (1964).

128. Villalobos, J., Jr., Steele, W. J., and Busch, H., Increased ribonuclease activity in nucleoli of livers of thioacetamide-treated rats. *Biochim. Biophys. Acta* **103**, 195–200 (1965).

129. Vincent, W. S., Isolation and chemical properties of nucleoli of starfish oocytes. *Proc. Natl. Acad. Sci. U. S.* **38**, 139–145 (1952).

130. Vincent, W. S., Structure and chemistry of nucleoli. *Intern. Rev. Cytol.* **5**, 269–299 (1955).

131. Vincent, W. S., Proteins of starfish nucleoli. *Biol. Bull.* **113**, 334 (1957).

132. Vincent, W. S., The nucleolus. *Proc. 11th Intern. Congr. Genet., The Hague, 1963* Vol. II, pp. 343–358. Pergamon Press, Oxford, 1965.

133. Waksman, S. A., Robinson, H., Metzger, H. J., and Woodruff, H. B., Toxicity of actinomycin. *Proc. Soc. Exptl. Biol. Med.* **47**, 261–263 (1941).

134. Waksman, S. A., and Woodruff, H. B., Bacteriostatic and bacteriocidal substances produced by a soil actinomyces. *Proc. Soc. Exptl. Biol. Med.* **45**, 609–614 (1940).

135. Weiss, S. B., Enzymatic incorporation of ribonucleoside triphosphate into the interpolynucleotide linkages of ribonucleic acid. *Proc. Natl. Acad. Sci. U. S.* **46**, 1020–1030 (1960).

136. Weiss, S. B., Biosynthesis of ribopolynucleotides. *Federation Proc.* **21**, 120–126 (1962).

137. Weiss, S. B., and Nakamoto, T., The enzymatic synthesis of RNA: Nearest-neighbor base frequencies. *Proc. Natl. Acad. Sci. U. S.* **47**, 140–145 (1961).

137a. Woods, P. S., and Zubay, G., Biochemical and autoradiographic studies of different RNA's: Evidence that transfer RNS is chromosomal in origin. *Proc. Natl. Acad. Sci. U. S.* **54**, 1705–1712 (1965).

137b. Wyatt, G. R., and Tata, J. R., The hybridization capacity of ribonucleic acid produced during hormone action. *Biochem. J.* **109**, 253–258 (1968).

138. Yamamoto, K., The origin and significance of the Amphinucleus studies on the oocytes in *Limax flavus. J. Nara Med. Assoc.* **17**, 497–535 (1966).

139. Yamamoto, K., and Yamada, M., Behavior of nucleoli in the growth period of egg cells of a slug (*Limax flavus*). *Proc. Japan. Histochem. Assoc.* **1**, 12–15 (1960).

139a. Yamana, K., and Shiokawa, K., Occurrence of an inhibitory factor of ribosomal RNA synthesis in embryonic cells of *Xenopus laevis. Exptl. Cell Res.* **44**, 283–286 (1966).

139b. Yamanaka, H., Furuya, N., Shimazaki, J., and Shida, K., Effects of testosterone administration on RNA synthesis in kidney of castrated rats. *Endocrinol. Japon.* **16**, 29–36 (1969).

140. Zirkle, R. E., Uretz, R. B., and Haynes, R. H., Disappearance of spindles and

pharagmoplasts after microbeam irradiation of cytoplasm. *Ann. N. Y. Acad. Sci.* **90,** 435–439 (1960).

General Reviews on the Nucleolus

Bernhard, W., and Granboulan, N., The fine structure of the cancer cell nucleus. *Exptl. Cell Res.* Suppl. 9, 19–53 (1962).

Birnstiel, M., The nucleolus in cell metabolism. *Ann. Rev. Plant Physiol.* **18,** 25–58 (1967).

Busch, H., Byvoet, P., and Smetana, K., The nucleolus of the cancer cell: A review. *Cancer Res.* **23,** 313–339 (1963).

Darnell, J. E., Jr., Ribonucleic acids from animal cells. *Bacteriol. Rev.* **32,** 262–290 (1968).

Gates, R. R., Nucleoli and related nuclear structures. *Botan. Rev.* **8,** 337–409 (1942).

Hay, E. D., Structure and function of the nucleolus in developing cells. *In* "The Nucleus" (A. J. Dalton and F. Haguenau, eds.), pp. 1–79. Academic Press, New York, 1968.

Kopac, M. J., and Mateyko, G. M., Nucleolar chromosomes: Structures, interactions, and perspectives. *Advan. Cancer Res.* **8,** 121–190 (1964).

Perry, R. P., The nucleolus and the synthesis of ribosomes. *Progr. Nucleic Acid Res. Mol. Biol.* **6,** 219–257 (1967).

Sirlin, J. L., The nucleolus. *Progr. Biophys. Biophys. Chem.* **12,** 25–66 (1962).

Sirlin, J. L., Jacob, J., and Birnstiel, M. L., Synthesis of Transfer RNA in the Nucleolus of *Smittia. Natl. Cancer Inst. Monograph* **23,** 255–270 (1966).

Stich, H., Bau und function der nukleolen. *Experientia* **12,** 7–14 (1956).

Swift, H., Studies on nucleolar function. *Symp. Mol. Biol.* pp. 266–303 (1959).

Swift, H., Cytochemical studies on nuclear fine structure. *Exptl. Cell Res.* Suppl. 9, 54–67 (1963).

Vincent, W. S., Structure and chemistry of nucleoli. *Intern. Rev. Cytol.* **4,** 269–298 (1955).

Vincent, W. S., and Miller, O. L., Jr., eds., "International Symposium on the Nucleolus, its Structure and Function, Montevideo," Natl. Cancer Inst. Monograph No. 23. U. S. Govt. Printing Office, Washington, D. C., 1966.

CHAPTER II

The Cytochemistry and Ultrastructure of the Nucleolus

I. Cytochemistry and Nucleolar Morphology

The first steps in the investigation of the nucleolar composition in regard to the morphological methods were the experiments distinguishing the nucleolar material and nucleoli from the substances forming the rest of the nucleus. Nucleoli were divided into two main groups according to their staining properties. The "real" nucleoli (plasmosomes) (*102*) were those nucleoli which could be stained with acidic dyes and which were, therefore, different from the rest of the nuclear material stained with basic dyes. The basophilic nucleoli or "karyosomes" (*102*) were "false nucleoli" (*97*) since they were composed of chromatin possessing similar staining properties. Jörgensen (*63*) used enzymatic digestion with pepsin acidified by hydrochloric acid for identification of nucleoli; he found that nucleoli were removed by digestion with pepsin while chromatin-containing structures remained intact. He suggested that nucleoli could be basophilic or acidophilic depending on cell growth and formation of proteins. The conclusion that nucleoli were basophilic was verified by staining nucleoli with a combination of two basic dyes (methyl green–pyronin) in the Unna-Pappenheim procedure (*120*). In this procedure, nucleoli were stained by the pyronin, i.e., they were "pyroninophilic," and chromatin stained green with methyl green. The description of a "chromatic layer" around the nucleolus by Heidenhein (*46*) was very likely the first real demonstration of the nucleolus-associated chromatin. The presence of a "chromatin layer" around the nucleolus which is connected with the "chromatin" network of the nucleus is in agreement with contemporary knowledge of nuclear morphology (*11, 24, 48*).

A. RNA

The improvement of light microscopic cytochemical methods and their application to research on the nuclear structures provided basic information on the composition of the nucleolus. The most exciting discovery in this field was the demonstration of the presence of RNA in the nucleolus by means of two completely different methods (Fig. 2.1a). Brachet (*18, 19*) showed that the pyroninophylic properties of the nucleolus were abolished by enzymatic digestion with ribonuclease. Caspersson and Schultz (*24*) found a UV-absorption spectrum typical for nucleic acid in the Feulgen-negative nucleolus. About 10 years later Estable and Sotelo (*33*) demonstrated the presence of trabecular structures, the "nucleolonemas," containing RNA in the nucleolus (Fig. 2.1b) and verified indirectly some previous findings describing network-like structures in the nucleoli (*118*).

With the toluidine blue–molybdate method, at least two types of ribonucleoproteins were found in the nucleoli (*81*). One type of ribonucleoprotein consists of multiple small solid or hollow spherules, the "nucleolini," or Love bodies. The other type of ribonucleoprotein constituted the

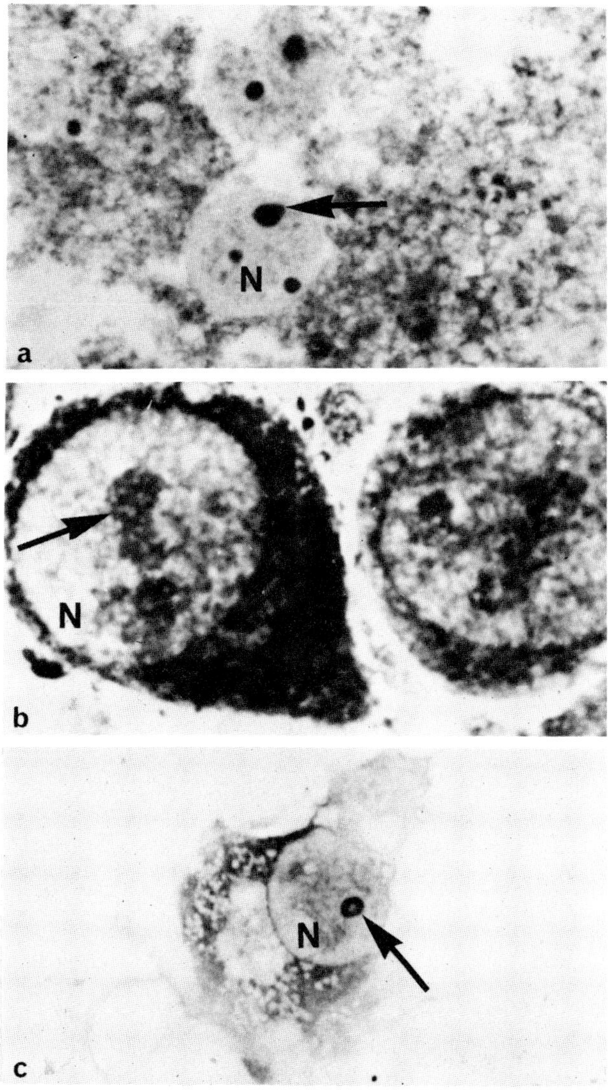

Fig. 2.1. Distribution of RNA-containing structures in nucleoli of smeared cells after staining with toluidine blue for RNA. N, nuclei, arrows, nucleoli. a. Nucleoli with a uniform distribution of RNA in liver cells, ×1800. b. Nucleoli with trabecular structures in Walker tumor cells, ×2800. c. Ring-shaped nucleolus in mature plasmacytes, ×2600.

main body of the nucleolus and was termed "pars amorpha" because of its amorphous appearance. As noted in Chap. I, the "nucleolini" are enlarged and variable in cancer cells. Recently, it has been found that trophoblast nucleolini are normally the same in size and shape but after *in vitro* culture they resemble those of tumor cells (*41a*).

Nucleoli of very unusual appearance in regard to the distribution of RNA were found in several cell types as, for example, some plant cells, nerve cells, in mature lymphocytes, plasmocytes, spermatocytes, and oocytes (*25, 31, 32, 61, 74, 113, 131, 138*). In these cells, nucleoli seem to have a ring-like appearance due to the localization of structures containing RNA only in their periphery (Fig. 2.1c).

The microphotometric methods facilitated quantitative studies on the content of RNA in nucleoli of various cells under physiological, experimental, and pathological conditions (*23, 150, 153, 161, 166*). In general, according to these studies, characteristic high RNA levels were associated with intense cellular activities connected with protein synthesis. However, the amounts of RNA in nucleoli and in chromosomes of some cells vary markedly depending on their functional activity (*153*).

B. DNA

The presence of structures containing DNA in the nucleolus was a subject of many controversial discussions in the light microscopic cytochemical literature (see Chapter IV). According to the majority of investigators (*48, 167*), nucleoli were Feulgen-negative and therefore they did not contain DNA. Thus, the cytochemical definition of the nucleolus as a Feulgen-negative nuclear structure was based on this conclusion (*48*). The chromatin structures containing DNA were present only around the nucleolus as nucleolus-associated chromatin. This perinucleolar structure has, however, a close relationship to the function of the nucleolus and it is actually a part of the nucleolar apparatus (*23, 48*). The fact that Feulgen reactions are negative in some nucleoli and the presence of chromatin structures in the nucleoli can be demonstrated by means of electron microscopic methods suggests that the amount of chromatin structures in some nucleoli is below the limit of detectability by light microscopic procedures (*132*). On the other hand, several investigators (*21, 27, 80, 98, 100, 101, 171*) demonstrated the presence of structures containing DNA in the nucleolar body even by light microscopic methods. According to Lettré and Siebs (*80*), the filamentous structures in nucleoli were Feulgen-positive and nucleolonemas therefore contained DNA. The light microscopic demonstration of DNA in nucleolonemas is still a subject of discussion (*21, 48*), because the trabecular

Feulgen-positive structures might be chromatinic septa-like structures penetrating from nucleolus-associated chromatin into the nuclear body, rather than being in the nucleolonemas proper. Such chromatin structures have now been observed not only by means of light microscopic methods (Fig. 2.2) (*134*), but also electron microscopically (see p. 94) (*40, 133*).

Basic proteins were found cytochemically in nucleoli of several cell types (*15, 23, 25, 33, 50, 90*). However, some observations questioned these findings (*129, 167*). Nevertheless, basic proteins (probably histones) were demonstrated by means of Alfert and Geschwind's procedure (*2*) in association with chromatin-like septa penetrating from the nucleolus-associated chromatin into the nucleolar body (*134*) (Fig. 2.3) or as fibrillar structures within the nucleolus (*171*). The uniform distribution of basic proteins in the nucleolus was also described (*25*). According to the cytochemical studies, the basic proteins can be associated not only with chromatin structures containing DNA, but also with structures containing RNA (*16, 25, 147*). The histone-like basic proteins have also been demonstrated in nucleoli by means of a very interesting fluorescence method (89). This procedure is probably based on the reaction of basic proteins with alpha globulin fractions of nonimmune serum. The nucleolar basic proteins not detectable with the Alfert and Geschwind procedure (*2*) seem to have a striking affinity to zinc (*145*). It was suggested that these proteins may represent a nucleolar product which functions in other parts of the cell.

C. Acidic Proteins

Only a few histochemical studies have been made on the presence and localization of acidic proteins in nucleoli (*78, 90, 109, 167, 168*) (see also Chapter VII). Acidic proteins seem to be more uniformly distributed in the nucleolar body (Fig. 2.4) than histones which are associated with DNA (*134*) (Fig. 2.3). Tryptophan-containing as well as tyrosine-rich proteins were also described in nucleoli (*78*).

Proteins have also been found in central light areas of nucleoli of nerve cells. Their presence in this area seems to be related to the synthesis of nucleolar protein (*31*). However, incorporation of ^{14}C-lysine was not found in this area of ring-shaped nucleoli (*55*) of other cell types. The sulfhydryl groups of nucleolar proteins were reported to be present in higher concentrations on the nucleolar surface than in the central portion of the nucleolus (*1*). According to cytophotometric studies the concentration of SH groups is higher in nucleoli than in the rest of the nucleus (*119*).

Spectrophotometric studies have shown that the protein bound to RNA

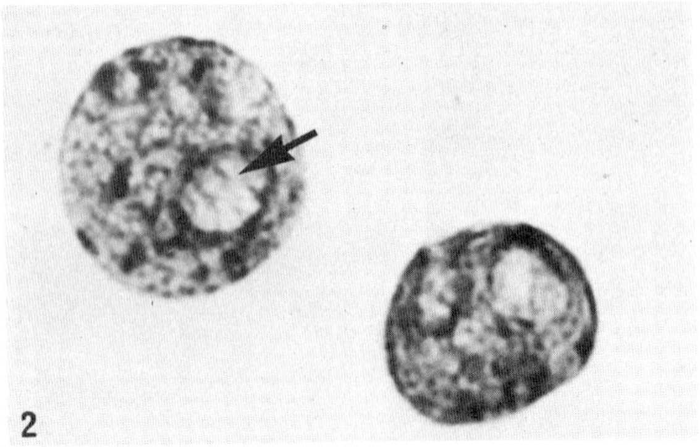

Fig. 2.2. Walker tumor cells. Chromatinlike septa penetrating from the nucleolus-associated chromatin into the nucleolus-arrow. Staining for DNA by a modified Feulgen method (*134*). ×2000.

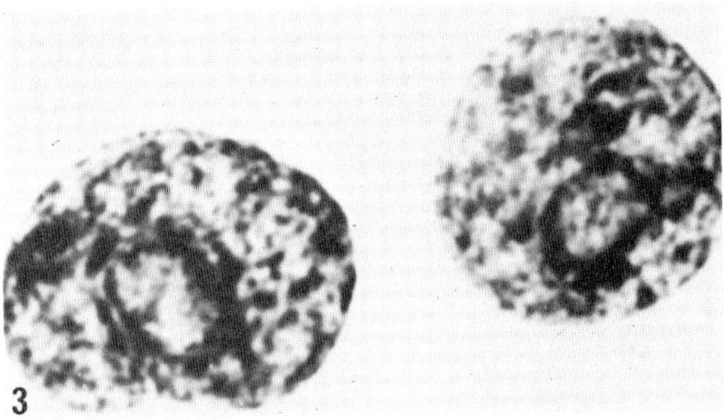

Fig. 2.3. Walker tumor cells stained by Alfert and Geschwind's method (*2*) for demonstration of histones (*134*). ×2200.

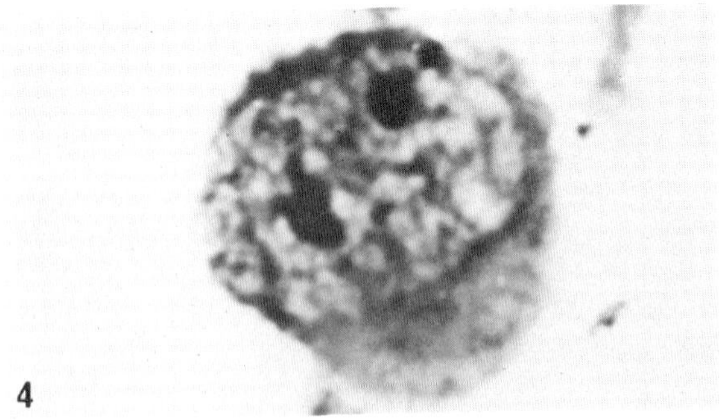

Fig. 2.4. Walker tumor cell stained for acidic proteins (*134*). Note the dense staining of nucleoli and interchromatin areas, particularly the large fibrils of nuclear ribonucleoprotein network. ×2200.

in the nucleolus differs from that bound to the cytoplasmic RNA (*49*). On the other hand, no qualitative differences in UV-absorption curves have been observed between nuclear and nucleolar proteins (*154*).

Liporibonucleoprotein complexes in the nucleolus were found by Zagury (*175*). These complexes contained sulfhydryl groups in the peripheral part of the nucleolus. The staining with Sudan black B for "bound lipids" has shown the presence of these substances in nucleoli of several cell types, particularly in the central regions of ringlike nucleoli in nerve cells (*31*) and in some oocytes (*32*) during their development. Similarly, phospholipids were observed in nucleolar central light areas of some oocytes (*103*) and they seemed to migrate to the peripheral part of the nucleolus during the maturation stages. Phospholipids found in nucleoli of cells from root meristems of various plants differed from those on the chromosomes (*71*). Droplets of neutral lipids which were found in nucleoli of rat liver cells after treatment with thioacetamide may have been derived from the cytoplasm by the invagination into nuclear structures (*160*).

The PAS* positivity indicating the presence of polysaccharides or other substances with 1,2 glycols groups has been noted in nucleoli of various cells (*25, 31, 33*). A few studies indicated also the presence of mucopolysaccharides in nucleoli (*79, 129*).

Although the cytochemistry of enzymes has improved greatly during past years, only a few reports deal with the localization of enzymes in the nucleoli. Among the enzymes detected cytochemically in nucleoli of various cells (Table 2.1) are nonspecific and specific alkaline phosphatases (*39, 76, 117, 144, 159a, 169, 169a, 176*), acid phosphatase (*30, 169, 177*), and succinic dehydrogenase (*28, 29*) and cholinesterase (*159a*). From the cytochemical point of view, however, the demonstration of enzymatic activity in nuclear structures by means of Gomori's type of procedures is a subject of controversy (*107*). For example, Gomori's method provided positive results in nucleoli of Walker tumor cells when azo-coupling procedures with Fast Blue RR were negative. Moreover, a positive nucleolar staining could be obtained without incubation with the substrate. Therefore, the staining with Gomori's method seemed to be due to the known affinity of nucleolar components to heavy metal ions (*99, 145, 157*).

On the other hand, some activity of alkaline phosphatase was noted in nucleoli of hyphae of *Neurospora* with the diazo-coupling method (*176*). Furthermore, the activity of alkaline phosphatase was demonstrated with Gomori's reaction in the same material after stratification of cell struc-

* PAS = periodic acid, Schiff.

TABLE 2.1

ENZYMES REPORTED TO BE PRESENT IN NUCLEOLI ON THE BASIS OF CYTOCHEMICAL STUDIES

Enzyme	Found in nucleoli of:	Reference
Alkaline phosphatases	Ascitic tumor (mouse)	*76*
(specific and nonspecific)	Hyphae of neurospora	*176*
	Onion (*Allium cepa*)	*117*
	Corn root tips *Zea mays*	
	Liver, duodenum (rat)	
	Kidney, skeletal muscle (rat)	
	Hepatoma, cirrhotic liver (mice)	*144*
	Nerve cells (rat)	*159a*
5′ Nucleotidase	Sarcoma (mouse)	*169*
	Ehrlich ascites carcinoma (mouse)	
	Nerve cells (squirrel monkey)	*54a*
ATPase	Sarcoma (mouse)	*169*
	Ehrlich ascites carcinoma (mouse)	*76, 169*
	Fibroblasts (granuloma pouch)	*169*
	Nerve cells (squirrel monkey)	*54a*
Glucose-6-phosphatase	Nerve cells (rat)	*159a*
Guanosine triphosphatase	Liver (rat)	*169a*
Succinic dehydrogenase	Liver (rat)	*28, 29*
	Mammary carcinoma (mouse)	
	Pancreas by pilocarpine	
	Regenerating liver (rat)	
	Nerve cells (squirrel monkey, rat)	*54a, 159a*
Cholinesterase	Nerve cells (rat)	*159a*

tures *in vivo* in the "supernatant" cytoplasm and nucleoli. Since cytoplasm and nucleoli were separated spatially, Zalokar (*176*) concluded that the product of cytoplasmic reactions did not migrate into the nucleolus. The presence of the enzyme in nucleoli depends on the type of cell investigated. For example, adenosine triphosphatase has been found in nucleoli of cells in rapidly growing tissues, including tumor cells (*76, 169*), but it was absent in nucleoli of muscle nuclei (*39*). However, some adenosine triphosphatase activity was observed in nucleoli of muscle nuclei even after denervation of these muscles (*39*). Similarly the presence and activity of enzymes in nucleoli can also depend on other factors as the aging of the cells (*76*).

In the earlier studies by Gersch (*37*), the methods for demonstration of iron and potassium provided negative results in nucleoli but the presence of phosphorus was noted in these cell structures. The orthophosphate ions detected in nucleoli seemed to be bound to lipids (*158*). The presence of zinc has been found in oocyte nucleolus of the starfish (*35*) and nucleoli of prostatic cancer cells (*42*). On the other hand, zinc was not detected in nucleoli of HeLa cells although the nucleolar proteins seemed to have a great affinity for this metal (*145*).

II. Ultrastructure of the Nucleolus

A. Methods and Introductory Remarks

Exploration of nucleoli by electron microscopy has been made possible by the great improvements in the techniques of ultramicrotomy and specimen preparation. The ultrastructural morphology of the nucleolus can be investigated by means of the usual electron microscopic procedures. The visualization of different nucleolar components depends, however, on the procedures used. Fixation with aldehydes contributed to the demonstration of chromatin structures in the nucleolus and around it since classical fixation with osmium tetroxide did not provide satisfactory results in the visualization of these structures (*11, 12*) (Fig. 2.5). Osmium tetroxide still remains a basic fixative in such procedures; the density of ribonucleoprotein structures in the nucleolus and nucleus is higher than that of the chromatin (Fig. 2.5) (*84, 85*). The combination of both fixatives, glutaraldehyde and osmium tetroxide, provides satisfactory results for the general demonstration of the nucleolar structures including its chromatin (Fig. 2.6). Formaldehyde (Fig. 2.7) although less frequently used, is a good fixative for ultrastructural cytochemical methods and also can be used for postfixation after fixation with osmium tetroxide (Fig. 2.8). In some studies of nucleolar structures uranyl acetate was used during dehydration in ethanol, not only to prevent extraction and polymerization artifacts, but also as a further postfixation. The epoxy mixtures and polyester Vestopal are most commonly used as the hydrophobic embedding media and glycolmethacrylate is recommended as water-soluble embedding medium (*7*). The contrast of observed structures can be improved by suitable impregnation ("staining") with heavy metals, for example, with uranyl acetate and lead (*26, 56, 57, 73, 84, 87, 151, 152, 155, 178*).

The development of the ultrastructural cytochemical methods provided the main information on the composition of nucleolar structures, particularly on the localization of nucleic acids and protein. Two main groups of ultrastructural cytochemical methods are used for the investigation of nucleoli. In the first group, the cytochemical procedures are carried out in blocks of tissues or frozen sections or cell suspensions. Such specimens are then usually postfixed and embedded in a hydrophobic plastic (*11, 152*). The isolated nucleoli can also be investigated by means of such procedures. In the second group the cytochemical procedures are performed in the ultrathin sections of specimens fixed in aldehydes and

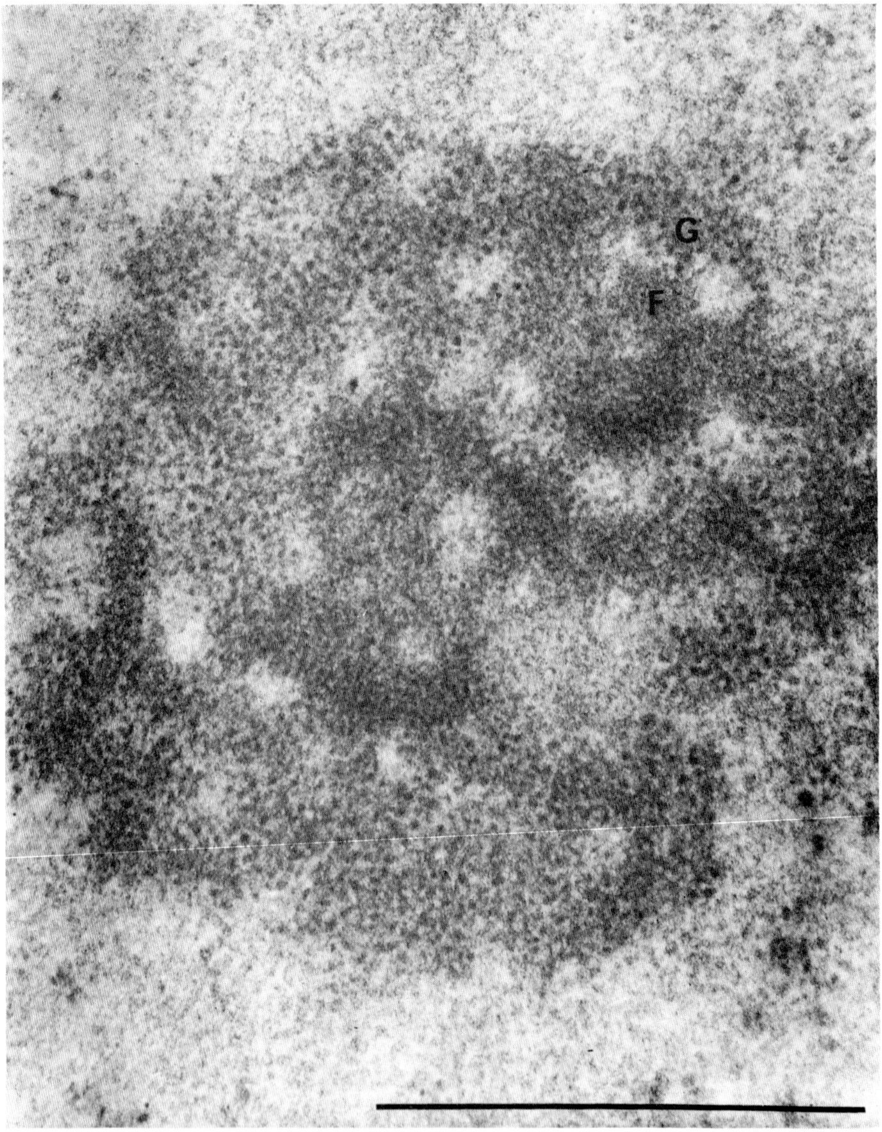

Fig. 2.5. Nucleolus of a liver cell. Fixation was carried out in osmium tetroxide. F, fibrillar components, G, granular components. ×65,000. In this and subsequent figures the marked line represents 1 μ unless specifically stated otherwise.

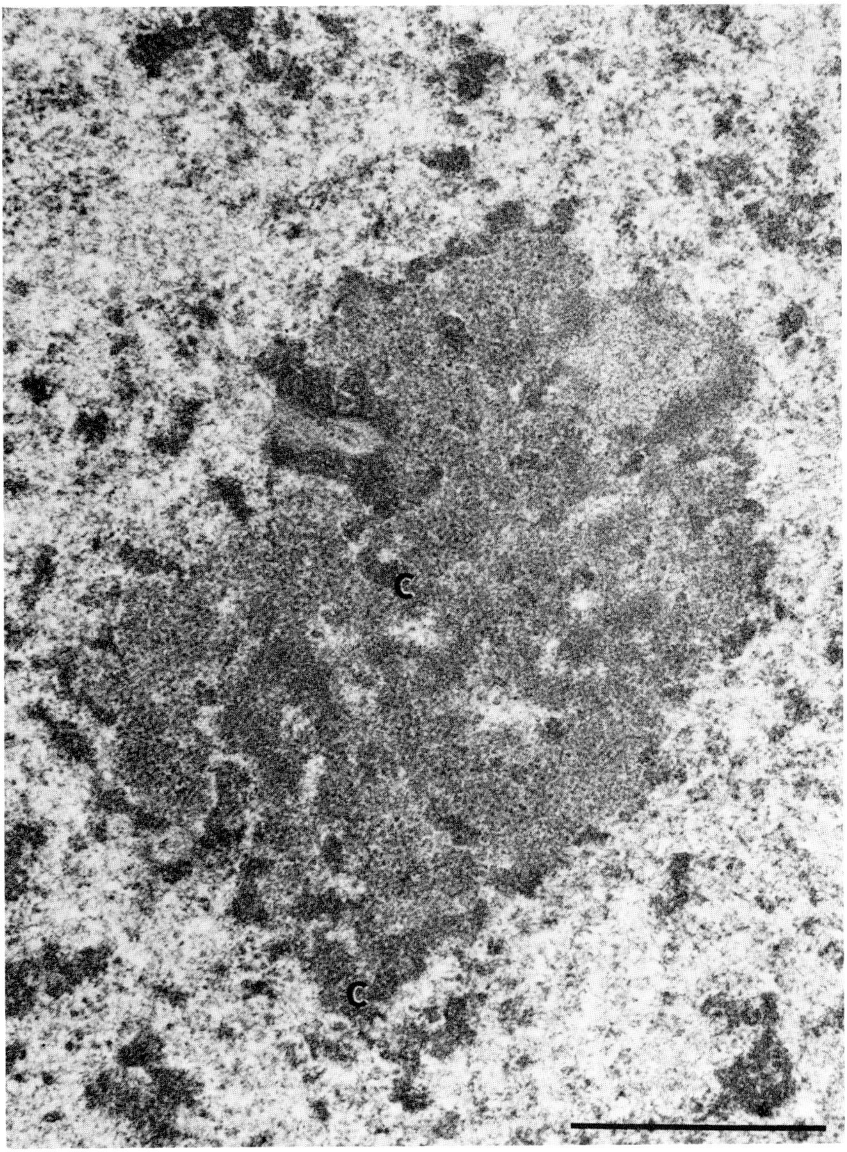

Fig. 2.6. Nucleolus of a Novikoff hepatoma cell, C, chromatin. Fixation was carried out in glutaraldehyde and postfixation in osmium tetroxide (*163*). ×34,000.

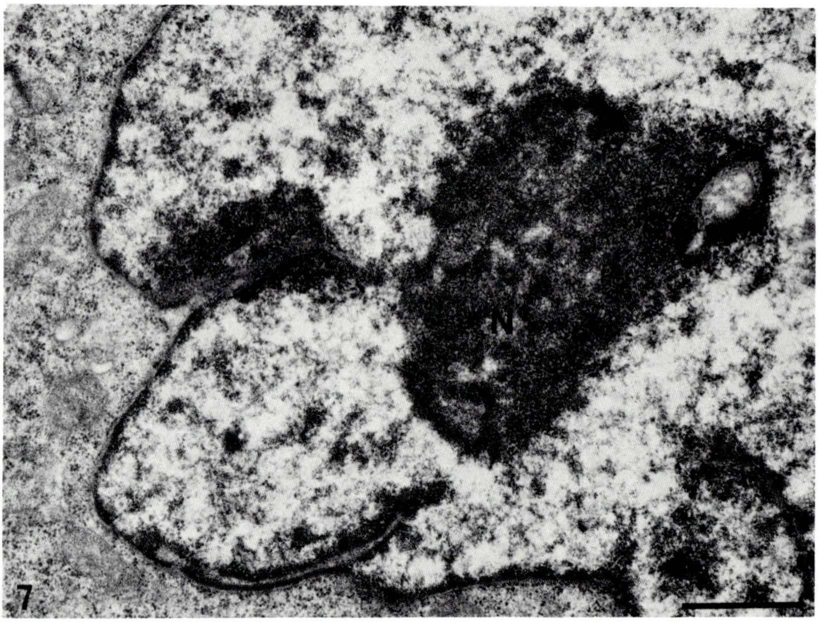

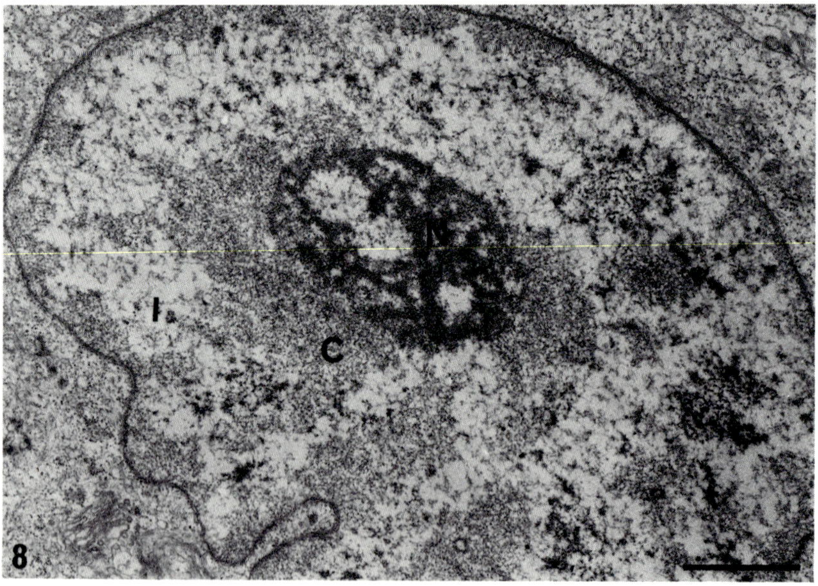

Fig. 2.7. Nucleolus of Novikoff hepatoma cell (N), fixed in formaldehyde and postfixed in osmium tetroxide. ×16,800.

Fig. 2.8. Nucleolus of a Walker tumor cell (fixed in osmium tetroxide and postfixed in formaldehyde) (*132*). N, nucleolus, C, chromatin, I, interchromatinic areas. ×14,500.

embedded in a water-soluble plastic (*11*). Fixation in glutaraldehyde or formaldehyde is recommended for such investigations.

The development of high resolution autoradiography facilitated the isotope incorporation studies in nucleolar structures (*38, 40, 41, 45*). These methods have also been combined with ultrastructural cytochemical studies in ultrathin sections when the specimens were embedded in the water-soluble medium.

The ultrastructural studies of nucleoli verified the majority of light microscopic observations on the morphology of nucleoli and they provided more information on the fine nucleolar morphology (*12, 43, 73*). Classical electron microscopic studies (*9, 13, 17, 51, 110*) have shown that the nucleolus is composed of trabecular structures—nucleolonemas—containing dense particles embedded in a matrix of lower electron density (Figs. 2.5, 2.8). The less well-defined pars amorpha or light spaces was present between nucleolonemas or on the periphery of the nucleolus. Andres (*4*) described the primarily fibrillar structures in nucleoli as component A, the primarily granular structures as the component B, the light areas as the component C, and cisterna-like areas as the component Z.

The dense structures in nucleoli were first described as granules or fibrils. Marinozzi (*85*) and other investigators have shown that both types of structures are present in the nucleolus. High magnification made it possible to distinguish further structures in these structural components of the nucleolus (Fig. 2.9) similar to those in cytoplasmic ribosomes (*54, 123, 132, 133, 173, 174*). The studies on the ultrastructure of different cell types demonstrated the existence of nucleoli without nucleolonemas. Such nucleoli were characterized by the uniform distribution of dense particles (see p. 79) (*11, 12, 14, 52a, 67, 82, 135, 156*) or by the absence of these particles in the central part of the nucleolus (see p. 83) (*31, 72, 88, 92, 135, 138, 172*).

Nucleoli without nucleolonemas were also noted in salivary glands of sciarids (*58*). At a certain stage of development, these nucleoli were composed of a dense, narrow periphery and a less dense central part with a sparser distribution of nucleolar structures. The peripheral part of the nucleolus can, however, be arranged in a reticulum with trabecular structures containing ribonucleoprotein granules (*142*). The absence of nucleolonemas was noted in nucleoli of protozoan *Tetrahymena pyriformis* (*22*).

Several studies with improved methods provided more information on the presence of chromatin structures in nucleoli (*40, 133*). According to these studies, chromatin structures penetrate from the nucleolus-associated chromatin into the nuclear "body" as septa-like structures (Fig. 2.6). Solitary chromatin fibers were also observed in the nucleolonemas (Fig. 2.9) (*106*).

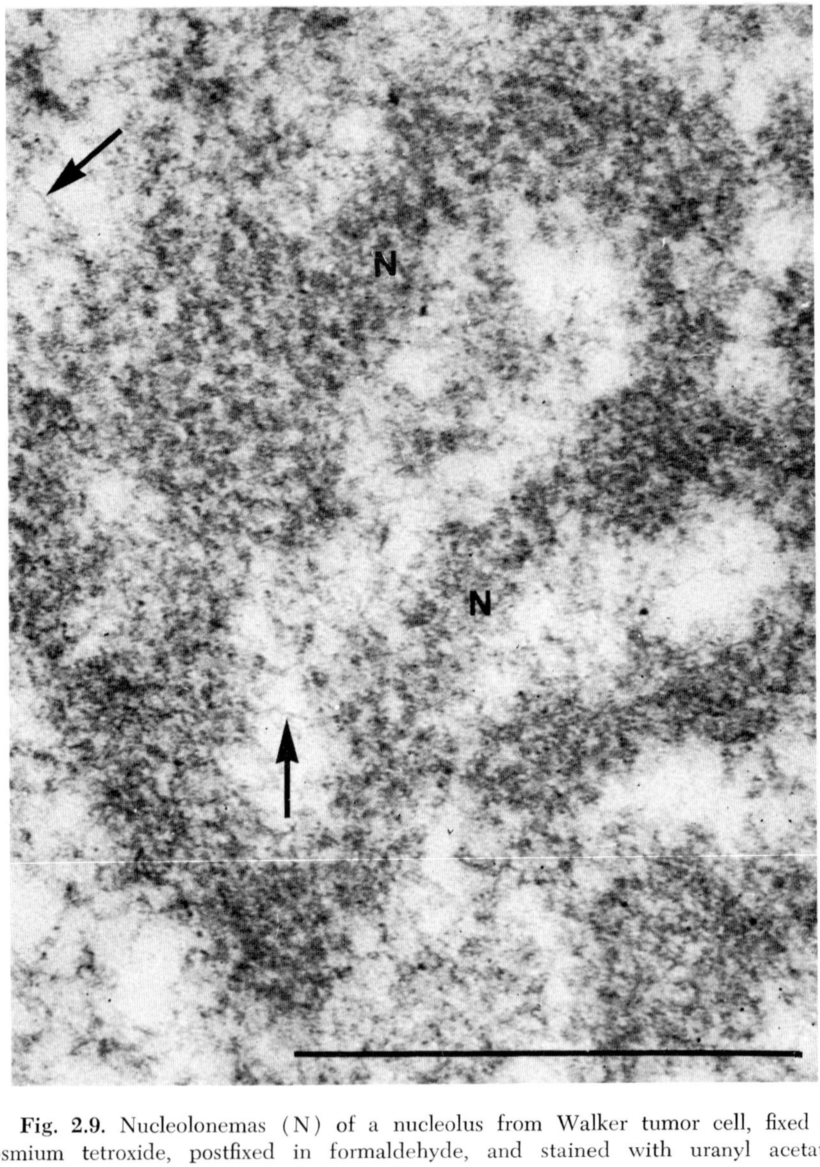

Fig. 2.9. Nucleolonemas (N) of a nucleolus from Walker tumor cell, fixed in osmium tetroxide, postfixed in formaldehyde, and stained with uranyl acetate. Chromatinlike fibrils, arrows. ×70,000.

B. Special Nucleolar Structures

Very unusual structures were noted in epidermal cells of the water planarian *Dugesia tigrina* (69). These structures present in the peripheral part of the nucleolus were composed of a central dense core 300–350 Å in width surrounded by a layer of dense granules 150 Å in diameter. A "nucleolar channel system" (Fig. 1.16) was observed in nucleoli of human endometrial secretory cells (*159*). This system is discussed in more detail in Chapter I (p. 37).

Atypical dense granules (Fig. 2.10a) were noted in nucleoli of some mammalian tumor cells (*122, 165*). Similar granules were observed also in nucleoli of plants (Fig. 2.10d; *73, 146*) and nucleoli of liver cells after partial hepatectomy (Fig. 2.10b) or 24 hours after administration of actinomycin D to the experimental animals (Fig. 2.10c). For demonstration of these granules in nucleoli of liver cells and Walker tumor in rats, it was necessary to fix the specimens in collidine-buffered osmium tetroxide.

C. The General Ultrastructure of the Nucleolus

Numerous studies of different cells have shown the existence of three types of nucleoli according to the distribution of ribonucleoprotein structures: nucleoli with nucleolonemas, compact nucleoli without nucleolonemas and with rather uniform distribution of ribonucleoprotein structures, and nucleoli with ribonucleoprotein structures present only in their peripheries.

D. Nucleoli with Nucleolonemas

Nucleoli found in most cells are characterized by the presence of nucleolonemas (Figs. 2.5–2.8). Nucleolonemas* (Fig. 2.9) are composed of structures containing RNA which are embedded in an amorphous matrix of lower electron density after usual fixation and "staining" procedures. Between the nucleolonemas there are light areas† of different size (Figs. 2.5–2.8).

The structures containing RNA appear to be granules or fibers at certain magnifications (Fig. 2.5). In some cells, the granules are in the

* According to the "Committee on the Nucleolar Nomenclature" (*24a*) nucleolonemas are threadlike elements about 0.1 μ in diameter or larger which are anastomosing and arranged either loosely or in a compact mass.

† Nucleolar light areas of lower density than surrounding structures without limiting membranes can be called vacuoles according to the Committee on the Nucleolar Nomenclature (*24a*). However, some participants of the symposium on the nucleolus objected to this term (*24a*, p. 52). Therefore, in the present volume the term "light areas" is used to express the morphological appearance of regions between nucleolonemas. The term "nucleolar interstitia" (*44*) is also a convenient description for the light areas. The term "nucleolar vacuoles" seems to be more convenient for describing enlarged nucleolar light areas or round light areas in more compact nucleoli as seen in nucleoli in Figs. 2.8 and 2.20.

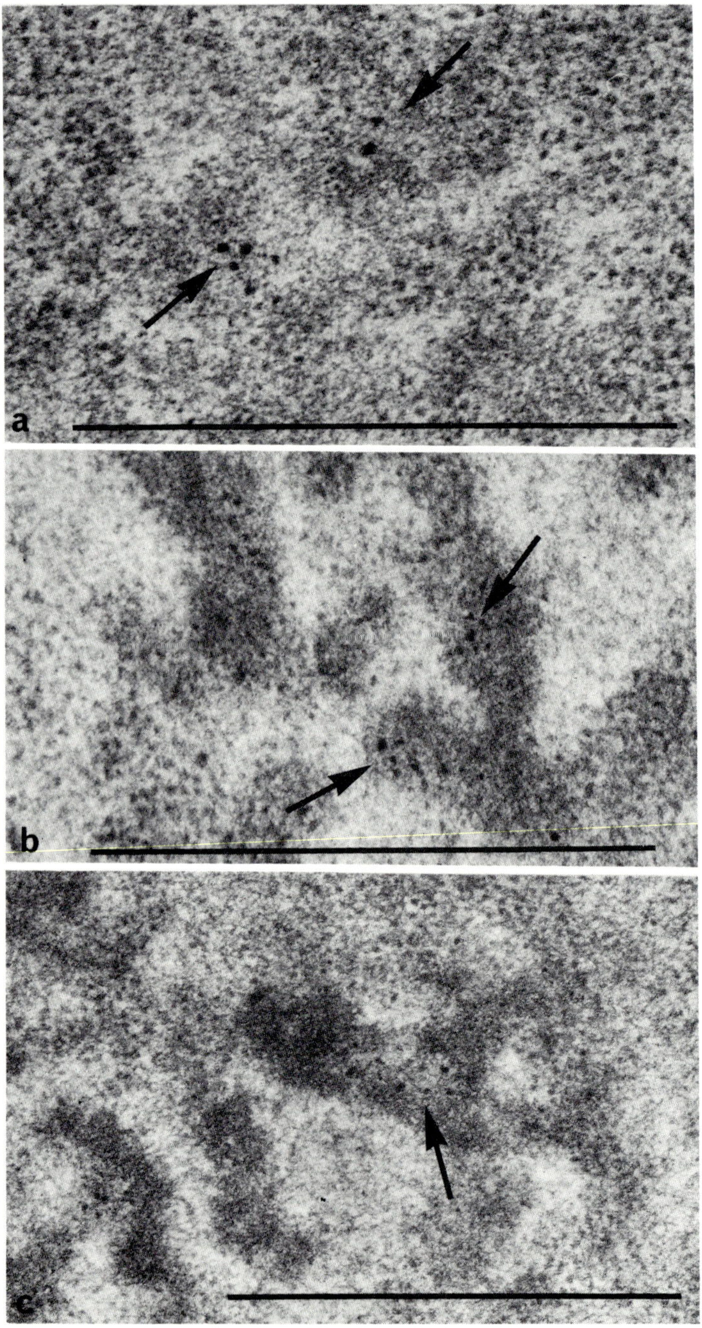

Fig. 2.10. a. Dense granules (arrows) in a nucleolus of Walker tumor cell (*122*),

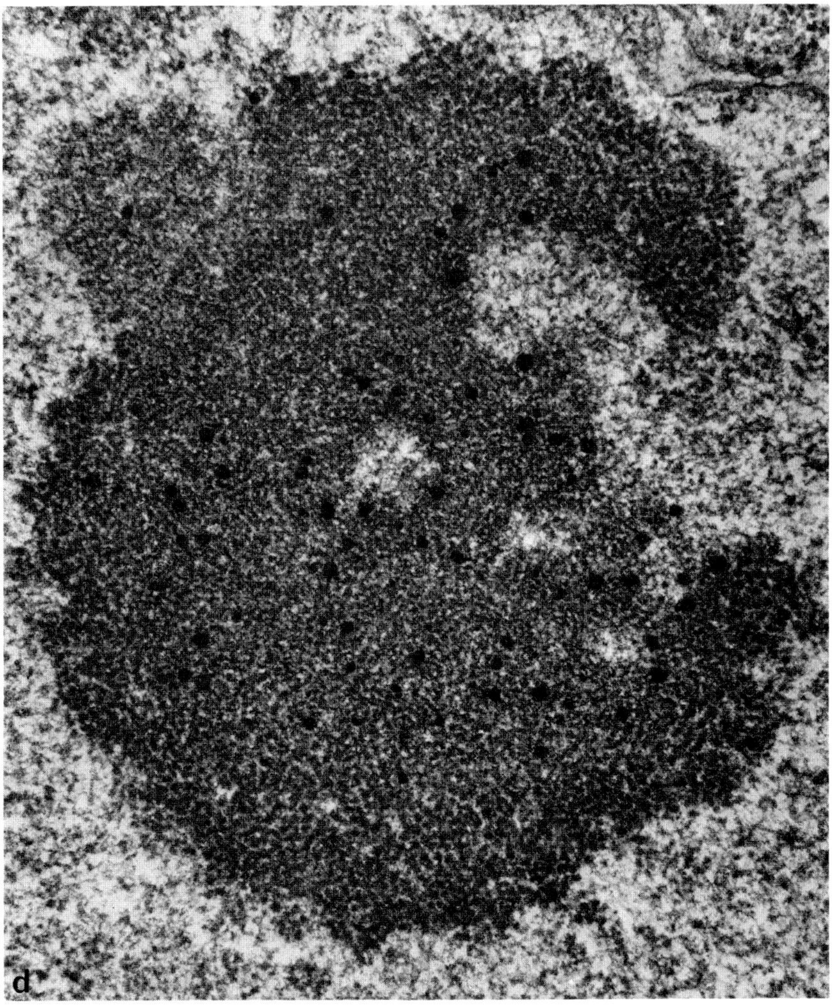

fixed in osmium tetroxide buffered with collidine. ×80,000. b. Dense granules in a cell from regenerating liver (arrows). Same fixation as in Fig. 2.10a. ×75,000. c. Nucleolus of liver cell 24 hours after administration of 150 μg actinomycin D to experimental animals. The fixation was the same as in Fig. 2.10a. Dense granules, arrows. ×60,000. d. Dense granules in nucleoli in the quiescent zone of the root tip of Plantago ovata. ×85,000 (*53a*). (Courtesy of Dr. Beal B. Hyde, Department of Botany, University of Vermont.)

peripheral part of the nucleoli (*9, 85*) while fibrillar structures can be observed in nucleolonemas located in the central regions. In many cells, however, fibrillar structures can also be found in the peripheral nucleolonemas.

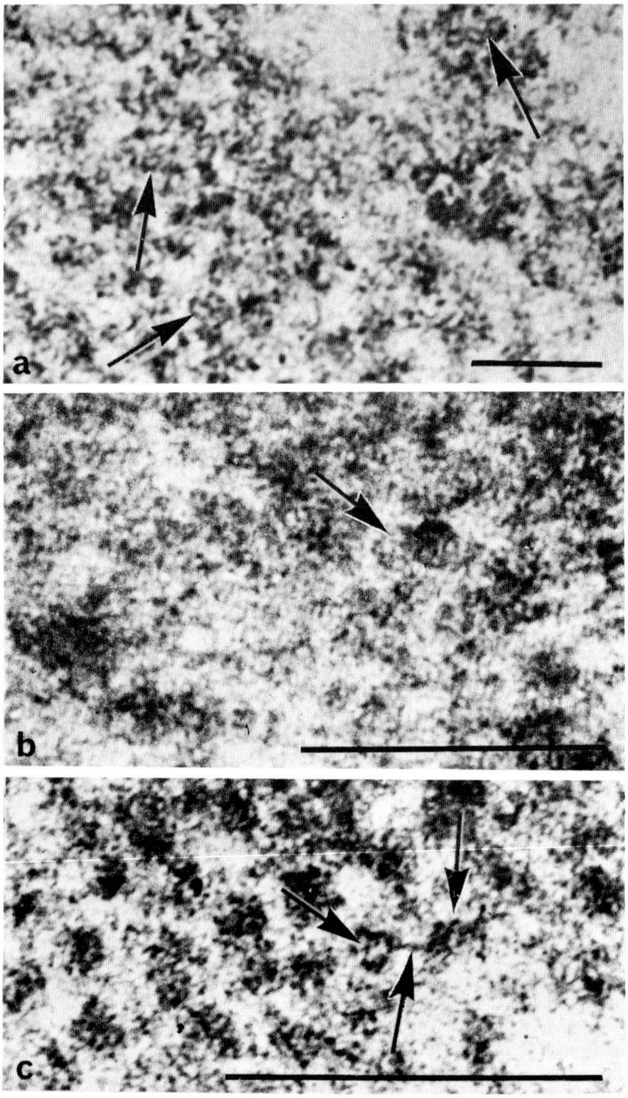

Fig. 2.11. a. Dense nucleolar ribonucleoprotein particles seem to be composed of a dense peripheral part (points) and a central light core. Polysome-like particles, arrow. ×185,000. The black line represents 0.1 μ. b. Dimeric particle (arrow) containing filaments 20 Å wide in a nucleolus of a Novikoff hepatoma cell (*141*). ×400,000. c. Nucleolar ribonucleoprotein particles containing fine filaments 20 Å in width (*141*). Note connection of two ribonucleoprotein particles (points) by a dense filament (arrow). ×500,000. The black line represents 0.1 μ.

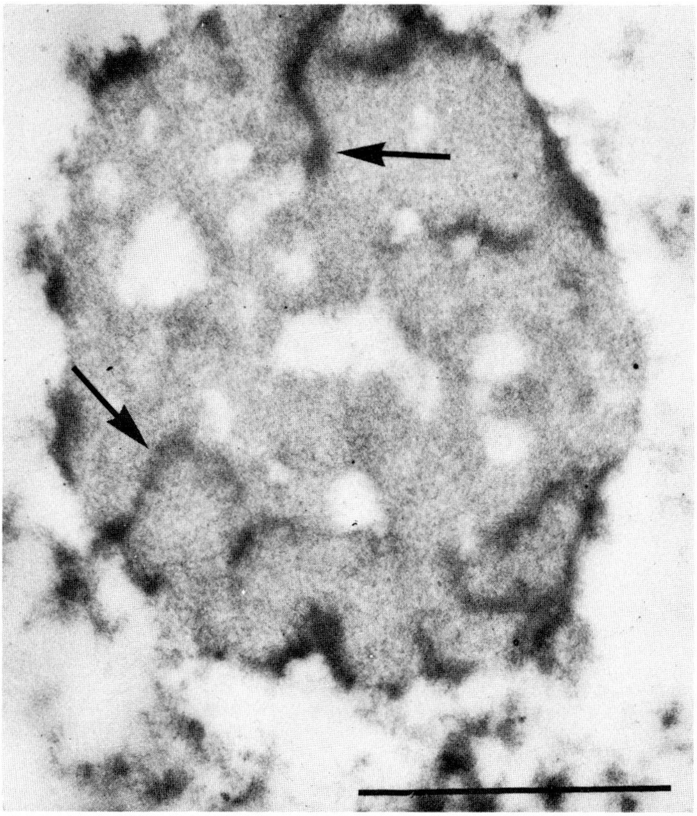

Fig. 2.12. Chromatinlike septal structures (arrows) penetrating from the perinucleolar nucleolus-associated chromatin into the isolated nucleolus of a liver cell. These isolated nucleoli were treated with RNase. ×42,000.

The granules are much like the cytoplasmic ribosomes in regard to their staining properties. Furthermore, dimeric particles and particles joined by filaments 20 Å wide were also found in the nucleolus (Fig. 2.11). These configurations of ribonucleoprotein particles are, however, very rare in the nucleolus. The size of nucleolar granules is about 150 Å in ultrathin sections. Their shape is polymorphic like that of the cytoplasmic ribosomes (*8*). At high magnifications of about 100,000, the granules seem to be composed of dense peripheral structures and central light cores (Fig. 2.11a) (*132, 133*). Therefore, these structures can appear as small vesicles or short tubules in ultrathin sections depending on their position (*65, 132, 133, 173*). This description is, however, only a strictly morphological expression of their appearance, because coiled filamentous structures can have a similar appearance in sections (*132*). At magnifications over 150,000, ribonucleoprotein particles (*71a, 137*) like ribonucleoprotein strands (*54, 123*) contain a thin dense filament 20 Å wide (Fig. 2.11a,c).

The fibrillar regions (Fig. 2.5) contain structures which have a similar affinity to heavy metals as granular components. The width of the fibrillar components ranges from 40 to 80 Å, and the length about 200 to 400 Å. The density of these structures is higher than that of chromatin fibers after osmium tetroxide fixation, while there is less contrast after aldehyde fixation (*8*).

The pars amorpha is less well defined (*8*). Many investigators have avoided the use of this term in the description of nucleoli. The light areas between the nucleolonemas (Fig. 2.5, 2.9) sometimes referred to as "nucleolar interstitia" (*44*), may contain RNA, DNA, and protein in other organized forms (*85, 132, 133*).

Chromatin structures are represented by extremely fine, solitary fibrils (Fig. 2.9) or by bundles and clusters of filaments (*3, 40, 44, 45, 75, 83, 85, 132, 133*). Chromatinic septa-like structures are very prominent in isolated nucleoli fixed in formaldehyde and postfixed in osmium tetroxide (Fig. 2.6, 2.12) (*133*).

E. Compact Nucleoli with Uniform Distributions of RNA-Containing Structures

These nucleoli are characterized by a compact appearance without developed nucleolonemas and by the uniform distribution of structures containing RNA (Fig. 2.13a). The morphology of the structures containing RNA is not different from that of nucleoli with nucleolonemas (Fig. 2.13b).

Nucleoli with uniform distributions of structures containing RNA were described in only a few cell types (*11, 12, 14, 22, 52a, 62, 67, 91, 136, 138, 156*). Under experimental conditions, compact nucleoli without nucleolonemas were observed after administration of thioacetamide and actinomycin D to experimental animals (*124*). In compact nucleoli studied after administration of actinomycin D, segregation of the fibrillar and granular components was usually noted (see Chapter X).

In compact nucleoli without nucleolonemas, persistent nucleoli, pronucleoli, or nucleoluslike bodies found in different cells during mitotic division (*20, 43, 52a, 143*), there is a uniform distribution of dense RNP particles. Practically nothing is known about the distribution of chromatin structures in compact nucleoli without nucleolonemas. However, chromatin structures seem to be dispersed as solitary fibrils in the whole nucleolar body (Fig. 2.14) when nucleoli were digested with ribonuclease.

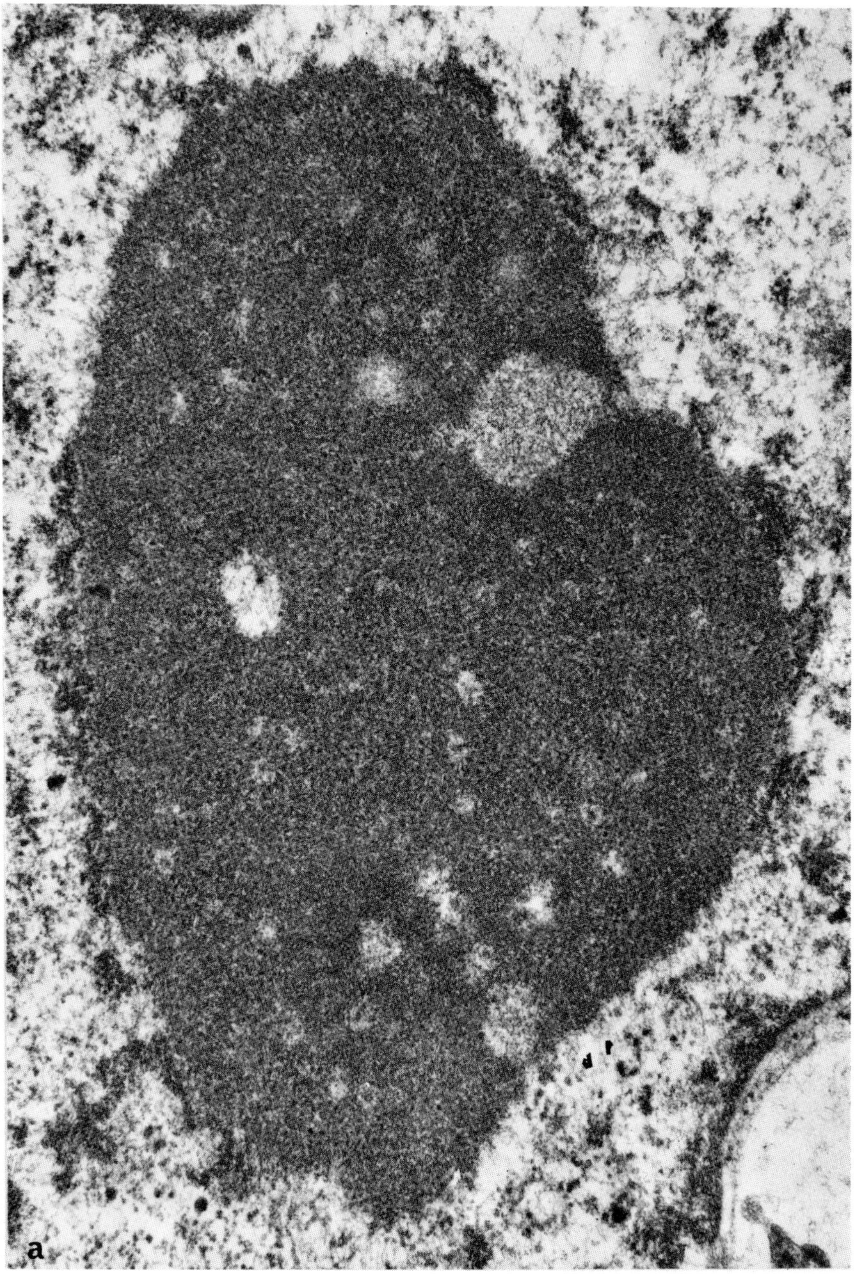

Fig. 2.13a. Compact nucleolus without nucleolonemas in a lymphoblast (*138*). ×45,700.

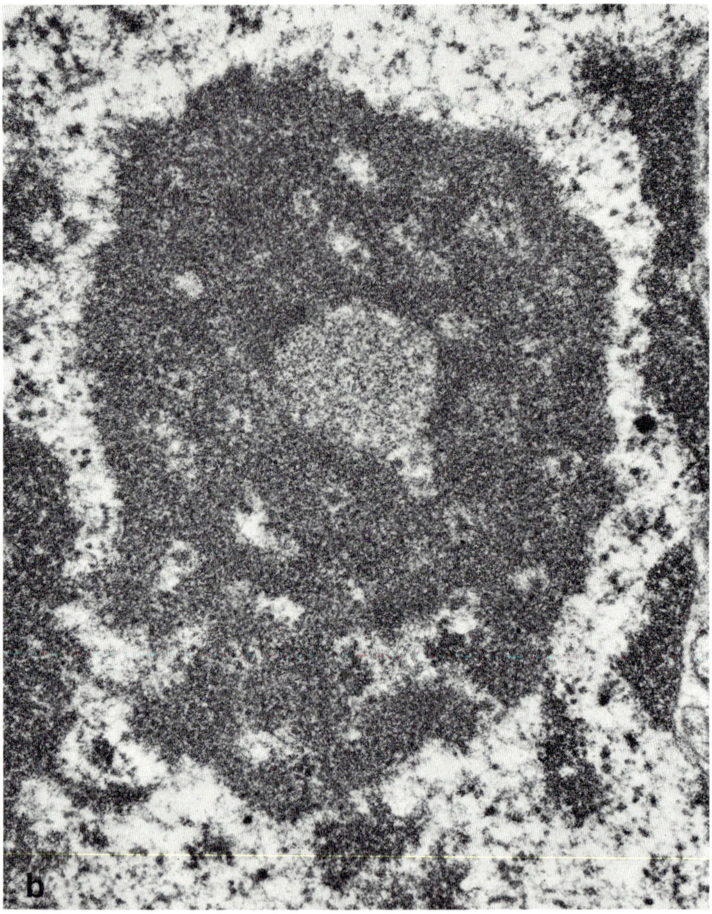

Fig. 2.13b. Transitional nucleolus showing microspherules between nucleolonemas and a central light core; this nucleolus is presumably an intermediate state in the development of nucleolonemas and ring-shaped nucleoli.

F. Ring-Shaped Nucleoli

Ring-shaped nucleoli are characterized by the presence of structures containing RNA only in the peripheral part of the nucleolus forming a shell or ring which determines its appearance (Figs. 2.15–2.17). In the peripheral ring containing ribonucleoprotein structures, residues of nucleolonemas have also been observed (Fig. 2.16) (*138*). The central part of the ring-shaped nucleoli seems to contain largely chromatinlike fibrils. In similar nucleoli of some plant cells, dispersed ribonucleoprotein par-

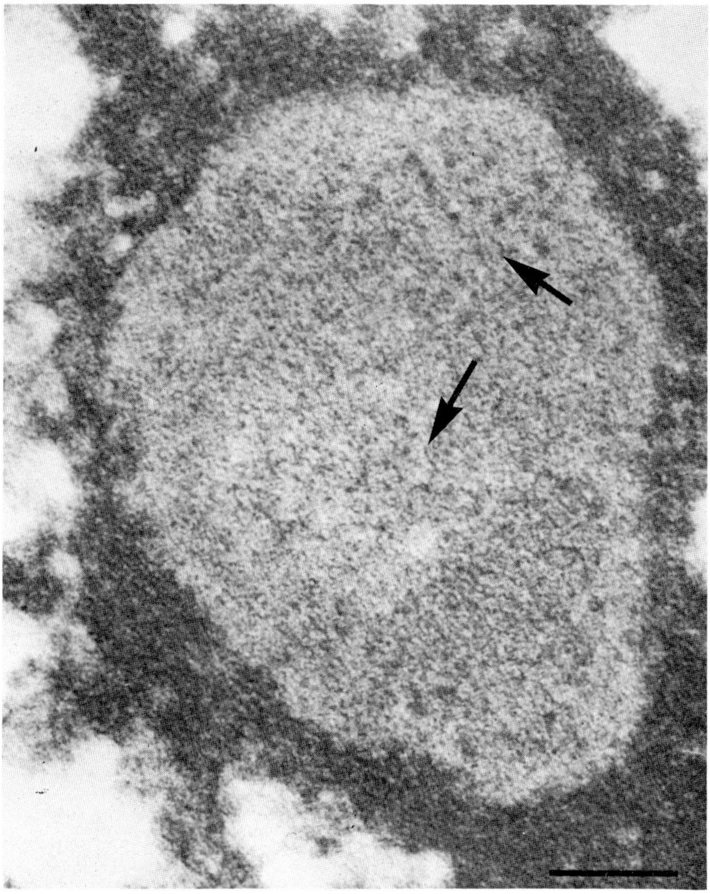

Fig. 2.14. Chromatin fibers in a compact nucleolus of a lymphoblast, fixed in formaldehyde and digested with ribonuclease. The chromatin fibers are those dense fibers which were not digested with ribonuclease. ×165,000. The black line represents 0.1 μ.

ticles were found beside chromatinlike fibrils in the central region of the nucleolus (*53, 74*).

In several studies, electron micrographs have shown the presence of ring-shaped nucleoli in mammalian cells (*82*), but the ultrastructural morphology of these cells was not described. This was probably due to the fact that ultrastructural studies were not correlated with simultaneous demonstration of RNA with light microscopic procedures. In the electron micrographs, ring-shaped nucleoli are sometimes really only cross sections of the peripheral portion of the nucleolus with prominent nucleolonemas and a light area between them. Light microscopic cytochemical

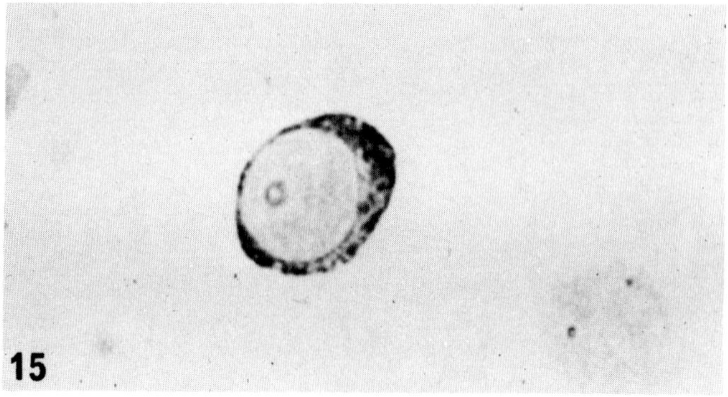

Fig. 2.15. Ring-shaped nucleolus in a leukemic lymphocyte (arrow) (*138*), stained with toluidine blue for RNA. ×2000.

studies (Fig. 2.15), however, have clearly shown the real existence of such nucleoli in which structures containing RNA were present only in the peripheral part of the nucleolus (*31, 32, 131, 138*). These light microscopic studies were also carried out in smear preparations to eliminate the possible error of interpretation due to the position of the nucleolus in the section (*138*).

The ultrastructure of ring-shaped nucleoli was investigated in only a few cell types and the presence of these nucleoli in some blood cells, spermatocytes (*20a*), and oocytes seemed to be interrelated with cell maturation and differentiation. Ring-shaped nucleoli were noted in cells after the administration of actinomycin D *in vitro* and *in vivo* (*64, 104, 139*) or chromomycin A3 *in vitro* (*112*). Starvation inducing a decrease of assimilation had a similar effect on some insect cells in regard to the formation of ring-shaped nucleoli (*172*). The transformation of ring-shaped nucleoli into compact nucleoli was noted in mixed leukocyte cultures or in short-term leukocyte cultures after stimulation of the RNA synthesis with phytohemagglutinin (*111*) or after antigenic stimulation (*137*). All these experiments seem to indicate that ring-shaped nucleoli described in some mammalian cells can reflect the reversible inhibition of the formation of new structures containing RNA. This conclusion is in agreement with some studies in which labeling with ^{3}H uridine was absent over ring-shaped nucleoli or over their central portions (*112, 112a*). Similarly, the incorporation of ^{14}C-labeled amino acids was absent in the central regions of such nucleoli (*55*). Moreover, cytoautoradiographic studies have also shown the cessation of RNA synthesis in mature lymphocytes (*162*) in which ring-shaped nucleoli occurred as the result of cell maturation and differentiation (*138*). Recently, ring-shaped

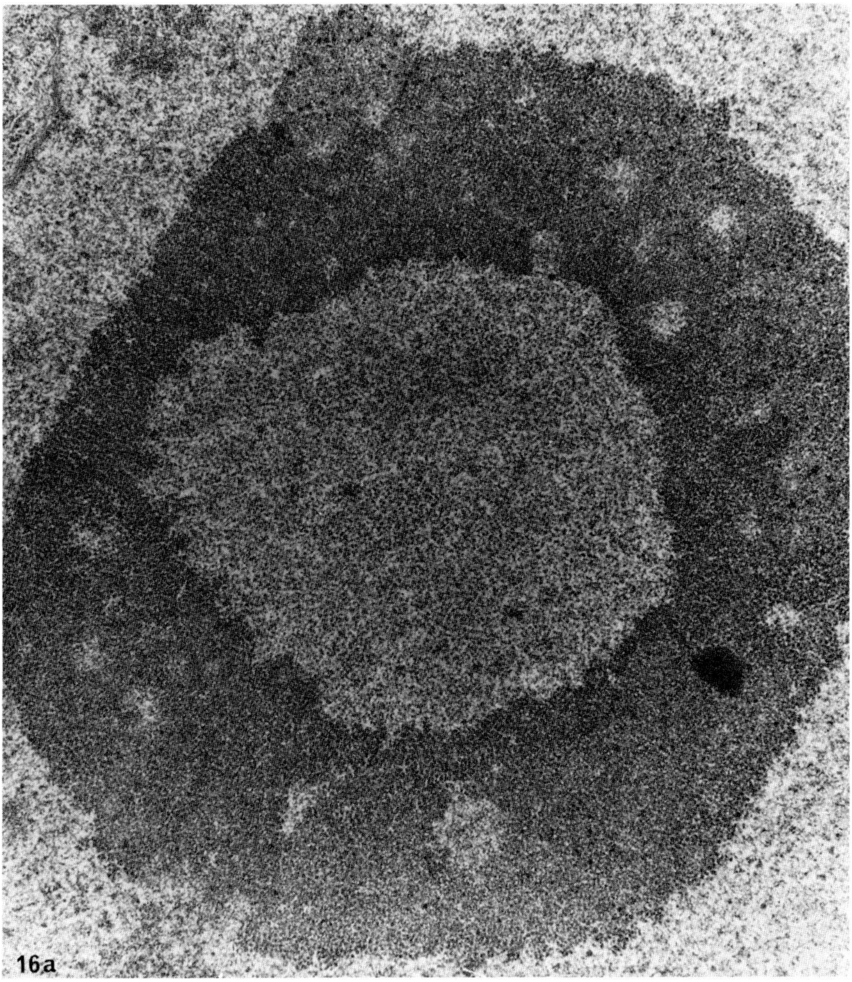

Fig. 2.16a. Nucleolus of a root tip (cortex) cell with large central lacuna containing chromatin structures as well as RNP particles. ×38,000 (*53a*). (Courtesy of Dr. Beal B. Hyde, Department of Botany, University of Vermont.)

nucleoli have been found in mature smooth muscle cells, endothelial cells, and lymphosarcoma cells.

In some cells, the ring-shaped appearance of nucleoli can also be due to the presence and orientation of a persistent chromosome which is surrounded by the ring of the nucleolar cortical material (*72, 108*). A similar nucleolar appearance has been also noted at telophase in Ehrlich ascites tumor cells. However, a structural difference between the outer and inner areas has not been observed (*173, 174*).

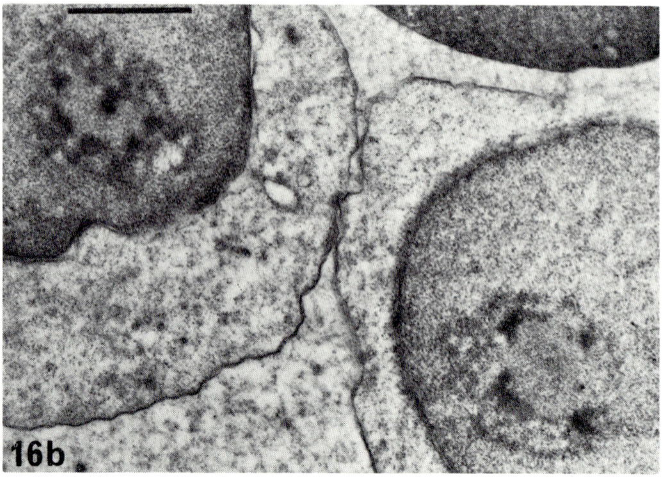

Fig. 2.16b. Ring-shaped nucleoli containing residues of nucleolonemas in leukemic lymphocytes (*138*). ×16,000.

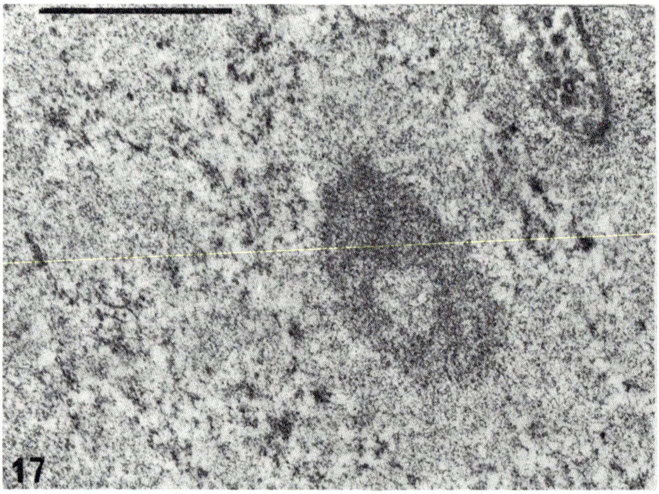

Fig. 2.17. Ring-shaped nucleolus of a leukemic lymphocyte (*137*). ×25,000.

G. RNA-Containing Fibrillar and Granular Components of the Nucleolus

The distribution and amount of the fibrillar and granular components in the nucleolus is very different in different cell types. The fibrillar and granular components are usually present in liver cells in small separated

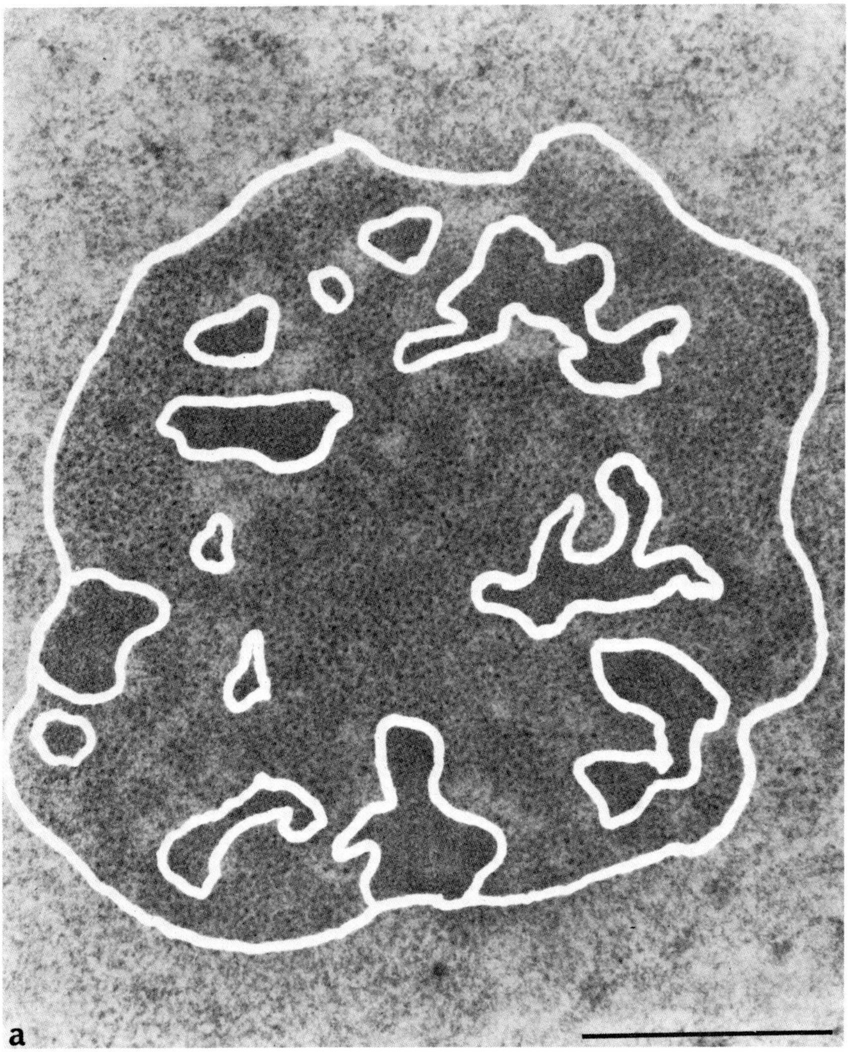

Fig. 2.18a. Nucleolus of liver cell with "maps" of fibrillar components (*140*). ×33,500.

areas (Fig. 2.18a,b). However, in some cells the fibrillar and granular components are mixed and separate areas composed either of granular or fibrillar components cannot be distinguished (*8, 11*). Regions containing both fibrillar and granular structures were described as fibrillogranular areas (*140*). The significance of both types of ribonucleoprotein structures is still a subject of much discussion. The granular com-

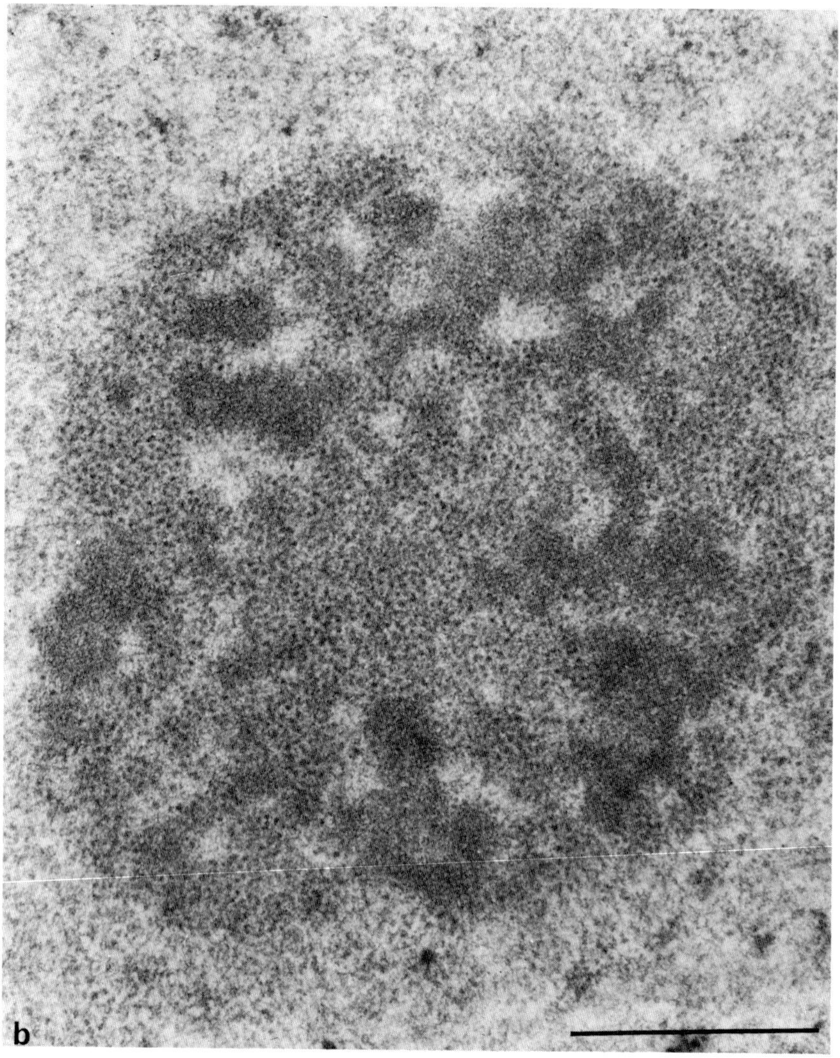

Fig. 2.18b. Same as Fig. 2.18a without mapping of fibrillar components.

ponents, according to the majority of ultrastructural studies including ultrastructural cytochemical observations (*85*) seem to be related to the formation of cytoplasmic ribosomes and may be their precursors. Jones (*62*) supported this concept by observations showing that nucleoli were compact and lacked granular components in cells in which there was no synthesis of ribosomal RNA. Some studies have suggested that the fibrillar components of the nucleolus may represent messenger RNA or that these structures represent precursors of granular components (*38, 85, 86, 142*).

H. Quantitative Analysis of Fibrillar and Granular Areas

To provide more information about the distribution of fibrillar or granular components and their relative amounts in nucleoli, areas containing these components were analyzed in nucleoli of liver cells subjected to different experimental conditions. The size of areas containing these components was measured with the aid of a planimeter (see Fig. 2.18a) and the percentage of the total nucleolar area composed of granular or fibrillar areas was determined (*140*). The reliability of planimetric studies was determined by means of evaluation of results obtained from planimetric analysis of serial sections. Similar ultrastructural studies were correlated with the sucrose density gradient sedimentation analysis of nucleolar RNA (*124*) (Table 2.2). These studies have shown that the fibrillar components comprised about 20% of the total nucleolar area in liver cells of control animals and a small decrease of fibrillar components was noted after stimulation of RNA synthesis in partial hepatectomy (Fig. 2.19). The relative amount of fibrillar components increased after administration of actinomycin D (Fig. 2.20) and decreased after administration of thioacetamide (Fig. 2.21). However, the size of the nucleoli increased considerably after partial hepatectomy or thioacetamide treatment (Chapter I). Corresponding changes were found in the amount of 4–7 S RNA. The nucleolar 4–7 S RNA increased (relatively) after treatment of experimental animals with actinomycin D and decreased in liver cells of animals treated with thioacetamide. When thioacetamide and actinomycin D were administered simultaneously, fibrillar structures were extruded from nucleoli as small spherules (Fig. 2.22a) leaving nucleoli composed practically only of granular components (Fig. 2.22b). In such nucleoli the 4–7 S RNA was almost absent and 28 S RNA was virtually the only species of RNA remaining. These observations provided evidence that 4–7 S RNA is present in the fibrillar components of the nucleolus and 28 S RNA is present in the granular structures (Chapter X).

I. Nucleolar Segregation

The phenomenon of the nucleolar segregation results from the redistribution of nucleolar components in separated areas containing fibrillar

TABLE 2.2

Percentage Composition of Nucleolar RNP Areas (*124*)

Tissue	No. nucleoli measured	Percentage of total nucleolar area containing: Fibrillar components	Percentage of total nucleolar area containing: Granular components	% Granular component area/% fibrillar component area	Percentage total RNA: 4–7 S RNA	Percentage total RNA: 28 S RNA	28 S RNA/ 4–7 S RNA	28 + >28 S RNA/ 4–7 S RNA
Normal liver	10	19	77	4	19	18.5	0.98	3.7
Regenerating liver 15 hours after partial hepatectomy	6	12	84	7				
Liver 1 hour after actinomycin D administration	29	48	50	1	34	36	1.06	1.9
Liver 9 days after thioacetamide administration	6	4.5	91	20	6.9	21	3.0	13
Liver treated with thioacetamide for 9 days, 1 hour after actinomycin D administration	55	<1	>99	∞	5.3	53.5	10.0	18

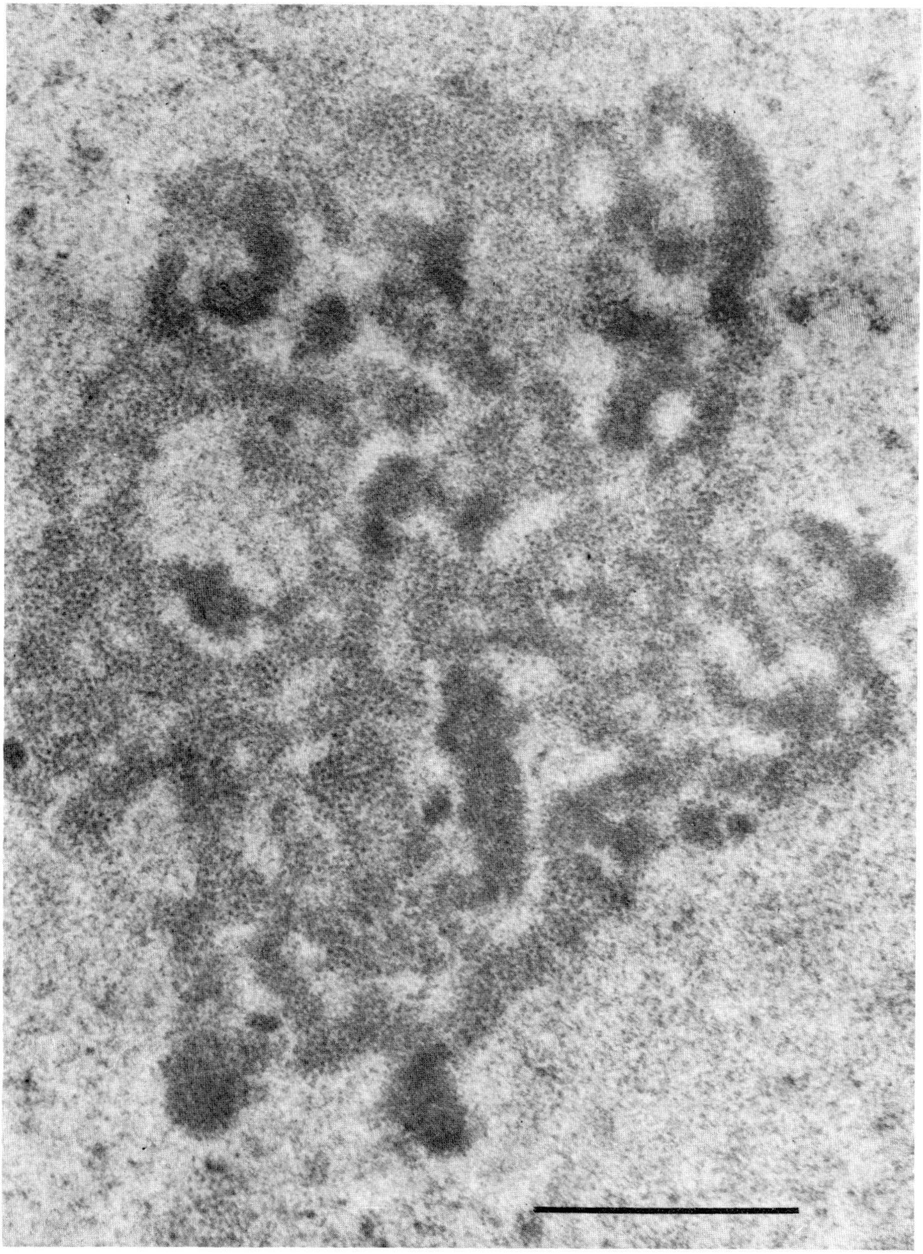

Fig. 2.19. Large nucleolus of a rat liver cell after partial hepatectomy. ×33,500.

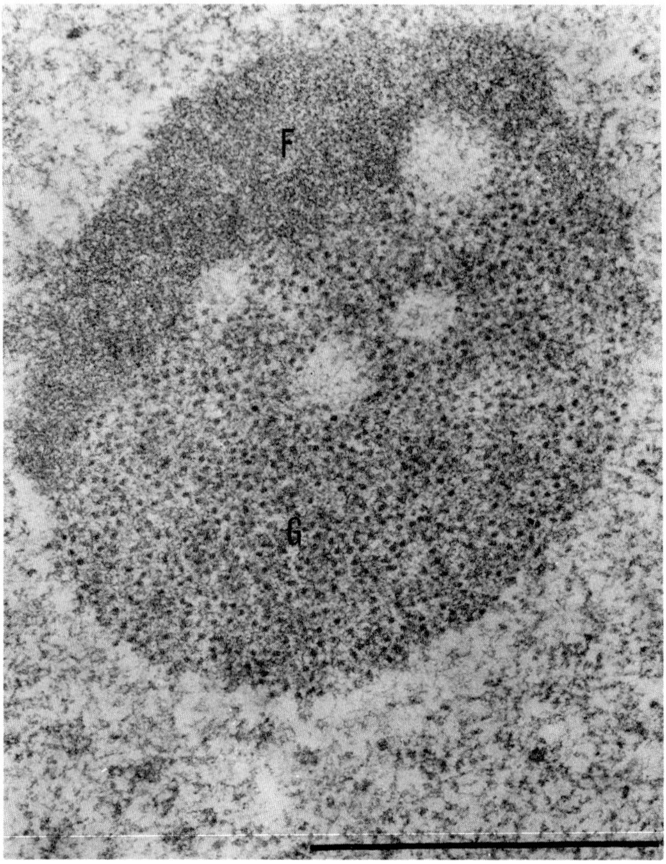

Fig. 2.20. Nucleolus of a rat liver cell after treatment with actinomycin D. Note segregation of granular (G) and fibrillar (F) components (*140*). ×46,000.

or granular ribonucleoprotein filaments or chromatin structures (Fig. 2.20, 2.23). The segregation of nucleolar components can be accompanied by the appearance of new structures of very high density (Fig. 2.23) (*142*) on the periphery of nucleoli or by the formation of caplike structures (*115, 116*). The increased density of nucleolar components, however, can be also due to the necrobiosis or autolysis (*5, 8, 38, 43, 52, 141a*).

The phenomenon of nucleolar segregation was described first after treatment of different cells with a variety of drugs (Chapter X) *in vitro* or *in vivo* (*59, 115, 121*). The changes of the relative size of the areas containing fibrillar or granular components were also quantitatively

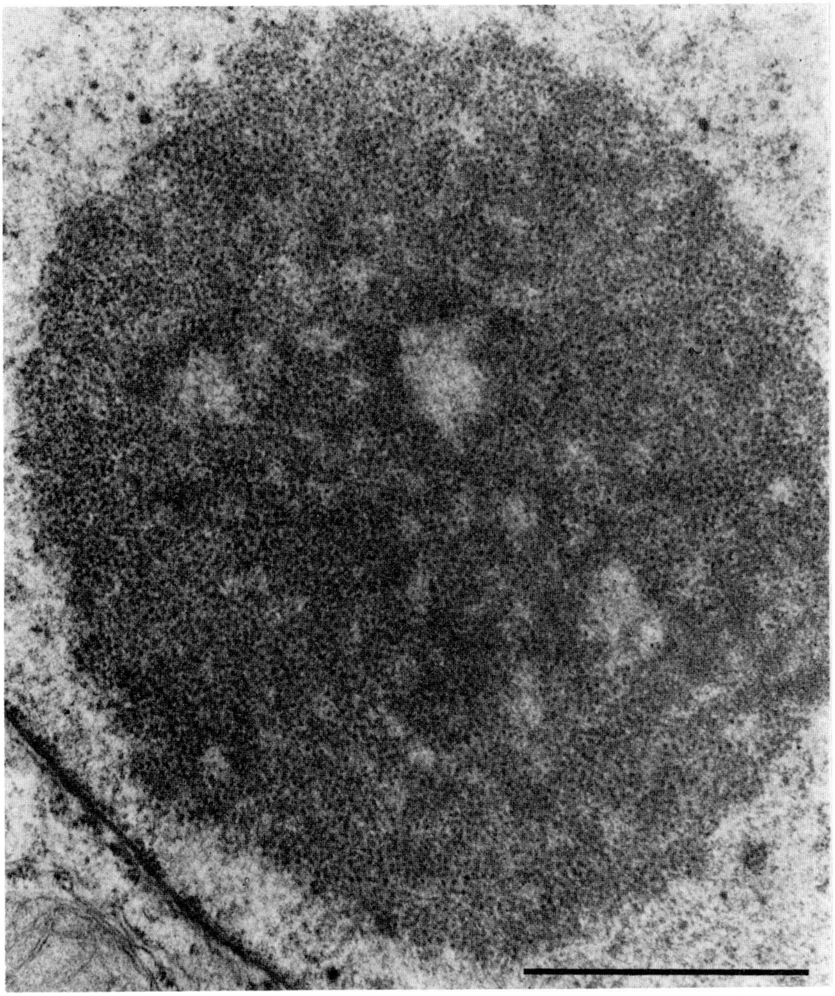

Fig. 2.21. Large nucleolus of a rat liver cell after treatment with thioacetamide (*125*). ×38,000.

determined during this process (*140*). Systematic studies of the effect of different antimetabolites on nucleoli have shown that the segregation phenomena can be induced, by antibiotics, carcinogens, physical factors, and such biological agents as viruses and mycoplasma (*34, 60, 70, 77, 95, 96, 105, 126, 127, 130, 149*). The phenomenon of nucleolar segregation has been described even in nucleoli of liver cells of animals treated with

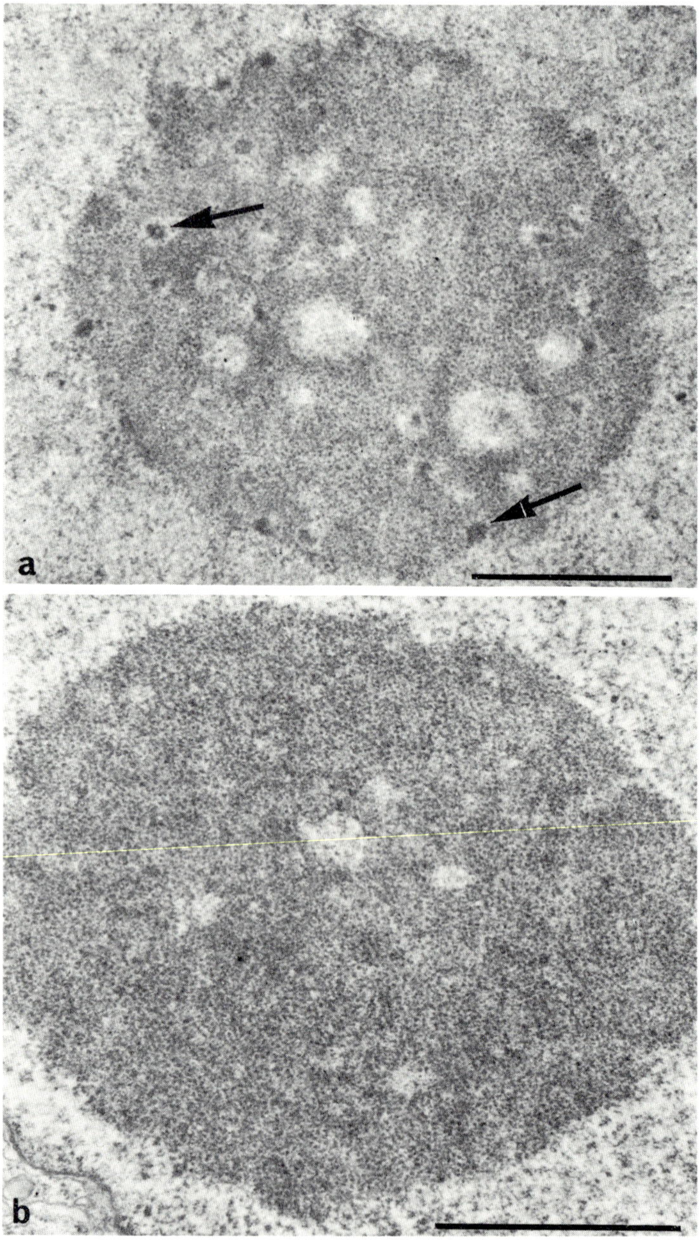

Fig. 2.22. a. Nucleolus of a rat liver cell for 9 days treatment with thioacetamide and then actinomycin D (*124*). Newly formed spherules (arrows) are present. ×27,000. b. Nucleolus of a rat liver cell after prolonged treatment with thioacetamide and then actinomycin D for 90 minutes (*124*). The spherules have moved from the nucleolus to the nucleoplasm. ×32,000.

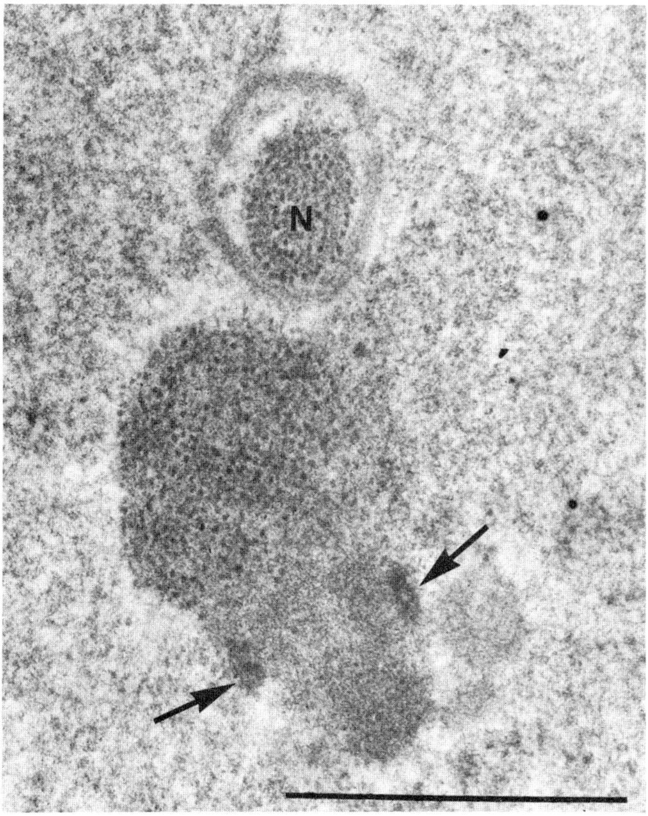

Fig. 2.23. Nucleolus of rat liver cell after treatment with actinomycin D. Note segregation of nucleolar components, formation of nucleoloid body (N), and appearance of dense homogeneous structures on the periphery of the nucleolus (arrows). ×45,000.

thioacetamide and then with actinomycin D or aflatoxin (*10, 124, 148*). The extrusion of segregated fibrillar components from nucleoli of liver cells was also noted in such experiments (*124*) and in Novikoff hepatoma cells after administration of high doses of actinomycin D (*163*). An unusual type of segregation has been described in liver nucleoli of experimental rats treated with tannic acid. This segregation, however, is not related to any significant changes in RNA or protein synthesis and the precise biochemical action of tannic acid on nucleoproteins is not known at present (*114*). Although the mechanism of segregation of nucleolar elements has not been clearly defined, it is now established that in each case there is an inhibition of nuclear RNA polymerase activity (*34*).

J. Ultrastructural Cytochemistry and High Resolution Autoradiography of the Nucleolus

Ultrastructural cytochemical studies of the nucleoli deal mainly with the localization of DNA, RNA, and proteins in nucleolar structures by means of digestion with deoxyribonuclease, ribonuclease, pepsin, trypsin, and pronase (*47, 85, 93, 94, 152*). These digestions were carried out in ultrathin sections or in blocks or cell suspensions after previous fixation and provided practically identical results. Although similar results have been obtained by digestion studies on isolated nucleoli or nuclei without previous fixation (*36, 54, 123, 125, 133*), it should be pointed out that fixation and also the post-treatment of specimens after enzymatic digestion can substantially influence the resulting appearance of the nucleolar structures. For example, treatment of digested samples with cold trichloroacetic acid (*152*) specifically enhances the effect of ribonuclease. A similar finding has been described in studies carried out on ultrathin sections of specimens embedded in water-soluble medium (*85*).

K. Digestion in Ultrathin Sections

According to the studies performed on ultrathin sections (*8, 84, 85*), dense granular components in nucleoli were partially or completely removed by treatment with ribonuclease and disappeared completely after ribonuclease treatment followed by treatment with pepsin. The dense fibrillar components can be removed with ribonuclease alone, and are most sensitive to the action of this enzyme (*85*). The protein matrix of low density in which the fibrillar or granular components are embedded is removable with pepsin. The density of the chromatin structures can be partially or completely removed with pepsin and deoxyribonuclease or trypsin (*8, 11, 85*).

L. Digestion in Fixed Tissue Blocks or Cell Suspensions

When the digestion procedures were carried out in blocks or cell suspensions after short fixation, both dense granular and fibrillar nucleolar ribonucleoprotein components could be removed with ribonuclease (*152*). The disappearance of nucleolar ribonucleoprotein structures after the treatment with ribonuclease is due to the sensitivity of the RNA filaments 20 Å wide present in these structures (*141*).

The studies of nucleoli after the digestion with ribonuclease provided more information on the relative size of areas containing chromatin structures associated with the nucleolus (Fig. 2.24) (*164*). The higher values obtained for the intranucleolar chromatin (Table 4.1) in Novikoff

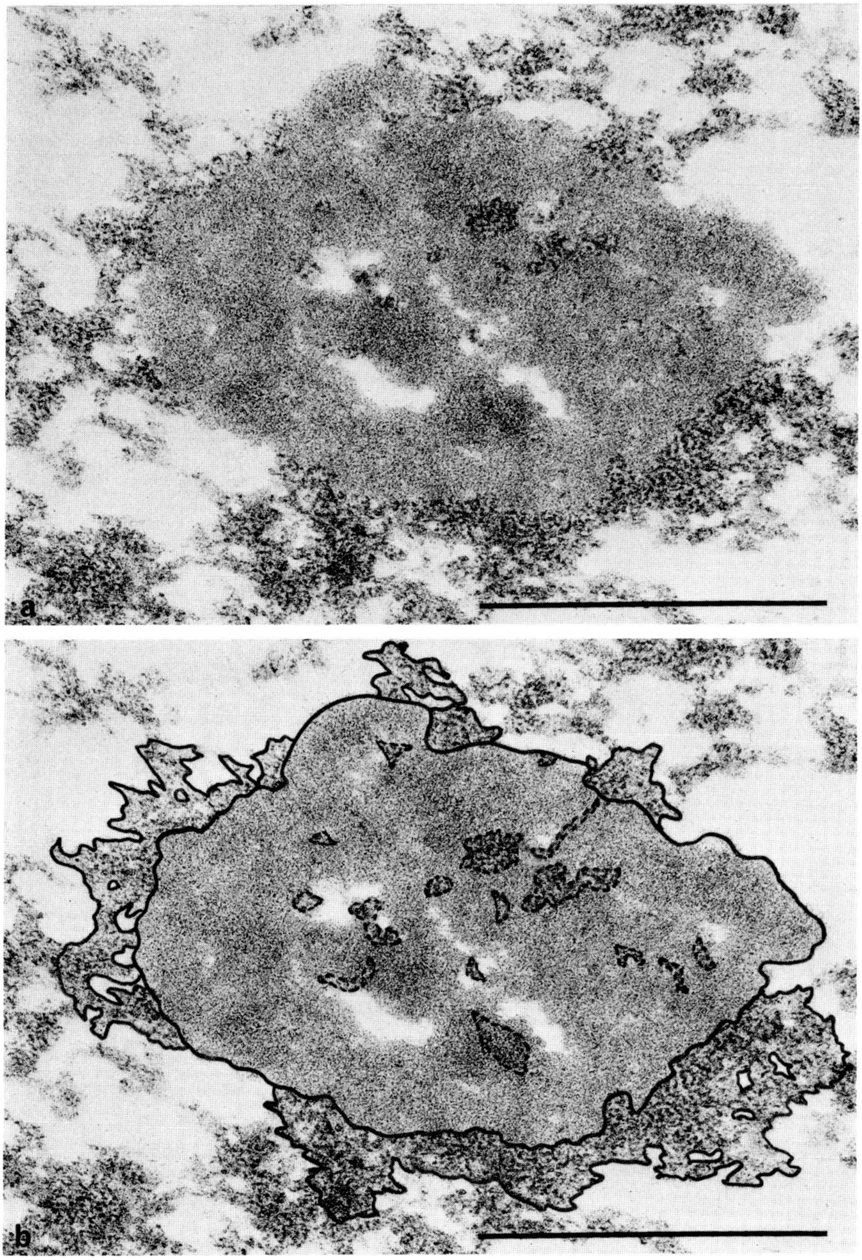

Fig. 2.24. a. Nucleolus of a rat liver cell, fixed in formaldehyde, digested with ribonuclease, and treated with trichloroacetic acid (*164*). ×46,000. b. Same specimen as in Fig. 2.24a, but with maps indicating the relative sizes of chromatin-containing areas (*164*). ×46,000.

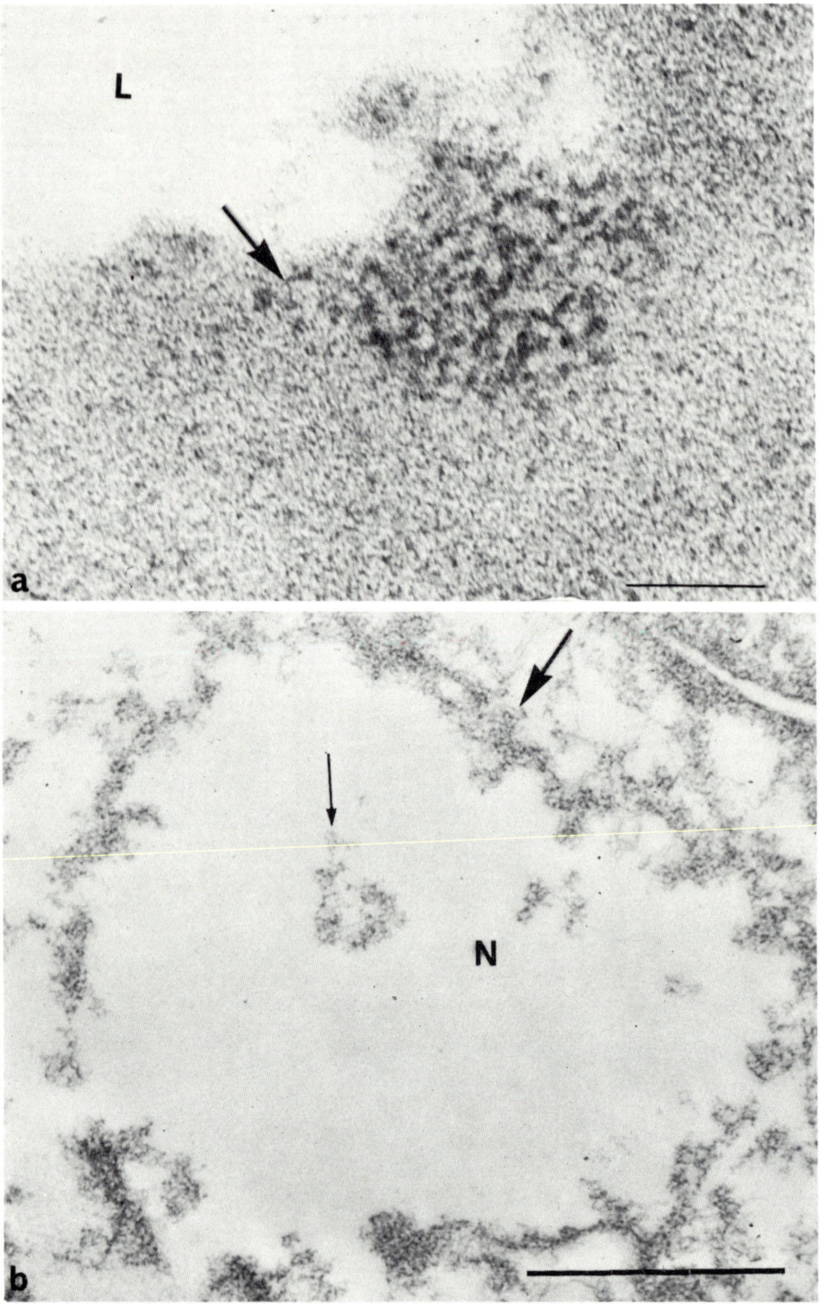

Fig. 2.25. a. Portion of nucleolus of a Novikoff hepatoma cell, fixed in formalde-

hepatoma than in normal rat liver suggested some differences in the function and organization of the intranucleolar chromatin in these two kinds of cells. However, no differences have been found in the ratios of all chromatin areas associated with the nucleolus to the whole nucleolar area between these cell types (Table 4.1). This may indicate that the ratio of the chromatinic areas to the nucleolar body is quite constant, but the organization of these structures is different and depends on the kind of cells observed. It should also be pointed out that solitary chromatin fibrils or filaments were not included in the planimetric measurements. These structures were studied after incomplete digestion with pepsin and complete digestion with ribonuclease (*141*). The chromatin structures are more prominent and the density of the fine filaments of the nucleolar matrix is reduced after such treatment (Fig. 2.25a). The chromatin fibrils 70–100 Å thick are composed of fine coiled filaments 20–25 Å wide (Fig. 2.25a). Uncoiled regions of these filaments were occasionally visible between the coiled regions (Fig. 2.25a). Since the isolated molecules of DNA observed after staining of uranyl acetate (*6, 68*) or precipitates of these molecules in ultrathin sections (*170*) are very similar to the fine chromatin filaments *in situ,* it seems to be likely that these structures represent extended, uncoiled molecules of DNA *in situ.* The presence of uncoiled chromatin filaments about 20 Å wide in chromatin structures associated with the nucleolus can be related to the activity of DNA molecules as a template. This possibility is supported by the fact that the nucleic acid precursors are incorporated into the nucleolar structures first in the regions of nucleolar chromatin (*40, 41, 66*).

Complete digestion with pepsin followed by ribonuclease or with ribonuclease followed by pepsin removed the fibrils of the nucleolar matrix beside ribonucleoprotein structures completely (Fig. 2.25b). After such treatment, the nucleoli appeared as empty spaces within the chromatin structure of the nucleus. The intranucleolar chromatin, however, was preserved in these nucleoli (Fig. 2.25b).

After digestion with pepsin the ribonucleoprotein structures of the nucleolus are more prominent (*47*), since the fine fibrillar background of the nucleolar matrix is removed and the dense RNA filaments present in ribonucleoprotein structures remain intact (Fig. 2.26) (*141*).

The chromatin structures disappeared or their density was reduced after digestion with deoxyribonuclease or double digestion with pepsin

hyde, digested with ribonuclease and pepsin, and post-treated with trichloroacetic acid (*137*). Light area between nucleolonemas, L; connection of coiled and uncoiled part of chromatin filament 20 Å wide, arrow. ×181,000. The black line represents 0.1 μ. b. Nucleolus of a Novikoff hepatoma cell, fixed with formaldehyde, digested with pepsin followed by ribonuclease, and post-treated in trichloroacetic acid. Nucleolus-associated chromatin, point; intranucleolar chromatin, arrow. ×34,500.

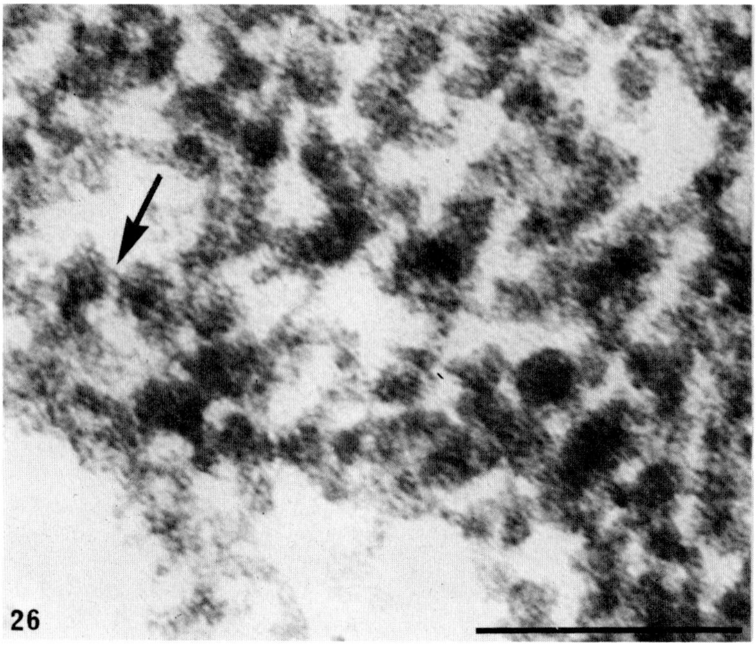

Fig. 2.26. Portion of nucleolus of Novikoff hepatoma cell, fixed with formaldehyde, digested with pepsin, and treated with trichloroacetic acid (*141*). Here the fine filaments 20 Å wide in ribonucleoprotein particles are preserved while the fine filamentous protein matrix of the nucleolus has been digested. Some particles are joined by a dense filament about 20 Å in width (arrow). ×350,000. The black line represents 0.1 μ.

and deoxyribonuclease (*47, 152, 163*). The ribonucleoprotein particles remained intact and dense RNA filaments could be observed in them (Fig. 2.27). Furthermore, the filaments connecting ribonucleoprotein particles were resistant both to treatment with deoxyribonuclease (Fig. 2.27b) and treatment with pepsin (Fig. 2.26). Thus, this filament represents a structure composed of RNA.

M. Extraction and Digestion of Isolated Nucleoli

In isolated nuclei or nucleoli the extraction procedures were used without previous fixation of the samples (*36, 54, 123, 133*). The nucleolar ribonucleoprotein structures were not destroyed by treatment with deoxyribonuclease or with 2 *M* NaCl (Fig. 2.28). The density of the ribonucleoprotein strands was reduced and the ribosome-like particles were not seen after Tris–saline extraction or after Tris–saline extraction followed by 2 *M* NaCl (Fig. 2.29) (*123*). Digestion with ribonuclease alone or after previous sequential extraction with Tris–saline and 2 *M*

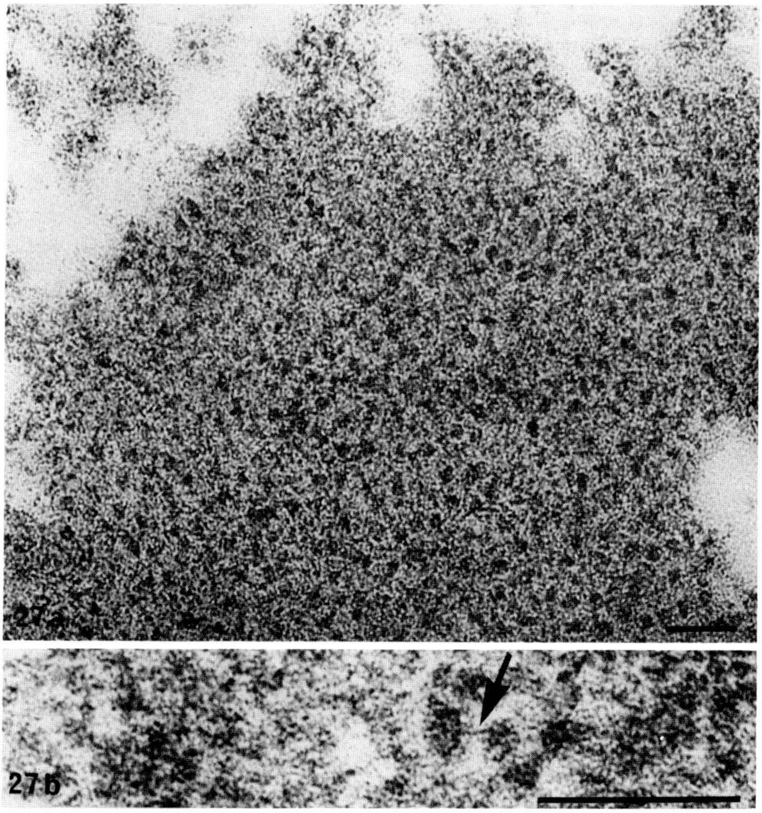

Fig. 2.27. a. Peripheral portion of nucleolus of Novikoff hepatoma cell treated with deoxyribonuclease (*141*). Note that the nucleolus-associated chromatin has disappeared. ×100,000. The black line represents 0.1 μ. b. Same specimen as in Fig. 2.27a. The filaments 20 Å wide in the ribonucleoprotein particles are preserved as well as the filaments connecting some of them (arrow). ×260,000. The black line represents 0.1 μ.

NaCl removed the structures containing RNA completely (*54, 123*) (Fig. 2.30). The treatment of samples with 2 *M* NaCl or sequential extraction with Tris–saline and 2 *M* NaCl or with Tris–saline, 2 *M* NaCl, and deoxyribonuclease removed most of the chromatin structures (Fig. 2.31). Figure 2.31 shows the residual fibrillar network of the nucleus. Similar observations have been made on the nucleoli of *Spirogyra* (*62a*). When isolated nucleoli were digested with deoxyribonuclease alone, thin chromatin fibrils disappeared, but a thin, dense layer remained on the nucleolar surface, particularly if the samples were postfixed in osmium tetroxide at high pH (Fig. 2.32). Since this dense layer could be ex-

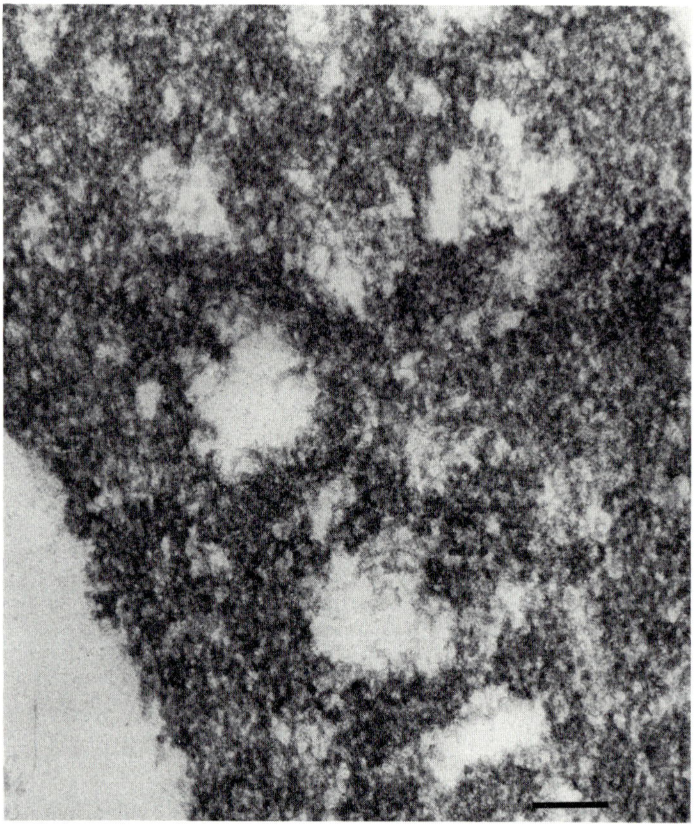

Fig. 2.28. Part of an isolated nucleolus of a Walker tumor cell after extraction with 2 *M* NaCl (*123*). ×97,500. The black line represents 0.1 μ.

tracted with 2 *M* NaCl, its occurrence after digestion with deoxyribonuclease may be due to the presence of histones or other proteins (*133*). The protein matrix of the nucleolonemas disappeared after digestion with pepsin.

N. Cytoautoradiography

Studies on the incorporation of tritiated thymidine in nucleolar structures (*114a*) have shown the presence of this labeled precursor in the nucleolus-associated chromatin and intranucleolar chromatin strips. Some differences in the labeling of different nucleoli were also noted since some nucleoli were not labeled in the same experiments. This finding has been explained by the variations of the nucleolar synthesis of DNA during the mitotic cycle (*8*).

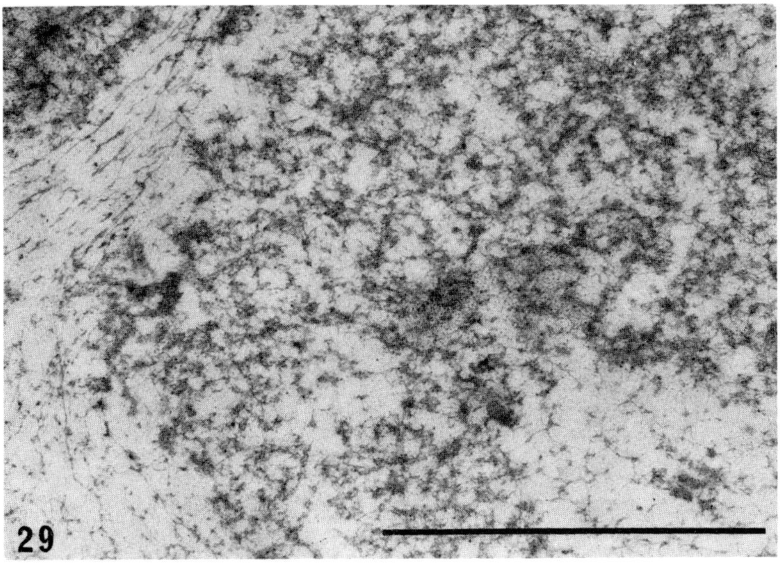

Fig. 2.29. Isolated tumor nucleoli (Walker 256 carcinosarcoma) first extracted with Tris–saline and then with 2 *M* NaCl. ×50,000 (*123*).

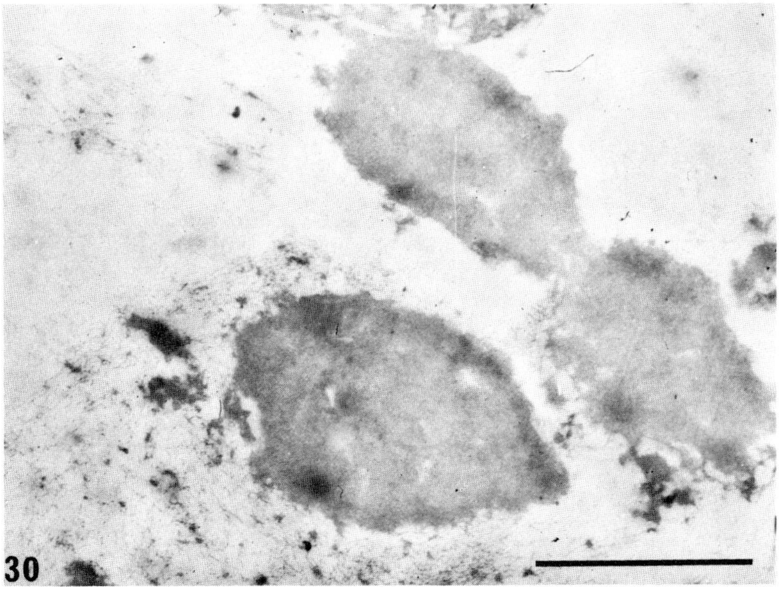

Fig. 2.30. Isolated tumor nucleoli (Walker 256 carcinosarcoma) after sequential extraction with Tris–salinc, and 2 *M* NaCl, and subsequent incubation with RNase. The long, thin fibrils seen also in Fig. 29 may be chromatin fibrils. ×28,000 (*123*).

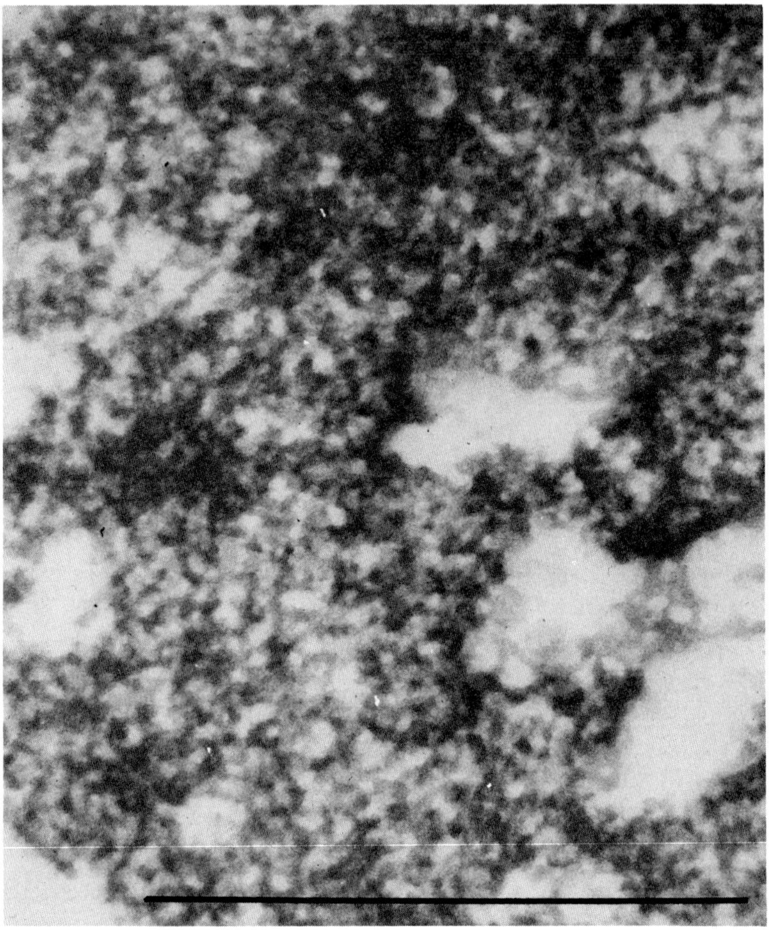

Fig. 2.31. Portion of an isolated tumor nucleolus (Walker 256 carcinosarcoma) after sequential extraction with Tris–saline, and 2 *M* NaCl, and subsequent incubation with DNase. ×80,000 (*123*).

The incorporation of tritiated uridine (*38, 40, 66*) was observed first in the region of the nucleolus-associated chromatin and its intranucleolar strips and fibrillar areas (Figs. 1.4, 1.15). The radioactivity was found in fibrillar regions and some silver grains were present over the areas containing granular components. The dispersed euchromatin areas were labeled at the same time. The whole nucleolar body, including granular components, was labeled later. The labeling of chromatin-containing structures with tritiated uridine can be explained by the synthesis of

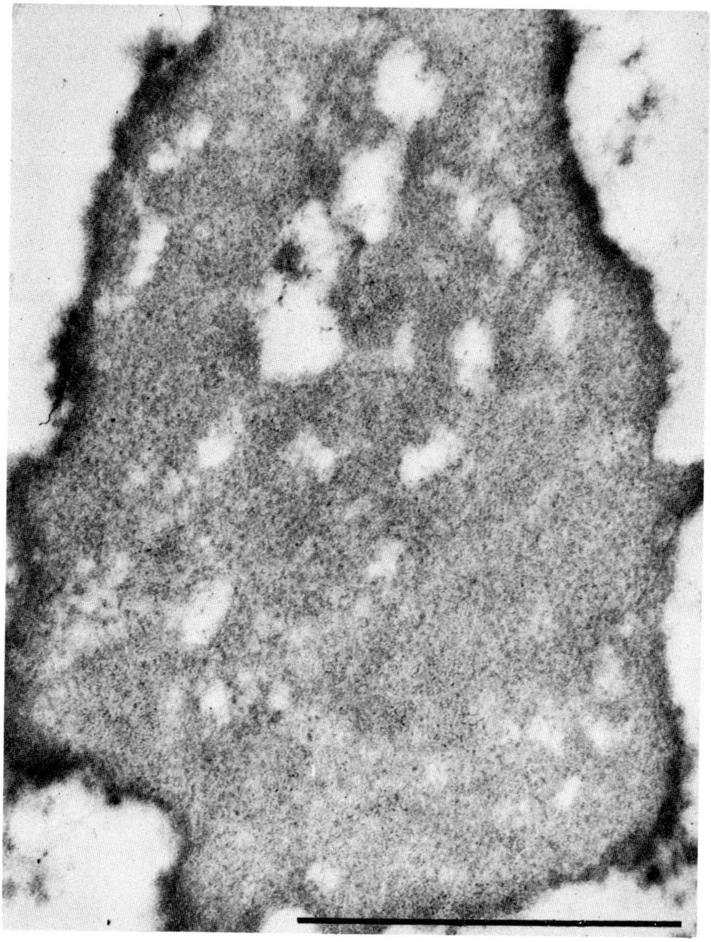

Fig. 2.32. Isolated nucleolus of the Walker tumor treated with deoxyribonuclease prior to fixation. A dense, thin chromatin layer remained on the surface of the nucleolus. ×51,000.

newly formed RNA transcribed on the DNA templates present in these structures. On the basis of differences in the labeling of granular and fibrillar components containing RNA, two alternative possibilities have been postulated. The first possibility is that two types of RNA exist in the nucleolus, one with a high turnover rate and the other with a low turnover rate (Chapters V, VI). The second possibility is that a single type of RNA migrates from the fibrillar to the granular components; this idea is supported by experiments (*38*) which showed that the areas containing fibrillar components contained only a negligible amount or no incorporation of tritiated uridine after inhibition of RNA synthesis with actinomycin D, although the areas containing granular components were still labeled. In the control experiments, the labeling was observed first in the regions containing fibrillar components and later also in areas with the granular ribonucleoprotein structures. Thermic shock induced a 90% reduction in the nucleolar incorporation of tritiated uridine by inhibiting RNA synthesis. The resulting nucleoli were composed only of fibrillar components since the granular components disappeared (*128*). These observations suggested that the granular components were reconverted to fibrillar structures by simple unraveling. The possibility that the ribonucleoprotein granular components had been released in the nucleus or cytoplasm can be excluded because no increased nuclear or cytoplasmic labeling of ribonucleoprotein structures has been observed (*128*).

III. Conclusions

Light microscopic cytochemical studies have provided basic information about the presence of RNA in the nucleolus. The demonstration of DNA in nucleoli in only a limited number of cell types is probably due to the presence of only small amounts of chromatin structures in these nucleoli which are below the limit of detectability by light microscopic procedures. The Feulgen positivity of trabecular structures in nucleoli seems to correspond to the presence of chromatin structures penetrating the nucleolar body as septalike structures from the nucleolus-associated chromatin. The basic proteins appear to have a similar localization, or they are present in structures containing RNA; the acidic proteins are distributed more uniformly in the nucleolus.

The ultrastructural cytochemical studies carried out on nucleoli *in situ* or after isolation have established the presence of nucleic acids and proteins in the nucleolar structures. The RNA is present in granular and fibrillar components of the nucleolus which both appear to be dense after fixation with osmium tetroxide. The DNA is present in chromatin structures that can be found in nucleoli as clumps of fibrils penetrating from nucleolus-associated chromatin into the nucleolus or as solitary fine fibrils. The proteins form a relatively homogeneous fine fibrillar matrix for the structures containing RNA. High resolution autoradiographic studies have shown the incorporation of tritiated precursors of both DNA and RNA into nucleolar structures.

With regard to general ultrastructure three main types of nucleoli have been found in different cells: nucleoli with nucleolonemas, compact nucleoli without nucleolonemas and with relatively uniform distributions of ribonucleoprotein structures, and ring-shaped nucleoli with ribonucleoprotein structures only in their periphery.

The structures containing RNA have a granular or fibrillar appearance at high magnifications. These structures can be present in separated areas or they are mixed in fibrillogranular regions. The relative amounts of fibrillar and granular components seem to depend on the cell type and can be changed by treatment of cells with some inhibitors of RNA synthesis or thioacetamide. The granular components are similar in some respects to the cytoplasmic ribosomes and they probably contain 28 S RNA. Correlative biochemical and ultrastructural studies provided evidence that some 4–7 S RNA is probably present in the fibrillar components. However, some fibrillar components may represent sites of synthesis of precursors for granular components or contain messenger RNA.

The phenomenon of the nucleolar segregation and redistribution of nucleolar components was observed not only after treatment of cells with some antimetabolites *in vitro* or *in vivo*, but also in cells infected with mycoplasma or *Herpes simplex* virus (see Chapter IX). The structures of enhanced density were also noted in nucleoli with segregation of nucleolar components. Extrusion of microspherules containing segregated fibrillar components was observed in liver cells of animals treated with thioacetamide and then with actinomycin D.

REFERENCES

1. Albertini, L., Etude cytophotometrique des proteins —SH nucléolaires de la racine jacinthe; réparation précise de ces composes au sein du nucléoli. *Compt. Rend.* **298,** 3976–3978 (1959).
2. Alfert, M., and Geschwind, T. T., A selective staining method for the basic proteins of cell nuclei. *Proc. Natl. Acad. Sci. U. S.* **139,** 991–999 (1953).
3. Altmann, H. A., Stöcker, E., and Thoenes, W., Uber chromatin und DNS synthese im nucleolus. *Z. Zellforsch. Mikroskop. Anat.* **59,** 116–133 (1963).
4. Andres, K. H., Untersuchungen über die Feinbau von spinal Ganglien. *Z. Zellforsch. Mikroskop. Anat.* **55,** 1–48 (1961).
5. Arcusoy, M., Smuckler, E. A., and Benditt, E. P., Unpublished data cited by Smuckler and Benditt (*141a*).
6. Beer, M., and Zobel, C. R., Electron strains. II. Electron microscopic studies on the visibility of stained DNA molecules. *J. Mol. Biol.* **3,** 717–726 (1961).
7. Bernhard, W., Inclusion de tissue en présence d'eau. *Proc. 3rd Reg. Conf. (Eur.) Electron Microscopy, Prague, 1964* Vol. B., pp. 9–10. Publ. House Czech. Acad. Sci., Prague, 1964.
8. Bernhard, W., Ultrastructural aspects of the normal and pathological nucleolus in mammalian cells. *Natl. Cancer Inst. Mongraph.* **23,** 913–938 (1966).
9. Bernhard, W., Bauer, A., Gropp, A., Haguenau, F., and Oberling, C., L'ultrastructure de nucléole de cellules normales et cancereuses. *Exptl. Cell Res.* **9,** 88–100 (1955).

10. Bernhard, W., Frayssinet, C., Lafarge, C., and LeBreton, E., Lesions nucléolaires précoces provoqué par l'aflatoxine dans le cellules hépatiques du rat. *Compt. Rend.* **261**, 1785–1788 (1965).
11. Bernhard, W., and Granboulan, N., The fine structure of the cancer nucleus. *Exptl. Cell Res.* Suppl. 9, 19–53 (1963).
12. Bernhard, W., and Granboulan, N., Electron microscopy of the nucleolus in vertebrate cells. *In* "The Nucleus" (A. J. Dalton and F. Haguenau, eds.), pp. 81–149. Academic Press, New York, 1968.
13. Bernhard, W., Haguenau, F., and Oberling, C., L'ultrastructure du nucléole de quelques cellules animales, révélée par le microscope électronique. *Experientia* **8**, 58–59 (1952).
14. Bessis, M., and Thiery, J. P., Etudes au microscope electroniquedes hémosarcomes humains. *Nouvelle Rev. Franc. Hematol.* **2**, 577–601 (1962).
15. Bloch, D., and Hew, H. C., Changes in nuclear histones during fertilization and early embryonic development in pulmonate snail. *J. Biophys. Biochem. Cytol.* **8**, 69–81 (1960).
16. Bolund, L., and Ringertz, N. R., Intracellular distribution of histone-like proteins in *Tetrahymena pyriformis*. *Exptl. Cell Res.* **44**, 606–613 (1966).
17. Borysko, E., and Bang, F. B., Structure of the nucleolus as revealed by the electron microscope. *Bull. Johns Hopkins Hosp.* **89**, 468–470 (1951).
18. Brachet, J., La détection histochimique des acides pentose nucléiques. *Compt. Rend. Soc. Biol.* **133**, 88–89 (1940).
19. Brachet, J., La localisation des acides pentose nucléiques dans les tissues animales et les oeufs d'amphibiens en voie de développement. *Arch. Biol.* (*Liege*) **53**, 207–257 (1942).
20. Brinkley, B. R., The fine structure of the nucleolus in mitotic divisions of Chinese hamster cells *in vitro*. *J. Cell Biol.* **27**, 411–422 (1965).
20a. Brinkley, B. R., Personal communication (1969).
21. Busch, H., Byvoet, P., and Smetana, K., The nucleolus of the cancer cell. A review. *Cancer Res.* **23**, 313–339 (1963).
22. Cameron, T., and Guile, E. E., Nucleolar and biochemical changes during unbalanced growth of *Tetrahymena pyriformis*. *J. Cell Biol.* **26**, 848–855 (1967).
23. Caspersson, T. O., "Cell Growth and Cell Function. A Cytochemical Study." Norton, New York, 1950.
24. Caspersson, T. O., and Schultz, J., Ribonucleic acids in both nucleus and cytoplasm and the function of the nucleolus. *Proc. Natl. Acad. Sci. U. S.* **26**, 507–515 (1940).
24a. Committee on the Nucleolar Nomenclature, *Natl. Cancer Inst. Monograph.* **23**, 573–574 (1966).
25. Cowden, R. R., Cytochemical studies of oocyte growth in the lancelet Branchiostoma caribaeum. *Z. Zellforsch. Mikroskop. Anat.* **60**, 399–408 (1963).
26. Dalton, A. J., and Haguenau, F., eds., "Ultrastructure in Biological Systems," Vol. 3, The Nucleus. Academic Press, New York, 1968.
27. Davis, J. M., The ultrastructure of the mammalian nucleolus. *In* "The Cell Nucleus" (J. S. Mitchell, ed.), pp. 3–14. Butterworth, London and Washington, D. C., 1960.
28. De, P., Chatterjee, R., and Mitra, S., Nucleolar localization of succinic dehydrogenase in normal rat hepatic cells. *Nature* **191**, 925 (1961).
29. De, P., Chatterjee, R., and Mitra, S., Nucleolar succinic dehydrogenase in mouse mammary carcinoma cells. *Nature* **192**, 286–287 (1961).

30. De Jong, D. W., Olson, A. C., and Jansen, E. F., Glutaraldehyde activation of nuclear acid phosphatase in cultured plant cells. *Science* **155,** 1672–1674 (1967).
31. Dutta, C., Siegesmund, K. A., and Fox, C. A., Light and electron microscope observation on an intranucleolar body in nerve cells. *J. Ultrastruct. Res.* **8,** 542–551 (1963).
32. Esper, H., Studies on the nucleolus vacuole in the oogenesis of *Arbacia punctulata. Exptl. Cell Res.* **38,** 85–96 (1965).
33. Estable, C., and Sotelo, J. R., Una nueva estrutura cellulare: El nucleolonema. *Publ. Inst. Invest. Sci. Biol.* **1,** 105–126 (1951).
34. Floyd, L., Unuma, T., and Busch, H., Effects of aflatoxin and other carcinogens upon nucleolar RNA of various tissues in the rat. *Exptl. Cell Res.* **51,** 423–438 (1968).
35. Fujii, T., Presence of zinc in nucleoli and its possible role in mitosis. *Nature* **174,** 1108 (1954).
36. Georgiev, G. P., and Chentsov, J. S., On the structural organization of nucleolo-chromosomal ribonucleoproteins. *Exptl. Cell Res.* **27,** 570–573 (1962).
37. Gersch, M., Untersuchungen über die Bedeutung der Nucleolen in Zellkern. *Z. Zellforsch. Mikroskop. Anat.* **30,** 484–528 (1940).
38. Geuskens, M., and Bernhard, W., Cytochimie ultrastructurale du nucléole. III. Axiom de l'actinomycine sur le métabolisme du RNA nucléolaire. *Exptl. Cell Res.* **49,** 579–598 (1966).
39. Golarz, M. N., and Bourne, G. H., Induction and accentuation of phosphatase activity in the nucleoli of muscle nuclei by denervation and injected nucleotides. *Exptl. Cell Res.* **25,** 691–693 (1961).
40. Granboulan, N., and Granboulan, P., Cytochimie ultrastructurale du nucleole. *Exptl. Cell Res.* **34,** 71–87 (1964).
41. Granboulan, N., and Granboulan, P., Cytochimie ultrastructurale du nucleole. *Exptl. Cell Res.* **38,** 604–619 (1965).
41a. Gwatkin, R. B. L., and Love, R., Nucleolini of mouse trophoblast cells *in vitro. Exptl. Cell Res.* **53,** 686–690 (1968).
42. Györkey, F., Min, K. W., Huff, J. A., and Györkey, P., Zinc and magnesium in human prostate gland, normal, hyperplastic and neoplastic. *Cancer Res.* **27,** 1348–1353 (1967).
43. Hay, E. D., Structure and function of the nucleolus in developing cells. *In* "The Nucleus" (A. T. Dalton and F. Haguenau, eds.), pp. 1–79. Academic Press, New York, 1968.
44. Hay, E. D., and Revel, J. P., The DNA component of the nucleolus, studied in autoradiographs viewed with the electron microscope. *Proc. 5th Intern. Congr. Electron Microscopy, Philadelphia, 1962* Vol. II, pp. 1–7. Academic Press, New York, 1962.
45. Hay, E. D., and Revel, J. P., The fine structure of the DNP component of the nucleus. An electron microscopic study utilizing autoradiography to localize DNA synthesis. *J. Cell Biol.* **16,** 29–51 (1963).
46. Heidenhein, M., "Plasma und Zelle." Fischer, Jena, 1907.
47. Heine, U., Langlois, A. J., and Beard, W. J., Ultrastructural alteration in avian leukemic myeloblasts exposed to actinomycin D *in vitro. Cancer Res.* **26,** 1847–1858 (1966).
48. Hertl, M., Zum nucleolus problem. *Z. Zellforsch. Mikroskop. Anat.* **46,** 18–51 (1957).

48a. Hertwig, G., Allgemeine mikroskopische Anatomie der lebenden masse. *In* "Handbuch der mikroskopischen Anatomie des Menchen" (W. Bargmann, ed.), Vol. 1, Part 1, pp. 1–420. Springer, Berlin, 1929.

49. Himes, M., A quantitative difference in ribonucleoprotein of the nucleolus and cytoplasm. *J. Histochem. Cytochem.* **6**, 392 (1958).

50. Horn, E. C., and Ward, C. L., The localization of basic proteins in the nuclei of larval *Drosophila* salivary glands. *Proc. Natl. Acad. Sci. U. S.* **43**, 776–779 (1957).

51. Horstmann, E., and Knoop, A., Zur struktur des nukleolus und des kernes. *Z. Zellforsch. Mikroskop. Anat.* **55**, 1–48 (1961).

52. Hruban, Z., Swift, H., and Slesers, A., Effect of azaserine on the fine structure of the liver and pancreatic acinar cells. *Cancer Res.* **25**, 708–724 (1965).

52a. Hsu, T. C., Arrighi, F. E., Klevecz, R. R., and Brinkley, B. R., The nucleoli in mitotic division of mammalian cells *in vitro*. *J. Cell Biol.* **26**, 539–553 (1965).

53. Hyde, B. B., Changes in nucleolar ultrastructure associated with differentiation in the root apex. *Natl. Cancer Inst. Monograph.* **23**, 39–51 (1966).

53a. Hyde, B. B., Changes in nucleolar ultrastructure associated with differentiation in the root tip. *J. Ultrastruct. Res.* **18**, 25–54 (1967).

54. Hyde, B. B., Shankarnarayan, K., and Birnstiel, M. L., Observations on fine structure in pea nucleoli *in situ* and isolated. *J. Ultrastruct. Res.* **12**, 652–667 (1965).

54a. Iijima, K., Shantha, T. R., and Bourne, G. H., Enzyme-histochemical studies on the hypothalamus with special reference to the supraoptic and paraventricular nuclei of squirrel monkey (*Saimiri sciureus*). *Z. Zellforsch.* **79**, 76–91 (1967).

55. Immers, J., Comparative study of the localization of incorporated ^{14}C labeled amino acids and 35 SO in the sea urchin ovary, egg and embryo. *Exptl. Cell Res.* **24**, 356–378 (1961).

56. Jacob, J., An electron microscope autoradiographic study of the site of initial synthesis of RNA in the nucleolus of Smittia. *Exptl. Cell Res.* **48**, 276–282 (1967).

57. Jacob, J., and Sirlin, J. L., Electron microscope studies on salivary gland cells. *J. Cell Biol.* **17**, 153–165 (1963).

58. Jacob, J., and Sirlin, J. C., Electron microscope studies on salivary gland cells. The nucleus of *Smittia Parthenogenetica* (*Chironomidae*) with special reference to the nucleus and effect of actinomycin thereon. *J. Ultrastruct. Res.* **11**, 315–328 (1964).

59. Jezequel, A. M., and Bernhard, W., Modifications ultrastructurales du pancréas exacrine de rat sous l'effet de l'actinomycin D. *J. Microscopie* **3**, 279–296 (1964).

60. Jezequel, A. M., Seeve, M. M., and Steiner, J. W., Segregation of nucleolar components in mycoplasma infected cells. *Lab. Invest.* **16**, 287–304 (1967).

61. Johnson, J. M., and Jones, L. E., Behaviour of nucleoli and contracting nucleolar vacuoles in tobacco cells growing in microculture. *Am. J. Botany* **54**, 189–198 (1967).

62. Jones, K. W., The role of the nucleolus in the formation of ribosomes. *J. Ultrastruct. Res.* **13**, 257–262 (1965).

62a. Jordan, E. G., and Godward, M. B. E., Some observations on the nucleolus in *Spirogyra*. *J. Cell Sci.* **4**, 3–15 (1969).

63. Jörgensen, M., Zellenstudien. I–III. *Arch. Zellforsch.* **10,** 1–202 (1913).
64. Journey, C. J., and Goldstein, M. N., Electron microscope studies on HeLa cell lines sensitive and resistant to actinomycin D. *Cancer Res.* **21,** 929–932 (1961).
65. Kalnins, V. J., Stich, H. F., and Bencosme, S. A., Fine structure of the nucleolar organizer of salivary gland chromosomes. *J. Ultrastruct. Res.* **11,** 282–291 (1964).
66. Karasaki, S., Electron microscopic examinations of the sites of nuclear RNA synthesis during amphibian embryogenesis. *J. Cell Biol.* **26,** 939–958 (1965).
67. Kessel, R. G., An electron microscope study of nuclear cytoplasmic exchange in oocytes of liona intestinalis. *J. Ultrastruct. Res.* **15,** 181–196 (1966).
68. Kiselev, N. A., and Nikolaeva, N. V., Electron microscopy of unradiated and irradiated deoxyribonucleic acid. *Biokhimiya* **26,** 956–960 (1961).
69. Krugelis, Mac R., Organized structures in the nucleolus. *J. Cell Biol.* **23,** 195–200 (1964).
70. Kume, F., Maruyama, S., D'agastino, A. N., and Chiga, M., Nucleolar changes produced by mithramycin in rat hepatic cells. *Exptl. Mol. Pathol.* **6,** 254–260 (1967).
71. LaCour, L. F., Chayen, J., and Gahan, P. S., Evidence for lipid material in chromosomes. *Exptl. Cell Res.* **14,** 469–485 (1958).
71a. LaCour, L. F., and Wells, B., The loops and ultrastructure of the nucleolus of *Ipheion uniflorum. Z. Zellforsch.* **82,** 25–45 (1966).
72. Lafontaine, J. G., Structure and mode of formation of the nucleolus in meristematic cells of *Vicia faba* and *Allium cepa. J. Biophys. Biochem. Cytol.* **4,** 777–784 (1958).
73. Lafontaine, J.-G., Structural components of the nucleus in mitotic plant cells. *In* "The Nucleus" (A. J. Dalton and F. Haguenau, eds.), pp. 151–196. Academic Press, New York, 1968.
74. Lafontaine, J. G., and Chouinard, L. A., A correlated light and electron microscope study on the nucleolar material during mitosis in *Vicia faba. J. Cell Biol.* **17,** 167–227 (1963).
75. Lafontaine, J. G., and Lord, F. A., Ultrastructure and mode of formation of the nucleolus in plant cells. *Natl. Cancer Inst. Monograph.* **23,** 67–75 (1966).
76. Landschütz, C., Zytochemischer nachweis der adenosintriphosphatase (ATPase) in der zellen des ascites tumor der maus. *Experientia* **6,** 232–234 (1950).
77. Lapis, K., and Bernhard, W., The effect of mitomycin-C on nuclear fine structure of KB cells in cell culture. *Cancer Res.* **25,** 628–646 (1965).
78. Lazarus, S. S., Vethamany, V. G., Shapiro, S. H., and Amsterdam, D., Histochemistry and fine structure of 4-nitroquinoline N-oxide induced nuclear inclusions. *Cancer Res.* **26,** 2229–2238 (1966).
79. Lazzarini, A. A., and Weissman, G., Metachromatic granules of mucopolysaccharides in human aortic intimal cells after hexavalent hexol nitrate. *Science* **131,** 1736–1737 (1960).
80. Lettré, R., and Siebs, W., Beobactungen zur struktur des nucleolus in normalen zellen sowie tumorzellen. *Z. Kresbforsch.* **60,** 564–580 (1954).
81. Love, R., and Bharadwaj, T. P., Two types of ribonucleoprotein in the nucleolus of mammalian cells. *Nature* **183,** 1453–1454 (1959).
82. Low, F. N., and Freeman, J. A., "Electron Microscopic Atlas of Normal and Leukemic Human Blood." McGraw-Hill, New York, 1958.
83. Maggio, R., Siekevitz, P., and Palade, G. E., Studies on isolated nuclei. II. Isolation and chemical characterization of nucleolar and nucleoplasmic subfractions. *J. Cell Biol.* **18,** 293–312 (1963).

84. Marinozzi, V., The role of fixation in electron staining. *J. Roy. Microscop. Soc.* [3] **81**, 141–154 (1963).
85. Marinozzi, V., Cytochemie ultrastructurale du nucleole RNA aux proteines intranucleolaires. *J. Ultrastruct. Res.* **10**, 433–456 (1964).
86. Marinozzi, V., and Bernhard, W., Présence dans le nucléole des deux types de ribonucléoprotéines morphologiquement distinctes. *Exptl. Cell Res.* **32**, 595–598 (1963).
87. Marinozzi, V., and Gautier, A., Fixations et colorations. *J. Ultrastruct. Res.* **7**, 936–951 (1962).
88. Mazanec, K., Differentiation of cell structures during the cleavage of the mammalian egg. *Acta Fac. Med. Univ. Brun.* (1969).
89. Meisel, J. C., and Lytle, R. I., Affinity of animal cell nucleoli for normal serum. *J. Cell Biol.* **29**, 461–473 (1966).
90. Menzies, D. W., Nucleolar definition in Hemalum and Eosin staining. *Stain Technol.* **37**, 41–42 (1962).
91. Miller, O. L., Structure and composition of peripheral nucleoli of salamander oocytes. *Natl. Cancer Inst. Monograph.* **23**, 53–66 (1966).
92. Millonig, G., and Guidice, G., Electron microscopic study of the reaggregation of cells dissociated from sea urchin embryos. *Develop. Biol.* **15**, 91–101 (1967).
93. Monneron, A., Utilisation de la pronase en cytochimie ultrastructurale. *J. Microscopie* **5**, 583–596 (1966).
94. Monneron, A., and Bernhard, W., Action de certaines enzymes sur des tissus inclus en epon. *J. Microscopie* **5**, 697–714 (1966).
95. Montgomery, P. O'B., Karney, D., Reynolds, R. C., and McClendon, D., Cellular and subcellular effects of ionizing radiations. *Am. J. Pathol.* **44**, 727–746 (1964).
96. Montgomery, P. O'B., and Reynolds, R. C., Cellular and subcellular responses to ultraviolet radiation. *Lab. Invest.* **14**, 1419–1427 (1964).
97. Montgomery, T. H., Comparative cytological studies with special regard to the nucleolus. *J. Morphol.* **15**, 265–269 (1898).
98. Mulnard, J., Présence d'inclusions Feulgen positives dans les nucléoles larvaires de quelques diptères. *Arch. Biol.* (*Liege*) **67**, 485–498 (1956).
99. Muramatsu, M., Smetana, K., and Busch, H., Quantitative aspects of isolation of nucleoli of the Walker carcinosarcoma and liver of the rat. *Cancer Res.* **23**, 510–518 (1963).
100. Nash, D., and Plaut, W., On the presence of DNA in larval salivary nucleoli in *Drosophila melanogaster. J. Cell Biol.* **27**, 682–686 (1965).
101. O'Donnell, E. H. J., Deoxyribonucleic acid structures in the nucleolus. *Nature* **191**, 1325–1326 (1961).
102. Ogata, M., Die veranderungen der pankreaszellen bei der sekretion. *Arch. Physiol.* (see Hertwig, *48a*).
103. Olivier, J., Chimie du nucléole de l'ovacyte au cours de l'ovogenise de la Némerte Lineus ruber (Müller). *Compt. Rend.* **D263**, 289–291 (1967).
104. Panteleakis, P. N., and Mateyko, G. M., Effects of actinomycin D on the nucleoli of different cells *in vitro. J. Cell Biol.* **31**, 154A (1966).
105. Parsons, D. F., An electron microscope study of radiation damage in the mouse oocyte. *J. Cell Biol.* **14**, 31–48 (1962).
106. Paweletz, N., Siebs, W., and Lettré, R., Untersuchungen zur argentataffin reaction des nukleolus. *Z. Zellforsch. Mikroskop. Anat.* **76**, 577–605 (1967).
107. Pearse, A. G. E., "Histochemistry." Churchill, London, 1960.

108. Peveling, E., Elektronenmikorskopische untersuchungen an zellkernen von cucumis sativus L. *Planta* **56,** 530–554 (1961).
109. Pollister, A. W., and Ris, H., Nucleoprotein determination in cytological preparations. *Cold Spring Harbor Symp. Quant. Biol.* **12,** 147–157 (1947).
110. Porter, R. K., Problems in the study of nuclear fine structure. *Proc. 4th Intern. Conf. Electron Microscopy, Berlin, 1958* Vol. 2, pp. 186–199. Springer, Berlin, 1960.
111. Possnerova, V., and Smetana, K., The morphology of nucleoli of lymphocytes in peripheral human blood after short time cultivation *in vitro. Folia Morphol.* **14,** 240–244 (1966).
112. Potmesil, M., and Smetana, K., The inhibitory effect of chromomycin A3 on the RNA synthesis and the formation of ring-shaped nucleoli in cell cultures. *Folia Biol.* (*Prague*) **14,** 132–139 (1968).
112a. Potmesil, M., and Smetana, K., Significance of ring-shaped nucleoli and micro-nucleoli in human lymphocytes. *Folia Biol.* **15,** 300–305 (1969).
113. Puza, V., and Hroch, M., Form and struktur der kernkorperthen lebender Zellen. *Sb. Ved. Praci, Lekar. Fak. Karlovy Univ. Hradci Kralove* **10,** 569–573 (1967).
114. Racela, A., Grady, H., and Svoboda, D., Ultrastructural nuclear changes due to tannic acid. *Cancer Res.* **27,** 1658–1671 (1967).
114a. Revel, J. P., and Hay, E. D., Autoradiographic localization of DNA synthesis in a specific ultrastructural component of the interphase nucleus. *Exptl. Cell Res.* **25,** 474–480 (1961).
115. Reynolds, R. C., Montgomery, P. O'B., and Hughes, B., Nucleolar "caps" produced by actinomycin D. *Cancer Res.* **24,** 1269–1278 (1964).
116. Reynolds, R. C., Montgomery, P. O'B., and Karney, D., Nucleolar "caps"—a morphologic entity produced by the carcinogen 4-nitroquinoline N-oxide. *Cancer Res.* **23,** 535–538 (1963).
117. Ross, M. H., and Ely, J. O., Alkaline phosphatases in fixed plant cells. *Exptl. Cell Res.* **2,** 339–348 (1951).
118. Ruzicka, V., Zur geschichte und kenntnis der feineren struktur der nucleolen centraler nervenzellen. *Anat. Anz.* **16,** 557–563 (1899).
119. Sandritter, W., and Krygier, A., Zytophotometrische bestimmungen von proteingebundenen thiolen in der mitose und interphase von HeLa-zellen. *Z. Krebsforsch.* **62,** 596–610 (1953).
120. Schneider, D. D., "Nucleolen": *in* "Encyklopädie der mikroskopischen Technik," Vol. II. Urban & Schwarzenberg, Munich, 1927.
121. Schoefl, G., The effect of actinomycin D on the fine structure of the nucleus. *J. Ultrastruct. Res.* **10,** 224–243 (1964).
122. Shankaranarayan, K., and Busch, H., Dense granules in nucleoli of Walker 256 carcinosarcoma cells of the rat. *Exptl. Cell Res.* **38,** 434–437 (1965).
123. Shankaranarayan, K., Muramatsu, M., Smetana, K., and Busch, H., Ultrastructural studies on RNA and DNA components of isolated nucleoli of Walker 256 carcinosarcoma. *Exptl. Cell Res.* **41,** 81–98 (1966).
124. Shankaranarayan, K., Steele, W. J., and Busch, H., Evidence that the granular and fibrillar components of nucleoli contain 28 and 6 S RNA, respectively. *Exptl. Cell Res.* **43,** 483–492 (1966).
125. Shankaranarayan, K., Steele, W. J., Smetana, K., and Busch, H., Ultrastructural aspects of the ribonucleoprotein network in nuclei of Walker tumor and rat liver. *Exptl. Cell Res.* **46,** 65–77 (1967).

126. Simard, R., Specific nuclear and nucleolar ultrastructural lesions induced by proflavin and similarly acting antimetabolites in tissue culture. *Cancer Res.* **26,** 2316–2328 (1966).
127. Simard, R., and Bernhard, W., Le phénomène de la segregation nucléolaire: Spécificité d'action de certains antimetabolites. *Intern. J. Cancer* **1,** 463–479 (1966).
128. Simard, R., and Bernhard, W., A heat-sensitive cellular function located in the nucleolus. *J. Cell Biol.* **34,** 61–76 (1967).
129. Sirlin, J. L., The nucleolus. *Progr. Biophys. Biophys. Chem.* **12,** 27–66 (1962).
130. Sirtori, C., and Bossisio-Bestetti, M., Nucleolar changes in KB tumor cells infected with *Herpes simplex* virus. *Cancer Res.* **27,** 367–376 (1967).
131. Smetana, K., A further contribution to the question of the incidence of nucleoli in the nuclei. *Folia Biol. (Prague)* **10,** 268–274 (1961).
132. Smetana, K., and Busch, H., On the ultrastructure of the Walker 256 carcinosarcoma. *Cancer Res.* **23,** 1600–1603 (1963).
133. Smetana, K., and Busch, H., Studies on the ultrastructure of the nucleoli of the Walker tumor and rat liver. *Cancer Res.* **24,** 537–557 (1964).
134. Smetana, K., and Busch, H., Studies on staining and localization of acidic nuclear proteins in the Walker 256 carcinosarcoma. *Cancer Res.* **26,** 331–337 (1966).
135. Smetana, K., Hermansky, F., and Possnerova, V., Nucleolar coefficient and glycogen content of lymphocytes in chronic lymphadenoses. *Neoplasma* **9,** 549–554 (1962).
136. Smetana, K., and Hermansky, F., A contribution to the knowledge of the ultrastructure of leukaemic lymphocytes. *Neoplasma* **10,** 405–411 (1963).
137. Smetana, K., and Ivanyi, D., Cytological analysis of mixed leukocyte culture reaction. *Folia Biol. (Prague)* **13,** 150–154 (1967).
138. Smetana, K., Lane, M., and Busch, H., Studies on nucleoli of leukemic agranulocytes and plasmacytes in multiple myeloma. *Exptl. Mol. Pathol.* **5,** 236–244 (1966).
139. Smetana, K., and Potmesil, M., Further note to the question of the incidence of ring-shaped nucleoli in liver cells of rats after treatment with actinomycin D. *Z. Zellforsch. Mikroskop. Anat.* **92,** 62–69 (1968).
140. Smetana, K., Shankaranarayan, K., and Busch, H., Quantitative analysis of ultrastructural components of nucleoli of the Walker tumor and liver. *Cancer Res.* **26,** 786–796 (1966).
141. Smetana, K., Unuma, T., and Busch, H., Ultrastructural studies on nucleic acids of granular components in Novikoff Hepatoma cells. *Exptl. Cell Res.* **51,** 105–122 (1968).
141a. Smuckler, E. A., and Benditt, E. P., The early effect of actinomycin on rat liver. Changes in ribosomes and polysomes. *Lab. Invest.* **14,** 1699–1709 (1965).
142. Stevens, B. J., The effect of actinomycin D on nucleolar and nuclear fine structure in salivary gland cell of *Chironomus thumni. J. Ultrastruct. Res.* **11,** 329–352 (1964).
143. Stevens, B. J., The fine structure of the nucleolus during mitosis in the grasshopper neuroblast cells. *J. Cell Biol.* **24,** 349–368 (1965).
144. Stowell, R. E., Lee, C. S., Tsuboi, K. K., and Villasana, A., Histochemical and microchemical changes in experimental cirrhosis and hepatoma formation in mice by carbon tetrachloride. *Cancer Res.* **11,** 345–354 (1951).
145. Studzinski, G. P., Selective binding of zinc by basic proteins of the HeLa cell nucleolus. *J. Histochem. Cytochem.* **13,** 365–375 (1965).

146. Sun, C. N., The occurrence of dense granules of unknown function in the nucleoli of certain plant cells. *Exptl. Cell Res.* **25**, 213–215 (1966).
147. Suskind, G., Autoradiographic and cytochemical evidence for synthesis of a lupine containing ribonucleoprotein in nucleoli inhibited by actinomycin D. *J. Cell Biol.* **24**, 309–316 (1965).
148. Suter, E. R., and Salomon, J. C., Effect de l'actinomycin D sur la structure fine du nucléole des hepatocytes de rats intoxiqués par la thioacetamide. *Exptl. Cell Res.* **43**, 248–251 (1966).
149. Svoboda, D., and Soga, J., Early effects of myrolizidine on the fine structures of rat liver cells. *Am. J. Pathol.* **48**, 347–374 (1966).
150. Swift, H., Quantitative aspects of nuclear nucleoproteins. *Intern. Rev. Cytol.* **2**, 1 (1953).
151. Swift, H., Studies on nuclear fine structure. *Brookhaven Symp. Biol.* **12**, 139–152 (1959).
152. Swift, H., Cytochemical studies on nuclear fine structure. *Exptl. Cell Res.*, Suppl. 9, 54–67 (1963).
153. Swift, H., Microphotometry in biologic research. *J. Histochem. Cytochem.* **14**, 842–852 (1966).
154. Swift, H., and Rasch, E. M., Nucleoproteins in *Drosophila* polytene chromosomes. *Exptl. Cell Res.* **2**, 456–457 (1954).
155. Swift, H., and Stevens, B. J., Nucleolar-chromosomal interactions in microspores of maize. *Natl. Cancer Inst. Monograph* **23**, 145–166 (1967).
156. Szollosi, D., Extrusion of nucleoli from pronuclei of the rat. *J. Cell Biol.* **25**, 545–562 (1965).
157. Tandler, C. J., An acid soluble component of the nucleolus; the cytochemical specificity of the lead acetate reaction. *J. Histochem. Cytochem.* **4**, 331–341 (1956).
158. Tandler, C. J., The detection of orthophosphate within the cell nucleus. *Exptl. Cell Res.* **25**, 246–250 (1961).
159. Terzakis, J. A., The nucleolar channel system of human endometrium. *J. Cell Biol.* **27**, 293–304 (1965).
159a. Tewari, H. B., and Bourne, G. H., The histochemistry of the nucleus and nucleolus with reference to nucleo-cytoplasmic relations in the spinal ganglion neuron of the rat. *Acta Histochem.* **13**, 323–350 (1962).
160. Thoenes, W., Fett im nucleolus. *J. Ultrastruct. Res.* **10**, 194–206 (1964).
161. Thorell, B., Studies on the formation of cellular substances during blood cell production. *Acta Med. Scand.* Suppl. 200, 1–120 (1947).
162. Torelli, U., Guaglino, D., Artussi, T., Emilia, G., Ferrari, G., and Mauri, C., An autoradiographic study of the RNA and protein metabolism of normal plasma cells and phytohemagglutinin stimulated lymphocytes. *Exptl. Cell Res.* **42**, 1–9 (1966).
163. Unuma, T., and Busch, H., Formation of microspherules in nucleoli of tumor cells treated with high doses of actinomycin D. *Cancer Res.* **27**, 1232–1242 (1967).
164. Unuma, T., Smetana, K., and Busch, H., A morphologic study on the chromatin areas associated with the nucleolus. *Exptl. Cell Res.* **48**, 665–671 (1967).
165. Uzman, B. G., Foley, G. E., Farber, S., and Lazarus, H., Morphologic variations in human leukemic lymphoblasts (CCRF-CEM Cells) after long term culture and exposure to chemotherapeutic agents. *Cancer* **19**, 1725–1742 (1966).
166. Vendrely, C., and Vendrely, R., Localisation de l'acide ribonucléique dans les différents tissus et organes de vertebres. *In* "Handbuch der Histochemie" (W.

Graumann and K. Neumann, eds.), Vol. 2, Part 3, pp. 84–283. Fischer, Jena, 1959.

167. Vincent, W. S., Structure and chemistry of nucleoli. *Intern. Rev. Cytol.* **4**, 269–298 (1955).

168. Vincent, W. S., "The Nucleolus. Genetics Today." Pergamon Press, Oxford, 1964.

169. Vorbrodt, A., Histochemically demonstrable phosphatases and protein synthesis *Exptl. Cell Res.* **15**, 1–20 (1958).

169a. Vorbrodt, A., Krzyzowska-Gruca, S., and Steplewski, Z., Cytochemical investigation on the localization of guanosine triphosphatase A in cell nuclei. *Bull. Ac. Pol. Sci.* **12**, 337–341 (1964).

170. Watson, M. L., Considerations of nucleic acid morphology in fixed tissues. *Proc. 5th Intern. Congr. Electron Microscopy, Philadelphia, 1962* Vol. II, Art. 0–5. Academic Press, New York, 1962.

171. Weissenfels, N., Struktur und Verhalten der Nukleolen von Huhnerherz Myoblasten in Gewebe Kultur wahrend des interphase Wachstums und der Mitose. *Z. Zellforsch. Mikroskop. Anat.* **62**, 667–700 (1964).

172. Wessing, A., Der Nucleolus und seine Beziehung zu den Ribosomen des Cytoplasmas. *Z. Zellforsch. Mikroskop. Anat.* **65**, 445–480 (1965).

173. Yasuzumi, Y., and Sugihara, R., The fine structure of nuclei as revealed by electron microscopy I. *Exptl. Cell Res.* **37**, 207–229 (1965).

174. Yasuzumi, Y., and Sugihara, R., The fine structure of nuclei as revealed by electron microscopy II. *Exptl. Cell Res.* **40**, 45–55 (1965).

175. Zagury, D., Existence d'un complexe liporibonucléoprotéique a groupement sulphydryle au sein du nucléole. *Compt. Rend.* **242**, 1825–1827 (1957).

176. Zalokar, M., Cytochemistry of centrifuged hyphae of neurospora. *Exptl. Cell Res.* **19**, 114–132 (1960).

177. Zjadela, F., and Morin, G. A., Etude cytoloque et histochemique du noyau hepatique de souris isole. *Rev. Hematol.* **7**, 628–651 (1952).

178. Zobel, C. R., and Beer, M., The use of heavy metal salts as electron stains. *Intern. Rev. Cytol.* **18**, 363–400 (1965).

CHAPTER III

Nucleolar Organizers, Nucleolar Size, Shape, and Number

I. Introduction

Nucleolar Organizer Concept

Only specific chromosomes in any given cell or species are involved in the formation of nucleoli. These specific DNA templates are not generally distributed throughout such chromosomes but rather are localized to special regions of these specific chromosomes. The chromosomes that carry these DNA templates are referred to as "nucleolar chromosomes." The specific regions containing the DNA templates involved in the synthesis of nucleolar structures are usually referred to as "nucleolus organizer regions (NOR's)," "SAT (sine acido thymonucleico) zones," or "secondary constrictions;" they have also been called nucleolar zones or olistherozones (*16, 18, 127, 129, 133*). The composition of the NOR is not completely defined but it is believed to be nucleolar DNA or rDNA (see Chapter IV) and a surrounding matrix containing special proteins and other components essential for synthesis of rRNA (ribosomal RNA). In some cells, as in oocytes, the elements for future nucleolar function are laid down and separated from the chromosomes. In others, the DNA is spread out as a fine filamentous structure separated from the main chromosomal mass (Chapter IV) (*71, 127, 129, 136, 137*). Regardless of its location or structure, the nucleolus organizer region is essential to the functional activity of cells. In its absence, as in specific lethal anucleolate mutants, the mutants fail to develop into full-grown members of the species (*30*); in some cases losses of the NOR may be reflected in abnormal morphology of the chromosomes (*101*).

II. Nucleolar Chromosomes

The evidence that there are special "nucleolar chromosomes" has emerged from studies on mutants or with drugs; these studies show in essence that nucleolar formation is inhibited by the loss of specific "nucleolar chromosomes" or by deletion of specific chromosome regions that presumably contain rDNA. On the other hand, losses of other chromosomal regions or other chromosomes do not result in either losses or alterations in the number or types of nucleoli. The "nucleolar chromosomes" vary in size and shape, and in certain micronuclei where there is chromosomal "pulverization" (*65*) there is uncertainty as to which chromosomal elements are involved.

A. Relation to Secondary Constrictions

The sites of synthesis of some nucleoli are identified in prophase or metaphase chromosomes (Fig. 3.1) as "secondary chromosomal constrictions" (*47, 88*). These sites of nucleolar synthesis are less-defined regions in some chromosomes, but in some cases they are at termini of chromosomes or at junctional regions of two chromosomes. These regions will be referred to as NOR's or nucleolus organizer regions, a term which indicates that the nucleolus originates in these regions. The chromosomal

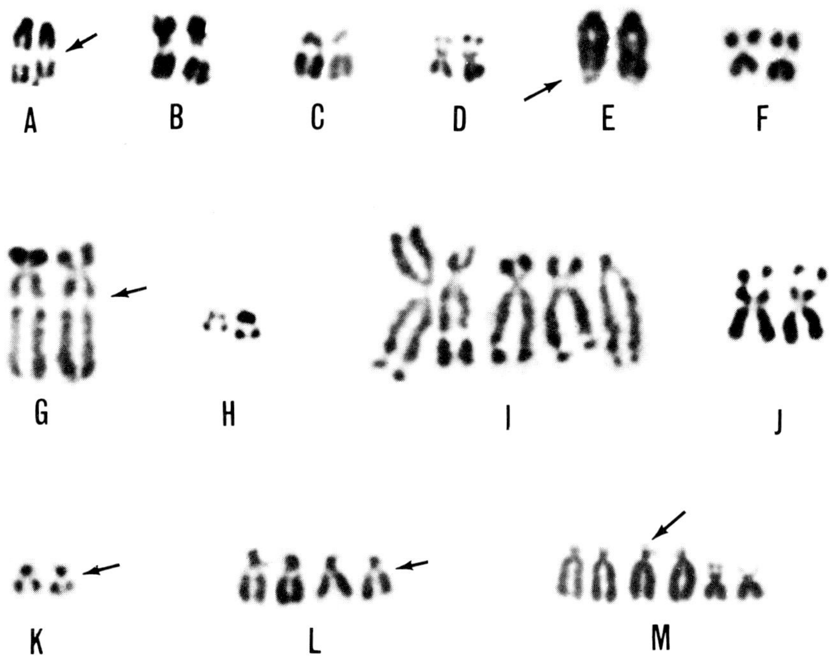

Fig. 3.1. Chromosomes of a number of species with secondary constrictions. The arrows represent the positions of the constrictions. Top row from left to right: A. *Spilogale putorius* (spotted skunk); B. *Mephitis mephitis* (striped skunk); C. *Mustela putorius* (ferret); D. *Felis catus* (domestic cat); E. *Cervus canadensis* (elk); F. *Sus scorfa* (domestic pig). Second row: G. *Carollia perspicillata* (fruit bat); H. *Pipistrellus subflavus* (Eastern pipistrelle); I. *Tamiasciurus hudsonicus* (red squirrel); J. *Chinchilla laniger* (chinchilla). Third row: K. *Tupaia glis* (tree shrew); L. *Alouatta caraya* (black howler); *Homo sapiens* (man). (Courtesy of Prof. T. C. Hsu, M. D. Anderson Hospital, Houston, Texas.)

constrictions reflect a lesser staining with Feulgen or other stains due to lesser condensation of DNA or greater concentrations of proteins in these regions. Evidence that DNA is not completely absent from these regions is shown in Fig. 1.3. Attempts have been made to correlate the amount of chromatin and the size of these regions with the number and size of the nucleoli, but, for reasons indicated below (88), these have not always been fruitful.

B. Basic Aspects of the Nature of Chromosomal Constrictions and Nucleolus Organizers

By definition, the "primary chromosomal constrictions" are those chromosomal regions that are the sites of attachment of the spindle and also the apparent junction between the long and short arms and the sister chromatids; this region is also referred to as the "kinetochore."

"Secondary chromosomal constrictions" are those chromosomal regions that are visualized along the arms of the chromosomes as narrowed or less dense regions because of the thinning of the deoxyribonucleoproteins in these regions (see Chapter I, Fig. 1.4). All of these regions are visible only in mitosis and their states in the interphase nucleus are not defined.

C. Chromosome Condensation and "Secondary" Chromosomal Constrictions

Although the position of the primary chromosomal constriction is both specific and very reproducible for individuals and species, the position, size, and degree of definition of the secondary chromosomal constrictions are apparently dependent upon the dynamic state of the chromosome at prophase. Depending upon the dynamics of the chromosomal condensation, there may be a large, small, or absent secondary constriction even in the regions where the secondary constriction is usually seen.

Since the nucleolus is a large structure which persists during prophase when there is a progressive condensation of the "nucleolar chromosomes," the process of formation of the secondary constriction may be viewed as an arrest of the condensation of the region bearing the nucleolus. At the critical time when the nucleolus disintegrates, there is insufficient time for complete condensation of the chromatin in the region of the nucleolus organizer. As a result, there is a region of extended or dispersed chromatin which "marks" the position of the nucleolus organizer. The relative difference in the rate of the condensation of the chromatin and the disappearance of the nucleolus is responsible for the persistence of the constriction. If the nucleolus disintegrates early; i.e., at a rate that is equal to or greater than the rate of chromosomal condensation, the nucleolus organizer region will not appear as a secondary constriction.

This is apparently the case in the rat, mouse, and some other rodents. In many species and in many cells, however, the obverse is true; i.e., the rate of disintegration of the nucleolus is slower than the rate of condensation of the chromatin; in these, "secondary constrictions" are readily identified.

In some species that appear to have intermediate rates of NOR condensation, nucleoli have occasionally been found to be associated, as in human chromosomes 13, 14, 15, 21, and 22 (Fig. 3.1). Generally these NOR's are not conspicuous. Some "satellited" chromosomes appear to have a long secondary constriction between the chromosomal mass and the "satellite."

The number of nucleoli is only partially related to the number of secondary constrictions in the chromosomes. For example, in some species, such as the rat, there are no secondary constrictions in the chromosomes (*51a*) and yet the nuclei may contain as many as seven nucleoli. Recently, Hsu *et al.* (*53*) analyzed the secondary chromosomal constrictions in cells of various mammals and noted that they contained 0–6 regions with secondary constrictions (Table 3.1). These constrictions appeared to have unique chromosomal localizations in any given species; the analyses of the width of the chromosomal fibers in these regions indicated that they were 50–70 Å in diameter as compared with 150–200 Å of the fibers in condensed chromosomal regions.

The position of the NOR is apparently variable from species to species along the arms, being anywhere between the primary constriction and the chromosomal ends. As will be noted (p. 120), the termini of the chromosomes may also contain NOR's which do not have significant differences in function from NOR's within chromosomes, but the question as to their structure has not been answered, particularly as to whether they

TABLE 3.1

SECONDARY CONSTRICTIONS (*53*)

Animal	Cell type[a]	Number of secondary constrictions
Green monkey	EL	0–2
Striped hyena	FL	0–2
Caracal	FL	0–2
Desert shrew	FL	0–2
Elk	FL	0–2
Fox squirrel	FL	0–5
Red squirrel	FL	0–6
Red squirrel	BM	0–3
Chinese hamster	FL	0–2

[a] EL = epithelial-like, FL = fibroblast-like, BM = bone marrow.

are similar to those of the secondary constrictions. Although secondary chromosomal constrictions were found in the species shown in Table 3.1, other species were not found to contain secondary chromosomal constrictions. Yerganian and Papoyan (*145*) reported that at metaphase, chromosomes of *Cricetulus migretorius* may associate at the ends; they concluded that such ends might serve as a source of the secondary constrictions of what would be functionally the equivalents of the secondary constrictions in the cells. Under conditions where the chromosomes are subjected to external influences, such as radiation or noxious drugs, these terminal secondary constrictions or NOR's may become visible as constrictions joining adjacent long chromosomes. Sometimes, there is a "fusion" of two chromosomes, possibly at the sites where they have NOR's.

Kopac and Mateyko (*71*) pointed out that secondary constrictions do not always elaborate nucleoli; there may be a lack of a nucleolar organizer in the constriction and a competition may exist between nucleolar organizers in the karyotype in which the "strongest" competitor is successful in organizing the nucleolus (*5*). In addition, secondary chromosomal constrictions may result from disease states or treatment of cells with drugs that interfere with DNA synthesis.

D. Nucleolar Chromosomes of Specific Species

In most mammalian species, the nucleolar chromosomes can be identified by the presence of secondary constrictions in mitotic metaphase (Table 3.2). In the rat kangaroo, the X chromosome is a nucleolar chromosome. The organizer is located at the proximal portion of the long

TABLE 3.2

Specific Nucleolar Chromosomes

Species	Chromosome number	Reference
Zea mays	6	*88, 91*
Bibio	One pair	*4, 21*
Rat lymphoblasts	Autosomes	*100*
Mouse	X, Y 6	*46, 113*
Rana pipiens	10	*27*
Gerbil	X, Y	*60*
Neurospora crassa	2	*38*
Triticum aestivum (wheat)	I, X, XIV, XVIII	*82*
Chironomus tentans	II, III, IV	*8, 9a, 67, 104*
Man	XIII, XIV, XV, XXI, XXII	*51a, 80a*

arm of the X chromosome. In cats (including lions, tigers), a small metacentric chromosome known as E_1 has a NOR. In humans, NOR's are located on ten chromosomes* (13, 14, 15, 21, and 22). Ohno and Kinosita (*100*) found that only a few autosomes were "nucleolus-formers" in rat lymphoblasts and the mesothelial cells of rat omentum. On the other hand, in mouse cells, the X and Y chromosomes were associated with the nucleolus and the organizer region was located near the centromeres (*99*). In the Chinese hamster, it is possible that the organizers are scattered. In the squirrels, three or four different pairs of chromosomes have nucleolus organizers. In some sublines of mouse plasma cell tumors (*146*), long telocentric chromosomes have been found that have secondary constrictions; they are called "SAT chromosomes" (see p. 116), but their role in nucleolar formation is not clear. One possibility suggested was that the constrictions result from breakage of the normal chromosomes.

E. NOR's of Other Species

Despite the evidence that implicates NOR regions as the sites of nucleolar synthesis, in many species it is not possible to identify the chromosomes involved. In the chicken, *Gallus domesticus*, Ohno *et al.* (*98*) noted that there are a number of microchromosomes that are associated with the nucleolus. Instead of the one chromosome with a specific NOR, in this case there are approximately 12 microchromosomes associated with the nucleolus; whether all 12 chromosomes participate in the organization of the nucleolus is not clear. Since there are many nucleolar protein and nucleic acid components (see Chapters I, V–VIII), this number would not be unrealistic if the functional elements involved in their synthesis were separated into individual entities.

In some species, such as the worm *Lumbricus*, the nucleoli are large enough to be visible at 320 magnifications. Hancock (*46*) found two nucleolar organizers on the periphery of one giant nucleolus. He suggested that there is endopolyploidy and possibly fusion of the nucleoli.

F. NOR's of Midges and Flies

In *Chironomus tentans*, each chromosome may be associated with one nucleolus with the exception of the chromosome I (*8, 67, 104*). The

* In human cells, studies have been made on the Barr body (*6*) which was once believed to be a "nucleolar satellite;" this dense X chromosome occurs only in female cells where it is present in 100% of the female cells (*132*). It is not related to nucleolar synthetic reactions. In the rat, a Barr body (*6*) is found that is a "chromocenter" or a region of high density.

nucleolus of chromosome IV is apparently only a secondary nucleolus formed under special circumstances which have not yet been defined. Beermann (*9a*) has made extensive studies on NOR number and the associated chromosomes in *Chironomus tentans* and *Chironomus pallidivittatus.* He obtained recombinant genotypes which ranged in nucleolar number from three to none. Thus, three NOR's are present, as indicated above. He also established a balanced heterozygous strain which regularly segregated approximately 25% anucleolate progeny which died as embryos. Instead of nucleoli, the nuclei of these anucleolate embryos contained numerous small spherical granules. There is also an accessory NOR in a mutant fourth chromosome of *C. tentans* and *C. pallidivittatus.* This nucleolus has a different structure from that of the normal nucleoli. The normal NOR's may be broken by irradiation into fragments of different lengths, each of which can form a nucleolus. These data and those of McClintock (*88, 89,* v.i.) provide evidence that even one partial organizer is sufficient to provide for normal nucleolar development.

Methods have been developed for the analysis of chromosome function in the *Drosophila* (*49, 71a*), but relatively few studies have been made on the "nucleolar chromosomes." McClintock studied the chromosomes of *Neurospora* (*89*) but the hope that some useful information on nucleolar chromosome function would emerge from this species has not been realized.

G. Plants

Extensive studies have been made on the chromosomes involved in nucleolar biosynthesis. The maidenhair tree was studied for nucleolar chromosomes by Ho (*50*) who noted that cells of young leaves of microsporangiate trees and megasporangiate trees have four satellites. These are on the short arms of the longest subterminally constricted pairs. However, there are fewer satellites or constrictions than the number of these chromosomes; i.e., chromosomes 3 and 12 are the only ones that contain constrictions and each of their chromatids contains one constriction. The total of four secondary constrictions which reflects the maximum number of possible nucleoli is accounted for by the total number of constrictions in these two chromosomes. Studies have been made on the mutations of the nucleolar chromosomes in a variety of plants. For example, in *Allium cepa,* Battaglia (*7*) found a variety of mutants; some have tandem satellites on M_7 and M_8, and others have a supernumerary centric fragment or a heteromorphic seventh pair. In some with two nucleolar pairs, one had terminal satellites and the other a secondary constriction in the submedian position.

Among the other studies that have been carried out in plants are those of Longwell and Svihla (*82*) and those of McClintock (*88, 90, 91*) who have made such extensive contributions that they are discussed in separate sections below. They showed that there are specific organizers that are localized to specific chromosomal regions which may or may not be associated with secondary constrictions in plants. Some of these regions contain "strong" and some "weak" nucleolus-forming segments.

H. Nucleolar Fusion and Number of NOR's

One problem in defining the total numbers of NOR's is that of fusion of the NOR's or nucleoli. Discrete NOR's may be seen as broad or fine secondary constrictions but in rodents they may not be seen and yet many nucleoli may be formed. Possibly these animals have terminal NOR's or NOR's that condense early (v.s.).

In some cells, there are as many nucleoli as there are constrictions or organizers; included in this group are cells of hyacinths, salamanders, axolotls, gerbils, and others (*71*). Synergism or fusion reduces the number of nucleoli in cells of *Drosophila, Pisum,* some spermatogonia and Sertoli cells, and in meiosis in many species (*71*). Lin (*81*) showed that there were as many as five nucleolar organizers in microsporocytes of maize and yet only one nucleolus was formed. This fact was interpreted to mean that in some cases, there is suppression of the development of the nucleoli by some external or internal forces.

I. Nucleolar Chromosomes in Meiosis

The NOR's have been mostly studied in metaphase chromosomes because they are distinguished in mitosis as "secondary chromosomal constrictions." In meiosis, they are distinguished as chromosomal knobs. Meiosis is that phase of chromosome formation in germinal cells in which chromosomes are first duplicated and separated as pairs in which the DNA is so closely approximated that it looks as if the cells are haploid. However, in the next division, there is no DNA replication and yet true haploid germ cells form. During the pachytene stage, when the chromosomes are completely paired, the nucleolus is large but is apparently completely inactive; the formation of ribosomes has ceased and the NOR is so condensed that it is referred to as a "knob" to which the inactive nucleolus is attached. Since the remainder of the chromosome is active, the chromatin is extended in most other regions except the NOR. Thus, in meiosis, there is a direct obverse situation to that found in mitosis; i.e., the NOR in mitosis is extended or dispersed and is detected as a "chromosomal constriction (secondary)" and in meiosis, the NOR is condensed and is seen as a knob. In each case, the attached nucleolus may be large and hence the region is readily identifiable.

III. Studies on Nucleolar Chromosomal Mutants in Plants

One of the classical experiments on the nucleolar chromosomes is that of McClintock (*88, 91*) on the corn *Zea mays* which was reported almost 35 years ago. A number of earlier workers in this field had shown the relationships of specific chromosomes to the nucleolus but Heitz (*47*) noted that satellite stalks and secondary constrictions in the chromosomes were functionally the same in that they served as nucleolus organizer regions. Heitz (*47*) referred to these regions as SAT regions; i.e., sine acid thymonucleinico; i.e., lacking DNA although it is now clear that some DNA is present. He indicated that the number of nucleoli was proportional to the number of these regions in the cell and moreover that differences in the sizes of the nucleoli could depend upon the activity of the different nucleolus-organizing regions.

The studies of McClintock (*88, 91*) were aided by the finding of a "satellited" chromosome (chromosome 6) in *Zea mays.* This chromosome has an NOR at one end, where a pale chromomere joins the "satellite" portion to the denser portion of chromosome 6. Chromosome 6 has a dense heteropycnotic "knob" referred to earlier as being condensed chromatin of the nucleolus in meiosis. Synthesis of the nucleolus occurs in a sister satellite chromosome as well. Since there is frequently fusion of the two nucleoli, the diploid cells of *Zea mays* contain either one or two nucleoli.

Interchange of chromatin was found to occur between chromosome 6 and chromosome 9. This interchange, whether X-radiation–induced or spontaneous, occurred at the NOR of chromosome 6; i.e., the long end of chromosome 9 interchanged with the short end of chromosome 6. As a result, part of the NOR was translocated to chromosome 9 and much of the mass of the long end of chromosome 9 was transferred to chromosome 6. Thus, 2 NOR's were formed, one remaining on chromosome 6 and the other on chromosome 9; both were active and yet differed in activity so that small and large nucleoli were produced. The number of nucleoli increased from two to as many as four in the progeny of these cells. The larger of the nucleoli was produced by the segment of the chromosome that migrated from its origin on chromosome 6 to chromosome 9 (the latter being called chromosome 9^6).

Since the portion of the NOR transferred to chromosome 9 was smaller than that left on chromosome 6, the smaller section clearly exhibited the greater activity; i.e., it was associated with the larger nucleolus. This result suggests that within a given NOR, the regions have different activities. The variation in the number and sizes of nucleoli was verified by the use of mutants that were heterozygous with respect to the chromosomes 6, 9, 6^9, and 9^6.

When normal female flowers were fertilized with X-ray–treated pollen whose NOR on chromosome 6 was deleted, the microspores contained one or no organizers. Those without NOR did not form normal nucleoli.

Thus, in these studies on *Zea mays,* McClintock (*88, 91*) showed that the "nucleolar chromosome" 6 contained the NOR as a "secondary constriction" in mitosis and a heteropycnotic knob in meiosis. If the "heteropycnotic knob" in the nucleolar organizer of the *Zea mays* is broken into unequal portions by X-ray, the two parts are capable of forming nucleoli of equal size. Since these nucleoli may be equal in size to control nucleoli, there seemed little doubt that the total number of DNA regions that code for the nucleoli are seldom read out. Instead there seems to be ample additional genomic material to provide for the synthesis of at least two normal-sized nucleoli.

A. Nucleolar Chromosomal Mutants of Wheat

In a species of wheat, *Triticum aestivum,* there are well-mapped chromosomal aberrations, some of which are associated with the numbers and types of nucleoli. In this species, four chromosomes have been reported to be associated with nucleoli; i.e., I, X, XIV, and XVIII. Tetrasomy of the non-nucleolar chromosomes, such as XI and XXI, did not produce significant changes in the number of nucleoli, but tetrasomic chromosomes I or X produced the formation of double the normal number; i.e., on the average there were three nucleoli per cell as compared with 1.5 for the control (*82*).

Although the result with the tetrasomic chromosomes I and X were not surprising, the results with the tetrasomic chromosomes XIV and XVIII were of interest in view of the previous concepts of "gene dosage" as being regulatory for cell function. When these chromosomes were tetrasomic, there was only an average of two nucleoli per cell; i.e., more than the usual number and yet not so many as when there was an increase in the number of the chromosomes I and X. These results supported the conclusion that there are "strong" (I and X) and "weak" (XIV and XVIII) nucleolar chromosomes. The mechanism of the effect associated with gene dosage has not been evaluated in molecular biological terms, but presumably would be related to the total number

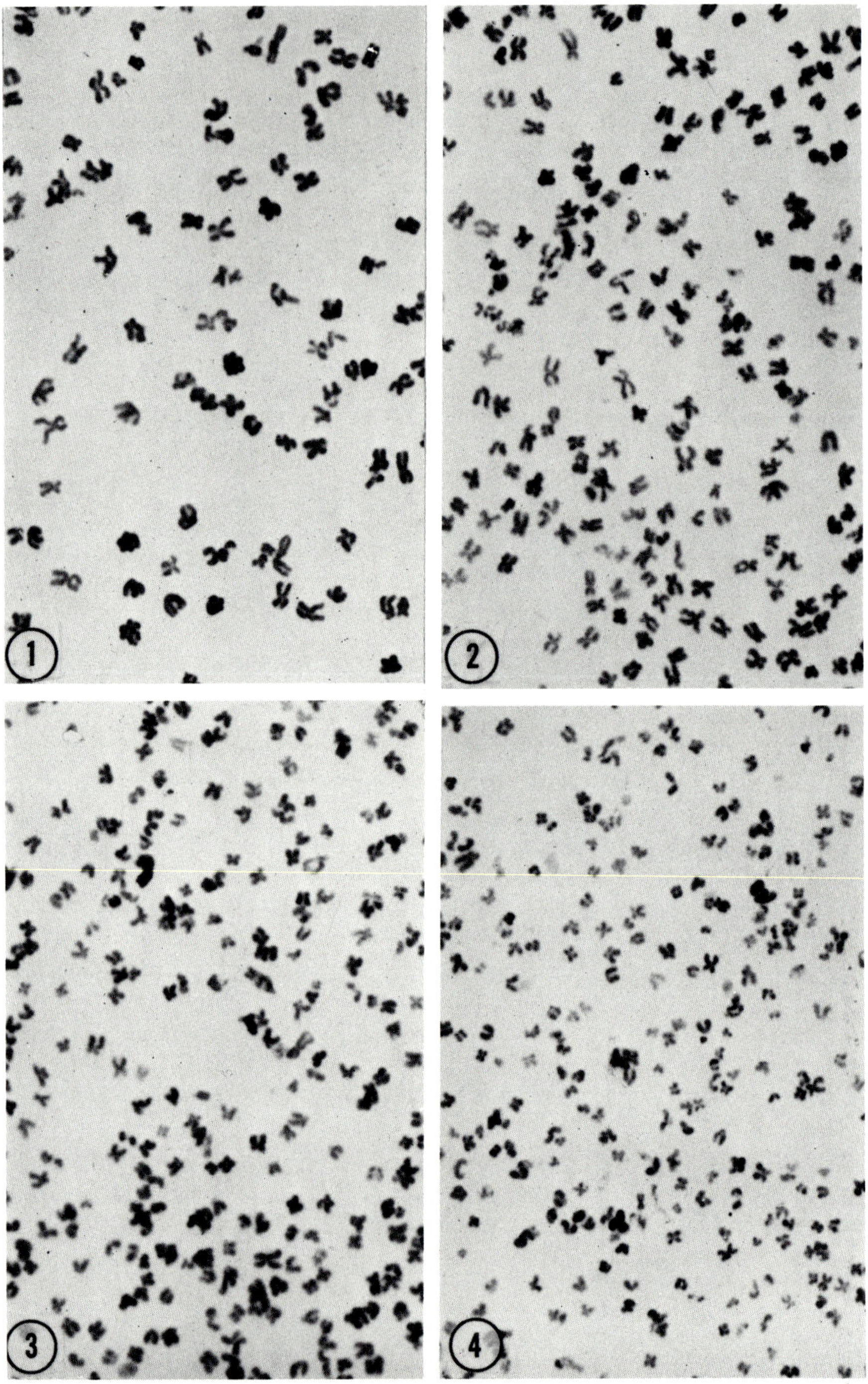
1
2
3
4

of rDNA templates that may prime for synthesis of ribosomal RNA in these species.

Studies on the relationship between chromosomal type, the increased number of nucleoli and cytoplasmic composition showed that there was an increase in cytoplasmic RNA and dry weight when there was an increase in the number of nucleoli. This result differs from those reported for *Drosophila* (*109–111*).

B. Micronucleoli

There are marked differences between nucleoli and "micronucleoli" (Fig. 1.13). The number of micronuclei or regions that look like nucleoli may approach thousands in some types of cells such as the salivary cells of *Chironomus* or *Rhynchosciara* angelae. Bauer (*8*) described the presence of micronucleoli ("Nebennukleolen") in *Chironomus tentans* in large numbers in addition to the two "macronucleoli" usually found. In virtually every band of the giant polytene chromosomes, there is a "micronucleolus" (*61, 62*). These micronucleoli may have the same functions as the nucleoli since their electron microscopic appearance is similar to that of the nucleoli. Whether their ribonucleoproteins are the same as those produced in the macronucleoli is uncertain at present because of the lack of satisfactory techniques for analysis of the RNA in the micronucleoli. Moreover, none of the genetics of these interesting structures have been clarified although there may be competition between the various genes for the synthesis of the RNP "droplets" or micronucleoli (*105*). If the micronucleoli do produce ribosomal RNA, at least in such cells, the distribution of rDNA is very widespread in such chromosomes.

C. Isolation of Nucleolar Chromosomes

It is now possible to isolate metaphase chromosomes (*55, 56, 86, 92*) (Figs. 3.2, 3.3) and the availability of such methods offers the hope that not only will it be possible to isolate the specific chromosomes responsible for formation of nucleoli, but also that the NOR's themselves will be isolated. Isolated chromosomes of the HeLa cells have been found to hybridize with 18 S and 28 S ribosomal RNA. In HeLa cells,

Fig. 3.2. Isolated and fractionated HeLa metaphase chromosomes. A small amount of a suspension in FM (fractionation medium) containing 30% glycerol of each pooled chromosome class from chromosome preparation I (see text) was spread on a clean glass slide and allowed to dry for 60 minutes. The half-wet chromosomes were then fixed in acetic acid–ethanol (3:2 v/v) for 10 minutes and stained in 1% orcein in lactic acid–acetic acid (1:1 v/v). Bright field. ×800 (*56*).

group symbol
corresponding human chromosomes
cell a
cell b
A
1-3
B
4-5
C
6-12,X
D
13-15
E
16-18
F
19-20
G
21-22

these RNA's hybridized best with the smaller chromosomes. On the other hand, the mRNA (presumably the template RNA for biosynthesis of the cytoplasmic proteins) hybridizes with all of the chromosomes. The isolated chromosomes were separated partially by sucrose density gradient centrifugation into groups coded A through G (Fig. 3.3). The ribosomal RNA hybridizes with chromosomes in regions D and G presumably because they contain the nucleolus organizers (see Figs. 3.2, 3.3).

IV. Functional State of the Nucleolar Organizers

Very little is understood about the state of the nucleolar organizer as it actually functions in cells in interphase. It is not known whether this portion of the genome is a counterpart of the operator gene of bacterial systems or actually serves as the structural gene on which there is synthesis of nucleolar RNA. It is also not known whether the organizer is a condensed structure or is divided into fibrillar regions in a number of loci where polycistronic readouts are produced (see Chapter IV).

One analogy that has been drawn for the possible functional state of the NOR or the structural nucleolar genes is that of the lampbrush chromosomes (*40, 41, 48, 73, 140*). These chromosomes (Fig. 3.4a) are best seen in vertebrate oocytes and cockroach or cricket cells. Usually the oocytes are the optimal sources of these interesting structures. In essence, these chromosomes are believed to be like very long hawsers whose loops are paid out from one end and at the other reform the end that is initially paid out. Another analogy drawn is that of a conveyor belt (Fig. 3.4b). Accordingly, there appears to be a stretched or extended portion at any given time that is the template for RNA being synthesized. Moreover, the genes are read in sequence. Whether this is the case for the nucleolus is not yet clear, but such a possibility would suggest that nucleolar products or ribosomal components would be pro-

---

Fig. 3.3. Karyotypes of two individual HeLa cells. Cells were blocked in metaphase with colchicine (5×10^{-6} M), suspended in 1% sodium citrate for 10 minutes to allow spreading of chromosomes, fixed in acetic acid–ethanol (3:2 v/v) for 10 minutes and then stained with 1% orcein in lactic acid–acetic acid (1:1 v/v). Cells suspended in stain solution were squashed by thumb pressure between a slide and a cover slip. Photographic enlargements were made of the metaphase chromosomes sets of single cells. The images of individual chromosomes were cut out of the enlargements, arranged as well as possible into morphological groups according to the Denver system, and rephotographed. Bright field. ×1300. "Group symbols" are the letter symbols commonly applied to the morphological groups of the Denver system. The numbers assigned by the Denver system to the human chromosomes in different groups are given for comparison.

Figures 3.2 and 3.3 were generously supplied by Prof. G. Attardi (*56*).

duced only at specified times in cell function. This seems to be the case for some bacteria as shown by the genetic maps (Chapter IV). In mammalian cells, such as liver cells, there is apparently continuous synthesis of ribosomal RNA in the nucleolus throughout the life of the animal.

V. Ultrastructure of the Nucleolus Organizer Region (NOR)

The nucleolus organizer region has been observed best in *chironomids* and *sciarids*. The general organization of this region is shown in Fig. 3.5 of Jacob and Sirlin (*58, 59a*). According to this diagram, the nucleolus organizer region is similar to other portions of the chromosome. Proximal to this region there develops an amorphous mass with some granular elements and expanding out from this region there is a larger mass which is surrounded by granular elements. These granular elements apparently emerge from the center of the nucleolar mass (see also Figs. 1.10, 1.11, and 1.12). The initial labeling of nucleolar RNA occurs in the proximal fibrillar regions of the nucleolus which may be the regions that contain dispersed or "euchromatin."

In *chironomids*, Kalnins *et al.* (*61, 62*) found that the nucleolus organizer regions (NOR) in the chromosome contain 100 Å fibers. The NOR is a disc-like region of these fibers and only 100 Å fibers are nucleolar organizer fibers (Figs. 1.10–1.12). The three-dimensional structure of the organizer region was suggested to be like an Indian arrowhead. Around this is a region they referred to as "pars amorpha" that separates the nucleolus proper and the organizer. The coils that interlace these regions may be parallel chromosomal fibers 100 Å in diameter; these fibers may be aggregates of DNA that are functional in biosynthesis of nucleolar elements.

In *Sciaridae*, the nuclei of the salivary gland cells contain nucleoli that are relatively small and are apparently not incorporated into the chromosomal structure (*39, 59*); these are referred to as a less dense internal structure that is fibrillar and resembles that of lymphocytes (see Chapter X); the fibrillar regions contain fibrils similar to those of the chromosomes. Larger granules, 150 Å in diameter, are found in nucleoli, chromosomes, and the nuclear sap. Initially, the "elementary" nucleoli are attached to chromosomal bands. Later a large nucleolus appears but whether this is formed *de novo*, by coalescence or by another mechanism is not at all clear. Unlike mammalian nucleoli, these nucleoli lack nucleolonemas.

In mammalian cells it is only rarely that the nucleolus organizer region can be visualized; for example, in some lymphocytes the extensions into

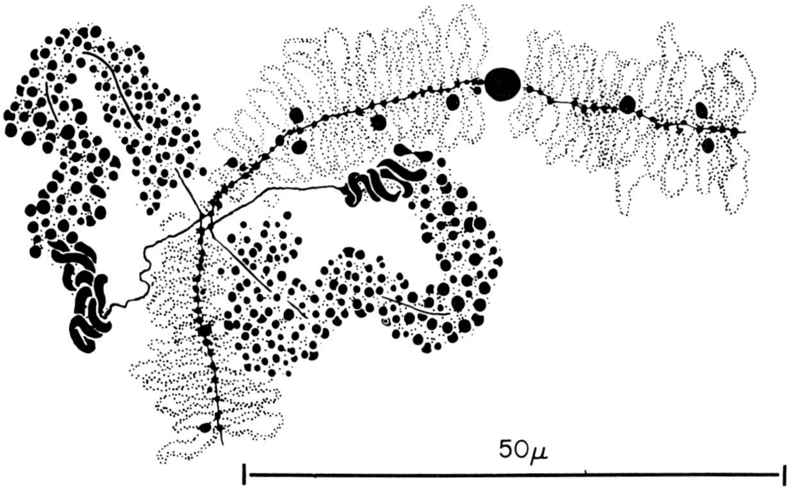

a

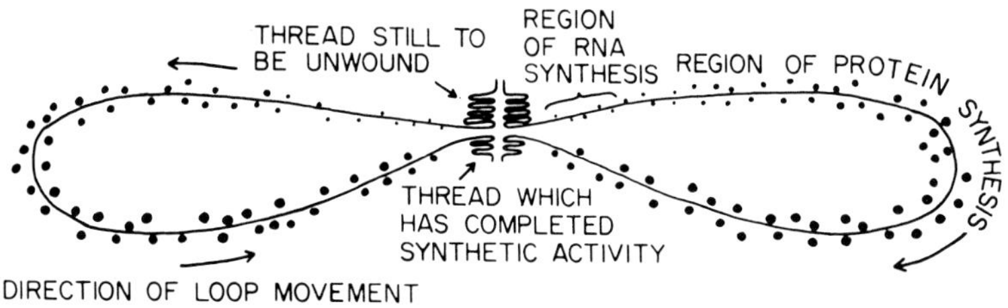

b

Fig. 3.4. a. The left end of chromosome XII from an oocyte of the newt *Triturus cristatus*. A large number of "normal" loop pairs are shown in addition to the "giant granular loops" characteristic of this chromosome region. All loops are asymmetric in the sense that one end is thinner than the other. In the giant granular loops, the thin insertion consists of a fine thread leading to a dense contorted region. The thick insertion, as well as the bulk of the loop, consists of a coarsely granular matrix surrounding a very delicate axis. (From Gall, J. G. and Callan, H. G.: *Proc. Natl. Acad. Sci. U. S.* **48**, 562–570, 1962.) b. Interpretation of labeling experiments of lampbrush chromosome XII in terms of a conveyor-belt model of the loops. (From Gall, J. G. and Callan, H. G.: *Proc. Natl. Acad. Sci. U. S.* **48**, 562–570, 1962).

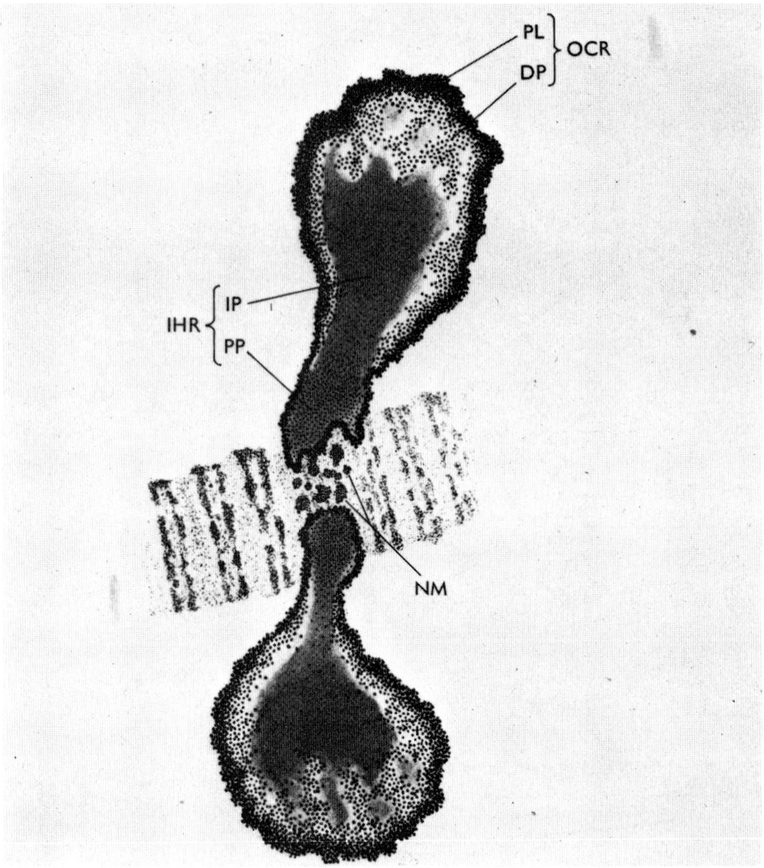

Fig. 3.5. This diagram summarizes the pattern of organization in the nucleolus of the larval salivary gland of the *chironomid, Smittia.* The parts shown are: nucleolar material (NM) within the chromatin of the organizer, proximal part (PP) of the nucleolus, intermediate part (IP), distal part (DP), and peripheral layer (PL). The proximal part and the intermediate part together constitute the inner homogeneous region (IHR) seen in low power electron micrographs, while the distal part and the peripheral layer comprise the outer coarse region (OCR) observed at low power. (From Jacob, J. and Sirlin, J. L.: *J. Ultrastr. Res.* **11**, 315–328, 1964.)

the nucleolus of DNA which is uncoiling can be seen as shown in Figs. 3.6 and 3.7. This uncoiling may be similar to that proposed for the lamp-brush chromosomes of various oocytes.

For mammalian cells, there is now an increasing amount of evidence that there are at least two kinds of organization of chromatin strands in the nucleolus aside from the chromatin of the perinucleolar "nucleolus-

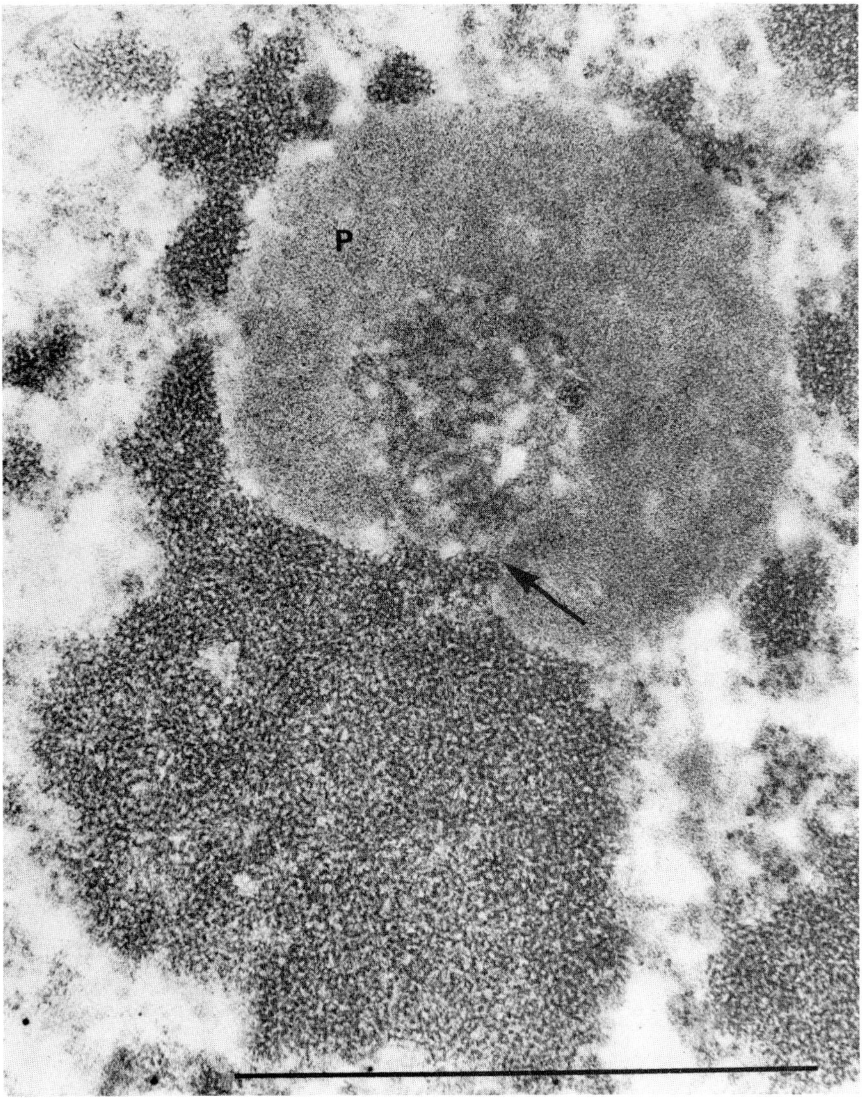

Fig. 3.6. Ring-shaped nucleolus of human lymphocyte with a chromatin pedicle in the peripheral ring (arrow). There is expansion, uncoiling, and separation of the chromatin filaments within the nucleolus. Chromatin filament (large point); peripheral nucleolar ring (P). ×78,000.

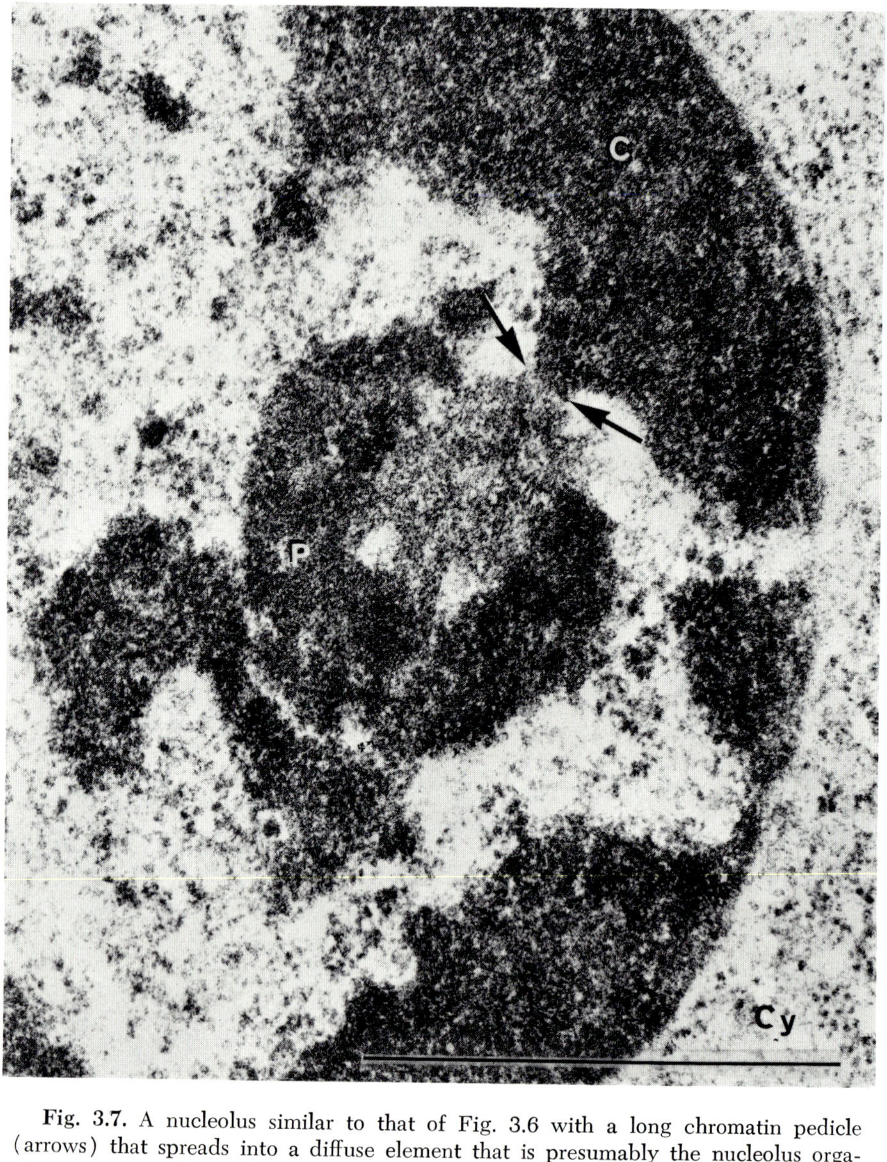

Fig. 3.7. A nucleolus similar to that of Fig. 3.6 with a long chromatin pedicle (arrows) that spreads into a diffuse element that is presumably the nucleolus organizer region or the nucleolar rDNA. Cy, cytoplasm; C, chromatin; P, peripheral nucleolar mass. ×64,000.

associated chromatin." One of these types of chromatin is shown in Fig. 3.8 which is an electron micrograph of the nucleolus after treatment of the section with ribonuclease so as to remove the RNA nucleolar granular elements without changing the chromosomal components. The nu-

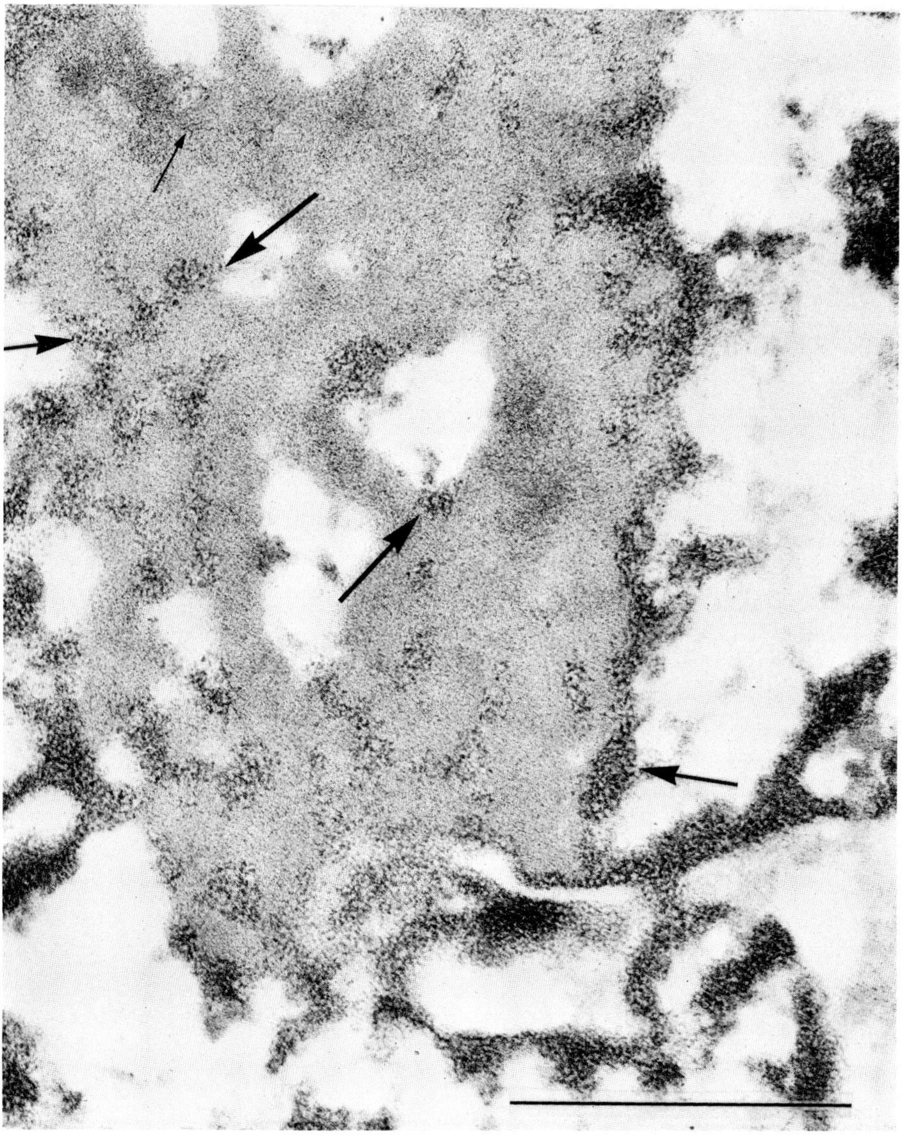

Fig. 3.8. A nucleolus after digestion with ribonuclease and pepsin. Chromatin strips are visualized within the nucleolus; these penetrate from the nucleolus-associated chromatin. ×46,000.

cleolar chromosomal components are either sections of chromatin that have entered the nucleolus in a trabecular formation (see Chapter II) or they represent the chromosomal elements described by O'Donnell and by Lettre and others as being distributed within the nucleolus (*38, 75, 77–80, 96, 97, 102, 135, 136*). These regions compose approximately 5% of the total nucleolar areas as seen in electron micrographs (*134*). Interestingly, DNA approximates 5 to 10% of the total nucleolar dry mass.

Another form of the nucleolar chromatin is that associated with very fine fibrillar elements; the diameter of these small strands approximates 20 Å (see Chapter IV). These small fibers are found in the nucleolar light spaces and also in high power pictures of the nucleolus; they are also visible after extractions with dilute and concentrated salt solutions (Fig. 2.25). Presumably, these fibrils are the chromosomal components involved in the biosynthesis of ribosomal precursors in the nucleolus rather than the dense chromatin masses.

VI. Number of Nucleoli per Cell

The number of nucleoli varies from cell to cell and possibly also in any given cell. The number of nucleoli in cells may encompass the very wide range of values from 0 to 2000 or more in individual cells of different species. From the point of view of characterizing cells, the number of nucleoli is of interest but it is not completely specific (Table 3.3). Even in as homogeneous a mammalian tissue as liver, there is a marked variation in the number of nucleoli per nucleus. The mean number of nucleoli per nucleus in normal liver is 2.7 but the range is 1 to 7 nucleoli per nucleus. Although there are relatively few cells with six or seven nucleoli per nucleus, most have either three or four.

The range over the animal and plant kingdom is much greater than that of such mammalian tissues as liver. In some cells, there are no nucleoli; i.e., bacteria and viruses lack organized nucleoli, as do some mutant forms of plants (*88, 91*) and some mutants of *Xenopus laevis* (*30, 36, 138, 139*). On the other hand, in tulips, as many as 13 nucleoli can be found per nucleus (*143*) and Montgomery (*95*) noted that the ova of *Doto*, a mollusc, had as many as 40 nucleoli per nucleus. In *Piscicola*, an annelid, there is a range of 12–400 nucleoli per nucleus in subcuticular gland cells (*95*).

In recent years, Miller (*93*), Callan (*17a*), and Gall (*40*) and their associates have been particularly concerned with the nucleoli of the developing amphibian oocyte and the products of these nucleoli. As many as 1000–2000 nucleoli have been estimated to form what may be a "nucleolar" layer in a shell just under or near to the nuclear envelope of

TABLE 3.3

AVERAGE NUMBER OF NUCLEOLI PER CELL IN TISSUES

Tissue	Size	Average number	Range	Reference
Walker tumor (rat)	Large	1.5	1–3	
Normal liver (rat)	Variable	2.7	1–7	
Mouse liver		2.3	1–6	*113*
Human lymphocytes		1.0		
Mouse lymphocytes	Small		1–3	
Leukemic lymphocytes		1.4		
Megakaryocytes		40		
Tulips			6–13	
Spermatids, spermatozoa (mouse)		0		
Piscicola (polychaete annelid)				
Subcuticular gland cells			12–400	*95*
Onion ganglion cells		1.0		
Muscle cells		12		
Amphibian oocyte			1000–2000	*40, 93*
Blastocyst (rabbit)			5–17	*45*
HeLa cell in culture				
In normal nutritional medium		1.7		*129*
In saline media		3.7	1–8	*129*
Doto (mollusc)			1–13	*95*
Ambystoma tigrinum		2	1–3	*26*
Heart cells in culture (*Triturus viridescens*)			1–4	*5*

these cells. Also, as noted previously, large numbers of micronucleoli have been found in the chromosomes of midges, mites, gnats, and small flies.

During development of oocytes the increased number of nucleoli is largely due to the formation of hundreds of free nucleoli. Callan (*17a*) has reported that the several hundred free nucleoli in oocytes are genetically related to the two nucleoli of diploid somatic interphases. It is not clear whether fully developed nucleoli detach from the organizer loci and add to the population of free nucleoli in oocytes or whether DNA-generators of free nucleoli detach from the organizer loci (Chap. IV).

In blastocyst nuclei, the nucleolar number ranges from 5 to 17 (*45*). In the rat pronucleus as many as 17 nucleoli have been found; with maturation the number of nucleoli diminishes.

In tissue cultures of fibroblasts of adult human bone marrow and human embryonic skin, Petersen and Therkelsen (*107, 132*) found that the usual number of nucleoli was two to four although five nucleoli were not uncommon. As many as eight nucleoli were found per cell. These authors noted that female cells contained more nucleoli than male cells possibly because they counted the "juxtanucleolar" Barr body.

In view of the fact that most of the mouse liver cells studied contained a maximum of six nucleoli, Shea and Leblond (*113*) concluded that the number of NOR's in the mouse cell was six (Table 3.3). Since the number was increased to 11 in tetraploid cells, they concluded that there was a doubling of the number of NOR's in these cells. On the other hand (*94*), there was a marked decrease in the number of nucleoli in regenerating liver cells; i.e., as few as one nucleolus per cell was commonly found. Irradiation of such cells virtually destroyed the synthesis of nucleolar components and as a result, there was a marked loss of nucleolar elements.

Variation in the Number of Nucleoli in a Given Cell

The data presented earlier on the number of the nucleoli in such cells as those of the normal liver or other mammalian cells, such as HeLa cells (*129*), indicate that the potentiality of these cells for the production of nucleoli is greater than would appear from the number of nucleoli usually present. Since many mammalian liver cells have five nucleoli and some have as many as seven, the potentiality of the cell for production of nucleoli is in excess of that normally expressed. The gene potential or "chromosome dosage" for production of nucleoli is thus generally inhibited and may be so repressed that only one nucleolus is formed. At present, there is no clear evidence on the nature of factors that influence the number of NOR's or gene sites that are operative in the biosynthesis of nucleoli although some secondary factors have been well defined.

In the studies of Swift (*129*) on HeLa cells in culture, it was noted that a conversion of the medium from the normal nutritional medium to one consisting only of salts resulted in the formation of as many as eight nucleoli. In saline media the average number of nucleoli per cell was 3.7 and the range was from 1 to 8. When the cells were incubated in a nutritionally adequate medium, the number of nucleoli decreased to 1.7 per cell. When the medium was adequate, the nucleoli had a more rounded shape and an increased size. The number of nucleoli per cell has been determined for specific cells as indicated in Table 3.3.

Although it has been suggested that there may be some relationship between the size of the nucleoli and their number, at least for the nucleoli of liver cells treated with thioacetamide (Fig. 1.2), this does not seem to be the case (*16*). After treatment with this drug, there is a marked compensatory increase in nucleolar RNA synthesis (*1, 2*). For the period of the first 48 hr after its administration, the size increases but the number of nucleoli per liver cell is not changed. Later, the number decreases but there is a great increase in the size of these nucleoli; the nucleolus may virtually fill the entire nucleus (*17, 68, 69, 109*). Another increase in the size of the nucleoli with a decrease in their number is found in regenerating liver in which both hypertrophy and the irregularity of these nucleoli may be related to the activity in production of new ribosomal precursors. In these cells there is essentially no change in the overall ultrastructural appearance of the nucleoli but instead there is a hypertrophic alteration in the components; i.e., there are more nucleolar elements but they do not differ either in type or in distribution.

VII. Nucleolar Number in Specific Mutants

Several types of mutants of animals have become available that contain differing numbers of nucleoli. One of these is a mutant of the toad *Xenopus laevis*. In the homozygous form, such progeny die as microcephalic young larvae which have shortened tails with crumpled tips that are bent ventrally; they also develop cardiac edema. At Oxford, Fischberg and Wallace (*36, 138*) found a mutant female whose eggs yielded heterozygotes with only one nucleolus in the larvae when they were fertilized by sperm from wild-type *Xenopus*. The crosses of the heterozygotes produced 25% wild-type offspring, 50% mutants of the same nucleolar types as the parent heterozygotes, and 25% lethal mutants which die as larvae; these had, respectively, two, one, and no nucleoli (*31a*).

The nuclei of the anucleolate mutants contained tiny RNA bodies one-eighth the size of normal nucleoli. In the anucleolate mutants, Hay and Gurdon (*46a*) found pseudonucleoli which have only a little granular element; i.e., a cap on the periphery of the fibrous element. Interestingly, the fibrillar element continues to be formed. The toads with two nucleoli were normal as were those with one. Elsdale *et al.* (*30*) predicted that such mutations would have use for marker studies, for transplantation studies, and for interphase nuclei, and as a source of material for experiments on the function of the nucleolus (see Chapter IV).

The presumption is that each of these nucleoli arises on one NOR. Normally there are two such NOR's per nucleus. If one nucleolus is present, it might result from either nucleolar fusion or some "cooperation" of the nucleolus-forming regions. Since the groups from heterozygotic parents contained two, one, and no nucleoli, respectively, it was concluded that the wild-type contain two NOR's, the heterozygotes, one NOR, and the lethal anucleolate mutants, none (*4*). Although the nucleolar number was reduced in the heterozygote, the amount of RNA in the nucleolus was not changed from the control (*4, 5*).

Support for a chromosomal difference between the anucleolate mutants and the wild type was obtained by Kahn who found that a secondary chromosomal constriction of the normal chromosomal complement was missing in the anucleolate mutants (*60*). These findings support the idea of Barr and Esper about the nature of the hereditary change (*5*).

Another kind of anucleolate lethal mutant is the t^{12}/t^{12} mutant of mice (*125*). The designation "t" indicates that the line is tailless. In this case, the recessive homozygote is also lethal. Although RNA synthesis is apparently markedly diminished in these mutant mice, thus far, this mutant has been much less studied than the *Xenopus* mutant.

A. The Bobbed Mutants of *Drosophila*

In *Drosophila*, there are both lethal and degenerative mutants that involve the nucleolus organizer region. Ritossa *et al.* (*110*) found that the "bobbed mutants" are deficient in NOR's; there is some controversy as to whether such partial deletions of the NOR's produce quantitative deficiencies in ribosomal RNA synthesis or whether other factors operate to repress nucleolar genes in *Drosophila* (*67a, 109a*). Analysis of DNA hybridizable with ribosomal RNA led to the suggestion that the hybridizable DNA in these mutant forms was proportionate to the number of NOR's (see Chapter IV). One fascinating aspect of these mutants was that if even one normal fly is present in the colony, the whole population becomes normal. If the weakened flies are kept as a series of mutants, eventually the nucleolar number and amount returns to normal suggesting that there is a tendency of the nucleolar genes to endoreduplicate for RNA synthesis. Ritossa has referred to this phenomenon as an unstable redundancy of genes for ribosomal RNA synthesis (*109b*), although there is a possibility that a few healthier flies simply multiply faster than the weaker mutants (*42a*).

B. Maximum Number of Nucleoli

Since there are a limited number of NOR's in the chromosomes of any cell, there is little doubt that the maximum number of nucleoli equals total number of NOR's. Since the nucleolar readouts of mammalian cells represent largely rRNA molecules rich in guanylic and cytidylic acids, the DNA on which the nucleolar RNA molecules are formed must be rich in deoxyguanylic and deoxycytidylic acids. However, DNA of the nucleolar organizer has been isolated and separated from the cells in only a few instances (*12*) (see also Chapter IV). The correlation of the amount of this DNA with the number of nucleoli has not been worked out thus far. In some cells, however, it seems that NOR's can endoreduplicate or produce "slave copies" which accounts for the huge number of nucleoli in oocytes. The mechanisms involved are unknown (*17a, 40, 41, 93*), but Callan (*17a*) has found a chromosomal locus that apparently controls the endoreduplication of the NOR copies.

To summarize the morphological and ultrastructural findings on nucleolus organizers (NOR's), they appear in some species as distinct "secondary chromosomal constrictions," although in others such constrictions are not demonstrable. Ultrastructural studies have shown that the DNA in the nucleolus is spread in a manner resembling that of lampbrush loops. Whether this is the DNA of the NOR or structural genes containing rDNA is not yet clear. The total number of NOR's varies considerably from species to species but the usual numbers are small ranging from 0 to 12. Elegant studies have been made on specific "nucleolar chromosomes" in wheat and corn with respect to the location of the nucleolus organizers and their activities in mutants.

Recently, chromosomes have been isolated from mammalian cells and localization of some nucleolus organizers in specific groups has been possible. The precise role of the nucleolus organizers is not established but presumably these regions either contain operator genes that are necessary for the activity of structural genes for synthesis of rRNA, or they are the regions that contain the DNA (rDNA) coding for ribosomal RNA.

In most instances the number of nucleoli in cells is smaller than the number of organizers which apparently represent a maximum number. However, in the amphibian oocytes (see Chapter IV), there is a large increase in the number of nucleoli by an obscure process referred to as endoreduplication. In mutants of the toad *Xenopus*, the nucleolar number is related numerically and precisely to the number of NOR's; i.e., two, one, none. Changes in size and shape of the nucleoli appear to be related to the state of nucleolar function.

VIII. Activation of the Nucleolar Organizers (Nucleolus Organizer Region)

At present, there is little understanding of the mechanisms of gene control for the biosynthesis of ribosomal RNA or other nucleolar products. It is of interest whether the Jacob-Monod (*57*) operon concept applies; i.e., is there a messenger RNA specific for activation of genes containing rDNA? Kasten *et al.* (*63, 64*) found that if DNA synthesis is inhibited by the incubation of cells in excessive concentrations of thymidine, the synthesis of RNA of the nucleolus is also blocked as is the release of RNA to the cytoplasm. It is clear that DNA synthesis is part of the continuing process of cell replication in tissue culture; the relationship of a block in its synthesis to inhibiton of nucleolar function is not defined. One possibility is that a messenger RNA is produced that is specific for the activation of the NOR's (*119*).

Evidence that the nucleolar cistrons for the biosynthesis of RNA differ in drug sensitivities (Chapter X) from that of the remainder of the cistrons has provided some support for the concept of selective activation of NOR's. The suggestion for differential activation emerges from the fact that actinomycin selectivity inhibits the nucleolar genes and TRB (trichlorobenzimidazole riboside) selectively inhibits extranucleolar gene readouts without inhibiting the activity of the nucleolar genes (*118*). These data do not support the concept that there is a "messenger" RNA that is "nucleolus specific" inasmuch as extranucleolar RNA synthesis was blocked by both di- and trichlorobenzimidazoles while that of the nucleolus organizer region did not show the inhibition. If the block on chromosomal RNA synthesis was complete, the substances responsible for the nucleolar synthetic reactions would not appear to be mRNA species.*

Other evidence for differences in nucleolar and extranucleolar biosynthetic actions comes from studies of Sirlin *et al.* (*120, 122*) on *Smittia;* they showed that the rates of labeling of RNA in various nuclear fractions were different; i.e., the labeling of RNA in the nucleolus proper proceeded faster than that of either the nucleolus organizer or the remainder of the chromosomes. Since both rates of labeling and the effects of inhibitors differ for nucleolar and extranucleolar RNA synthesis (see Chapters V, VI), it is questionable whether extranucleolar RNA serves to activate or "drive" nucleolar synthetic reactions.

Studies on the cell cycle show that early in telophase and late in prophase, the NOR's are active yet the remainder of the chromatin is relatively inactive at this time; this is an additional type of evidence for the relative "independence" of nucleolar biosynthetic reactions. In any event, it would seem that some signal to the NOR's comes from either genetic elements in the nucleus or functional elements in the cytoplasm. All of the responses made by the nucleoli are telologically related to their role in the biosynthesis of the ribosomes; i.e., they are activated during normal growth, abnormal growth, protein synthesis, special responses to hormones, and other types of predictable responses to increased cellular need for protein.

* These data do not rule out the possible existence of a nucleolus-specific messenger whose synthesis is sensitive to actinomycin and resistant to benzimidazole.

The precise nature of the "message" that stimulates the synthesis of nucleolar products to the exclusion of other types of products has been the object of speculation without definition. The only evidence presently available about the chemical nature of genetic regulators or "repressors" suggests that they are probably proteins and are likely to be acidic proteins (*43, 108*). One model system for studies on nucleolar regulation appears to be the thioacetamide-treated liver (Fig. 1.2). In this instance, there is an enormous increment in the biosynthesis of nucleolar RNA; i.e., up to 30 fg of RNA per nucleolus per minute (ten times that of the normal liver nucleolus). This increased biosynthesis occurs as the levels of ribosomes and their associated RNA and protein fall in the cytoplasm. Thioacetamide produces damage to lysosomes with an increased level of free RNase and probably proteases in the cytoplasm; destruction of the ribosomes occurs (*19*, see also Chapter VII).

The fall in the ribosome content of the cytoplasm is coupled with an increased synthesis of nucleolar products; it is tempting to suggest that the rate-limiting factor in the activity of the nucleolus is the level of cytoplasmic ribosomes or some feedback from them. An increased re-entry of free ribosomes into the nucleus and/or the nucleolus or a product which inhibits nucleolar biosynthetic reactions might be an effective control. Recently, evidence has been provided that cytoplasmic factors influence gene readouts (*132a*) and other evidence suggests that ribonucleoproteins may function as gene derepressors (*54a*). Accordingly, ribosomal ribonucleoproteins may function in a similar manner to derepress nucleolar genes (see also p. 46).

IX. Timing of Nucleolar Synthesis in the Cell Cycle

A. The Synthesis of Nucleolar Products in Telophase and Interphase

The cessation of nucleolar and nuclear RNA synthesis during metaphase may be a physical effect due to chromosome condensation or the cell may accumulate enough repressors to block chromosomal functions (*13, 87*). In telophase and early interphase, there is a large burst of synthesis of nucleolar products presumably because the NOR's are derepressed early or differ physically from other chromosomal regions (*26, 47*). At this time, the accumulating nucleolar products push apart the chromosomal threads and thereby decrease the visibility of the chromatin in the NOR (*9, 19–21, 26, 103*).

In plant cells (*Allium*), Das (*25*) noted that the rate of synthesis of

RNA decreased in condensing chromosomes during prophase but as long as the nucleoli persisted there was RNA synthesis. RNA synthesis stopped during the phase of chromosome condensation and resumed during telophase.

There is uncertainty about the precise state of the NOR at the time of activation. Although under exceptional circumstances the nucleolus may persist in cells during metaphase, persistence is a rare phenomenon (see p. 146). Although there is little reason to believe that the reformation is not completely *de novo* in daughter cells, in some cases small pieces of the disintegrated nucleolus may coalesce and re-form some parts of a new nucleolus (*126;* see also Chapter I). For example, in the grasshopper there is a loss of nucleolar substance and dispersal of the granules in prophase; at anaphase, there appear to be clumps of nucleolar material in small layers on chromosome surfaces. Stevens (*126*) noted that, in anaphase, ribonucleoprotein was laid down on some chromosome surfaces; in telophase, there was coalescence of these nucleolar products. Thus, some granular nucleolar precursors or products appear to be grouped together during telophase. In mammalian cells, there is also a large increase in the perichromosomal RNA which appears to form an envelope around the metaphase chromosomes (*124*). Because this ribonucleoprotein layer is so rich in acidic nucleolar proteins, it is readily visible with the improved staining methods for these proteins.

Errera and Brunfaut (*31*) concluded that some residual RNA in HeLa cells may coalesce to initiate nucleolar synthesis or possibly serve as a focus for nucleolar synthesis. During prophase in these cells, there is migration of nucleolar RNA to the cytoplasm. After mitosis, RNA synthesis resumes rapidly and in late telophase, radioactivity is associated with the chromatin. Because there were nonradioactive basophilic regions in the nucleoli, they suggested that some of the unlabeled nucleolar RNA must have been derived from coalescence of the RNA of the nucleus. A similar concept was presented by Feinendegen and Bond (*33*) and Feinendegen *et al.* (*34*).

Lafontaine and Chouinard (*73a, 74, 74a*) noted that in *Vicia faba* during early prophase, the nucleolus was round; in mid- and late-prophase, the nucleolus was irregular in shape. This fact may account in part for the irregularity of morphology noted in tumor cells that are rapidly growing as compared with those of the Morris hepatoma series (*16, 70*). In late prophase and anaphase, the nucleolar material merged imperceptibly into the nucleoplasm. Projections in the regions of the adjacent chromosomes were noted. However, these workers suggested that the nucleoli were formed *de novo* on specific chromosomal regions and not by coalescence.

B. Control of the Nucleolus Organizer

Is the shut-off mechanism for the NOR or the biosynthesis of nucleolar components the same as that which modulates the NOR in interphase? In Chapter V, the biosynthetic activity of the nucleolus will be presented in greater detail but there is little doubt in such cells as normal liver, the NOR potential is repressed at least 90%. One hypothesis for bacterial systems (*44*), is that the level of ribosomal components themselves can be rate-limiting for rRNA synthesis (v.s.). This type of "feedback control" could be important in two states other than mitosis; (a) the metabolic states of bacteria in which the organisms have no nucleoli and the DNA is in proximity to all of the metabolic events of the cell which usually take place rapidly (division in bacteria occurs as often as every 10–20 minutes), and (b) in the normal state of interphase in either functional or resting cells.

The possibility that "feedback control" is involved in the biosynthesis of ribosomal precursors in nuclei during mitosis is not very likely since there is probably little difference in the functional states of the cytoplasm of these cells just before, during, and after mitosis; i.e., protein synthesis continues. The generally condensed states of the chromosomes and the virtually complete loss of ribonucleoprotein from metaphase chromosomes suggests that other mechanisms may be operative to produce generalized blocks in the biosynthetic activities of the nucleolar genes. General repressors may function at this time as compared with the more discrete and discriminating repressors normally operative in liver cells and similarly "modulated" cells. Such controls may include histones (*15*), acidic proteins (*37, 43, 108*), specific alterations in the state of permeability of the nuclear membrane with the entry of special enzymes into the nucleus, or enzymes and proteins involved in coiling or super-coiling of chromosomes (*112*). As yet, no particular substance (*5, 14, 102, 106, 115–117, 123, 144*) has been proved to be involved in control of the NOR's.

C. Nucleolar Extrusion

Two major hypotheses have been presented on the fate of the nucleoli during prophase; i.e., disintegration and extrusion. In preprophase and prophase, there is an increased content of nuclear components, including RNA, acidic proteins, DNA, and histones (*15, 141, 142*) as might logically be expected to be required by the dividing cells. However, from the onset of metaphase, the amount of nuclear ribonucleoprotein decreases. In some instances in tissue culture, whole nucleoli have been observed to migrate from the nucleus into the cytoplasm; i.e., the nucleolus is "extruded." In other instances in tissue culture, the nucleolar contents are extruded into the cytoplasm; i.e., the nucleolus appears to release a sac-load of components in much the same way it is envisioned that a phage releases the virus into the cell. Some workers have suggested that this is the normal mechanism by which the nucleolar products move into the cytoplasm and this finding supported the old idea that the nucleolus is a "collecting station" for nuclear products. However, others have suggested that extrusion of the nucleolus is an agonal phenomenon, a morphological curiosity, or the result of the effects of toxic drugs (*3, 25a, 54, 66, 77–80, 114, 128*). Other evidence has been provided that nucleolar products are released into the cytoplasm as organized ribonucleoprotein particles (*36, 131;* see also Chapter I). In prophase, Love has noted that most of the parachromatin (ribonucleoprotein) diffuses into the cytoplasm and it is not reincorporated into the nucleus or nucleolus (*35, 83–85*). He has suggested that there is an exchange of ribonucleoproteins of the cytoplasm and the nucleus during the process of mitosis.

An alternative concept indicates that organized nucleolar and other ribonucleoproteins normally move through a nuclear network to the nuclear membrane and then into the cytoplasm (see Chapter VIII).

At present, the mechanisms of extrusion of the nucleolar products are not defined, but it is doubtful that whole nuclear extrusion is a normal interphase process. It is more likely that there is a mass extrusion of the nucleolar components into the cytoplasm during cell division.

X. Nucleolar Persistence

Not all nucleoli disappear in mitosis. Hsu and his associates (*54*) noted that after the treatment of cells in tissue culture with fluorodeoxyuridine or other drugs, some nucleoli persisted in the nucleus or in the cytoplasm throughout mitosis. Similar effects were observed by Studzinski (*128*) after treatment of cells with puromycin. Inasmuch as the nucleoli persist, presumably nucleolus-associated DNA or another special chromosomal mass keeps them functional or protects them against dissolution. No precise role of these "persistent nucleoli" is established but in another report, Hsu *et al.* (*52, 54*) showed that nucleolar persistence through mitosis was more common in some cell lines in tissue culture than in others. For example, in the Dede strain, there was 98% nucleolar persistence; in an LM and Lagothrix strain there was only a 4–5% persistence. Persistent nucleoli have been found in *Psilotum nudum,* a plant, by Fabbri (*32*). In most cells, these aberrant nucleoli would be completely destroyed by cytoplasmic components as was found by Hoskins and Montgomery (*51*).

XI. Protein Synthesis in the NOR

Whether the protein that is required for the formation of nucleolar ribonucleoproteins is produced in or near the chromosomes or whether it is a product of the cytoplasm that migrates to the nucleolus, Balbiani rings and puffs is a topic that has not yet been satisfactorily approached experimentally (see Chapter VII). Some labeling of the proteins in the regions of the chromosomes was reported by Sirlin and Knight (*121*). After administration of labeled leucine and/or methionine to larvae of *Drosophila* or *Smittia*, there was localization of the label along the polytene chromosomes (Fig. 3.9). Since the period of labeling was 2–6 hr, it was not certain whether the isotope in the chromosomes reflected direct protein synthesis on the chromosomes or movement of protein from the cytoplasm into the nucleus. They suggested that the synthesis occurred on the chromosomes and that the protein synthesized consisted of many products formed by the action of many genes.

Evidence that the synthesis of protein is important in the reformation of nuclear structures after mitosis was provided by Cummins *et al.* (*24*) in slime molds. If protein synthesis was blocked just prior to dissolution of the nucleolus in prophase by actidione or cycloheximide, there was a failure of reconstruction of the nucleus. The reactions involved may be similar to the ones that were investigated by Gaulden and Perry (*42*) in their studies on the effects of ultraviolet microbeam irradiation on the role of the nucleolus in mitosis. In cells with irradiated nucleoli, there was a failure of cell division.

To summarize the findings on the activity of the NOR's, apparently they do not function after late prophase, in anaphase, or metaphase in most cells. The mechanism for the shutoff of these genes is not established but among the functional genes, the nucleolar genes apparently are among the last to cease functioning in mitosis and among the earliest to begin again. The fate of the nucleolus varies during mitosis; i.e., it may persist, or it may be extruded although it generally disintegrates. Persistence is an occasional phenomenon in most cell types, but there are some cell lines where it occurs more frequently. Although the nucleolus usually disintegrates in mitosis and its products migrate to the cytoplasm, the fate of the nucleolar residues is not defined; i.e., they may be destroyed or serve as a base for the resynthesis of the nucleoli. The mechanism of activation of the newly formed nucleolus organizers in

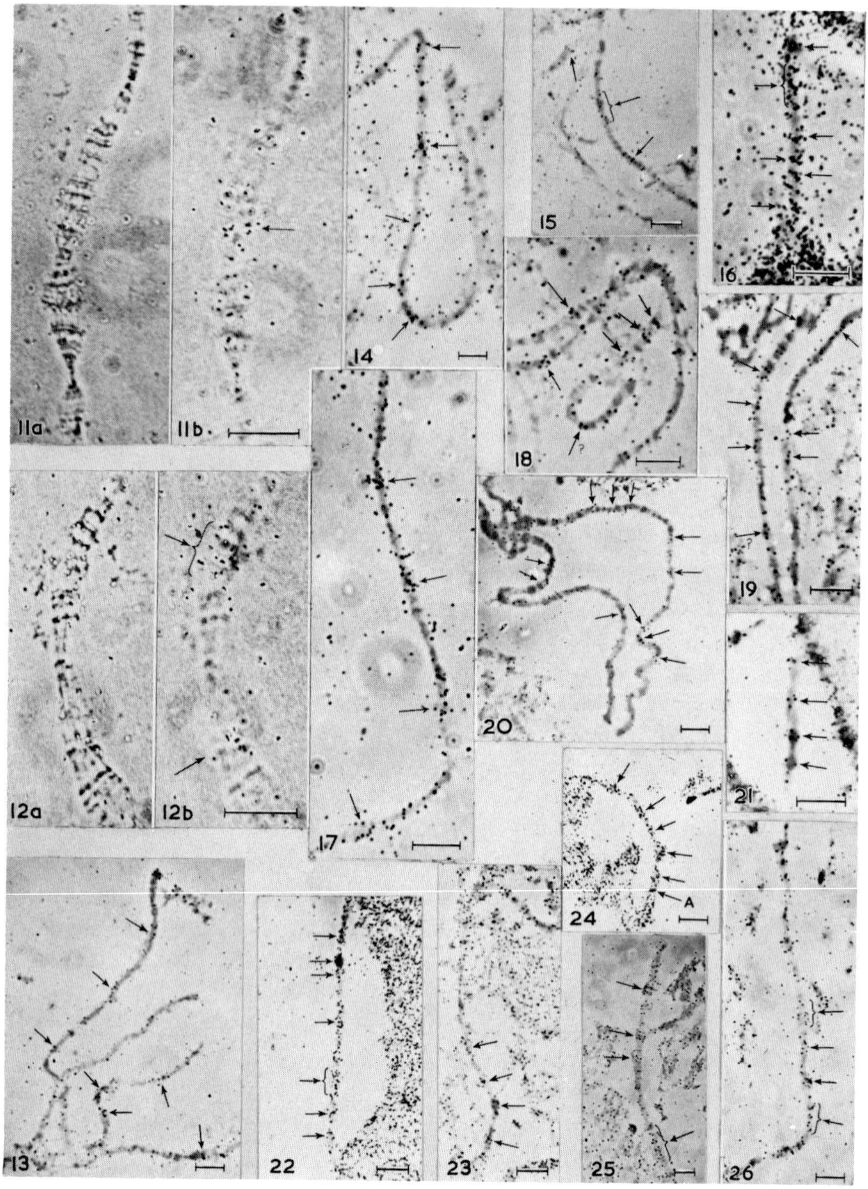

Fig. 3.9. Uptake of labeled amino acids on the chromosomes of *Drosophila*. The dots represent reduced silver grains and the inserts—methionine-^{35}S—24.5 hr (inserts 11–12), methionine-^{14}C, 2 hr (inserts 13–15) or 6 hr (inserts 16 and 17), leucine-^{3}H, 2 hr (inserts 18–21), 4 hr (inserts 22, 23, and 27) or 6 hr (inserts 24–26). In insert 24 A = dark bands without reduced silver grains. (From Sirlin and Knight, *121*).

telophase is unknown but is believed to be different from the quantitative modulation exercised during interphase which enables the nucleolus to produce greater or lesser quantities of product as required for cellular function. As will be noted in Chapter X, the activity of the nucleolus is sensitive to the effects of a variety of toxic agents and physiological demands of the cell but their precise effects on the NOR's remains to be defined.

XII. The Relationship of Nucleoli to Puffs and Balbiani Rings

In the polytene chromosomes of the *Drosophila* and other flies, midges, and insects, the individual chromomeres become activated at different times, under different conditions of hormonal influence and other physiological changes. There are variations in puff sizes and locations, but nucleoli appear to be localized in well-defined regions or bands of the chromosomes which account for 0.5% of the total genome. In *Chironomus tentans,* chromosomes II and III have NOR's (see p. 121); the NOR's have been located with relative ease in nuclei of the malpighian tubules of the adult forms of *Chironomus tentans* (*8, 10, 11, 104, 105*). An "accessory nucleolus" may develop on chromosome IV, generally at the end of the chromosome.

Inasmuch as the formation of puffs and Balbiani rings is associated with the production of ribonucleoproteins, it might seem that there is some similarity of these events with the formation of nucleoli. However, the similarity would appear to be rather limited since the composition of the components of the Balbiani rings and puffs is not clearly defined but they do not serve for the formation of ribosomal precursors which is one of the main functions of the nucleoli (*29a*). Moreover, nucleoli are present in these cells throughout their functional stages while the Balbiani rings and puffs are uniquely sensitive to the formation of specific secretory granules (*10, 11*), molting stages (*23*), hormones, particularly juvenile hormone and ecdysone (*22, 23, 72*), and specific stages of cell function (*76*) dependent upon the particular organism studied.

These puffs are found at specific places on the chromosomes and contain both RNA and protein as is the case with the nucleoli. Swift (*130*) have suggested that these puffs also contain DNA and histone but there are difficulties in making precise biochemical studies on such small amounts of material as can be obtained from these insect substructures. Although the possibility of denaturation of the proteins is serious and hydrolytic cleavage of RNA and DNA may occur in these structures as they are prepared for isolation, the microanalytical techniques are of considerable interest.

There are few studies that are more intriguing than those on the Balbiani rings of the salivary glands of the *Chironomus* and those on ecdysone, a hormone that induces molting in *Drosophila* and other Diptera (*72*). In *Chironomus thummi,* there are marked changes in morphological appearance during the last instar which is the prepupal stage. These changes are associated with increases and decreases in the sizes of the ribonucleoprotein containing puffs of the chromosomes. Moreover, these changes are apparently quite specific in the sense that the puffs affected are localized in specific places within the chromosomes. Ecdysone, the molting hormone, is the product of the prothoracic gland and the juvenile hormone determines the nature of the molt; both produce enlargement of specific puffs in the chromosomes. Kroeger (*72*), utilized transplantation to produce some puffing changes that were quite similar to those produced with purified hormones.

Other substances may mimic the effects of ecdysone and juvenile hormone (*72*). Solutions of $ZnCl_2$, $CdCl_2$, chloroform, butanol, and urethane all had effects similar to those of ecdysone when given in relatively high doses. Similarly, the effects of juvenile hormone could be duplicated by the wounding of the cells, X-radiation, or oxygen poisoning. From these results, it was concluded that the effects of these substances may not occur directly on the genes but rather on some cytoplasmic input elements that later produced altered gene outputs. Conceivably the hormones also act indirectly through some form of cytoplasmic messenger that communicates the need for specific readouts to the genome. This concept is similar to the one postulated for nucleolar synthesis of ribonucleoproteins (p. 141).

One of the decisive experiments that was carried out on the phenotypic development of cells and the activity of the Balbiani rings was that of Beermann (*10*). He related the production of specific secretory granules in the salivary glands to a specific chromosomal locus. In the *Chironomus,* there are only four cells which are active in the production of the normal salivary secretions. These cells contain the biochemical apparatus for production of the secretory granules referred to as SZ granules. Mutants were obtained which lack the ability to produce the SZ granules, either because the cells are inactive or absent; the salivary secretions of this type of fly differs from that of the normal flies.

The production of the secretory granules has been related to the activity of one specific Balbiani ring on chromosome IV; in the absence of this specific puff, these cells are not functional. The conclusion has been drawn that the BR_4 or the Balbiani ring on chromosome number IV is a normally activated locus which controls a specific function. Thus, it seems possible that each Balbiani ring or perhaps several of these rings acting in concert may produce specific biochemical activities in cells and thus may represent the intermediaries by virtue of which the chromosomes express genetic information.

To the extent that specific drugs, such as actinomycin and puromycin, exert inhibitory effects on the formation and function of the various puffs and Balbiani rings (*22, 23*) there is little doubt that the processes of protein synthesis and RNA synthesis are similar to those of the normal nucleolar synthetic reactions as indicated in Chapters V and VII.

One of the main points of similarity of nucleolar function and those of the synthesis of Balbiani rings and puffs is with respect to the specificity of the regions involved. For the nucleoli, only specific chromosomal regions are definitely related to nucleolar synthesis and only specific kinds of DNA seem to serve as templates for the transcription of RNA. In the case of the Balbiani rings, the situation seems similar although specific loci probably produce specific messenger or template RNA molecules.

In the flies and gnats, there is a very marked similarity of the appearance and functional activity of the nucleoli to those of the mammalian cells (see Chapter I), but there is one major point of difference. As will be indicated in the chapter on nucleolar RNA (Table 5.3a,b) the nucleolus of mammalian cells produces an RNA rich in guanylic (G) and cytidylic (C) acids; i.e., a GC-rich RNA. However, in the cells of *Drosophila* and in *Chironomus tentans* the nucleolar RNA is AU-rich; i.e., A 30, U 27, G 20, C 22 (*11, 28, 29, 111;* see also Chapter IV). It is not possible to evaluate the significance of this difference which is of interest because in mammalian cells, messenger RNA is generally reported to be AU-rich.

Acknowledgment

The authors express their sincere appreciation to Drs. T. C. Hsu and Tae Suk Ro-Choi for their very kind cooperation in the preparation of parts of this chapter.

REFERENCES

1. Adams, H. R., and Busch, H., Labeling of nucleolar RNA with orotic acid-2-C^{14} in livers of normal and thioacetamide-treated rats. *Biochem. Biophys. Res. Commun.* **9**, 578–583 (1962).
2. Adams, H. R., and Busch, H., Effects of thioacetamide on incorporation of orotic acid-2-C^{14} into RNA fractions in liver. *Cancer Res.* **23**, 576–582 (1963).
3. Albanese, M. P., Passagio di nucleoli attraverso la membrana nucleare negli ovociti in accrescimento di Patella coerulea L. *Experientia* **20**, 550–551 (1964).
4. Barr, H. J., Problems in the developmental cytogenetics of nucleoli in *Xenopus. Natl. Cancer Inst. Monograph* **23**, 411–424 (1966).

5. Barr, H. J., and Esper, H., Nucleolar size in cells of *Xenopus laevis* in relation to nucleolar competition. *Exptl. Cell Res.* **31**, 211–214 (1963).
6. Barr, M. L., and Bertram, E. G., A morphological distinction between neurones of the male and female, and the behaviour of the nucleolar satellite during accelerated nucleoprotein synthesis. *Nature* **163**, 676–677 (1949).
7. Battaglia, E., Mutazioni coinvolgenti cromosomi nucleolari in *Allium cepa L. Caryologia* **16**, 405–429 (1963).
8. Bauer, H., The structure of chromosomes of the salivary glands of chironomus thummikiefer. *Z. Zellforsch. Mikroskop. Anat.* **23**, 280–313 (1935).
9. Bauer, H., and Caspersson, T. O., Cytochemical observations on nucleolus formation in *Chironomus. Hereditas* Suppl. 35 (1948).
9a. Beermann, W., The nucleolus as a component of the cell nucleus essential to life. *Chromosoma* **11**, 263–296 (1960).
10. Beermann, W., Ein Balbiani-ring als locus einer speicheldrusen mutation. *Chromosoma* **12**, 1–25 (1961).
11. Beermann, W., Cytological aspects of information transfer in cellular differentiation. *Am. Zoologist* **3**, 23–32 (1963).
12. Birnstiel, M. L., Wallace, H., Sirlin, J. L., and Fischberg, M., Localization of the ribosomal DNA complements in the nucleolar organizer region of *Xenopus laevis. Natl. Cancer Inst. Monograph* **23**, 431–448 (1966).
13. Brachet, J., "Biochemical Cytology." Academic Press, New York, 1957.
14. Bucknall, R. A., and Carter, S. B., A reversible inhibitor of nucleic acid synthesis. *Nature* **213**, 1099–1101 (1967).
15. Busch, H., "Histones and Other Nuclear Proteins." Academic Press, New York, 1965.
16. Busch, H., Byvoet, P., and Smetana, K., The nucleolus of the cancer cell: A review. *Cancer Res.* **23**, 313–339 (1963).
17. Busch, H., Muramatsu, M., Adams, H. R., Smetana, K., Steele, W. J., and Liau, M. C., Isolation of nucleoli. *Exptl. Cell Res.* Suppl. 9, 150–163 (1963).
17a. Callan, H. G., Chromosomes and nucleoli of the axolotyl, *Ambystoma Mexicanum. J. Cell Sci.* **1**, 85–108 (1966).
18. Caspersson, T. O., "Cell Growth and Cell Function. A Cyto-Chemical Study." Norton, New York, 1950.
19. Chakravorty, A. K., and Busch, H., Effect of thioacetamide on the activity of alkaline ribonuclease and ribonuclease inhibitor in rat liver. *Biochem. Pharmacol.* **16**, 1711–1718 (1967).
20. Chen, T. T., Observations on mitosis in opalinids (protozoa ciliata). I. The behaviour and individuality of chromosomes and their significance. *Proc. Natl. Acad. Sci. U. S.* **22**, 594–602 (1936).
21. Chen, T. T., Observations on mitosis in opalinids (protozoa ciliata). II. The association of chromosomes and nucleoli. *Proc. Natl. Acad. Sci. U. S.* **22**, 602–607 (1936).
22. Clever, U., Genaktivatäten in den riesenchromosomen von Chironomus tentans und ihre beziehungen zur entwicklung. IV. Das verhalten der puffs in der larvenhäutung. *Chromosoma* **14**, 651–675 (1963).
23. Clever, U., Induction and repression of a puff in *Chironomus tentans. Develop. Biol.* **14**, 421–438 (1966).
24. Cummins, J. E., Blomquist, J. C., and Rusch, H. P., Anaphase delay after inhibition of protein synthesis between late prophase and prometaphase. *Science* **154**, 1343–1344 (1966).

25. Das, N. K., Chromosomal and nucleolar RNA synthesis in root tips during mitosis. *Science* **140**, 1231–1233 (1963).
25a. Das, N. K., Luykx, P., and Alfert, M., The nucleolus and RNA metabolism in *Urechis* eggs. *Develop. Biol.* **12**, 72–78 (1965).
26. Dearing, W. H., The material continuity and individuality of the somatic chromosomes of *ambyostoma tigrinum* with special reference to the nucleolus as a chromosomal component. *J. Morphol.* **56**, 157–179 (1934).
27. Di Berardino, M. A., The karyotype of *Rana pipiens* and investigation of its stability during embryonic differentiation. *Develop. Biol.* **5**, 101–126 (1962).
28. Edström, J. E., The base composition of RNA from nucleoli formed at two different organizers in *Chironomus tentans*. *Biochem. Biophys. Res. Commun.* **18**, 341–344 (1965).
29. Edström, J. E., Composition of the nucleolus as a basis for views on its function. *Natl. Cancer Inst. Monograph* **23**, 223–234 (1966).
29a. Edström, J.-E., and Daneholt, B., Sedimentation properties of the newly synthesized RNA from isolated nuclear components of *Chironomus tentans* salivary gland cells. *J. Mol. Biol.* **28**, 331–343 (1967).
30. Elsdale, T. R., Fischberg, M., and Smith, S., A mutation that reduces nucleolar number in *Xenopus laevis*. *Exptl. Cell Res.* **14**, 642–643 (1958).
31. Errera, M., and Brunfaut, M., Observations of mitotic figures in pulse labeled HeLa cells. *Exptl. Cell Res.* **33**, 105–111 (1964).
31a. Esper, H., and Barr, H. J., A study of the developmental cytology of a mutation affecting nucleoli in *Xenopus* embryos. *Develop. Biol.* **10**, 105–121 (1964).
32. Fabbri, F., Contributo per l'interpretazione della persistenza nucleolare durante la mitosi in *Psilotum nudum* (L.) Beauv. *Caryologia* **13**, 297–337 (1960).
33. Feinendegen, L. E., and Bond, V. P., Observations on nuclear RNA during mitosis in human cancer cells in culture (HeLa-S_3), studied with tritiated cytidine. *Exptl. Cell Res.* **30**, 393–404 (1963).
34. Feinendegen, L. E., Bond, V. P., Shreeve, W. W., and Painter, R. B., RNA and DNA metabolism in human tissue culture cells studied with tritiated cytidine. *Exptl. Cell Res.* **19**, 443–459 (1960).
35. Feldherr, C. M., Nucleocytoplasmic exchanges during cell division. *J. Cell Biol.* **31**, 199–203 (1966).
36. Fischberg, M., and Wallace, H., A mutation which reduces nucleolar number in *Xenopus laevis*. *In* "The Cell Nucleus" (J. S. Mitchell, ed.), pp. 30–33. Butterworth, London and New York, 1960.
37. Franklin, R. M., and Baltimore, D., Patterns of micromolecular synthesis in normal and virus-infected mammalian cells. *Cold Spring Harbor Symp. Quant. Biol.* **27**, 175–195 (1962).
38. Furieri, P., Chromatina associata al nucleoli in alcuni mammiferi. *Boll. Soc. Ital. Biol. Sper.* **33**, 260–261 (1957).
39. Gabrusewycz-Garcia, N., and Kleinfeld, R. G., A study of the nucleolar material in *Sciara coprophila*. *J. Cell Biol.* **29**, 347–359 (1966).
40. Gall, J. G., Nuclear RNA of the salamander oocyte. *Natl. Cancer Inst. Monograph* **23**, 475–488 (1966).
41. Gall, J. G., and Callan, H. G., H^3 uridine incorporation in lampbrush chromosomes. *Proc. Natl. Acad. Sci. U. S.* **48**, 562–570 (1962).
42. Gaulden, M. E., and Perry, R. P., Influence of the nucleolus on mitosis as revealed by ultraviolet microbeam irradiation. *Proc. Natl. Acad. Sci. U. S.* **44**, 553–559 (1958).

42a. Gersh, E. S., Mutants at the bobbed locus in *Drosophila melanogaster:* Relation to ribosomal RNA synthesis? *Science* **162**, 1139 (1968).
43. Gilbert, W., and Müller-Hill, B., Isolation of the lac repressor. *Proc. Natl. Acad. Sci. U. S.* **56**, 1891–1898 (1966).
44. Gros, F., Gilbert, W., Hiatt, H. H., Attardi, G., Spahr, P. F., and Watson, J. D., Molecular and biological characterization of messenger RNA. *Cold Spring Harbor Symp. Quant. Biol.* **26**, 111–132 (1961).
45. Hancock, R. L., Nucleolar organizers of blastocyst nuclei. *Growth* **28**, 251–256 (1964).
46. Hancock, R. L., Giant chromocentres and nucleoli in *Lumbricus. Nature* **205**, 1337–1338 (1965).
46a. Hay, E. D., and Gurdon, J. B., Fine structure of the nucleolus in normal and mutant *Xenopus* embryos. *J. Cell Sci.* **2**, 151–162 (1967).
47. Heitz, E., Uber totale und partielle somatische heteropyknose, sowie strukturelle geschlechtschromosomen bie *Drosophila funebris. Z. Zellforsch. Mikroskop. Anat.* **19**, 720–742 (1933).
48. Hennig, W., Untersuchungen zur struktur und funktion des lampenbürsten-Y-chromosoms in der spermatogenese von *Drosophila. Chromosoma* **22**, 294–357 (1967).
49. Hess, O., Struktur-differenzierungen im Y-chromosom von *Drosophila hydei* und ihre bezieungen zu gen-aktivitäten. *Chromosoma* **16**, 222–248 (1965).
50. Ho, T., The nucleolar chromosomes of the maiden-hair tree. *J. Heredity* **54**, 67–74 (1963).
51. Hoskins, G. C., and Montgomery, P. O'B., Nucleolar phagocytosis by HeLa and Chang liver cells in tissue culture as studied by time lapse phase contrast cinematography. *Exptl. Cell Res.* **26**, 534–540 (1962).
51a. Hsu, T. C., Personal communication (1969).
52. Hsu, T. C., Arrighi, F. E., Klevecz, R. R., and Brinkley, B. R., The nucleoli in mitotic divisions of mammalian cells *in vitro. J. Cell Biol.* **26**, 539–553 (1965).
53. Hsu, T. C., Brinkley, B. R., and Arrighi, F. E., The structure and behavior of the nucleolus organizers in mammalian cells. *Chromosoma* **23**, 137–153 (1967).
54. Hsu, T. C., Humphrey, R. M., and Somers, C., Persistent nucleoli in animal cells following treatments with fluorodeoxyuridine and thymidine. *Exptl. Cell Res.* **33**, 74–77 (1964).
54a. Huang, R. C. C., and Huang, P. C., Effect of protein-bound RNA associated with chick embryo chromatin on template specificity of the chromatin. *J. Mol. Biol.* **39**, 365–378 (1969).
55. Huberman, J. A., and Attardi, G., Isolation of metaphase chromosomes from HeLa cells. *J. Cell Biol.* **31**, 95–105 (1966).
56. Huberman, J. A., and Attardi, G., Studies of fractionated HeLa cell metaphase chromosomes. I. The chromosomal distribution of DNA complementary to 28 S and 18 S ribosomal RNA and to cytoplasmic messenger RNA. *J. Mol. Biol.* **29**, 487–505 (1967).
57. Jacob, F., and Monod, J., Genetic regulatory mechanisms in the synthesis of proteins. *J. Mol. Biol.* **3**, 318–356 (1961).
58. Jacob, J., An electron microscope antoradiographic study of the site of initial synthesis of RNA in the nucleolus of *Smittia. Exptl. Cell Res.* **48**, 276–282 (1967).
59. Jacob, J., and Sirlin, J. L., Electron microscope studies on salivary gland cells. I. The nucleus of *Bradysia mycorum* frey with special reference to the nucleolus. *J. Cell Biol.* **17**, 153–165 (1963).

59a. Jacob, J., and Sirlin, J. L., Electron microscope studies on salivary gland cells. IV. The nucleus of *Smittia parthenogenetica* (Chironomidae) with special reference to the nucleolus and the effects of actinomycin thereon. *J. Ultrastruct. Res.* **11**, 315–328 (1964).
60. Kahn, J., The nucleolar organizer in the mitotic chromosome complement of Xenopus laevis. *Quart. J. Microscop. Sci.* **103**, 407–409 (1962).
61. Kalnins, V. I., Stich, H. F., and Bencosme, S. A., Fine structure of the nucleolar organizer of salivary gland chromosome of Chironomids. *J. Ultrastruct. Res.* **11**, 282–291 (1964).
62. Kalnins, V. I., Stich, H. F., and Bencosme, S. A., Fine structure of nucleoli and RNA-containing chromosome regions of salivary gland chromosomes of Chironomids and their interrelationship. *Can. J. Zool.* **42**, 1147–1155 (1964).
63. Kasten, F. H., and Strasser, F. F., Nucleic acid synthetic patterns in synchronized mammalian cells. *Nature* **211**, 135–140 (1966).
64. Kasten, F. H., Strasser, F. F., and Turner, M., Nucleolar and cytoplasmic ribonucleic acid inhibition by excess thymidine. *Nature* **207**, 161–164 (1965).
65. Kato, H., and Sandberg, A. A., Chromosome pulverization in human cells with micronuclei. *J. Natl. Cancer Inst.* **40**, 165–179 (1968).
66. Kessel, R. G., and Beams, H. W., Nucleolar extrusion in oocytes in *Thyone briareus*. *Exptl. Cell Res.* **32**, 612–615 (1963).
67. Keyl, H. G., Chromosomenevolution bei chironomus I. Strukturabwandlungen an speicheldrusenchromosomen. *Chromosoma* **12**, 26–47, (1961).
67a. Kiefer, B. I., Dosage regulation of ribosomal DNA in *Drosophila melanogaster*. *Proc. Natl. Acad. Sci. U. S.* **61**, 85–89 (1968).
68. Kleinfeld, R. G., Altered patterns of RNA metabolism in liver cells following partial hepatectomy and thioacetamide treatment. *Natl. Cancer Inst. Monograph* **23**, 369–378 (1966).
69. Kleinfeld, R. G., and von Haam, E., Effect of thioacetamide on rat liver regeneration. II. Nuclear RNA in mitosis. *J. Biophys. Biochem. Cytol.* **6**, 393–398 (1959).
70. Kopac, M. J., and Mateyko, G. M., Malignant nucleoli: Cytological studies and perspectives. *Ann. N. Y. Acad. Sci.* **73**, 237–282 (1958).
71. Kopac, M. J., and Mateyko, G. M., Nucleolar chromosomes: Structures, interactions and perspectives. *Advan. Cancer Res.* **8**, 121–190 (1964).
71a. Krivshenko, J., New evidence for the homology of the short euchromatic elements of the X and Y chromosomes of *Drosophila busckii* with the microchromosome of *Drosophila melanogaster*. *Genetics* **44**, 1027–1040 (1959).
72. Kroeger, H., Experiments on the extranuclear control of gene activity in Dipteran polytene chromosomes. *Symp. Macromol. Aspects Cell Cycle, Oak Ridge, 1963*, pp. 45–59, *J. Cell Comp. Physiol.* **62**[1] (1963).
73. Kunz, W., Lampenbürstenchromosomen und multiple nukleolen bei orthopteren. *Chromosoma* **21**, 446–462 (1967).
73a. Lafontaine, J. G., Structural components of the nucleus in mitotic plant cells. *In* "The Nucleus" (A. J. Dalton and F. Haguneau, eds.), pp. 151–196, Academic Press, New York, 1968.
74. Lafontaine, J. G., and Chouinard, L. A., A correlated light and electron microscope study of the nucleolar material during mitosis in *Vicia faba*. *J. Cell Biol.* **17**, 167–201 (1963).
74a. Lafontaine, J. G., Ultrastructure and mode of formation of the nucleolus in plant cells. *Natl. Cancer Inst. Monograph* **23**, 67–75 (1966).

75. Laird, A. K., Nuclear changes induced in rat liver cells by thioacetamide. *Arch. Biochem. Biophys.* **46**, 119–127 (1953).

76. Lara, F., and Hollander, F., Changes in RNA metabolism during the development of *Rhynchosciara angelae*. IEG No. 7, Memo No. 766 (1966).

77. Lettré, R., and Siebs, W., Beobachtungen am nucleolus *in vitro* gezuchteter zellen. *Z. Krebsforsch.* **60**, 19–20 (1954).

78. Lettré, R., and Siebs, W., Beobachtungen zur struktur des nucleolus in normalen zellen sowie in tumorzellen. *Z. Krebsforsch.* **60**, 564–580 (1955).

79. Lettré, R., and Siebs, W., Film sur des études du nucléole de cellules cultivées *in vitro*. *Chemotherapia* **2**, 223–225 (1961).

80. Lettré, R., and Siebs, W., Some studies of the nucleolus of cells cultivated *in vitro*. *Pathol. Biol., Semaine Hop.* [N. S.] **9**, 819–823 (1961).

80a. Lima-de-Faria, A., Reitalu, J., and O'Sullivan, M. A., Replication of autosomal heterochromatin in man. *Chromosoma* **16**, 152–161 (1965).

81. Lin, M., Chromosomal control of nucleolar composition in maize. *Chromosoma* **7**, 340–370 (1955).

82. Longwell, A. C., and Svihla, G., Specific chromosomal control of the nucleolus and of the cytoplasm in wheat. *Exptl. Cell Res.* **20**, 294–312 (1960).

83. Love, R., Interchange of ribonucleic acid between the nucleus and cytoplasm during mitosis of tumor cells. *Acta Unio Intern. Contra Cancrum* **16**, 1150–1152 (1960).

84. Love, R., Differences in the internal structure of nucleoli of diploid and non-diploid transformed or neoplastic cells *in vitro*. *Exptl. Cell Res.* **40**, 188–192 (1965).

85. Love, R., Anisonucleolinosis in mammalian cell cultures. *Natl. Cancer Inst. Monograph* **23**, 167–180 (1966).

86. Maio, J. J., and Schildkraut, C. L., Isolated mammalian metaphase chromosomes. I. General characteristics of nucleic acids and proteins. *J. Mol. Biol.* **24**, 29–39 (1967).

87. Mazia, D., Mitosis and the physiology of cell division. *In* "The Cell" (J. Brachet and A. E. Mirsky, eds.), Vol. 3, p. 373. Academic Press, New York, 1961.

88. McClintock, B., The relation of a particular chromosomal element to the development of the nucleoli in Zea mays. *Z. Zellforsch. Mikroskop. Anat.* **21**, 294–328 (1934).

89. McClintock, B., Neurospora. I. Preliminary observations of the chromosomes of neurospora crassa. *Am. J. Botany* **32**, 671–678 (1945).

90. McClintock, B., Controlling elements and the gene. *Cold Spring Harbor Symp. Quant. Biol.* **21**, 197–216 (1956).

91. McClintock, B., The relation of a particular chromosomal element to the development of the nucleoli in *Zea mays*. *Am. Naturalist* **95**, 265–328 (1961).

92. Mendelsohn, J., Moore, D. E., and Salzman, N. P., Separation of isolated Chinese hamster metaphase chromosomes into three size groups. *J. Mol. Biol.* **32**, 101–112 (1968).

93. Miller, O. L., Jr., Structure and composition of peripheral nucleoli of salamander oocytes. *Natl. Cancer Inst. Monograph* **23**, 53–66 (1966).

94. Mironescu, S., and Dragomir, C., Nucleolar behavior in regenerating liver of normal and whole-body-irradiated rats. *Cancer Res.* **27**, 1819–1830 (1967).

95. Montgomery, T. H., Comparative cytological studies with special regard to the nucleolus. *J. Morphol.* **15**, 265–564 (1898).

96. Mulnard, J., Présence d'inclusions feulgen positives dans les nucléoles larvaires de quelques dipteres. *Arch Biol.* (*Liege*) **67**, 485–498 (1956).
97. O'Donnell, E. H. J., Deoxyribonucleic acid structures in the nucleolus. *Nature* **191**, 1325–1326 (1961).
98. Ohno, S., Christian, L. C., and Stenius, C., Nucleolus-organizing microchromosomes of *Gallus domesticus*. *Exptl. Cell Res.* **27**, 612–614 (1962).
99. Ohno, S., Kaplan, W. D., and Kinosita, R., Heterochromatic regions and nucleolus organizers in chromosomes of the mouse, *Mus musculus*. *Exptl. Cell Res.* **13**, 358–364 (1957).
100. Ohno, S., and Kinosita, R., The primary and secondary constriction on the chromosomes of the rat lymphoblast. *Exptl. Cell Res.* **8**, 558–562 (1955).
101. Ohno, S., Trukillo, J. M., Kaplan, W. D., and Kinosita, R., Nucleolus organizers in the causation of chromosomal anomalies in man. *Lancet* **II**, 123–126 (1961).
102. Ohno, S., Weiler, C., and Stenius, C., A dormant nucleolus organizer in the guinea pig, *Cavia cobaya*. *Exptl. Cell Res.* **25**, 498–503 (1961).
103. Pätau, K., SAT-chromosomen und spiralstruktur der chromosomen der extrakapsularen korper. *Cytol. Fujii Jubilee* pp. 667–680 (1937).
104. Pelling, C., Chromosomal synthesis of ribonucleic acid as shown by incorporation of uridine labelled with tritium. *Nature* **184**, 655 (1959).
105. Pelling, C., and Beermann, W., Diversity and variation of the nucleolar organizing regions in chironomids. *Natl. Cancer Inst. Monograph* **23**, 393–410 (1966).
106. Person, S., Phillips, S., Funk, F., and Osborn, M., Suppression of nonsense by revertant bacteria. *IEG* No. 7, Sci. Memo No. 279 (1966).
107. Petersen, G. B., and Therkelsen, A. J., Number of nucleoli in female and male human cells in tissue culture. *Exptl. Cell Res.* **28**, 590–592 (1962).
108. Ptashne, M., Specific binding of the λ phage repressor to λ DNA. *Nature* **214**, 232–234 (1967).
109. Rather, L. J., Experimental alteration of nuclear and cytoplasmic components of the liver cell with thioacetamide. I. Early onset and reversibility of volume changes of the nucleolus, nucleus and cytoplasm. *Bull. Johns Hopkins Hosp.* **88**, 38–58 (1951).
109a. Ritossa, F. M., Nonoperative DNA complementary to ribosomal RNA. *Proc. Natl. Acad. Sci. U. S.* **59**, 1124–1131 (1968).
109b. Ritossa, F. M., Unstable redundancy of genes for ribosomal RNA. *Proc. Natl. Acad. Sci. U. S.* **60**, 509–516 (1968).
110. Ritossa, F. M., Atwood, K., Lindsley, D., and Spiegelman, S., On the chromosomal distribution of DNA complementary to ribosomal and soluble RNA. *Natl. Cancer Inst. Monograph* **23**, 449–472 (1966).
111. Ritossa, F. M., Atwood, K. C., and Spiegelman, S., On the redundancy of DNA complementary to amino acid transfer RNA and its absence from the nucleolar organizer region of *Drosophila melanogaster*. *Genetics* **54**, 663–676 (1966).
112. Schultz, J., Perspectives for the conference on the nucleolus. *Natl. Cancer Inst. Monograph* **23**, 1–9 (1966).
113. Shea, J. R., Jr., and Leblond, C. P., Number of nucleoli in various cell types of the mouse. *J. Morphol.* **119**, 425–434 (1966).
114. Shimizu, N., and Ishii, S., Electron microscopic observations on nucleolar extrusion in nerve cells of the rat hypothalamus. *Z. Zellforsch. Mikroskop. Anat.* **67**, 367–372 (1965).
115. Shiokawa, K., and Yamana, K., Pattern of RNA synthesis in isolated cells of *Xenopus laevis* embryos. *Develop. Biol.* **16**, 368–388 (1967).

116. Shiokawa, K., and Yamana, K., Inhibitor of ribosomal RNA synthesis in *Xenopus laevis* embryos. *Develop. Biol.* **16**, 389–406 (1967).
117. Shiokawa, K., and Yamana, K., Ribonucleic acid (RNA) synthesis in dissociated embryonic cells of *Xenopus laevis*. III. Effects of uridine and cytidine on inhibition of ribosomal RNA synthesis by a cytoplasmic factor. *Proc. Japan Acad.* **43**, 148–151 (1967).
118. Sirlin, J. L., and Jacob, J., Sequential and reversible inhibition of synthesis of ribonucleic acid in the nucleolus and chromosomes: Effect of benzamide and substituted benzimidazoles on dipteran salivary glands. *Nature* **204**, 545–547 (1964).
119. Sirlin, J. L., Jacob, J., and Kato, K.-I., The relation of messenger to nucleolar RNA. *Exptl. Cell Res.* **27**, 355–359 (1962).
120. Sirlin, J. L., Kato, K., and Jones, K. W., Synthesis of ribonucleic acid in the nucleolus. *Biochim. Biophys. Acta* **48**, 421–423 (1961).
121. Sirlin, J. L., and Knight, G. R., Chromosomal synthesis of protein. *Exptl. Cell Res.* **19**, 210–219 (1960).
122. Sirlin, J. L., Tandler, C. J., and Jacob, J., The relationship between the nucleolus organizer and nucleolar RNA. *Exptl. Cell Res.* **31**, 611–615 (1963).
123. Sirtori, C., and Bosisio-Bestetti, M., Nucleolar changes in KB tumor cells infected with *Herpes simplex* virus. *Cancer Res.* **27**, 367–376 (1967).
124. Smetana, K., and Busch, H., Studies on staining and localization of acidic nuclear proteins in the Walker 256 carcinosarcoma. *Cancer Res.* **26**, 331–337 (1966).
125. Smith, S. J., A morphological and histochemical investigation of a preimplantation lethal (t^{12}) in the house mouse. *J. Exptl. Zool.* **132**, 51–79 (1956).
126. Stevens, B. J., The fine structure of the nucleolus during mitosis in the grasshopper neuroblast cell. *J. Cell Biol.* **24**, 349–368 (1965).
127. Stich, H., Bau und function der nukleolen. *Experientia* **12**, 7–14 (1956).
128. Studzinski, G. P., Nucleolus-like inclusions in the cytoplasm of HeLa cells treated with puromycin. *Nature* **203**, 883–884 (1964).
129. Swift, H., Studies on nucleolar function. *Symp. Mol. Biol.* 266–293 (1960).
130. Swift, H., The histones of polytene chromosomes. *In* "The Nucleohistones" (J. Bonner and P. Ts'o, eds.), pp. 169–183. Holden Day, San Francisco, California, 1964.
131. Swift, H., and Stevens, B. J., Nucleolar-chromosomal interaction in microspores of maize. *Natl. Cancer Inst. Monograph* **23**, 145–166 (1966).
132. Therkelsen, A. J., and Petersen, G. B., Frequency of sex-chromatin-positive cells in the logarithmic and post-logarithmic growth phases of human cells in tissue culture. *Exptl. Cell Res.* **28**, 588–590 (1962).
132a. Thompson, L. R., and McCarthy, B. J., Stimulation of nuclear DNA and RNA synthesis by cytoplasmic extracts *in vitro*. *Biochem. Biophys. Res. Comm.* **30**, 166–172 (1968).
133. Tobias, P. V., "Chromosomes, Sex-cells and Evolution in a Mammal." Lund, Humphries & Co., London, 1956.
134. Unuma, T., Smetana, K., and Busch, H., A morphological study on the chromatin areas associated with the nucleolus. *Exptl. Cell Res.* **48**, 665–671 (1967).
135. Vincent, W. S., Isolation and chemical properties of nucleoli of starfish oocytes. *Proc. Natl. Acad. Sci. U. S.* **38**, 139–145 (1952).
136. Vincent, W. S., Structure and chemistry of nucleoli. *Intern. Rev. Cytol.* **5**, 269–299 (1955).

137. Vincent, W. S., Baltus, E., Lovlie, A., and Mundell, R. E., Proteins and nucleic acids of starfish oocyte nucleoli and ribosomes. *Natl. Cancer Inst. Monograph* **23,** 235–254 (1966).
138. Wallace, H., Autoradiographic studies of RNA and protein synthesis in anucleolate *Xenopus* embryos. *Natl. Cancer Inst. Monograph* **23,** 425–430 (1966).
139. Wallace, H., and Birnstiel, M. L., Ribosomal cistrons and the nucleolar organizer. *Biochim. Biophys. Acta* **114,** 296–310 (1966).
140. Whitehouse, H. L. K., A cycloid model for the chromosome. *J. Cell Sci.* **2,** 9–22 (1967).
141. Woodard, J., Gelber, B., and Swift, H., Nucleoprotein changes during the mitotic cycle in *Paramecium aurella. Exptl. Cell Res.* **23,** 258–264 (1961).
142. Woodard, J., Rasch, E., and Swift, H., Nucleic acid and protein metabolism during the mitotic cycle in *Vicia faba. J. Biophys. Biochem. Cytol.* **9,** 445–462 (1961).
143. Woods, M. W., The nucleolus in tulipa. *Am. J. Botany* **24,** 528–536 (1937).
144. Yamana, K., and Shiokawa, K., Occurrence of an inhibitory factor of ribosomal RNA synthesis in embryonic cells of *Xenopus laevis. Exptl. Cell Res.* **44,** 283–286 (1966).
145. Yerganian, G., and Papoyan, S., Isomorphic sex chromosomes, autosomal heteromorphism and telomeric associations in the grey hamster of Armenia, *Cricetulus migratorius* Pall. *Hereditas* **52,** 307–319 (1964).
146. Yoshida, T. H., Potter, M., and Imai, H. T., Comparative study of karyotypes in sublines producing and non-producing γ-globulin in mouse plasma cell tumors. *Natl. Inst. Genet. (Japan)* **17,** 13–14 (1966).

CHAPTER IV

*Nucleolar DNA**

* The authors are most appreciative to Dr. Tae Suk Ro-Choi for her aid in preparation of some of the tabular material.

I. Introduction

Several types of DNA, including the rDNA* cistrons and other DNA's, are associated with the nucleolus. Presumably, the rDNA cistrons are located intranucleolarly, but other types of DNA are probably also present within the nucleolar mass or the "nucleolar body," which is the structure contained within the nucleolus-associated chromatin (*6, 7*). In addition there is a dense perinucleolar chromatin mass which may contain a small amount of rDNA but probably contains mainly other types of DNA.

Morphologically, the two forms of DNA associated with the nucleolus are the intranucleolar DNA which may exist in dense masses or extended strands as shown in Figs. 4.1–4.2 (*4, 5*) and the dense perinucleolar nucleolus-associated DNA from which the intranucleolar DNA penetrates the nucleolus (Figs. 4.3, 2.12). Whether the perinucleolar nucleolus-associated DNA is another type of DNA than the intranucleolar DNA or simply represents an inactive nucleolar DNA (possibly rDNA) is not yet clear. Most workers believe that the nucleolar products are formed from within the nucleolar mass although it is still possible that some nucleolar synthesis is centripetal.

There is some evidence that the nucleolus is involved in a variety of biosynthetic reactions other than the synthesis of rRNA (see Chapters V, X). The presence of special proteins in the nucleolus (Chapter IV, *32*) shows that a number of cistrons must be functional for their synthesis and others must be functional for synthesis of specialized nucleolar structures such as the channels noted in Chapter I (p. 43). It seems possible that some of these syntheses occur in the nucleolus, but the number of active cistrons other than those for rRNA synthesis is not defined. Nonetheless, the evidence suggests that other cistrons are functional (see Chapter V).

II. Evidence for the Presence of Intranucleolar DNA

A. Light Microscopic Evidence for the Presence of Intranucleolar DNA

Very early studies on the nucleolus by means of either the Feulgen stain (Fig. 4.3) or methyl green (both of which stain for DNA) demon-

*The following abbreviations are used in this chapter: rDNA, that DNA which contains the cistrons coding for ribosomal RNA (rRNA); AU, adenylic (A) and uridylic (U) acids or deoxyadenylic (A) and deoxyuridylic (U) acids; GC, guanylic (G) and cytidylic (C) acid or deoxyguanylic (G) and deoxycytidylic (C) acids.

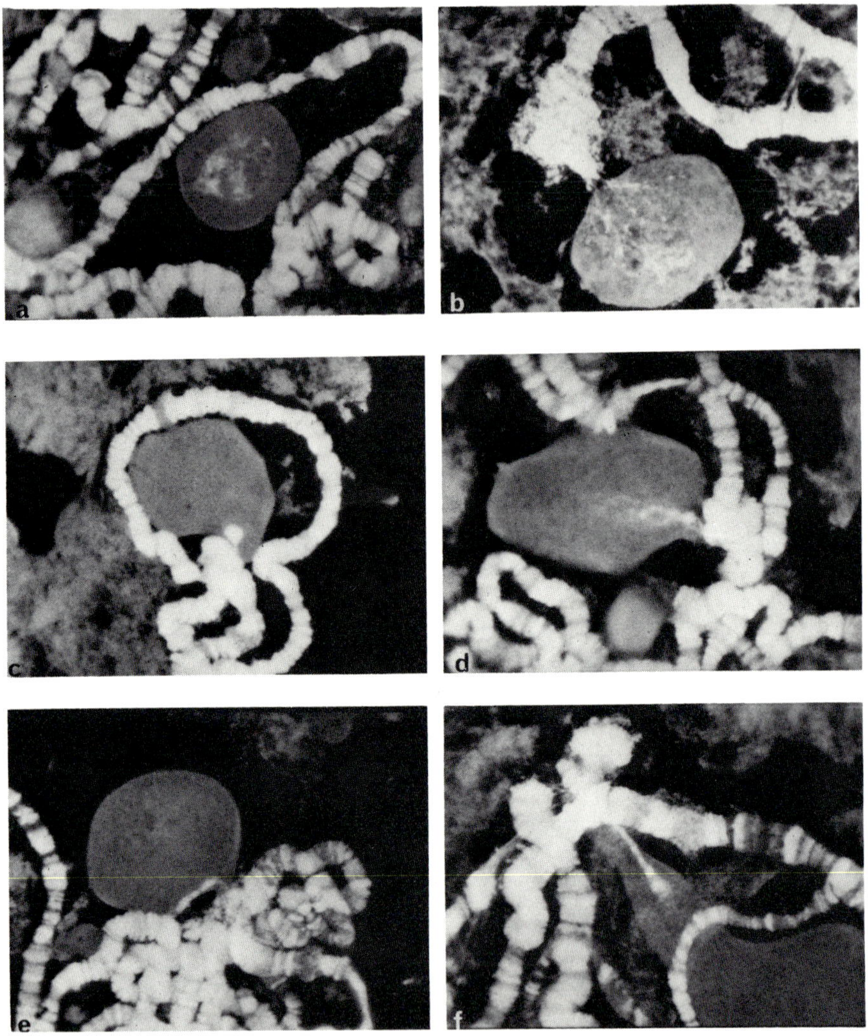

Figures 4.1 and 4.2 show the marked morphological variations of the intranucleolar DNA of various types of *Drosophila* species.

Fig. 4.1. a. *D. mercatorum mercatorum.* Salivary-gland cells showing DNA within nucleoli, but no morphological continuity between this DNA and the chromosomes. Stained with Coriphosphine-O and photographed at ×540. b. *D. peninsularis.* Salivary-gland cells showing nucleolar DNA which appears to be morphologically continuous with a terminal heterochromatin region of a chromosome. Stained with Coriphosphine-O and photographed at ×540 and ×324, respectively. c–f. *D. fulvimaculoides.* Salivary-gland cells, stained with Coriphosphine-O and photographed at ×540, showing some of the forms nucleolar DNA can take in this species. In Figs. 4.1e and f, note the banded appearance of the nucleolar DNA. Since such structures are found in both male and female larvae, they are most likely not Y chromosomes (courtesy of Dr. H. J. Barr, Department of Anatomy, University of Wisconsin).

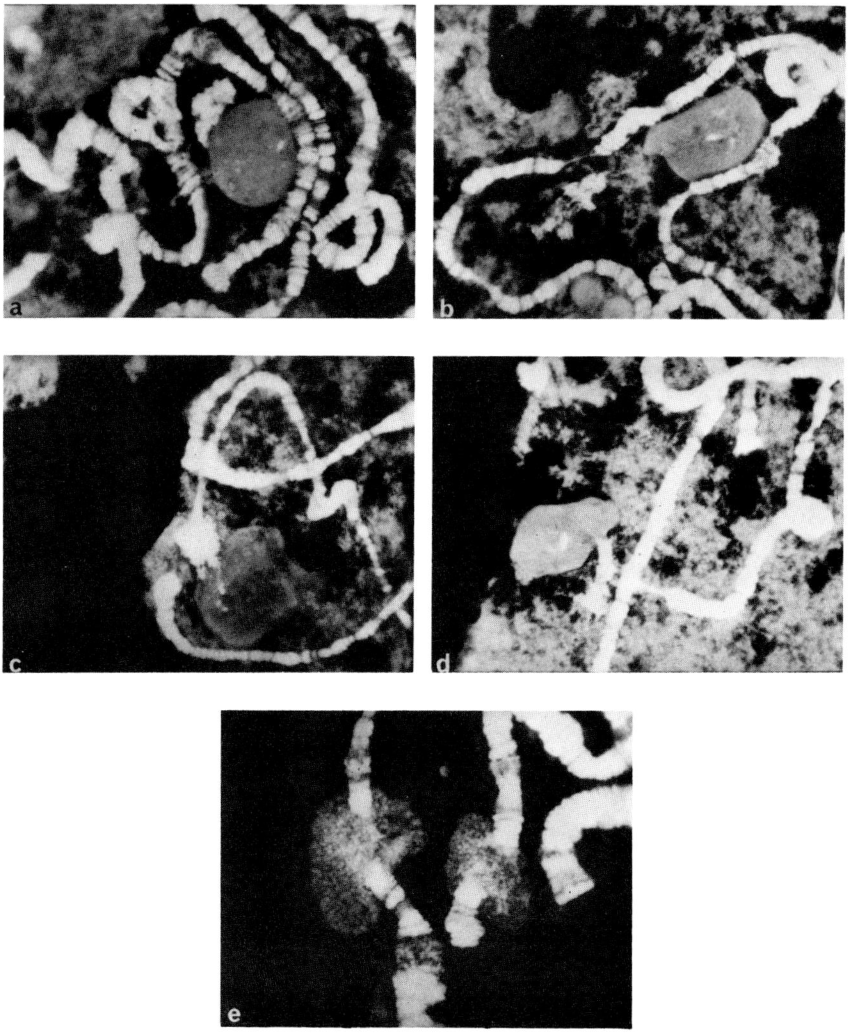

Fig. 4.2. a–d. *D. tumiditarsus*. Salivary-gland cells, stained with Coriphosphine-O and photographed at ×540. In Figs. 4.2a and b, note that the heterochromatic dot chromosome appears associated with the nucleolus. In Figs. 4.2c and d, a structure associated with the nucleolus and resembling the heterochromatic dot chromosome appears to be connected to a chromosome end. e. *Sciara coprophila*. Salivary-gland chromosomes showing DNA puffs. Stained with Coriphosphine-O and photographed at a magnification of ×540 (courtesy of Dr. H. J. Barr, Department of Anatomy, University of Wisconsin).

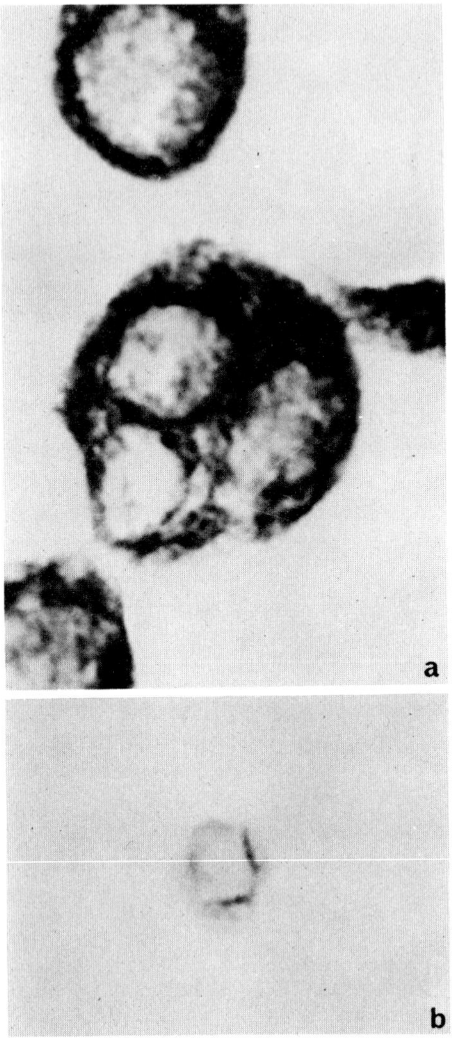

Fig. 4.3. a. Isolated nuclei of livers of rats treated with thioacetamide (50 mg/kg) for 9 days. The specimens were stained with a modified Feulgen stain for DNA (toluidine blue after treatment with 1 *N* HCl at 60°C). The enlarged nucleoli were surrounded by dense nucleolus-associated chromatin. A few strands within the nucleolar region appear to be intranucleolar chromatin. ×2600. b. Isolated nucleolus of a rat liver treated with thioacetamide for 9 days. The specimen was stained as for Fig. 4.3a. Perinucleolar nucleolus-associated chromatin is visible around the nucleolus but intranucleolar chromatin is less clearly defined. ×2600. From Unuma *et al.* (*76*).

strated the presence of a dense perinucleolar shell of chromatin referred to as the "nucleolus-associated chromatin." Caspersson (*14*) called this chromatin "heterochromatin" because of its density. Subsequently it was suggested that this chromatin might be the most active cellular chromatin because of its proximity to the nucleolus and hence its ability to add its own genetic readout to the nucleolar product (see Chapters I, III). Since polysomes are produced by the combined action of genes that are read for the ribosomal RNA and proteins, and the genes read for messenger RNA, the morphological suggestion is that those DNA strands located proximally to the nucleolus could have the best opportunity to add their messengers to the developing ribosomes (Chapter I, p. 10).

The demonstration of intranucleolar DNA (Figs. 4.1–4.3) was a problem that seems less important now, but was at one time a difficult one. By light microscopy (*40, 43, 54*) DNA containing fibrillar structures were found coursing through the nucleolus; the question whether these deoxyribonucleoprotein strands were above or below the large mass of ribonucleoprotein in the main "nucleolar body" was finally resolved by electron microscopy (*29a, 68c, 69, 69a*).

Light microscopic evidence (Figs. 4.1 and 4.2) of Barr and Plaut (*4, 5*) shows not only the presence of intranucleolar DNA, but also that there are differences in morphology of the nucleolar DNA in various species of *Drosophila* (*53*). The DNA ranges in form from microdroplets to extended masses and sprays of fibrils similar to those noted in Chapter III. In *Drosophila,* Wolstenholme (*81*) found that DNA in puffs and nucleoli was continuous with the DNA along the chromosomes. In rat liver cells, Altmann *et al.* (*1*) showed that there are penetrations of the perinucleolar chromatin into the nucleolus. Since both nucleolar DNA and perinucleolar DNA were labeled with thymidine, Altmann *et al.* (*1*) concluded that DNA synthesis occurred both in the nucleolus and in the perinucleolar chromatin. Evidence for the penetration of the DNA from the perinucleolar chromatin was also obtained by light microscopic studies on *Spirogyra* (*54*) and Walker tumor (*69b,* see also Fig. 2.12).

B. Electron Microscopic Studies

At present, the electron microscopic evidence has superseded the evidence obtained with most other types of analytical methods that have suggested the presence of DNA in the nucleolus (*1, 24, 29a, 33, 34, 37, 69, 69a, 71–74, 76, 77;* see also Chapter II). As shown in Fig. 4.4, dense chromatin fibrils persist in the nucleolar matrix after treatment of the nucleoli with RNase. The density of these fibrils is the same as that of the perinucleolar chromatin (Fig. 4.4; see Figs. 2.12, 2.14, 3.6, 3.7, 3.8).

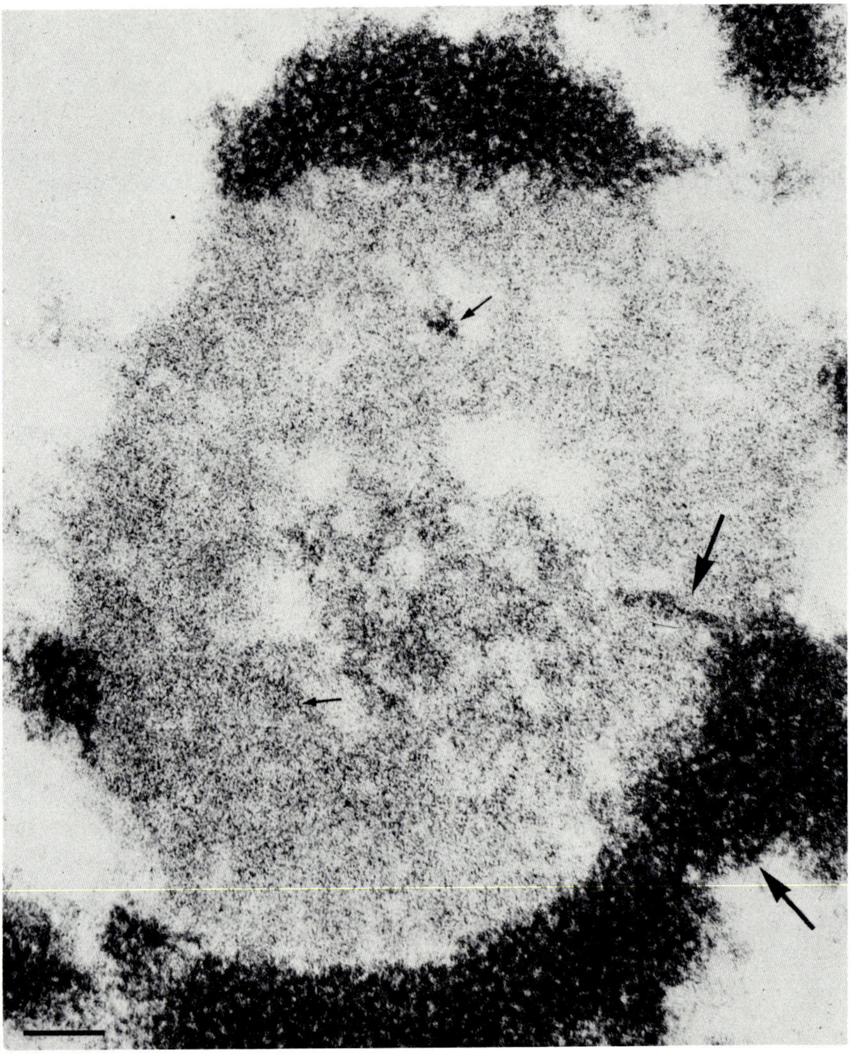

Fig. 4.4. Ring-shaped nucleolus of a human lymphocyte showing various morphological forms of nucleolar chromatin. The sample was completely digested with RNase and subsequently with pepsin. A chromatin strip (arrow) penetrates from the nucleolus-associated chromatin (large point) into the nucleolus. The intranucleolar condensed chromatin (small point) and individual fibrils (small arrow) are visible. The protein filaments in the nucleolus are less dense than DNA filaments. The marker line is 0.1 μ. $\times$100,000.

After treatment with DNase, these intranucleolar chromatin densities disappear.

These chromatin fibers are sometimes highly coiled (condensed) and at other times apparently stretched out (dispersed), or uncoiled (*24, 25, 69c*), possibly because they differ in their activity states (*24–26*). In any event, the presence of these fibers in ultrathin sections and their continuity with the chromatin fibers that are extranucleolar supports the idea that the intranucleolar fibers are simply continuations of the extranucleolar fibers (*1, 29a, 69a*). The functional DNA fibers of the nucleolus are not yet identified, i.e., even thinner DNA or chromatin strands have been demonstrated in the nucleoli.

C. Autoradiographic Studies

Following the technique utilized in this laboratory (*67*), efforts have been made to demonstrate selectively nucleolar DNA by means of studies with labeled actinomycin D (*68e*), an antibiotic that theoretically should localize in the nucleolus (Chapter X). Ebstein (*22a*) found that the labeled actinomycin was bound to chromosomes of the mature oocyte including a lampbrush chromosome and also labeled the oocyte nucleoli in their various forms (see below). These studies provide evidence that nucleolar DNA in oocyte nucleoli is a detached loop (*22*).

III. DNA of Isolated Nucleoli

Although one approach to the problem of the study of nucleolar DNA would be the isolation of the nucleolar chromosomes (see Chapter III) and NOR's (*2, 3, 46*), little progress has been made in this area. Two other approaches have been fruitful; i.e., studies on DNA of isolated nucleoli and studies on "satellite DNA" (v.i.).

For reasons that are not understood, the perinucleolar chromatin shell is "hardened" by calcium or other divalent ions (*12, 13, 19, 20, 52*, see Chapter XI). This property permits the isolation of intact nucleoli.* If this outer shell is either removed or altered by DNase or 2 *M* NaCl, the nucleolus swells, loses its microscopic definition (*13, 52*), and cannot be isolated by the usual techniques. In isolated nucleoli (see Chapter XI), most of the DNA is in the perinucleolar nucleolus-associated chromatin and the remaining DNA is in the intranucleolar chromatin (*6, 7, 33, 34, 40–42, 54, 69, 73*). The term "nucleolar body" developed by Bernhard and

* As in a number of instances, oocyte nuclei are exceptions to the rule; i.e., some of their nucleoli seem to retain their morphology without a layer of perinucleolar chromatin (*7b*).

TABLE 4.1

STUDIES ON THE NUCLEOLAR CHROMATIN AREAS (77)

	Percentage of nucleolar area composed of intranucleolar chromatin	Ratio of the area of nucleolus-associated chromatin to that of intranucleolar chromatin	Volume ratio	Ratio of the area of nucleolar body to that of total nucleolar chromatin[b,c]	Volume ratio[a′]
Liver nucleoli: Control					
In situ	6.0	6.5	9[a]	2.1	1.5
Isolated by sonication method	4.1	4.8	—	4.1	1.0
Isolated by press method	7.2	3.7	—	3.4	2.5
Thioacetamide-treated liver nucleoli					
Isolated by sonication method	1.3	5.2	—	17.5	—
Treated by DNase[d]	1.0	0.8	—	57.5	—
Novikoff hepatoma nucleoli					
In situ	12.0	2.9	3.9	2.2	1.7

[a] The ratio of the volume of the nucleolus-associated chromatin to that of the intranucleolar chromatin was calculated by the following formulas:

$$(1)\ \frac{r_2^3 - r_1^3}{r_1^3} \times \frac{100}{c} \qquad (2)\ \frac{a_2b_2^2 - a_1b_1^2}{a_1b_1^2} \times \frac{100}{c}$$

[a′] The ratio of the volume of the nucleolar body to that of the total nucleolar chromatin was calculated by the following formulas:

$$(3)\ \frac{r_1^3}{\frac{100}{100 - c} r_2^3 - r_1^3} \qquad (4)\ \frac{a_1b_1^2}{\frac{a100}{100 - c} a_2b_2^2 - a_1b_1^2}$$

(1,3 are formulae for the spherical nucleolus; 2,4 are formulae for the ellipsoidal nucleolus; r_1 = radius of the nucleolar body, r_2 = radius of the nucleolus including nucleolus-associated chromatin; a_1b_1 = major and minor semiaxis; a_2b_2 = major and minor semiaxis of the ellipsoidal nucleolus including nucleolus-associated chromatin (the ellipse was rotated about its major axis); or percentage of the area of the intranucleolar chromatin to the nucleolar area).

[b] Nucleolus without the nucleolus-associated chromatin or the areas of intranucleolar chromatin.

[c] Total of the areas of both nucleolus-associated chromatin and intranucleolar chromatin.

[d] Conditions for DNase treatment were 10 μg/ml of DNase in the presence of Mg^{2+} at 4°C for 40 minutes.

his associates (*6, 7*) is useful to separate conceptually these types of chromatin. The nucleolar body is defined as that part of the nucleolus within the confines of the perinucleolar chromatin; within the nucleolar body, the intranucleolar chromatin is readily identifiable although it is small in amount (*77*).

A. Nucleolar Chromatin Areas

Mapping studies of the type shown in Chapter II indicate that the area of the perinucleolar chromatin in electron micrographs is six times

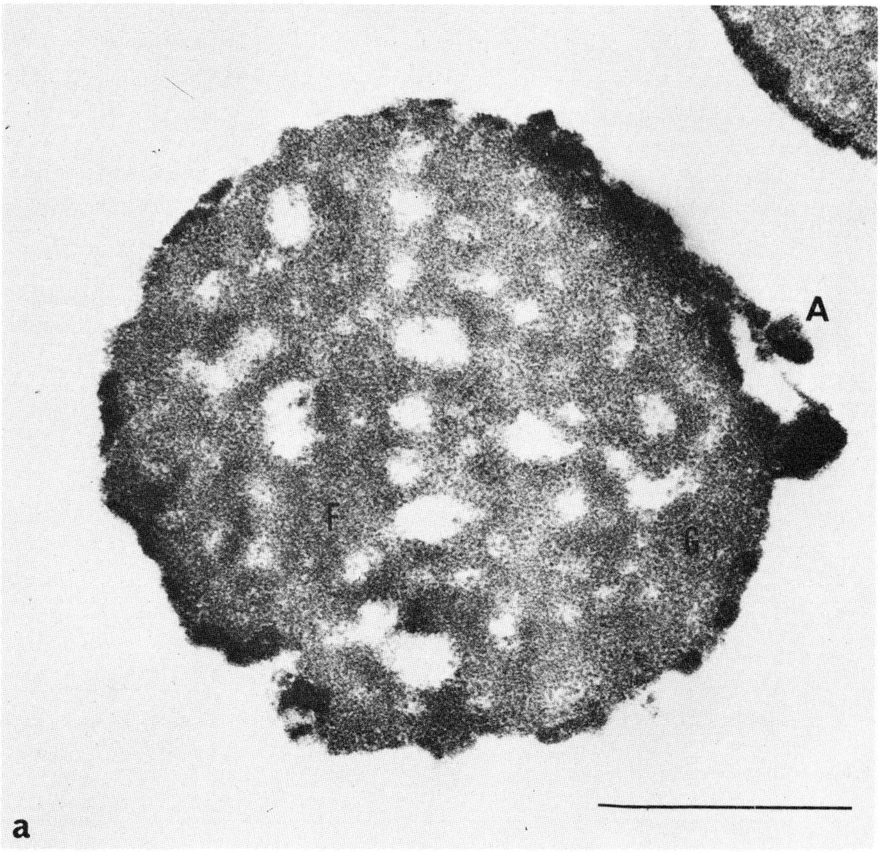

Fig. 4.5–4.9. Isolated nucleoli of thioacetamide-treated rat livers. 4.5a. Granular (G) and fibrillar (F) components of the nucleolus are preserved. Clumps of perinucleolar nucleolus-associated chromatin (A) are present around the nucleolus. The intranucleolar chromatin is less clearly defined. ×34,000. b. RNase treatment removed the granular and fibrillar RNP components of these nucleoli. The perinucleolar nucleolus-associated chromatin and intranucleolar chromatin are more dense than nucleolar matrix. ×34,000. From Unuma *et al.* (*76*).

that of the intranucleolar chromatin in liver nucleoli and about three times that of the intranucleolar chromatin of the Novikoff hepatoma (Table 4.1). These data suggest the intranucleolar DNA is only 13–25% of the total nucleolar DNA.

B. Dispersed Intranucleolar Chromatin

The dispersed form of intranucleolar chromatin is not readily visible because it is less dense than intranucleolar DNA masses shown in Fig. 4.4 (see also Fig. 3.6). When the large nucleoli of livers of thioacetamide-treated rats (Fig. 4.5a, 4.5b) are treated with DNase under carefully controlled conditions there is selective reduction of the amount of perinucleolar DNA (Fig. 4.6a). Increased concentrations of DNase also removed the intranucleolar chromatin (Fig. 4.6b). Fusion of the nucleolar residues occurred when RNase was also added (Fig. 4.7).

After treatment with DNase in low concentrations, about 30% of the DNA remained in the nucleoli instead of the 16% calculated (Table 4.1).

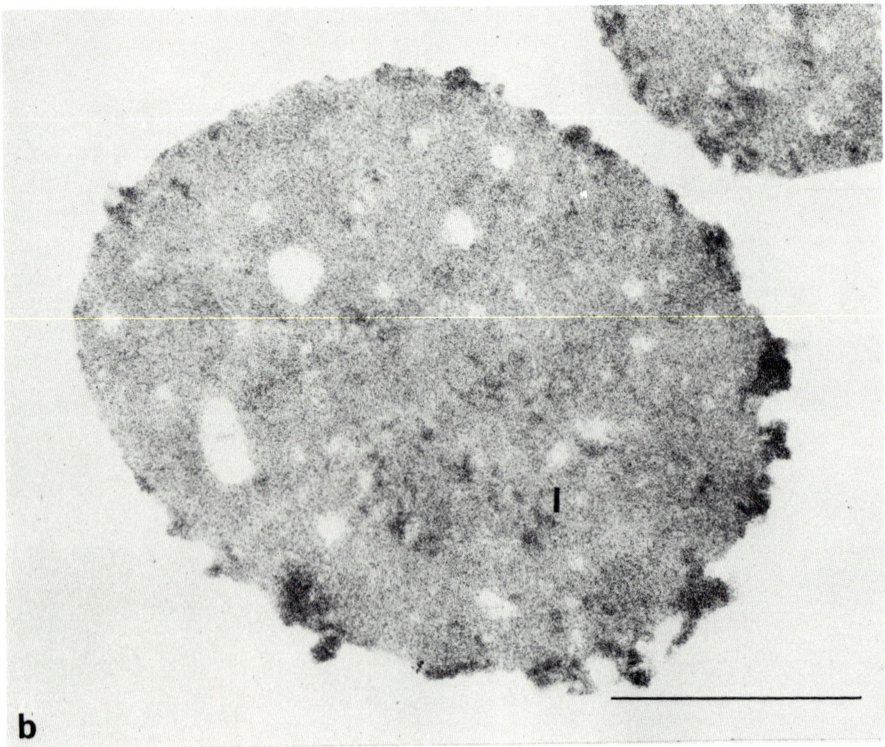

Fig. 4.5 (*continued*).

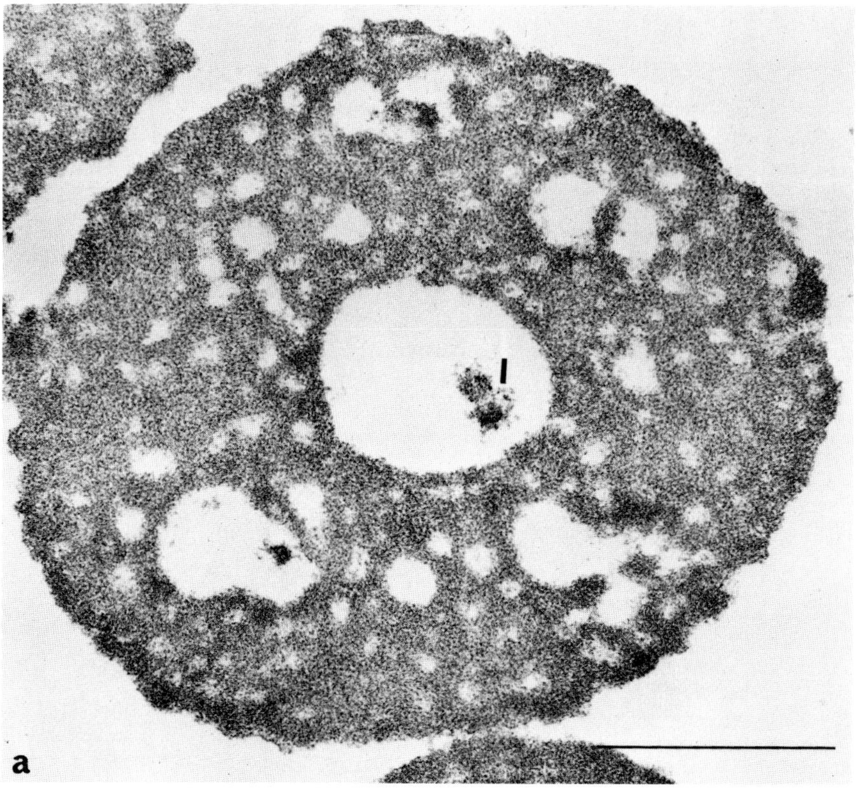

Fig. 4.6. a. DNase treatment, 10 μg/ml with Mg^{2+} (7.5 m*M*) for 40 minutes. Nucleolus-associated chromatin was removed, but the intranucleolar chromatin (I) was preserved. ×32,000. b. DNase treatment, 100 μg/ml with Mg^{2+} (7.5 m*M*) for 1 hour. Both nucleolus-associated chromatin and intranucleolar chromatin were removed. An arrow shows a small clump of granular components of the nucleolus. G = granular components, F = fibrillar components. ×32,000. From Unuma *et al.* (*76*).

Thus, the dispersed fine fibrils of intranucleolar DNA (Chapter II) probably account for as much DNA as the denser intranucleolar masses. As will be noted in Chapter VIII, a dense ribonucleoprotein fraction is not extracted from nucleoli either by treatment with 0.15 *M* NaCl or with 2 *M* NaCl, alone or successively and the fine fibrils may well be this fraction.

In summary of the problems of defining the localizations of nucleolar DNA, in any study it is necessary to differentiate: perinucleolar DNA, intranucleolar condensed DNA, and intranucleolar dispersed DNA. Ultimately, the information available on the amounts of nucleolar DNA should differentiate these physical states and localizations of DNA.

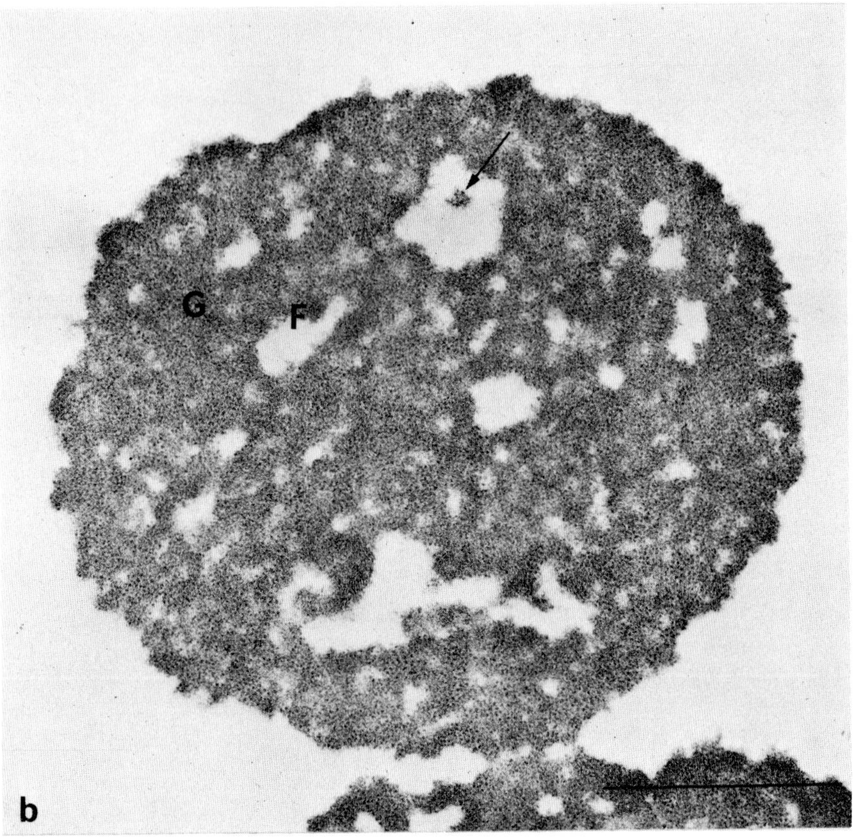

Fig. 4.6 (*continued*).

C. Amount of Nucleolar DNA

Table 4.2 shows the amounts of nucleolar DNA and RNA in various types of cells. The DNA and RNA contents of the plant cells are higher than those of mammalian cells, but the methods employed were different so the results cannot be precisely compared. The orders of magnitude are similar. The results reported for *Vicia faba* were similar to those of *Pisum sativum*. Studies were also made on *Phaseolus vulgaris, Allium cepa, A. ampeloprasum, A. babintonii, Tradescantia ohioensis* and *Zea mays* (*49*).

Since the total DNA per nucleolus was approximately 1.3 pg in Novikoff hepatoma cells and 0.38 pg in liver cells (Table 4.2) and the morphological analyses (Table 4.1) showed that 25% of the nucleolar chromatin in tumors is intranucleolar as compared with 13% in the liver cells,

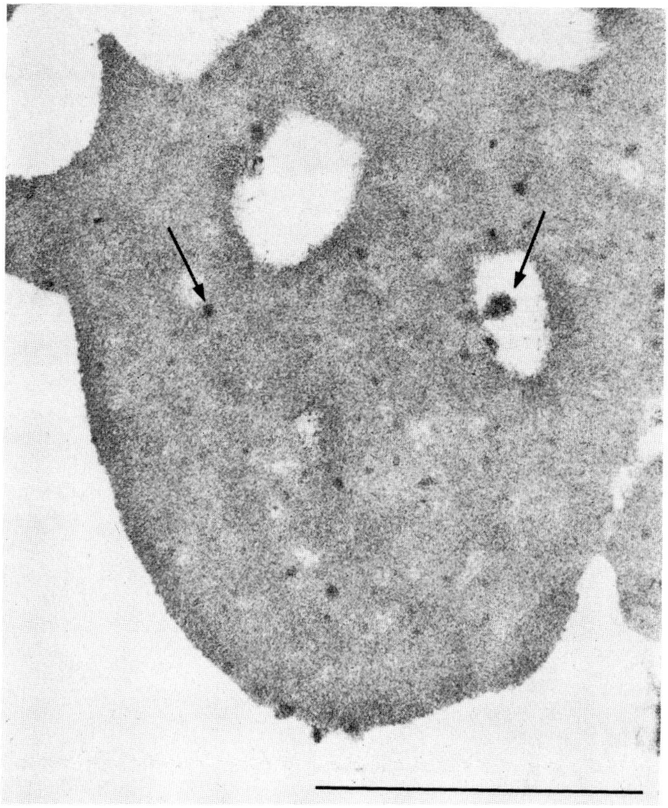

Fig. 4.7. Preservation of intranucleolar DNA in samples like those of Fig. 6a digested with DNase in low concentration and subsequently digested with RNase. The nucleolus-associated chromatin was virtually completely removed but the dense masses of intranucleolar chromatin are still visible (arrows). ×44,800. From Unuma *et al.* (*76*).

the intranucleolar chromatin should be 0.05–0.3 pg per nucleolus in liver cells and in Novikoff hepatoma cells.

1. *Amount of rDNA*

Presumably nucleolar DNA consists of both rDNA and other types of DNA. The amount of rDNA that specifically codes for 18 S and 28 S RNA in the mammalian cells is small as a part of the total; i.e., about 0.005–0.01% (*3, 48, 70a*). Attardi *et al.* (*2, 3*) concluded that perhaps the total rDNA is 0.008% of the total DNA with 0.002–0.003% coding for 18 S RNA and 0.003–0.005% coding for 28 S RNA.* The values obtained by Attardi

* This number is subject to error inasmuch as some overlap was found for the hybridization of the 18 S and 28 S RNA for the same site (*2, 3*).

TABLE 4.2

DNA AND RNA CONTENT OF NUCLEOLI[a]

	DNA/No	RNA/No	Reference
Vicia faba	10.3	5.1	*49, 50*
Walker tumor	0.91	1.22	*19a*
	1.70	1.90	*52*
Xenopus laevis (oocytes)	30		*56*
Rat liver	0.38	0.38	*52*

[a] The values are picograms (10^{-12} gm)/nucleolus. The total nucleolar nucleic acid/nucleus is calculated by multiplying the values by the nucleolar number (see Chapter III), *except for Xenopus;* this value is the total picograms of nucleolar DNA. No = nucleolus.

et al. (*2, 3*) are supported by other results; i.e., McConkey and Hopkins (*48*) derived a value of approximately 0.005% from their studies on HeLa cells. The amount of DNA in the mammalian cell is usually considered to be 10×10^{-12} gm or 6.0×10^{12} daltons. If 0.01% of this DNA is involved in the synthesis of ribosomal RNA, 6×10^{8} daltons of single-stranded DNA would be involved (12×10^{8} daltons of double-stranded DNA). If the molecular weight of one initial product, 45 S nucleolar RNA, is 4×10^{6}, there would be 150 genes or cistrons coding for the 45 S RNA precursor of ribosomal RNA in such cells. Mohan *et al.* (*51c*) have recently reported that in the rat the amount of rDNA is constant in rat liver and brain, i.e., about 0.04% of the genome (see also p. 262).

By comparison with the low value reported for mammalian cells, there is a much higher percentage of rDNA of the genome of other species including *Drosophila, R. pipiens,* tobacco plants, and bacteria (*3, 62, 74a, 78a, 82*). Values ranging from 0.16 to 0.25% of the DNA have been reported to be hybridized with the RNA of the ribosomes. In studies on the rDNA content per cell, a wide range in the total number of genes has been found for varying species; i.e., 6 for *E. coli* (*82*), 440 for HeLa cells (*3*), 1600 for *Xenopus laevis* (*78a*), and 200 for chicken cells (*62*). In *Xenopus laevis* the cells with one nucleolus would have only 800 rDNA molecules per cell.

In the chicken (*51*), the saturation level that was reached for ribosomal RNA was 0.02% of the DNA. In comparison, the value of 0.2% was found for *E. coli.* The input ribosomal RNA had a base composition of A 25, U 20, G 31, and C 24, and the hybridized RNA had a base composition of A 25, U 18, G 35, and C 22 for *E. coli.* In the chick the input RNA had the base composition: A 19, U 17, G 35, and C 29, and the hybridized RNA had a base composition of A 27, U 14, G 37, and C 22 (*51*).

The series of "bobbed" mutants of *Drosophila* has a number of differences from that of the normal flies in terms of phenotype and also has a marked difference in the number of genes from the diploid number of 260 coding for ribosomes. The essence of the data obtained by Ritossa and colleagues (*61–66*) is that the *Drosophila* bobbed locus is in the nucleolar organizer region. In the "bobbed" mutants, there are marked variations in the number of cistrons that contain rDNA. In these flies, lethality or survival for a given mutant is dependent upon the number of genes containing rDNA (*60a, 60b*). If the gene set has in excess of about 50 genes for production of the ribosomes, survival is possible. However, if there are less than 50 genes, the flies do not survive.

IV. Satellite DNA

Although direct efforts to isolate nucleolar DNA from either chromosomes or nucleolar masses have had only limited success thus far, the whole field of isolation of DNA that codes for rRNA has been brightened and clarified by the findings relating to the "satellite DNA" of the toad, *Xenopus laevis*, and more recently of other species.

Birnstiel *et al.* (*7a, 8a*) reported that from the total DNA of *Xenopus laevis*, a DNA "satellite" fraction could be isolated which hybridized to a much greater extent with the ribosomal RNA than did the remainder of the DNA. This rDNA fraction was present in greatest amount in wild type *Xenopus laevis*, less in the mutant heterozygous with respect to NOR's (see Chapter III) and was virtually absent in the anucleolate mutants of the *Xenopus laevis*. In studies on amphibian oocytes, a variety were found by Brown and David (*10a*) to contain satellite DNA.

After DNA from red cells of the *Xenopus laevis* was banded in cesium chloride (CsCl), fractions were collected and tested for their ability to hybridize with the 18 S and/or 28 S ribosomal RNA. The ribosomal RNA was made radioactive by pulse labeling the newly hatched tadpoles with radioactive sodium carbonate. DNA–RNA hybrids were formed by annealing as indicated by the resistance of these products to RNase, their high buoyant density, and sharp melting temperatures (T_m). The RNA of the hybrid had a content of 62% guanylic and cytidylic acids. The DNA hybridizing with 18 S and 28 S rRNA was localized to the same regions of the CsCl gradient (*8*). In more recent studies, this DNA has been isolated in highly purified form (*8*) as shown in Fig. 4.8.

A. Hybrid Formation with DNA Derived from the Mutants

Although the DNA from erythrocytes of wild-type *Xenopus laevis* readily hybridized with ribosomal RNA, DNA from the mutants that lacked nucleoli did not hybridize to detectable levels and presumably

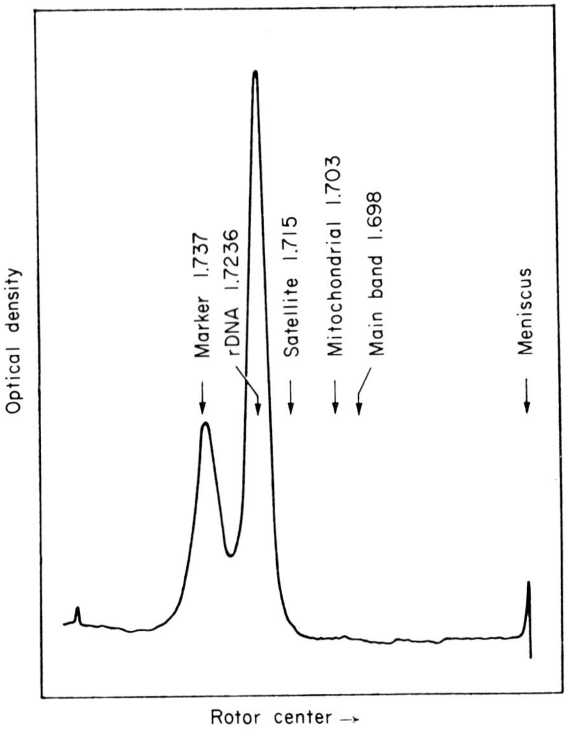

Fig. 4.8. Analytical centrifugation of the purified, isolated rDNA. rDNA (2 μg) was analyzed in the Spinco analytical centrifuge. Denatured *Pseudomonas aeroginosa* DNA was included as a density marker (1.737 gm/cm^3) (from 8).

lacks rRNA. As shown in Fig. 4.9, the annealing with the nucleoli of toads with 1 nucleolus (1n) was intermediate between that of the normal controls (2n) and that of the DNA of the toads that had no nucleoli (0n).

B. Sedimentation of Xenopus DNA

The DNA was fractionated in part by virtue of the different buoyant density of the GC-rich DNA which would be the logical template on which ribosomal GC-rich RNA should be synthesized. As shown in Figs. 4.8 and 4.10, the hybridization of the ribosomal RNA was maximal for the DNA with a buoyant density of 1.724 gm/cm^3. The bulk of *Xenopus* DNA had a buoyant density of 1.698–1.700 gm/cm^3. Another satellite DNA had a buoyant density of 1.715 gm/cm^3. The satellite rDNA composed 0.15–0.2% of the total cellular DNA and was absent in the homozygous mutant that contained no nucleoli (Fig. 4.11).

The amount of total cell DNA accounted for in these "satellite" frac-

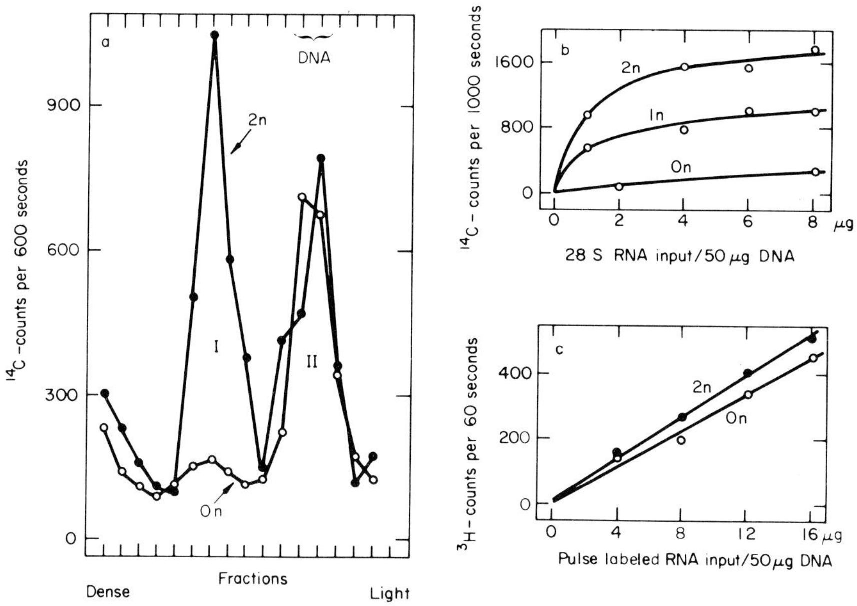

Fig. 4.9. Hybrid formation between ribosomal RNA or "messenger" RNA and DNA from the three genotypes of *Xenopus*. a: Twenty micrograms 28 S ribosomal RNA was annealed to either 2n or 0n DNA. Fractionation on a CsCl gradient reveals that 0n DNA produces only negligible high density hybrids (peak I), unlike 2n DNA. b: Saturation curves for annealing of 28 S ribosomal RNA with 2n, 1n, and 0n DNA. Only the high density hybrids have been scored. Contrary to pulse-labeled RNA, ribosomal RNA saturates the relevant DNA stretches at relatively low RNA levels. c: Hybrid formation between pulse-labeled "messenger" RNA and DNA is similar in both 0n and 2n preparations. Pulse-labeled RNA was prepared by incubating explants of tadpoles with 100 μCi uridine-^{3}H/ml (2×10^{3} μCi/μmole) for 45 minutes at room temperature and subsequent extraction with hot phenol and detergent. From Birnstiel *et al.* (*8a*).

tions is in the order of 0.08–0.14% for the 28 S rDNA and 0.05–0.08% for the 18 S rDNA. This total percentage of DNA is similar to that which hybridizes with ribosomal RNA from bacteria (*82*), pea seedlings (*15*), and fruit flies (*61–63, 65;* see p. 174).

C. DNA Satellites of Plants

In pumpkins and Chinese cabbage, there is a satellite rDNA with a buoyant density of 1.707 gm/cm^3 compared to the buoyant density of 1.695 gm/cm^3 for the bulk of the DNA. The satellite DNA of the pumpkins is greatly enriched in cistrons for ribosomal RNA (Table 4.3). In comparing the number of cistrons that hybridize ribosomal RNA in

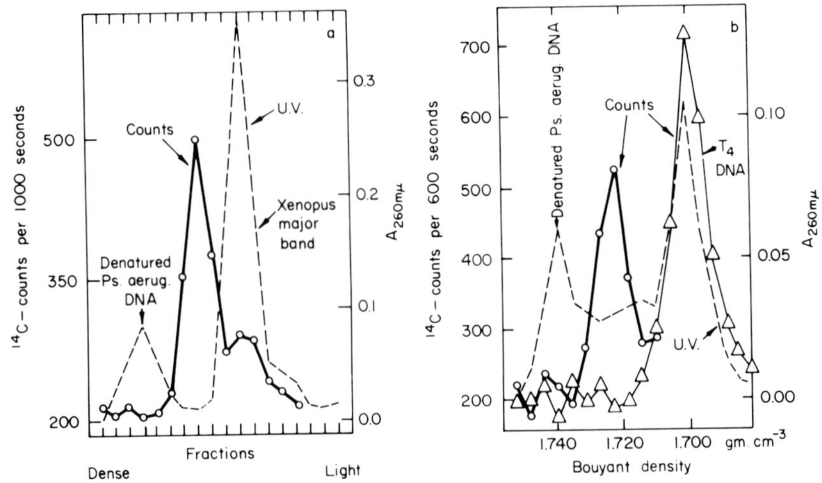

Fig. 4.10. Detection of a DNA component base-complementary to 28 S RNA. a: *Xenopus* erythrocyte DNA (mol wt 2×10^7) was fractionated by a first CsCl density gradient centrifugation and each DNA fraction annealed with ribosomal RNA. The *Xenopus* DNA complementary to *Xenopus* 28 S RNA is present mainly in a region intermediate between denatured *Ps. aeruginosa* DNA and the major *Xenopus* DNA mass. b: The DNA fractions, which had previously (a) been shown to be complementary to 28 S RNA, were collected from 3–6 gradients and redistributed on a CsCl gradient by a second centrifugation. DNA fractions possessing a buoyant density of 1.722–1.724 gm/cm^3 annealed to 28 S ribosomal ^{14}C-RNA. T_4DNA that serves as a marker and is also radioactive prevents the accurate measurement of the hybridization in the lighter fractions. DNA fractions of *Xenopus* of lesser buoyant densities are relatively inert, as shown in a. From Birnstiel *et al.* (*8a*).

DNA's of plant nuclei (*47, 68d*) tobacco nuclei have 1500 cistron copies for rRNA; the nuclei of Chinese cabbage and pumpkin have about ten times this number of cistrons. The pumpkin nuclear DNA satellite was found to rapidly renature by comparison with the remainder of the DNA,

TABLE 4.3

Hybridization of Tobacco Ribosomal RNA to Different DNA Fractions from Pumpkin (*47*)

DNA source	Hybridization (% of DNA)
Tobacco	0.084
Whole pumpkin	0.39
Main band, pumpkin	0.22
Satellite band, pumpkin	3.6

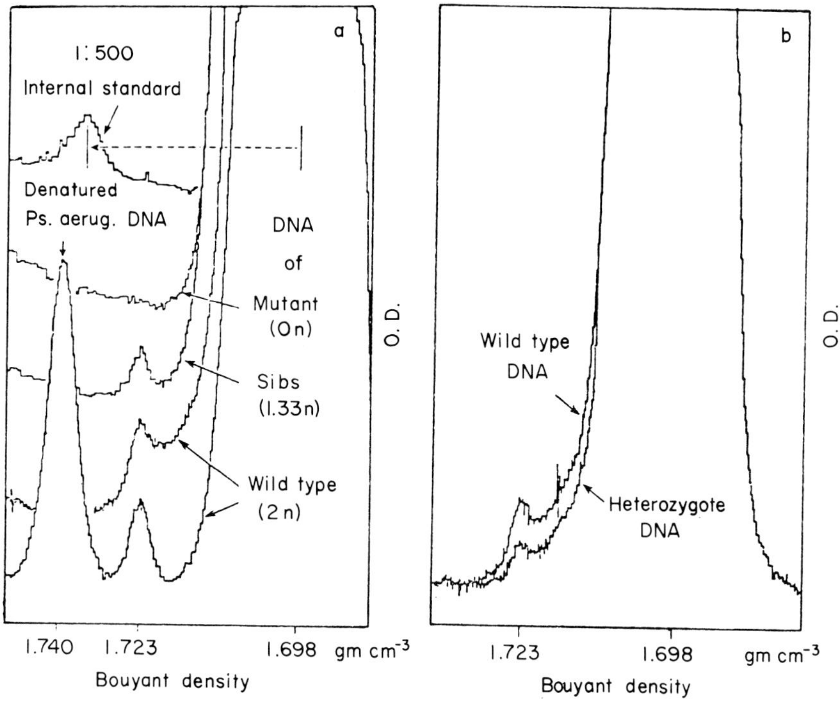

Fig. 4.11. Tracings of photographic records obtained from analytical centrifugation of DNA samples. a: The CsCl gradient (at a mean density of 1.720 gm/cm^3) contained the following amounts of DNA (mol wt 10–50 $\times$ 10^6): homozygous mutant: 87 μg; phenotypically normal sibs of the same matings: 69 μg; wild type (without marker): 102 μg; wild type with denatured *Ps. aeruginosa* as a density marker: 95 μg. As an internal standard for the determination of the satellite mass, a 500-fold dilution of the wild-type DNA (without marker) was used in parallel experiments. The major DNA band (1.698 gm/cm^3) shows a considerable spread owing to the fact that it is present in 100-fold excess. The tracings have been displaced vertically to facilitate comparison. Similarly, the internal standard has been shifted to a region of high density. A satellite at 1.723 gm/cm^3 is present in all preparations at high DNA input, with the exception of the homozygous mutant DNA. (A further *Xenopus* component at 1.715 gm/cm^3 also becomes apparent, especially in DNA samples of a molecular weight of about 30 $\times$ 10^6 (not shown).) b: CsCl gradient centrifugation of homozygous and heterozygous DNA. The DNA input was 105 μg in each case. The samples were run simultaneously in the analytical centrifuge which was equipped with an alternator for two-cell operation. The heterozygous DNA contains only half of the 1.723 satellite mass. The DNA material at 1.715 gm/cm^3 is probably also reduced in the mutant. From Birnstiel *et al.* (*8a*).

probably because it contained fewer species of DNA than did the bulk of the DNA (Table 4.3, 4.4b).

In studies on the satellite DNA's of higher plants, including *Nicotiana tabacum, Tagetes patula, Antirrhinum majus, Dianthus caryophilus,*

TABLE 4.4a

Satellite DNA of Animal Tissues

Organism	Major DNA Density	Major DNA Origin	Major DNA Hybrid formation	Satellite DNA Density	Satellite DNA Origin	Satellite DNA Hybrid formation	Reference
Mouse							
Liver, lung	1.702	Nucleus		1.691	Nucleus		*9, 23*
Liver, lung, brain, kidney, spleen	1.701			1.690			*39a′*
Thymus	1.700	Nucleus (including nucleolus)	10 S RNA 28 S RNA	1.690	Nucleus (excluding nucleolus)	16 S RNA 28 S RNA	*31*
Myeloma tumors (MPC11 MPC_2, BJ)	1.702	Nucleus		1.691	Nucleus		*15*
Lymphoblast tumor	1.700	Nucleus		1.690	Nucleolus		*68b*
Chick liver	1.701	Nucleus		1.708	Mitochondria		*9*
	1.698	Nucleus		1.707	Mitochondria		*59a*
	1.701	Nucleus		1.707	Mitochondria		*16a*
heart	1.698	Nucleus		1.707	Mitochondria		*59b*
Guinea pig							
liver	1.697			1.703			*39a′*
	1.700	Nucleus		1.705	Nucleus		*16a*
kidney, spleen	1.697			1.703			*39a′*
Sheep heart	1.704	Nucleus		1.714	Mitochondria		*16a*
Xenopus laevis							

Somatic cell	1.700	Nucleus	mRNA	1.724	Nucleolar organizer region	rRNA	*8a*
				1.715	Mitochondria		*8a*
	1.699	Chromosomal		1.724	Nucleoli	rRNA	*10a*
Oocytes	1.700	Chromosomal		1.729	Peripheral	rRNA	*27a*
	1.699	Chromosomal		1.729	Nucleoli	rRNA	*10a*
Siredon Mexicanum							
Somatic cell	1.704	Chromosomal		1.718	Nucleoli	rRNA	*10a*
Oocytes	1.704	Chromosomal		1.725	Peripheral	rRNA	*10a*
N. maculosus oocytes	1.707	Chromosomal		1.723	Peripheral	rRNA	*10a*
Sea urchin egg (*Lytechinus pictus*)	1.703	Mitochondria		1.694	Nucleoli Nucleus		*56a*

TABLE 4.4b

SATELLITE DNA OF PLANTS

Organism	Major DNA Density	Major DNA Origin	Satellite DNA Density	Satellite DNA Origin	Reference
Yeast	1.700	Nuclear	1.685	Mitochondria	*16a*
Neurospora crassa	1.712	Nuclear	1.701	Mitochondria	*45a*
Blastocladiella emersonii (aquatic fungus)	1.725	Nuclear (chromatin and nucleolar)	1.710	Nuclear (only in the chromatin)	*16*
Euglena gracilis (algal flagellate)	1.707	Nuclear	1.685	Chloroplast	*59d*
	1.707	Nuclear	1.690	Mitochondria	*59d, 59e*
	1.707	Nuclear	1.686	Chloroplast	*22b*
	1.707	Nuclear	1.691	Mitochondria	*22b*
Chlamydomonas (green alga)	1.728	Nuclear	1.702	Chloroplast	*68a*
Chinese cabbage	1.695	Nuclear	1.707	Nuclear	*47*
Pinto bean	1.695	Nuclear	1.707	Nuclear	*47*
Pumpkin	1.695	Nuclear	1.707	Nuclear[a]	*47*
Spinacea oleracea (spinach)	1.695	Nuclear	1.705	Chloroplast	*15b*
Beta vulgaris (beet leaves)	1.695	Nuclear	1.719	Mitochondria	*15b*
Tagetes patula (marigold)	1.692	Nuclear	1.702 1.707	Chloroplast	*30*
Antirrhinum majus (snapdragon)	1.697	Nuclear	1.709	Chloroplast	*30*
Nicotiana tabaccum (tobacco)			1.706	Chloroplast	*30a*
	1.696	Nuclear	1.710	Mitochondria	*30b*
	1.690	Nuclear	1.703	Chloroplast	*68d*
	1.697	Nuclear	1.702	Chloroplast	*74a*
Ipomoea batatas (sweet potato)	1.692	Nuclear	1.706	Mitochondria	*70b*
Phaseolus aureus (mung bean)	1.691	Nuclear	1.706	Mitochondria	*70b*
Brassica rapa (turnip)			1.700	Chloroplast	
	1.692	Nuclear	1.706	Mitochondria	*70b*
Dianthus caryophilus (carnation)	1.695	Nuclear	1.706	Chloroplast	*30*
Beta vulgaris, var. *cicla* (swiss chard)			1.700	Chloroplast	
	1.690	Nuclear	1.705	Mitochondria	*39c*
Ranunculus repens (buttercup)	1.697	Nuclear	1.703	Chloroplast	*30*
Allium cepa (onion)	1.688	Nuclear	1.706	Mitochondrial	*70b*
Triticum vulgaris (wheat)	1.707		1.716		*36a*
Equisetum sp. (horsetails)	1.697		1.713	Chloroplast	*30*

[a] This satellite has been shown to be complementary to rRNA.

Ranunculus repens, and *Equisetum* (Table 4.4b), Green and Gordon (*30*) found that there were two satellite DNA's, one a constituent of chloroplasts and the other which composes 2% of the total DNA and is labeled faster than is nuclear DNA. The GC contents were greater for these satellites than for the bulk of the nuclear DNA's. It is possible that the 2% of the DNA may be a nucleolar satellite but this is not yet established.

D. DNA Satellites of Other Species

In studies on myeloma tumor satellite DNA, Greenberg and Uhr (*31*) found a mitochondrial satellite DNA with a bouyant density of 1.692 gm/cm^3 and another high specific-activity satellite DNA that hybridized more specifically with 16 S rRNA than did the bulk of the DNA. This satellite may also code for 28 S rRNA but further evidence is required.

In other forms, somewhat confusing results have been obtained. In *Blastocladiella emersonii* (*16*), the major DNA component had 66% GC and the minor component had 51% GC. The nucleolar RNA in this species was reported to have a high content of uridylic acid (27–31%) and adenylic acid (23–25%). Thus, it would be possible that the DNA coding for nucleolar RNA would be more AU rich than would the remainder of the DNA and further that this satellite was, in fact, the one that coded for nucleolar RNA. However, the more GC-rich component was suggested to be the nucleolar component despite the high AU content of the nucleolar RNA of this species. Although other satellites have been found, their relationships to ribosomal formation or function in nuclei or mitochondria are not clear at present (*9a, 39d*).

E. Nucleolar Satellite DNA of Mammalian Cells

Schildkraut and Maio (*68b*) isolated a "nucleolar satellite" DNA with a bouyant density of 1.690, a lower T_m (84°) than that of whole nuclear DNA (86°), and a lower GC content (35%) than whole nuclear DNA (41%). Since rDNA must have a high GC content, the role of this nucleolar satellite remains to be defined. Satellite DNA's have been studied in other species including the mouse (*9*). At this time, however, it is not possible to conclude whether those satellites obtained represent nucleolar, mitochondrial, or other types of DNA (Table 4.4a,b).

V. Base Compositions of Nucleolar DNA

The base compositions of nucleolar DNA are presented in Table 4.5a,b. The total nucleolar DNA has been analyzed but the base compositions of

TABLE 4.5a

Base Composition of DNA of Isolated Nuclei and Nucleoli[a]

Tissue	Base (% total residues) Adenine	Thymine	Cytosine	5-Methyl cytosine	Guanine	Ref.
Tumor (Walker, Novikoff)						
Nucleus	28.3	27.4	20.8		23.2	*13*
Nucleolus	28.1	27.2	21.3		23.2	*13*
Liver						
Nucleus	28.3	29.2	21.0		21.5	*13*
Nucleolus	28.6	28.9	21.4		21.1	*13*
Vicia faba						
Nucleus	30.0	30.0	14.9	5.7	19.1	*50*
Nucleolus	29.6	31.4	15.4	5.7	17.9	*50*

[a] Determined by paper chromatography of the bases.

the intranucleolar DNA of mammalian cells have not been determined. The satellite DNA from *Xenopus laevis* may be intranucleolar DNA (*8a, 78*), but this is not yet proved. Its buoyant density shows that it has a high content of deoxyguanylic and deoxycytidylic acids as the primer should for the synthesis of GC-rich rRNA (*79, 80,* Table 4.5b).

Surprisingly, no significant difference was found in the base composition of whole nucleolar DNA and the whole nuclear DNA (Tables 4.5a, 4.5b) in plant and animal cells (*13, 19, 20*). Although a rather random sample of the total nuclear DNA may be associated with isolated nucleoli, the composition of the nucleolar DNA may still represent only specific nucleolus-associated cistrons.

TABLE 4.5b

GC Contents of Nucleolar Satellite DNA

Tissue	Method: Buoyant density (CsCl)	Tm	Other	Ref.
Xenopus laevis satellite	65%	55%		*10*
Drosophila melanogaster[a]	40%			*62*
Pumpkin	48%	55.7%		*47*
Mouse lymphoblast			36%	*68b*

[a] This rDNA has same density as other nuclear DNA. GC is the total of deoxyguanylic and deoxycytidylic acids.

When methods become available for isolation of the intranucleolar DNA, its base composition may reflect that of a more restricted sample of total cellular DNA, and GC-rich components should be identifiable.

VI. Synthesis of Nucleolar DNA

In the S phase of the cell cycle, when DNA is being synthesized, the synthesis of extranucleolar DNA proceeds earlier and/or faster than that of nucleolar DNA (*15a, 15c, 28, 33, 34, 44, 60*).* Studies of this type in which labeling of DNA with tritiated thymidine has been analyzed by autoradiography or after the isolation of DNA (*18a*) from isolated nucleoli showed that euchromatin or extranucleolar chromatin was labeled earlier than the intranucleolar chromatin or the nucleolus-associated perinucleolar "heterochromatin." Since the chromosomes associated with the nucleolus are among the last to cease functioning and hence to separate as condensed metaphase chromosomes (see Chapters I, III), it is not surprising that they are late replicating and late labeling (*39*).

Although most studies suggest that nucleolar DNA is synthesized later than that of the other cellular DNA, Lima-DeFaria *et al.* (*45*) have shown that human metaphase chromosomes 21 and 22 apparently replicate early. These have short arms and satellites and are probably involved in formation of the nucleoli (Chapter III; p. 117, 129).

A. Nucleolar Gene Amplification

Nucleolar DNA synthesis is not limited in all species to the S phase. By a process of "gene amplification" or "endoreduplication" some DNA "self-duplicates." In the fly, *Tipula orescena,* a "DNA body" develops in oocytes and composes as much as 59% of the total DNA (*5a, 5b, 44, 56*). This body contains a multiplicity of genes for biosynthesis of nucleolar products in this fly; i.e., it apparently serves to markedly augment the biosynthesis of ribosomal precursors.

In the polytene chromosomes of *Drosophila melanogaster,* Plaut *et al.* (*57*) found a large number of replicating regions which they referred to as DNA "replicons;" in at least 30 sites (less than 15% of the total length) of the chromosome, DNA synthesis seems to occur more or less autonomously. Some parts are more active than others and some parts are active

* One possible factor that may affect such studies emerges from the findings of Hill (*36*) that labeled donor DNA of lymphocytes produces labeling in nucleoli of some cells. This result suggests that products of DNA may migrate selectively into the nucleoli and this may also be the case for DNA precursors such as thymidine (*36*).

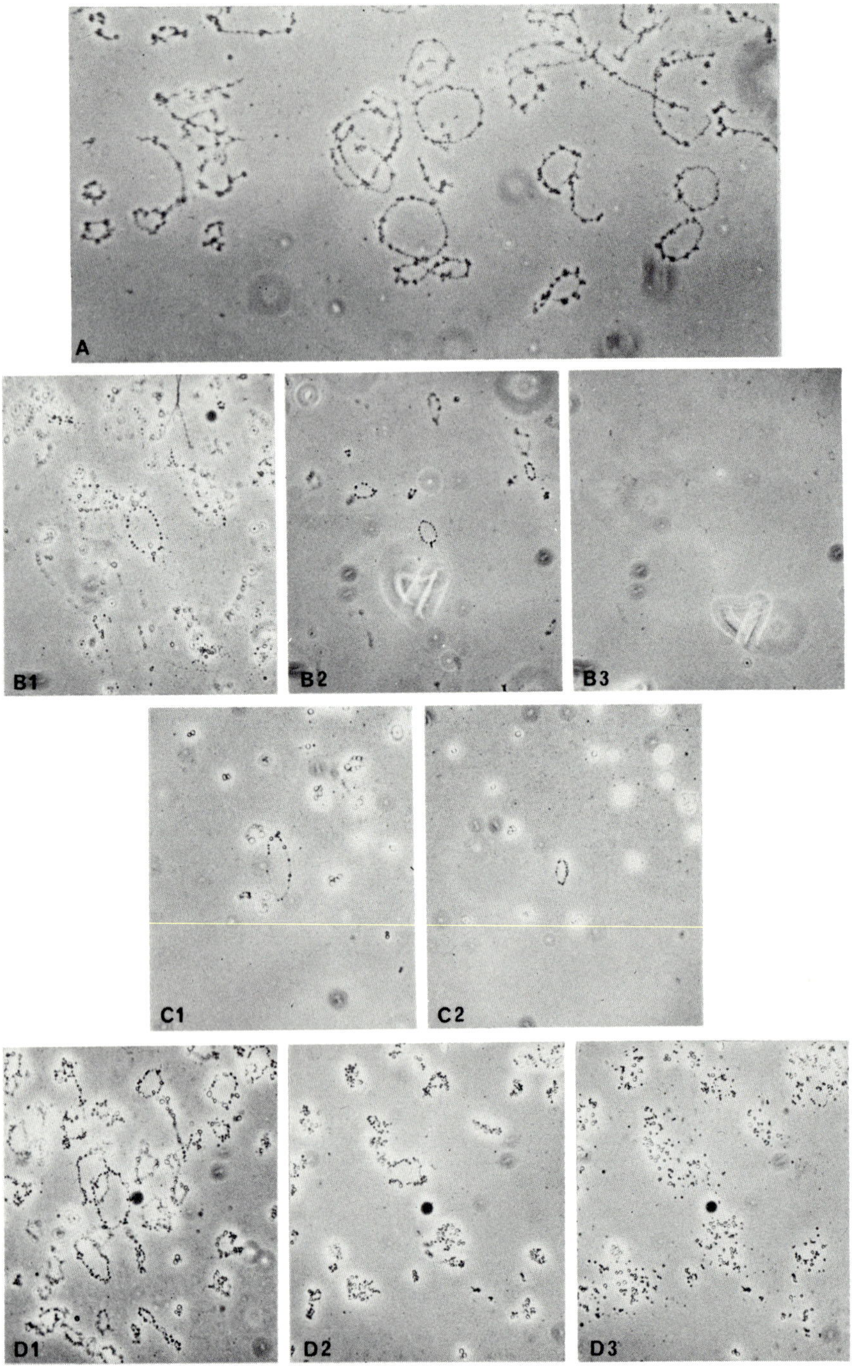

Fig. 4.12. A. Nucleolar cores in circular form derived from cluster nucleoli during late stages of oogenesis by isolation in low molarity saline. Largest circles are

longer than others. Clearly this autonomous behavior in specific gene segments results in a "metabolic DNA" which is involved in the "gene amplification" or "endoreduplication" such as was reported by Heinonen and Halkka in cricket oocytes (*35*). During vitellogenesis, some cistrons produce massive increases in new RNA that is responsible for the tremendous increase in the volume of the oocyte. Studies on *Sciara coprophila* have confirmed the asynchronous replication of DNA in polytene structures (*27*). Similar gene amplifications have been found in water beetles (*27c, 39a*). Recently, Gall has summarized the information on the many forms of extra-chromosomal DNA in oocyte nuclei (*27b*).

B. Nucleolar "Gene Amplification" in Amphibian Oocytes

The number of nucleoli is, in general, related to the number of NOR's (Chapter III). Usually, the number of NOR's per genome is characteristic, and DNA from various tissues contains the same proportion of ribosomal genes (*62*). However, oocytes and particularly those of amphibia show interesting deviations from this general rule.

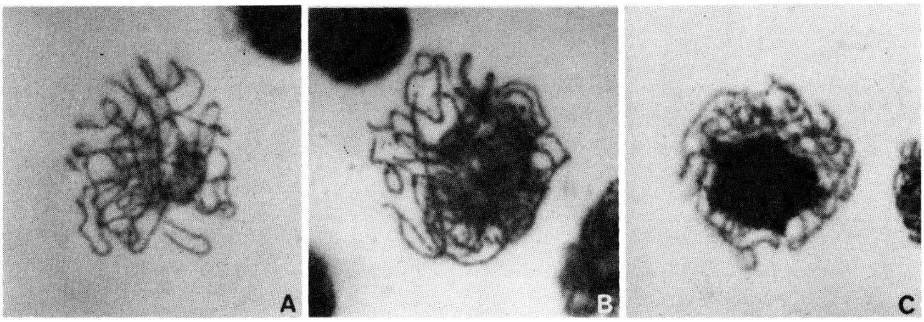

Fig. 4.13. *Xenopus* oocyte nuclei during the period of nucleolar DNA synthesis. Feulgen stain. ×1700 (*27a*). A. Very early stage containing normal pachytene chromosomes with the 4C amount of DNA. Arrow indicates nucleolus. B. Excess DNA begins to accumulate as granules around the nucleolus. Such nuclei show intense thymidine incorporation. C. After synthesis the nuclear DNA content has more than tripled and the excess DNA appears as a compact mass. Subsequently, the mass will break up into hundreds of granules which become associated with the multiple nucleoli (courtesy of J. G. Gall, Yale University).

---

approximately 165 μ in circumference. Linear cores presumably result from opening of the circles by mechanical breakage during isolation. Phase contrast. ×240. B. Sequential stages in the digestion of circular nucleolar cores isolated into a medium containing trypsin. Phase contrast. ×255. C. Same as Fig. 4.12B except cores isolated into medium containing ribonuclease. ×255. D. Same as Fig. 4.12B except cores isolated into medium containing deoxyribonuclease. ×255 (courtesy of Dr. O. L. Miller and Dr. B. R. Beatty, Oak Ridge National Laboratory).

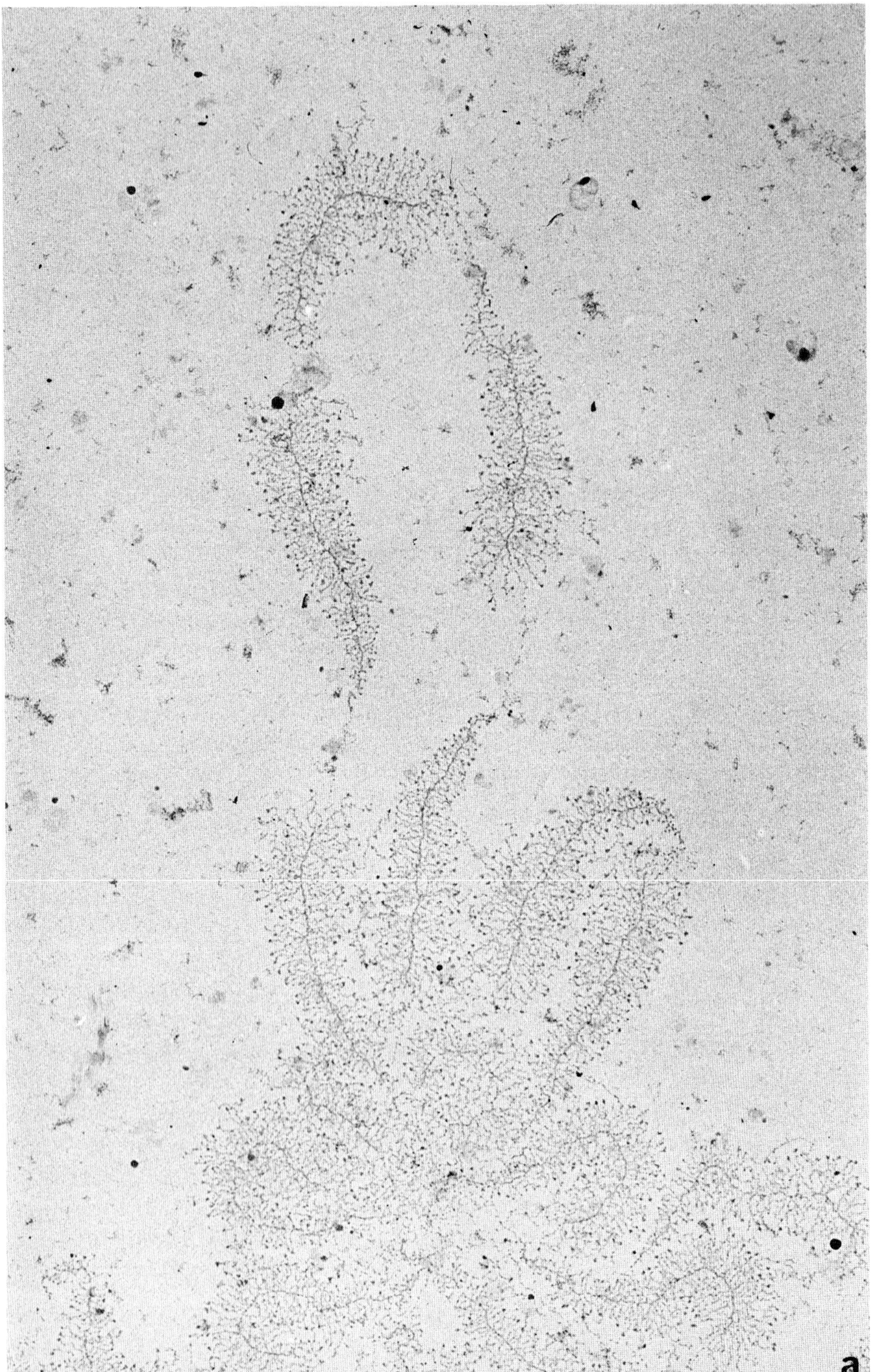
a

Cytological studies show that the nucleolus organizer undergoes differential replication in amphibian oocytes. For example, 600, 1000, and 1600 "peripheral" nucleoli per nucleus were found in oocytes for *Triturus* (*45b*), *Siredon mexicanum* (*13a*) and *Xenopus laevis* (*56*). Since the growing oocyte persists in the first meiotic prophase for an extended period (*59c*), the cell is tetraploid and, therefore, would be expected to contain four nucleoli. This huge increase in nucleolar rDNA genes and in nucleoli is referred to as "gene amplification."

The "peripheral" nucleoli of oocytes (*51a, 51b*) have the same characteristics of morphology and chemical constituents as somatic nucleoli; i.e., they contain both granular and fibrillar elements. However, they are located just within the nuclear envelope and are separated from the chromosomes (*39b, 45c, 51a*). Like other nucleoli, they rapidly incorporate uridine and other precursors into RNA and this synthesis is inhibited by actinomycin (*36b*). The RNA of these nucleoli, especially in the newt *Triturus*, is similar in base composition to ribosomal RNA (*22c*).

Each of these nucleoli contains a small amount of DNA (*9b, 39f, 55b*), which is released as DNP cores in the shape of beaded circles or necklaces in hypotonic solutions (*39b, 51b*) (Fig. 4.12). These oocytes have a greater percentage of satellite DNA of bouyant density 1.724 gm/cm^3 than normal cells (*27a*). In the *Xenopus laevis*, the total amount of high density DNA of the oocyte nucleus is 25 pg, or 5000 times the haploid amount of nucleolar DNA. This value is very close to the amount measured by Perkowska *et al.* (*56*) which was 30 pg of amplified DNA and 5200 nucleolar cores in the same organism. The ribosomal RNA (both 18 and 28 S rRNA but not 4 S or 5 S RNA) was specifically hybridized to this high density DNA. Enrichment of rDNA in oocytes has been also shown by the greater hybridizability with rRNA of unfractionated egg DNA compared to somatic DNA (*10a*).

The origin of this nucleolar DNA was first studied by King in 1908 (*39b*). She showed that the extra "chromatin" forming in the oocytes of the toad *Bufo* during the pachytene stage of meiosis is associated with

---

Fig. 4.14. a. Portion of isolated core of nucleolus from *Xenopus laevis* oocyte-presumptive rRNA cistrons. Uranyl acetate stain. The DNA fibers are visualized at the centers of elongated periodic masses. Along these masses there are RNA fibers of progressively increasing chain length. The distance between these RNA fibers presumably represents spaces occupied by RNA polymerase or other synthetic elements. ×20,000. The DNA in the spaces between the synthetic clusters has been referred to as "spacer DNA" but its structure and function are unknown. b. Portion of isolated cortex (granular component) of nucleolus from *Xenopus* oocyte. All granules appear to be attached to well-defined fibers. Uranyl acetate stain (courtesy of Drs. O. L. Miller and B. R. Beatty, Oak Ridge National Laboratory). ×27,500.

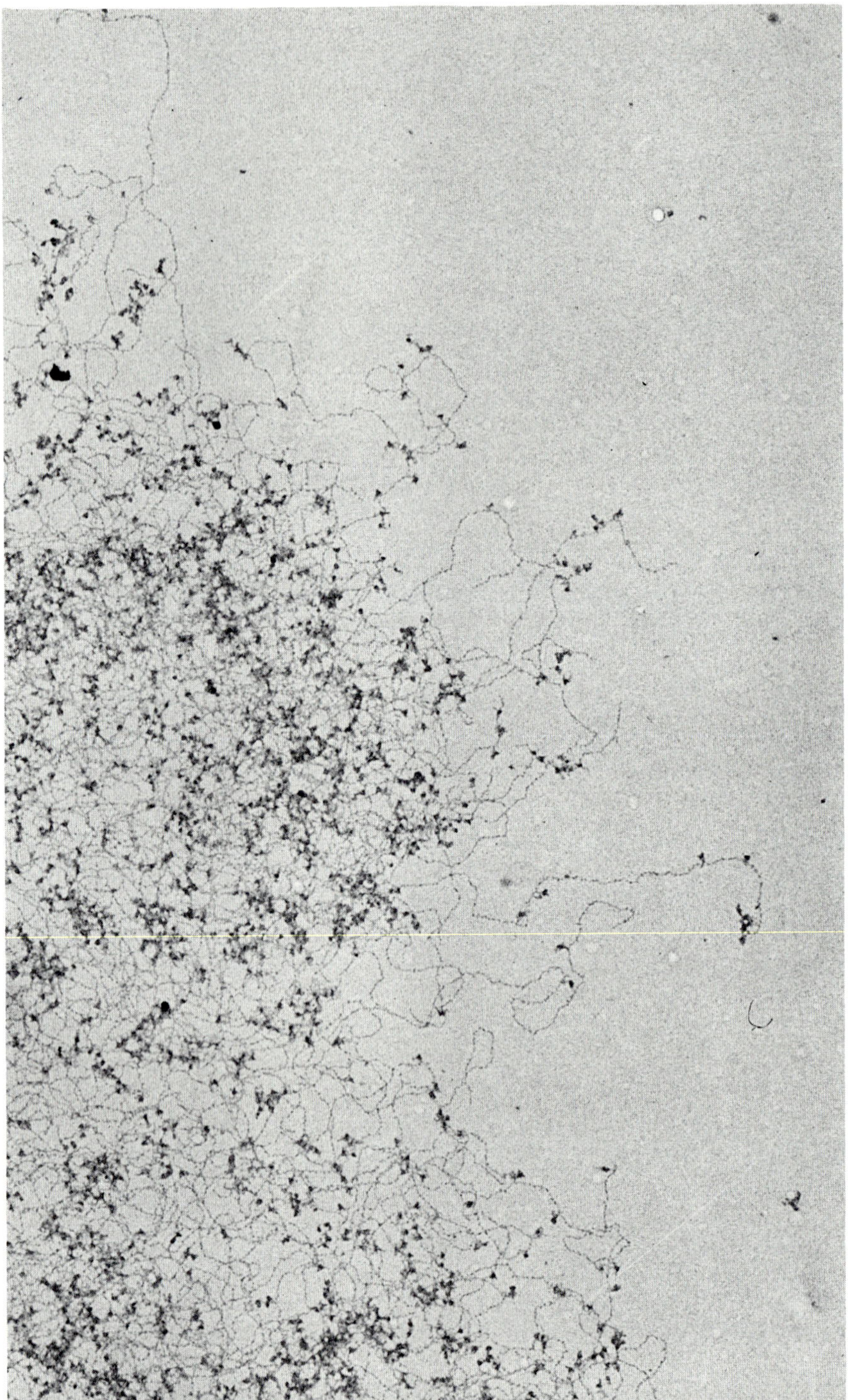

Fig. 4.14 (*continued*).

the forming nucleoli (Fig. 4.13). Later Painter and Taylor (*55b*) used the Feulgen reaction to demonstrate that the material observed by King was in fact DNA. Photometry showed that in *Xenopus* oocytes the nucleolar DNA synthesis resulted in a 3.5-fold increase in whole nuclear DNA (*56*); the "extra" DNA was in the form of minute granules or fibers that separated from the chromosomes (Fig. 4.12) (*27a*).

During the period of DNA "amplification," the oocyte nuclei exhibited a marked uptake of thymidine-^{3}H into a "nucleolar cap," while the chromosomes remained unlabeled (*27a*). The analysis of ^{3}H thymidine-labeled DNA on CsCl equilibrium centrifugation confirmed that up to 30% of the total radioactivity was in the higher density DNA (*27a*).

The sequence of events suggested by Gall (*27a*) is: (a) during the pachytene stage, specific replication of the nucleolus organizer region takes place; (b) the extra organizers, probably circular DNA molecules, migrate to the nuclear envelope, where they serve as the templates for the synthesis of rRNA precursors (Fig. 4.14); (c) multiple nucleoli form; (d) the synthesis of extra DNA is limited to the pachytene stage, but the DNA itself is metabolically stable and persists throughout oogenesis; (e) when the egg is mature and ready for ovulation, all RNA synthesis stops; (f) the germinal vesicle breaks down at the first meiotic reduction division, and the multiple nucleoli disappear (*10, 39a, 45d*).

After meiosis the extra rDNA no longer functions; there is no rRNA synthesis in eggs or embryos until the onset of gastrulation (*10b*). Although the extra rDNA is present in unfertilized eggs, it is not replicated during cleavage. Thus, hybridization experiments show that DNA from gastrulas composed of about 30,000 cells, contains the same proportion of rDNA and 5 S DNA as does adult somatic DNA (*10c*). These extrachromosomal copies of rDNA are apparently used for rRNA synthesis only during oogenesis and are lost at the first meiotic reduction division (*27a*).

Although the presence of multiple nucleoli is exceptional in amphibian germinal vesicles, both urodele and anuran (individual oocytes of most animals contain a single, prominent nucleolus), the patterns of oocyte maturation are remarkably similar in widely different species (*59b, 80a*). Brown *et al.* (*10a*) have shown rDNA amplification in the eggs of the echiuroid worm, *Urechis caupo,* and the surf clam, *Spisula solidissima,* although only a single nucleolus is present in these oocytes.

C. Control and Mechanism of Increased Synthesis of DNA

The large number of extrachromosomal DNA-containing nucleoli in amphibian oocytes results from an amplification of the ribosomal RNA cistrons without concurrent duplication of the rest of the genome in the same cell. Although the amphibia studied (*10a*) have widely different contents of DNA per cell and widely different numbers of rDNA replicas in their somatic genomes, they accumulate about the same amount of extra rDNA per oocyte (*56*). The rDNA content in oocyte nuclei isolated from heterozygous (1-No) females is the same as that found in the wild-type (2-No) oocytes, although the somatic nuclei of the heterozygous animals have lost half their NOR's or rDNA complement (*78a*).

One of the most fascinating questions in nucleolar physiology is that of the mechanisms by virtue of which DNA can be induced to replicate locally. Gene control of biosynthetic events through protein (*11, 29, 58, 58a*), "chromosomal controls," or other mechanisms (*38*) (see Chapter III) has been discussed, but how chromosomal receptors may respond to the cytoplasmic need for RNA of the ribosomes is a most interesting question. Activation or stimulation of the structural genes suffices to meet the ribosomal RNA synthesis in most mature cells. When these activities are inadequate to meet the demands of the cells, as in the case of the oocytes, apparently auxiliary genes are activated that not only stimulate maximal ribosomal RNA production but also produce activation of duplicase reactions to reproduction of rDNA. As yet, there is no apparent parallel for this remarkable phenomenon for other genes.

D. Detachment of Ribosomal DNA Templates from the Chromosomes

"Amplification" of rDNA in oocytes is accompanied by detachment of "nucleolar granules" or "rDNA granules" from the chromosomal "bouquet" or "nuclear cap."

The mechanism by which the free nucleoli are separated from the chromatin and spaced next to the nuclear envelope remains obscure (*8a, 62*), particularly since the pachytene-stage cells, which give rise to the larger oocytes with peripheral nucleoli, contain the normal number of nucleolar organizers. The rDNA cistrons are detached from the chromosomes without impairing the genome for future cell generations; mechanistically, it is possible that: (a) NOR's on the chromosomes are repetitively duplicated and the copies are detached and dispersed or (b) a single copy (or limited number of copies) is made, detached, and reduplicated. A choice between these alternatives is not yet possible (*13a*).

VII. Gene Dosage and the Synthesis of Ribosomal RNA

The linear relationship of the DNA satellite mass of *Xenopus* with the nucleolar number of the mutants (*8a*) is one example of "gene dosage."

As noted earlier (Chapter III), there is approximately half of the satellite mass in the total DNA of the *Xenopus* mutant with one nucleolus as there is in the *Xenopus* wild-type toads that have two nucleoli. In the mutants that have no nucleoli, there is no detectable satellite DNA (*8a*).

Ritossa *et al.* (*61–66*) have made extensive studies on relationship between the amount of RNA capable of hybridizing with DNA and the number of nucleolus organizers in *Drosophila melanogaster.* In the wild type, it was found that 0.27% of the total DNA hybridizes with labeled ribosomal RNA. By contrast, the total extent of hybridization was 10–20% when the "messenger" RNA was utilized for hybridization. They estimated that there are 130 complementary sites for the synthesis of ribosomal RNA in a haploid chromosomal set and 260 sites for biosynthesis of RNA in the diploid type of set of chromosomes of *Drosophila.*

In *Drosophila melanogaster,* both the X and Y chromosomes possess a single nucleolus organizer region (NOR) in the heterochromatic region of the X and in the short arm of the Y chromosome that is heterochromatic (*59a, 60a, 60b, 65, 66*). In studies on "bobbed" mutant loci, an association was found of the mutation in the sc^8–sc^4 regions of these chromosomes with the rDNA genes (*66*); i.e., this locus for "bobbed" character is closely associated with the nucleolus organizer regions. In all cases, the "bobbed" mutants had a lower hybridization with ribosomal RNA but there was marked variation (see Table 4.6). Reversion of bobbed mutants to a wild state occurred spontaneously, probably by crossing over, but total reversion did not occur. The conclusion was drawn that, although in these mutants the "machinery for ribosomal RNA formation was depleted," but it spontaneously adjusted to virtual normalcy over a number of generations. The mechanisms involved, possibly local gene amplification, remain to be defined (*60a, 60b*).

Employing mutants on the X and Y chromosomes, populations were obtained with NOR's ranging from 1 to 4. A precisely linear proportionality of the number of NOR's and the hybridization with rRNA was obtained as shown in Figs. 4.15 and 4.16.

Inasmuch as each haploid genome contains approximately 0.2–1 pg of DNA, they calculated that the total genome contains 0.2×10^{-12} gm $\times$ 6×10^{23} daltons (= 1 gm) or 1.2×10^{11} daltons of DNA. Since 0.27% of the total DNA codes for the ribosomal RNA, the rDNA contains 3.2×10^8 daltons. The total molecular weight of 28 S and 18 S RNA approximates 2.4×10^6 and accordingly, there are about 130 cistrons ($3.2 \times 10^8/2.4 \times 10^6$) for the synthesis of ribosomal RNA in the total haploid genome; about 260 cistrons are present in the total diploid genome.

There is an interesting correlation between this value and that of the number of microspherules that have been formed in the nucleoli of

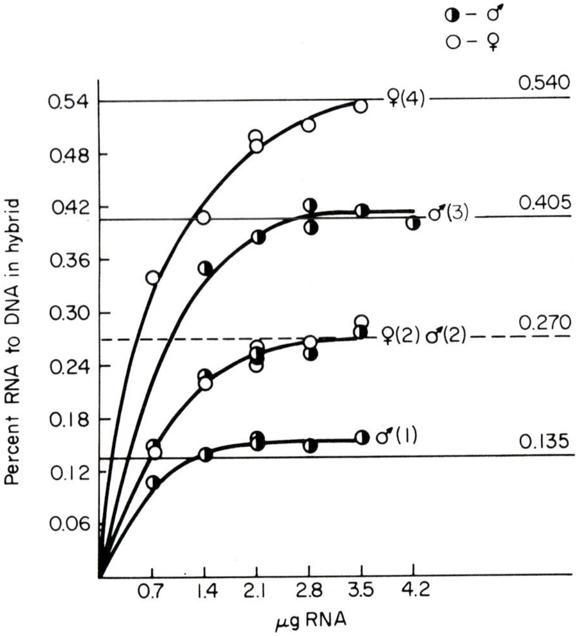

Fig. 4.15. Saturation levels of DNA containing different dosages of nucleolar organizer (NOR) regions with rRNA. Dosage of NOR is indicated by the number in parentheses. Dotted horizontal line at 0.27 is assumed to be a correct estimate for a dosage of 2, and solid horizontal lines represent predicted plateaus for dosages of 1, 3, and 4, respectively. Numerical values of the plateaus are given on right (*62*).

mammalian cells after treatment with actinomycin D (*75*). As noted in Chapter X, the number of fibrillar dense microspherules was estimated by planimetry to average 220 microspherules per nucleolus of the Novikoff hepatoma with a range of 150–310. However, this may not be the upper limit of rDNA cistrons in these cells since it is possible that not all of the potential cistrons for synthesis of rRNA are present even in such a large nucleolus. In any cell, many of the cistrons for synthesis of rRNA are inactive at any given time or indeed even during the whole life of the cell; some types of red cells and sperm do not synthesize rRNA and yet have the same gene dosage of rDNA as cells with two nucleoli (*62*). Although the precise mechanisms involved in the control of nucleolar genes have not been elucidated, it appears from studies on inactive rDNA (*60b*) that one operator gene may control large segments of rDNA readouts. Ritossa (*60b*) found a mutant in which substantial segments of rDNA were nonfunctional. Since this mutant probably arose as a result of mutation in one locus, the limited number of ribosomal RNA molecules

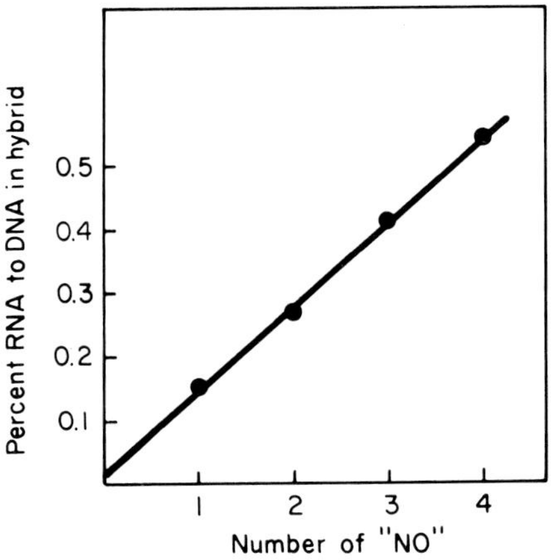

Fig. 4.16. Relationship between number of nucleolar organizers and percentage of rRNA to DNA in hybrid at saturation (*62*). NO = nucleolus organizers.

produced apparently results from inactivity of a large DNA segment containing a number of rDNA genes (*60b*).

VIII. Genetic Maps of Ribosomal Cistrons

The preparation of genetic maps has made it possible to define the sequential interrelationships of the various genetic activities in a variety of phages, microbial and mold mutants, and other species (*29, 70*). Genetic maps have been constructed for the loci in a variety of cell types that pertain to the ribosomal cistrons. In almost all of these there is close agreement that the genes for the synthesis of 18 S and 28 S rRNA of mammalian cells (or some insects) and those of the bacterial 16 and 23 S RNA are closely approximated. Many techniques employing analyses similar to those used for rII mutants of T_4 phage (*68*) and the "bobbed" mutants of *Drosophila* (*62*) have indicated that the two cistrons for ribosomal RNA are in close approximation, i.e., regardless of refinement of the degree of mutation for various readouts these genes are not separated.

Although studies of Dubnau, Smith, and Marmur (*21*) have shown that 4, 16, and 23 S RNA are produced from cistrons that are closely clustered on the gene map, much evidence from other sources has shown

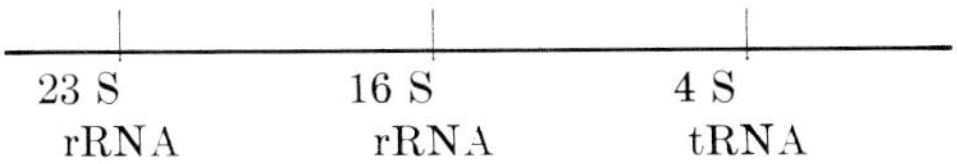

Gene map of cistrons for rRNA and tRNA (*21*)

that at least in plants (*80a*), amphibia (*10*), and *Drosophila* (*61*) the genes for synthesis of the transfer RNA and for the synthesis of ribosomal RNA are not closely clustered. In the *Drosophila melanogaster* employed by Ritossa *et al.* (*61–66*) 0.27% of the total DNA* is rDNA. Quagliarotti and Ritossa (*59a*) have shown (Table 4.6) that in these regions the genes for synthesis of 18 S rRNA and 28 S rRNA are closely approximated. On the other hand, the t-RNA genes that compose about 0.015% of the DNA are not included in the NOR regions. Oishi *et al.* (*55, 55a*) reported that the t-RNA genes are closely clustered to the ribosomal RNA genes in the *B. subtilis* chromosome between two adenine loci:

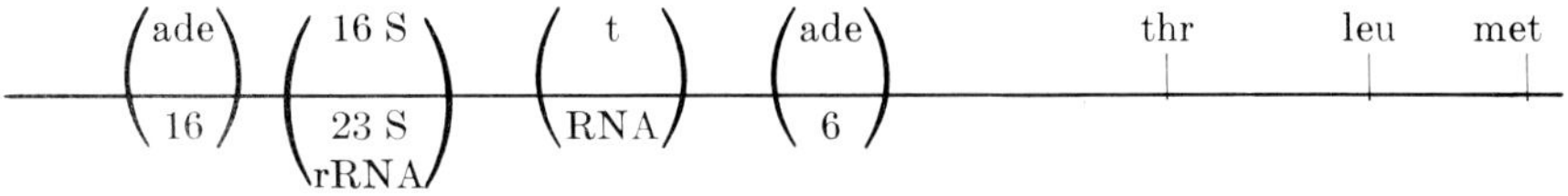

Interestingly, they noted competition of 23 S and 16 S rRNA with transfer RNA in their hybridization studies. Recently Oishi (*54a*) has isolated an H strand of DNA from *B. subtilis* which contains the rDNA as shown by its exclusive hybridization with the 16 S and 23 S rRNA.

In a recent study on localization of the genes for rRNA synthesis, Cutler and Evans (*17, 18*) made a fascinating study on Hfr (high frequency replicating) mutants of *E. coli.* In these cells, at least two separate gene sites are involved in the biosynthesis of ribosomal RNA. Their elegant system for "staircase" mapping indicated that these chromosomal sites are separate and distinct. In their regions 8–9 and 13–14, the cistrons for synthesis of ribosomal RNA and tRNA were most active. In both sites, there was geographical separation of the genes for synthesis of the transfer RNA and ribosomal RNA; their data agree with the model of Oishi *et al.* (*55*). In addition, Schweizer *et al.* (*68b′*) have reported that in yeast the genes for 18 S and 26 S rRNA are redundant, i.e., there are 140 cistrons for each of these RNAs.†

Another type of evidence for segregation of genes that are involved in the synthesis of transfer RNA and rRNA was derived by Brown (*10*) in studies on the *Xenopus* mutants that lack cistrons for nucleolar synthesis (see Chapter III). In the anucleolate mutant, neither the synthesis of 4 S transfer RNA or DNA-like RNA, which usually occurs after gene amplification, was affected by the presence or absence of the nucleolus.

* This number may vary in different species and values of 0.38% have been found in other strains (*59*).

† In an interesting experiment on cell hybrids of human and mouse cells, it was found that only the mouse type of 28 S RNA was produced, probably because the transcriptions of human ribosomal genes was suppressed. The implication of this experiment remains to be fully assessed (*22d*).

TABLE 4.6

RATIO OF DNA COMPLEMENTARY 28 AND 18 S rRNA IN THE WILD TYPE AND IN DIFFERENT MUTANTS

Type[a]	a rRNA/DNA (% in hybrid)	b 28 S rRNA/DNA (% in hybrid)	c 18 S rRNA/DNA (% in hybrid)	$\frac{a}{c}$	$\frac{a}{b}$	$\frac{b}{c}$
Wild type	0.373 ± 0.013	0.247 ± 0.012	0.122 ± 0.002	3.05 ± 0.15	1.51 ± 0.12	2.02 ± 0.13
yw bb S1	0.165 ± 0.009	0.111 ± 0.004	0.053 ± 0.003	3.11 ± 0.28	1.48 ± 0.13	2.09 ± 0.23
yw bb S2	0.205 ± 0.006	0.138 ± 0.004	0.066 ± 0.003	3.10 ± 0.24	1.48 ± 0.09	2.09 ± 0.18
car bb S1	0.254 ± 0.006	0.170 ± 0.003	0.082 ± 0.003	3.09 ± 0.18	1.49 ± 0.06	2.07 ± 0.11
car bb S2	0.168 ± 0.006	0.109 ± 0.005	0.053 ± 0.002	3.17 ± 0.23	1.54 ± 0.13	2.05 ± 0.17
UCO3 bb S1	0.185 ± 0.006	0.123 ± 0.004	0.060 ± 0.002	3.08 ± 0.20	1.50 ± 0.10	2.05 ± 0.13

Data from Quagliarotti and Ritossa (*59a*).

[a] *bb* = bobbed mutants.

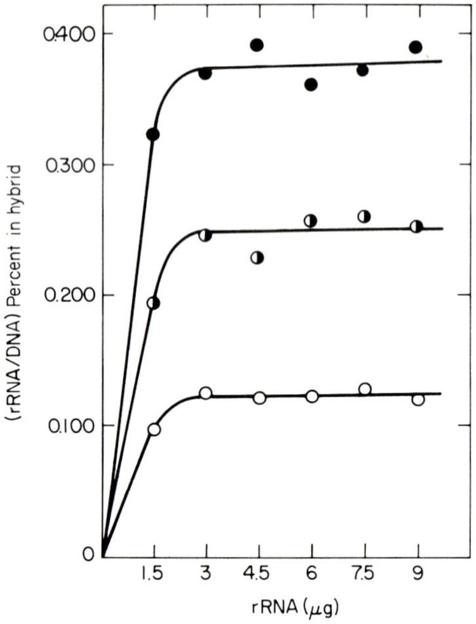

Fig. 4.17. Saturation curves of hybridization experiments between wildtype DNA and ^{32}P-labeled ribosomal RNA from *D. melanogaster*. The upper curve (—●—●—) was obtained using a mixture of hot 28 S and 18 S rRNA. The middle curve (—◑—◑—) was obtained with labeled 28 and cold 18 S in the proportion of 2:1. The lowest curve (—○—○—) was obtained with cold 28 S and labeled 18 S in the proportion of 2:1. (From Quagliarotti and Ritossa, *59a*.)

In the normal form of this oocyte, there is rapid synthesis of the rDNA through the mechanism of gene amplification or endoreduplication (see pages 187), after which there is a rapid proliferation of the ribosomes; there then follows a rapid synthesis of DNA-like RNA and tRNA. These temporal relationships and the lack of effect of the nucleolar deletions on other synthetic rDNA reactions indicate a segregation of nucleolar genes containing rDNA from the other gene loci.

IX. Spatial Relationships of the Genes for Synthesis of 18 S and 28 S RNA

The development of methods for the isolation of the satellite rDNA from *Xenopus laevis* (Fig. 4.8) has permitted a much more precise analysis of the rRNA cistrons than has hitherto been possible (*8, 10c, 10d, 59a*). In view of the large differences in the base compositions and oligonucleotide frequency distributions of 18 S rRNA, and 28 S rRNA, it is clear (see Chapter V), that these two types of RNA are formed on different cistrons. With the purified satellite rDNA and other analytical methods (*10b, 59a*), it was shown that both the 18 S rRNA and 28 S rRNA hybridized independently and reached individual plateaus (Fig. 4.17); the saturation level for 28 S rRNA was 2.0–2.3 times that of 18 S rRNA.

When rDNA was subjected to scission, it was shown that rDNA of "average" molecular weight of 1.6×10^6 readily hybridized with both the 28 S and 18 S in approximately the same ratios as unsheared rDNA.

When the rDNA was cleaved further to molecular weights (average) of 0.5 and 0.3×10^6, there was selective hybridization of the 28 S rRNA with the piece of higher bouyant density and 18 S rRNA with the DNA of lower bouyant density. The clear implication is that there is a close spatial relationship between the genes for synthesis of 18 S and 28 S RNA. These findings are supported by the results of Brown and Weber (*10c,d*) and Quagliarotti and Ritossa (*59a*) who carried out similar studies on *Xenopus laevis* and *Drosophila melanogaster*, respectively.

The most probable model for the arrangement of the cistrons for rDNA is shown in Fig. 4.18. There are alternating 18 S and 28 S rDNA cistrons and interspersed segments coding for other RNA's which presumably would be cleaved away from the rRNA's. These results pose an interesting question in the light of the analytical data for nucleoli which suggests that although both 18 S and 28 S rDNA are present (*70a*) and in a 1:1 ratio (*59a*), the nucleolar 28 S RNA product has not been chemically differentiated from nucleolar 35 S and 45 S RNA and 18 S rRNA has not yet been directly demonstrated in the nucleolus (see p. 255).

One interesting fact is that in the *Drosophila melanogaster* the compositions of 18 S rRNA and 28 S rRNA are markedly different from those of the mammalian cells as shown in Table 4.7 (Chapter V, p. 227). Thus in *Drosophila*, the rDNA cistrons must differ completely in base

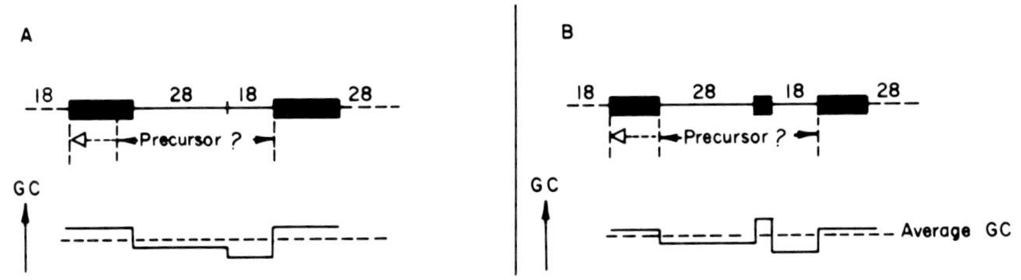

Fig. 4.18. Arrangement of the 28 S and 18 S rRNA and nonribosomal sequences on the rDNA. A, the alternating, partially integrated model; B, the alternating, fully integrated model. From (*8*).

TABLE 4.7

BASE COMPOSITION OF *Drosophila melanogaster* RNA FRACTIONS (*62*)[a]

	C	A	U(T)	G
28 S rRNA	17.5	30.4	30.1	22.0
18 S rRNA	17.8	28.1	31.0	23.1
tRNA	25.9	20.2	24.6	29.3
DNA	20.0	30.0	30.0	20.0

[a] ^{32}P-labeled 28 S and 18 S fractions were purified on sucrose density gradient (12 hours at 25,000 rpm in an SW 25 (Spinco head). tRNA was purified by MAK column to release it from free ^{32}P. RNA's were digested 18 hours at 30°C in 0.35 *N* NaOH and the nucleotides were separated by a column of Dowex 1-X8.

composition from the rDNA of mammalian cells. Moreover, in these insects, the 18 S rRNA and 28 S rRNA have virtually the same compositions; in other cell types, these rRNA's have very different compositions. Nonetheless, saturation-hybridization experiments (Fig. 4.17) showed that, as expected from the molecular weight, the hybridization for the 28 S rRNA was twice that of 18 S rRNA and the two were additive. Accordingly, the conclusion was made that for this species as well as *B. megaterium* there is essentially no overlap of the genes for the biosynthesis of the two ribosomal RNA's (Table 4.6).

In the studies on the tRNA of *Drosophila* it was also found that there were few if any overlaps of rDNA and cistrons for tRNA biosynthesis (*62*). In addition, the redundancy is less for tDNA.

X. Summary

The information currently available on DNA of the nucleolus is limited compared to that which will be required to ultimately characterize this DNA; i.e., to learn the sequences of bases in nucleolar DNA and the total number of types of DNA present. Nucleolar DNA consists of a variety of morphological types which may have multiple functions; rDNA is one portion of total nucleolar DNA for which the function is well defined at present.

The greatest progress made thus far has been in the isolation of rDNA which now has been obtained from bacterial sources (*39e*) and from amphibians (*8*) because of the fortunate circumstance that the rDNA has a different buoyant density from the whole DNA. The evidence thus far indicates that the rDNA consists of alternating stretches of segments coding for 18 S rRNA and 28 S rRNA in animal cells and 16 S and 23 S rRNA in bacteria. However, precise definition of the oligonucleotide sequences is necessary to establish this point and to learn what other types of RNA products may be produced. There is considerable uncertainty about some of the RNA products of the nucleolus as will be indicated in Chapters V and VI.

The availability of purified rDNA should permit more definitive studies on operator genes affecting nucleolar DNA (*60b*), and the control mechanisms that may affect the rate of these readouts. Much more needs to be learned about perinucleolar nucleolus-associated DNA and intranucleolar DNA that is not rDNA. The selective degradation procedures may offer an approach to the study of these different types of molecules.

REFERENCES

1. Altmann, H.-W., Stöcker, E., and Thoenes, W., Uber chromatin und DNS-synthese im nucleolus: Elektronenmikroskopische, autoradiographische und lichtmikroskopische untersuchungen an leberzellen von ratten. *Z. Zellforsch. Mikroskop. Anat.* **59**, 116–133 (1963).
2. Attardi, G., Huang, P. C., and Kabat, S., Recognition of ribosomal RNA sites in DNA. I. Analysis of the *E. coli* system. *Proc. Natl. Acad. Sci. U. S.* **53**, 1490–1498 (1965).
3. Attardi, G., Huang, P. C., and Kabat, S., Recognition of ribosomal RNA sites in DNA. II. The HeLa cell system. *Proc. Natl. Acad. Sci. U. S.* **54**, 185–192 (1965).
4. Barr, H. J., and Plaut, W., Nucleolar DNA in *Drosophila*. *J. Cell Biol.* **31**, 10A (1966).
5. Barr, H. J., and Plaut, W., Comparative morphology of nucleolar DNA in *Drosophila*. *J. Cell Biol.* **31**, C17–C22 (1966).

5a. Bauer, H., Die histologie des ovars von *Tipula paludosa. Meig.* *Z. Zool.* **143**, 53–76 (1932).

5b. Bayreuther, K., Die oogenese der *tipuliden*. *Chromosoma* **7**, 508–557 (1956).

6. Bernhard, W., Ultrastructural aspects of the normal and pathological nucleolus in mammalian cells. *Natl. Cancer Inst. Monograph* **23**, 13–38 (1966).
7. Bernhard, W., and Granboulan, N., The fine structure of the cancer cell nucleus. *Exptl. Cell Res.* Suppl. 9, 19–53 (1962).

7a. Birnstiel, M., Some experiments relating to the homogeneity and arrangement of the ribosomal RNA genes of *Xenopus laevis*. *Ciba Found. Symp., Cell Differentiation* pp. 178–191 (1967).

7b. Birnstiel, M., Personal communication (1969).

8. Birnstiel, M., Speirs, J., Purdom, I., and Jones, K., Properties and composition of the isolated ribosomal DNA satellite of *Xenopus laevis*. *Nature* **219**, 454–463 (1968).

8a. Birnstiel, M., Wallace, H., Sirlin, J. L., and Fischberg, M., Localization of the ribosomal DNA complements in the nucleolar organizer regions of *Xenopus laevis*. *Natl. Cancer Inst. Monograph* **23**, 431–448 (1966).

9. Bond, H. E., Flamm, W. G., Burr, H. E., and Bond, S. B., Mouse satellite DNA, further studies on its biological and physical characteristics and its intracellular localization. *J. Mol. Biol.* **27**, 289–302 (1967).

9a. Borst, P., Ruttenberg, G. J. C. M., and Kroon, A. M., Mitochondrial DNA. I.

Preparation and properties of mitochondrial DNA from chick liver. *Biochim. Biophys. Acta* **149**, 140–155 (1967).

9b. Brachet, J., La localisation de l'acide thymonucléique pendant l'oogenese et la maturation chez les amphibiens. *Arch. Biol.* (*Liege*) **51**, 151–165 (1940).

10. Brown, D. D., The nucleolus and synthesis of ribosomal RNA during oögenesis and embryogenesis of *Xenopus laevis*. *Natl. Cancer Inst. Monograph* **23**, 297–310 (1966).

10a. Brown, D. D., and David, I. B., Specific gene amplification in oocytes. *Science* **160**, 272–280 (1968).

10b. Brown, D. D., and Littna, E., RNA synthesis during the development of *Xenopus laevis*, the South African clawed toad. *J. Mol. Biol.* **8**, 669–687 (1964).

10c. Brown, D. D., and Weber, C. S., Gene linkage by RNA-DNA hybridization. I. Unique DNA sequences homologous to 4 S RNA, 5 S RNA and ribosomal RNA. *J. Mol. Biol.* **34**, 661–680 (1968).

10d. Brown, D. D., and Weber, C. S., Gene linkage by RNA-DNA hybridization. II. Arrangement of the redundant gene sequences for 28 S and 18 S ribosomal RNA. *J. Mol. Biol.* **34**, 681–697 (1968).

11. Busch, H., "Histones and Other Nuclear Proteins." Academic Press, New York, 1965.

12. Busch, H., Byvoet, P., and Smetana, K., The nucleolus of the cancer cell: A review. *Cancer Res.* **23**, 313–339 (1963).

13. Busch, H., Muramatsu, M., Adams, H. R., Smetana, K., Steele, W. J., and Liau, M. C., Isolation of nucleoli. *Exptl. Cell Res.* Suppl. 9, 150–163 (1963).

13a. Callan, H. G., Chromosomes and nucleoli of the axolotl, *Ambystoma mexicanum*. *J. Cell Sci.* **1**, 85–108 (1966).

14. Caspersson, T. O., "Cell Growth and Cell Function. A Cytochemical Study." Norton, New York, 1950.

15. Chipchase, M. I. H., and Birnstiel, M. L., On the nature of nucleolar RNA. *Proc. Natl. Acad. Sci. U. S.* **50**, 1101–1107 (1963).

15a. Chun, E. H. L., and Littlefield, J. W., The replication of the minor DNA component of mouse fibroblasts. *J. Mol. Biol.* **7**, 245–248 (1963).

15b. Chun, E. H. L., Vaughan, M. H., Jr., and Rich, A., The isolation and characterization of DNA associated with chloroplast preparations. *J. Mol. Biol.* **7**, 130–141 (1963).

15c. Church, K., Replication of chromatin in mouse mammary epithelial cells grown *in vitro*. *Genetics* **52**, 843–849 (1965).

16. Comb, D. G., Brown, R., and Katz, S., The nuclear DNA and RNA components of the aquatic fungus *Blastocladiella emersonii*. *J. Mol. Biol.* **8**, 781–789 (1964).

16a. Corneo, G., Moore, C., Sanadi, D. R., Grossman, L. I., and Marmur, J., Mitochondrial DNA in yeast and some mammalian species. *Science* **151**, 687–689 (1966).

17. Cutler, R. G., and Evans, J. E., Isolation of selected segments from the genome of Hfr *Escherichia coli*. *J. Mol. Biol.* **26**, 81–90 (1967).

18. Cutler, R. G., and Evans, J. E., Relative transcription activity of different segments of the genome throughout the cell division cycle of *Escherichia coli*. The mapping of ribosomal and transfer RNA and the determination of the direction of replication. *J. Mol. Biol.* **26**, 91–105 (1967).

18a. Desjardins, R., Grogan, D. E., Arendell, J. P., and Busch, H., Effects of antitumor agents on the synthesis of nucleolar DNA. *Cancer Res.* **17**, 159–164 (1967).

19. Desjardins, R., Smetana, K., and Busch, H., A large scale procedure for isolation of nucleoli from rat liver. *Exptl. Cell Res.* **40**, 353–360 (1965).
19a. Desjardins, R., Smetana, K., Grogan, D., Higashi, K., and Busch, H., A large scale procedure for isolation of nucleoli from Walker tumor. *Cancer Res.* **26**, 97–104 (1966).
20. Desjardins, R., Smetana, K., Steele, W. J., and Busch, H., Isolation of nucleoli of the Walker carcinosarcoma and liver of the rat following nuclear disruption in a French pressure cell. *Cancer Res.* **23**, 1819–1823 (1963).
21. Dubnau, D., Smith, I., and Marmur, J., Gene conservation in Bacillus species. II. The location of genes concerned with the synthesis of ribosomal components and soluble RNA. *Proc. Natl. Acad. Sci. U. S.* **54**, 724–730 (1965).
22. Ebstein, B. S., The distribution of DNA within the nucleoli of the amphibian oocyte as demonstrated by tritiated actinomycin D radioautography. *J. Cell Sci.* **5**, 27–44 (1969).
22a. Ebstein, B. S., Tritiated actinomycin D as a cytochemical label for small amounts of DNA. *J. Cell Biol.* **35**, 709–713 (1967).
22b. Edelman, M., Schiff, J. A., and Epstein, H. T., Studies of chloroplast development in *Euglena* XII. Two types of satellite DNA. *J. Mol. Biol.* **11**, 769–774 (1965).
22c. Edström, J. E., and Gall, J. G., The base composition of ribonucleic acid in lampbrush chromosomes, nucleoli, nuclear sap and cytoplasm of *Triturus* oocytes. *J. Cell Biol.* **19**, 279–284 (1963).
22d. Eliceiri, G. L., and Green, H., Ribosomal RNA synthesis in human-mouse hybrid cells. *J. Mol. Biol.* **41**, 253–260 (1969).
23. Flamm, W. G., Bond, H. E., Burr, H. E., and Bond, S. B., Satellite DNA isolated from mouse liver: Some physical and metabolic properties. *Biochim. Biophys. Acta* **123**, 652–654 (1966).
24. Frenster, J. H., Nuclear polyanions as de-repressors of synthesis of ribonucleic acid. *Nature* **206**, 680–683 (1965).
25. Frenster, J. H., A model of specific de-repression within interphase chromatin. *Nature* **206**, 1269–1270 (1965).
26. Frenster, J. H., Localized strand separations within deoxyribonucleic acid during selective transcription. *Nature* **208**, 894–896 (1965).
27. Gabrusewycz-Garcia, N., and Kleinfeld, R. G., A study of the nucleolar material in *Sciara coprophila. J. Cell Biol.* **29**, 347–359 (1966).
27a. Gall, J. G., Differential synthesis of the genes for ribosomal RNA during amphibian oogenesis. *Proc. Natl. Acad. Sci. U. S.* **60**, 553–560 (1968).
27b. Gall, J. G., The genes for ribosomal RNA during oögenesis. *Genetics,* Suppl. **61**, 121–132 (1969).
27c. Gall, J. G., MacGregor, H. C., and Kidston, M. E., Gene amplification in the oocytes of dytiscid water beetles. *Chromosoma* (Berlin) **26**, 169–187 (1969).
28. Galton, M., and Holt, S. F., Late replication of mouse sex chromosomes. *Exptl. Cell Res.* **37**, 111–116 (1965).
29. Gilbert, W., and Muller-Hill, B., The lac operator is DNA. *Proc. Natl. Acad. Sci. U. S.* **58**, 2415–2421 (1967).
29a. Granboulan, N., and Granboulan, P., Cytochimie ultrastructurale du nucléole. I. Mise en évidence de chromatine à l'intérieur du nucléole. *Exptl. Cell Res.* **34**, 71–87 (1964).
30. Green, B. R., and Gordon, M. P., The satellite DNA's of some higher plants. *Biochim. Biophys. Acta* **145**, 378–390 (1967).

30a. Green, B. R., and Gordon, M. P., Satellite DNA's of tobacco. *Federation Proc.* **24**, 539 (1965).
30b. Green, B. R., and Gordon, M. P., Replication of chloroplast DNA of tobacco. *Science* **152**, 1071–1074 (1966).
31. Greenberg, L. J., and Uhr, J. W., Myeloma tumor satellite DNA: A role in ribosomal RNA synthesis. *Biochem. Biophys. Res. Commun.* **27**, 523–528 (1967).
32. Grogan, D. E., Desjardins, R., and Busch, H., Nucleolar proteins of rat liver and Walker tumor. *Cancer Res.* **26**, 775–779 (1966).
33. Hay, E., and Revel, J. P., The DNA component of the nucleolus studied in autoradiographs viewed with the electron microscope. *Proc. 5th Intern. Conf. Electron Microscopy, Philadelphia, 1962* Vol. II, pp. 1–7. Academic Press, 1962.
34. Hay, E. D., and Revel, J. P., The fine structure of the DNP component of the nucleus. *J. Cell Biol.* **16**, 29–51 (1963).
35. Heinonen, L., and Halkka, O., Early stages of oogenesis and metabolic DNA in the oocytes of the house cricket, *Acheta domesticus* (L.). *Ann. Med. Exptl. Biol. Fenniae (Helsinki)* **45**, 101–109 (1964).
36. Hill, M., Translocation of donor DNA into nucleolar associated chromatin of recipient cell *in vitro*. *Experientia* **22**, 726 (1966).
36a. Hotta, Y., Basel, A., and Stern, H., Nuclear DNA and cytoplasmic DNA from tissues of higher plants. *J. Cell Biol.* **27**, 451–457 (1965).
36b. Izawa, M., Allfrey, V. G., and Mirsky, A. E., The relationship between RNA synthesis and loop structure in lampbrush chromosomes. *Proc. Natl. Acad. Sci. U. S.* **49**, 544–551 (1963).
37. Jacob, J., Intranucleolar deoxyribonucleic acid components in insect cells as revealed by electron microscopy. *Nature* **211**, 36–38 (1966).
38. Jain, H. K., Chromosomal control of nucleolar synthesis. *Experientia* **20**, 156–157 (1964).
39. Kasten, F. H., and Strasser, F. F., Nucleic acid synthetic patterns in synchronized mammalian cells. *Nature* **211**, 135–140 (1966).
39a. Kato, K., Cytochemistry and fine structure of elimination chromatin in *Dytiscidae*. *Exptl. Cell Res.* **52**, 507–522 (1968).
39a′. Kezer, J., Cited in W. J. Peacock, "Chromosome Replication." *Natl. Cancer Inst. Monograph* **18**, 101 (1965).
39b. King, H. D., The oogenesis of *Bufo lentiginosus*. *J. Morphol.* **19**, 369–438 (1908).
39c. Kislev, N., Swift, H., and Bogorad, L., Nucleic acids of chloroplasts and mitochondria in Swiss chard. *J. Cell Biol.* **25**, 327–344 (1965).
39d. Kit, S., Species differences in animal deoxyribonucleic acids as revealed by equilibrium sedimentation in density gradients. *Nature* **193**, 274–275 (1962).
39e. Kohne, D. E., Isolation and characterization of bacterial ribosomal RNA cistrons. *Biophys. J.* **8**, 1104–1118 (1968).
39f. Lane, N. J., Spheroidal and ring nucleoli in amphibian oocytes. *J. Cell Biol.* **35**, 421–434 (1967).
40. Lettré, R., and Siebs, W., Beobachtungen zur struktur des nucleolus in normalen zellen sowie in tumorzellen. *Z. Krebsforsch.* **60**, 564–580 (1955).
41. Lettré, R., and Siebs, W., Some studies of the nucleolus of cell cultivated *in vitro*. *Pathol. Biol. Semaine Hop.* [N.S.] **9**, 819–823 (1961).
42. Lettré, R., and Siebs, W., Behavior of nucleoli of cells *in vitro*. *Proc. Intern. Union Physiol. Sci., 22nd Intern. Congr. Leiden, 1962,* pp. 837–840. Excerpta Med. Found., Amsterdam, 1962.

43. Lettré, R., Siebs, W., and Paweletz, N., Morphological observations on the nucleolus of cells in tissue culture, with special regard to its composition. *Natl. Cancer Inst. Monograph* **23**, 107–124 (1966).
44. Lima-De-Faria, A., Differential uptake of tritiated thymidine into hetero and euchromatin in *melanophus* and *secale*. *J. Biophys. Biochem. Cytol.* **6**, 457–466 (1959).
45. Lima-De-Faria, A., Reitalu, J., and O'Sullivan, M. A., Replication of autosomal heterochromatin in man. *Chromosoma* **16**, 152–161 (1965).
45a. Luck, D. J. L., and Reich, E., DNA in mitochondria of *neurospora crassa*. *Proc. Natl. Acad. Sci. U. S.* **52**, 931–938 (1964).
45b. MacGregor, H. C., The role of lampbrush chromosomes in the formation of nucleoli in amphibian oocytes. *Quart. J. Microscop. Sci.* **106**, 215–228 (1965).
45c. MacGregor, H. C., Pattern of incorporation of ^{3}H uridine into RNA of amphibian oocyte nucleoli. *J. Cell Sci.* **2**, 145–150 (1967).
45d. MacGregor, H. C., Nucleolar DNA in oocytes of *Xenopus laevis*. *J. Cell Sci.* **3**, 437–444 (1968).
46. Maio, J. J., and Schildkraut, C. L., Isolated mammalian metaphase chromosomes. I. General characteristics of nucleic acids and proteins. *J. Mol. Biol.* **24**, 29–39 (1967).
47. Matsuda, K., and Siegel, A., Hybridization of plant ribosomal RNA to DNA: The isolation of a DNA component rich in ribosomal RNA cistrons. *Proc. Natl. Acad. Sci. U. S.* **58**, 673–680 (1967).
48. McConkey, E. H., and Hopkins, J. W., The relationship of the nucleolus to the synthesis of ribosomal RNA in HeLa cells. *Proc. Natl. Acad. Sci. U. S.* **51**, 1197–1204 (1964).
49. McLeish, J., Deoxyribonucleic acid in plant nucleoli. *Nature* **204**, 4953 (1964).
50. McLeish, J., Chemical and autoradiographic studies of intranucleolar DNA in *Vicia Faba*. *Exptl. Cell Res.* (1969) (in press).
51. Merits, I., Schulze, W., and Overby, L. R., Ribosomal ribonucleic acid sites in chicken deoxyribonucleic acid. *Arch. Biochem. Biophys.* **115**, 197–205 (1966).
51a. Miller, O. L., Studies on the ultrastructure and metabolism of nucleoli in amphibian oocytes. *Proc. 5th Intern. Conf. Electron Microscopy, Philadelphia, 1962* Vol. II, Art. NN8. Academic Press, New York.
51b. Miller, O. L., Structure and composition of peripheral nucleoli of salamander oocytes. *Natl. Cancer Inst. Monograph* **23**, 53–66 (1966).
51c. Mohan, J., Dunn, A., and Casola, L., Ribosomal DNA in the rat. *Nature* **223**, 295–296 (1969).
52. Muramatsu, M., Smetana, K., and Busch, H., Quantitative aspects of isolation of nucleoli of the Walker carcinosarcoma and liver of the rat. *Cancer Res.* **23**, 510–518 (1963).
53. Nash, D., and Plaut, W., On the presence of DNA in larval salivary gland nucleoli in *Drosophila melanogaster*. *J. Cell Biol.* **27**, 682–686 (1965).
54. O'Donnell, E. H. J., Deoxyribonucleic acid structures in the nucleolus. *Nature* **191**, 1325–1326 (1961).
54a. Oishi, M., The transcribing strands of *Bacillus subtilis* DNA for ribosomal and transfer RNA. *Proc. Natl. Acad. Sci. U. S.* **62**, 256–262 (1969).
55. Oishi, M., Oishi, A., and Sueoka, N., Location of genetic loci of soluble RNA on *Bacillus subtilis* chromosome. *Proc. Natl. Acad. Sci. U. S.* **55**, 1095–1103 (1966).
55a. Oishi, M., and Sueoka, N., Location of genetic loci of ribosomal RNA on *Bacillus subtilis* chromosome. *Proc. Natl. Acad. Sci. U. S.* **54**, 483–491 (1965).

55b. Painter, T. S., and Taylor, A. N., Nucleic acid storage in the toad's egg. *Proc. Natl. Acad. Sci. U. S.* **28**, 311–317 (1942).
56. Perkowska, E., MacGregor, H. C., and Birnstiel, M. L., Gene amplification in the oocyte nucleus of mutant and wildtype *Xenopus laevis*. *Nature* **217**, 649–650 (1968).
56a. Pikó, L., Blair, D. G., Tyler, A., and Vinograd, J., Cytoplasmic DNA in the unfertilized sea urchin egg: Physical properties of circular mitochondrial DNA and the occurrence of catenated forms. *Proc. Natl. Acad. Sci. U. S.* **59**, 838–845 (1968).
57. Plaut, W., Nash, D., and Fanning, T., Ordered replication of DNA in polytene chromosomes of *Drosophila melanogaster*. *J. Mol. Biol.* **16**, 85–93 (1966).
58. Ptashne, M., Specific binding of the λ phage repressor to λ DNA. *Nature* **214**, 232–234 (1967).
58a. Ptashne, M., Isolation of the λ phage repressor. *Proc. Natl. Acad. Sci. U. S.* **57**, 306–313 (1967).
59. Quagliarotti, G., Personal communication (1969).
59a. Quagliarotti, G., and Ritossa, F., On the arrangement of genes for 28 S and 18 S ribosomal RNAs in *Drosophila melanogaster*. *J. Mol. Biol.* **36**, 57–69 (1968).
59b. Rabinowitz, M., Sinclair, J., DeSalle, L., Haselkorn, R., and Swift, H., Isolation of deoxyribonucleic acid from mitochondria of chick embryo heart and liver. *Proc. Natl. Acad. Sci. U. S.* **53**, 1126–1133 (1965).
59c. Raven, C. P., "Oogenesis: The Storage of Developmental Information." Pergamon Press, Oxford, 1962.
59d. Ray, D. S., and Hanawalt, P. C., Properties of the satellite DNA associated with the chloroplasts of *Euglena gracilis*. *J. Mol. Biol.* **9**, 812–824 (1964).
59e. Ray, D. S., and Hanawalt, P. C., Satellite DNA components in *Euglena gracilis* cells lacking chloroplasts. *J. Mol. Biol.* **11**, 760–768 (1965).
60. Ribas-Mundo, M., DNA replication patterns of normal human leukocyte cultures. Time sequence of DNA synthesis in relation to the H^3-thymidine incorporation over the nucleolus. *Blood* **28**, 891–900 (1966).
60a. Ritossa, F. M., Unstable redundancy of genes for ribosomal RNA. *Proc. Natl. Acad. Sci. U. S.* **60**, 509–516 (1968).
60b. Ritossa, F. M., Nonoperative DNA complementary to ribosomal RNA. *Proc. Natl. Acad. Sci. U. S.* **59**, 1124–1131 (1968).
61. Ritossa, F. M., and Atwood, K. C., Unequal proportions of DNA complementary to ribosomal RNA in males and females of *Drosophila simulans*. *Proc. Natl. Acad. Sci. U. S.* **56**, 496–499 (1966).
62. Ritossa, F. M., Atwood, K. C., Lindsley, D. L., and Spiegelman, S., On the chromosomal distribution on DNA complementary to ribosomal and soluble RNA. *Natl. Cancer Inst. Monograph* **23**, 449–472 (1966).
63. Ritossa, F. M., Atwood, K. C., and Spiegelman, S., On the redundancy of DNA complementary to amino acid transfer RNA and its absence from the nucleolar organizer region of *Drosophila melanogaster*. *Genetics* **54**, 663–676 (1966).
64. Ritossa, F. M., Atwood, K. C., and Spiegelman, S., A molecular explanation of the bobbed mutants of *Drosophila* as partial deficiencies of "ribosomal" DNA. *Genetics* **54**, 819–834 (1966).
65. Ritossa, F. M., and Spiegelman, S., Localization of RNA complementary to ribosomal RNA in the nucleolus organizer region of *Drosophila melanogaster*. *Proc. Natl. Acad. Sci. U. S.* **51**, 1197–1204 (1964).
66. Ritossa, F. M., and Spiegelman, S., Localization of DNA complementary to ribo-

somal RNA in the nucleolus organizer region of *Drosophila melanogaster*. *Proc. Natl. Acad. Sci. U. S.* **53**, 737–745 (1965).

67. Ro, T. S., Narayan, K. S., and Busch, H., Effects of actinomycin D on base composition and nearest neighbor frequency of nucleolar RNA of the Walker tumor and liver. *Cancer Res.* **26**, 780–785 (1966).

68. Ruger, W., and Bautz, E. K. F., Non-repeating nucleotide sequences in the genome of bacteriophage T4. *J. Mol. Biol.* **31**, 83–90 (1968).

68a. Sager, R., and Ishida, M. R., Chloroplast DNA in chlamydomonas. *Proc. Natl. Acad. Sci. U. S.* **50**, 725–730 (1963).

68b. Schildkraut, C. L., and Maio, J. J., Studies on the intranuclear distribution and properties of mouse satellite DNA. *Biochim. Biophys. Acta* **161**, 76–93 (1968).

68b′. Schweizer, E., MacKechnie, C., and Halvorson, H. O., The redundancy of ribosomal and transfer RNA genes in *Saccharomyces cerevisiae*. *J. Mol. Biol.* **40**, 261–277, 1969.

68c. Shankaranarayan, K., Muramatsu, M., Smetana, K., and Busch, H., Ultrastructural studies on RNA and DNA components of isolated nucleoli of Walker 256 carcinosarcoma. *Exptl. Cell Res.* **41**, 81–89 (1966).

68d. Shipp, W., Kieras, F. J., and Haselkorn, R., DNA associated with tobacco chloroplasts. *Proc. Natl. Acad. Sci. U. S.* **54**, 207–213 (1965).

68e. Simard, R., The binding of actinomycin D-^{3}H to heterochromatin as studied by quantitative high resolution radioautography. *J. Cell Biol.* **35**, 716–722 (1967).

69. Smetana, K., and Busch, H., On the ultrastructure of the Walker 256 carcinosarcoma. *Cancer Res.* **23**, 1600–1603 (1963).

69a. Smetana, K., and Busch, H., Studies on the ultrastructure of the nucleolus of the Walker tumor and the rat liver. *Cancer Res.* **24**, 537–557 (1964).

69b. Smetana, K., and Busch, H., Studies on staining and localization of acidic nuclear proteins in the Walker 256 carcinosarcoma. *Cancer Res.* **26**, 331–337 (1966).

69c. Smetana, K., Unuma, T., and Busch, H., Ultrastructural studies on nucleic acids of nucleolar granular components in Novikoff hepatoma cells. *Exptl. Cell Res.* **51**, 105–122 (1968).

70. Snyder, L., and Geiduschek, E. P., *In vitro* synthesis of T4 late messenger RNA. *Proc. Natl. Acad. Sci. U. S.* **59**, 459–466 (1968).

70a. Steele, W. J., Localization of deoxyribonucleic acid complementary to ribosomal ribonucleic acid and preribosomal ribonucleic acid in the nucleolus of rat liver. *J. Biol. Chem.* **243**, 3333–3341 (1968).

70b. Suyama, Y., and Bonner, W. D., DNA from plant mitochondria. *Plant Physiol.* **41**, 383–388 (1966).

71. Swift, H., Cytochemical techniques for nucleic acids. *In* "The Nucleic Acids" (E. Chargaff and J. N. Davidson, eds.), Vol. 2, pp. 51–92. Academic Press, New York, 1955.

72. Swift, H., Studies on nucleolar function. *Symp. Mol. Biol., Univ. Chicago,* pp. 266–303 (1959).

73. Swift, H., Cytochemical studies on nuclear fine structure. *Exptl. Cell Res.* Suppl. 9, 54–67 (1963).

74. Swift, H., and Stevens, B. J., Nucleolar-chromosomal interaction in microspores of maize. *Natl. Cancer Inst. Monograph* **23**, 145–166 (1966).

74a. Tewari, K. K., and Wildman, S. G., Chloroplast DNA from tobacco leaves. *Science* **153**, 1269–1271 (1966).

75. Unuma, T., and Busch, H., Formation of microspherules in nucleoli of tumor cells treated with high doses of actinomycin D. *Cancer Res.* **27,** 1232–1242 (1967).
76. Unuma, T., Floyd, L., and Busch, H., Selective removal of perinucleolar nucleolus associated chromatin from isolated nucleoli of livers of thioacetamide-treated rats. *Exptl. Cell Res.* **52,** 101–111 (1968).
77. Unuma, T., Smetana, K., and Busch, H., A morphological study on the chromatin areas associated with the nucleolus. *Exptl. Cell Res.* **48,** 665–671 (1967).
78. Wallace, H., Autoradiographic studies of RNA and protein synthesis in anucleolate *Xenopus* embryos. *Natl. Cancer Inst. Monograph* **23,** 425–430 (1966).
78a. Wallace, H., and Birnstiel, M. L., Ribosomal cistrons and the nucleolar organizer. *Biochim. Biophys. Acta* **114,** 296–310 (1966).
79. Watson, J. D., and Crick, F. H. C., A structure for deoxyribosenucleic acid. *Nature* **171,** 737–738 (1953).
80. Watson, J. D., and Crick, F. H. C., Genetic implications of the structure of deoxyribonucleic acid. *Nature* **171,** 964–967 (1953).
80a. Wilson, G. B., "The Cell in Development and Heredity." MacMillan, New York, 1928.
81. Wolstenholme, D. R., Direct evidence for the presence of DNA in interbands of *Drosophila* salivary gland chromosomes. *Genetics* **53,** 357–360 (1966).
82. Yankofsky, S. A., and Spiegelman, S., Distinct cistrons for the two ribosomal RNA components. *Proc. Natl. Acad. Sci. U. S.* **49,** 538–544 (1963).

CHAPTER V

Nucleolar RNA of High Molecular Weight

* In this chapter, nRNA = nucleolar RNA; rRNA = ribosomal RNA.

The evidence of Brachet (*17–25*) and Caspersson (*46–48*) and their associates that RNA is concentrated in the nucleolus and the methods employed for the cytophotometric and staining techniques have been previously reviewed in this volume and elsewhere (*161, 227, 228*). Moreover, the importance of nucleolar and ribosomal RNA and their relationships to one another and to the physiology of the cell have been discussed extensively in recent years (*13, 41*) as the functions of the various types of RNA became apparent and the role of the nucleolus in the synthesis of ribosomal RNA was verified (Chapter I, Table 1.1). As the roles of 18 S, 28 S, and other types of rapidly sedimenting RNA became defined with respect to their structural and template functions, questions developed such as: (1) does the nucleolus synthesize only ribosomal precursor RNA or is messenger or template RNA synthesized in the nucleolus and (2) does the nucleolus synthesize all the ribosomal precursors?

In view of the presence in nuclei of multiple NOR's (Chapter III), another question is whether there are several kinds of nucleolar rRNA precursors. As noted earlier (*38*), these problems have been difficult to approach experimentally because of the limitations of methods for the separation of large RNA molecules, despite the remarkable progress (*5a, 105, 118*) made with respect to small RNA molecules (see Chapter VI). Techniques have become readily available that permit separation of RNA molecules of various size classes (*1, 7, 8, 62, 77, 97–99, 119, 132, 171–173, 204, 210, 215*); i.e., 4–7 S, 10–12 S, 18 S, 28 S, 35 S, 45 S, 55 S, 65 S, and 85 S RNA molecules; satisfactory fractionation has been obtained only for very different AU-rich and GC-rich RNA of the 18 S to 85 S-size classes. Probably many types of RNA are present in the various sedimentation classes of nuclear RNA.

It is of considerable importance to define the mechanism of the assembly of cytoplasmic elements in the nucleus and in the nucleolus. Although it would seem logical that whole ribosomes would be formed in one place in the cell and that the nucleolus might subserve this function completely, the evidence has not been adequate to establish this point unequivocally (*110, 128, 130, 195, 215–218*). All of the data available support the idea that the nucleolus plays a vital role in the synthesis of the ribosomal precursors but whether this includes the proteins (see Chapter VII) and all types of ribosomal RNA is uncertain (*45, 58*).

In this connection, one puzzling feature about the nucleus is the apparent absence of free ribosomes and polysomes (*148*). In most of the methods used for the preparation of nucleoli and nuclei, these structures sediment rather rapidly owing to the addition of calcium or other divalent ions to the homogenizing medium (*51;* see Chapter XI). However, there are other RNA-containing structures adhering to the nucleus, including the perinuclear ribosomes present in the outer layer of the nuclear envelope. When these are removed from the nucleus by Triton X or citric acid, there is no evidence for the presence in the nucleus of free ribosomes. As noted in Chapter II, ribosomelike structures are visualized in the nucleolus by electron microscopy but their function is not defined.

The nucleolus contains other RNA besides ribosomal precursor RNA (see Chapter VI). The evidence which indicates that in normal liver the rapidly sedimenting RNA's in the nucleolus differ from that present both in anaplastic and minimally deviated tumors (Morris tumors), suggests that more than one type of rapidly sedimenting product may be present; it is possible that some polyriboadenylic acid derived from nicotinamide dinucleotide is also present in normal liver nucleoli (*50, 142, 143*). In addition, the nucleolus contains a number of low molecular weight RNA's, (*54, 71, 140a, 221, 226*) some of which are rich in uridylic acid (Chapter VI).

I. Content of Nucleolar RNA

The amount of RNA of the nucleoli has been measured by a number of techniques each of which has some limitations. In essence, the optimal procedure would be the utilization of the ultramicrotechniques which are capable of determining with precision the amount of RNA in a single nucleolus. However, at this point, no such procedure is readily available and the techniques of microspectrophotometry do not have the required precision.

The procedures employed in this and other laboratories are based upon the determination of RNA in aliquots of nucleolar preparations followed by direct counting of the nucleoli in hemacytometers. The limitations of this procedure are (a) the possibility that the sample of nucleoli may not be completely representative, (b) the analysis by hemacytometric procedures may produce aberrations in the result due to aggregation of nucleoli, (c) the possibility that some nucleolar RNA is lost, and (d) the possibility that some "adventitious" RNA is present. In studies in this

TABLE 5.1

CONTENTS OF NUCLEOLI[a]

Tissue	DNA	RNA	Protein	Reference
Liver				
Sonication	0.38	0.38	3.8	*44, 140*
Compression–decompression	0.22	0.55	6.5	*44, 59*
Walker tumor				
Sonication	1.7	1.9	21.6	*44, 140*
Compression–decompression	1.6	2.2	9.8	*44, 59*

[a] Expressed as pg/nucleolus.

laboratory it was found that isolated nucleoli had a tendency to aggregate and as a result low values for nucleolar numbers were obtained unless a mixture of 10% sodium citrate and 0.1% azure C in 0.25 *M* sucrose (7:3) was employed for staining; with this mixture little aggregation was observed (*136, 140*).

Depending upon the procedure for isolation of the nucleoli and the tissue from which the nucleoli are isolated, the values for the content of RNA and DNA per nucleolus exhibited some variation (Table 5.1). In comparison with the values obtained by the compression–decompression procedure, the nucleoli obtained by sonication contained somewhat less RNA and more DNA. Although these differences were not very marked there were greater differences in the protein contents of these nucleoli.

Marked differences were found in the values obtained for the Walker tumor and the liver (*42*). The nucleoli of the Walker tumor had about five times as much RNA and DNA as those of the normal liver in agreement with the results noted earlier with respect to the sizes and volumes of the nucleoli of tumors. In the tumors, the nucleolar coefficient or the number of nucleoli per cell was 1.7 as compared with 2.7 for the liver so that the amount of nucleolar RNA per nucleus in the liver is about 1 pg as compared with approximately 3.4 pg per nucleus in the Walker tumor.

Although these values represent quantitative differences between the tumors and other tissues based upon direct biochemical analysis of isolated nucleoli, other studies have been made on the basis of dry weight. Table 5.2 shows the percentage composition of isolated nucleoli from a number of tissues based on the dry weight.

The comparative composition of nuclei and nucleoli has also been evaluated. In recent studies by Amano (*2, 3*), the nucleolus has been shown to be different from the nucleus in its higher RNA to DNA ratio; i.e., 2 in nucleoli as opposed to 0.16 in nuclei. The protein content of the

TABLE 5.2

COMPOSITION OF ISOLATED NUCLEOLI FROM VARIOUS SOURCES (% dry weight)

Sample	RNA	DNA	Protein	RNA/DNA	Reference
Rat liver	11.0	8.0	78.0	1.4	*59, 60, 135, 140*
Walker 256 carcinosarcoma	10.4	8.5	81.0	1.2	*59, 60, 135, 140*
Novikoff hepatoma	8.0	7.5	84.5	1.1	*127*
Regenerating liver (6 hr)	7.6	4.6	87.8	1.7	*135*
Regenerating liver (18 hr)	15.5	5.4	79.1	2.9	*135*
Guinea pig liver	4.1	9.5	86.4	0.43	*129*
Pea stem (4 days)	15.4	10.6	74.0	1.5	*14*
Pea seedlings (36 hr)	16.7	6.4	76.9	2.6	*14*
Pea seedlings (48 hr)	11.0	5.0	84.4	2.2	*13*

nucleolus and nucleus was approximately the same; i.e., 69 and 66%, respectively.

In nerve cells (*147*), the volumes of the nucleoli range from 2–60 μ^3 and in some fish, there are nucleoli that have a volume of 6000 μ^3. Edström (*64a–67a*) has reported that when there is increased motor activity, the nucleolar volume of the associated nerve cells more than doubles. In nerve nucleoli, the weight per unit volume is 0.4 pg per μ^3. The volume of the "spherical" nucleolus is about $4/3\ \pi\ r^3$ and estimating that the diameter is about 5 μ for tumor nucleoli, the total volume is 65.4 μ^3. If this is the case and the value for the nerve nucleoli is correct, then about 26 pg (40%) would be organic matter. Since about 10–20% of the organic matter is RNA, and assuming it has a density of 1, the value of 2.5–5.0 pg is in good agreement with the value of 1.7–3.3 pg found by direct chemical analysis. Values of 1.7 pg of RNA per nucleolus have been found also for cells in the supraoptic nucleus. Such data are of interest in comparisons from tissue to tissue.

II. Types of RNA Present in the Nucleolus

METHODS

In isolation of rapidly sedimenting nuclear and nucleolar RNA, most of the problems have been resolved by improvements in the experimental methods employed. The earliest refinement was the development of the phenol extraction procedure (*116, 117*). The use of sodium dodecyl sulfate (SDS) was introduced by Gilbert (*81*) and modified by Hiatt *et al.* (*101*) for studies on animal tissues. The employment of sucrose gradients (*26*) permits the isolation of the RNA in good yield and in an undegraded state so that there is little problem in the reproduction of the results from laboratory to laboratory. The employment of large-scale techniques, particularly large-scale homogenization and zonal ultracentrifugation (*103, 104*), has added to the opportunities for isolation, purification, and analysis of the structure of these RNA species (*39, 104*). The most critical factor in studies on high molecular weight RNA is the prevention of its degradation either by accompanying RNases or by inadvertent addition of extrinsic RNases (*49*). With appropriate precautions, high molecular weight RNA is stable for months in the cold (*73a, 103a*).

For isolation of nucleolar RNA, the most commonly employed procedure has been extraction of the RNA with sodium dodecyl sulfate and phenol buffered with sodium acetate and acetic acid to pH 5.1 (*76–78, 138, 173, 200*). The sample is generally heated to 63° following the thermal fractionation procedure of Georgiev *et al.* (*76–79*). The RNA is precipitated from the aqueous phase with ethanol (2–2.5 volumes). The RNA is then dissolved in water and subjected to further analytical procedures.

Treatment of various fractions with phenol at low and high temperatures to provide a "thermal" fractionation was utilized by Sibatani *et al.* (*181, 181a, 181b*) and Georgiev *et al.* (*77, 137*). Under the conditions employed (*136*), the RNA extracted from the nucleus with cold phenol (aRNA) is largely ribosomal RNA and products of low molecular weight; i.e., 4 S–28 S RNA. At higher temperatures (60–65°), the RNA that is extracted has a high molecular weight and is richer in AU (*77, 132, 181*). Georgiev *et al.* (*77*) utilized a more refined "thermal fractionation procedure" in which ranges of temperature employed were: room temperature, 45–50°, 50–55°, 55–63°, and 68–85°. As the temperature increased, the turnover rates of the RNA also increased. Because phenol functions as a protein denaturant, the RNA released at the higher temperatures is that most tightly bound to protein, or to DNA. Since most of the RNA in the cell is protein bound, some protein denaturant is required to free the RNA from its protective or structurally associated proteins.

Two methods have been of value in chemical characterization of the RNA of the nucleolus; i.e., UV nucleotide analysis and ^{32}P nucleotide analysis (*211–213*). The results obtained with these two procedures have not agreed completely because the former measures the composition of the total RNA in the mass of the nucleolus and the latter because it has been used for the analysis of the "rapidly-turning-over" or "newly synthesized" fraction of nucleolar RNA. Two methods for fractionation of RNA have been particularly useful for the study of nucleolar RNA; i.e., sucrose density gradient sedimentation (Fig. 5.1a) and the chromatography of RNA on Sephadex, DEAE-cellulose and other types of absorbents (27). Sucrose density gradient sedimentation has also been used for purification of nucleolar RNA sedimentation classes such as 45 S nRNA (Fig. 5.1b).

III. Sucrose Density Sedimentation Patterns of Nucleolar RNA

Figure 5.1 shows there are three major sedimentation classes of rapidly sedimenting nucleolar RNA which have approximate sedimentation

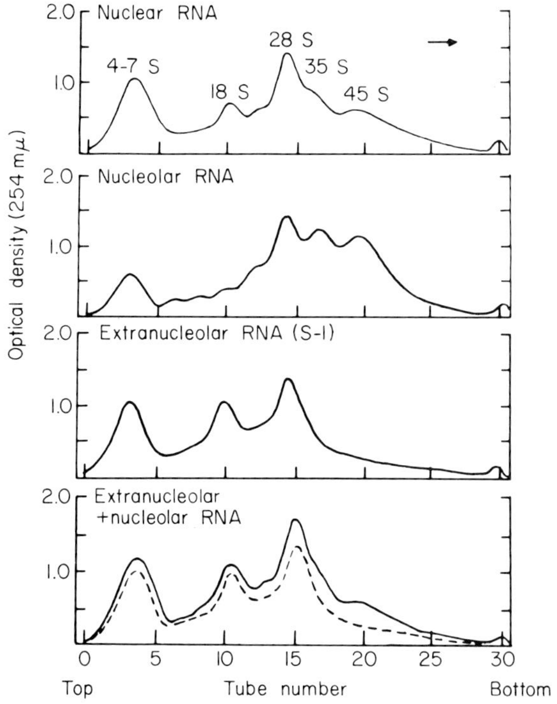

Fig. 5.1a. a. Sucrose density-gradient patterns of rat liver nuclear RNA (top), nucleolar RNA (second from top) and extranucleolar RNA (second from bottom). The sums of the two middle RNA preparations are shown in the bottom graph. The direction of sedimentation is indicated by the arrow and the approximate sedimentation coefficients are indicated above the peaks and shoulders of the top gradient pattern. The concentration of 18 S RNA is low in the nucleolar RNA and the concentrations of 28, 35, and 45 S RNA are almost all equally high. The only other regions in the nucleolus that have substantial amounts of RNA are the 4–7 S region and the pre-28 S region referred to as 23 S RNA (from *137*). b. Purification of nucleolar 45 S RNA of Novikoff hepatoma ascites cells by repeated sucrose density-gradient sedimentations.

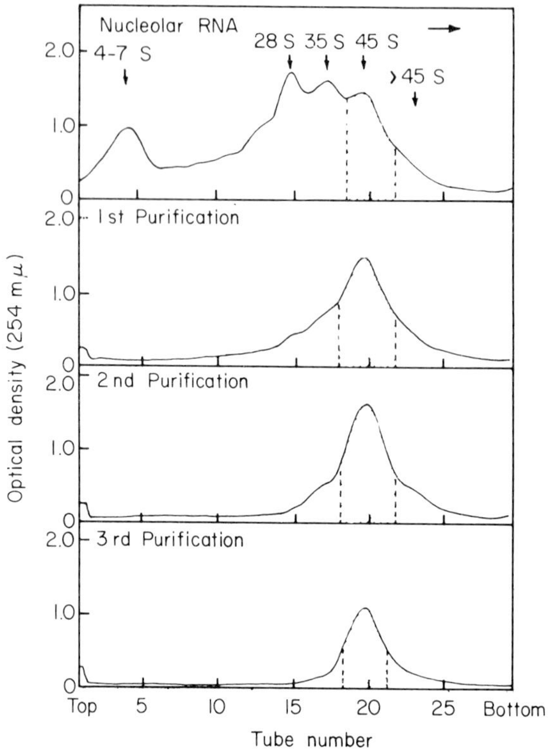

Fig. 5.1b. (*continued*).

coefficients of 28 S, 35 S, and 45 S (*136, 153–156, 173, 196*). The smaller peak in the 4–7 S region will be discussed more fully in Chapter VI.

The sedimentation classes of nucleolar RNA differ in some respects from sedimentation classes of RNA of the cytoplasm and the nucleus (Fig. 5.1a, 5.2). In the cytoplasm there are three major classes of RNA; i.e., 4–7 S, 18 S, and 28 S. It is now clear that there are many subspecies of 4–7 S RNA, including the various types of transfer RNA and the recently characterized 5 S RNA's (*36, 37, 53, 168, 219, 220*) which may be binding factors for the 30 S and 50–60 S ribonucleoprotein subunits of the ribosomes. If there are rapidly sedimenting species of RNA in the cytoplasm related to the more rapidly sedimenting nuclear and nucleolar RNA, they have not been isolated; in part this may reflect the problem of degradation of cytoplasmic RNA by homogenization and cytoplasmic RNases.

In the nucleus of the normal liver cell (Fig. 5.1) and most other tissues studied, each of the sedimentation classes of RNA of the cytoplasm is present along with the more rapidly sedimenting RNA classes found in the nucleoli; i.e., 35 S, 45 S, and 55 S RNA. Although the sedimentation classes are the same, there probably are different types of RNA in each of these fractions, some of which are precursors of ribosomal RNA and others of messenger or template RNA.

The chief differences in the pattern of RNA of the nucleolus from those of the other cellular fractions are that the nucleolus contains larger amounts of the rapidly sedimenting 45 S RNA and 35 S RNA by comparison with the remainder of the cell and, conversely, the nucleolus has little or no 18 S RNA which is uniformly present in the other fractions.

IV. On the Nature of the Gene Readout in the Nucleolus

Recently, electron micrographs of the apparent process of gene readout of the nucleolar products have been produced. The process of lengthen-

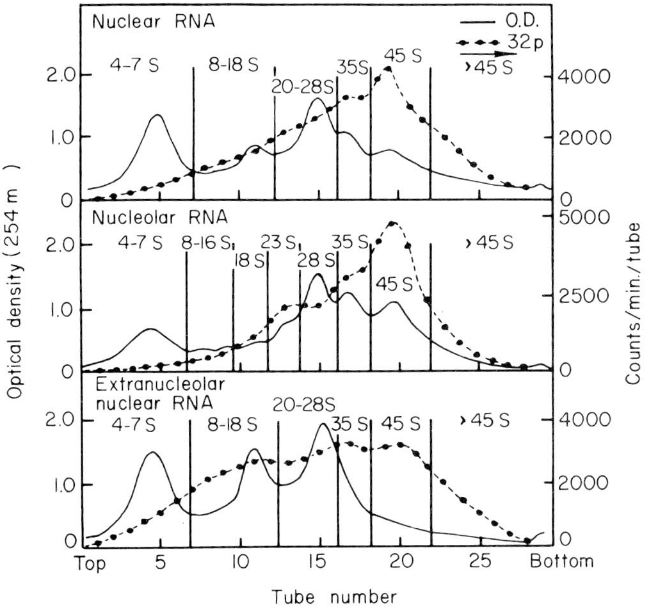

Fig. 5.2. Pulse-labeling (15–30 minutes) patterns with ^{14}C- or ^{32}P-labeled precursors are shown for the whole nuclear (top), nucleolar (middle), and extranucleolar nuclear (bottom) liver RNA fractions. The maximal labeling initially occurred in the 45 S or greater than 45 S region of the gradient. In the extranucleolar nuclear region, the labeling was not always found in this rapidly sedimenting fraction presumably because of the artifacts introduced by the short- or long-term sonication procedure which may have destructive effects on the nuclear RNA as well as the release of RNases which produce hydrolytic reactions that result in hetero- or polydispersity of the broken RNA (from *138*).

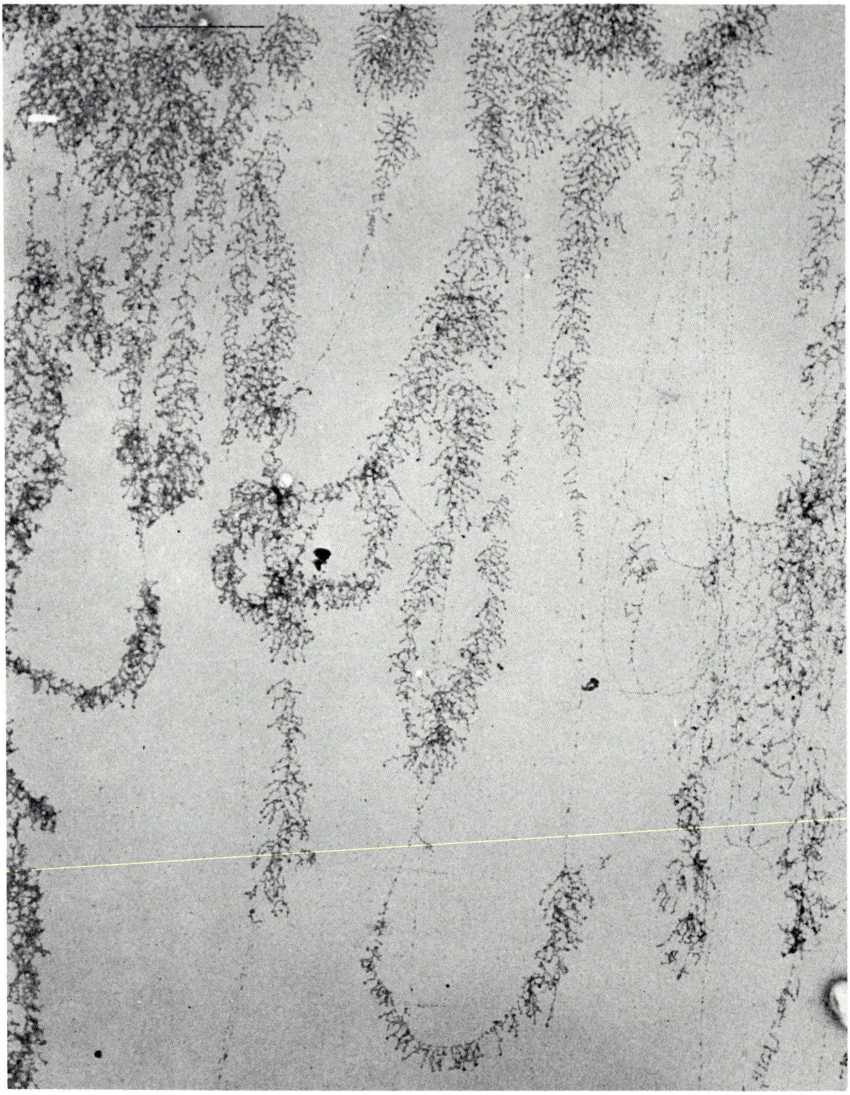

Fig. 5.3. High magnification (×29,600) of nucleolar core showing periodic gene readouts. Along these DNA fibers are seen RNA elements with progressively longer chain lengths. The marker line is 1 μ. Although it is tempting to assume that the increasing length of the fibers on the DNA chains reflect progressive synthesis, it is interesting to note that the largest molecules (45 S RNA and 55 S RNA) are apparently the first labeled products found in nucleolar RNA pools; possibly, the fiber lengths may be decreasing along the DNA chain as the RNA is being processed to form rRNA (courtesy of Drs. O. L. Miller, Jr. and B. Beatty, Oak Ridge National Laboratory).

ing of RNA fibers along DNA chains has been visualized in nucleolar cores of oocytes (Fig. 5.3, see also pp. 188–190).

Electron Microscopic Analysis of Fiber Lengths

Figures 5.4–5.7 show RNA fibers of varying lengths found in rapidly sedimenting nuclear and nucleolar RNA. Some short fragments were found (Fig. 5.4, top) which may be artifacts induced by the procedure, or may represent pieces of intermediate size in the course of synthesis. The mean length of the fibers of the 28 S RNA is approximately 1.0–1.1 μ (Fig. 5.4, middle). On the other hand, the fibers of 35 S RNA had mean lengths of 2.2 and 3.1 μ which suggests that these may be dimers and trimers of 28 S RNA [Figs. 5.4 (bottom), 5.5, 5.6] or just longer chains. The 45 S RNA contained fibers of approximately 4 and 5 μ in length that accounted for the majority of the total molecules present (Figs. 5.6–5.8). In addition to these fibers, there were also much longer fibers; i.e., 6 μ or even longer (Fig. 5.6). It is possible that these longer fibers are oligomers of the nucleolar 45 S RNA, or long-chain messenger RNA in these preparations (*40*).

Recently an extensive study on lengths of RNA fibers has been published by Granboulan and Scherrer (91a) who found marked variations in the lengths of such fibers in all of the RNA classes including 18 S and 28 S rRNA and higher molecular weight nuclear RNA's. Their mean length for 28 S RNA was 1.2 μ as found in studies in the authors' laboratory. Their 32 S and 45 S RNA's had modal lengths of 1.7 and 3.5 μ which are somewhat below the modal lengths obtained in this laboratory. They also found fibers that were much longer in both the 32 S and 45 S RNA, i.e., up to approximately 6 μ.

V. Content and Types of Ribosomal RNA

Since rRNA is the major nucleolar product, its composition must be considered in any analysis of nucleolar RNA. Mammalian ribosomal RNA is rich in guanylic and cytidylic acids (Table 5.3a). However, there are substantial differences between the compositions of 18 S rRNA and 28 S rRNA; 18 S rRNA of rat liver is richer in both adenylic and uridylic acids and poorer in guanylic and cytidylic acid than is the 28 S rRNA and accordingly, the ratio r(r = AMP + UMP/GMP + CMP) is 0.72 for the 18 S RNA as compared to 0.53 for the 28 S rRNA. Values for rRNA of other mammalian tissues and tumors are presented in Table 5.3a.

There is serious difficulty in direct transfer of information on rRNA from species to species since the types of products found in one species

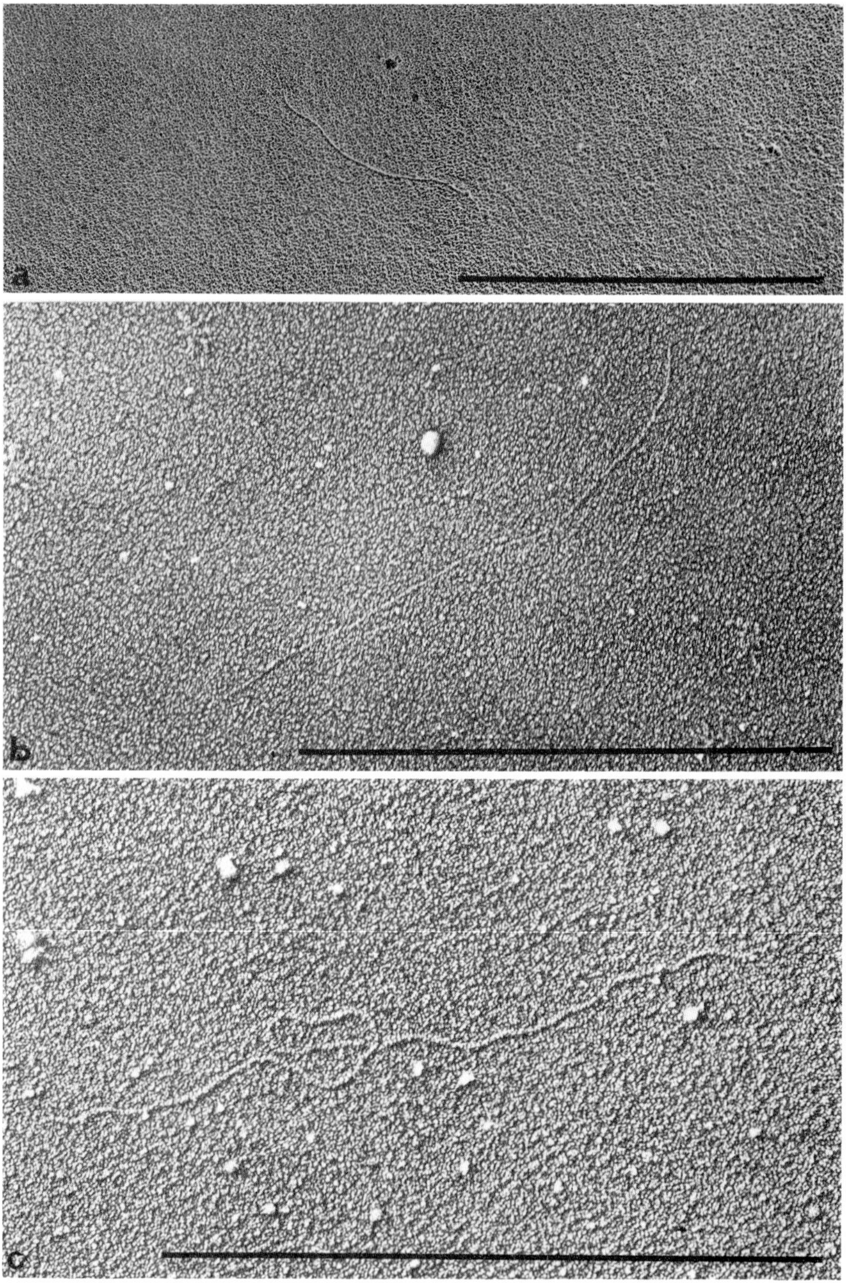

Fig. 5.4. a. A filament of 28 S nuclear RNA about 0.6 μ in length. ×50,000. b. A filament of 28 S nuclear RNA 1.1 μ in length. ×72,000. c. A filament of 35 S nucleolar RNA. 1.6 μ in length. ×87,000. The measured lines are 1 μ in this and subsequent figures.

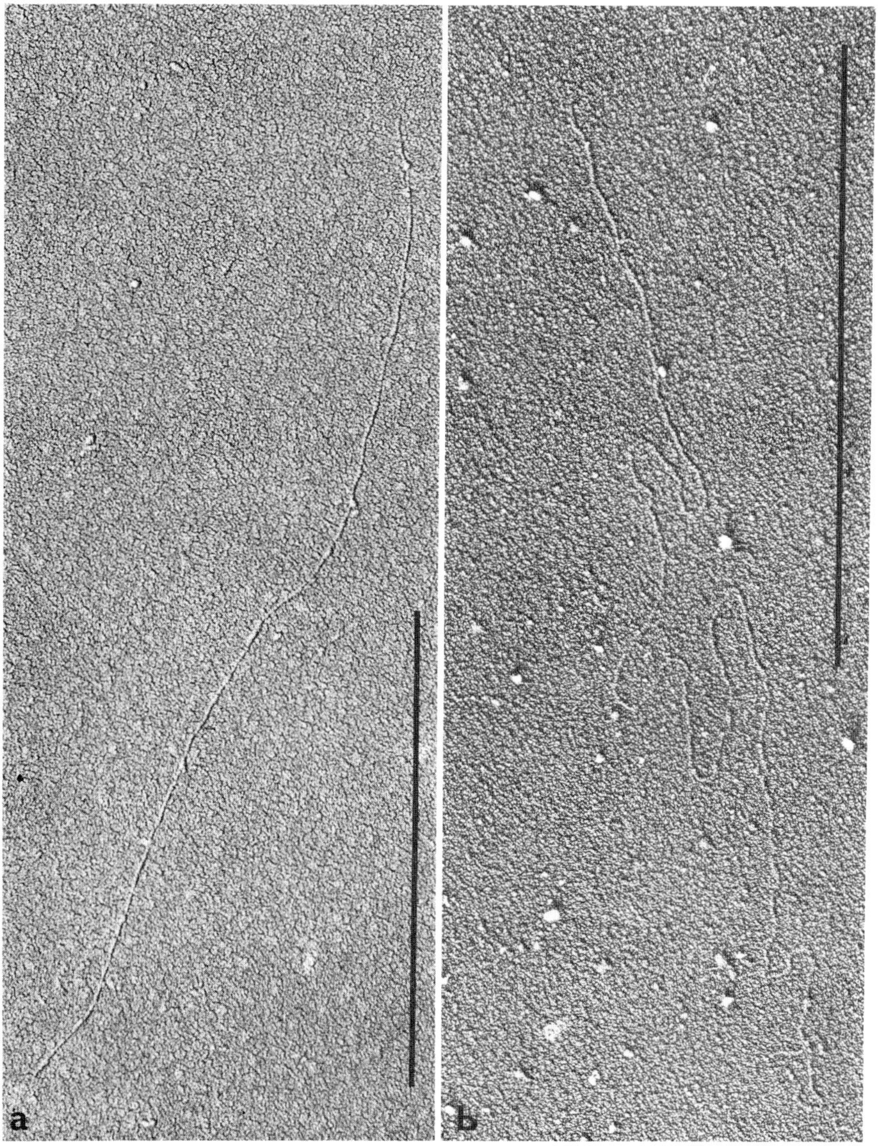

Fig. 5.5. a. A filament of 35 S nucleolar RNA 2.2 μ in length. $\times$63,000. b. A filament of nucleolar 35 S RNA 3.2 μ in length. $\times$82,000.

differ markedly from those in others (Table 5.3b). One example is the remarkable difference in the ribosomal RNA of the flies and those of the mammalian systems (see Chapter IV).

Another problem that exists relates to the method used for extraction

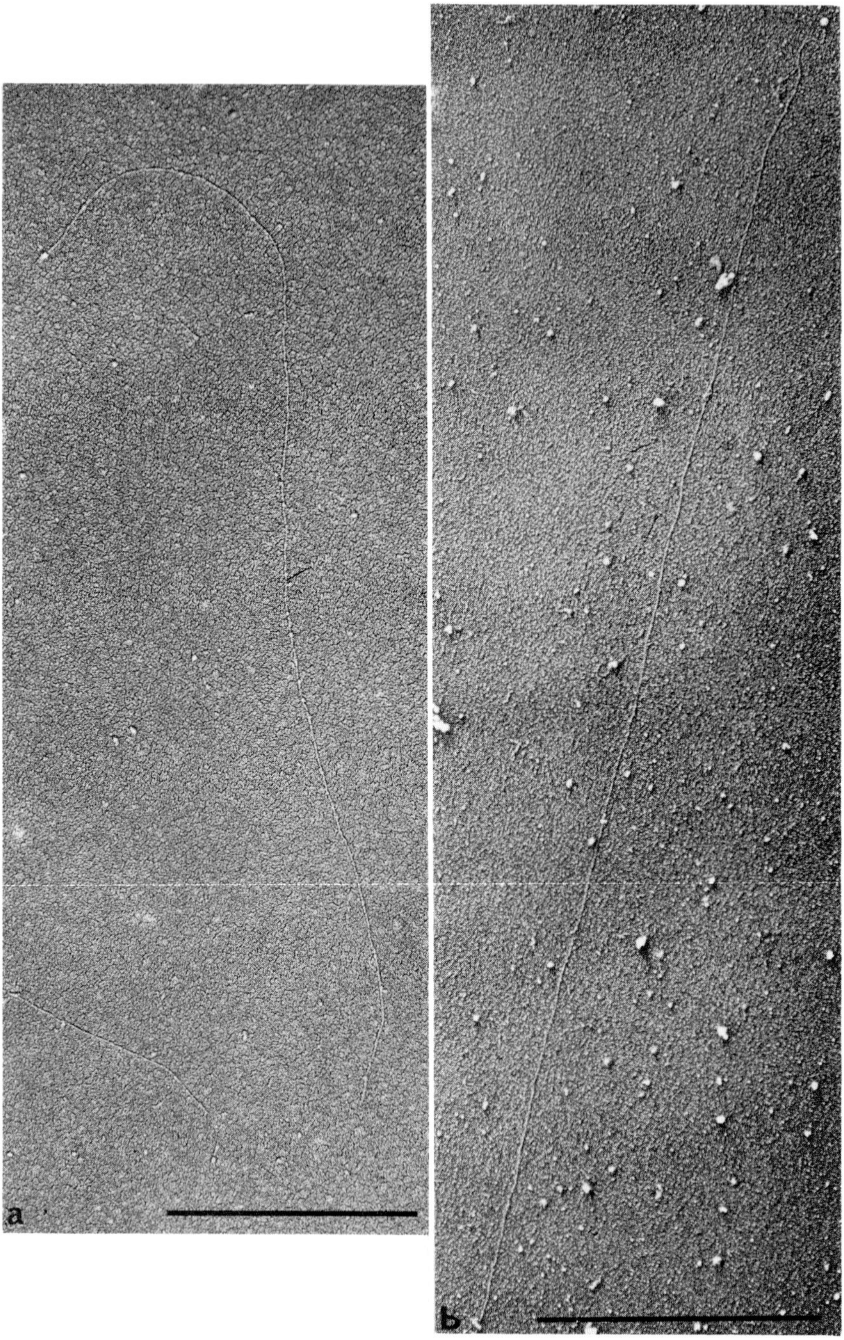

Fig. 5.6. a. A filament of 35 S nucleolar RNA. ×34,000. b. A filament of nucleolar 45–55 S RNA. ×38,000. a, 4 μ in length; b, 5 μ.

Fig. 5.7. A 6-μ long filament of nuclear 45–55 S RNA. $\times$68,000.

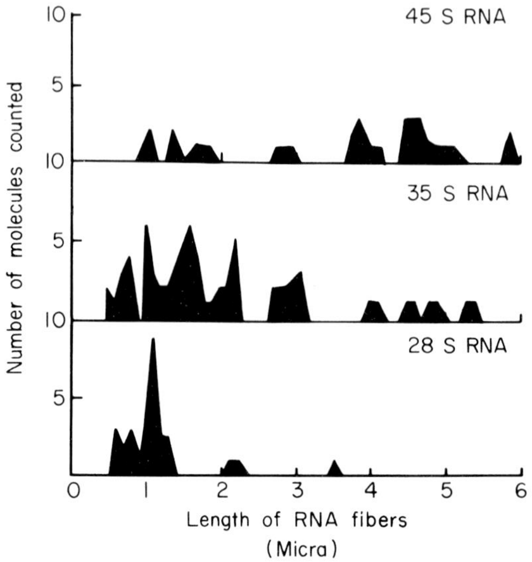

Fig. 5.8. Summary of data on the distribution of lengths of RNA in 28, 35, and 45 S sedimentation classes (see also *91a*).

of the RNA. Giudice and Mutola (*83*) found that in the sea urchin, ribosomal RNA had the base composition A-25, U-36, G-23, and C-15 after extraction with hot phenol and A-23, U-47, G-18, and C-11 after extraction with cold phenol. Whether U-rich messenger RNA is present in the preparations obtained by extraction with cold phenol or whether several types of ribosomes are formed is not clear from the data.

VI. Nucleolar 45 S RNA

The most important property of the rapidly sedimenting RNA and that which attracted the attention of so many laboratories is its rapid rate of labeling or turnover (Fig. 5.2). In both nuclear and nucleolar RNA, the labeling rates of 45 S RNA and even more rapidly sedimenting RNA are the highest of the RNA fractions.*

A. Isolation

In the nucleus, there are at least two types of 45 S RNA (Table 5.4), one of which is nucleolar and is the precursor of ribosomal RNA and

* Recently, in the authors' laboratory, Dr. E. J. Hidvegi and Dr. A. W. Prestayko have shown that the rapidly sedimenting nucleolar RNA contains 65 S and 85 S RNA as well as 45 S RNA. These longer fibers were found after careful use of cold conditions. It is also possible that fibers of even greater size are present although the amounts are quite small. Interestingly, the more rapidly labeled RNA classes were the 65 S and 85 S RNA's.

TABLE 5.3a

Base Composition of Mammalian 18 S and 28 S Ribosomal RNA[a]

Tissue	Method[b]	18 S RNA A	18 S RNA U	18 S RNA G	18 S RNA C	18 S RNA $\frac{A+U}{G+C}$	28 S RNA A	28 S RNA U	28 S RNA G	28 S RNA C	28 S RNA $\frac{A+U}{G+C}$	Reference
Rat liver ribosomes	UV	22	20	30	28	0.72	18	19	33	30	0.59	*133*
Rat liver	UV	20	18	32	30	0.61	18	17	33	32	0.53	*116*
Rat liver	$^{32}P_i$	22	21	31	26	0.76	21	18	34	27	0.64	*96*
Rat liver	UV	22	20	32	26	0.72	18	16	35	31	0.52	*93a*
Krebs 2 ascites	UV	21	22	29	28	0.74	16	17	36	31	0.50	*131*
BHK-21 (baby hamster kidney)	$^{32}P_i$	22	22	31	26	0.77	18	18	34	30	0.56	*35*
Stice (pig kidney)	$^{32}P_i$	23	22	31	25	0.80	18	19	36	28	0.58	*35*
HLM (human liver)	$^{32}P_i$	23	23	29	25	0.86	17	18	35	30	0.54	*35*
HeLa (human epithelium)	$^{32}P_i$	22	23	29	25	0.83	17	17	35	31	0.51	*170*
HeLa (human epithelium)	$^{32}P_i$	23	23	29	25	0.83	17	18	35	30	0.54	*35*
HeLa (human epithelium)	$^{32}P_i$	23	23	25	28	0.87	21	19	28	32	0.66	*173*
HeLa cells	$^{32}P_i$	21	22	31	27	0.64	16	17	35	32	0.49	*105a, 195*
L (mouse fibroblasts)	$^{32}P_i$	24	22	30	25	0.83	20	18	35	28	0.61	*35*
FLS ascites (mouse)	$^{32}P_i$	21	21	31	27	0.72	17	18	35	30	0.52	*94, 95*

[a] Compositions are in terms of moles per cent of the given nucleotide or are percentages of the total ^{32}P in the nucleotides. A = adenylic acid; U = uridylic acid; G = guanylic acid; C = cytidylic acid.

[b] Method: UV = ultraviolet analysis of nucleotides; $^{32}P_i$ = incorporation of ^{32}P-orthophosphate.

another is part of the "polydisperse" RNA which is "messenger" (*28, 102*). Barondes *et al.* (*11*) reported that the rapidly sedimenting RNA of the nucleus had the highest stimulatory activity for the incorporation of amino acids into the ribosomal system; however, there is little evidence to support the view that this RNA is nucleolar since similar studies with rapidly sedimenting purified nucleolar RNA have not shown that it stimulates high levels of amino acid incorporation. Some confusion in this subject has emerged from studies of Brentani *et al.* (*29, 30*) who reported that RNA derived from the residue fraction of nuclei (nuclear elements insoluble in 0.15 or 2 *M* NaCl) had high levels of stimulatory activity for amino acid incorporation. This RNA is composed partly of nucleolar RNA but also contains the RNA of the nuclear ribonucleoprotein network (Chapter VIII) which is bound to protein and nuclear dense granules; it is not readily extracted. When nucleoli were obtained with the standard sonication procedure (Chapter XI), the activity of their RNA as messenger RNA was remarkably lower.

TABLE 5.3b

BASE COMPOSITION OF 16–18 S AND 23–28 S RIBOSOMAL RNA OF OTHER CELLS (*1a*)[a]

Tissue	16–18 S RNA					23–28 S RNA					Reference
	A	U	G	C	$\frac{A+U}{G+C}$	A	U	G	C	$\frac{A+U}{G+C}$	
Psittacosis organisms (*Meningopneumonitis* organism)	26.4	24.3	32.2	17.1	1.02	28.3	24.4	30.6	16.7	1.11	*1a*
Bacteria (*E. coli*, *A. aerogenes*, *B. subtilis*, *B. emersonii*, *P. aeruginosa*, *P. vulgaris*)	25.0	21.1	31.3	22.6	0.85	26.0	20.9	31.3	21.8	0.88	*1a*
Mitochondria (*Neurospora*)	31.8	31.7	20.4	16.0	1.74	33.9	31.9	19.1	15.0	1.91	*1a*
Fungi (yeast, *Neurospora*, *Coprinus*, *Psalliota*)	25.2	26.0	28.9	19.9	1.02	26.5	24.0	28.9	20.6	1.01	*1a*
Dicotyledons (potato, pea, cauliflower, cabbage, parsnip, celery)	24.0	22.9	30.7	22.4	0.89	23.7	20.0	32.5	23.8	0.78	*1a*
Monocotyledons (*Zea*)	22.0	20.2	32.8	25.0	0.73	21.2	16.9	33.9	28.0	0.62	*1a*

Insects (*Drosophila*, *Chironomus*)	28.8	27.4	23.5	20.3	1.28	30.8	27.1	22.5	19.6	1.37	*1a*
Echinoderms (*Arbacia*)	23.1	23.2	29.9	23.8	0.87	19.9	19.3	34.9	26.0	0.65	*1a*
Amphibians (*Xenopus*)	24.1	22.9	28.9	24.1	0.88	19.7	17.4	35.0	27.9	0.59	*1a*
Birds (chick, duck)	23.3	22.1	28.8	25.8	0.82	18.0	18.8	34.7	28.6	0.58	*1a*
Chicken embryo (7 days)	24.0	22.0	29.0	25.0	0.85	19.0	18.0	34.0	28.0	0.60	*126*
Chicken erythroblastic cells	21.0	21.0	30.0	28.0	0.73	17.0	17.0	36.0	30.0	0.51	*94*
Xenopus laevis	22.0	18.0	31.0	29.0	0.67	17.0	16.0	37.0	30.0	0.49	*52*
Potato tuber	25.4	25.2	27.2	22.2	1.02	25.1	21.2	31.7	22.0	0.86	*52*
Pea seedlings	23.7	25.1	31.1	20.1	0.96	23.6	21.6	32.1	22.6	0.82	*52*
Cauliflower	25.6	21.0	32.9	20.6	0.87	25.7	18.4	33.1	22.6	0.79	*52*
Escherichia coli	24.8	21.5	31.0	22.7	0.86	25.4	19.6	33.5	21.5	0.82	*52*
Bacillus subtilis	26.5	21.6	29.6	22.3	0.93	26.5	19.3	32.0	22.5	0.84	*52*
Pseudomonas aeruginosa	25.1	20.5	32.8	21.6	0.84	26.3	21.3	31.2	21.2	0.91	*52*
Arbacia punctulata	22.8	20.9	31.2	24.9	0.78	21.0	18.1	34.7	26.0	0.65	*202a*

[a] Compositions are in terms of moles per cent of the given nucleotide or are the percentages of the total ^{32}P in the nucleotides. A = adenylic acid; U = uridylic acid; G = guanylic acid; C = cytidylic acid.

TABLE 5.4a

UV AND ^{32}P-ORTHOPHOSPHATE NUCLEOTIDE COMPOSITIONS OF VARIOUS NUCLEAR RNA FRACTIONS

	UV-Base composition of RNA					% Distribution of ^{32}P						
Fraction	A	U	G	C	$\frac{A+U}{G+C}$	A	U	G	C	$\frac{A+U}{G+C}$	Pulse time (min.)	Reference
Whole nuclear												
Calf thymus	19.3	24.2	30.4	26.1	0.77							*146*
Calf thymus	20.4	18.3	32.9	28.1	0.63							*181*
Rat liver						14.3	24.4	34.7	26.6	0.63	40	*101*
Rat liver	20.2	24.6	25.7	29.5	0.81							*68*
Rat liver	20.9	24.8	29.5	24.9	0.84	27.1	24.8	22.0	26.1	1.08	15	*200*
Rat liver	17	29	29	24	0.87							*3*
HeLa	19.0	16.4	30.2	34.2	0.55	30.0	25.2	24.0	20.8	1.23	30	*100*
HeLa						22.7	23.6	31.8	22.0	0.86	60	*100*
HeLa						19.5	25.3	28.8	26.6	0.81	75	*170*
Nucleolar RNA												
Rat liver	15.9	21.7	35.8	27.9	0.61	21.0	17.2	37.3	24.5	0.62		*166*
Walker tumor	14.2	20.4	35.4	28.7	0.56	13.0	21.1	34.7	31.3	0.52		*166*
Guinea pig liver	12.3	22.4	50.8	14.5	0.53	21.8	22.4	25.0	30.8	0.79	60	*129*
Starfish oocyte (*Asterias rubens*)	21.7	14.7	36.3	27.3	0.57							*67a*
Spider egg (*Tehenavia domestica*)	25.2	22.2	29.8	22.9	0.90							
Starfish oocytes (*Asterias rubens*)	23.7	18.5	33.4	24.3	0.74							*67a*

Nuclear sap												
Calf thymus	19.2	19.3	34.5	27.0	0.63							
Calf thymus	20.4	18.6	32.7	28.2	0.64							*181*
Guinea pig liver	23.9	26.4	34.9	14.8	1.01	22.2	21.9	20.8	35.2	0.79	60	*129*
Rat liver	18.5	20.3	35.8	25.4	0.63	22.4	24.1	30.4	23.2	0.87	15	*200*
Rat liver	20	32	28	22	0.98							*3*
Nuclear residue												
Calf thymus	19.8	25.5	29.9	24.8	0.83							*146*
Calf thymus	21.8	23.2	30.5	25.7	0.80							*181*
Fr I	18.9	20.6	34.2	26.3	0.65							*181*
Fr II	26.3	28.7	25.8	19.2	1.22							*181*
Ehrlich ascites	16.9	21.3	32.4	29.3	0.62	16.6	20.0	32.5	31.0	0.58	30–60	*78*
Rat liver	16.3	20.7	33.0	29.9	0.59	18.3	20.3	31.7	29.7	0.63	30–60	*78*
Rat liver	17.5	21.7	33.9	26.9	0.65	23.4	20.4	32.2	23.9	0.78	15	*20*
Deoxyribonucleoprotein												
Calf thymus	19.6	24.5	29.6	26.3	0.79							*181*
Guinea pig liver	22.6	28.7	33.3	14.9	1.05	20.9	20.6	19.5	33.0	0.79	60	*129*
Rat liver	21.4	22.6	32.7	23.3	0.79	30.8	21.3	25.9	22.0	1.09	15	*200*
Rat liver	17	34	30	19	1.06							*3*
DNA–RNA												
FLS ascites						27.7	25.8	23.2	23.3	1.15	30	*94*
Rat liver	23.3	24.4	31.8	20.6	0.91	29.6	21.0	27.5	22.0	1.02	15	*200*
HeLa						35.1	25.4	18.5	21.0	1.53	15	*122*
Regenerating rat liver (18 hr)	15.1	20.8	35.5	29.4	0.55	19.7	16.8	37.8	25.6	0.58		*135*

TABLE 5.4b

NUCLEOTIDE COMPOSITION OF RNA OF SUCROSE DENSITY GRADIENT FRACTIONS OF NUCLEAR RNA[a]

Source	Approximate S value	UV-nucleotide composition of RNA					% Distribution of ^{32}P					Pulse time (min.)	Reference
		A	U	G	C	$\frac{A+U}{G+C}$	A	U	G	C	$\frac{A+U}{G+C}$		
Rat liver	55	20.8	28.4	30.7	20.1	0.97	26.1	22.6	26.5	24.8	0.95	10	*145a*
Rat liver	55						25.7	23.7	26.3	24.3	0.98	30	*137*
Walker tumor	55	17.5	23.6	34.1	24.8	0.70	17.6	24.0	30.0	28.5	0.71	10	*145a*
Rat liver	45	19.1	25.7	31.0	24.2	0.81	24.9	21.2	29.5	24.4	0.86	10	*145a*
Rat liver	45						25.1	22.1	28.5	24.3	0.89	30	*137*
Walker tumor	45	16.5	20.5	35.6	27.3	0.59	15.3	22.7	32.3	29.7	0.61	10	*145a*
HeLa	45						19.8	20.1	29.3	29.8	0.67	15–60	*173*
FLS ascites	45						21.5	21.5	30.4	26.7	0.75	30	*94*
Rat liver	35						24.7	22.1	29.1	24.1	0.88	30	*137*
HeLa	35						19.9	20.7	28.8	30.4	0.69	15–60	*173*
FLS ascites	35						21.8	23.4	29.0	25.8	0.82	30	*94*
Rat liver	20–28						25.2	23.5	27.6	23.7	0.95	30	*137*
Rat liver	8–18						25.6	23.9	26.2	24.3	0.98	30	*137*
Rat liver	4–6						24.3	24.5	26.2	25.0	0.95	30	*137*
FLS ascites	4–6						20.6	21.3	23.1	34.9	0.72	30	*94*
L cells	4						18.1	24.4	28.8	28.7	0.74	30	*155*

[a] The values for base composition are moles per cent of the given nucleotide as determined by ultraviolet analysis of nucleotides obtained by ion-exchange chromatography; the values for pulse-labeled RNA are percentages of the total ^{32}P in nucleotides obtained by ion-exchange chromatography.

B. Methods for the Study of Nucleolar 45 S RNA

Nucleolar 45 S RNA was found first in the authors' laboratory when analyses were made for the presence of rapidly sedimenting nucleolar RNA in normal growing, thioacetamide-treated, and neoplastic cells (*43, 124, 138, 139, 194, 199*). Confirmatory reports have been presented from other laboratories (*2, 3, 136a*). Although the question was raised as whether this RNA was simply an aggregate, 45 S RNA is a unique species of RNA as shown both in studies on its failure to disaggregate and on its preservation under a variety of conditions (*103a*); nucleolar RNA is most rapidly labeled in the 45 S region and with time its label is transferred to the 35 S and 28 S RNA.

One of the critical hopes of the workers in this field has been that a series of methods could be developed for fractionation of the rapidly sedimenting RNA into individual molecular species. Utilizing the techniques of polyacrylamide gel electrophoresis (*221*), similar although somewhat sharper peaks of 45 S RNA were demonstrated in "nucleolar" preparations obtained by the extraction technique of Penman (*149, 150*). Chromatographic techniques that have been employed include analysis on ion-exchange chromatography, calcium phosphate gels, countercurrent distribution (*114, 115, 194*), and hybridization of the RNA (*82, 198*). None of these methods has provided a satisfactory chemical product from the point of view of demonstrated purity. Accordingly, the best presumption at the present time is that studies on both UV nucleotide

TABLE 5.5

UV Nucleotide Compositions of Nucleolar RNA[a]

	Adenylic acid (A)	Uridylic acid (U)	Guanylic acid (G)	Cytidylic acid (C)	$\frac{A+U}{G+C}$
Normal rat liver (*138*)					
45 S RNA	14.5	20.8	35.3	29.4	0.55
35 S RNA	14.3	19.8	35.1	30.8	0.52
28 S RNA	15.0	19.9	34.9	30.2	0.54
Walker tumor (*194*)					
55 S RNA	15.3	19.2	36.7	28.8	0.53
45 S RNA	14.6	20.5	35.1	29.7	0.54
35 S RNA	15.4	19.0	35.0	30.6	0.52
28 S RNA	16.0	19.0	35.3	29.6	0.54

[a] Ultraviolet nucleotide composition of nucleolar RNA components. Each purified RNA fraction was subjected to alkaline hydrolysis followed by chromatography on Dowex-1 column. The values for each purine or pyrimidine nucleotide are averages of the percentage of total purine and pyrimidine bases in the RNA fraction determined by ultraviolet absorption at the specific wave length for each nucleotide. The nucleotide compositions are percentages of the total nucleotides.

analysis and ^{32}P distribution in nucleotides simply reflect distributions in population series of molecules of high molecular weight (see p. 264).

Table 5.4a presents the UV and ^{32}P nucleotide compositions of various fractions of whole nuclear RNA. The ^{32}P nucleotide compositions reflect the synthesis of an AU-rich nuclear RNA in the whole nuclear RNA fraction. In the nucleolar RNA, however, the newly synthesized nucleolar RNA is GC-rich (Table 5.4a). The compositions of the whole nuclear fractions are reflected in the composition of the RNA sedimentation classes (Figs. 5.1, 5.2) as shown in Table 5.4b. Liver nuclear 55 S RNA is more AU-rich than that of the Walker tumor, but its composition does not differ markedly from that of other fractions obtained from rat liver. Studies on nucleolar RNA sedimentation classes (Table 5.5) showed a striking agreement of UV nucleotide compositions of 28 S RNA with the 35 S and 45 S RNA; i.e., they do not differ significantly from one another in the Walker tumor or the normal liver.

Although the UV nucleotide compositions of the various fractions of the rapidly sedimenting RNA are virtually the same, they all differ sharply from those of the 4–7 S RNA of the nucleoli (Chapter VI). In the 4–7 S RNA, there is little transfer activity, but there is a substantial amount of other components; i.e., RNA species with a very high content of uridylic acid (*134, 140a, 221*).

TABLE 5.6a

^{32}P NUCLEOTIDE COMPOSITION OF 45 S NUCLEAR RNA (*43*)[a]

	A	U	G	C	r
Hepatoma 9618A	16.9	21.1	31.7	30.4	0.61
Hepatoma 9121	16.8	20.6	33.5	29.0	0.60
Hepatoma 9108	17.4	21.7	31.1	29.0	0.66
Hepatoma 9098	15.8	22.6	31.5	30.1	0.61
Hepatoma 8995	16.0	21.3	31.9	30.8	0.59
Hepatoma 7800	21.6	25.3	29.8	23.5	0.88
Hepatoma 7787	20.0	22.4	30.2	27.4	0.74
Hepatoma 3924A	18.3	22.2	31.5	28.0	0.68
Hepatoma R7	17.0	21.7	31.9	29.1	0.63
Hepatoma R3B	19.5	24.6	30.8	25.1	0.79
Walker tumor (*145a*)	15.3	22.7	32.3	29.7	0.61
Normal liver (*104, 138, 145a*)	26.6	23.0	25.8	24.3	1.00
Regenerating liver (*135*)	21.8	18.5	34.7	24.9	0.68
Thioacetamide-treated liver	25.6	23.0	26.2	25.1	0.95

[a] ^{32}P nucleotide composition of 45 S nuclear RNA after pulse labeling with orthophosphate-^{32}P. The values are percentages of the total ^{32}P in each of the four nucleotides The values for each of the tumors are averages of two to four determinations on individual samples. A, adenylic acid; C, cytidylic acid; G, guanylic acid, U, uridylic acid; r, ratio of A + U/G + C.

TABLE 5.6b

^{32}P Nucleotide Composition of Nucleolar 45 S RNA (*43*)

	A	U	G	C	r
Hepatoma 9618A	14.7	19.5	34.4	31.4	0.52
Hepatoma 9618B	14.4	20.3	34.3	30.6	0.54
Hepatoma 9121	14.0	18.6	35.4	32.0	0.47
Hepatoma 9098	12.7	20.3	34.1	32.9	0.49
Hepatoma 7787	14.8	19.4	35.0	29.5	0.56
Novikoff hepatoma (*145a*)	12.1	21.3	33.5	33.1	0.50
HeLa cells (*145a*)	13.1	17.2	36.6	33.2	0.44
H-50 cells SV40 transformed (*145a*)	13.4	22.4	35.4	28.8	0.56
Walker tumor (*194*)	13.2	20.6	34.4	31.7	0.51
Ehrlich ascites (*145a*)	12.6	22.1	33.2	32.0	0.53
Regenerating liver (*136*)	19.3	17.9	38.8	24.0	0.59
Normal liver	19.3	16.6	39.4	24.7	0.56
Thioacetamide-treated liver (2 days)	17.1	16.1	38.1	28.4	0.49
Thioacetamide-treated liver (9 days)	18.5	16.4	37.6	26.1	0.55
Monkey kidney-tissue culture (*145a*)	13.6	33.3	32.2	21.0	0.88

Extensive studies have been made on the ^{32}P nucleotide composition of 45 S RNA in both whole nuclear and nucleolar preparations (Table 5.6). Interestingly, the ^{32}P nucleotide compositions of the nucleolar RNA of normal and regenerating liver were virtually identical (*43, 135*). This finding suggests that there is little change in the nucleolar gene readouts when the liver changes from a normal resting state to one of active cell division. This conclusion, made on the basis of nucleotide compositions of the newly synthesized nucleolar RNA, is supported by data on the nearest neighbor frequencies of nucleolar RNA of regenerating liver (*163, 164*) and that obtained by hybridization studies (*64*).

On the other hand, there is a striking difference in the values obtained for nucleotide compositions of newly synthesized nucleolar RNA of a number of tumors and those of the normal and regenerating liver (Table 5.6b). In each case, the newly synthesized nucleolar RNA of the tumors has a lower content of adenylic acid (A) and a higher content of cytidylic acid (C) than does the RNA of either normal or regenerating liver (*43, 136b, 141*). These data are of interest not only because they are quite consistent from tumor to tumor but also because of the differences in the origins of the tumors.

C. Morris Hepatomas

Of late, considerable interest has been evoked by the Morris hepatomas which have been found to differ only "minimally" from the normal liver

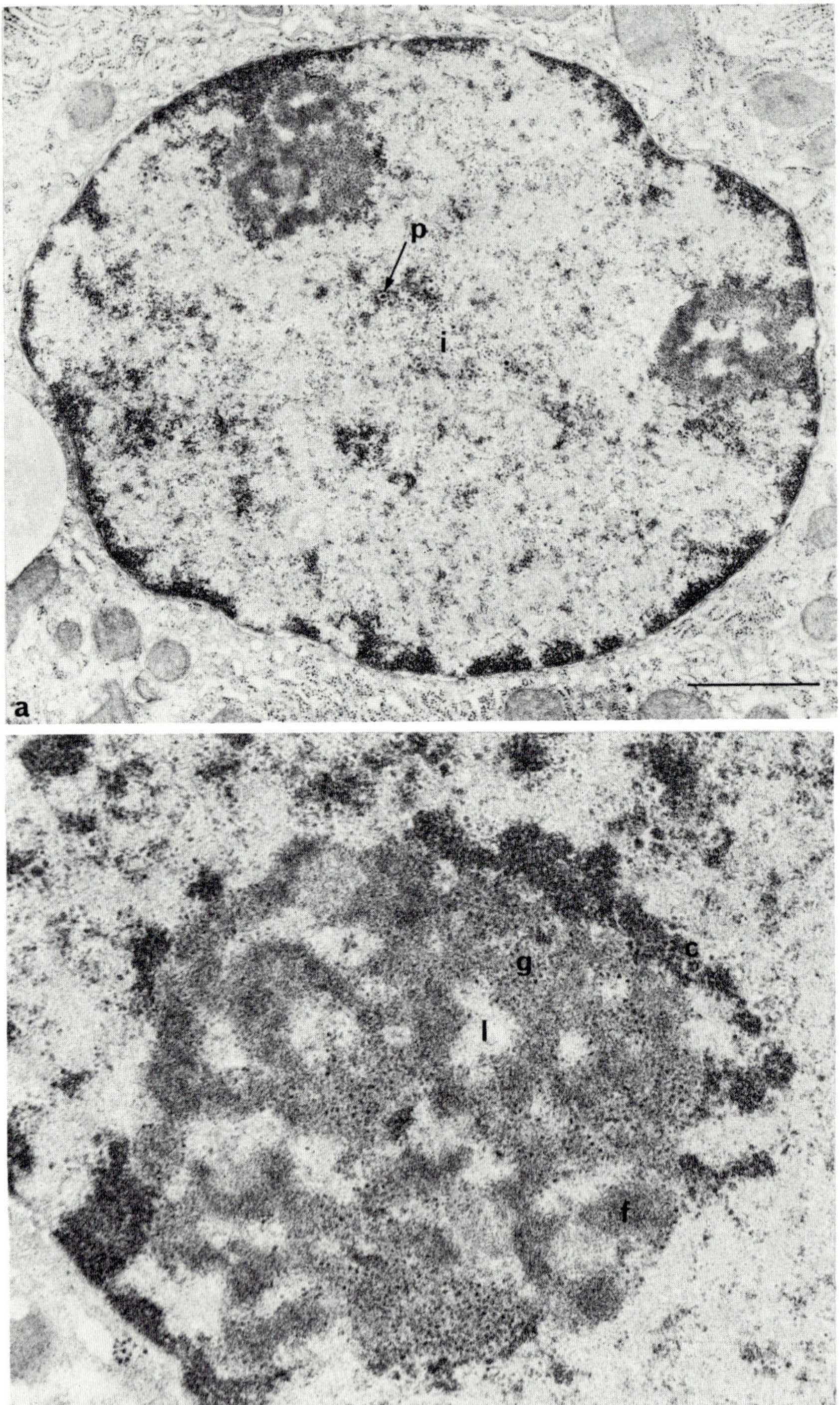
p
i
a
c
g
l
f
b

as compared to anaplastic tumors such as Novikoff hepatomas. Without delving into the extensive studies on the biochemical phenotypes of these tumors, suffice it to say that some are very similar to liver in that they produce glycogen, bile, and a number of cytoplasmic structural and functional components of the liver. Karyological studies by Nowell *et al.* (*144*) showed that among these tumors, there are two, the 9618A and the 9633 that have perfectly normal karyotypes or chromosome compositions. Presumably, the DNA of these chromosomes would either be the same (or only slightly different) from that of the normal liver cells from which the tumors were derived. Accordingly, the greatest interest from a karyological point of view centers around the readouts in the nucleoli of the 9618A tumor. This tumor, like the others studied, however, has the ^{32}P nucleotide composition common to the neoplasms rather than that of the normal liver (Table 5.6b).

The nucleolus of the 9618A tumor is more compact (Fig. 11.5) and less similar to that of the normal liver than the nucleolus of the hepatoma 7787 which cannot be distinguished by electron microscopic criteria from that of the normal liver (Fig. 5.9, 5.10). The nucleolar readout as shown by nucleotide composition of the newly synthesized nucleolar RNA of hepatoma 7787 is similar to that of the other tumors, rather than that of the liver (Table 5.6b). In short, all of the tumors are more like one another than like either other tissues or the phenotypically differentiated tissue of the host as would be expected from Greenstein's "convergence" theory of cancer (*93*). Because of some of the uncertainties of ^{32}P nucleotide compositions, however, it is difficult to be certain of the meaning of some of these results until structures can be defined for these RNA's.

From the point of view of the effects of carcinogens that produce alterations in the nucleoli, the results with thioacetamide-treated livers (p. 8) are noteworthy because of their similarity to results obtained with the normal liver (Table 5.6b) despite the large size changes produced by thioacetamide (Chapter X). These nucleoli had only a slightly lower content of adenylic acid in their newly synthesized RNA than those of normal liver and a slightly higher content of cytidylic acid. Whether this effect reflects specific hyperactive synthesis of one nucleolar component, a tendency toward the neoplastic readouts or some other effect remains to be worked out.

Fig. 5.9. a. Nucleus of a liver cell with two nucleoli. p, perichromatinic granule; i, interchromatinic granules. ×18,000. b. Nucleolus of a liver cell with well-separated nucleolonemas. g, granular components; f, fibrillar components; l, intranucleolar light areas. ×39,000.

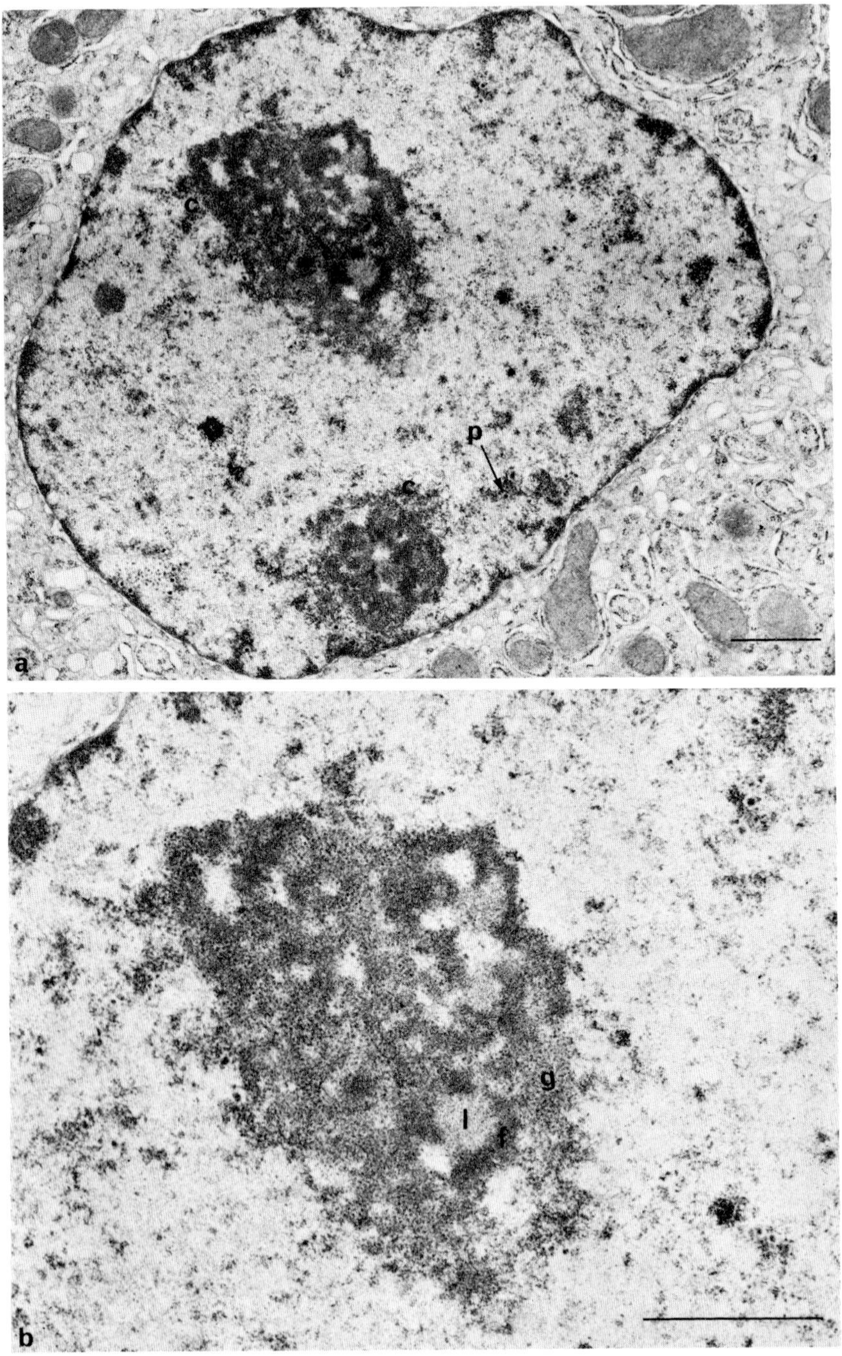
c
p
c
a
g
l
f
b

VII. Rates of Synthesis of Nucleolar RNA

The rates of synthesis of 45 S RNA have been of interest for two reasons, the first being that this RNA is the most rapidly synthesized of the readily isolated nucleolar RNA's and second, it is this species that is rate-limiting for the formation of the other nucleolar RNA fractions since it either is unfolded or cleaved to yield the subspecies. Thus far, it has not been possible to directly calculate or analyze the rates of synthesis of this RNA since there is no known mechanism for direct shutting off its synthesis and turning it on again. The approximations that have been derived are based upon the demonstrated rapid termination of the synthesis on nucleolar RNA in the presence of actinomycin D.

A. Use of Inhibitors in the Study of Synthesis of 45 S RNA

After the discovery of Journey and Goldstein (*112*) that actinomycin D, an antibiotic derived from *Streptomyces griseus*, had an inhibitory effect on nucleolar function, Perry (*151–156*) and subsequently many other workers (*41*) utilized this antibiotic to block nucleolar synthetic reactions. This antibiotic and subsequently a host of others that have been studied (Chapter X) specifically block nucleolar metabolism when administered in low doses. In higher doses, they exert general inhibitory effects on nuclear functions and in very high doses they produce a series of nonspecific effects, apparently related to dissolution of lysosomes (*201*). Inhibitors that produce similar effects are 2-mercapto-1-(β-4-pyridethyl)benzimidazole, 8-azaguanine, aflatoxin B_1 (*63, 113, 175*), and others listed in Table 10.2.

Morphologically, these agents produce a segregation of the granular and fibrillar regions of the nucleolus (Chapters II, X). This segregation appears to be simply a continuation of normal nucleolar processes that are unmasked when new synthesis fails to keep up with the transport of completed nucleolar products. This effect is like an assembly line in which the input was cut off earlier and only the outgoing products are visible. The ability of the inhibitors to shut off new synthesis is useful for determining rates of export of products and rates of degradation or processing. Assuming that there is little abnormality introduced into the processes with low doses, these processing rates are probably equal to rates of synthesis in the uninhibited system.

Fig. 5.10. a. Nucleus of a Morris hepatoma 7787 cell. c, nucleolus-associated chromatin; p, perichromatinic granule. ×12,000. b. Nucleolus of a Morris hepatoma 7787 cell with distinct nucleolonemas. g, granular components; f, fibrillar components; l, intranucleolar light areas separating nucleolonemas. ×24,000.

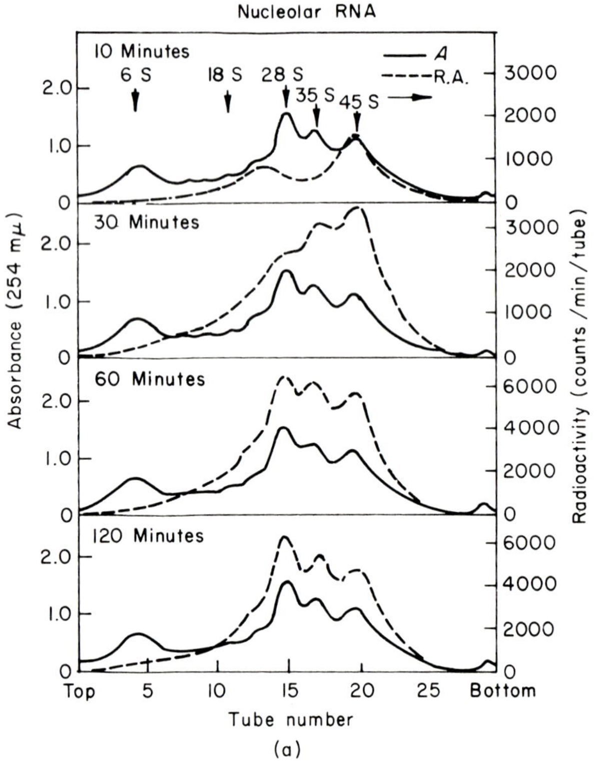

Fig. 5.11. a. Five microcuries per rat of [6-^{14}C] orotic acid were injected intravenously. At specified time points, the animals were killed and the liver nucleolar preparation was isolated. The time after injection of the tracer is shown on each pattern. b. Ten minutes after intravenous injection of 6.3 μCi of [6-^{14}C] orotic acid, 750 μg/kg of actinomycin D were injected intravenously. At specified time points, the animals were killed and liver nucleoli were isolated. The time after injection of actinomycin D is shown on each pattern (from *139*).

In the usual studies, after the termination of pulse-labeling period (10–20 min) for the nuclear and nucleolar RNA, either with a ^{14}C- or ^{32}P-labeled precursor, actinomycin D is introduced into the system in concentrations sufficient to suppress further formation of the labeled precursors of other RNA molecules. Under these circumstances, the series of gradients in Fig. 5.11–5.12 develops in normal liver. After the actinomycin block is developed, there is a loss of both label and RNA from the 45 S and 55 S regions followed by a loss of 35 S RNA, leaving only 28 S RNA. From such data, the conclusion has been drawn that in the nucleolus:

$$45\text{ S RNA} \rightarrow 35\text{ S RNA} \rightarrow 28\text{ S RNA}$$

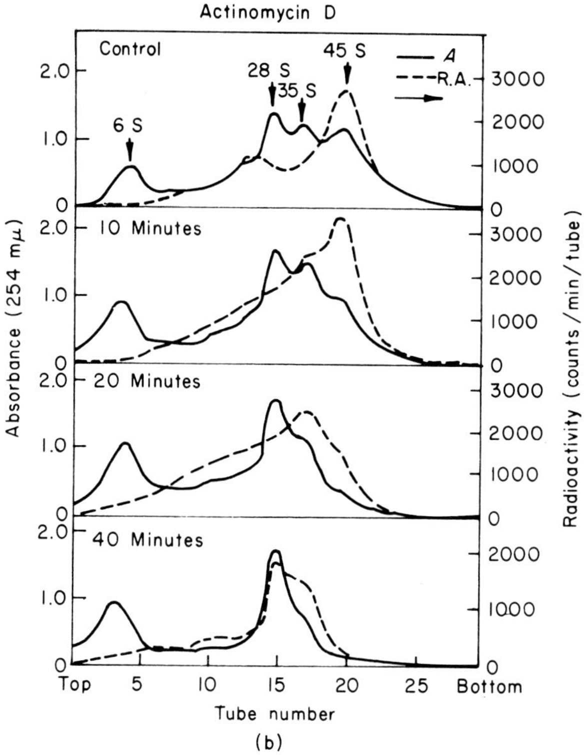

Fig. 5.11 (*continued*).

At this point, it is necessary to mention two other assumptions, (a) at the doses used neither actinomycin nor its effects degrade RNA and (b) the quantitative recovery of radioactivity in nuclear and nucleolar RNA is unaltered by actinomycin treatment. If actinomycin D would enhance the activity of RNases or would otherwise promote the destruction of RNA, the validity of the results obtained would be questionable.

Until recently, a model system for investigating the validity of the analytical systems that have been employed in this and other laboratories for studies on the rates biosynthesis of nucleolar RNA was not available but now studies have been made on the Walker tumor in its logarithmic growth phase that validate the results that have been obtained by the "indirect method" (*107*). The "indirect method" is based upon planimetric analysis of the gradient patterns employing the Burrell-Ott planimeter. With this device, it is possible to calculate the area under the gradient pattern occupied by any of the peaks (Fig. 5.12). Since the amount of RNA, either labeled or unlabeled, is directly related to the

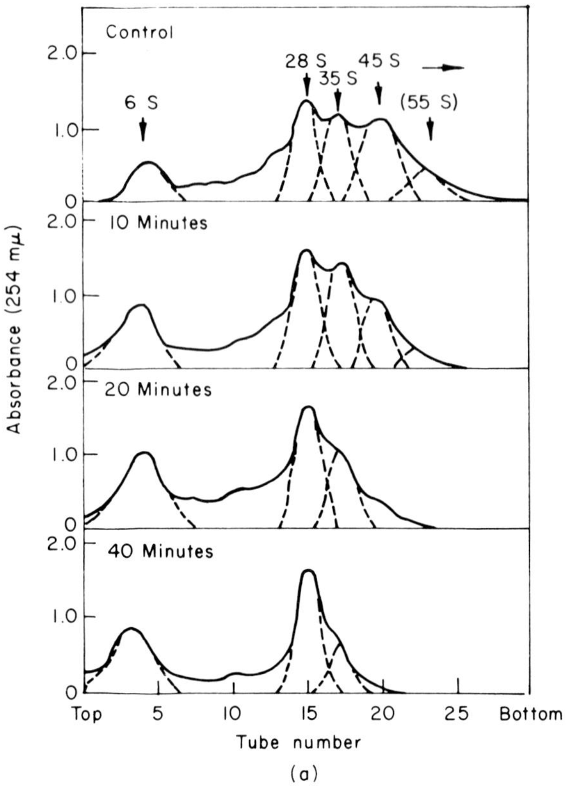

Fig. 5.12. a. Procedure for determination of the area under each peak. The areas were determined with a planimeter. b. The ratio of 45 S RNA to 4–7 S RNA is plotted on logarithmic scale against time after actinomycin D injection. The amount of 45 S RNA was calculated from the sucrose density-gradient profiles (from *139*) and its half-life in this experiment is approximately 8 minutes.

size of the absorbancy peak, the values obtained by planimetry reflect the actual amounts of a given sedimentation class of RNA that is present in a preparation, or a peak. Accordingly, the changes in peak heights can not only be calculated with respect to one another but also on an absolute basis when the total amounts of RNA on the gradients have been defined (Fig. 5.12).

With this information, regression curves have been plotted in which the logarithm of the area under the 45 S peak have been plotted against the period of treatment with actinomycin D. From the linear character of these plots (Fig. 5.12b), it has been possible to calculate the half-life times of 45 S RNA as shown in Table 5.7. From the half life of the 45 S RNA of the Walker tumor, the following calculations were made with respect to turnover and synthesis of nucleolar RNA:

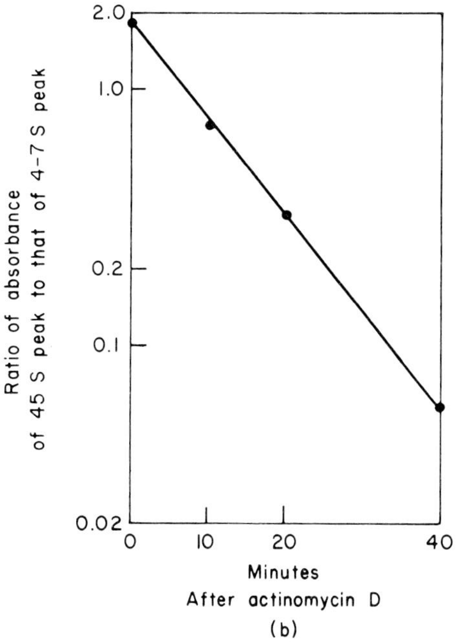

Fig. 5.12 (*continued*).

1. The half life of 45 S RNA in the nucleolus of the Walker tumor = 6.7 minutes.
2. Of the total 1.9 pg of RNA in the Walker tumor nucleolus, 23%, or 0.44 pg, is 45 S RNA.
3. Therefore, 0.22 pg is lost from the nucleolus in 6.7 minutes and under steady-state conditions, 0.22 pg are synthesized in 6.7 minutes.

TABLE 5.7

HALF-LIVES OF 45 S RNA OF NUCLEAR AND NUCLEOLAR RNA OF REGENERATING LIVER (18 HR), AND OF NUCLEOLAR RNA OF THE WALKER TUMOR

	Half-life of 45 S RNA (min) by		
		Radioactivity	
Tissue	Optical density	Orotic acid-^{14}C	^{32}P-Orthophosphate
Normal liver nucleoli (*139*)	10–12	10	
Regenerating liver nuclei (*107*)	8	4.1	
Regenerating liver nucleoli (*107*)	4.7	3.7	
Walker tumor nucleoli (*107*)	6.7		6.8

4. Since 0.22 pg is 220 fg (10^{-15} gm), the rate of synthesis of 45 S RNA in these nucleoli is 32 fg/minute.
5. On this basis, the nucleolus synthesizes 1.9 pg of 45 S RNA per hour and 45.6 pg per day; since the nucleolar number averages 1.5, 68 pg would be produced per day in these nucleoli.
6. Previous studies from this laboratory (*134*) have shown that in the Walker tumor, the nucleolus synthesized approximately 60–70% of total nuclear RNA and a total of 104 pg of RNA would be produced per cell per 24 hours (the generation time of these cells).

In view of the fact that whole amount of RNA might not remain intact, it is likely that this amount of 104 pg RNA is a minimal estimate of synthesis. Since the total cellular RNA is approximately 96 fg, this amount of RNA would necessarily be reproduced in 24 hours because that is the approximate generation time of these cells.

B. Rates of Synthesis of 45 S RNA in Other Tissues

In normal liver, total nucleolar RNA is approximately 0.33 pg (per nucleolus) and of this approximately 60 fg are nucleolar 45 S RNA. The half life of the nucleolar 45 S RNA is approximately 10 minutes in the normal liver as measured by the actinomycin regression curves, and accordingly, approximately 30 fg of 45 S nucleolar RNA would be lost per nucleolus per 10 minutes or approximately 3 fg would be lost per minute. Assuming a "steady state" of the liver, there is a synthesis of 3 fg per minute of 45 S RNA in the liver nucleolus.

In the liver, there are approximately 2.5 mg of ribosomal RNA per gram; since there are about 10^8 cells per gram, there are about 25 pg of ribosomal RNA per cell. This RNA turns over with a half life of approximately 5.5 days; i.e., about 12.5 pg turns over in 5.5 days, 2.3 pg per day, and 0.1 pg per hour. Based upon the data that 3 fg are synthesized per minute, the total 45 S RNA synthesized would be sufficient to account for approximately 180 fg of RNA per hour. Assuming there are 2.7 nucleoli per nucleus, the total would be sufficient to account for about 500 fg per hour or a fivefold excess over the total actually required to replace the lost ribosomal RNA. By comparison with the process which occurs in the Walker and Novikoff tumors, this would seem to be an excessive amount of synthesis. However, as found recently (*49*), there is a significantly greater content of nucleolar RNase in the normal liver than in the tumors and in addition, the concentration of ribonuclease inhibitor in the liver is apparently much lower than that of the tumors that have been studied (*169*). As will be indicated in Chapter VII, the difference in the concentration of the ribonuclease inhibitors in the tumors and other tissues is unaccounted for in terms of the cellular functions but may be related to the differences in turnover rates.

By comparison with the normal liver, the nucleolus of the regenerating liver produces RNA at a substantially increased rate; i.e., some five times as much RNA is produced per nucleolus per minute in the regenerating liver as in the normal liver (*73, 107*). The half-life time of nucleolar RNA is 3.7 minutes in the nucleoli of regenerating liver as measured by turnover of radioactivity and 4.7 minutes as measured by optical density. These values compared with 10 and 12 minutes respectively for the normal liver and reflect the fact that the regenerating liver is not only producing proteins for the normal functions of the liver but is also producing new proteins for the biosynthesis of new cells (Table 5.7).

VIII. Methylation of Nucleolar RNA

Although its significance is still not clear, there is good evidence that methylation of nucleolar RNA occurs directly in the nucleolus and hence specific methylases may be present there (see Chapter VII). The first studies made on this subject were those of Jacob, Sirlin, and their colleagues from 1961–1963 (*184–193*). Utilization of ^{14}C- or ^{3}H-methyl-methionine; i.e., methionine labeled in the transferable methyl group, permitted autoradiographic and biochemical analysis of the labeling of the RNA. Subsequently a series of studies have been reported from different laboratories that showed that methylation occurred on the 45 S RNA presumably either when it was being synthesized or shortly thereafter (*92, 136a, 229*). In the studies of Greenberg and Penman (*92*), the RNA with the highest specific initial activity was the 45 S "nucleolar" RNA; as the RNA was processed further, the label appeared in nucleolar 35 S and 28 S RNA in exactly the same way as shown in Fig. 5.11. When the RNA was hydrolyzed, the label was found in dinucleotides since alkaline hydrolysis does not hydrolyze dinucleotides with a 2′-*O*-methyl group on the nucleotide on the 5′ side of the oligonucleotide. Greenberg and Penman (*92*) also suggested that the polydisperse RNA (the extranucleolar RNA) was not methylated appreciably and presumably the methylated RNA represented only ribosome precursors.

The methylation occurs mainly on the hydroxyl groups; i.e., it is 2′-*O*-methylation rather than *N*-methylation of the bases; *N*-methylation also occurs. Vaughan *et al.* (*209, 210*) reported that there is 1.9 and 1.3% methylation in 18 S and 28 S RNA, respectively. In cells deprived of methionine, Vaughan *et al.* (*209*) found that although nucleolar ribosomal precursor RNA was formed, there was a failure of formation of the ribosomes. It is not clear whether this resulted from a lack of methionine for protein synthesis or whether the processing of RNA was impeded.

Thus far, the most searching study on methylation of nucleolar RNA has been carried forward by Wagner *et al.* (*213a*) who not only found that most of the methylation was on the 2′-hydroxyl of the ribose but also that alkali-resistant dinucleotides differed for 18 S and 28 S rRNA.

They also identified a number of alkali-resistant dinucleoside phosphates; i.e., 2′-*O*-methyl AU, 2′-*O*-methyl AG, 2′-*O*-methyl GU and 2′-*O*-methyl GG and 2′-*O*-methyl UG, or 2′-*O*-methyl GU. 2′-*O*-methyl GU or UG were present in highest amounts. On the basis of the methylation patterns, the authors concluded that the methylation of 45 S RNA was similar to an equimolar mixture of 18 S and 28 S rRNA. The methylation pattern of 32 S RNA was similar to that of 28 S RNA. They concluded that 32 S RNA was the precursor of 28 S RNA in the nucleolus. Methylation of nucleolar RNA also occurs subsequent to its initial formation according to Zimmerman (*228a*). For example, the dimethylaminopurine associated with 18 S rRNA is not found in 35 S nucleolar RNA. This methylation may occur at the time of cleavage. Extensive studies on the methylated dinucleotides and basis of 16 S and 28 S RNA have been made by Lane and Tamaoki (*124a, 203a*).

IX. Other Types of Rapidly Sedimenting RNA

A. Newly Synthesized Bacterial RNA

Although it has been suggested that in mammalian species newly synthesized RNA has a high molecular weight while that of bacteria is of low molecular weight, it is now clear that this RNA is probably newly formed as 23 S–28 S or other high molecular weight RNA (*6, 72*). The RNA was originally found to have a low molecular weight because the RNase levels were high in the original preparations of bacteria that were studied. The same is apparently true in the preparations of RNA obtained from the *Drosophila* and other insects.

B. Other Types of RNA Synthesized—Poly A

In recent months, substantial interest has been occasioned by the finding of a series of reactions that convert nicotinamide adenine dinucleotide (NAD) into a polyriboadenylic acid that has a molar ratio of adenine = 1, ribose = 2, phosphate = 2 and periodate consumption of 1 (*50, 73b, 100a, 142, 143*). The synthesis of this interesting polymer is catalyzed by an enzyme that may be an RNA polymerase. This enzyme may be found in nucleoli or may be present in extranucleolar elements (*142, 143*).

Although very likely unrelated, a similar homopolymer has been reported to be present in reovirus by Bellamy and Joklik (*12*). This A-rich RNA contains 85% adenylic acid and is synthesized in the presence of progressively more active infection. This RNA lacks guanylic acid; uridylic acid accounts for 10.5% of the total residues and 0.5% is cytidylic acid. These small molecules are only 60 nucleotides in length and in each virion of the reovirus, there are 50–100 such molecules.

X. Independence of Nucleolar RNA Synthesis

In the course of studies on isolated nucleoli in this laboratory, two biochemical observations were made which have established that the nucleolar RNA synthesis is a local process and that it proceeds "semi-independently" from the biosynthesis of RNA in the remainder of the nucleus. The kinetics of labeling of two species of RNA, one extractable with cold phenol (a-RNA) and the other with hot phenol (i-RNA) were analyzed (*139*). On the basis of counts per minute per milligram (Fig. 5.13a) the nucleolar a-RNA was labeled faster and to a greater extent in a 2-hour period than was the extranucleolar a-RNA or the total nuclear a-RNA. This result led to the conclusion that at least the nucleolar a-RNA was a nucleolar product. Analysis of the UV nucleotide composition of RNA showed that the rapidly labeled nucleolar RNA had a much different nucleotide composition (Table 5.4) from that of the remainder of the nucleus (*137*). With ^{14}C-orotic acid as a tracer, nucleolar RNA and nuclear RNA are labeled with its product, ^{14}C-uridylic acid, and hence the specific activities should be calculated on the basis of specific activity of the uridylic acid in the nucleolus (Fig. 5.13b). On this basis, the values for the specific activities of the nucleolar RNA were equal to or greater than those of the nuclear RNA. These data rule out the possibility that the nucleolar RNA is synthesized at a distance from the nucleolus and instead show that nucleolar RNA is biosynthesized "semi-independently" from that of the remainder of the nucleolus.

Another concept was disposed of by these studies; i.e., that the nucleolus serves a storage function. If this were the case, the specific activity of the nucleolar RNA should be relatively low depending on the pool size of the product in the nucleolus. Since, in fact, the specific activity of the nucleolar RNA was equal to or greater than that of the whole nuclear RNA, it is clear that the nucleolus is really the site of synthesis of its own RNA's.

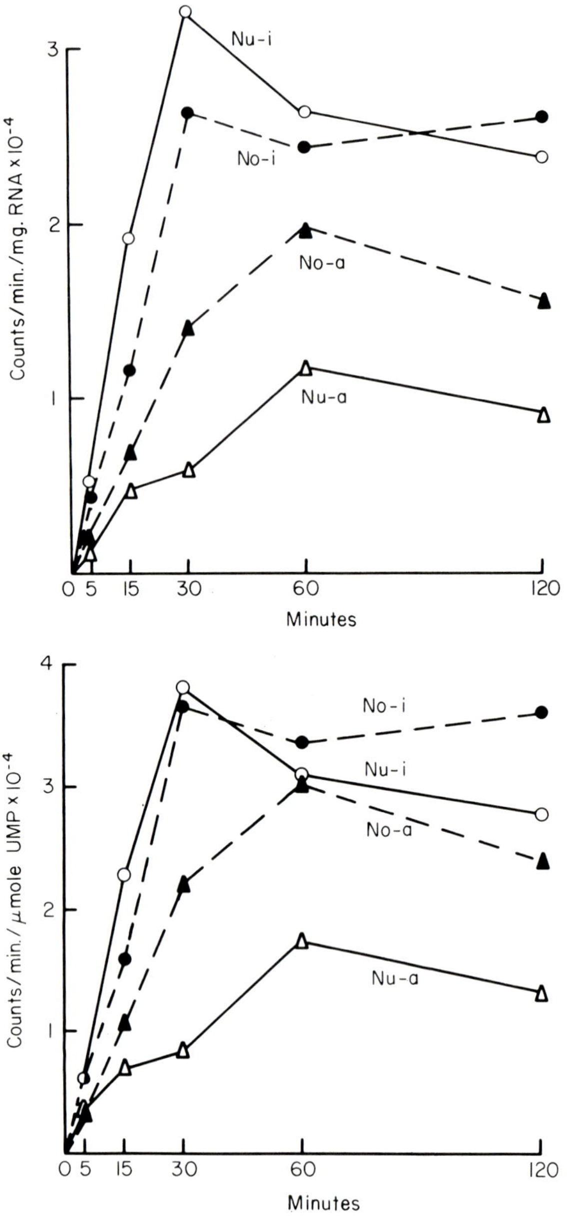

Fig. 5.13. a. Kinetics of labeling of a- and i-RNA of the rat liver nucleus (Nu) and of the nucleolus (No) after intravenous injection of 6-^{14}C orotic acid. The specific activity of RNA was plotted against the time after injection of the tracer. Each point represents an average of 2–6 experiments using eight rats for each experiment. b. Kinetics of labeling of a- and i-RNA calculated on the basis of UMP content of each RNA fraction. The specific activity of UMP was plotted against the time after injection of the tracer. Nu, nucleus and No, nucleolus (from *137*).

TABLE 5.8

In Vitro INCORPORATION OF UTP-^{14}C INTO ISOLATED NUCLEOLI[a,b]

	Specific activity (cpm/mg RNA)	$\mu\mu$M incorporated per mg RNA
Complete system	15,000	1,630
-ATP	7,800	850
-GTP	4,600	500
-CTP	5,400	590
-GTP-CTP	2,500	270
-Mg^{2+}	4,200	460
-ATP-PK-PEP	3,600	390
-PEP-PK	11,700	1,270
-Mg^{2+} $\pm$ Mn^{2+} 2 μmoles	18,800	2,040
5 μmoles	6,600	660
10 μmoles	2,700	290
DNase-treated	680	70

[a] From Ro and Busch (*163*).

[b] The reaction mixture (complete system) contained the following in final volume of 2.0 ml: 20 μg pyruvic kinase (PK), 10 μmoles phosphoenol pyruvate (PEP), 2 μmoles ATP, 0.25 μmoles each of GTP and CTP, 5 μmoles $MgCl_2$, 50 μmoles Tris–HCl, pH 8.2, 0.2 μCi UTP 2-^{14}C (11.3 mCi/mM) and nucleoli obtained from 5 gm of rat liver. The incubation time was 15 minutes. Radioactivity was determined on PCA-precipitable residues.

A. Nucleolar RNA Polymerase

Further evidence for a direct role of the nucleolus in the biosynthesis of RNA was obtained in studies on the RNA polymerase of the nucleoli (*163–165*). As was previously reported (*222–223a*), all four nucleotide triphosphates were required for the catalytic activity of this enzyme. In addition, both an energy source and a divalent ion, either Mg^{2+} or Mn^{2+}, were required (Table 5.8). The overall reaction catalyzed by this enzyme is:

$$\text{NTP} \xrightarrow[\text{RNA polymerase}]{Mg^{2+},\ \text{DNA}} \text{RNA}$$

NTP is nucleoside triphosphate; DNA is required as a primer from which the transcription of DNA can occur by virtue of the activity of the enzyme, RNA polymerase which functions as a "transcriptase." As might be anticipated from the requirement for DNA, if the DNA is either hydrolyzed by DNase or is removed in other ways, the system is inactivated.

One of the interesting points with respect to this enzyme is its high concentration in the nucleolus (*109, 182, 183*). As shown in Table 5.9 the

TABLE 5.9

Distribution of RNA Polymerase in Nuclear Fractions[a]

	Activity	% Nuclei	cpm/mg RNA/20 min	cpm/mg Protein/20 min
Nuclei	574	100	9,900	140
Nuclei, disrupted	503	88	8,400	157
Nuclear sap	82	14	1,500	37
Chromatin, broken nucleoli	316	55	53,900	263
Nucleoli	263	45	49,700	3,300

[a] From Siebert *et al.* (*183*).

fractions obtained from nuclei that contained nucleoli also had most of the activity of the nuclear RNA polymerase of liver.*

The liver nucleoli contain a minimum of 45% of the total nuclear RNA polymerase and also contain the most active nuclear RNA polymerase. The ratio of the activity in the nucleoli to that of the overall nuclei was approximately 23 with respect to the "specific activities" of the enzyme (counts per milligram protein per 20 minutes) and only 2 for the mixed chromatin and nucleolar fraction. For other enzymes studied (see Chapter VII), including ribonuclease, ATPase, and NAD synthetase, the ratios of activity of the nucleolus were approximately 6, 3, and 2, respectively (*183*).

B. "Nearest-Neighbor Frequencies" of the Product of Nucleolar RNA Polymerase

The precise relationship of the product produced by nucleolar RNA polymerase *in vitro* and the product synthesized *in vivo* has not yet been evaluated. Analysis has been made of nearest-neighbor frequencies of the nucleolar product. The method of nearest-neighbor analysis was used by Weiss and Nakamoto (*223a*) in their studies on the RNA polymerase as it was by Josse *et al.* (*111*) for analysis of DNA synthesized as a means for avoiding the difficulties of dealing with the whole DNA molecule. The idea is that a labeled nucleoside triphosphate with ^{32}P in the phosphate proximal to the nucleoside will enter the RNA in the position specific for that nucleotide. On hydrolysis of the RNA with alkali, the RNA is cleaved so that the phosphate moves to the 3′-phosphate of the

* A number of recent findings have indicated that the optimal conditions are different for the RNA polymerases in nucleoli and extranucleolar chromatin (*106*). The ionic strength, the types of divalent cations present and the method of preparing the enzyme produce marked variation in their activities. The precise activities of these RNA polymerases in the nuclei *in vivo* are not yet definable.

nucleotide adjacent to the nucleotide used as a tracer. Thus, the labeled phosphate-^{32}P is on the "nearest-neighbor" nucleotide to the nucleotide utilized as the tracer. This procedure offers a basis for comparison of the newly synthesized products of different systems and in addition, some idea of the frequency of nucleotide pairs in a molecule of nucleic acid.

In the usual procedure for the study of the nearest-neighbor frequency of the newly synthesized RNA, the incubation period is short; i.e., 15 or 30 minutes. The RNA is then isolated and the distribution of the ^{32}P in the four nucleotides is determined. Differences in the products obtained for the normal liver and the Walker tumor in these isolated nucleolar systems are shown in Table 5.10.

TABLE 5.10

PERCENTAGE OF TOTAL ^{32}P FROM ATP-U-^{32}P IN THE NUCLEOTIDES OF NUCLEOLAR RNA SYNTHESIZED *in Vitro* (*163*)

	Adenylic acid	Uridylic acid	Guanylic acid	Cytidylic acid
Liver	30.0	13.8	35.4	20.7
Walker tumor	33.3	17.3	24.0	25.3

Thus, after the addition of proximally labeled ATP to the nucleolar system of liver, 30 and 35.4% of the labeled nucleotide was incorporated in positions adjacent to adenylic and guanylic acids respectively. On the other hand, in the tumor, only 24% was incorporated into guanylic acid residues adjacent to adenylic acid. In the tumor, more isotope was incorporated into uridylic and cytidylic acids residues adjacent to adenylic acid and accordingly, the tumor should have less G-A linkages and more C-A and U-A linkages. Another form of expression of this data is in Table 5.11 which shows the ratios of incorporated nucleotides relative to

TABLE 5.11

RATIO OF ^{32}P FROM ATP-U-^{32}P IN NUCLEOTIDE TO ^{32}P IN ADENYLIC ACID IN NUCLEOLAR RNA SYNTHESIZED *in Vitro* (*163*)

	Adenylic acid	Uridylic acid	Guanylic acid	Cytidylic acid
Liver	1.0	0.46	1.18	0.60
Walker tumor	1.0	0.52	0.72	0.76

the A-A (adenylic acid: adenylic acid) linkages. That such differences are found with another precursor was shown with labeled guanosine triphosphate as a precursor (Table 5.12). In this case, the guanylic acid is mainly adjacent to guanylic and cytidylic acid residues, presumably

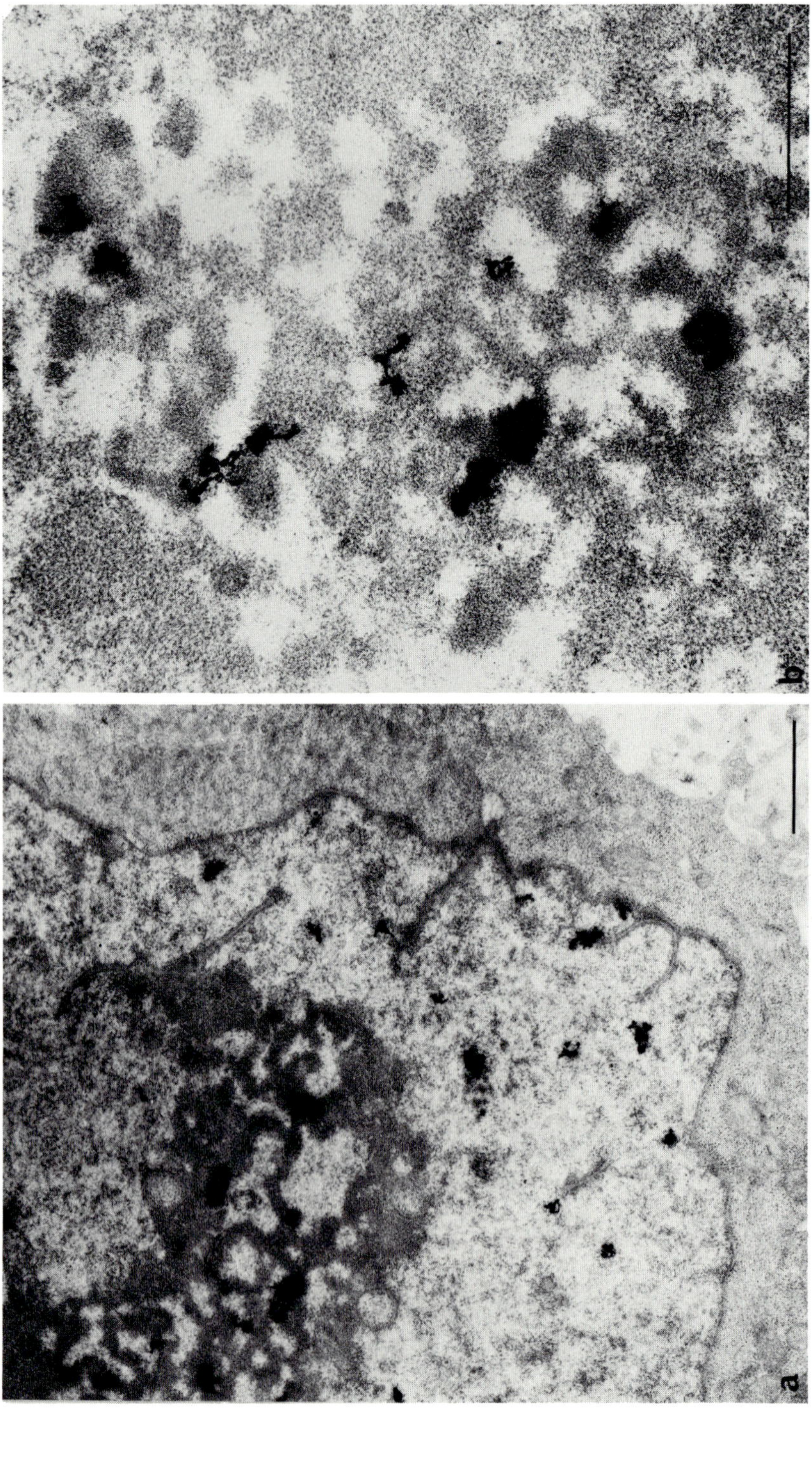

Fig. 5.14. Novikoff hepatoma cells labeled with ^{3}H-5-uridine *in vivo*. Markers lines in the figures denote 1 μ. Specimens were fixed with glutaraldehyde and osmium tetroxide, embedded in Epon-Araldite, and stained with uranyl acetate and lead citrate (*207a*). a. Nucleus and cytoplasm, 15-minute pulse. Practically no silver grains are found in the cytoplasm. $\times$15,000. b. Nucleolus, 5-minute pulse. Most of the silver grains are located over the fibrillar components of the nucleolus. $\times$25,000.

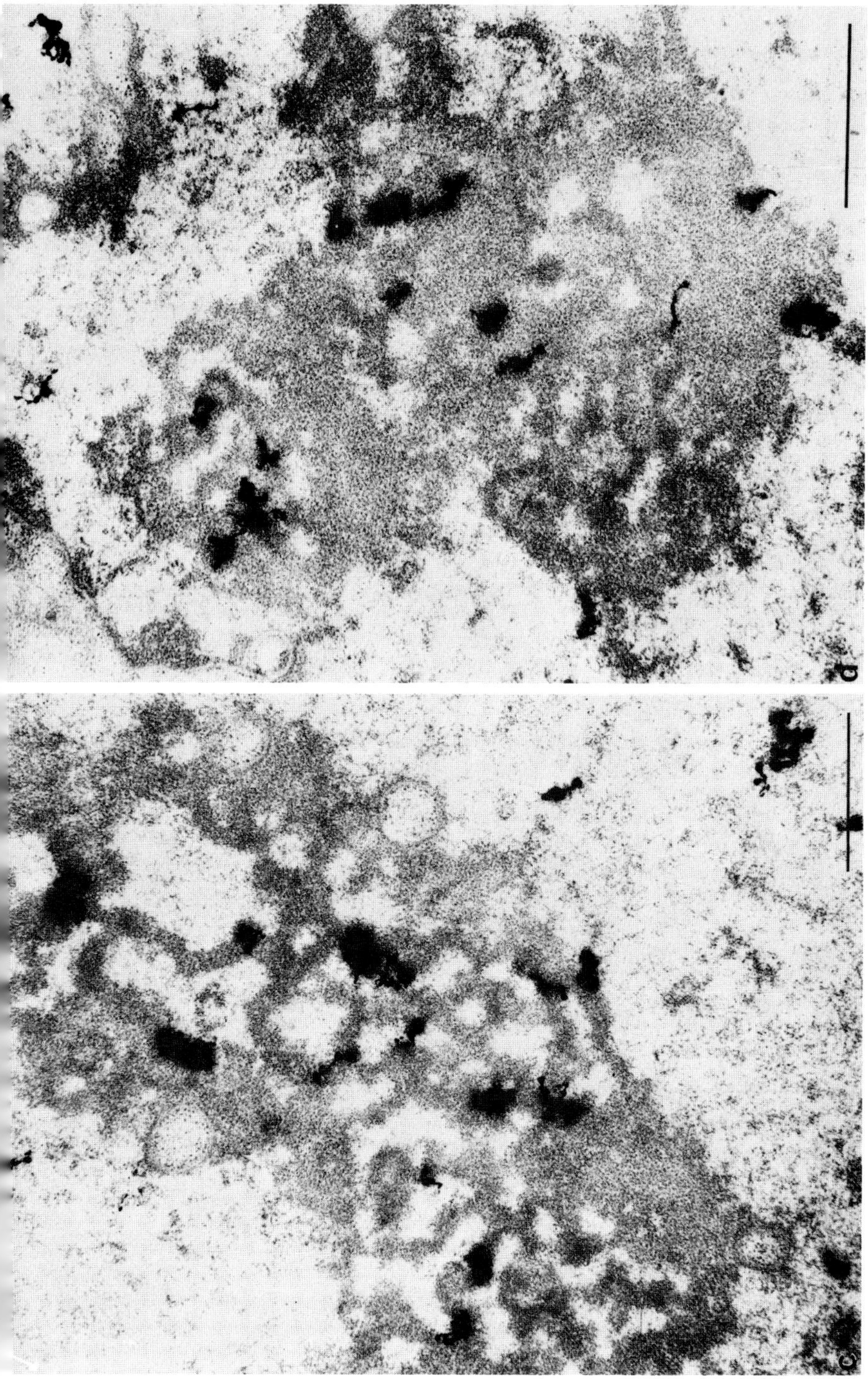

Fig. 5.14. c. Nucleolus, 15-minute pulse. Some of the grains are located over the granular components, but most of them are over the fibrillar components of the nucleolus. ×21,000. d. Nucleolus, 60-minute pulse. Most of the grains are over the granular components. ×25,000 (from *207a*).

TABLE 5.12

PERCENTAGE OF TOTAL ^{32}P FROM GTP-α-^{32}P IN INDIVIDUAL NUCLEOTIDES OF NUCLEOLAR RNA SYNTHESIZED *in Vitro* (*163*)

	Adenylic acid	Uridylic acid	Guanylic acid	Cytidylic acid
Liver	11.0	11.8	43.8	33.4
Walker tumor	13.0	13.5	37.9	35.7

indicating that there is a large number of linear sequences of guanylic residues and/or cytidylic acid residues in these nucleotide sequences of nucleolar RNA. The values for the tumor differ from those for the liver only to a small extent for nucleotides other than guanylic acid. These data along with those for the adenylic acid indicate that the sequences of newly synthesized RNA of the nucleoli of the liver and tumor have significant differences. The differences must be defined in terms of the linear sequences of nucleotides in these RNA molecules.*

XI. Autoradiographic Studies on the Synthesis of RNA in the Nucleolus

Goldstein and his associates (*84–89*) who initially made brilliant use of autoradiography for the analysis of traffic in and out of the nucleus, established that most of the RNA of the cytoplasm was synthesized in the nucleus of protozoa. In addition, they showed that some of the cytoplasmic products migrated back into the nucleus. They applied the technique of quantitative autoradiography to the analysis of the origins of nucleolar RNA; although their results were equivocal, they concluded initially that nucleolar RNA was derived from chromosomal activity that was extranucleolar. Later they retracted this idea, basing the newer concept on a number of types of analysis (*14, 85, 108*).

Extensions of the work on synthesis of nucleolar RNA involved the efforts of many cytologists whose conclusion was that in many cells, particularly those that grew rapidly or produced much protein, nucleolar labeling preceded that of the cytoplasm or even the remainder of the nucleus. The concept developed that the nucleolar product emerged into the cytoplasm (*4, 5, 9, 10, 16, 55–57, 69, 69a, 74, 75, 80, 125, 126a, 130a, 145, 158, 159, 161a, 174, 197, 205, 206, 224a, 224b, 225, 225a*).

Autoradiographic studies also showed that there was more label over the nucleolus in tumors than in resting tissues such as the normal liver. Using Zirkle's technique (*230*), Perry *et al.* (*151–159*) showed that after

* Studies on the purification of nuclear and nucleolar RNA polymerases are in progress and will be discussed more fully in Chapter VII (*54a*).

microirradiation of nucleoli there was a drop in the labeling (grains) over nucleolar RNA; they could not differentiate between an intra- or perinucleolar origin of nucleolar RNA in view of the problems of ascertaining kinetics of change in specific activity and the possibility of differential sensitivity of special genes to the effects of irradiation.

In addition to these types of studies, autoradiography was of great value in the assessment of the effects of drugs, particularly actinomycin and puromycin, on the fate of labeled precursors of RNA (*150, 153, 203*). Of more significance to the sites of RNA synthesis have been the studies combining the use of electron microscopy and autoradiography. With this combination, Firket and Granboulan (*70*) and Granboulan and Granboulan (*90, 91*) localized the sites of initial labeling of nucleolar RNA to the fibrillar regions of the nucleolus although quantitative studies have shown that an appreciable portion of the total isotope taken up into RNA became incorporated into the granular elements as well (*207a*).

Kinetic studies on the labeling of RNA in granular and fibrillar com-

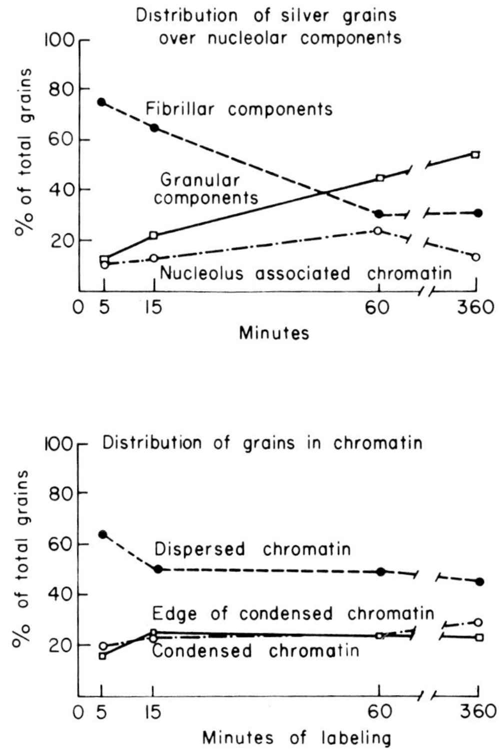

Fig. 5.15. Kinetic studies on quantitative distribution of silver grains in nuclear components (from *207a*).

ponents of the nucleoli (*207a*) have indicated that initially the labeling takes place in the fibrillar elements of the nucleolus, and apparently there is a transfer in a relatively short time of the isotope to the granular elements (Fig. 5.14, 5.15). The possibility that most of the synthetic reactions occur in the looser framework of the fibrillar elements (*70, 90, 91*) is in general agreement with the idea that the dispersed chromatin has a higher biosynthetic activity than the condensed chromatin.

The composition of the fibrillar elements to the nucleolus is of interest but has proved difficult to define completely. LaCour and Crawley (*123*) have suggested that in plants there is an intense nucleolar labeling that disappears on treatment with RNase. Utilizing this method, they have shown that labeling occurs in skeins that are centrally located in these nucleoli; they referred to these skeins as "nucleolonemas" (see Chapter II). Combined analytical and ultrastructural studies carried out in this laboratory (*175a, 176*) have suggested that the low molecular weight nucleolar RNA (see Chapter VI) is limited in localization to the fibrillar regions of the nucleolus.

XII. Processing of the 45 S Nucleolar RNA: Formation of 28 S rRNA and 18 S rRNA

As indicated in Chapter IV, convincing evidence has emerged from genetic and hybridization studies that the nucleolar DNA contains the codes for both 28 S rRNA and 18 S rRNA (*11, 12a–15, 31–34, 160, 162, 177, 208, 214*). A number of lines of evidence obtained by biochemical analysis of nucleolar RNA permitted an extension of these conclusions and, in addition, a further insight into the structures of the molecules that are intermediates in the process.

One interesting aspect of the biosynthetic reactions in the nucleolus is that the initial product has a higher sedimentation coefficient; i.e., 45 S, than the final 28 S product. Evidence obtained by polyacrylamide gel electrophoresis supports the conclusion that the 45 S nucleolar RNA has a higher molecular weight than either 35 S or 28 S nucleolar RNA. In addition, the electron microscopic analyses (Figs. 5.4–5.8) also suggest that fibers of longer lengths are present in the more rapidly sedimenting nuclear RNA. It is not clear why the initial nucleolar product is such a large RNA strand.

Among the methods employed to study the processing of nucleolar RNA have been: (a) UV-nucleotide compositions, (b) ^{32}P nucleotide compositions, (c) oligonucleotide compositions, (d) hybridization studies, (e) analyses of the 3′ and 5′ terminals of the RNA, (f) analyses of the split products of the RNA's on polyacrylamide gels, and (g) electron microscopy.

On the basis of all of the types of data available at this point, the current concept is that oligomers of 45 S nucleolar RNA are formed initially; these have approximate sedimentation coefficients of 65 S, 85 S, 100 S and possibly 120 S (*176a*).* These then are cleaved to 45 S nucleolar RNA which then undergoes the cleavage reactions postulated in Fig. 5.16.

A. Chemical Similarity of 45 S, 35 S and 28 S Nucleolar RNA

Table 5.5 shows that the nucleotide compositions are similar for 55 S RNA, 45 S RNA, 35 S RNA, and 28 S RNA in nucleoli of the Novikoff hepatoma and the normal liver. Similar results for ^{32}P nucleotide compositions within a given tissue are presented in Table 5.13. These data suggested the possibility that the higher size classes of the newly synthesized nucleolar RNA's were oligomers; i.e., dimers, trimers, tetramers, or pentamers of 28 S RNA or that the 55 S RNA, 45 S RNA, and 35 S RNA have conformational differences from 28 S RNA.

TABLE 5.13

^{32}P Nucleotide Composition of Various Sedimentation Classes of Nucleolar High Molecular Weight RNA[a]

	A	U	G	C	$\frac{A + U}{G + C}$
Walker tumor (*194*)					
28 S	12.5	24.7	32.1	30.8	0.59
35 S	13.8	20.1	34.6	31.6	0.51
45 S	13.2	20.6	34.4	31.7	0.51
55 S	12.7	20.5	34.7	31.9	0.50
Novikoff hepatoma (*141*)					
28 S	11.5	23.3	32.8	32.3	0.53
35 S	12.1	22.7	32.8	32.3	0.53
45 S	12.1	21.3	33.5	33.1	0.50
55 S	12.0	21.3	33.4	33.2	0.50
Normal liver (*138*)					
28 S	20.5	17.2	36.3	26.0	0.61
35 S	20.3	17.1	39.1	23.5	0.60
45 S	19.3	16.6	39.4	24.7	0.56
>45 S	20.3	16.6	37.1	26.0	0.58

[a] Nucleotide composition of newly synthesized RNA was determined by the distribution of ^{32}P in the four nucleotides.

* E. J. Hidvegi, A. W. Prestayko, and H. Busch, unpublished data.

B. Oligonucleotide "Frequencies" of Nucleolar RNA

These "frequencies" have been determined on either whole nuclear 35 S and 45 S precursor RNA or "nucleolar ribosomal precursor RNA" independently, by Roberts and D'Ari (*167*), by Jeanteur *et al.* (*109*), and in the authors' laboratory.

The oligonucleotide "frequencies" are usually determined by analysis of radioactivity distributions in the small oligonucleotides produced by pancreatic RNase (Table 5.14). After a defined period of labeling, usually with ^{32}P orthophosphate, the RNA is purified, hydrolyzed, and

TABLE 5.14

Radioactivity Distribution in Mono- and Oligonucleotides of Ribosomal, Nucleolar, and Nuclear RNA after Labeling with ^{32}P-Orthophosphate[a]

	RNA species							
	Ehrlich cells (*167*)		Nucleolar Novikoff ascites (*226a*)			Ehrlich cells (*167*)		
Sequence	35 S Nuclear RNA	45 S Nuclear RNA	28 S nRNA	35 S nRNA	45 S nRNA	18 S rRNA	28 S rRNA	18 S + 28 S rRNA
C	26.4	25.5	28.7	29.0	27.7	19.0	23.2	21.8
ψ	0.5	0.5	0.8	0.7	0.8	1.2	0.8	0.9
U	14.1	14.1	13.6	13.2	13.7	15.0	12.7	13.5
AC	4.2	4.2	4.2	3.9	4.4	6.0	4.5	5.0
GC	13.0	12.6	14.4	14.6	15.0	9.0	12.3	11.2
AU	2.7	2.7	2.9	2.7	2.9	5.5	3.1	3.9
GU	9.9	10.0	9.3	9.4	9.8	6.0	9.2	8.1
AAC	1.2	1.3	1.8	1.7	1.7	2.8	1.4	1.9
(AG)C	4.7	4.6	5.0	5.5	4.9	6.4	5.9	6.1
GGC, A_3U	6.5	6.4	3.4	3.7	4.4	5.8	6.7	6.4
AAU	0.9	1.0	1.4	1.5	1.3	2.2	1.1	1.5
(AG)U	3.7	3.6	2.7	3.4	3.6	6.8	4.2	5.1
GGU	4.8	5.0	3.1	2.7	2.6	3.7	3.9	3.9
A_3C	0.6	0.5	1.4	1.3	0.9	0.6	0.8	0.7
$(A_2G)C$	2.0	2.0	2.4	2.6	1.7	2.2	2.7	2.5
$(AG_2)C$	1.8	2.0	1.4	1.6	1.5	2.4	2.1	2.2
$(A_2G)U$	1.6	1.5	1.6	1.4	1.6	2.8	1.9	2.2
$(AG_2)U$	2.1	2.1	1.4	1.0	1.0	2.3	2.5	2.4

[a] Figures give the percentage of the total radioactivity from each species found in a given sequence. Each number represents an average of at least five determinations on separate samples of RNA. Calculated from the 18 S and 28 S columns; these values are for equal molar amounts of 18 S and 28 S RNA and assume that the molecular weight of 28 S RNA is twice that of 18 S RNA. Average of the standard deviations calculated for the spots in each of the four RNA species.

the oligonucleotides are subjected to a "finger-printing procedure" (*167*). The percentage of the total radioactivity recovered in any given oligonucleotide is a measure of the frequency of its occurrence in the total RNA molecule. This method depends upon the random formation of nucleotide precursors of RNA and insertion of the nucleotides into their specific positions in the RNA from pools that are adequately equilibrated. The results obtained by all three laboratories are in reasonably good (if not perfect) agreement, and establish that the oligonucleotides do not differ significantly for 45 S and 35 S nucleolar RNA in the tumor cell types studied (Table 5.14). By contrast with the similarities for the nuclear and nucleolar high molecular weight RNA's, the radioactivity distribution in the 18 S and 28 S rRNA is quite different, particularly with respect to the radioactivity in C, GC, GU, AAU, AGU, U, AU, and A_2GU (Table 5.14).

C. Structural Analysis of Nucleolar High Molecular Weight RNA

In this laboratory some new approaches have been made to the analysis of the structures of nucleolar and ribosomal RNA's. The first of these is shown in Table 5.15 which presents the results of studies on the 5′ terminal nucleotides and the alkali stable (methylated) dinucleotides of nucleolar 45 S, 35 S, and 28 S RNA.

The chain length of 45 S nucleolar RNA (Table 5.15) is approximately 12,800 nucleotides; the chain lengths of 35 S nucleolar RNA and 28 S nucleolar RNA are approximately 7800 and 5700 nucleotides, respectively. The major 5′ terminal in these molecules is the methylated dinucleotide pCmpUp (Table 5.15) which is common to nucleolar 45 S, 35 S, and

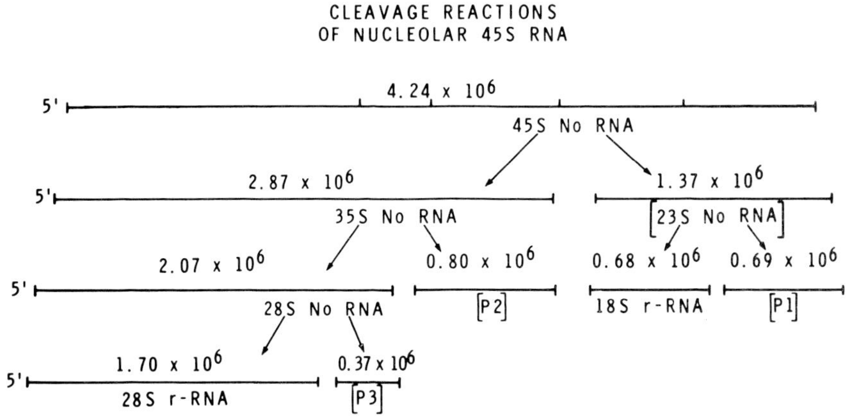

Fig. 5.16. No = nucleolar; r = ribosomal; P1, P2, and P3 = postulated polynucleotide intermediates that have not yet been identified. From (*161b*).

TABLE 5.15

5′ Termini and Alkali Stable Dinucleotides of 45 S Nucleolar RNA and Subcomponents *(51b)*

5′-Terminal	Alkali stable dinucleotides[a]	Mono-nucleotides	Alkali stable dinucleotides	Mono-nucleotides	Alkali stable dinucleotides	Mono-nucleotides
	3 AmpAp		0 AmpAp		3 AmpAp	
	6 AmpUp		1 AmpUp		4 AmpUp	
	12 AmpGp		3 AmpGp		4 AmpGp	
pCmpUp(80%)	+ GmpAp		+ GmpAp		+ GmpAp	
	6 AmpCp		1 AmpCp		3 AmpCp	
	2 UmpAp		0 UmpAp		2 UmpAp	
pAp(10%)	1 UmpUp		0 UmpUp		1 UmpUp	
	4 UmpGp		1 UmpGp		1 UmpGp	
	5 UmpCp	5700 Np	0 UmpCp	2100 Np	2 UmpCp	5000 Np
	+ CmpUp		+ CmpUp		+ CmpUp	
	6 GmpUp		2 GmpUp		2 GmpUp	
	10 GmpGp		3 GmpGp		4 GmpGp	
pCp(10%)	1 GmpCp		1 GmpCp		1 GmpCp	
	2 CmpAp		1 CmpAp		2 CmpAp	
	4 CmpGp		1 CmpGp		1 CmpGp	
	4 CmpCp		2 CmpCp		1 CmpCp	
	Total 66		Total 16		Total 31	

←——— Nucleolar 28 S RNA ———→ (columns 1–3)

←——— Nucleolar 35 S RNA ———→ (columns 1–5)

←——— Nucleolar 45 S RNA ———→ (columns 1–7)

[a] The alkali stable trinucleotide, UmpUmpCp, is not included.

28 S RNA. The presence of this common terminal dinucleotide shows that the nucleolar 28 S RNA is on the 5′ end of the 45 S nucleolar RNA molecule (Fig. 5.16). Recently, pCmpUp was also found to be the 5′ terminal of 28 S rRNA.* It is postulated that in the conversion of nucleolar 45 S RNA to nucleolar 35 S RNA, a 23 S RNA fragment (5000 nucleotides) is released and this fragment is subsequently cleaved to form the 18 S rRNA and the polynucleotide intermediate P1 (Fig. 5.16). The remainder of the 35 S nucleolar RNA molecule undergoes further cleavage (Fig. 5.16) to nucleolar 28 S RNA and a polynucleotide P2.

D. Other Common Fragments of High Molecular Weight Nucleolar RNA

There are other common fragments of nucleolar 45 S, 35 S, and 28 S RNA. In addition to the 5′ terminal pCmpUp, two other common fragments have been found in the nucleolar 45 S, 35 S, and 28 S RNA, i.e., the alkali stable trinucleotide UmpUmpCmp (Table 5.15) and a large oligonucleotide fragment (the B_3 fragment) derived by T_1 RNase digestion. Interestingly, the gel patterns (Fig. 5.17) obtained by partial hydrolysis with T_1 RNase have indicated that the large digestion products produced from nucleolar 45 S, 35 S, and 28 S RNA and 28 S rRNA are virtually indistinguishable (*224*). The fragment referred to as the B_3 fragment was isolated in an electrophoretically homogeneous state (Fig. 5.18). This fragment has an approximate molecular weight of 200,000, a fairly sharp T_m (melting temperature) at 75° and a nucleotide composition: AMP, 4.6; UMP, 14; GMP, 41; and CMP, 41. This fragment is particularly interesting because it suggests that a portion of the nucleolar 45 S, 35 S, and 28 S RNA and 28 S rRNA is highly hydrogen bonded and may be double stranded like some 2 S rRNA fragments recently analyzed (*195a, 206a*). Like the evidence from analysis of alkali resistant oligonucleotides and the 5′ terminal analysis, these data provide further evidence for the presence of common residues in the nucleolar 45 S, 35 S, and 28 S RNA's and 28 S rRNA.

E. Hybridization Studies

Hybridization studies of two types, i.e., saturation studies (Fig. 5.19) and competition studies (Fig. 5.20), have been carried out in this laboratory (*161b*) on nucleolar RNA of Novikoff hepatoma. Saturation analysis provides a basis for comparisons of the percentage of the total nucleolar DNA hybridized with a variety of types of nucleolar and

* In a variety of plants the 3′ termini of 18 S and 28 S rRNA have been reported to be adenosine and uridine respectively (*93b*).

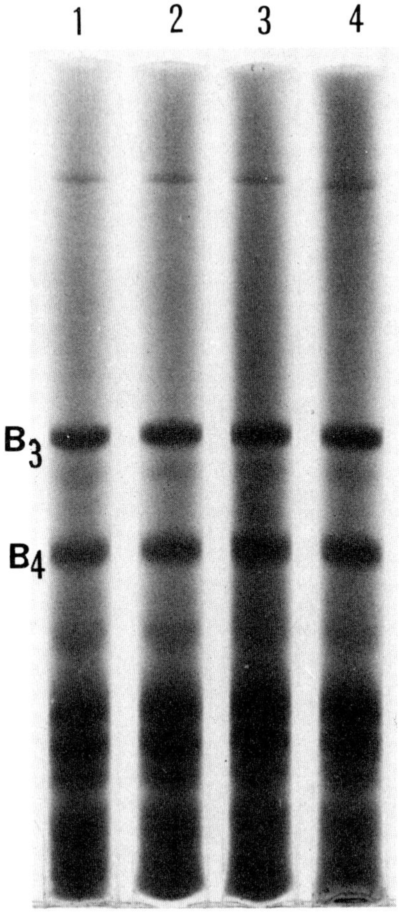

Fig. 5.17. Polyacrylamide gels of digestion products of RNA incubated 30 min at 37° with 20 U T_1 RNase/mg RNA for nucleolar 28 S RNA (slot 1), ribosomal 28 S RNA (slot 2), nucleolar 35 S RNA (slot 3), and nucleolar 45 S RNA (slot 4). From (*224*).

ribosomal RNA's (Table 5.16). Competition experiments were then carried out between labeled 45 S, 65 S, and 85 S nucleolar RNA and other nucleolar and ribosomal RNA's (Fig. 5.20). On the basis of these results and the assumption that the molecular weight of 28 S rRNA is 1.7×10^6 (*127a*), molecular weights for the various fragments were determined as shown in Table 5.16. These results (Table 5.16, Fig. 5.16) provide evidence that 18 S and 28 S rRNA together compose approximately 60% of the total nucleotides of the 45 S nucleolar RNA. They also indicate that 28 S nucleolar RNA is a longer chain than 28 S rRNA (Fig. 5.16). Previously, it had been reported (*198*) that both 18 S and 28 S rRNA inhibited the hybridization of nucleolar 28 S, 35 S, and 45 S RNA.

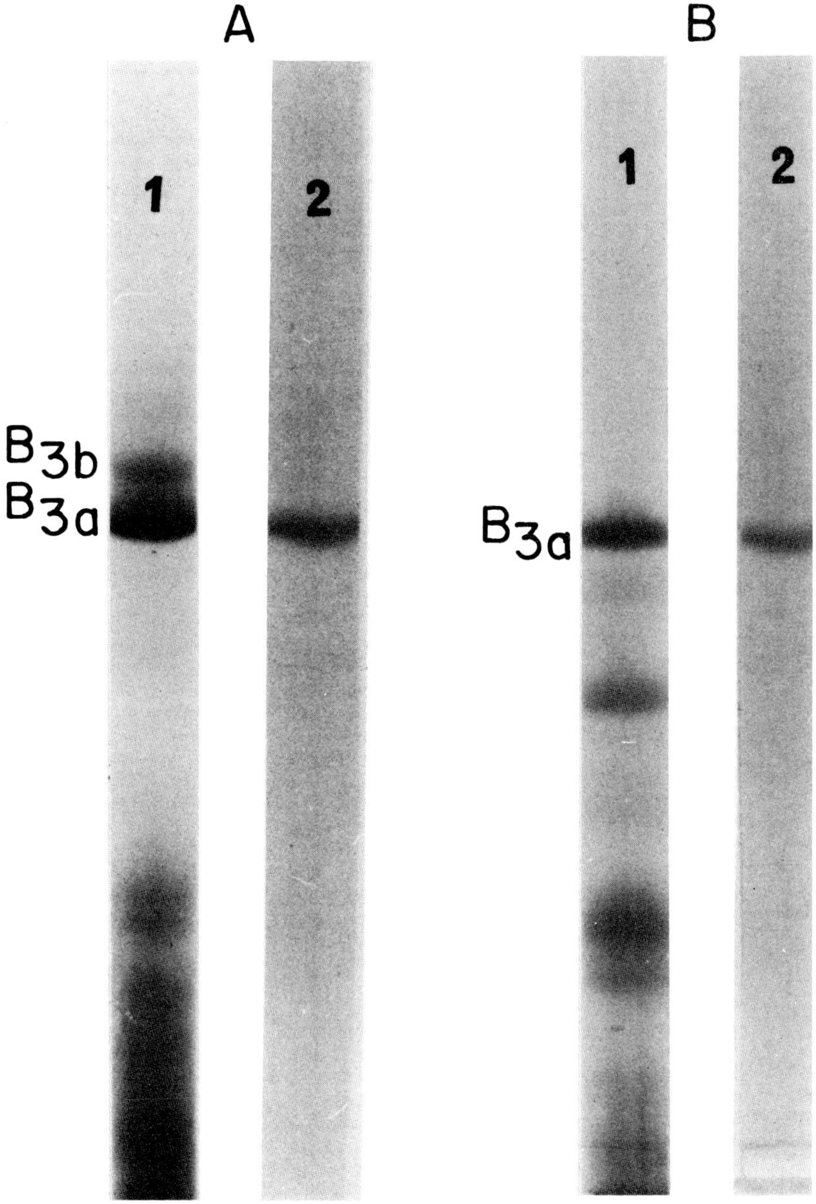

Fig. 5.18. Polyacrylamide gel electrophoresis of digestion products of 28 S rRNA (A_1) after digestion with 20 U T_1 RNase/mg RNA at 4° for 30 min and (B_1) after digestion with 20 U T_1 RNase/mg RNA at 37° for 30 min. The B_3 bands (A_2 and B_2) extracted from the corresponding gels had identical R_f values and were electrophoretically homogeneous. From (*224*).

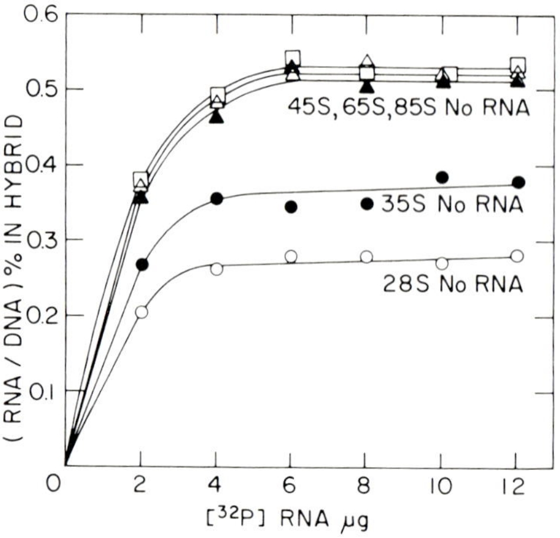

Fig. 5.19. Saturation curves of nucleolar DNA and ^{32}P-labeled nucleolar 28 S, 35 S, 45 S, 65 S, and 85 S RNA; the specific activities of the labeled RNAs were 5.2×10^3, 4.9×10^3, 5.7×10^3, 5.6×10^3 cpm/μg, respectively. No = nucleolar. From (*161b*).

F. The Final Cleavage of Nucleolar RNA to 28 S rRNA and 18 S rRNA

The precise sites of final cleavages for formation of 18 S rRNA and 28 S rRNA is not yet established, but it can be stated with some certainty that this cleavage does not occur in the nucleolus. The data shown in Table

TABLE 5.16

Percentage of the Nucleolar 45 S RNA Contained in Various Nucleolar and Ribosomal RNA's (*161b*)[a]

	I. Saturation analysis		II. Competition		
	% of 45 S Nucleolar RNA	Molecular weight	% of 45 S Nucleolar RNA	Molecular weight	Mean of I, II
18 S rRNA	15.4	0.69×10^6	17	0.67×10^6	0.68×10^6
28 S rRNA	37.5	1.70×10^6	43	1.70×10^6	1.70×10^6
28 S Nucleolar RNA	48.0	2.17×10^6	50	1.97×10^6	2.07×10^6
18 S + 28 S rRNA	53.1	2.40×10^6	62	2.45×10^6	2.42×10^6
35 S Nuclcolar RNA	67.3	3.05×10^6	68[b]	2.70×10^6	2.87×10^6
45 S Nucleolar RNA	100	4.53×10^6	100	3.95×10^6	4.24×10^6

[a] Percentages were determined by the saturation and competition analyses. The conversion of these percentages to molecular weights was made on the assumption that the molecular weight of 28 S rRNA is 1.7×10^6.

[b] This value was obtained indirectly from competition experiments between labeled 35 S nucleolar RNA and unlabeled 28 S nucleolar RNA.

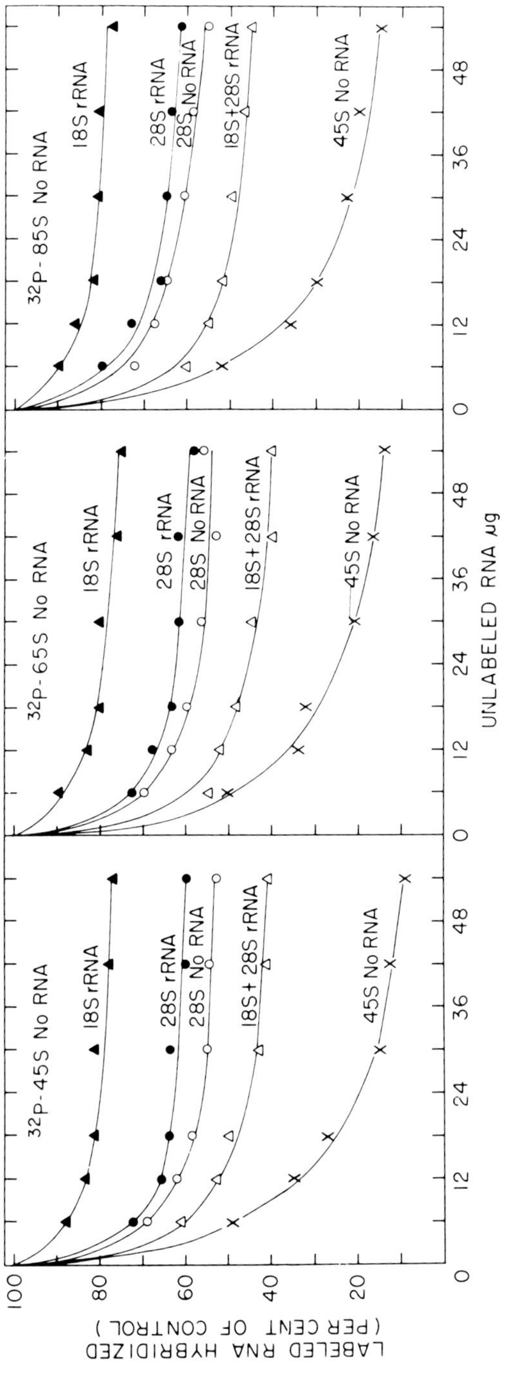

Fig. 5.20. Competition experiments between ^{32}P-labeled 45 S, 65 S, and 85 S nucleolar RNAs and unlabeled 18 S rRNA, 28 S rRNA, 18 S + 28 S rRNA, 28 S nucleolar RNA, and 45 S nucleolar RNA; 6 μg of labeled nucleolar RNA were used for each point. No = nucleolar. From (*161b*).

5.16 indicates that 28 S nucleolar RNA undergoes some further cleavage and as suggested in Fig. 5.16 nucleolar 23 S RNA may well be the source of 18 S rRNA. Whether these final conversions occur in the nuclear ribonuclear protein network (Chapter VIII) or at the nuclear membrane remains to be determined.

G. Other Suggested Mechanisms for Formation of rRNA

Other authors have made a variety of suggestions for the processing of nucleolar RNA to form rRNA as noted below:

1. 45 S nRNA → 18 S rRNA + 28 S rRNA + "X" (*12a, 153–156, 173*)
2. 45 S nRNA → 18 S rRNA + 32 S nRNA (*153–156, 209, 210*)
3. 45 S nRNA → 35 S nRNA → 18 S rRNA + 28 S rRNA (*1b, 109*)
4. 45 S nRNA → 41 S nRNA → 36 S nRNA → $\left.\begin{matrix}\text{28 S rRNA}\\ \text{18 S rRNA}\end{matrix}\right\}$ (*210a, 221*)

Although the basic concept in most of these is in agreement with the formulation presented above, there are some differences in each of these mechanisms. As noted on page 267, the critical problem in completely defining these mechanisms is the lack of specific information on structure. A number of authors have proposed that a large piece of the 45 S nucleolar RNA is not included in 18 S or 28 S rRNA (*1b, 109*). The reason is that there is no explanation at present for the lower content of adenylic acid in the whole nucleolar 45 S RNA; i.e., 15% (Table 5.5) and 20% and 18%, respectively, in 18 S and 28 S rRNA (Table 5.3a). This difference in composition represents an important unsolved difference between these molecules.

Warner and Soeiro (*215a*) pointed out that a 7 S RNA is formed in the nucleolus, possibly as a product of higher molecular weight RNA. However, this fragment does not account for the differences in these molecules (see Chapter VI).

It should also be pointed out that Zimmerman and Holler (*229*) suggested that there was only a conformational difference between 45, 35, and 28 S nRNA, but this seems unlikely at this point.

H. Evidence that the Nucleolus May Produce High Molecular Weight RNA's Other Than rRNA Precursors

Two types of evidence that the nucleolus may produce RNA other than ribosomal precursor RNA have been developed. The first of these emerged from studies in which the Novikoff hepatoma ascites cells are treated with actinomycin D and an RNA was found with the nucleotide

TABLE 5.17

^{32}P-NUCLEOTIDE COMPOSITION OF NEWLY SYNTHESIZED NUCLEOLAR 45 S RNA (*51a*)[a]

	AMP	UMP	GMP	CMP
Control	12	22	34	31
Actinomycin treated	25	25	25	25

[a] Nucleotide composition of newly synthesized 45 S RNA of control and actinomycin D-treated nucleoli.

composition shown in Table 5.17 which is compared with that of the usual nucleolar RNA (*51a*).

In addition, analysis of the oligonucleotide composition of these RNA's has shown that the overall oligonucleotide compositions of the product obtained after actinomycin D treatment differs significantly from that obtained in the control RNA and more importantly from that of the RNA of the chromatin fraction (Table 5.18). Thus, perhaps 10% of the total nRNA may represent RNA that is neither rRNA nor like that of whole nuclei. Such RNA may be messenger RNA linked to the ribosomal products that move to the cytoplasm in the form of polysomes. This possible "proximity effect" of the chromatin adjacent to the nucleolus has been alluded to in Chapters I and IV (p. 10, 165).

XIII. Comparative Analysis of Nucleolar RNA of Tumors and Other Tissues

As indicated in Table 5.6b, the ^{32}P nucleotide compositions of the nucleolar RNA of a variety of tumors differ markedly from those of normal, regenerating, and thioacetamide-treated liver (*120, 121, 136b, 199, 202*) and monkey kidney in tissue culture. To ascertain whether these differences represented total differences in sequences or simply the variations in a relatively few oligonucleotides, the isotope distributions in the oligonucleotides of 28 S, 35 S, and 45 S nRNA's were determined for Novikoff hepatoma, Morris hepatoma, and normal liver (Table 5.19).

There are many similarities in the isotope distribution in the oligonucleotides; i.e., similar values were found for C, U, AC, AU, (AG)U, A_2U, A_3C, $(AG_2)C$, $(A_2G)U$, and $(AG_2)U$. On the other hand, as Table 5.19 indicates, there are greater amounts of isotope in the dinucleotides, GC and GU, of the tumors. These data indicate that the differences in the nucleolar gene readouts of the tumors and normal liver may reside in relatively limited regions of nRNA molecules. Whether in both normal and neoplastic cells a variety of genes are read for nRNA is not yet known and hence it is not clear whether different gene sets are read

TABLE 5.18

PERCENTAGE DISTRIBUTION OF RADIOACTIVITY IN NUCLEOTIDES AND OLIGONUCLEOTIDES OBTAINED BY RIBONUCLEASE DIGESTION OF ^{32}P-LABELED RNA (*51a*)[a]

	Control nucleolar 45 S RNA	Actinomycin D-treated nucleolar 45 S RNA	Extranucleolar chromatin 45 S RNA
C	27.9	19.4	19.0
U	14.8	19.3	22.7
ψ	0.6	0.5	0.4
AC	4.5	12.4	6.9
AU	3.0	3.7	10.3
GC	16.3	12.8	7.3
GU	12.5	11.6	10.3
AAC	1.2	2.2	1.9
AAU	1.9	1.6	2.4
AGC	5.0	4.5	3.9
AGU	3.5	4.1	4.3
GGC + AAAU	3.2	1.7	2.6
GGU	2.3	2.7	2.0
AAAC	0.4	1.2	1.0
AAGC	1.6	1.5	1.7
AAGU	1.4	0.9	2.0
AGGC	0.9	1.2	1.5
AGGU	0.8	0.8	0.5
AC/GC	0.3	1.0	1.0
AU/GU	0.2	0.3	1.0

[a] Percentage distribution of radioactivity in the nucleotides of newly synthesized 45 S RNA (20-min pulse) obtained from control nucleoli, actinomycin D-treated nucleoli, and the extranucleolar chromatin fraction. The oligonucleotides of 4 mg RNA (10,000–100,000 counts/minute 45 S RNA and yeast RNA as carrier), obtained after complete digestion with 166 μg pancreatic ribonuclease at pH 7.1, were separated by electrophoresis at pH 2.7 (0.5% formic acid–ammonium hydroxide) for 18 hours at 6 V/cm. After electrophoresis, chromatography was performed for 38–40 hours with a mixture of *tert*-butanol and electrophoretic buffer (adjusted to pH 3.8) 55:45, v/v. The nucleotide spots were eluted with 0.01 *M* HCl. Radioactivity was determined in an automatic liquid-scintillation spectrophotometer. The values were calculated from the total sum of the oligonucleotides recovered.

in the tumors or simply quantitative differences exist in those read in normal liver from those read in the tumors. Supportive evidence for the concept that there are significant differences in the RNA of the tumors and other tissues has recently come from hybridization and other types of studies (*51′, 51c, 167a*).

TABLE 5.19

RADIOACTIVITY DISTRIBUTION IN MONO- AND OLIGONUCLEOTIDES IN NUCLEOLAR RNA AFTER 6-HOUR LABELING WITH ^{32}P-ORTHOPHOSPHATE[a] (*226a*)

	Normal liver			Morris hepatoma 9618A			Novikoff hepatoma		
Oligo-nucleotides	28 S	35 S	45 S	28 S	35 S	45 S	28 S	35 S	45 S
C	26.7	27.8	27.7	28.8	28.6	28.0	28.7	29.0	27.7
U	13.4	13.1	12.6	14.7	14.2	14.5	13.6	13.2	13.7
Pseudo-U	1.0	0.6	1.3	0.8	0.6	0.6	0.8	0.7	0.8
Total mono	41.1	41.5	41.7	44.3	43.4	43.1	43.1	42.9	42.2
AC	4.3	4.3	4.6	5.2	5.3	5.1	4.2	3.9	4.4
GC	13.2	14.0	13.5	18.3	18.5	19.9	14.4	14.6	15.0
AU	2.5	2.5	3.1	1.9	1.9	1.8	2.9	2.7	2.9
GU	5.8	5.9	5.8	6.3	6.4	6.4	9.3	9.4	9.8
Total di	25.8	26.7	27.0	31.7	32.1	32.4	30.8	30.6	32.1
AAC	2.1	2.5	2.8	2.0	2.0	1.8	1.8	1.7	1.7
(AG)C	6.5	6.5	5.6	7.5	7.2	7.3	5.0	5.5	4.9
GGC/A_3U	5.2	5.0	4.6	2.5	2.2	2.4	3.4	3.7	4.4
AAU	1.8	2.1	1.5	1.5	1.7	1.5	1.4	1.5	1.3
(AG)U	3.8	3.0	3.6	2.2	2.7	2.5	2.7	3.4	3.6
GGU	2.6	2.1	2.7	1.0	1.0	0.9	3.1	2.7	2.6
Total tri	22.0	21.2	20.8	16.7	16.9	16.4	17.4	19.5	18.5
A_3C	1.3	1.5	1.5	1.1	1.4	1.1	1.4	1.3	0.9
(AAG)C	2.8	2.9	2.9	2.6	2.6	3.0	2.4	2.6	1.7
(AGG)C	3.1	2.6	2.0	1.4	1.1	1.1	1.4	1.6	1.5
(AAG)U	2.7	2.2	2.8	1.4	1.5	2.3	1.6	1.4	1.6
(AGG)U	1.2	1.3	1.7	0.6	0.8	0.6	1.4	1.0	1.0
Total tetra	11.2	10.5	10.9	7.1	7.4	8.1	8.2	7.9	6.7

[a] The values for each sequence are the percentages of the total radioactivity from each RNA found in a given sequence. The abbreviations are those used by Roberts and D'Ari (*167*).

FURTHER STRUCTURAL ANALYSIS OF THE NUCLEOLAR RNAs

It is clearly a gigantic task of the future of science to determine the entire linear sequence of the oligonucleotides in nucleolar RNA and the nucleolar RNA products of which rRNA has been highlighted in this chapter (*61, 69b, 207*). Little progress has been made in such structural analysis and yet, the need for such analysis is plain, in view of the differences between tumors and other tissues and also because of the uncertainties about the precise differences between nucleolar RNA, 18 S and 28 S rRNA.*

Another difficult problem in carrying out sequential studies on nucleolar RNA's is that 28 S nRNA (mol wt ca. 1.5×10^6) contains approximately 5000 nucleotide residues. None of the methods employed for studies on the viral RNA's of approximately this size have yet provided a satisfactory sequence. Much better systems for analysis of fragments are required to answer such questions as the mechanism of cleavage of nRNA's, the presence of spacers, the relationships of nRNA and rRNA, the nature of the messengers in the nucleolar product, and the difference between nRNA's of tumors and other tissues.

XIV. Regulation of Nucleolar Function

Although little is established mechanistically about the controls of the synthesis of nucleolar RNA, it is obvious from the discussion on nucleolar organizers (Chapter III) and nucleolar DNA (Chapter IV) that there is a dynamic balance in the cell between cell requirement, nucleolar synthesis, and responsiveness of the nucleolar organizer regions to changes in cell need. Some of the best examples of these responses are seen in connection with the effects of a variety of drugs on the nucleolar products.

Increased interest in these control mechanisms developed with the findings of Shiokawa and Yamana (*178–180*) that there are apparently factors in the cytoplasm that specifically inhibit the synthesis of ribosomes and that synthesis of ribosomes can be continued if uridine and cytidine are added back to their culture system.

Their inhibitor is of interest in view of its heat stability and the fact that it is dialyzable. It was found only in the midblastulae cells and not in neurula cells in which there is rapid ribosomal RNA synthesis. The precise nature of the compounds is yet to be defined but for the present two types of material seem to be present, one in the soluble and the other in the residue fraction of the 15,000 *g* centrifugate.

Although the mechanism of the inhibition of ribosomal RNA synthesis is not yet clear, it is apparently reversible. The inhibitor occurs in highest concentrations in the cells that have the lowest activity in the synthesis of ribosomes, so that it is possible that the inhibitor represents a physiologically active principle. In mammalian cells, it seems clear that there

* It would be most desirable if pure RNA's were available from nucleoli, but thus far purification methods have not provided completely satisfactory products. It would seem from the effects of actinomycin and other effects that the nucleolar product may be 90% pure rRNA precursor and satisfactory sequence analysis might be approached even in the presence of a 10% contamination by other RNA's.

are functional regulatory mechanisms, but thus far, there is no information on these mechanisms.*

XV. Summary

In the course of evolution a variety of types of mechanisms appear to have been developed for the synthesis of high molecular weight ribosomal RNA's which vary markedly in composition from species to species. In higher organisms, including mammalian cells, a huge precursor is apparently produced which is then broken down to a long RNA chain with a sedimentation coefficient of approximately 45 S. This fragment undergoes a series of cleavage reactions which are not yet completely understood because all the products and their functions have not been defined. Among these products are nucleolar 28 S RNA and 23 S RNA which probably undergo other conversion reactions in the nuclear ribonucleoprotein network (Chapter VIII) to the final 18 S and 28 S rRNA. Similar mechanisms appear to operate in other species including *Xenopus laevis* (*124'*) and in *Chironomus* (*162a*). Currently studies are proceeding on the structural analysis of nucleolar 45 S RNA and its products which are of particular interest because of the differences demonstrated between these molecules in normal liver and a number of types of tumors. Evidence that nucleoli can produce other types of high molecular weight RNA has been obtained but the functions of these RNA's are not known at present.

REFERENCES

1. Altmann, H., Dolejs, I., and Fetter, F., The separation of nucleic acids on a modified methylated albumin-kieselguhr (MAK) column. *Anal. Biochem.* **21**, 477–480 (1967).

1a. Amaldi, F., Non-random variability in evolution of base compositions of ribosomal RNA. *Nature* **221**, 95–96 (1969).

1b. Amaldi, F., and Attardi, G., Partial sequence analysis of ribosomal RNA from HeLa cells. I. Oligonucleotide pattern of 28 S and 18 S RNA after pancreatic ribonucleae digestion. *J. Mol. Biol.* **32**, 737–755 (1968).

2. Amano, M., Metabolism of RNA in the liver cells of the rat. I. Isolation and chemical composition of nucleus, nucleolus, chromatin, nuclear sap and cytoplasm. *Exptl. Cell Res.* **46**, 169–179 (1967).

3. Amano, M., Metabolism of RNA in the liver cells of the rat. II. Chemical analysis of RNA in cellular components. *Exptl. Cell Res.* **46**, 180–190 (1967).

* Although the nucleolus is clearly the site of synthesis of 18 S and 28 S rRNA, present evidence indicates that the other species of ribosomal RNA, i.e., 5 S rRNA is probably synthesized extranucleolarly (*32–34, 159a*). Possibly the new technique of John *et al.* (*109a*) which permits intracellular hybridization of RNA to DNA will define more satisfactorily the localization of the genes for synthesis of 5 S rRNA.

4. Amano, M., and Leblond, C. P., Comparison of the specific activity time curves of ribonucleic acid in chromatin, nucleolus and cytoplasm. *Exptl. Cell Res.* **20**, 250–253 (1960).
5. Amano, M., Leblond, C. P., and Nadler, N. J., Radioautographic analysis of nuclear RNA in mouse cells revealing three pools with different turnover times. *Exptl. Cell Res.* **38**, 314–340 (1965).
5a. Apgar, J., Holley, R. W., and Merrill, S. H., Purification of the alanine-, valine-, histidine-, and tyrosine-acceptor ribonucleic acids from yeast. *J. Biol. Chem.* **237**, 796–802 (1962).
6. Artman, M., Silman, N., and Engelberg, H., Pulse labelled RNA associated with ribosomes in *Escherichia coli*. *Nature* **213**, 39–42 (1967).
7. Attardi, G., Huang, P. C., and Kabat, S., Recognition of ribosomal RNA sites in DNA. I. Analysis of the *E. coli* system. *Proc. Natl. Acad. Sci. U. S.* **53**, 1490–1498 (1965).
8. Attardi, G., Parnas, H., Huang, M.-I. H., and Attardi, B., Giant size rapidly labeled nuclear ribonucleic acid and cytoplasmic messenger ribonucleic acid in immature duck erythrocytes. *J. Mol. Biol.* **20**, 145–182 (1966).
9. Bal, A. K., and Gross, P., Asynchronous synthesis of RNA in nucleoli of root meristem. *Science* **143**, 808–810 (1964).
10. Bannasch, P., and Thoenes, W., Zum problem der nucleolaren stoffabgabe. Elektronenmikroskpische untersuchungen am pankreas der weissen maus. *Z. Zellforsch. Mikroskop. Anat.* **67**, 674–692 (1965).
11. Barondes, S. H., Dingman, C. W., and Sporn, M. B., *In vitro* stimulation of amino-acid incorporation into protein by liver nuclear RNA. *Nature* **196**, 145–147 (1962).
12. Bellamy, A. R., and Joklik, W. K., Studies on the A-rich RNA of reovirus. *Proc. Natl. Acad. Sci. U. S.* **58**, 1389–1395 (1967).
12a. Birnstiel, M., The nucleolus in cell metabolism. *Ann. Rev. Plant Physiol.* **18**, 25–68 (1967).
12b. Birnstiel, M. L., Chipchase, M.-I. H., and Hayes, R. J., Incorporation of L-(^{14}C) leucine by isolated nuclei. *Biochim. Biophys. Acta* **55**, 728–733 (1962).
13. Birnstiel, M. L., Chipchase, M.-I. H., and Hyde, B. B., The nucleolus, a source of ribosomes. *Biochim. Biophys. Acta* **76**, 454–462 (1963).
13a. Birnstiel, M. L., and Hyde, B. B., Protein synthesis by isolated pea nucleoli. *J. Cell Biol.* **18**, 41–50 (1963).
14. Birnstiel, M. L., Rho, J. H., and Chipchase, M.-I. H., Fractionation of isolated pea nuclei. *Biochim. Biophys. Acta* **55**, 734–740 (1962).
15. Birnstiel, M. L., Wallace, H., Sirlin, J. L., and Fischberg, M., Localization of the ribosomal DNA complements in the nucleolar organizer region of *Xenopus laevis*. *Natl. Cancer Inst. Monograph* **23**, 431–448 (1966).
16. Bogoroch, R., and Siegel, B. V., Some metabolic properties of the nucleolus as demonstrated by recent radioisotope experiments. *Acta Anat.* **45**, 265–287 (1961).
17. Brachet, J., La détection histochimique des acides pentosenucléiques. *Compt. Rend. Soc. Biol.* **133**, 88–89 (1940).
18. Brachet, J., La localisation des acides pentosenucléiques dans les tissus animaux et les oeufs d'amphibiens en voie de développement. *Arch. Biol. (Liege)* **53**, 207–257 (1942).
19. Brachet, J., Biochemical and physiological interrelations between nucleus and cytoplasm during early development. *Growth* **11**, 309–324 (1947).

20. Brachet, J., The role of the cell nucleus in oxidations and phosphorylations. *Biochim. Biophys. Acta* **9**, 221–222 (1952).
21. Brachet, J., "Biochemical Cytology." Academic Press, New York, 1957.
22. Brachet, J., Ribonucleic acids and the synthesis of cellular proteins. *Nature* **186**, 194–199 (1960).
23. Brachet, J., "The Biological Role of Ribonucleic Acids." Elsevier, Amsterdam, 1960.
24. Brachet, J., Nucleocytoplasmic interactions in unicellular organisms. *In* "The Cell" (J. Brachet and A. E. Mirsky, eds.), Vol. 2, pp. 696–745. Academic Press, New York, 1961.
25. Brachet, J., Chantrenne, H., and Vanderhaeghe, F., Recherches sur les interactions biochimiques entre le noyau et le cytoplasme chez les organismes unicellulaires. *Biochim. Biophys. Acta* **18**, 544–563 (1955).
26. Brakke, M. K., Photometric scanning of centrifuged density gradient columns. *Anal. Biochem.* **5**, 271–283 (1963).
27. Brakke, M. K., and Daly, J. M., Density gradient centrifugation: Non-ideal sedimentation and the interaction of major and minor components. *Science* **148**, 387–389 (1965).
28. Brawerman, G., Gold, L., and Eisenstadt, J., A ribonucleic acid fraction from rat liver with template activity. *Proc. Natl. Acad. Sci. U. S.* **50**, 630–638 (1963).
29. Brentani, M., Brentani, R., and Raw, I., Role of nucleolar ribonucleic acid in incorporation of ribosomal amino-acid. *Nature* **201**, 1130 (1964).
30. Brentani, R., Brentani, M., and Raw, I., Messenger activity of purified RNA from rat liver nuclei. *Nature* **214**, 1122–1123 (1967).
31. Brown, D. D., The nucleolus and synthesis of ribosomal RNA during oogenesis and embryogenesis of *Xenopus laevis*. *Natl. Cancer Inst. Monograph* **23**, 297–310 (1966).
32. Brown, D. D., and Gurdon, J. B., Absence of ribosomal RNA synthesis in the anucleolate mutant of *Xenopus laevis*. *Proc. Natl. Acad. Sci. U. S.* **51**, 139–146 (1964).
33. Brown, D. D., and Littna, E., The synthesis and accumulation of low molecular weight RNA during embryogenesis of *Xenopus laevis*. *J. Mol. Biol.* **20**, 95–112 (1966).
34. Brown, D. D., and Weber, C. S., Gene linkage by RNA-DNA hybridization. II. Arrangement of the redundant gene sequences for 28 S and 18 S ribosomal RNA. *J. Mol. Biol.* **34**, 681–697 (1968).
35. Brown, F., and Martin, S. J., Base composition of ribosomal ribonucleic acid fractions from five membrane tissue-culture strains. *Biochem. J.* **97**, 20C–22C (1965).
36. Brownlee, G. G., and Sanger, F., Nucleotide sequences from the low molecular weight ribosomal RNA of *Escherichia coli*. *J. Mol. Biol.* **23**, 337–353 (1967).
37. Brownlee, G. G., Sanger, F., and Barrell, B. G., The sequence of 5 S ribosomal ribonucleic acid. *J. Mol. Biol.* **34**, 379–412 (1968).
38. Busch, H., "An Introduction to the Biochemistry of the Cancer Cell." Academic Press, New York, 1962.
39. Busch, H., Isolation and purification of nuclei. *Methods Enzymol.* **12**, Part A, Sect. III, 421–464 (1967).
40. Busch, H., Arendell, J., Higashi, K., Nakamura, T., Neogy, R., Schwartz, S., and Smith, S., RNA of nucleoli of Walker tumor and liver cells. *2nd Intern.*

Symp. Cell Chem., Ohtsu, Japan, 1966 pp. 73–90. Pergamon Press, Oxford, 1967.

41. Busch, H., Byvoet, P., and Smetana, K., The nucleolus of the cancer cell: A review. *Cancer Res.* **23**, 313–339 (1963).
42. Busch, H., Desjardins, R., Grogan, D., Higashi, K., Jacob, S. T., Muramatsu, M., Ro, T. S., and Steele, W. J., Composition of nucleoli isolated from mammalian cells. *Natl. Cancer Inst. Monograph.* **23**, 193–212 (1966).
43. Busch, H., Hodnett, J. L., Morris, H. P., Neogy, R., Smetana, K., and Unuma, T., ^{32}P nucleotide composition of nuclear and nucleolar RNA of Morris hepatomas. *Cancer Res.* **28**, 672–683 (1968).
44. Busch, H., Muramatsu, M., Adams, H. R., Smetana, K., Steele, W. J., and Liau, M. C., Isolation of nucleoli. *Exptl. Cell Res.* Suppl. 9, 150–163 (1963).
45. Carneiro, J., and Mota, M. A. G., Time sequence of leucine-H^3 incorporation into cytoplasm, chromatin and nucleolus during protein deficiency in the rat. *Pathol. Biol. Semaine Hop.* [N.S.] **12**, 594–601 (1964).
46. Caspersson, T. O., "Cell Growth and Cell Function. A Cytochemical Study." Norton, New York, 1950.
47. Caspersson, T. O., Farber, S., Foley, G. E., and Killander, D., Cytochemical observations on the nucleolus-ribosome system. Effects of actinomycin D and nitrogen mustard. *Exptl. Cell Res.* **32**, 529–552 (1963).
48. Caspersson, T. O., and Schultz, J., Ribonucleic acids in both nucleus and cytoplasm and the function of the nucleolus. *Proc. Natl. Acad. Sci. U. S.* **26**, 507–515 (1940).
49. Chakravorty, A. K., and Busch, H., Effect of thioacetamide on the activity of alkaline ribonuclease and ribonuclease inhibitor in rat liver. *Biochem. Pharmacol.* **16**, 1711–1718 (1967).
50. Chambon, P., Weill, J. D., and Mandel, P., Nicotinamide mononucleotide activation of a new DNA-dependent polyadenylic acid synthesizing nuclear enzyme. *Biochem. Biophys. Res. Commun.* **11**, 39–43 (1963).
51. Chauveau, J., Moule, Y., and Rouiller, C., Isolation of pure and unaltered liver nuclei, morphology and biochemical composition. *Exptl. Cell Res.* **11**, 317–321 (1956).

51′. Chiarugi, V. P., Changes of nuclear RNA in hepatomas as revealed by RNA DNA hybridization. *Biochim. Biophys. Acta* **179**, 129–135 (1969).

51a. Choi, Y. C., and Busch, H., Effects of actinomycin D on the oligonucleotide composition of nucleolar 45 S RNA of Novikoff hepatoma ascites cells. *Biochim. Biophys. Acta* **174**, 766–769 (1969).

51b. Choi, Y. C., and Busch, H., Structural analysis of nucleolar precursors of ribosomal RNA: Studies on the 5′-terminal and alkali resistant dinucleotides of nucleolar high molecular weight RNA. *J. Biol. Chem.* **245**, April (1970).

51c. Church, R. B., Luther, S. W., and McCarthy, B. J., RNA synthesis in taper hepatoma and mouse-liver cells. *Biochim. Biophys. Acta* **190**, 30–37 (1969).

52. Click, R. E., and Hackett, D. P., Evidence that the two ribosomal RNA species of plant tissue are synthesized on different genetic loci. *J. Mol. Biol.* **16**, 279–284 (1966).
53. Comb, D. G., and Katz, S., Studies on the biosynthesis and methylation on transfer RNA. *J. Mol. Biol.* **8**, 790–800 (1964).
54. Comb, D. G., and Sarkar, N., The binding of 5 S ribosomal ribonucleic acid to ribosomal subunits. *J. Mol. Biol.* **25**, 317–330 (1967).

54a. Cunningham, D. D., and Steiner, D. F., Extraction of RNA polymerase from rat liver nuclei in a soluble form. *Biochim. Biophys. Acta* **145**, 834–836 (1967).

55. Das, N. K., Chromosomal and nucleolar RNA synthesis in root tips during mitosis. *Science* **140**, 1231–1233 (1963).

56. Das, N. K., and Alfert, M., Nucleolar RNA synthesis during mitotic and meiotic prophase. *Natl. Cancer Inst. Monograph.* **23**, 337–352 (1966).
57. Das, N. K., Luykx, P., and Alfert, M., The nucleolus and RNA metabolism in Urechis eggs. *Develop. Biol.* **12**, 72–78 (1965).
58. De, P., Chatterjee, R., and Mitra, S., Nucleolar succinic dehydrogenase in mouse mammary carcinoma cells. *Nature* **192**, 285–286 (1961).
59. Desjardins, R., Smetana, K., and Busch, H., A large scale procedure for isolation of nucleoli from rat liver. *Exptl. Cell Res.* **40**, 127–137 (1965).
60. Desjardins, R., Smetana, K., Steele, W. J., and Busch, H., Isolation of nucleoli of the Walker tumor and liver following nuclear disruption in a French pressure cell. *Cancer Res.* **23**, 1819–1823 (1963).
61. Diemer, J., McLennan, B., and Lane, B. G., Studies of the primary structure of 18 S + 28 S ribonucleates. *Biochim. Biophys. Acta* **114**, 191–194 (1966).
62. DiGirolamo, A., Henshaw, E. C., and Hiatt, H. H., Messenger ribonucleic acid in rat liver nuclei and cytoplasm. *J. Mol. Biol.* **8**, 479–488 (1964).
63. DiMarco, A., Silvestrini, R., Dasdia, T., and DiMarco, S., Nucleolar and chromosomal origin of nuclear RNA and its passage to cytoplasm. *Riv. Istochim.* **11**, 211–220 (1965).
64. Drews, J., and Brawerman, G., Alterations in the nature of ribonucleic acid synthesized in rat liver during regeneration and after cortisol administration. *J. Biol. Chem.* **242**, 801–808 (1967).
64a. Edström, J.-E., Extraction, hydrolysis and electrophoretic analysis of ribonucleic acid from microscopic tissue units (microphoresis). *J. Biophys. Biochem. Cytol.* **8**, 39–43 (1960).
65. Edström, J.-E., Chromosomal RNA and other nuclear RNA fractions. *In* "The Role of Chromosomes in Development" (M. Locke, ed.), pp. 137–152. Academic Press, New York, 1964.
66. Edström, J.-E., The base composition of RNA from nucleoli formed at two different organizers in Chironomus tentans. *Biochem. Biophys. Res. Commun.* **18**, 341–344 (1965).
67. Edström, J.-E., and Gall, J. G., The base composition of ribonucleic acid in lampbrush chromosomes, nucleoli, nuclear sap and cytoplasm of Triturus oocytes. *J. Cell Biol.* **19**, 279–284 (1963).
67a. Edström, J.-E., Grampp, W., and Schor, N., The intracellular distribution and heterogeneity of ribonucleic acid in starfish oocytes. *J. Biophys. Biochem. Cytol.* **11**, 549–557 (1961).
68. Elson, D., Trent, L. W., and Chargaff, E., The nucleotide composition of pentose nucleic acids in different cellular fractions. *Biochim. Biophys. Acta* **17**, 362–366 (1955).
69. Errera, M., Hell, A., and Perry, R. P., The role of the nucleolus in ribonucleic acid and protein synthesis. II. Amino acid incorporation into normal and nucleolar inactivated HeLa cells. *Biochim. Biophys. Acta* **49**, 58–63 (1961).
69a. Feinendegen, L. E., Bond, V. P., Shreeve, W. W., and Painter, R. B., RNA and DNA metabolism in human tissue culture cells studied with tritiated cytidine. *Exptl. Cell Res.* **19**, 443–459 (1960).
69b. Fellner, P., and Sanger, F., Sequence analysis of specific areas of the 16 S and 23 S ribosomal RNAs. *Nature* **219**, 236–238 (1968).
70. Firket, H., and Granboulan, P., Electron microscope autoradiographs of tritiated uridine incorporation. *J. Roy. Microscop. Soc.* [3] **81**, 227–228 (1963).
71. Forget, B. G., and Weissman, S. M., Low molecular weight RNA components from KB cells. *Nature* **213**, 878–882 (1967).

72. Fry, M., and Artman, M., Sedimentation behaviour of rapidly labeled RNA from *Escherichia coli. Nature* **217**, 661–664 (1968).
73. Fujioka, M., Koga, M., and Lieberman, I., Metabolism of ribonucleic acid after partial hepatectomy. *J. Biol. Chem.* **238**, 3401–3406 (1963).
73a. Fujisawa, T., and Muramatsu, M., Studies on the sedimentation properties of nuclear, nucleolar and extranucleolar nuclear RNA. *Biochim. Biophys. Acta* **169**, 175–187 (1968).
73b. Futai, M., Mizuno, D., and Sugimuro, T., Hydrolysis of the polymer formed from NAD with rat liver phosphodiesterase yielding nucleoside 5′mono-phosphate. *Biochem. Biophys. Res. Commun.* **28**, 395–399 (1967).
74. Gall, J. G., Uptake of nucleic acid precursors by oocytes of the newt. *Records Genet. Soc. Am.* **28**, 72 (1959).
75. Gavosto, F., Pileri, A., Pegoraro, L., and Gabutti, W., Richerche *in vivo* sul metabolismo dello RNA negli elementi di leucemia acuta umana. *Tumori* **51**, 419–431 (1965).
76. Georgiev, G. P., Ribonucleic acid of the nucleolo-chromosomal apparatus. *Biokhimiya* **26**, 1095–1107 (1961).
77. Georgiev, G. P., The nucleolus and ribosomal RNA synthesis in mammalian cells. *Natl. Cancer Inst. Monograph* **23**, 513–526 (1966).
78. Georgiev, G. P., Samarina, O. P., Lerman, M. I., Smirnov, M. N., and Seretzov, A. N., Biosynthesis of messenger and ribosomal ribonucleic acids in the nucleolochromosomal apparatus of animal cells. *Nature* **200**, 1291–1294 (1963).
79. Georgiev, G. P., Samarina, O. P., Mantieva, V. L., and Zbarsky, I. B., Further studies on the nuclear ribonucleoproteins. *Biochim. Biophys. Acta* **46**, 399–400 (1961).
80. Geuskens, M., Accumulation nucleolaire d'acide ribonucleique (RNA) dan l'oocyte d'asterie. *Exptl. Cell Res.* **30**, 322–330 (1962).
81. Gilbert, W., Polypeptide synthesis in Escherichia coli. II. The polypeptide chain and s-RNA. *J. Mol. Biol.* **6**, 389–403 (1963).
82. Gillespie, D., and Spiegelman, S., A quantitative assay for DNA-RNA hybrids with DNA immobilized on a membrane. *J. Mol. Biol.* **12**, 829–842 (1965).
83. Giudice, G., and Mutola, V., Synthesis of ribosomal RNA during Sea-Urchin development. *Biochim. Biophys. Acta* **138**, 276–285 (1967).
84. Goldstein, L., Cailleau, R., and Crocker, T. T., Nuclear-cytoplasmic relationships in human cells in tissue culture. II. The microscopic behavior of enucleate human cell fragments. *Exptl. Cell Res.* **19**, 332–342 (1960).
85. Goldstein, L., and Eastwood, J. M., On the primary site of nuclear RNA synthesis. A retraction. *J. Cell Biol.* **31**, 195–199 (1966).
86. Goldstein, L., and Micou, J., On the primary site of nuclear RNA synthesis. *J. Biophys. Biochem. Cytol.* **6**, 301–304 (1959).
87. Goldstein, L., and Micou, J., Nuclear-cytoplasmic relationships in human cells in tissue culture. III. Autoradiographic study of interrelation of nuclear and cytoplasmic ribonucleic acid. *J. Biophys. Biochem. Cytol.* **6**, 1–6 (1959).
88. Goldstein, L., Micou, J., and Crocker, T. T., Nuclear-cytoplasmic relationships in human cells in tissue culture. IV. A study of some aspects of nucleic acid and protein metabolism in enucleate cells. *Biochim. Biophys. Acta* **45**, 82–86 (1960).
89. Goldstein, L., and Prescott, D. M., Proteins in nucleocytoplasmic interactions. II. Turnover and changes in nuclear protein distribution with time and growth. *J. Cell Biol.* **36**, 53–61 (1968).

90. Granboulan, N., and Granboulan, P., Cytochimie ultrastructurale du nucléole. I. Mise en évidence de chromatine à l'intérieur du nucléole. *Exptl. Cell Res.* **34**, 71–87 (1964).
91. Granboulan, N., and Granboulan, P., Cytochemie ultrastructurale du nucléole. II. Etude des sites de synthèse du RNA dans le nucléole et le noyau. *Exptl. Cell Res.* **38**, 604–619 (1965).
91a. Granboulan, N., and Scherrer, K., Visualisation in the electron microscope and size of RNA from animal cells. *European J. Biochem.* **9**, 1–20 (1969).
92. Greenberg, H., and Penman, S., Methylation and processing of ribosomal RNA in HeLa cells. *J. Mol. Biol.* **21**, 527–535 (1966).
93. Greenstein, J. P., "Biochemistry of Cancer," 2nd rev. ed. Academic Press, New York, 1954.
93a. Hadjivassiliou, A., and Brawerman, G., Template and ribosomal ribonucleic acid components in the nucleus and the cytoplasm of rat liver. *Biochemistry* **6**, 1934–1940 (1967).
93b. Hallöin, J. M., Pollard, C. J., Becker, V. E., and Fowlks, E. R., The 5′-linked termini of plant ribosomal ribonucleic acids. *Biochemistry,* **8**, 1940–1947 (1969).
94. Harel, J., Harel, L., Lacour, F., Boer, A., and Imbenotte, J., Fractions with differing base composition in RNA from malignant cells of mouse. *J. Mol. Biol.* **7**, 645–651 (1963).
95. Harel, J., Harel, L., Lacour, F., Boer, A., and Imbenotte, J., *In* "Cellular Control Mechanisms and Cancer" (P. Emmelot and O. Mühlbock, eds.), p. 37. Elsevier, Amsterdam, 1964.
96. Harel, L., Harel, J., Boer, A., Imbenotte, J., and Carpeni, N., Persistance d'une synthèse de D-RNA dans le foie de rat traité par l'actinomycine. *Biochim. Biophys. Acta* **87**, 212–218 (1964).
97. Harris, H., Turnover of nuclear and cytoplasmic ribonucleic acid in two types of animal cell, with some further observations on the nucleolus. *Biochem. J.* **73**, 362–369 (1959).
98. Harris, H., Nuclear ribonucleic acids. *Progr. Nucleic Acid Res.* **2**, 19–59 (1963).
99. Harris, H., The ribonucleic acids in the nucleus and cytoplasm of animal cells. *Endeavour* **24**, 50–56 (1965).
100. Harris, H., Fisher, H. W., Rodgers, A., Spencer, T., and Watts, J. B., An examination of the ribonucleic acids in the HeLa cell with special reference to current theory about the transfer of information from nucleus to cytoplasm. *Proc. Roy. Soc.* **B157**, 177–198 (1963).
100a. Hasegawa, S., Fujimura, S., Shimizu, Y., and Sugimura, T., The polymerization of adenosine 5′-diphosphate ribose moiety of NAD by nuclear enzyme II. Properties of the reaction product. *Biochim. Biophys. Acta* **149**, 369–376 (1967).
101. Hiatt, H. H., A rapidly labeled RNA in rat liver nuclei. *J. Mol. Biol.* **5**, 217–229 (1962).
102. Hiatt, H. H., Henshaw, E., and DiGirolamo, A., Messenger ribonucleic acid and protein synthesis in rat liver. *Natl. Cancer Inst. Monograph* **13**, 131–139 (1963).
103. Higashi, K., Adams, H. R., and Busch, H., Base composition of RNA fractions of nuclei of Walker tumor isolated with the citric acid procedure. *Cancer Res.* **26**, 2196–2201 (1966).

103a. Higashi, K., and Busch, H., Stability of nucleolar 45 S RNA. *Biochim. Biophys. Acta* **138,** 76–79 (1967).
104. Higashi, K., Shankaranarayan, K., Adams, H. R., and Busch, H., Utilization of the citric acid procedure and zonal ultracentrifugation for mass isolation of nuclear RNA from Walker 256 carcinosarcoma. *Cancer Res.* **26,** 1582–1590 (1966).
105. Holley, R. W., Apgar, J., Doctor, B. P., Farrow, J., Marini, M. A., and Merrill, S. H., A simplified procedure for the preparation of tyrosine and valine-acceptor fractions of yeast "soluble ribonucleic acid." *J. Biol. Chem.* **236,** 200–202 (1961).
105a. Houssais, J.-F., and Attardi, G., High molecular weight nonribosomal-type nuclear RNA and cytoplasmic messenger RNA in HeLa cells. *Proc. Natl. Acad. Sci. U. S.* **56,** 616–623 (1966).
106. Ittel, M. E., Wintzerith, M., and Mandel, P., Localisation des synthèses des RNA nucléaires du foie de rat. *Bull. Soc. Chim. Biol.* **49,** 13–23 (1967).
107. Jacob, S. T., Steele, W. J., and Busch, H., Turnover of 45 S RNA of regenerating liver and Walker tumor. *Cancer Res.* **27,** 52–60 (1967).
108. Jain, H. K., Chromosomal control of nucleolar synthesis. *Experientia* **20,** 156–157 (1964).
109. Jeanteur, P., Amaldi, F., and Attardi, G., Partial sequence analysis of ribosomal RNA from HeLa cells. II. Evidence for nonribosomal-type sequences in 45 S and 32 S ribosomal RNA precursors. *J. Mol. Biol.* **33,** 757–775 (1968).
109a. John, H. A., Birnstiel, M. L., and Jones, K. W., RNA-DNA hybrids at the cytological level. *Nature* **223,** 582–587 (1969).
110. Jones, K. W., The role of the nucleolus in the formation of ribosomes. *J. Ultrastruct. Res.* **13,** 257–262 (1965).
111. Josse, J., Kaiser, A. D., and Kornberg, A., Enzymatic synthesis of deoxyribonucleic acid. VIII. Frequencies of nearest neighbor base sequence in deoxyribonucleic acid. *J. Biol. Chem.* **236,** 864–875 (1961).
112. Journey, L. J., and Goldstein, M. N., Electron microscope studies on HeLa cell lines sensitive and resistant to actinomycin D. *Cancer Res.* **21,** 929–932 (1961).
113. Kasten, F. H., Strasser, F. F., and Turner, M., Nucleolar and cytoplasmic ribonucleic acid inhibition by excess thymidine. *Nature* **207,** 161–164 (1965).
114. Kidson, C., and Kirby, K. S., Fraction of DNA. I. Countercurrent distribution of native and degraded DNA. *Biochim. Biophys. Acta* **36,** 624–627 (1963).
115. Kidson, C., Kirby, K. S., and Ralph, R. K., Isolation characteristics of rapidly labeled RNA from normal rat liver. *J. Mol. Biol.* **7,** 312–315 (1963).
116. Kirby, K. S., Isolation and characterization of ribosomal ribonucleic acid. *Biochem. J.* **96,** 266–269 (1965).
117. Kirby, K. S., Deoxyribonucleic acids and cancer. *Methods Cancer Res.* 3, 1–26 (1967).
118. Kirby, K. S., Hastings, J. R. B., and O'Sullivan, M. A., Countercurrent distribution of ribonucleic acids. IV. *Biochim. Biophys. Acta* **61,** 978–979 (1962).
119. Kitazume, Y., Ycas, M., and Vincent, W. S., Metabolic properties of a ribonucleic acid fraction in yeast. *Proc. Natl. Acad. Sci. U. S.* **48,** 265–282 (1962).
120. Kleinfeld, R. G., Altered patterns of RNA metabolism in liver cells following partial hepatectomy and thioacetamide treatment. *Natl. Cancer Inst. Monograph.* **23,** 369–378 (1966).
121. Kleinfeld, R. G., and von Haam, E., Effect of thioacetamide on rat liver regeneration. II. Nuclear RNA in mitosis. *J. Biophys. Biochem. Cytol.* **6,** 393–398 (1959).

122. Krsmanovic, V., Kanazir, D., and Errera, M., A DNA-pulse-labelled RNA complex from HeLa cell nuclei. *Biochim. Biophys. Acta* **130,** 544–548 (1965).
123. LaCour, L. F., and Crawley, J. W. C., The site of rapidly labeled ribonucleic acid in nucleoli. *Chromosoma* **16,** 124–132 (1965).
124. Lamar, C., Jr., Munger, W. L., and Pitot, H. C., Studies on a 32 S component of nuclear RNA. *Arch. Biochem. Biophys.* **119,** 98–104 (1967).
124′. Landesman, R., and Gross, P. R., Patterns of macromolecule synthesis during development of *Xenopus laevis* II. Identification of the 40 S precursor to ribosomal RNA. *Develop. Biology* **19,** 244–260 (1969).
124a. Lane, B. G., and Tamaoki, T., Methylated bases and sugars in 16 S and 28 S RNA from L cells. *Biochim. Biophys. Acta* **179,** 332–340 (1969).
125. Leblond, C. P., and Amano, M., Symposium: Synthetic processes in the cell nucleus. IV. Synthetic activity in the nucleolus as compared to that in the rest of the cell. *J. Histochem. Cytochem.* **10,** 162–174 (1962).
126. Lerner, A. M., Bell, E., and Darnell, J. E., Ribosomal RNA in the developing chick embryo. *Science* **141,** 1187–1188 (1963).
126a. Lettré, R., and Siebs, W., Film sur des études du nucléole de cellules cultivées *in vitro. Chemotherapia* **2,** 223–225 (1961).
127. Liau, M. C., Hnilica, L. S., and Hurlbert, R. B., Regulation of RNA synthesis in isolated nucleoli by histones and nucleolar proteins. *Proc. Natl. Acad. Sci. U. S.* **53,** 626–633 (1965).
127a. Loening, U. E., Molecular weights of ribosomal RNA in relation to evolution. *J. Mol. Biol.* **38,** 355–365 (1968).
128. Love, R., Improved staining of the nucleoproteins of the nucleolus. *J. Histochem. Cytochem.* **10,** 227 (1962).
129. Maggio, R., Siekevitz, P., and Palade, G. E., Studies on isolated nuclei. II. Isolation and chemical characterization of nucleolar and nucleoplasmic subfractions. *J. Cell Biol.* **18,** 292–313 (1963).
130. Marinozzi, V., and Bernhard, W., Présence dans le nucléole de deux types de ribonucléoprotéines morphologiquement distinctes. *Exptl. Cell Res.* **32,** 595–598 (1963).
130a. McMaster-Kaye, R., Symposium: Synthetic processes in the cell nucleus. III. The metabolism of nuclear ribonucleic acid in salivary glands of *Drosophila repleta. J. Histochem. Cytol.* **10,** 154–161 (1961).
131. Montagnier, L., and Bellamy, A. D., 18 S and 30 S fractions from the RNA of Krebs-2 ascites cells with differing base composition. *Biochim. Biophys. Acta* **80,** 157–160 (1964).
132. Morrison, W. W., and McCluer, R. H., The temperature-dependent extraction of two rapidly-labeled, DNA-like RNA fractions from rat liver. *Biochim. Biophys. Acta* **138,** 649–651 (1967).
133. Munro, A. J., Structural ribonucleic acid of rat liver ribosomes. *Biochem. J.* **91,** 21C–22C (1964).
134. Muramatsu, M., and Busch, H., Studies on nucleolar RNA of the Walker 256 carcinosarcoma and the liver of the rat. *Cancer Res.* **24,** 1028–1034 (1964).
135. Muramatsu, M., and Busch, H., Studies on the nuclear and nucleolar ribonucleic acid of regenerating rat liver. *J. Biol. Chem.* **240,** 3960–3966 (1965).
136. Muramatsu, M., and Busch, H., Isolation, composition and function of nucleoli of tumors and other tissues. *Methods Cancer Res.* **4,** 303–359 (1967).
136a. Muramatsu, M., and Fujusawa, T., Methylation of ribosomal RNA precursor and tRNA in rat liver. *Biochim. Biophys. Acta* **157,** 476–492 (1968).

136b. Muramatsu, M., Higashinakagawa, T., Ono, T., and Sugano, H., The nuclear RNA of minimal deviation hepatomas. *Cancer Res.* **28,** 1126–1137 (1968).

137. Muramatsu, M., Hodnett, J. L., and Busch, H., Studies on the "Independence" of nucleolar ribonucleic acid synthesis. *Biochim. Biophys. Acta* **91,** 592–597 (1964).

138. Muramatsu, M., Hodnett, J. L., and Busch, H., Base composition of fractions of nuclear and nucleolar ribonucleic acid obtained by sedimentation and chromatography. *J. Biol. Chem.* **241,** 1544–1550 (1966).

139. Muramatsu, M., Hodnett, L., Steele, W. J., and Busch, H., Molecular properties of RNA of isolated nucleoli of rat liver. I. Synthesis of 28 S RNA in the nucleolus. *Biochim. Biophys. Acta* **123,** 116–125 (1966).

140. Muramatsu, M., Smetana, K., and Busch, H., Quantitative aspects of isolation of nucleoli of the Walker carcinosarcoma and liver of the rat. *Cancer Res.* **23,** 510–518 (1963).

140a. Nakamura, T., Prestayko, A. W., and Busch, H., Studies on nucleolar 4 to 6 S ribonucleic acid of Novikoff hepatoma cells. *J. Biol. Chem.* **243,** 1368–1375 (1968).

141. Nakamura, T., Rapp, F., and Busch, H., Common features of the base composition of rapidly labeled RNA of nucleoli in a number of experimental tumors. *Cancer Res.* **27,** 1084–1091 (1967).

142. Nishizuka, Y., Ueda, K., Nakazawa, K., and Hayaishi, O., Studies on the polymer of adenosine diphosphate ribose. I. Enzymic formation from nicotinamide adenine dinucleotide in mammalian nuclei. *J. Biol. Chem.* **242,** 3164–3171 (1967).

143. Nishizuka, Y., Ueda, K., Nakazawa, K., Reeder, R. H., Honjo, T., and Hayaishi, O., Poly adenosine diphosphate ribose synthesis and nicotinamide adenine dinucleotide transglycosidases. *J. Vitaminol.* (*Kyoto*) **14,** 143–152 (1968).

144. Nowell, P. C., Morris, H. P., and Potter, V. R., Chromosomes of "minimal deviation" hepatomas and some other transplantable rat tumors. *Cancer Res.* **27,** 1565–1579 (1967).

145. Oehlert, W., Autoradiographische untersuchungen zur ribonukleinsaure-synthese in den verschiedenen strukturen der zelle. *Beitr. Pathol. Anat. Allgem. Pathol.* **124,** 311–350 (1961).

145a. Okamura, N., and Busch, H., Base composition of high molecular weight nuclear RNA of the Walker tumor and liver of the rat. *Cancer Res.* **25,** 693–697 (1965).

146. Osawa, S., Takata, K., and Hotta, Y., Nuclear and cytoplasmic ribonucleic acids of calf thymus. *Biochim. Biophys. Acta* **28,** 271–277 (1958).

147. Page, E., and Quastel, J. H., The nucleolus. *In* "Neurochemistry" (K. A. C. Elliott, I. H. Page, and J. H. Quastel, eds.), p. 337, Thomas, Springfield, Illinois, 1962.

148. Penman, S., Scherrer, K., Becker, Y., and Darnell, J. E., Polysomes in normal polio virus-infected HeLa cells and their relationship to messenger-RNA. *Proc. Natl. Acad. Sci. U. S.* **49,** 654–661 (1963).

149. Penman, S., Smith, I., and Holtzman, E., Ribosomal RNA synthesis and processing in a particulate site in the HeLa cell nucleus. *Science* **154,** 786–789 (1966).

150. Penman, S., Smith, I., Holtzman, E., and Greenberg, H., RNA metabolism in the HeLa cell nucleus and nucleolus. *Natl. Cancer Inst. Monograph* **23,** 489–512 (1966).

151. Perry, R. P., On the nucleolar and nuclear dependence of cytoplasmic RNA synthesis in HeLa cells. *Exptl. Cell Res.* **20**, 216–220 (1960).
152. Perry, R. P., The cellular sites of synthesis of ribosomal and 4 S RNA. *Proc. Natl. Acad. Sci. U. S.* **48**, 2179–2186 (1962).
153. Perry, R. P., Selective effects of actinomycin D on the intracellular distribution of RNA synthesis in tissue culture cells. *Exptl. Cell Res.* **29**, 400–406 (1963).
154. Perry, R. P., Role of the nucleolus in ribonucleic acid metabolism and other cellular processes. *Natl. Cancer Inst. Monograph.* **14**, 73–89 (1964).
155. Perry, R. P., The nucleolus and the synthesis of ribosomes. *Natl. Cancer Inst. Monograph.* **18**, 325–340 (1965).
156. Perry, R. P., On ribosome biogenesis. *Natl. Cancer Inst. Monograph* **23**, 527–546 (1966).
157. Perry, R. P., and Errera, M., The influence of nucleolar ribonucleic acid metabolism on that of the nucleus and cytoplasm. *In* "The Cell Nucleus" (J. S. Mitchell, ed.), pp. 24–29. Butterworth, London and Washington, D. C., 1960.
158. Perry, R. P., Errera, M., Hell, A., and Durwald, H., Kinetics of nucleoside incorporation into nuclear and cytoplasmic RNA. *J. Biophys. Biochem. Cytol.* **11**, 1–13 (1961).
159. Perry, R. P., Hell, A., and Errera, M., The role of the nucleolus in ribonucleic acid and protein synthesis. I. Incorporation of cytidine into normal and nucleolar inactivated HeLa cells. *Biochim. Biophys. Acta* **49**, 47–57 (1961).
159a. Perry, R. P., and Kelley, D. E., Persistent Synthesis of 5 S RNA when Production of 28 S and 18 S Ribosomal RNA is Inhibited by Low Doses of Actinomycin D. *J. Cell Physiol.* **72**, 235–246 (1968).
160. Perry, R. P., Srinivasan, P. R., and Kelley, D. E., Hybridization of rapidly labeled nuclear ribonucleic acids. *Science* **145**, 504–507 (1964).
161. Prescott, D. M., Nuclear function and nuclear-cytoplasmic interactions. *Physiol. Rev.* **40**, 17–44 (1960).
161a. Prescott, D. M., and Bender, M. A., Synthesis of RNA and protein during mitosis in mammalian tissue culture cells. *Exptl. Cell Res.* **25**, 260–268 (1962).
161b. Quagliarotti, J., Hidvegi, E. J., Wikman, J., and Busch, H., Structural analysis of nucleolar precursors of ribosomal RNA: Comparative hybridizations of nucleolar and ribosomal RNA with nucleolar DNA. *J. Biol. Chem.* **245**, April (1970).
162. Quagliarotti, G., and Ritossa, F., On the arrangement of genes for 28 S and 18 S ribosomal RNA's in *Drosophila melanogaster. J. Mol. Biol.* **36**, 57–69 (1968).
162a. Ringborg, U., personal communication.
163. Ro, T. S., and Busch, H., *In vitro* labeling of RNA in isolated nucleoli of the Walker tumor and liver. *Cancer Res.* **24**, 1630–1633 (1964).
164. Ro, T. S., and Busch, H., Studies on RNA polymerase of nuclear fractions of regenerating liver. *Biochim. Biophys. Acta* **134**, 184–187 (1967).
165. Ro, T. S., Muramatsu, M., and Busch, H., Labeling of RNA of isolated nucleoli with UTP ^{14}C. *Biochem. Biophys. Res. Commun.* **14**, 149–155 (1964).
166. Ro, T. S., Narayan, K. S., and Busch, H., Effects of actinomycin D on base composition and nearest neighbor frequency of nucleolar RNA of the Walker tumor and liver. *Cancer Res.* **26**, 780–785 (1966).
167. Roberts, W. K., and D'Ari, L., Base sequence differences between the ribo-

somal and "ribosomal precursor" ribonucleic acids from Ehrlich ascites cells. *Biochemistry* **7**, 592–600 (1968).

167a. Roche, J. G., Rosenau, W., and Goldberg, M. L., Differences in nuclear ribonucleic acid of human neoplastic and normal tissue. *Proc. Soc. Exptl. Biol. Med.* **131**, 465–468 (1969).

168. Rosset, R., Julien, J., and Monier, R., Ribonucleic acid composition of bacteria as a function of growth rate. *J. Mol. Biol.* **18**, 308–320 (1966).

169. Roth, J. S., Some observations on the assay and properties of ribonucleases in normal and tumor tissues. *Methods Cancer Res.* **3**, 153–242 (1968).

170. Salzman, N. P., Shatkin, A. J., and Sebring, E. D., The synthesis of a DNA-like RNA in the cytoplasm of HeLa cells infected with vaccinia virus. *J. Mol. Biol.* **8**, 405–416 (1964).

171. Samarina, O. P., Incorporation of labeled amino acids into protein fractions of the cell nuclei of liver and Ehrlich ascitic carcinoma. *Biokhimiya* **26**, 61–69 (1961).

172. Scherrer, K., and Darnell, J. E., Sedimentation characteristics of rapidly labeled RNA from HeLa cells. *Biochem. Biophys. Res. Commun.* **7**, 486–490 (1962).

173. Scherrer, K., Latham, H., and Darnell, J. E., Demonstration of an unstable RNA and of a precursor to ribosomal RNA in HeLa cells. *Proc. Natl. Acad. Sci. U. S.* **49**, 240–248 (1963).

174. Schultze, B., Oehlert, W., and Maurer, W., Uber eine allgemeine beziehung zwischen der umsatzrate der ribonukleinsaure und des eiweisses im organismus von maus und ratte. *Biochim. Biophys. Acta* **49**, 35–46 (1961).

175. Semmel, M., and Huppert, J., Interaction *in vitro* entre l'actinomycine D et le RNA ribosomal. *Biochim. Biophys. Acta* **103**, 702–704 (1965).

175a. Shankaranarayan, K., Muramatsu, M., Smetana, K., and Busch, H., Ultrastructural studies on RNA and DNA components of isolated nucleoli of Walker 256 carcinosarcoma. *Exptl. Cell Res.* **41**, 81–98 (1966).

176. Shankaranarayan, K., Steele, W. J., and Busch, H., Evidence that the granular and fibrillar elements of nucleoli contain 28 S and 6 S RNA, respectively. *Exptl. Cell Res.* **43**, 483–492 (1966).

176a. Sharma, O. K., Hidvegi, E. J., Marks, F., Prestayko, A. W., Smetana, K., and Busch, H., Isolation and characterization of 60 S and 85 S RNA from nuclear fractions of Novikoff hepatoma cells. *Physiol. Chem. Physics* **1**, 185–209 (1969).

177. Shearer, R. W., and McCarthy, B. J., Evidence for ribonucleic acid molecules restricted to the cell nucleus. *Biochemistry* **6**, 283–289 (1967).

178. Shiokawa, K., and Yamana, K., Pattern of RNA synthesis in isolated cells of *Xenopus laevis* embryos. *Develop. Biol.* **16**, 368–388 (1967).

179. Shiokawa, K., and Yamana, K., Inhibitor of ribosomal RNA synthesis in *Xenopus laevis* embryos. *Develop. Biol.* **16**, 389–406 (1967).

180. Shiokawa, K., and Yamana, K., Ribonucleic acid (RNA) synthesis in dissociated embryonic cells of *Xenopus laevis*. III. Effects of uridine and cytidine on inhibition of ribosomal RNA synthesis by a cytoplasmic factor. *Proc. Japan Acad.* **43**, 148–151 (1967).

181. Sibatani, A., deKloet, S. R., Allfrey, V. G., and Mirsky, A. E., Isolation of a nuclear RNA fraction resembling DNA in its base composition. *Proc. Natl. Acad. Sci. U. S.* **48**, 471–477 (1962).

181a. Sibatani, A., Yamana, K., Kimura, K., and Okagaki, H., Fractionation with phenol of ribonucleic acid of animal cells. *Biochim. Biophys. Acta* **33**, 590–591 (1959).

181b. Sibatani, A., Yamana, K., Kimura, K., and Takahashi, T., Fractionation of two metabolically distinct classes of ribonucleic acids in animal cells and its bearings on cancer. *Nature* **186**, 215–217 (1960).

182. Siebert, G., Enzymes of cancer nuclei. *Exptl. Cell Res.* Suppl. 9, 389–418 (1963).

183. Siebert, G., Villalobos, J., Ro, T. S., Steele, W. J., Lindenmayer, G., Adams, H., and Busch, H., Enzymatic studies on isolated nucleoli of rat liver. *J. Biol. Chem.* **241**, 71–78 (1966).

184. Sirlin, J. L., The nucleolus of the cell nucleus. *Endeavour* **20**, 146–153 (1961).

185. Sirlin, J. L., and Jacob, J., Cell function in the ovary of drosophila. II. Behaviour of RNA. *Exptl. Cell Res.* **20**, 283–293 (1960).

186. Sirlin, J. L., and Jacob, J., Function, development and evolution of the nucleolus. *Nature* **195**, 114–117 (1962).

187. Sirlin, J. L., and Jacob, J., Sequential and reversible inhibition of synthesis of ribonucleic acid in the nucleolus and chromosomes: Effect of Benzamide and substituted Benzimidazoles on dipteran salivary glands. *Nature* **204**, 545–547 (1964).

188. Sirlin, J. L., Jacob, J., and Birnstiel, M. L., Synthesis of different species of nucleolar ribonucleic acid. *Biochim. Biophys. Acta* **108**, 716–718 (1965).

189. Sirlin, J. L., Jacob, J., and Birnstiel, M. L., Synthesis of transfer RNA in the nucleolus of *Smittia*. *Natl. Cancer Inst. Monograph* **23**, 255–270 (1966).

190. Sirlin, J. L., Jacob, J., and Kato, K.-I., The relation of messenger to nucleolar RNA. *Exptl. Cell Res.* **27**, 355–359 (1962).

191. Sirlin, J. L., Jacob, J., and Tandler, C. J., Transfer of the methyl group of methionine to nucleolar ribonucleic acid. *Biochem. J.* **89**, 447–452 (1963).

192. Sirlin, J. L., Kato, K., and Jones, K. W., Synthesis of ribonucleic acid in the nucleolus. *Biochim. Biophys. Acta* **48**, 421–423 (1961).

193. Sirlin, J. L., Tandler, C. J., and Jacob, J., The relationship between the nucleolus organizer and nucleolar RNA. *Exptl. Cell Res.* **31**, 611–615 (1963).

194. Smith, S. J., Higashi, K., and Busch, H., Studies on nucleolar RNA fractions of the Walker and Novikoff tumors. *Cancer Res.* **27**, 849–954 (1967).

195. Soeiro, R., Birnboim, C., and Darnell, J., Rapidly labeled HeLa cell nuclear RNA. II. Base composition and cellular localization of a heterogeneous RNA fraction. *J. Mol. Biol.* **19**, 362–372 (1966).

195a. Spencer, M., Pigram, W. J., and Littlechild, J., Studies on ribosomal RNA structure I. The isolation of crystallizable fragments. *Biochim. Biophys. Acta* **179**, 348–359 (1969).

196. Sporn, M. B., and Dingman, C. W., The fractionation and characterization of nuclear ribonucleic acid from rat liver. *Biochim. Biophys. Acta* **68**, 389–400 (1963).

197. Srinivasan, P. R., Miller-Faurès, A., Brunfaut, M., and Errera, M., Kinetics of pulse-labeling of ribonucleic acid in HeLa cells. *Biochim. Biophys. Acta* **72**, 209–216 (1963).

198. Steele, W. J., Localization of deoxyribonucleic acid complementary to ribosomal ribonucleic acid and preribosomal ribonucleic acid in the nucleolus of rat liver. *J. Biol. Chem.* **243**, 3333–3341 (1968).

199. Steele, W. J., and Busch, H., Increased content of high molecular weight RNA

fraction in nuclei and nucleoli of livers of thioacetamide-treated rats. *Biochim. Biophys. Acta* **119**, 501–509 (1966).

200. Steele, W. J., Okamura, N., and Busch, H., Effects of thioacetamide on the composition and biosynthesis of nuclear and nucleolar RNA. *J. Biol. Chem.* **240**, 1742–1749 (1965).

201. Stewart, G. A., and Farber, E., The rapid acceleration of hepatic nuclear ribonucleic acid breakdown by actinomycin but not by ethionine. *J. Biol. Chem.* **243**, 4479–4485 (1968).

202. Stöcker, E., and Klärner, P., Autoradiograpische untersuchungen über die zelluläre nukleinsaure- und eiweiss-synthese in der niere unter thioacetamid. *Naturwissenschaften* **52**, 498–499 (1965).

202a. Sy, J., and McCarty, K. S., Ribosomal RNA of *Arbacia punctulata*. *Biochim. Biophys. Acta* **166**, 571–574 (1968).

203. Tamaoki, T., The particulate fraction containing 45 S RNA in L cell nuclei. *J. Mol. Biol.* **15**, 624–639 (1966).

203a. Tamaoki, T., and Lane, B. G., Methylation of sugars and bases in ribosomal and rapidly labeled ribonucleates from normal and puromycin-treated L cells. *Biochemistry* **7**, 3431–3440 (1968).

204. Tamaoki, T., and Mueller, G. C., Effect of puromycin on RNA synthesis in HeLa cells. *Biochem. Biophys. Res. Commun.* **11**, 404–410 (1963).

205. Tandler, C. J., Detection and origin of nucleolar components: A model for nucleolar RNA function. *Natl. Cancer Inst. Monograph* **23**, 181–190 (1966).

206. Tandler, C. J., and La Torre, J. L., The removal of uracil residues from RNA in tissue preparations. *Life Sci.* **3**, 1431–1435 (1964).

206a. Thomas, Jr., G. J., and Spencer, M., Studies on ribosomal RNA structure II. Secondary structures in solution of rRNA and crystallizable fragments. *Biochim. Biophys. Acta* **179**, 360–368 (1969).

207. Timasheff, S. N., Witz, J., and Luzzati, V., The structure of high molecular weight ribonucleic acid in solution. A small-angle X-ray scattering study. *Biophys. J.* **1**, 525–537 (1961).

207a. Unuma, T., Arendell, J., and Busch, H., High resolution autoradiographic studies of the uptake of ^{3}H-5-uridine into condensed and dispersed chromatin of nuclei and granular and fibrillar components of nucleoli of Novikoff hepatoma ascites cells. *Exptl. Cell Res.* **52**, 429–438 (1968).

208. Unuma, T., Floyd, L., and Busch, H., Selective removal of perinucleolar nucleolus associated chromatin from isolated nucleoli of livers of thioacetamide-treated rats. *Exptl. Cell Res.* **52**, 101–111 (1968).

209. Vaughan, M. H., Jr., Soeiro, R., Warner, J. R., and Darnell, J. E., Jr., The effects of methionine deprivation of ribosome synthesis in HeLa cells. *Proc. Natl. Acad. Sci. U. S.* **58**, 1527–1534 (1967).

210. Vaughan, M. H., Jr., Warner, J. R., and Darnell, J. E., Jr., Ribosomal precursor particles in the HeLa cell nucleus. *J. Mol. Biol.* **25**, 235–251 (1967).

210a. Vesco, C., and Penman, S., The fractionation of nuclei and the integrity of purified nucleoli in HeLa cells. *Biochim. Biophys. Acta* **169**, 188–195 (1968).

211. Vincent, W. S., and Baltus, E., The ribonucleic acids of nucleoli. *In* "The Cell Nucleus" (J. S. Mitchell, ed.), pp. 18–23. Butterworth, London and Washington, D. C., 1960.

212. Vincent, W. S., Bensam, B., and Bensam, A., Synthesis of ribonucleic acid by nucleoli. *Biol. Bull.* **115**, 342 (1958).

213. Volkin, E., Synthesis and function of the DNA-like RNA. *Federation Proc.* **21,** 112–119 (1962).
213a. Wagner, E. K., Penman, S., and Ingram, V. M., Methylation patterns of HeLa cell ribosomal RNA and its nucleolar precursors. *J. Mol. Biol.* **29,** 371–387 (1967).
214. Wallace, H., and Birnstiel, M. L., Ribosomal cistrons and the nucleolar organizer. *Biochim. Biophys. Acta* **114,** 296–310 (1966).
215. Warner, J. R., The assembly of ribosomes in HeLa cells. *J. Mol. Biol.* **19,** 383–398 (1966).
215a. Warner, J. R., and Soeiro, R., Nascent ribosomes from HeLa cells. *Proc. Natl. Acad. Sci. U. S.* **58,** 1984–1990 (1967).
216. Warner, J. R., and Soeiro, R., The involvement of RNA in protein synthesis. *New Engl. J. Med.* **276,** 563–570 (1967).
217. Warner, J. R., and Soeiro, R., The involvement of RNA in protein synthesis. *New Engl. J. Med.* **276,** 613–617 (1967).
218. Warner, J. R., and Soeiro, R., The involvement of RNA in protein synthesis. *New Engl. J. Med.* **276,** 675–680 (1967).
219. Watson, J. D., and Ralph, R. K., The nature of 7 S RNA in mammalian cells. *J. Mol. Biol.* **22,** 67–78 (1966).
220. Watson, J. D., and Ralph, R. K., Stability of 5 S ribonucleic acid in mammalian cells. *J. Mol. Biol.* **26,** 541–544 (1967).
221. Weinberg, R. A., Loening, U., Willems, M., and Penman, S., Acrylamide gel electrophoresis of HeLa cell nucleolar RNA. *Proc. Natl. Acad. Sci. U. S.* **58,** 1088–1095 (1967).
222. Weiss, S. B., Enzymatic incorporation of ribonucleoside triphosphate into the interpolynucleotide linkages of ribonucleic acid. *Proc. Natl. Acad. Sci. U. S.* **46,** 1020–1030 (1960).
223. Weiss, S. B., Biosynthesis of ribopolynucleotides. *Federation Proc.* **21,** 120–126 (1962).
223a. Weiss, S. B., and Nakamoto, T., The enzymatic synthesis of RNA: Nearest-neighbor base frequencies. *Proc. Natl. Acad. Sci. U. S.* **47,** 140–145 (1961).
224. Wikman, J., Howard, E., and Busch, H., Studies on the primary structure of ribosomal 28 S RNA and its nucleolar precursors. Isolation of an electrophoretically homogeneous fragment rich in guanylic and cytidylic acids. *J. Biol. Chem.* **244,** 5471–5480, 1969.
224a. Willems, M., Wagner, E., Laing, R., and Penman, S., Base composition of ribosomal RNA precursors in the HeLa cell nucleolus: Further evidence of non-conservative processing. *J. Mol. Biol.* **32,** 211–220 (1968).
224b. Woodard, J., Rasch, E., and Swift, H., Nucleic acid and protein metabolism during the mitotic cycle in *Vicia faba*. *J. Biophys. Biochem. Cytol.* **9,** 445–462 (1961).
225. Woods, P. S., RNA in nuclear-cytoplasmic interaction. *Brookhaven Symp. Biol.* **12,** 153–174 (1959).
225a. Woods, P. S., and Taylor, J. H., Studies of ribonucleic acid metabolism with tritium-labeled cytidine. *Lab. Invest.* **8,** 309–318 (1959).
226. Woods, P. S., and Zubay, G., Biochemical and autoradiographic studies of different RNA's: Evidence that transfer RNA is chromosomal in origin. *Proc. Natl. Acad. Sci. U. S.* **54,** 1705–1712 (1965).
226a. Yazdi, E., Ro-Choi, T. S., Wikman, J., Choi, Y. C., and Busch, H., ^{32}P distribu-

tions in oligonucleotides of rapidly sedimenting nucleolar RNA's of hepatomas and normal rat liver. *Cancer Res.* **29,** 1755–1762 (1969).

227. Zalokar, M., Nuclear origin of ribonucleic acid. *Nature* **183,** 1330 (1959).
228. Zalokar, M., Sites of protein and ribonucleic acid synthesis in the cell. *Exptl. Cell Res.* **19,** 559–576 (1960).

228a. Zimmerman, E. F., Secondary methylation of ribosomal ribonucleic acid in HeLa cells. *Biochemistry* **7,** 3156–3164 (1968).

229. Zimmerman, E., and Holler, B., Methylation of 45 S ribosomal RNA precursor in HeLa cells. *J. Mol. Biol.* **23,** 149–161 (1967).
230. Zirkle, R. E., Uretz, R. B., and Haynes, R. H., Disappearance of spindles and pharagmoplasts after microbeam irradiation of cytoplasm. *Ann. N. Y. Acad. Sci.* **90,** 435–439 (1960).

CHAPTER VI

4–7 S RNA of the Nucleolus

I. Introduction

As noted in the sucrose density gradient sedimentation patterns shown in Chapter V, an RNA fraction that is relatively slowly sedimenting is present in nucleolar RNA of both liver and tumors. This low molecular weight RNA is of interest for two reasons: (1) its rate of labeling is slow by comparison with the more rapidly sedimenting nucleolar RNA and (2) its nucleotide composition differs from that of the more rapidly sedimenting RNA. In addition, the nucleolar low molecular weight RNA

TABLE 6.1

NUCLEOTIDE COMPOSITIONS OF 4–7 S RNA

	AMP	UMP	GMP	CMP	$\frac{AMP + UMP}{GMP + CMP}$
Cytoplasmic RNA's					
4 S RNA (*28, 62*)	18	21	32	27	0.66
5 S RNA (*20–22, 52, 62*)	18	23	32	27	0.71
Novikoff nucleolar					
4–7 S RNA (*47*)	19	26	32	22	0.85
Walker nucleolar					
4–7 S RNA (*47*)	19	22	34	25	0.70
Liver nucleolar					
4–7 S RNA (*46*)	20	29	30	21	0.96
Liver nuclear					
4–7 S (*28, 46*)	22	27	31	21	0.95

had a different overall nucleotide composition in the fractions obtained from the tumor and the liver nucleoli (Table 6.1). Accordingly, it was of interest to isolate and fractionate the 4–7 S nucleolar RNA (*28, 45, 46*).

The remarkably successful research programs on the isolation, fractionation, and structural analysis of amino acid acceptor or transfer RNA (s- or t-RNA) and the more recent studies on 5 S RNA represent major contributions of fundamental biochemistry to the understanding of life processes, particularly in the field of protein synthesis (*1, 2, 5, 6, 13, 14, 16–18, 18a, 20–24, 29–32, 36, 39–44, 50, 61, 63–66*). Moreover, these studies have set an important pattern in the use of physical and biochemical methods for analysis of the structure of several transfer RNA's (Fig. 6.1) and 5 S RNA (*4–8, 17, 18*). As noted in Chapter V, in addition to studies on structures of other types of low molecular weight RNA (*7, 8, 10, 11, 37*), current goals include the application of similar techniques to the determination of the structure of the 18 S and 28 S rRNA (*15, 25, 53, 54*) and the higher molecular weight nuclear and nucleolar RNA's (*12, 67*), with particular respect to the relationship of their structures to the orientation of their accompanying proteins which provide the functional mosaics for protein synthesis (*19, 25*). Recent studies have shown that some low molecular weight RNA's are specifically localized to cell fractions, including cytoplasmic membranes (*3, 35*), and others to the nucleus (*28, 33, 36, 40, 49, 52, 56, 62, 66a*) and the nucleolus (*9, 46, 47, 52, 66a*).*

* Some species of 4–7 S RNA's have been isolated from *E. coli* (*27*) and these are presumably being characterized by the group at Cambridge. Approximately 3 S RNA rich in dihydrothymidylic acid has been reported to be present in nuclei of mammalian cells (*11a*); however, these reports have not yet been confirmed.

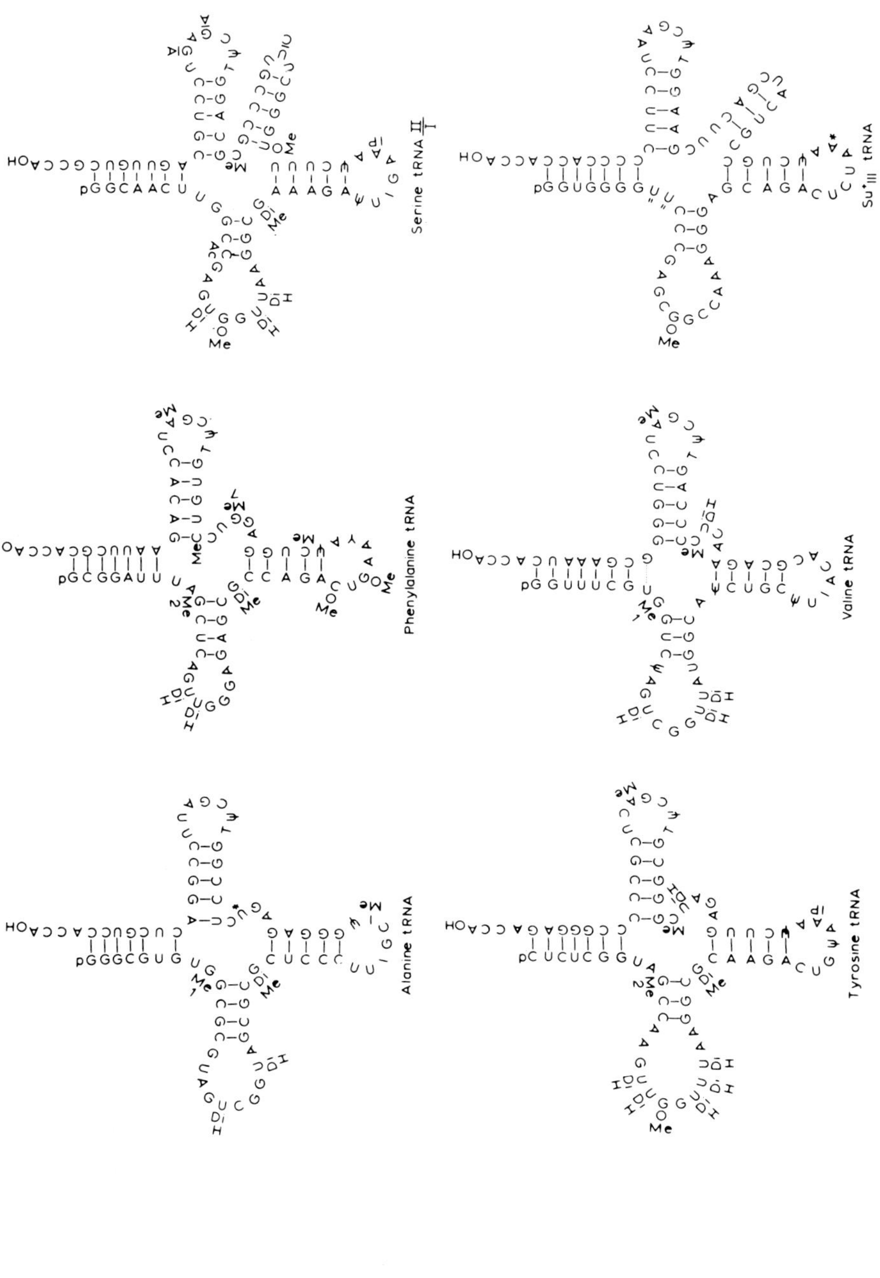

Fig. 6.1. Structures of several natural tRNA's and one suppressor tRNA (Su^+_{III}) (courtesy of Dr. James Madison).

For such studies on nucleotide sequences and conformations, highly purified RNA must be obtained.

With nucleoli, there are critical limitations on the amount of low molecular weight RNA available in the small quantities of the nucleoli usually isolated. The newer methods for mass isolation of nuclei and nucleoli (see Chapter XI) have made it possible to isolate and analyze sufficient RNA to show that a number of subspecies of 4–7 S RNA are localized to nuclei and that some of these are limited in localization to the nucleoli (*47, 52, 66a*).

II. Polyacrylamide Gel Electrophoresis of Nuclear and Nucleolar Low Molecular Weight RNA

One of the most important of the methods currently in use for the study of the various types of nucleolar and nuclear low molecular weight RNA species is polyacrylamide gel electrophoresis (*28, 45, 47, 49, 52, 66a*). This method can be employed either as an analytical procedure or as a larger scale method for isolation of purified electrophoretically homogeneous types of RNA. Figure 6.2 presents the analytical gels for low molecular weight RNA from various fractions of liver cells. Whole nuclear RNA contains the greatest variety of these RNA's of low molecular weight and the cytoplasmic sap contains the smallest number of mobility bands. The fast-moving RNA in the cytoplasmic sap is largely composed of transfer or amino acid acceptor RNA which has the highest mobility of any of the types of low molecular weight RNA in these groups. The low molecular weight RNA of the ribosomal fraction contains at least two bands in the 4 S region, a dark band of 5 S RNA that has a slower migration rate and another dark band that follows the 5 S RNA (Fig. 6.2). There are also fainter bands, one immediately behind the dense 5 S RNA band and others that follow the more dense bands at some distance. Probably, there are only three main types of low molecular weight RNA in the ribosomes; i.e., 4 S transfer RNA (with a trailing band), 5 S RNA (with a trailing band), and a third component which will be referred to as U1a or 7 S (28 S) RNA (*50*).

In the whole nuclear RNA, as obtained either from Novikoff hepatoma, normal liver, or other tissues (*28, 45, 66a*), there is a dense band of 4.5 S RNA between the 4 S and 5 S bands, and a distinct region containing two dense bands referred to as U1b and U1c that follow the U1a band (referred to previously). These closely approximated two bands are followed by a single dense band referred to as U2. The smaller U3 band that follows the U2 band is concentrated in the nucleolus.

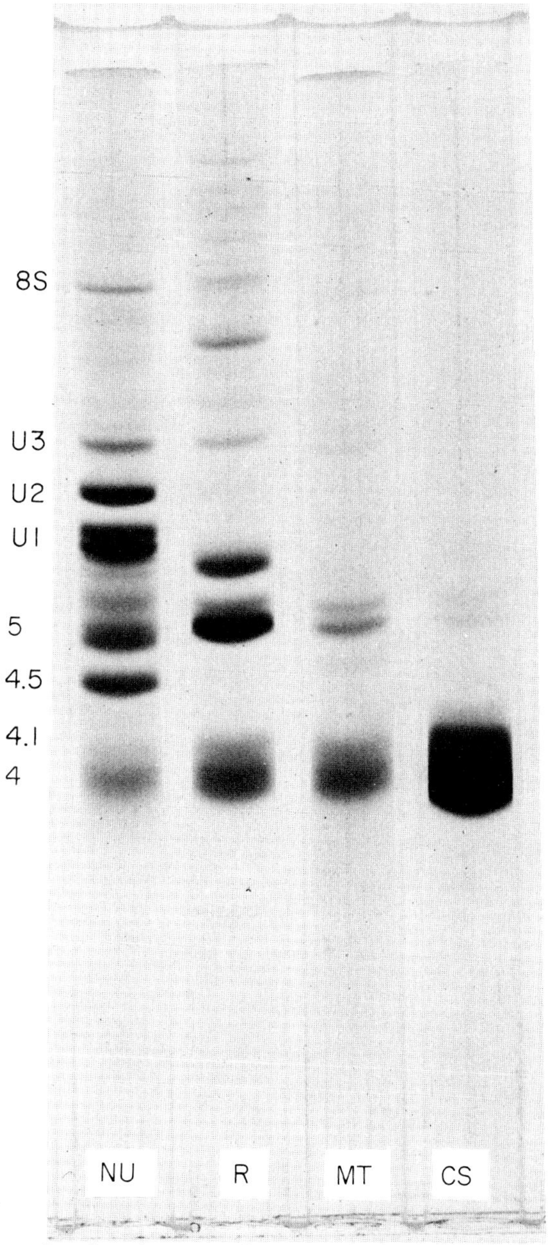

Fig. 6.2. Polyacrylamide gel patterns for nuclear RNA (Nu), ribosomal RNA (R), mitochondrial RNA (MT), and RNA of the cytoplasmic sap (CS). The dense band in the cytoplasmic sap is largely transfer RNA. In the nuclear RNA this is followed by a band of 4.5 S RNA and bands of 5 S RNA (which are also seen in the RNA of the ribosomal and mitochondrial fractions). The U1 RNA is a dense double band, the U2 RNA is a single dense band, and the U3 and 8 S RNA's are much less dense (courtesy of Dr. T. S. Ro-Choi).

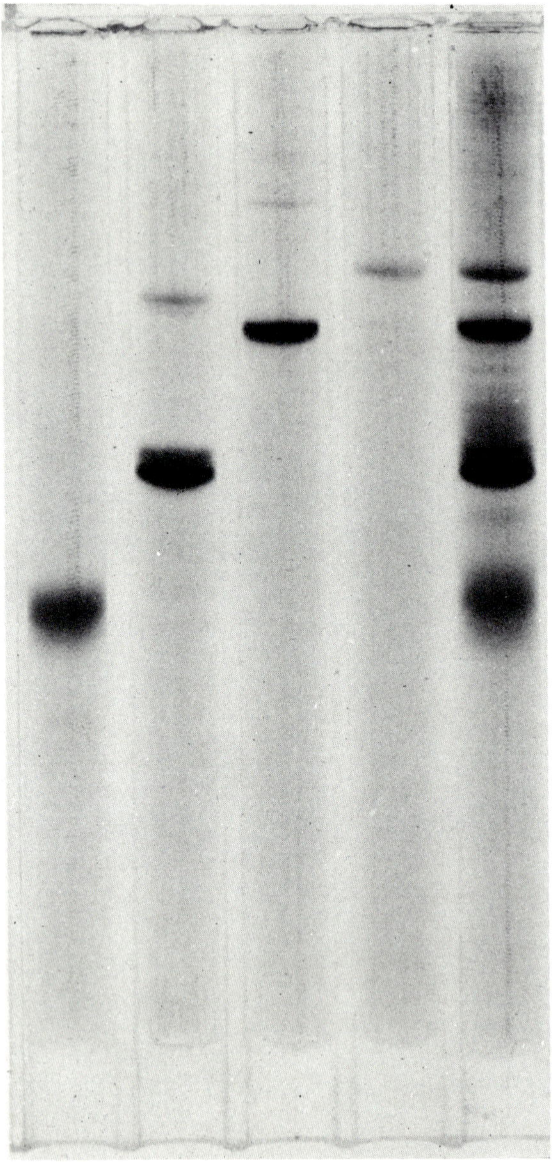

Fig. 6.3. Polyacrylamide gel patterns of nucleolar 4–8 S RNA and its subfractions. The right-hand slot (No) is whole nucleolar 4–8 S RNA. From left to right individual fractions are 4 S, 5 S, U3, and 8 S RNA (courtesy of Dr. A. W. Prestayko).

Nucleolar low molecular weight RNA has a much simpler pattern on gel electrophoresis than whole nuclear RNA (Fig. 6.3). Like the ribosomal RNA, the nucleolar RNA has only three major electrophoretic components; however, there are differences in these components; i.e., the nucleolus has 4 S RNA, 5 S RNA, and a slower moving U3 RNA.

Are the Low Molecular Weight RNA's of the Nucleus and Nucleolus Individual Molecular Species or Some Type of Artifact?

In any new field it is of critical importance that the components identified be shown to be definitive natural products than either aggregates or breakdown products of other known molecules (Table 6.2). One evidence

TABLE 6.2

Evidence that Nuclear Low Molecular Weight RNA's Are Native Molecular Species (*28, 45*)

1. The polyacrylamide gel patterns of nuclear low molecular weight RNA's are reproducible for individual cell types and are similar from cell type to cell type including liver, spleen, fibroblasts, ovary, kidney, and tumors (*28, 45, 52a*).
2. The nucleotide compositions of specific RNA fractions are constant and differ from those of high molecular weight RNA.
3. The RNase content of Novikoff hepatoma nuclear preparations is low and the residual RNase activity is inhibited at the low pH of citric acid used for these nuclear isolations.
4. When labeled 18 S and 28 S rRNA were added to the preparations, no degradation was observed during isolation of nuclear RNA either with hot or cold phenol extractions.
5. The nuclear RNA bands are found in preparations of RNA obtained from whole cells so they cannot be artifacts of nuclear isolation techniques.
6. The special nuclear RNA bands are equally well extracted with either hot or cold phenol.
7. The labeling patterns of the nuclear RNA with either ^{14}C-sodium formate or ^{14}C-methyl methionine differ from those of tRNA, 5 S RNA, or high molecular weight RNA.

for the natural existence of these molecules in cells is their existence in the RNA extracted from the whole Novikoff hepatoma (Figs. 6.4, 6.5) as well as in the nuclei obtained by the citric acid procedure (Chapter XI). Although the whole cell contains a greater preponderance of 4 S and 5 S RNA, each of the other low molecular weight nuclear RNA types is visible in whole cell RNA.

The evidence that the low molecular weight nuclear RNA is probably not a derivative of the high molecular weight RNA is derived from studies in which labeled rRNA was added to preparations of nuclei or whole cells before they were homogenized or the RNA was extracted.

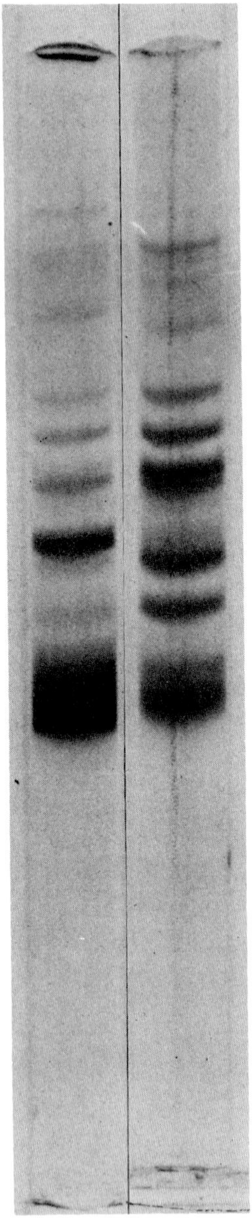

Fig. 6.4. Polyacrylamide gel patterns of 4–7 S RNA obtained from whole Novikoff cells (left) and nuclei (right) isolated with the citric acid procedure (Chapter XI). Each of the bands in the nuclear fraction, including the dense U-rich bands, is seen in the extract of the whole cell (*45*).

When the labeled rRNA was added to the whole cells, insufficient quantities of labeled RNA appeared in the nuclear preparations to permit detection even with high specific activity precursor RNA. When the labeled rRNA was added directly to nuclear preparations (Fig. 6.6) the isotope did not distribute into the 4–7 S RNA. These data and the other results presented in Table 6.2 indicate that the 4–7 S RNA is a nuclear constituent which is not a degradation product.

III. Isolation and Purification of Nucleolar and Nuclear 4–7 S RNA

Both the 4–7 S nuclear and nucleolar RNA's have been extracted from the purified organelles by means of the cold or hot phenol extraction procedures mentioned in Chapter V, usually as a component of the total

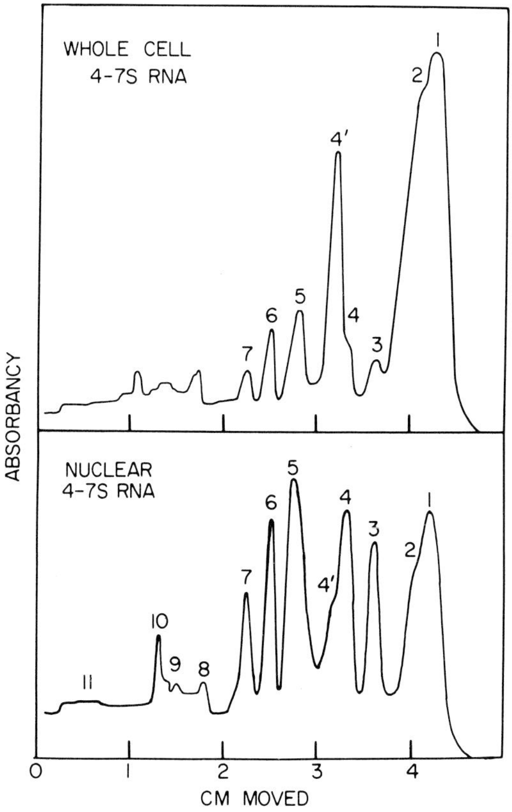

Fig. 6.5. Chromoscan patterns of the gels shown in Fig. 6.4. Peaks 5, 6, and 7 are the U-rich bands and peak 3 is the 4.5 S RNA (*45*).

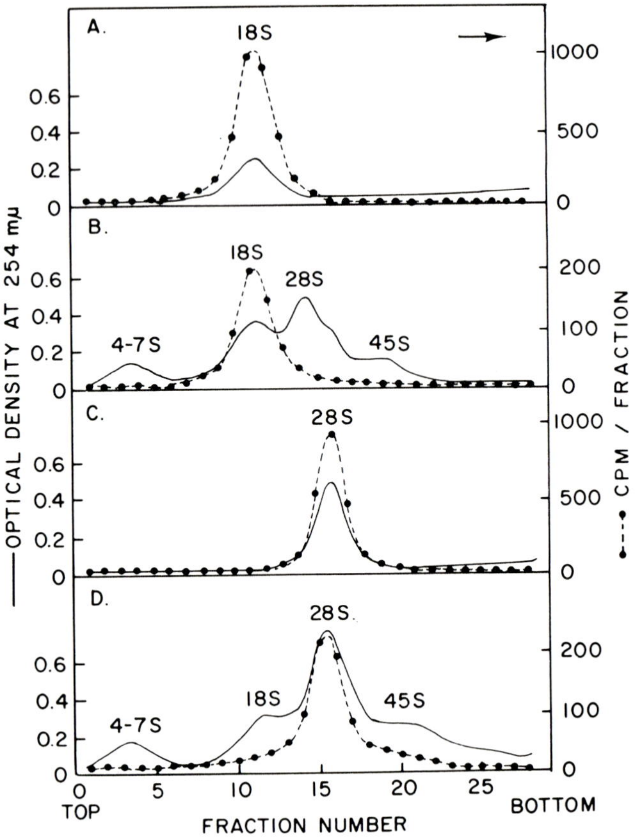

Fig. 6.6. Sucrose density gradient sedimentation profiles of Novikoff ascites hepatoma nuclear RNA extracted from nuclear preparations to which ^{14}C-labeled purified ribosomal 18 S or 28 S RNA (0.4 mg) was added. The RNA was extracted with the dodecyl SO_4-hot phenol procedure (from *45*). (A) ^{14}C-labeled ribosomal 18 S RNA alone; (B) RNA from nuclei to which 18 S RNA (a) had been added; (C) ^{14}C-labeled ribosomal 28 S RNA; (D) RNA from nuclei to which labeled 28 S RNA (C) had been added. Optical density (———); radioactivity (—●—●—) From (*45*).

RNA of the fraction. The whole RNA of the fraction is then subjected to sucrose density gradient centrifugation as shown in Fig. 6.7 for Novikoff hepatoma nuclear 4–7 S RNA (shaded portion, upper graph A; purified portion, bottom graph C). The amount of this RNA in the nucleus and the nucleolus is not large; i.e., about 14% of the total in Novikoff hepatoma nuclei, about 20% in normal liver nuclei, and about 6% in the nucleoli of Novikoff hepatoma cells. The overall amounts of this RNA were much greater in the nuclei of Novikoff hepatomas; i.e., about 90 mg/kg tumor cells as compared to about 15 mg/kg of normal liver nuclei. The anaplastic tumors, such as HeLa cells and Morris hepatoma cells, are a much richer source of whole nuclear 4–7 S RNA than normal liver cells (*26*).

For nucleolar RNA, the yields are far more favorable for the tumors than for liver. The reasons are apparent when the histology of these cells is considered: (a) the nucleus is a much smaller part of the total cell volume in normal tissues than in rapidly growing transplantable tumors in which the cytoplasm constitutes only about one half of the cell volume; (b) the nucleolus is much greater in size in the tumors; i.e., each nucleo-

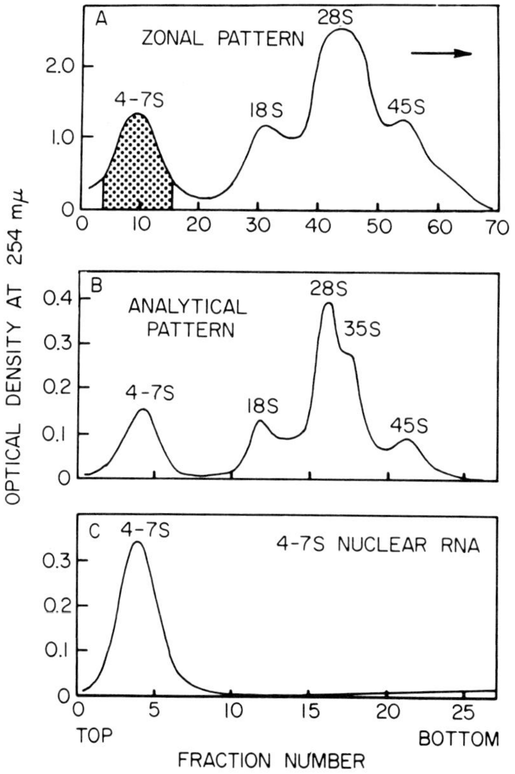

Fig. 6.7. Sucrose density gradient centrifugation profiles of Novikoff ascites hepatoma nuclear RNA (from *45*). (A) Pattern for 120 mg of RNA from a rate zonal centrifugation in the zonal ultracentrifuge; the pooled shadowed portion was the 4–7 S RNA fraction subjected to later fractionation; (B) analytical sucrose density gradient centrifugation profile for 0.2 mg of the nuclear RNA used in (A); (C) analytical scale profile of the 4–7 S RNA obtained from (A).

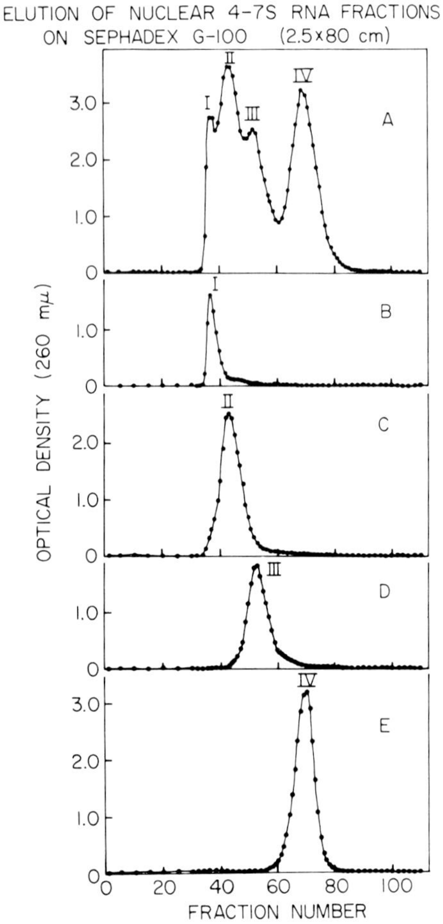

Fig. 6.8. a. Sephadex G-100 column chromatography of Novikoff ascites hepatoma nuclear 4–7 S RNA. (A) Whole nuclear 4–7 S RNA; (B) to (E) rechromatography of RNA from peaks I to IV obtained from (a) (from *45*). B. Elution pattern for nucleolar 4–7 S RNA, ribosomal 4–7 S RNA, and soluble RNA of the Novikoff hepatoma from Sephadex G-100. RNA was eluted with 0.05 *M* ammonium acetate, pH 5.1, at a flow rate of 3 ml per cm^2 per hour. Each fraction contained 4 to 5 ml of eluant. U-V, ultraviolet. (From *45*).

lus contains 1.5–3.5 pg of RNA in the tumors and only about 0.2–0.3 pg of RNA in the normal liver. The nucleolar 4–7 S RNA composes approximately 5–10% of the whole nucleolar RNA. Accordingly the overall yield of RNA per gram of tissue is much greater in the tumors by comparison with the normal liver.

A. Separation of Low Molecular Weight RNA on Sephadex Columns

These 4–7 S RNA bands of differing mobilities were partially separated by chromatography of whole nuclear and nucleolar 4–7 S RNA on Sephadex G-100 (Fig. 6.8a,b) by the technique originally described by Galibert and his associates (*20–22*). The nucleolar RNA of peak II and the nuclear RNA of peak III were similar in mobility to that of ribosomal 5 S RNA. As shown in Figs. 6.3 and 6.9, the nuclear and nucleolar RNA's were not completely separated by chromatography on Sephadex G-100. Although none of the nuclear RNA's was obtained as a homogeneous band (Fig.

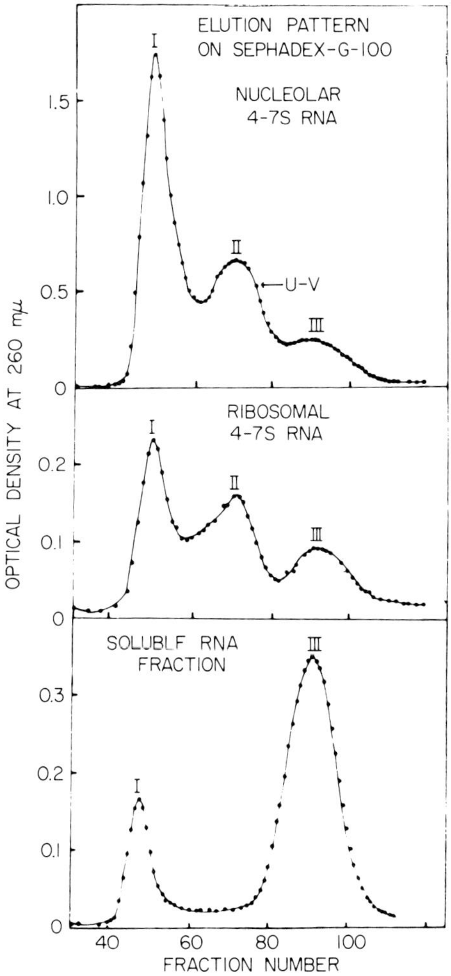

Fig. 6.8 (*continued*).

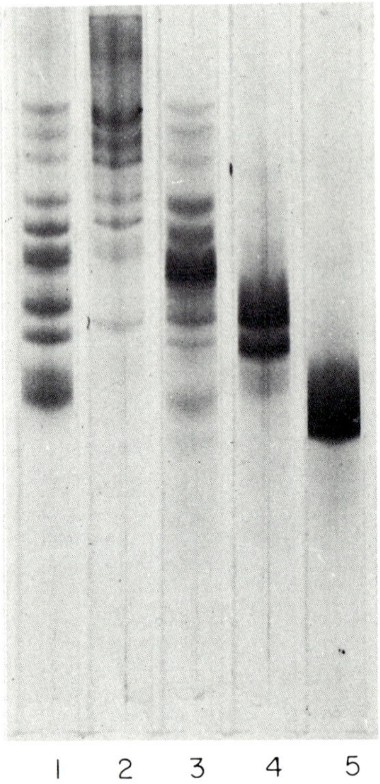

Fig. 6.9. Polyacrylamide gel electrophoresis patterns of 4–7 S RNA fractions obtained by Sephadex G-100 column chromatography (Fig. 6.8a). Slot 1: total 4–7 S RNA, slots 2 to 5: peaks I to IV, respectively. From (*45*).

6.9), peaks III and IV contained predominantly 5 S RNA and 4 S RNA, respectively.

B. Purification of the Nuclear 4–7 S RNA's by Preparative Gel Electrophoresis

Although chromatography on Sephadex G-100 was useful in separating the nuclear and nucleolar 4–7 S RNA into a group of size classes, this technique has not yet provided complete resolution of the nuclear or nucleolar RNA's into electrophoretically homogeneous single bands (Fig. 6.9). Two approaches to this important problem have yielded successful results. The first is preparative disc gel electrophoresis. Utilizing a product purified on columns of Sephadex G-100 (Fig. 6.8), the RNA was separated by preparative gel electrophoresis into zones containing U1 and U2 RNA (Table 6.3) as shown in Fig. 6.10. The U2 RNA isolated from the shaded portion is an electrophoretically homogeneous single band (Fig. 6.11) and the U1 RNA was purified as a double band (Fig. 6.11). Although these RNA's both contain large concentrations of uridylic acid, their nucleotide compositions differ as shown in Table 6.3.* Preparative gel electrophoresis was also used for the isolation of electrophoretically homogeneous 4.5 S nuclear RNA as shown in Fig. 6.12.

C. Purification of Nuclear 4–7 S RNA's by DEAE Sephadex

Another procedure that has provided electrophoretically homogeneous single RNA bands has been chromatography of 4–7 S RNA on DEAE-

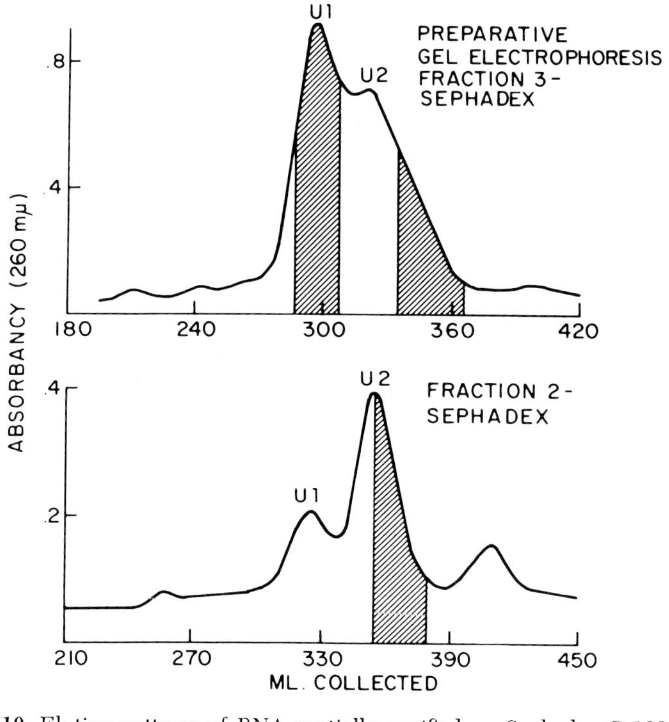

Fig. 6.10. Elution patterns of RNA partially purified on Sephadex G-100 columns and fractionated by preparative gel electrophoresis (with an 8% polyacrylamide gel, 4 cm in height in a Canalco PD 2/320 upper column). A 2-mg sample was subjected to electrophoresis at 30 mA constant current for 6 to 7 hours. The flow rate of flush buffer was 1 ml per minute and each fraction contained 3.0 ml. The fastest moving component is at the left. The shaded areas were pooled and contained the U1 and U2 fractions in high purity (Fig. 6.11). From (*28*).

* It is not clear whether the differences in the values for the liver and tumor RNA's reflect differences in about 3% of the total bases or whether they reflect the accuracy of the method employed for nucleotide analyses.

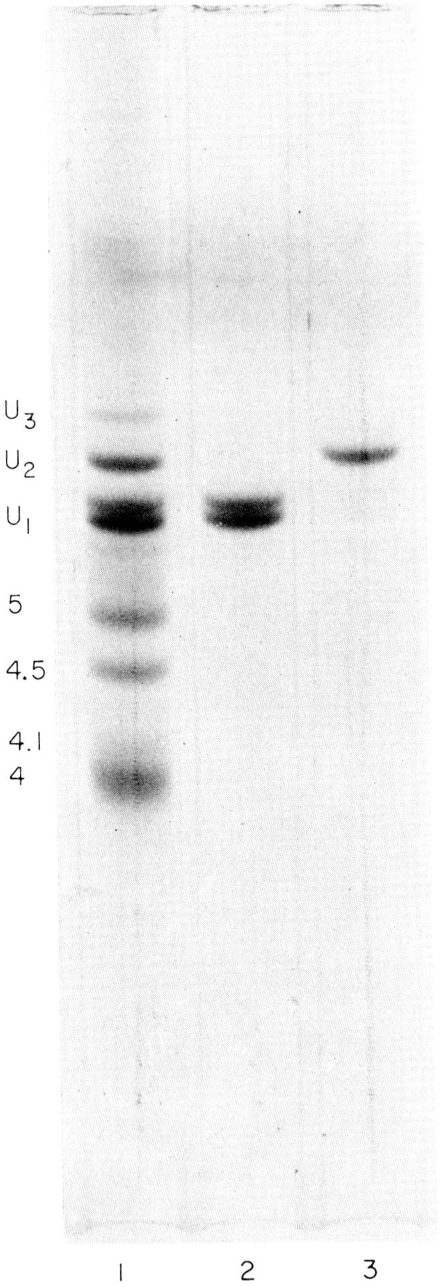

Fig. 6.11. Polyacrylamide gel electrophoresis of the individual 4–7 S RNA fractions obtained by preparative gel electrophoresis of Sephadex purified fractions. From left to right, the RNA samples are (1) whole rat liver nuclear 4–7 S RNA, (2) the U1 RNA fraction, and (3) the U2 RNA fraction from preparative gel electrophoresis (Fig. 6.10). The direction of migration is from top to bottom (from 28).

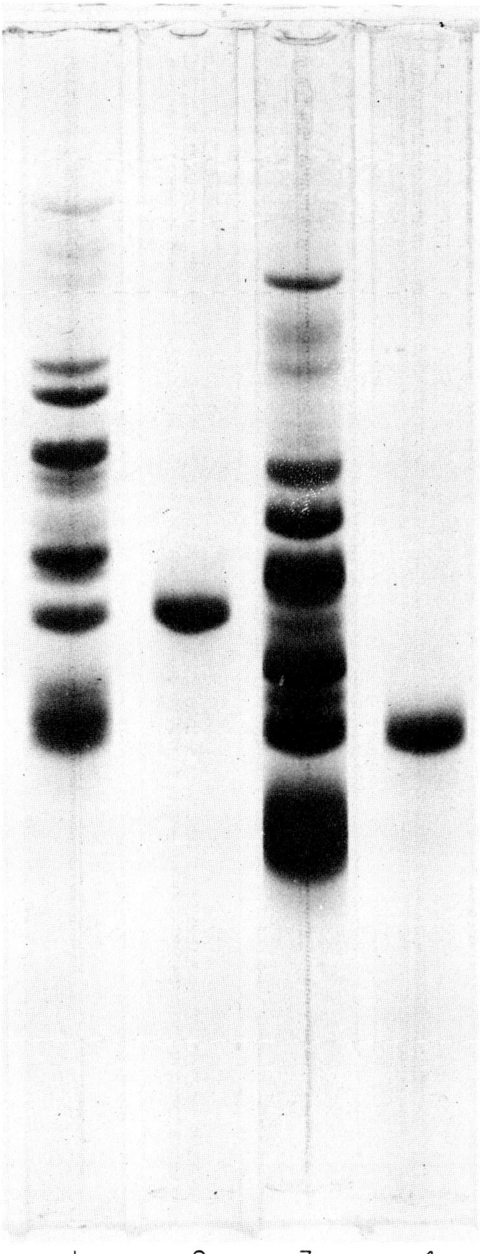

Fig. 6.12. Purification of nuclear 4.5 S RNA of preparative polyacrylamide gel electrophoresis. The two left-hand gels were 10% gels run at pH 7.2 and the two right-hand gels were 8% gels run at pH 5.1. Gels 1 and 3 represent whole nuclear 4–7 S RNA and gels 2 to 4 represent the purified 4.5 S RNA (courtesy of Dr. T. S. Ro-Choi).

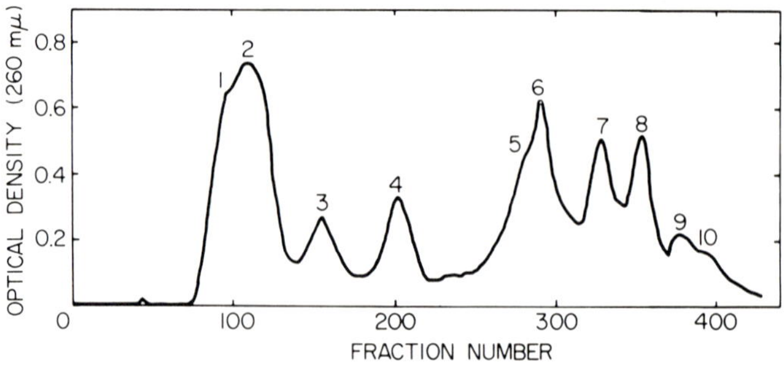

Fig. 6.13. Chromatography of nuclear 4–7 S RNA on DEAE-Sephadex. Approximately 40 mg of nuclear 4–7 S RNA was applied to a 1.50 × 100 cm column. The RNA was eluted with a linear NaCl gradient (0.45–0.6 *M*) in 0.02 *M* Tris-HCl buffer (pH 7.6). From (*45a*).

TABLE 6.3

ULTRAVIOLET NUCLEOTIDE COMPOSITIONS OF NUCLEAR U-RICH RNA FRACTIONS PURIFIED BY PREPARATIVE GEL ELECTROPHORESIS

	Fraction	AMP[a]	UMP	GMP	CMP	(AMP + UMP)/(GMP + CMP)
Liver (*28*)	U 1	19	27	31	23	0.85
	U 2	22	28	32	18	1.00
Novikoff hepatoma (*45*)	U 1	19	26	33	23	0.80
	U 2	22	26	33	19	0.94

[a] The abbreviations used are: AMP, adenylic acid; UMP, uridylic acid; GMP, guanylic acid; CMP, cytidylic acid.

Sephadex (A-50) columns and elution of the RNA with increasing concentrations of NaCl. Variables that are employed include modifications of the salt gradients, elutions at different pH values, and variations in rates of increases of salt gradients. Other variables that are being analyzed include urea gradients and addition of urea to the whole samples. With this technique the type of fractionation obtained is shown in Fig. 6.13; peak 7 contained an electrophoretically homogeneous U2 RNA as shown by polyacrylamide gel electrophoresis in Fig. 6.14.

IV. Nucleotide Composition of Nuclear and Nucleolar 4–7 S RNA Fractions

Table 6.4 shows the UV nucleotide compositions of 4–7 S RNA fractions of the Novikoff hepatoma nucleoli (Fig. 6.8b). The nucleolar fractions had a high content of uridylic acid, by comparison with the reported

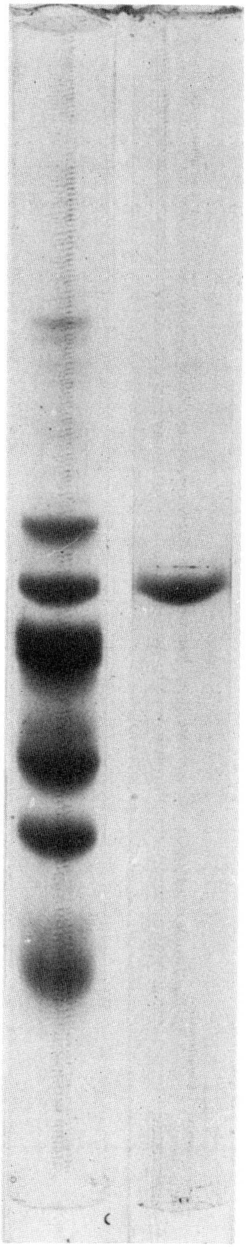

Fig. 6.14. Polyacrylamide gel electrophoresis pattern showing purification of U2 RNA on DEAE-Sephadex (Fig. 6.13). Left: total 4–7 S RNA; Right: the RNA from peak 7 in Fig. 6.13 (courtesy of Dr. Y. Moriyama).

TABLE 6.4

UV-Nucleotide Composition of Nucleolar 4–7 S RNA Fractions Obtained by Chromatography on Sephadex G-100[a]

Peak	AMP	UMP	GMP	CMP	AMP + UMP / GMP + CMP	Reference
Whole 4–7 S RNA	19	27	31	23	0.86	
Peak I	19	27	31	23	0.85	
Peak II	21	29	29	21	1.01	
5 S RNA	18	25	31	26	0.74	*(20–22)*
Peak III	22	26	31	21	0.92	
tRNA	17	22	31	25	0.78	*(20–22)*

[a] The values of Galibert *et al.* *(20–22)* for cytoplasmic 5 S and tRNA of liver are presented for comparison; the values for uridylic acid do not include pseudouridylic acid, which composed 4.8% of the total nucleotides in their tRNA and 0.26% of the total nucleotides in their 5 S RNA. The values obtained for the base composition of ribosomal 5 S RNA in this laboratory agree closely with those reported by Galibert *et al.* *(20–22)*.

values for either 4 S or 5 S RNA obtained from mammalian systems. Although the RNA in peak II (Fig. 6.8b) was similar in mobility to 5 S RNA, the nucleotide composition was markedly different inasmuch as uridylic acid constituted 28.8% of the nucleotides in the nucleolar RNA and only 25% in the ribosomal 5 S RNA. In peak I of Fig. 6.8b, the uridylic acid content was also high but because the content of adenylic acid was lower, the AMP + UMP/GMP + CMP ratio was lower than the value of 1.0 for the RNA in peak II. The nucleotide composition showed that the RNA in peak III had a high content of adenylic acid but it contained less uridylic acid than that of peak II. Peak III contained most of the amino acid acceptor activity of the nucleolar RNA (see p. 306), but its low overall activity reflects the presence of several kinds of RNA other than transfer RNA.

When liver nuclear 4–7 S RNA was fractionated on a large scale on Sephadex G-100 columns, the pattern shown in Fig. 6.15 was obtained and the samples were grouped into eight fractions. The nucleotide compositions of these RNA fractions (Table 6.5) indicate that the transfer RNA is concentrated in Fraction 8; the U-rich RNA is concentrated in Fractions 3–5. The RNA in Fraction 7 consists mostly of the 4.5 S RNA; its nucleotide composition is of interest because of the relatively equal distribution of the nucleotides in this RNA. This RNA is being purified (Fig. 6.12) and now has been found to consist of at least three components. The UV nucleotide compositions should be compared with ^{32}P nucleotide compositions obtained by extraction of RNA from individual polyacrylamide gels (Table 6.6). Fractions 1, 2–4, 7, and 8 of Table 6.5

TABLE 6.5

ULTRAVIOLET BASE COMPOSITIONS OF LIVER NUCLEAR 4–7 S RNA FRACTIONS OBTAINED BY EXCLUSION CHROMATOGRAPHY ON SEPHADEX G-100

Fraction	% Total 4–7 S RNA	AMP[a]	UMP	GMP	CMP	$\frac{\text{AMP} + \text{UMP}}{\text{GMP} + \text{CMP}}$
1	15	21	26	29	24	0.88
2	8	22	27	29	23	0.95
3	25	21	27	29	23	0.93
4	10	22	28	28	23	0.99
5	10	23	28	27	22	1.07
6	4	23	25	28	24	0.92
7	3	24	25	27	25	0.95
8	25	20	20	34	27	0.65

[a] The abbreviations used are: AMP, adenylic acid; UMP, uridylic acid; GMP, guanylic acid; CMP, cytidylic acid.

appear to contain predominantly components of the U3, U2, 4.5, and 4 S RNA bands of Table 6.6 (see also *40a*). Recently it has been found (*52b*) that the 4.5 S RNA consists of two major components (Table 6.6); the first (4.5 S_I RNA) has a 5′-terminal pGp and its 3′-terminal is uridine. For the other major component (4.5 S_{III}), the 5′-terminal is pAp and the 3′-terminal is 2′(3′)-*O*-methyluridine. In studies currently in progress it has also been found that the 5 S nuclear RNA contains two or three components (*52b*).

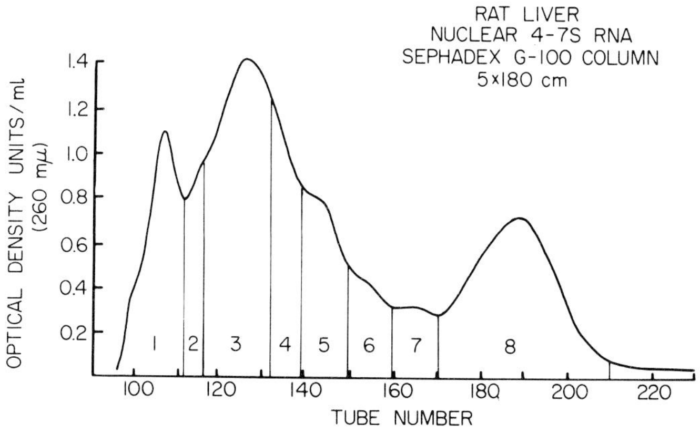

Fig. 6.15. Elution pattern of rat liver nuclear 4–7 S RNA from a 5 × 180-cm column of Sephadex G-100. A 45-mg sample was eluted with 0.05 *M* ammonium acetate, pH 5.1, at a flow rate of 2.1 ml per cm^2 per hour. Each fraction contained approximately 14 ml of eluant (from *28*).

TABLE 6.6

^{32}P NUCLEOTIDE COMPOSITIONS OF 4–7 S NUCLEAR RNA FRACTIONS[a]

	AMP	UMP	GMP	CMP	$\frac{A + U}{G + C}$	Number of methyl groups / Molecule (*66a*)	Major 3′ terminal nucleoside (*52b*)
U3	21	27	29	23	0.93	3	A
U2	22	28	28	21	1.01	6	A
U1b, U1c	20	26	28	26	0.86	3	G
U1a	19	23	31	27	0.73	0	
5 S 1	18	24	32	26	0.73	0	U
5 S 2	20	24	30	26	0.79		U
4.5 S_I	25	24	26	25	0.97	4	U
4.5 S_{III}	27	25	29	19	1.04	?	U-Me

[a] These nucleotide compositions are a composite from studies made in this laboratory (*28, 45, 47, 52*) and those of Weinberg and Penman (*66a*).

V. Amino Acid Acceptor Activity of Nucleolar and Nuclear 4–7 S RNA

Table 6.7 shows the amino acid acceptor activities of the whole nuclear and nucleolar 4–7 S RNA and the activities of various fractions obtained by chromatography of the RNA on Sephadex G-100. In each case the last fraction to emerge had the highest amino acid acceptor activity. The nuclear RNA in Fraction 8 (Fig. 6.15) had an activity greater than that of commercial yeast tRNA. In the nucleolar RNA, the highest amino acid acceptor activity was found in the RNA in peak III (Fig. 6.8b). However, this activity was only about one-fifth that of commercial tRNA and about one-eighth that of whole nuclear 4 S RNA. This result along with the marked difference in the nucleotide composition of this RNA by comparison with that of the tRNA (Table 6.7) suggests that the RNA in this fraction of the nucleolar RNA is probably different from transfer RNA. However, neither its structure nor the number of species present is known.

TURNOVER OF NUCLEAR AND NUCLEOLAR 4–7 S RNA

Figure 6.16 shows a sucrose density gradient centrifugation profile of Novikoff hepatoma nuclear RNA labeled with (^{14}C) sodium formate for 1 hour. The marked incorporation of ^{14}C into 35 S and 45 S nuclear RNA was similar to that reported previously with other tracers (see Chapter V). The specific activity in cpm/mg was approximately 100,000 for 4–7 S and 18 S, 80,000 for 28 S, 320,000 for 35 S, and 660,000 for 45 S RNA, respectively (*45–48*).

TABLE 6.7

ACCEPTOR ACTIVITY OF FRACTIONS OF NUCLEAR AND NUCLEOLAR 4–7 S RNA OBTAINED BY CHROMATOGRAPHY ON SEPHADEX G-100

RNA fractions	Acceptor activity cpm/mg RNA[a]
Nucleolar	
Whole 4–7 S RNA	6,100
Peak I	390
Peak II	1,000
Peak III	39,600
Nuclear	
Whole 4–7 S RNA	130,000
Fraction 1	0
Fraction 3	0
Fraction 7	6,000
Fraction 8	325,000
tRNA (yeast)	180,000

[a] The results are counts per minute of ^{14}C-amino acids incorporated per milligram of RNA. In these experiments, 0.5 μmole of total protein hydrolyzate (^{14}C-labeled) was used as the source of amino acids.

The 4–7 S RNA of Fig. 6.16 was collected and analyzed by polyacrylamide preparatory gel electrophoresis (Fig. 6.17). Six peaks were found. The specific activity of first peak (tRNA) was about 30,000 to 60,000 cpm/mg RNA. A higher specific activity of 100,000 to 140,000 was found in peaks 3 and 4 which contain the 4.5 S and 5 S RNA, respectively. In peaks 5 and 6 which contain the U1 and U2 RNA's, respectively, the specific activity of the RNA was only 14,000–16,000 cpm/mg. This low specific activity indicates a low turnover of some of the low molecular weight nuclear RNA's and, hence, they have been referred to as being "metabolically stable" (*45, 66a,b*).

VI. Methylation

Since methylation is an outstanding feature of a number of bases and nucleotides of transfer RNA, it was of interest to determine whether there was methylation of various types of nuclear and nucleolar 4–7 S RNA (*58–60*). Figure 6.18 shows the density gradient sedimentation profile of nuclear RNA labeled with (methyl-^{14}C) methionine. Radioactivity distribution profiles indicate that the labeling rate of 4–7 S RNA is somewhat less than that of 45 S and 35 S RNA. The incorporation of ^{14}C into 4–7 S RNA continued up to 24 hours after injection of (methyl-^{14}C)methionine; the labeling was decreased in the high molecular weight RNA at this

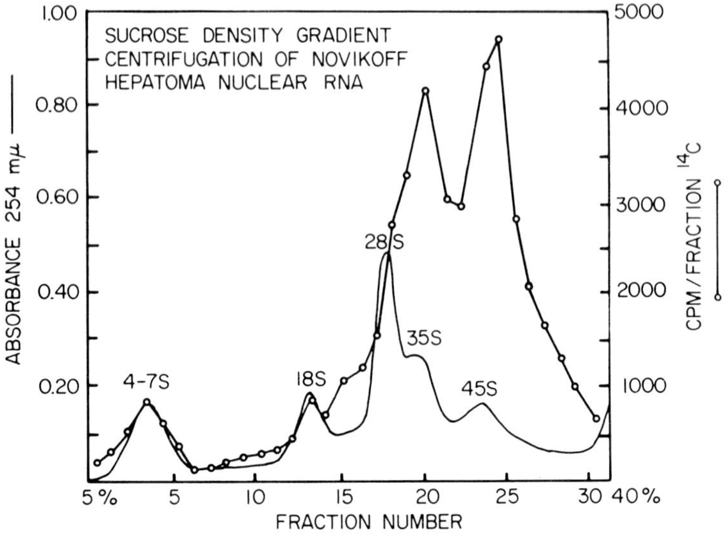

Fig. 6.16. Sucrose density gradient sedimentation profile of Novikoff hepatoma nuclear RNA labeled with [^{14}C] sodium formate. Absorbancy (———); radioactivity (—○—○—) (from *45*).

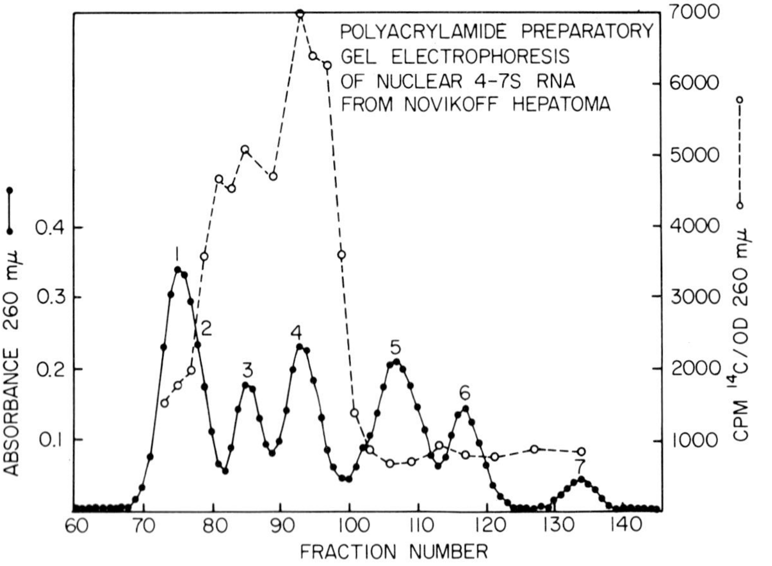

Fig. 6.17. Electrophoretic pattern on a preparative polyacrylamide gel of Novikoff hepatoma nuclear 4–7 S RNA labeled with [^{14}C] sodium formate. The 4–7 S RNA was obtained from the nuclear RNA of Fig. 6.16. Marked differences are apparent in the specific activities for the individual RNA peaks. Absorbancy (—●—●—); radioactivity (- -○- -○- -) (from *45*).

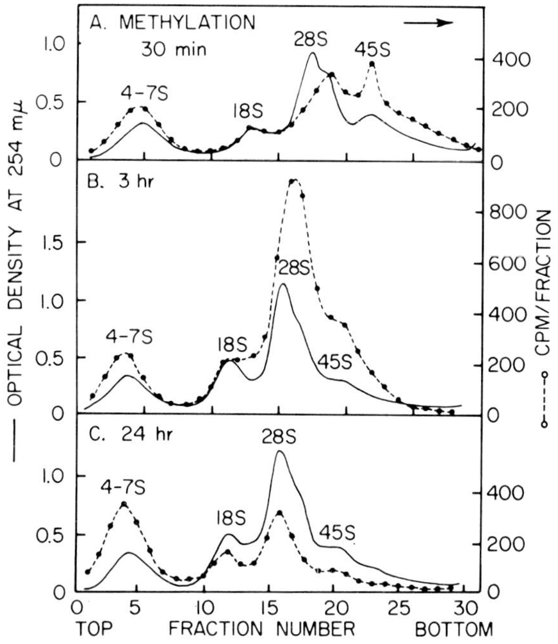

Fig. 6.18. Sucrose density gradient sedimentation profiles of Novikoff ascites hepatoma nuclear RNA labeled with L-[methyl-^{14}C]methionine. Rats were killed at 30 minutes (a), at 3 hours (b), and at 24 hours (c) after injection of L-[methyl-^{14}C]methionine. Optical density (———); radioactivity (--●--●--) (from *45*).

time. The turnover rate of the methyl groups was fastest in the 45 S RNA (see Chapter V).

The corresponding chromoscans of polyacrylamide gels after electrophoresis of 4–7 S nuclear RNA labeled with (methyl-^{14}C)methionine are shown in Fig. 6.19. At 30 minutes after injection of the tracer, ^{14}C was found only in the tRNA region of chromoscan peaks 1 and 2. At 3 hours, a radioactive peak appeared in chromoscan peaks 5 and 6. After 24 hours, the radioactivity distribution pattern was similar to that at 3 hours except that more of the radioactivity was localized to chromoscan peak 6.

All types of nucleolar 4–7 S RNA seem to be methylated as shown in Fig. 6.20 although the greatest rates of methylation occur on the newly synthesized 45 S and other rapidly sedimenting RNA (see Chapter V). In the nucleolus as well as in the nucleus, the lowest specific activities with respect to methylation were found in the RNA of peak II which was richest in uridylic acid. As determined by specific activity, the rates of methylation of the RNA in peak III were not significantly different from those of the cytoplasmic transfer RNA (Table 6.8). On the basis of comparative uptakes of labeled methyl groups of methionine and ^{32}P-ortho-

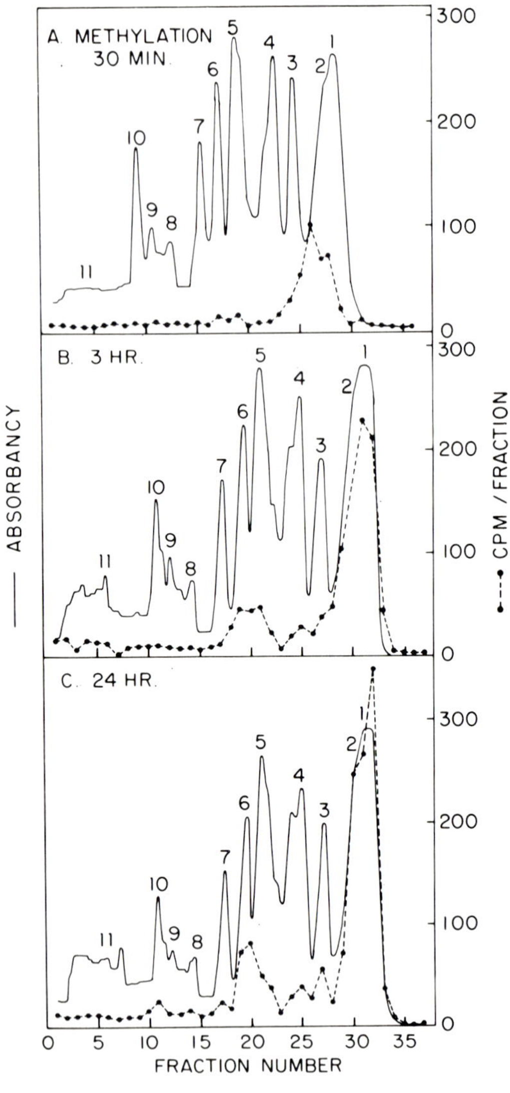

Fig. 6.19. Densitometric tracings of the polyacrylamide gel electrophoresis patterns of 4–7 S RNA labeled with L-[methyl-^{14}C]methionine (Fig. 6.18). The gels were sliced and the radioactivity in the fractions was determined in a scintillation counter. Absorbancy (———); radioactivity (- -●- -●- -) (from *45*).

phosphate into the various types of nuclear 4–7 S RNA, Weinberg and Penman (*66a*) calculated that these molecules contain from 0–6 methyl groups (see Table 6.6). They noted that transfer RNA contains an average of 6.5 methyl groups and the uridylic acid-rich RNA's contain

TABLE 6.8

EFFECT OF ALKALINE TREATMENT ON RELEASE OF METHIONINE-METHYL-^{14}C BOUND TO NUCLEOLAR RNA

Fraction	Specific activities: Before incubation	Specific activities: After incubation	Decrease after incubation (%)
Peak I	16,900	8,800	48
Peak II	10,900	5,040	54
Peak III	14,500	9,840	32
tRNA (cytoplasm)	12,400	7,540	39

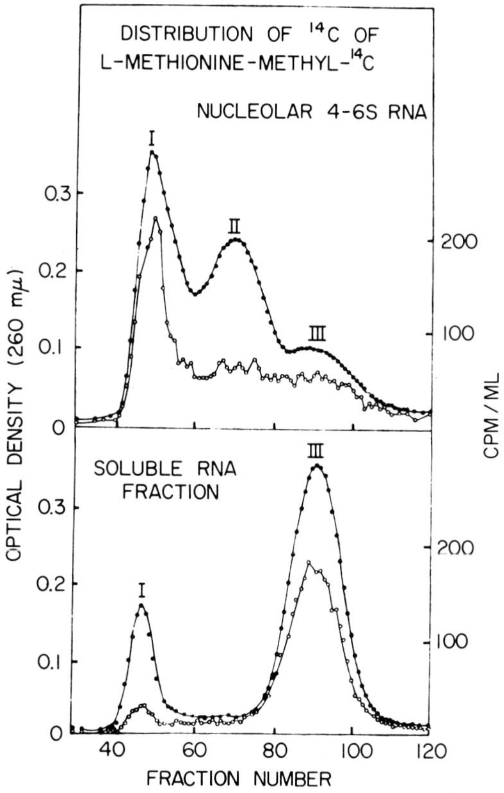

Fig. 6.20. Elution pattern from Sephadex G-100 of nucleolar 4–7 S RNA and soluble cytoplasmic RNA of the Novikoff hepatoma labeled with L-methionine-methyl-^{14}C.

3–6. Positions of the methyl groups in these 4–7 S RNA molecules are not defined at present.

One of the interesting aspects of the methylation of nucleolar RNA is shown in Table 6.8; i.e., the effect of alkaline treatment on release of methionine-methyl-^{14}C that is bound to various nucleolar RNA fractions. When methyl-^{14}C-methionine is linked to transfer RNA in ester linkage of the whole amino acid to the 3′-terminal adenosine, alkaline hydrolysis completely removes the methylated amino acid. However, the methyl groups bound to either the sugar hydroxyls or the base nitrogens should not be released under these conditions. Accordingly, the result obtained after incubation of the nucleolar RNA with alkali was surprising in that less label was released in the transfer RNA and the RNA of nucleolar peak 3 than in nucleolar RNA's of peaks 1 and 2 (Fig. 6.20). These results suggest that the methyl group in the nontransfer RNA is bound in a linkage that is even more labile than that of the transfer RNA. Presumably much of the isotope bound in the transfer RNA reflects *O*-methylation and *N*-methylation.

A. Localization of 4–7 S RNA in the Nucleolus

Although it is not completely clear where the 4–7 S RNA is topographically localized in the nucleolus, the evidence obtained by combined inhibitor studies noted in Chapters I, II, and X in which thioacetamide and actinomycin D were used concurrently to produce increases in nucleolar size and segregation of nucleolar fibrillar elements suggests that the nucleolar 4–7 S RNA is localized in the fibrillar elements of the nucleolus (55). This possibility is of particular interest in view of the fact that only one type of nuclear 4–7 S RNA, the U3 RNA, seems to be localized to the nucleolus. This RNA is moderately rich in uridylic acid and is readily isolated as a single band by electrophoresis. Whether this band represents some specific rDNA (NOR)-activating cofactor or "gene gate" remains to be learned in future studies. The fact that the U1 and U2 types of RNA are apparently localized in the chromatin fraction (52) suggests that external to the nucleolus there may be some functional gene gates or activators that differ from those within the nucleolus, either because different types of DNA are activated or because special nucleolar products are produced.

None of the functions of the nucleolar and nuclear 4–7 S RNA's are completely defined. For example, it is not yet certain whether the nuclear tRNA's function for protein synthesis or simply are products that migrate from the nucleolus and nucleus to the cytoplasm. It is now clear that tRNA is not synthesized in the nucleolus but is synthesized in the extranucleolar chromatin (*34, 38, 51, 57, 58, 68, 69*).

B. Numbers of Low Molecular Weight Nuclear and Nucleolar RNA Per Nucleus

On the basis of estimated molecular weights of 24,000 (80 nucleotides) to 55,000 (180 nucleotides) for 4–7 S RNA and an estimate of 6 pg of 4–7 S RNA per nucleus in rat liver cells, it was calculated that there were approximately 10^6–10^7 molecules of the 4–7 S RNA per nucleus (*52*). Weinberg and Penman (*66a*) have extended the calculation for each of the types of RNA per HeLa cell (Table 6.9). Although these values are

TABLE 6.9

Estimated Numbers of Molecules of Various Types of RNA in Nuclei, the Nucleoplasm, and the Cytoplasm

Species of RNA	Estimated No. of molecules/cell			Estimated No. of nucleotides
	Cytoplasm	Nucleoplasm	Nucleolus	
U3			2×10^5	180
U2		4×10^5	1×10^5	165
U1b, U1c		1×10^6	7×10^4	150
(28 sA)U1a	5×10^6	5×10^4	1×10^4	140
5 S	5×10^6	3×10^5	3×10^5	121
4.5 S		2×10^5	1×10^5	100
4 S tRNA	1×10^8	2×10^5		80
18 S	5×10^6	6×10^4		1.8×10^3
28 S	5×10^6	7×10^4		4.9×10^3
35 S			4×10^4	6.8×10^3
45 S			1×10^4	1.3×10^4

Modified from Weinberg and Penman (*66a*).

contingent on the presumption that there are no losses of these RNA molecules from the nuclei during isolation, they do provide an order of magnitude for the numbers of these molecules per cell which agrees with the numbers suggested above (*52*). Since these numbers of nuclear 4–7 S RNA's approach the total numbers of potentially functional genes in cells, it is interesting to speculate on the possibility that 4–7 S RNA's may be associated with genes on a one-to-one basis and may have a role in gene modulation. The association of the U-rich RNA of the U1–U2 groups with the chromatin fraction is another circumstantial evidence which favors this idea, as does the metabolic stability alluded to earlier (*28, 45, 47, 52, 66a*).

In summary, a new and exciting chapter has recently opened in the field of RNA chemistry and particularly that of the biochemistry of nucleolar and nuclear RNA; i.e., the uniquely localized low molecular weight RNA. The nucleolus contains four different types of RNA, 4 S, 5 S,

U3, and 8 S RNA and in addition to these, the whole nucleus contains 4.5 S, U1b, U1c, U2, and probably 10–12 S RNA's that are yet to be characterized. Their functions are not known but they are certainly not export products since their turnover rates are very slow and they are not found in the cytoplasm. Presumably they are part of either the synthetic or control mechanisms of the nucleolus and nucleus. It is important to note that the electrophoretic studies only provide a picture of the minimal number of 4–7 S species; i.e., already three components have been found in the nuclear 4.5 S RNA and at least two and possibly three components are in the nuclear 5 S RNA.

REFERENCES

1. Apgar, J., Everett, G. A., and Holley, R. W., Isolation of large oligonucleotide fragments from the alanine RNA. *Proc. Natl. Acad. Sci. U. S.* **53**, 546–548 (1965).
2. Apgar, J., Holley, R. W., and Merrill, S. H., Purification of the alanine-, valine-, histidine-, and tyrosine-acceptor ribonucleic acids from yeast. *J. Biol. Chem.* **237**, 796–802 (1962).
3. Attardi, B., and Attardi, G., A membrane-associated RNA of cytoplasmic origin in HeLa cells. *Proc. Natl. Acad. Sci. U.. S.* **58**, 1051–1058 (1967).
4. Bachvaroff, R. J., and Tongur, V., 5 S ribonucleic acid in ribosomes from mammalian tissues. *Nature* **211**, 248–250 (1966).
5. Brownlee, G. G., and Sanger, F., Nucleotide sequences from the low molecular weight ribosomal RNA of *Escherichia coli*. *J. Mol. Biol.* **23**, 337–353 (1967).
6. Brownlee, G. G., Sanger, F., and Barrell, B. G., The sequence of 5 S ribosomal ribonucleic acid. *J. Mol. Biol.* **34**, 379–412 (1968).
7. Cantor, C. R., Possible conformations of 5 S ribosomal RNA. *Nature* **216**, 513–514 (1967).
8. Cantor, C. R., The extent of base pairing in 5 S ribosomal RNA. *Proc. Natl. Acad. Sci. U. S.* **59**, 478–483 (1968).
9. Choi, Y. C., Smetana, K., and Busch, H., Studies on RNA isolated from nucleolar fractions of Novikoff hepatoma cells by sequential extractions with dilute and concentrated salt solutions. *Exptl. Cell Res.* **53**, 582–602 (1968).
10. Comb, D. G., and Katz, S., Studies on the biosynthesis and methylation on transfer RNA. *J. Mol. Biol.* **8**, 790–800 (1964).
11. Comb, D. G., and Sarkar, N., The binding of 5 S ribosomal ribonucleic acid to ribosomal subunits. *J. Mol. Biol.* **25**, 317–330 (1967).

11a. Dahmus, M. E., and McConnell, D. J., Chromosomal ribonucleic acid of rat ascites cells. *Biochemistry* **8**, 1524–1534 (1969).

12. deWachter, R., and Fiers, W., Studies on the bacteriophage MS2. IV. The 3′-OH terminal undecanucleotide sequence of the viral RNA chain. *J. Mol. Biol.* **30**, 507–527 (1967).
13. Doctor, B. P., Apgar, J., and Holley, R. W., Fractionation of yeast amino acid-acceptor ribonucleic acids by counter-current distribution. *J. Biol. Chem.* **236**, 1117–1120 (1961).
14. Doctor, B. P., and Connelly, C. M., Separation of yeast amino acid-acceptor ribonucleic acids by countercurrent distribution in modified Kirby's system. *Biochem. Biophys. Res. Commun.* **6**, 201–204 (1961).

15. Fellner, P., and Sanger, F., Sequence analysis of specific areas of the 16 S and 23 S ribosomal RNAs. *Nature* **219**, 236–238 (1968).
16. Forget, B. G., and Weissman, S. M., Low molecular weight RNA components from KB cells. *Nature* **213**, 878–882 (1967).
17. Forget, B. G., and Weissman, S. M., Nucleotide sequence of KB cell 5 S RNA. *Science* **158**, 1695–1699 (1967).
18. Forget, B. G., and Weissman, S. M., Oligonucleotides produced by digestion of KB cell ribosomal 5 S ribonucleic acid with specific nucleases. *J. Biol. Chem.* **243**, 5709–5723 (1968).
18a. Forget, B. G., and Weissman, S. M., The nucleotide sequence of ribosomal 5 S ribonucleic acid from KB cells. *J. Biol. Chem.* **244**, 3148–3165 (1969).
19. Fuller, W., and Hodgson, A., Conformation of the anticodon loop in tRNA. *Nature* **215**, 817–821 (1967).
20. Galibert, F., Larsen, C. J., Lelong, J. C., and Boiron, M., Recherches sur un acide ribonucléique ribosomique de faible poids moléculaire isolé des cellules de mammifères. *Bull. Soc. Chim. Biol.* **48**, 21–36 (1966).
21. Galibert, F., Lelong, J. C., Larsen, C. J., and Boiron, M., Comparative kinetics of synthesis of transfer RNA and 5 S RNA in bacteria and mammalian cells. *J. Mol. Biol.* **21**, 385–390 (1966).
22. Galibert, F., Lelong, J. C., Larsen, C. J., and Boiron, M., Position of 5 S RNA among cellular ribonucleic acids. *Biochim. Biophys. Acta* **142**, 89–98 (1967).
23. Gardner, J. A. A., and Hoagland, M. B., A unique ribonucleic acid of low molecular weight from rat liver microsomes. *J. Biol. Chem.* **243**, 10–19 (1968).
24. Goodman, H. M., Abelson, J., Landy, A., Brenner, S., and Smith, J. D., Amber suppression: A nucleotide change in the anticodon of a tyrosine transfer RNA. *Nature* **217**, 1019–1024 (1968).
25. Gould, H. J., The nature of high molecular weight fragments of ribosomal RNA. *J. Mol. Biol.* **29**, 307–313 (1967).
26. Higashi, K., Adams, H. R., and Busch, H., Base composition of RNA fractions of nuclei of Walker tumor isolated with the citric acid procedure. *Cancer Res.* **26**, 2196–2201 (1966).
27. Hindley, J., Fractionation of ^{32}P-labelled ribonucleic acids on polyacrylamide gels and their characterization by fingerprinting. *J. Mol. Biol.* **30**, 125–136 (1967).
28. Hodnett, J. L., and Busch, H., Isolation and characterization of the uridylic acid-rich RNA of rat liver nuclei. *J. Biol. Chem.* **243**, 6336–6344 (1968).
29. Holley, R. W., Experimental approaches to the determination of the nucleotide sequences of large oligonucleotides and small nucleic acids. *Progr. Nucleic Acid Res. Mol. Biol.* **8**, 37–47 (1968).
30. Holley, R. W., Apgar, J., Doctor, B. P., Farrow, J., Marini, M. A., and Merrill, S. H., A simplified procedure for the preparation of tyrosine and valine-acceptor fractions of yeast "soluble ribonucleic acid." *J. Biol. Chem.* **236**, 200–202 (1961).
31. Holley, R. W., Apgar, J., Everett, G. A., Marquisee, M., Merrill, S. H., Penswick, J. R., and Zamir, A., Structure of a ribonucleic acid. *Science* **147**, 1462–1465 (1965).
32. Holley, R. W., Doctor, B. P., Merrill, S. H., and Saad, F., Countercurrent distribution of rat-liver "soluble" fraction ribonucleic acids. *Biochim. Biophys. Acta* **35**, 272–273 (1959).
33. Huang, R. C., and Bonner, J., Histone-bound RNA, a component of native nucleohistones. *Proc. Natl. Acad. Sci. U. S.* **54**, 960–967 (1965).

34. Jacob, J., and Sirlin, J. L., The incorporation of pseudouridine into nucleolar transfer RNA. *J. Zool. Soc. India* **14**, 1–3 (1962).
35. King, H. W. S., and Fitschen, W., Characterisation of RNA from the smooth endoplasmic reticulum of rat liver. *Biochim. Biophys. Acta* **155**, 32–37 (1968).
36. Knight, E., Jr., and Darnell, J. E., Distribution of 5 S RNA in HeLa cells. *J. Mol. Biol.* **28**, 491–502 (1967).
37. Lake, J. A., and Beeman, W. W., On the conformation of yeast transfer RNA. *J. Mol. Biol.* **31**, 115–125 (1968).
38. Lal, B. M., and Burdon, R. H., Maturation of low molecular weight RNA in tumour cells. *Nature* **213**, 1134–1135 (1967).
39. Larsen, C. J., Galibert, F., and Lelong, J. C., Influence of conditions of extraction on the chromatographic property of 5 S RNA extracted from KB cells. *Compt. Rend.* **265**, 1337–1340 (1967).
40. Larsen, C. J., Galibert, F., Lelong, J. C., and Boiron, M., Charactérisation dans la fraction nucléaire de cellules KB d'un ARN métaboliquement stable de faible poids moléculaire. *Compt. Rend.* **264**, 1523–1526 (1967).
40a. Larsen, C. J., Galibert, F., Hampe, A., and Boiron, M., Etude des RNA nucleaires de faible poids moleculaire de la cellule KB. *Bull. Soc. Chim. Biol.* **51**, 649–668 (1969).
41. Madison, J. T., Primary structure of RNA. *Ann. Rev. Biochem.* **37**, 131–148 (1968).
42. Madison, J. T., Everett, G. A., and Kung, H., Nucleotide sequence of a yeast tyrosine transfer RNA. *Science* **153**, 531–534 (1966).
43. McPhie, P., Hounsell, J., and Gratzer, W. B., The specific cleavage of yeast ribosomal ribonucleic acid with nucleases. *Biochemistry* **5**, 988–993 (1966).
44. Morell, P., and Marmur, J., Association of 5 S ribonucleic acid to 50 S ribosomal subunits of *Escherichia coli* and *Bacillus subtilis*. *Biochemistry* **7**, 1141–1152 (1968).
45. Moriyama, Y., Hodnett, J. L., Prestayko, A. W., and Busch, H., Studies on the nuclear 4–7 S RNA of the Novikoff hepatoma. *J. Mol. Biol.* **39**, 335–349 (1969).
45a. Moriyama, Y., Ip, P., and Busch, H., Chromatography of nuclear 4–7 S RNA on DEAE–Sephadex columns. *Biochim. Biophys. Acta,* in press.
46. Muramatsu, M., Hodnett, J. L., and Busch, H., Base composition of fractions of nuclear and nucleolar ribonucleic acid obtained by sedimentation and chromatography. *J. Biol. Chem.* **241**, 1544–1550 (1966).
47. Nakamura, T., Prestayko, A. W., and Busch, H., Studies on nucleolar 4 to 6 S ribonucleic acid of Novikoff hepatoma cells. *J. Biol. Chem.* **243**, 1368–1375 (1968).
48. Nakamura, T., Rapp, F., and Busch, H., Common features of the base composition of rapidly labeled RNA of nucleoli in a number of experimental tumors. *Cancer Res.* **27**, 1084–1091 (1967).
49. Peacock, A. C., and Dingman, C. W., Resolution of multiple ribonucleic acid species by polyacrylamide gel electrophoresis. *Biochemistry* **6**, 1818–1827 (1967).
50. Pene, J. J., Knight, E., Jr., and Darnell, J. E., Jr., Characterization of a new low molecular weight RNA in HeLa cell ribosomes. *J. Mol. Biol.* **33**, 609–623 (1968).
51. Perry, R. P., The cellular sites of synthesis of ribosomal and 4 S RNA. *Proc. Natl. Acad. Sci. U. S.* **48**, 2179–2186 (1962).
52. Prestayko, A., and Busch, H., Low molecular weight RNA of the chromatin fraction from Novikoff hepatoma and rat liver nuclei. *Biochim. Biophys. Acta* **169**, 338–349 (1968).

52a. Rein, A., and Penman, S., Species specificity of the low molecular weight nuclear RNA's. *Biochim. Biophys. Acta* **190,** 1–9 (1969).
52b. Ro-Choi, T. S., Moriyama, Y., Choi, Y. C., and Busch, H., Isolation and purification of a nuclear 4.5 S ribonucleic acid of the Novikoff hepatoma. *J. Biol. Chem.* **245,** April, 1970.
53. Rushizky, G. W., and Knight, C. A., An oligonucleotide mapping procedure and its use in study of tobacco mosaic virus nucleic acid. *Virology* **11,** 236–249 (1960).
54. Rushizky, G. W., and Sober, H. A., The preparation and characterization of large oligoribonucleotides. *Progr. Nucleic Acid Res. Mol. Biol.* **8,** 171–195 (1968).
55. Shankarnarayan, K., Steele, W. J., and Busch, H., Evidence that the granular and fibrillar elements of nucleoli contain 28 S and 6 S RNA, respectively. *Exptl. Cell Res.* **43,** 483–492 (1966).
56. Shearer, R. W., and McCarthy, B. J., Evidence for ribonucleic acid molecules restricted to the cell nucleus. *Biochemistry* **6,** 283–289 (1967).
57. Sirlin, J. L., Synthesis of 4 S RNA in the nucleolus of *Smittia*. *In* "The Cell Nucleus—Metabolism and Radiosensitivity" (J. A. Cohen, chairman), pp. 87–96. Taylor & Francis, London, 1966.
58. Sirlin, J. L., Jacob, J., and Tandler, C. J., Transfer of (^{14}C-methyl) methionine to nucleolar ribonucleic acid. *Biochem. J.* **87,** 37P (1963).
59. Sirlin, J. L., Jacob, J., and Tandler, C. J., Transfer of the methyl group of methionine to nucleolar ribonucleic acid. *Biochem. J.* **89,** 447–452 (1963).
60. Smith, J. D., and Dunn, D. B., The occurrence of methylated guanines in ribonucleic acids from several sources. *Biochem. J.* **72,** 294–301 (1959).
61. Staehelin, M., Rogg, H., Baguley, B. C., Ginsberg, T., and Wehrli, W., Structure of a mammalian serine tRNA. *Nature* **219,** 1363–1365 (1968).
62. Steele, W. J., and Busch, H., RNA: Isolation and fractionation. *Methods Cancer Res.* **3,** 61–152 (1967).
63. Stephenson, M. L., and Zamecnik, P. C., Purification of valine transfer ribonucleic acid by combined chromatographic and chemical procedures. *Proc. Natl. Acad. Sci. U. S.* **47,** 1627–1635 (1961).
64. Watson, J. D., and Ralph, R. K., Template activity of 7 S RNA isolated from mammalian cells. *Biochem. Biophys. Res. Commun.* **24,** 257–263 (1966).
65. Watson, J. D., and Ralph, R. K., The nature of 7 S RNA in mammalian cells. *J. Mol. Biol.* **22,** 67–78 (1966).
66. Watson, J. D., and Ralph, R. K., Stability of 5 S ribonucleic acid in mammalian cells. *J. Mol. Biol.* **26,** 541–544 (1967).
66a. Weinberg, R. A., and Penman, S., Small molecular weight monodisperse nuclear RNA. *J. Mol. Biol.* **38,** 289–304 (1968).
66b. Weinberg, R., and Penman, S., Metabolism of small molecular weight monodisperse nuclear RNA. *Biochim. Biophys. Acta* **190,** 10–29 (1969).
67. Weith, H. L., and Gilham, P. T., Structural analysis of polynucleotides by sequential base elimination. The sequence of the terminal decanucleotide segment of the ribonucleic acid from bacteriophage f2. *J. Am. Chem. Soc.* **89,** 5473–5474 (1967).
68. Woods, P. S., Autoradiographic studies of transfer RNA metabolism in *Vicia faba* root-tip cells. *Natl. Cancer Inst. Monograph* **23,** 271–284 (1966).
69. Woods, P. S., and Zubay, G., Biochemical and autoradiographic studies of different RNA's: Evidence that transfer RNA is chromosomal in origin. *Proc. Natl. Acad. Sci. U. S.* **54,** 1705–1712 (1965).

CHAPTER VII

Nucleolar Structural Proteins and Enzymes

I. Introduction

Interest in the nucleolar proteins is increasing because of the evidence that there are a number of important types of proteins involved in enzymatic, structural, and possibly regulatory functions for nucleolar and nuclear activity. Of the synthetic enzymes or groups of enzymes in the nucleolus, reference has been made in Chapter V to both the aggregate nucleolar RNA polymerase and the nucleolar methylases. For reasons that are not entirely clear the nucleolus also contains NAD pyrophosphorylase or synthetase, a poly A synthetase, and some ATPases. As noted in the discussion of the current concept of nucleolar RNA synthesis, the nucleolus also must contain cleavage enzymes of which some species of RNase may be important, and, in addition, there may be other special cleavage enzymes (Chapter V).

In addition to the enzymes, there are in the nucleolus a substantial number of proteins that are "ribosomelike" in amino acid composition (*12*). These proteins may either be synthesized in the nucleolus, or arrive from extranucleolar sources to be incorporated into the nucleolar ribosomal precursors. The process of the formation and maturation of the nucleolar ribonucleoprotein masses are apparently as complex as they are fascinating.

Among the more exciting recent findings are those showing the presence in the nucleolus of special "nucleolus-specific" proteins (*4, 26a, 30, 31*), some of which may be hormone sensitive (*11*). The evidence for this idea has been developed from studies with antinucleolar antibodies (Fig. 7.1). Although the field is clearly in its infancy, it is manifest that the nucleolar proteins offer an interesting challenge to research in both chemistry and biology from an analytical and functional point of view as well as for studies on pathology. Recently Ritchie (*41′*) has indicated that it is possible to distinguish by immunological techniques three major antigenic structures within nucleoli of rat liver, i.e., the pars amorpha, the nucleolonemas, and the perinucleolar regions. This result supports the concept that nucleolini (nucleolar vacuoles) contain distinctive protein elements (*29a′*).

Stains for Nucleolar Proteins

As noted in Chapter II, techniques for histochemical demonstration of nucleolar acidic proteins have now become available (*34, 45a*). Although these techniques are not completely specific, they do permit visualization of acidic nucleolar proteins (Fig. 7.2) presumably because of their associations with the dyes under conditions (pH 4.9) when the binding groups of other proteins are unavailable.

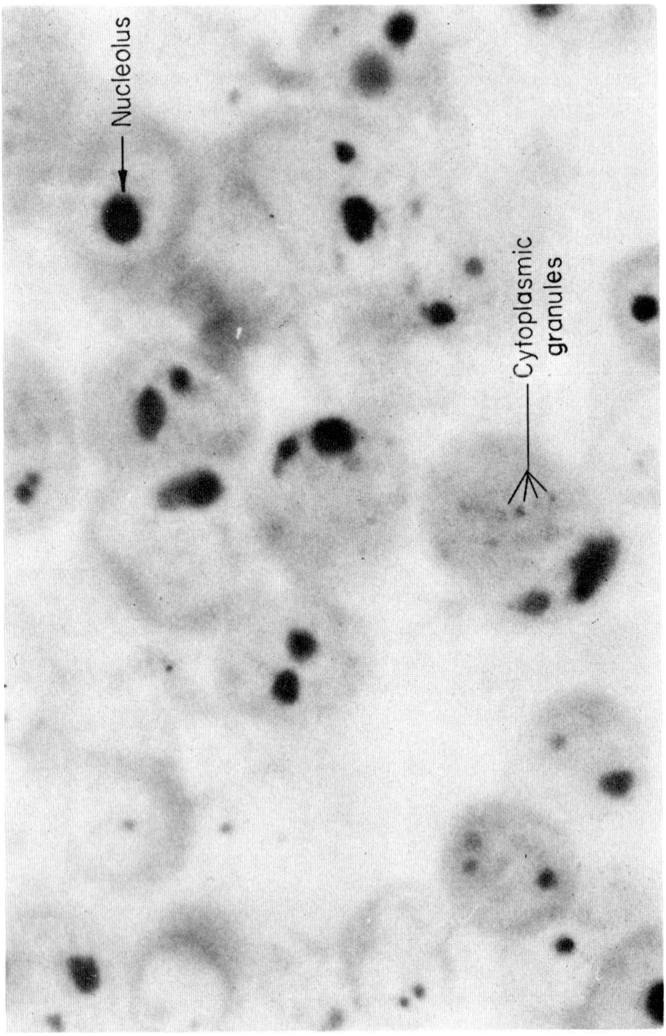

Fig. 7.1. Selective staining of nucleoli with antinucleolar antibodies, stained with fluorescein.

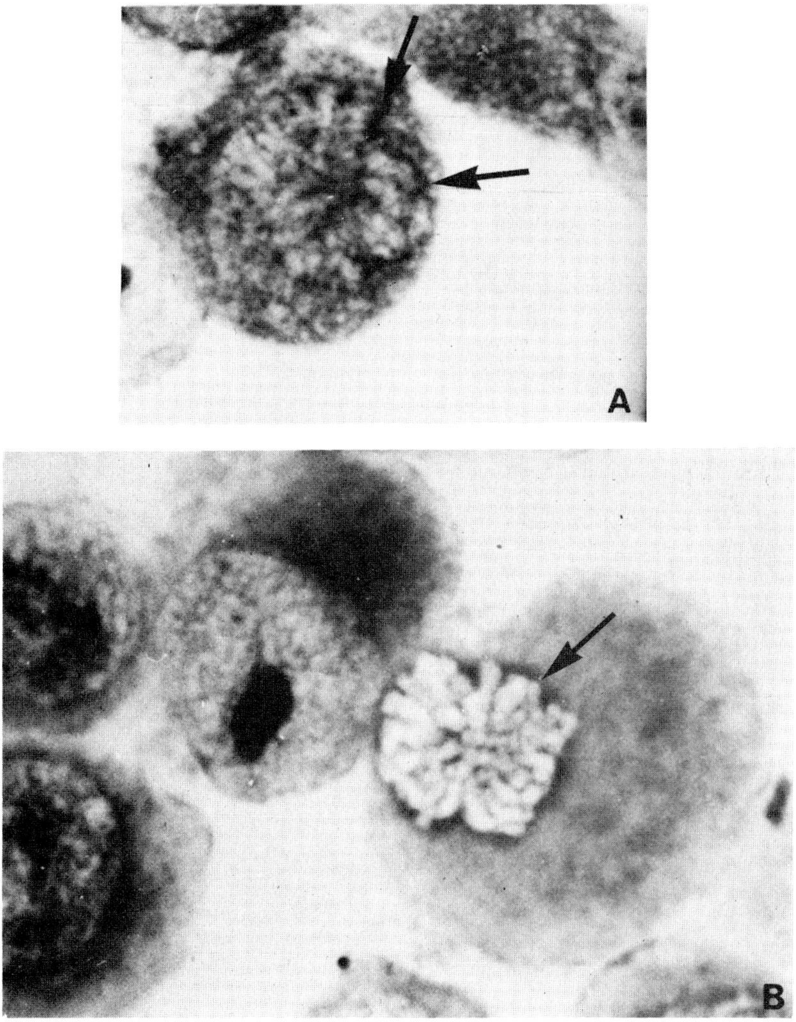

Fig. 7.2. The toluidine blue stain for acidic nuclear proteins (*45a*) showing the density of nucleolar proteins and the lesser density in the nuclear ribonucleoprotein network. (A) Acidic proteins in the nuclear ribonucleoprotein network. (B) One cell (left) has a densely stained nucleolus. The adjacent cell shows staining of acidic proteins in a mitotic figure (arrow); in the peri- and interchromosomal area a network of acidic proteins is visible.

II. Nucleolar Proteins as Ribosomal Precursor Proteins

In view of the evidence provided in preceding chapters that the nucleolus is the site of synthesis of ribosomal RNA the logical question is whether the nucleolus is the site of assembly of the ribosomes themselves in either a complete or partially completed state (*40*). If this is the case, one would anticipate the presence in the nucleolus of ribosomal proteins and the existing evidence supports the view that the vast bulk of the nucleolar mass is composed of these or precursor proteins. In earlier studies from the authors' laboratory (Tables 7.1, 7.2), it was pointed out

TABLE 7.1

Amino Acid Composition of Nuclear Proteins Soluble in 0.14 *M* NaCl (*12*)[a]

	Nuclear ribosomes			Globulins		Nuclear sap		Microsomes	
Amino acid	Calf thymus	Liver	Tumor	Liver	Tumor	Liver	Tumor	Liver	Tumor
Alanine	7.4	7.5	7.9	7.8	7.9	9.8	9.1	8.1	8.0
Arginine	7.8	5.6	6.3	5.1	5.2	2.9	3.6	4.9	5.8
Aspartic acid	9.1	9.8	9.6	10.0	10.4	10.5	10.6	10.0	9.6
Cystine	0.5	—	—	—	0.9	0.8	0.5	—	0.6
Glutamic acid	10.3	11.2	11.5	11.9	12.3	9.2	10.7	12.1	11.6
Glycine	9.3	10.8	9.7	9.4	7.1	8.4	8.5	8.4	8.7
Histidine	2.5	2.3	2.2	2.6	2.9	4.0	3.6	2.2	2.4
Isoleucine	5.2	3.9	4.5	4.3	4.7	3.8	4.0	4.4	4.7
Leucine	8.7	7.8	8.5	8.6	9.0	9.9	9.0	9.6	8.8
Lysine	10.0	8.7	9.1	8.5	9.0	8.3	9.1	7.0	9.1
Methionine	2.0	1.4	1.1	1.1	1.4	1.3	1.2	1.9	1.4
Phenylalanine	3.6	4.3	3.7	4.0	3.6	4.4	3.8	5.1	3.6
Proline	4.8	5.5	5.2	5.2	5.0	5.5	5.4	6.0	5.2
Serine	4.5	6.7	6.0	6.3	6.0	6.2	6.4	6.7	5.8
Threonine	4.8	4.8	5.0	4.9	5.2	5.1	5.2	5.1	5.2
Tyrosine	3.0	2.5	2.2	3.4	2.4	2.2	2.0	2.9	2.3
Valine	6.6	5.1	6.0	6.7	7.1	7.6	7.3	6.5	7.1

[a] The values are percentages of total moles of amino acids recovered. The tumor studied was the Walker 256 carcinosarcoma.

(*12*) that the amino acid compositions of the microsomal proteins obtained from the liver and the Walker tumor was not only similar for these two tissue types but also was similar to that of the "nuclear ribosomes" (probably largely derived from the outer layer of the nuclear envelope). The overall compositions of the nucleolar acidic proteins (*7, 12, 21, 22*) are like those of the microsomal proteins (Table 7.1). Even in those early studies, the evidence suggested that the nucleolus was composed largely of "ribosomal" or "ribosomal precursor" proteins.

TABLE 7.2

Amino Acid Composition of Nucleolar Acidic Proteins (*12*)[a]

Amino acid	Walker tumor	Rat liver
Alanine	7.0	7.0
Arginine	6.0	5.5
Aspartic acid	10.0	9.7
Glutamic acid	13.8	13.7
Glycine	7.6	7.3
Histidine	2.4	2.5
Isoleucine	4.1	4.6
Leucine	8.9	9.3
Lysine	7.3	7.7
Methionine	2.1	2.3
Phenylalanine	3.6	3.7
Proline	5.3	6.2
Serine	7.6	7.2
Threonine	5.1	5.3
Tyrosine	2.4	2.2
Valine	6.0	5.6

[a] The values are percentages of total moles of amino acids in the proteins. The acid-soluble proteins were previously extracted with 0.25 *N* HCl.

On the basis of a ratio of 8:1 of the dry mass of the precursor proteins to RNA, it would appear that the nucleolus has a prior pool of such proteins that waits for the synthesis of RNA. At this point, it is not possible to define the precise amount of precursor protein relative to the amount of RNA because so little is known about the various nucleolar proteins. However, about 40–60% of the nucleolar proteins would appear to fall into this "precursor" category on the basis of analyses made by Grogan *et al.* (*22*).

A. The Complex of Ribosomal Proteins

One of the problems that was not readily approached in the early studies on nucleolar proteins was the definition of the number of types of proteins present (*5, 12*). It is only recently that elegant techniques have been developed that permit fractionation and amino acid analysis of these proteins. In addition, reassembly or reconstitution of the ribosomes has been virtually completed in the presence of appropriately sized RNA fibers.

The work of a number of groups, including that of Moore *et al.* (*32*), Traub and Nomura (*53*), Tissieres and his colleagues (*51, 52, 54, 55, 58*) and Fogel and Sypherd (*16, 47*), has provided new methodology for

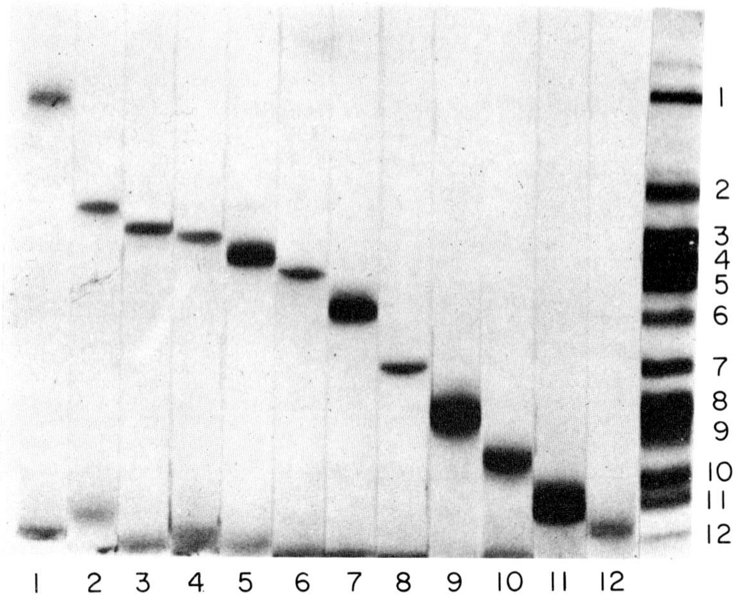

Fig. 7.3. Disc electrophoresis of individual ribosomal proteins in 7.5% acrylamide gel. The gels were made up at pH 4.3 in 8 *M* urea. The gel at the far right shows the unfractionated *E. coli* 30 S ribosomal protein. Migration was from anode (top) to cathode (bottom) (from *16*).

extraction of the ribosomal proteins and separation of these proteins on polyacrylamide gels. Figure 7.3 shows the type of separations that can be achieved in polyacrylamide gels buffered at pH 8.3 and containing 8 *M* urea. As noted in Fig. 7.3, the 30 S ribosomal particle contains approximately 12 components and each of these has been separated into electrophoretically homogeneous elements by this method. As shown in Table 7.3 (*16*), the constitution of the proteins in each of these bands varies greatly. Band 11 is very rich in lysine and alanine, band 2 contains the highest concentration of aspartic acid, band 3b contains the highest concentration of glutamic acid, and band 5, the highest concentration of valine. Bands 7, 8a, 8b, and 9 are interesting because of their high arginine concentrations; the values are similar to those of some of the histones (*12*). These results are in good agreement with those obtained for ribosomal proteins in *E. coli* by the group of Moore *et al.* (*32*) and others who carried out their extractions with acetic acid. They found 14 electrophoretic bands which were initially separated into groups by filtration on Sephadex, chromatography on carboxymethylcellulose, and separation on polyacrylamide gels. They prepared tryptic digests of these proteins and showed their differences in peptide fingerprints. The molecu-

TABLE 7.3

AMINO ACID COMPOSITION OF THE PROTEINS OF THE 30 S RIBOSOMAL PARTICLE OF *E. coli* (*16*)

	2	3a	3b	4a	4b	5	6	7	8a	8b	9	10	11	12	30 S
Lys	8.1	9.9	6.0	8.5	6.8	7.8	7.5	7.7	8.6	7.3	10.3	12.4	14.0	10.3	8.5
His	1.9	1.5	2.9	1.6	2.3	1.6	2.4	1.2	2.1	3.4	2.4	3.7	3.0	1.7	2.1
Arg	6.1	9.1	8.3	8.8	7.7	9.8	9.8	13.1	11.4	11.9	11.7	9.5	8.5	9.6	9.1
Asp	11.5	7.7	10.1	8.3	9.6	7.3	8.9	6.4	8.2	8.9	7.6	7.2	8.8	7.0	9.0
Thr	5.4	5.0	5.3	4.0	5.1	3.4	6.1	4.3	5.6	5.2	6.3	6.0	3.5	5.1	4.9
Ser	5.1	4.6	4.3	5.3	4.6	7.7	5.1	5.4	5.7	5.2	5.9	5.3	6.3	8.1	5.0
Glu	10.2	11.5	15.1	11.8	13.3	10.6	11.5	11.9	7.6	9.8	9.6	7.4	9.5	10.5	11.4
Pro	2.7	4.6	3.5	3.3	3.8	3.7	4.2	3.1	4.8	2.1	3.3	5.4	2.3	3.2	3.3
Gly	7.6	8.8	6.0	9.4	7.5	6.6	7.5	9.6	9.3	7.1	6.5	8.1	5.7	7.1	8.0
Ala	10.6	10.9	11.1	10.7	11.3	12.7	10.5	8.9	11.1	10.5	7.8	9.0	17.4	13.2	10.3
Val	8.2	7.8	8.6	8.4	8.6	10.1	7.2	8.1	7.9	8.1	9.8	8.7	4.6	4.7	8.5
Met	3.1	1.5	3.7	3.2	3.7	2.8	2.1	2.2	1.4	1.2	0.9	1.7	2.9	1.3	2.0
Iso	5.5	7.0	5.3	4.4	5.2	3.2	5.8	5.5	7.0	6.4	6.7	4.2	6.8	3.7	5.4
Leu	8.8	6.2	4.9	7.9	5.7	8.0	7.5	6.7	5.8	7.7	7.0	6.9	4.1	5.4	7.7
Tyr	1.0	1.7	2.2	2.3	2.0	2.1	1.4	3.0	1.2	1.7	1.2	1.3	0.2	0.9	1.8
Phe	4.4	2.2	2.8	2.1	2.7	2.8	2.6	3.0	2.3	3.6	2.9	3.1	2.3	2.2	2.9

TABLE 7.4

AMINO ACID COMPOSITIONS OF PROTEINS OF THE 30 S RIBOSOMAL PARTICLES OF *E. coli* (*32*)

Protein	Lys	His	Arg	Asp	Thr	Ser	Glu	Pro	Gly	Ala	Val	Met	Ile	Leu	Tyr	Phe
	Moles/100 moles of amino acids															
1	16.0	2.2	8.3	9.4	3.2	4.4	9.9	2.0	3.3	21.9	3.1	2.1	7.3	3.9	1.1	2.3
2a	14.0	5.2	9.0	6.6	6.0	5.5	6.0	7.1	8.3	7.3	7.7	1.4	4.5	7.6	1.4	4.5
3	11.3	3.4	9.5	7.6	7.0	6.9	10.2	2.9	5.6	3.8	13.1	1.2	7.9	6.0	1.0	2.5
4a	5.8	2.4	12.1	10.2	3.8	4.2	9.5	2.6	7.4	13.4	9.8	1.0	7.3	4.9	1.2	4.6
4b	7.9	4.1	11.7	8.4	5.9	5.8	11.7	1.6	7.1	9.1	6.4	0.7	4.1	11.0	2.2	2.4
5	8.0	1.2	13.5	6.1	4.2	5.4	12.7	2.5	10.1	8.5	7.6	2.0	5.3	7.1	3.0	3.0
6	5.7	1.9	10.3	10.1	7.3	2.8	12.7	4.9	5.4	9.8	8.9	0.8	8.1	9.4	0.0	2.1
7	8.4	1.3	11.1	7.8	3.7	6.2	12.3	4.0	5.5	12.2	10.6	1.9	3.1	7.3	1.8	2.5
8a	6.9	2.2	5.9	8.5	5.0	4.1	10.5	3.0	12.4	11.9	11.1	4.0	5.2	5.2	1.1	2.4
10a	5.0	4.2	9.2	11.0	5.0	3.8	16.0	3.8	4.5	10.9	8.0	4.0	5.1	4.3	2.4	3.3
10b	8.3	1.7	8.3	8.4	5.2	4.2	11.4	3.2	8.7	11.5	9.9	1.4	7.4	6.6	1.7	2.3
11	8.9	0.1	6.0	7.5	5.0	5.4	11.8	3.5	9.2	11.5	9.5	3.3	6.7	7.9	1.5	2.2
13	7.5	1.7	5.5	11.8	4.6	4.7	14.2	2.0	9.7	8.7	12.0	0.6	5.1	8.3	0.6	3.1
Total 30 S	8.6	2.0	7.8	8.6	4.9	4.5	10.6	3.7	7.9	10.3	9.2	2.5	5.8	7.5	1.9	2.9

lar weights of these proteins varied from 3000 to 28,000 and the amino acid analyses of proteins are shown in Table 7.4 (*32*). Although the values of the amino acids were not in complete agreement with those of Fogel and Sypherd (*16*), both groups found proteins that were rich in lysine and alanine, proteins rich in both arginine and lysine, like the histones, and proteins relatively rich in glutamic acid.

Recently, studies similar to those reported above on the composition of ribosomal 30 S particles that contain 16 S RNA have been reported for the 50–60 S ribosomal particle that contains 23 S RNA. These studies have again shown that there are many proteins in the particle and further established the substantial range of molecular weights, basicity, and features similar to those of the proteins in the smaller particle. As reported for the smaller particle, reconstitution studies have been made. Moore *et al.* (*32*) showed that in smaller 30 S particles, there is one and only one copy of each protein in the particle. Accordingly, these protein–RNA complexes are highly organized and definitely arranged mosaics in which the position of each protein element is precisely ordered. Since the reconstituted particle has the same ribosomal activity as the control, it would appear that bonding forces rather than specific energy sources are crucial to the localization of these components. In addition, it would seem that each of these individual proteins has a definite function in the overall activity of the ribosome, whether it serves as a binding site, an enzyme, or as a specific structural element of the particle.

One might speculate that the definite order which seems to exist must be related to sequences on the RNA molecule, inasmuch as the absence of the RNA does not produce the same ordered organization that is found when it is present. Inasmuch as the proteins are present in the nucleolus, at least in an overall sense, it may be suggested that there are rather unorganized structures in the nucleolus that lie in wait for the synthesis of the crucial organizing RNA fibers generated by the nucleolar organizer and its associated synthetic systems (Chapter V).

Little is known about specific nucleolar proteins in the degree of refinement known for the highly purified ribosomal proteins of *E. coli*. Studies similar to those made with these refined systems are in progress with isolated nucleoli of liver, Novikoff hepatoma and the Walker tumor (Fig. 7.4). Polyacrylamide gel electrophoretic patterns of nucleolar proteins of normal liver reveal two major bands and several minor bands (Fig. 7.4). Similar patterns from the nucleoli of Novikoff hepatoma revealed five major bands. The polyacrylamide gel patterns of the Walker tumor contain four major bands. Bands 3, 4, 8, and 9 of the tumor nucleolar proteins are present in very low concentration or are absent from the nucleolar protein patterns of normal liver. On the other hand,

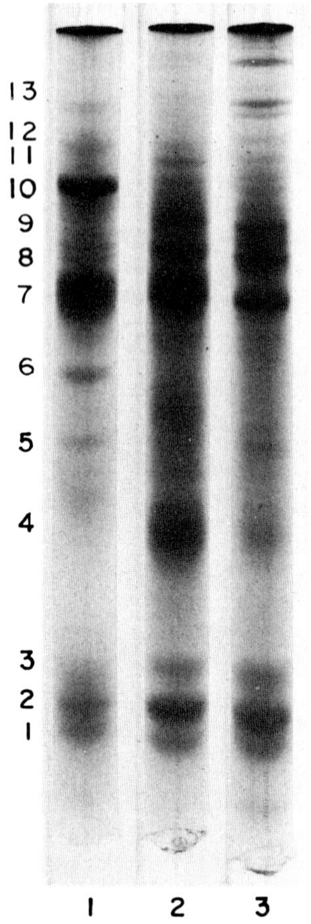

Fig. 7.4. Polyacrylamide gel electrophoresis patterns of nucleolar proteins of rat liver (1), Novikoff hepatoma (2), and Walker tumor (3). These proteins were extracted with the following buffer: Tris-acetate, 0.01 *M* (pH 7.6); magnesium acetate, 0.01 *M*; EDTA, 1 m*M*; and β-mercaptoethanol, 1 m*M* in 8 *M* urea. The polyacrylamide gel electrophoresis runs were made at pH 4.3 for 3 hours. The patterns for the proteins of the two tumors are remarkably similar and are distinctive from those of the normal liver. Protein bands 4, 8, and 9 are present in the tumors and are in low concentration or absent in the normal liver nucleoli (courtesy of Dr. James E. Dworak).

band 10 is dense in normal liver and is in low concentration in the nucleolar proteins of the two tumors. These differences in nucleolar protein patterns may be as important to the functional state of the nucleoli of the tumor cells as are the differences found in the newly synthesized 45 S nucleolar RNA (Chapter V).

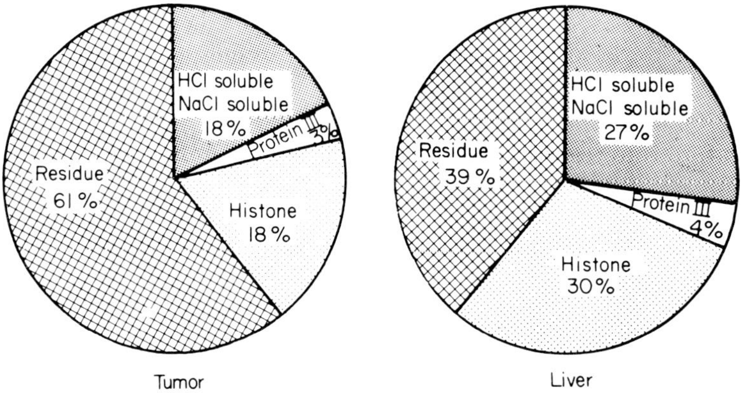

Fig. 7.5. Distribution of nucleolar proteins on the basis of solubility (from *22*).

B. Distribution of the Nucleolar Proteins

As shown in the accompanying diagram (Fig. 7.5), the nucleolar proteins can be separated into classes that are soluble in dilute saline solution, those that are soluble in acid and insoluble in dilute saline solution, and those that are neither soluble in acid nor in dilute saline solution (*21, 22*). Those soluble in dilute saline solution are generally also soluble in the dilute HCl so that workers isolating histones from nucleoli or nuclei with mineral acids should first treat the cells or the tissues with dilute saline solution or otherwise clearly differentiate between their product and the "true" histones; i.e., those basic proteins free of contaminating acidic protein elements that represent the non-histone acid-soluble elements derived from the nucleolus and possibly other nuclear sites (*12*).

Mundell (*33a*) has suggested that like some ribosomal proteins, some of the nucleolar proteins are basic and these may be proteins that complex with newly synthesized RNA to form the basis of the ribosome in a "nucleolar core." He also suggests that a "cortical region" of the nucleolus contains more mature preribosomal products.

C. The Saline-Soluble Nucleolar Proteins

The amino acid compositions (*12, 21*) show that most of the nucleolar proteins are acidic (Tables 7.2, 7.5). Aspartic and glutamic acids compose 21–25% of the total amino acids in these proteins and the basic amino acids, including lysine, arginine, and histidine, compose approximately 15% of the total amino acids of this series of proteins. The amino acid compositions of the proteins of the nucleolar "residue" are very similar

TABLE 7.5

Amino Acid Composition of Nucleolar Protein Fractions (*7*, *21*, *22*)

	Nucleolar									Nuclear acid-soluble (histone)[a]			Cytoplasmic ribosomal[a]	
	"Soluble"[a]			Acid-soluble[a]			Residual[a]							
	RL	WT	PS	RL	WT	PS	RL	WT	PS	RL	WT	PS	RL	PS
Alanine	9.1	9.5	6.7	11.9	11.0	6.7	7.9	8.0	8.3	11.8	10.8	10.6	8.6	8.4
Arginine	3.9	4.4	6.3	8.1	8.5	8.0	5.5	5.1	6.4	7.4	8.0	6.8	7.0	6.3
Aspartic acid	10.2	9.9	12.0	6.0	6.5	10.5	9.1	9.2	10.2	6.3	6.9	9.7	8.5	9.5
½ cysteine	1.9	1.0	—	Tr	Tr	—	1.4	0.7	—	Tr	Tr	—	1.0	—
Glutamic acid	11.3	12.8	15.9	9.7	10.4	14.6	12.8	13.0	12.8	9.9	10.5	11.0	9.2	10.2
Glycine	9.0	8.3	6.8	9.0	8.9	7.7	8.6	8.8	7.6	8.7	8.4	6.5	7.6	8.1
Histidine	2.9	1.7	1.8	1.8	1.6	2.0	2.1	1.7	2.0	1.6	1.6	1.8	2.2	1.4
Isoleucine	4.3	4.1	4.9	4.3	4.1	4.9	4.4	4.2	4.9	4.4	4.3	4.7	5.3	5.7
Leucine	9.7	8.5	8.8	7.9	7.7	7.7	9.5	9.1	8.6	8.0	7.9	6.9	9.4	8.7
Lysine	8.1	9.3	7.2	13.5	13.7	7.3	7.8	7.8	9.2	13.3	13.7	13.9	9.0	9.1
Methionine	1.0	1.7	0.7	0.9	0.5	1.0	1.4	1.8	1.3	1.0	0.6	0.8	2.2	1.6
Phenylalanine	4.6	2.7	4.3	2.3	2.3	5.6	3.5	3.5	3.9	2.4	2.3	2.9	3.9	4.1
Proline	5.7	5.0	4.8	5.1	5.3	5.0	5.8	5.4	4.7	5.4	5.4	5.8	5.1	5.0
Serine	6.6	6.4	7.4	6.1	6.0	7.0	7.7	6.4	6.5	5.9	5.9	5.0	5.5	5.9
Threonine	4.6	5.2	3.9	4.8	5.4	4.3	4.4	5.1	4.6	5.6	5.6	5.3	5.3	5.6
Tyrosine	1.7	1.9	2.7	2.1	1.8	2.1	1.6	1.6	2.2	1.8	1.6	2.4	3.3	3.0
Valine	7.3	6.2	6.8	6.5	6.8	5.6	6.5	6.1	6.7	6.3	6.3	5.9	6.9	7.4
Acidic/basic	1.4	1.5	1.8	0.7	0.7	1.5	1.4	1.5	1.3	0.7	0.7	0.9	1.0	1.2

[a] RL = rat liver; WT = Walker 256 carcinosarcoma; PS = pea seedlings; Tr = trace.

to those of the proteins extracted with dilute or concentrated saline solutions or with dilute HCl, although the content of aspartic and glutamic acids is slightly lower. Since studies on extraction of these proteins with urea-containing solutions and fractionation on polyacrylamide gels are only in their early stages it is not possible to describe further the properties of these proteins.

Analytical studies have been carried forward on the saline-soluble nucleolar proteins. In this laboratory, studies have been made (*22*) on the fractionation of the nucleolar saline-soluble proteins on DEAE-cellulose columns using salt gradients as indicated in Fig. 7.6. After a breakthrough peak, six peaks were eluted with increasing concentrations of NaCl and two additional fractions obtained by washes with 0.2 *N* HCl and 0.2 *N* NaOH, respectively. Slightly more than half the protein emerged in the fractions eluted with 0.1, 0.2, and 0.4 *M* NaCl. As noted in Table 7.6, the fraction eluted with 0.1 *M* NaCl contained the largest amount of serine in any of the fractions; i.e., 14.9%. This "serine-rich" protein may be of interest in view of one reaction suggested by Taborsky (*47a*); he found that serylphosphates may be an energy-transfer component of proteins when the 2,3-carbon-to-carbon-bond becomes a double

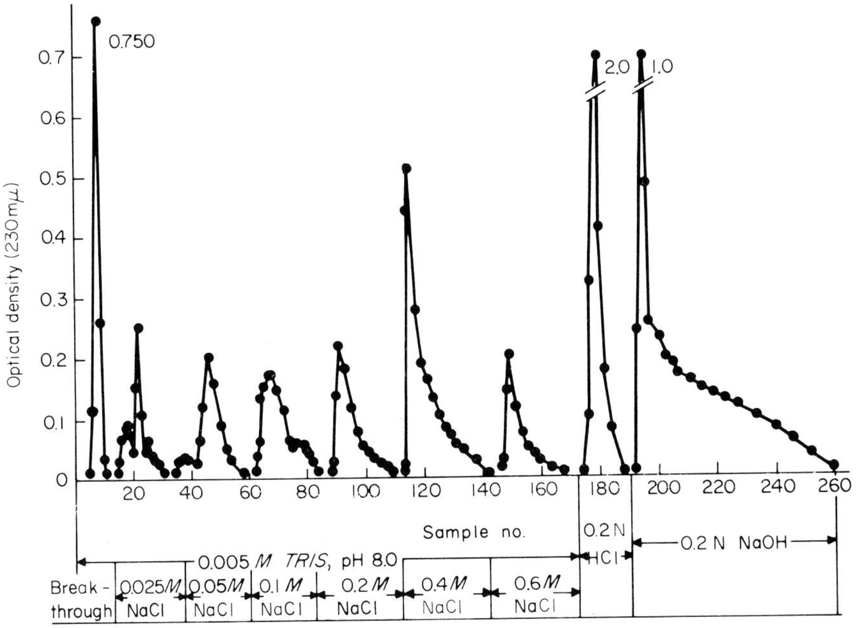

Fig. 7.6. The elution pattern of saline-soluble nucleolar proteins on DEAE-cellulose (*21*).

TABLE 7.6

Amino Acid Content of the Proteins Fractionated on DEAE–Cellulose[a]

Fraction	Tris-insoluble	Break-through	0.025 *M* NaCl	0.05 *M* NaCl	0.1 *M* NaCl	0.2 *M* NaCl	0.4 *M* NaCl	0.6 *M* NaCl	0.2 *N* HCl	0.2 *N* NaOH
Alanine	7.7	8.4	7.2	8.4	8.6	8.9	5.9	6.4	7.4	7.5
Arginine	5.9	4.8	3.3	2.7	3.4	2.4	3.5	3.7	5.8	5.1
Aspartic acid	10.5	7.9	9.9	9.8	10.5	11.5	17.8	8.8	10.1	9.7
½ Cystine	1.2	1.6	1.1	1.8	1.8	1.3	0.9	1.9	Trace	Trace
Glutamic acid	12.4	13.2	16.3	14.0	17.5	19.9	21.3	17.3	14.2	12.8
Glycine	8.5	9.9	14.4	12.1	10.2	6.3	6.9	11.6	7.0	8.8
Histidine	2.6	1.8	1.3	2.0	2.2	1.5	1.0	0.5	2.0	1.8
Isoleucine	4.1	3.1	3.2	2.2	2.2	2.7	2.0	1.8	3.5	4.6
Leucine	8.7	5.7	3.7	4.0	3.4	6.5	5.6	2.5	8.3	9.1
Lysine	8.7	10.6	9.3	8.7	8.2	11.2	8.6	11.9	8.6	7.2
Methionine	1.4	1.2	1.2	0.8	0.6	1.2	0.9	Trace	2.1	2.2
Phenylalanine	3.5	2.1	1.6	1.8	1.7	2.1	1.9	2.2	3.0	3.4
Proline	4.6	8.0	5.2	5.1	4.7	6.0	4.1	3.1	4.4	5.6
Serine	7.3	9.0	14.0	14.1	15.9	12.8	7.7	20.0	9.4	7.8
Threonine	5.2	6.1	5.2	5.8	5.6	4.7	3.1	4.3	5.8	5.6
Tyrosine	2.1	0.7	0.7	0.3	1.2	0.5	0.8	1.1	1.7	1.5
Valine	6.4	5.5	4.4	4.5	5.7	4.9	4.6	3.3	6.5	6.9
Acidic/basic	1.3	1.2	1.9	1.8	2.2	2.1	3.0	1.6	1.5	1.6

[a] The values are percentages of total moles of amino acid recovered and are averages of four experiments. The value for glycine was corrected for nucleic acid present. From Grogan and Busch (*21*).

bond by dehydrogenation; this reaction converts what is a simple phosphate ester bond to an "energy-rich bond."

Glutamic and aspartic acid-rich proteins were found in the fractions eluted with 0.2 and 0.4 *M* NaCl. In the fraction eluted with 0.2 *M* NaCl, which composed approximately 21% of the total nucleolar acid and saline-soluble proteins, the total number of glutamic and aspartic residues composed 31.5% of the amino acid residues in the protein. In the fraction eluted with 0.4 *M* NaCl, the glutamic and aspartic residues composed 39.1% of the total residues in the protein. This protein would then contain more glutamic and aspartic residues than have been previously reported for any mammalian species. Since an assessment of the total number of amide groups has not been made yet, it is not clear whether these proteins are acidic or basic. Lysine, histidine, and arginine altogether compose only approximately 15% of the amino acid residues and most of these proteins are probably acidic even if some of the glutamic and aspartic acid residues are the amide form glutamine and asparagine.

D. Functions of the Nucleolar Acidic Proteins

The precise functions of these proteins probably will not be evaluated until more information is available on those proteins that are enzymatically important (see p. 340), those that have special intranucleolar functions, and those that are export products. In view of their similarities in amino acid composition, it seems likely that at least some of the nucleolar proteins are precursors of ribosomal proteins although the sites of their synthesis are not yet clear. Although as will be noted in the section on synthesis of nucleolar RNA (p. 336), both the saline- and acid-insoluble "residue" proteins seem to be important in the synthesis of RNA; it is not yet clear why nucleolar RNA should be laid down on acidic rather than basic proteins or why the residue fraction should be the one where the most rapidly formed RNA should be localized. A simple idea is that such proteins may serve to protect the RNA from RNases or that they might serve as subunits for sizing-out of the various types of RNA's that are formed in the spooling-out of oligomers (Chapter V).

The concept that the nucleolar proteins are simply import products would suggest that the specific activities of the saline-soluble proteins which might contain such precursors would be greater after pulse labeling than the specific activity of the proteins of the residue fraction (*15, 36*). Table 7.7 shows that this idea may be correct for the Walker tumor but the converse is true for the normal liver. In view of this type of conflicting data which may reflect the activity of the tissue, this possi-

TABLE 7.7

LABELING OF NUCLEAR PROTEINS WITH L-LYSINE-U-^{14}C (*36*)

Fraction	Tumor (counts/min./mg)[a]	Liver (counts/min./mg)[b]
Acid-soluble proteins		
Nucleus	890 (853–914)	208 (164–280)
Nucleolus	1075 (977–1145)	265 (240–306)
Acid-insoluble proteins		
Nucleus	650 (632–674)	291 (275–300)
Nucleolus	310 (254–371)	357 (312–398)

[a] Values are averages for three experiments with pooled tissue from five rats for each experiment; 60-minute pulse.

[b] Values are averages for four experiments with pooled tissue from five rats for each experiment. The ranges are presented in parentheses; 60-minute pulse.

bility is difficult to prove (*21, 22, 36*). Presumably the improving methods which are being developed at present will answer some of these difficult questions.

One point is clear now, however, and that is that the nucleolar protein does not represent the completed microsomal product. The enzyme glucose-6-phosphatase is not found in the nucleolus but it is a microsomal protein that is presumably added in the cytoplasm, possibly directly on the membranes of the endoplasmic reticulum (*45*).

E. NUCLEOLAR HISTONES

The nucleolar histones isolated from the preparations of nucleoli obtained by the methods employed in Chapter XI were virtually identical to the histones present in the whole nuclei as indicated by amino acid analysis (Table 7.8) (*12, 39*) and by mobility of the different histone components in polyacrylamide gels. For those workers interested in demonstrating that the histones have a major role in control of gene activity, these results were quite disappointing, inasmuch as they did not show a difference between one of the most active parts of the nucleus and the less active total nucleus (*12*).

However, the portion of the genome found in the total nucleolar preparation constitutes approximately 10% of the total cellular DNA in the case of the Walker tumor and about 2.5% of the total cellular DNA in the liver nucleoli. Thus, much of the "inactive genome" must be included along with the 0.25% of the genome or less that codes for the nucleolar products. Accordingly, it is conceivable that the DNA at the "active sites" of transcription is associated with different histones or may not be associated with histones at all (*12*, see also Chapter IV).

TABLE 7.8

Amino Acid Content of the Histones of Liver and Tumor Nucleoli and Nuclei (*22*)[a]

Fraction	Nucleoli Liver	Nucleoli Tumor	Nuclei Liver	Nuclei Tumor
Alanine	11.9	11.0	11.8	10.8
Arginine	8.1	8.5	7.4	8.0
Aspartic acid	6.0	6.5	6.3	6.9
½ Cystine	Trace	Trace	Trace	Trace
Glutamic acid	9.7	10.4	9.9	10.5
Glycine	9.0	8.9	8.7	8.4
Histidine	1.8	1.7	1.6	1.6
Isoleucine	4.3	4.1	4.4	4.3
Leucine	7.9	7.7	8.0	7.9
Lysine	13.5	13.7	13.3	13.7
Methionine	0.9	0.5	1.0	0.6
Phenylalanine	2.3	2.3	2.4	2.3
Proline	5.1	5.3	5.4	5.4
Serine	6.1	6.0	5.9	5.9
Threonine	4.8	5.4	5.6	5.6
Tyrosine	2.1	1.8	1.8	1.6
Valine	6.5	6.8	6.3	6.3

[a] The values are percentages of total moles of amino acid recovered.

III. Synthesis of Nucleolar Proteins

One of the difficult questions confronting research in the area of the nucleolus and the nucleus is the origin of the nuclear and nucleolar proteins (*15, 29b, 41, 42*). In recent experiments, Robbins and Borun (*42*) suggested that the source of the nuclear proteins is the cytoplasm. Although both kinetic and other types of experimental data argue against the likelihood that all nuclear proteins are synthesized in the peripheral cytoplasm, their evidence suggests that the outer layer of the nuclear envelope may be the site of synthetic activity for these proteins. Fortunately, methods are now being developed in our laboratory and others (*26d, 43a, 45b, 64*) for the isolation of these membranes and it is hoped that analyses of the synthetic functions of this structure will soon be made.

A. Nucleolar Synthesis of Proteins

In early studies on this subject, Birnstiel *et al.* (*6–10*) analyzed the incorporation of labeled amino acids into the nucleolar and chromatin fractions of pea seedlings. These fractions were contaminated with other elements but the data did show that most of the isotope in the nucleolar fraction was incorporated into the nucleolar residue fraction and that this fraction had the highest specific activity of any of the protein fractions studied. Since similar data were obtained for the proteins of the chromatin, these results were a bit difficult to assess (*10*). However, labeling of the nucleolar protein was greater in all fractions than was that of the protein of the chromatin, and the authors concluded that the nucleolus is the main site of synthesis of nuclear proteins.

Unfortunately, the nucleolus of the mammalian cell had not been a satisfactory subject of studies on the synthesis of the proteins; i.e., neither net synthesis of proteins nor high levels of labeling of proteins *in vitro* was demonstrated despite repeated efforts (*29b*). Although evidence has been presented that the proteins may be synthesized outside of the nucleolus and transported into the nucleolus, recently Zimmerman *et al.* (*64b″*) have indicated that nucleolar protein synthesis can occur. At least two nucleolar proteins separated by polyacrylamide gel electrophoresis were found to be labeled at rates different from corresponding nucleoplasmic proteins. However, neither of these was a ribosomal precursor protein and, hence, it would appear that they still might be synthesized in another cellular compartment, possibly the ribosomes themselves (*27a, 29a″*).

B. Synthesis of Nucleolar Proteins in Thioacetamide-Treated Systems

In a sense, the thioacetamide-treated nucleolus represents the most "driven" of all the liver nucleolar systems inasmuch as its synthetic activity is apparently maximal, approaching 30 fg of RNA per minute per nucleolus (*1*). As noted earlier, this activity is essentially normal from the point of view of base composition of the newly synthesized RNA. Under these conditions there is a more rapid synthesis of the nucleolar acid proteins of the residue fraction in which the specific activity was found to exceed that of the other nucleolar and nuclear protein fractions. Although these results could mean that there is active biosynthesis of protein in such nucleoli, clearer data are required on purified individual proteins rather than the mixtures usually studied.

IV. Nucleolar Proteins and Nucleolar RNA Synthesis

It is well established that little of the RNA in the cell is free of protein under normal conditions (*35*). In the nucleolus, the vast bulk of the dry

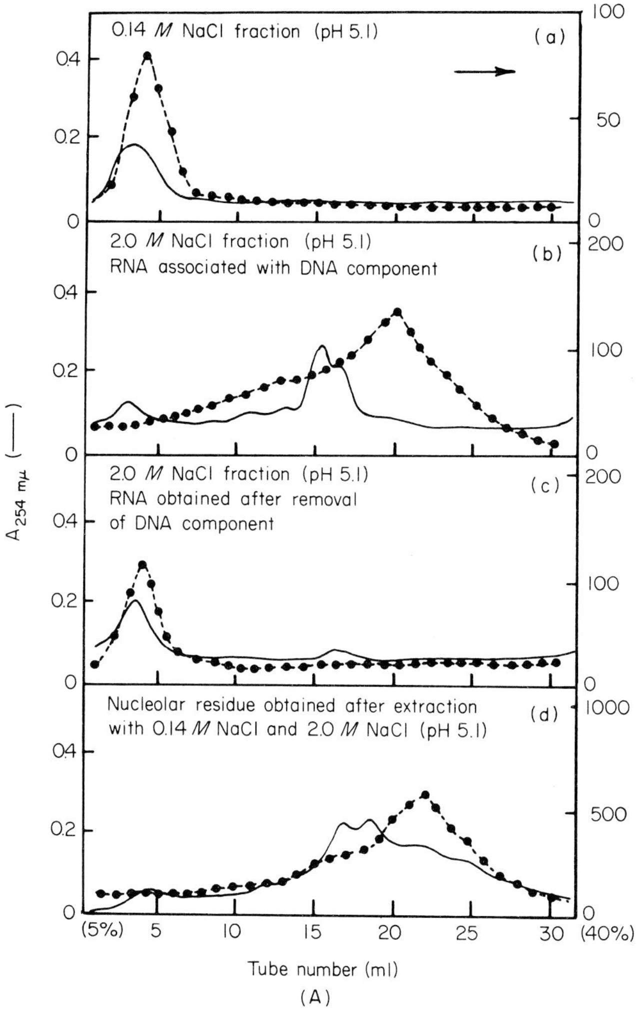

Fig. 7.7. A. Sedimentation profiles and distributions of radioactivity of RNA from the nucleolar fractions obtained at pH 5.1. The arrow indicates the direction of sedimentation. The numbers above the peaks indicate approximate sedimentation coefficients. a, 0.14 *M* NaCl fraction (pH 5.1); b, 2.0 *M* NaCl fraction (pH 5.1) RNA associated with DNA component; c, 2.0 *M* NaCl (pH 5.1) RNA obtained after removal of DNA component; d, nucleolar residue obtained after extraction with 0.14 *M* NaCl and 2.0 *M* NaCl (pH 5.1). B. Sedimentation profiles and distributions of radioactivity of whole nucleolar RNA and of RNA fractions obtained by differential salt extraction at pH 7.6. The arrow indicates the direction of sedimentation. The numbers above the peaks indicate approximate sedimentation coefficients. a, Nucleolar RNA; b, 0.14 *M* NaCl fraction (pH 7.6); c, 2.0 *M* NaCl fraction (pH 7.6) RNA obtained after removal of DNA component; d, nucleolar residue obtained after extraction with 0.14 *M* NaCl and 2.0 *M* NaCl (pH 7.6). Abscissa: Tube number (ml); ordinate: (left) A254 mμ (———); (right) cpm/ml (● - - ●). (From 12c).

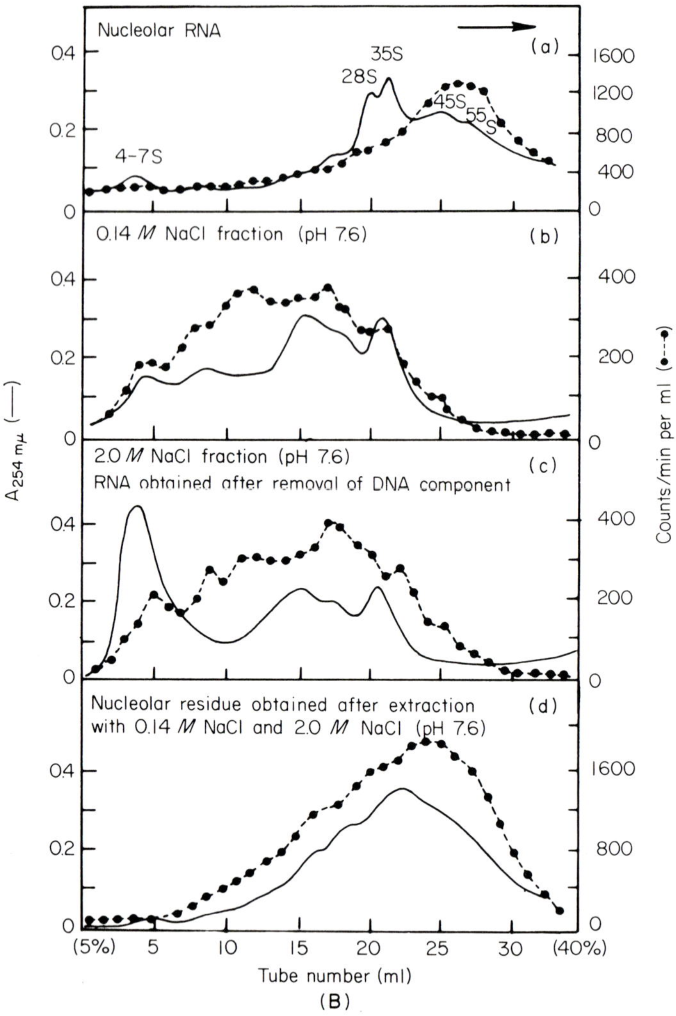

Fig. 7.7 (*continued*).

mass is protein and DNA and RNA only account for 15–25% of the total dry mass (Chapters I and V). The possibility that newly synthesized RNA is bound to the chromatin complex in its formative state by protein (*12a*) is supported by recent studies on the synthesis of 45 S nucleolar RNA. In the subfractions obtained by salt-extraction procedures, the bulk of the labeled RNA was insoluble in either dilute or concentrated salt solutions (Fig. 7.7). These findings support the view that when RNA is formed, it is laid down in intimate association with the very difficultly soluble nucleolar "residue" proteins which may form a matrix for its synthesis. Whether such products form precursors for the "informosomes" of Georgiev *et al.* (*18, 19, 27, 38*) cannot be defined at present.

In a most informative study on this subject, Tamaoki (*48*) and Tamaoki and Mueller (*49, 50*) found that there is a 45 S ribonucleoprotein in the nuclei of HeLa cells that contains 45 S RNA and newly synthesized protein. This particle sediments similarly to ribosomal particles and may actually be a portion of the ribosomes. Initially their report was that the fraction obtained was derived from the chromatin fraction of the nucleus and that its RNA with base compositions of A 21, U 19, G 28, and C 32 for the 28 S fraction and A 23, U 23, G 25, and C 28 for the 16 S fraction resembled ribosomal RNA.

After treatment of the chromatin fraction with deoxycholate and electrophoresis of the fraction on a Geon column they found that a 53 S ribonucleoprotein was present that contained the heavy RNA. They further concluded that DNase aided in the release of this ribonucleoprotein fraction. Their evidence supported the idea presented above that RNA is seldom free of protein in the cell.

Tamaoki (*48*) later developed the method for isolation of these ribonucleoprotein particules employing successive treatment with 0.5% DOC (deoxycholate), in 0.001 *M* Mg^{2+}, treatment of the lysate with DNase, and then treatment with sodium dextran sulfate. The particulate elements are released by this treatment; these studies (*48*) indicated that the ribonucleoproteins obtained contained 45 S RNA. Treatment of the cells with actinomycin D (see Chapter X) produced a disappearance of the 45 S ribonucleoprotein component and, moreover, caused the label to accumulate in 16 and 28 S ribosomal RNA. It would be of interest to ascertain whether there is a significant decrease in the formation of these particles in systems treated with puromycin or other inhibitors of protein synthesis. In the studies of Vaughan, Penman, and Soiero (*46, 55, 61–63a*), it became apparent that inhibition of protein synthesis produced a failure of the transport of the RNA to the cytoplasm (see Chapter VIII).

A. Recent Studies on Nucleolar Ribonucleoproteins

In recent months, major new findings have been made in studies on the biosynthesis of nucleolar ribonucleoproteins (*26b′, 26b″, 29, 40″, 55a*). Studies of several groups (*26b, 29, 55a*) have indicated that cleavages of RNP elements of high molecular weight can occur in nucleoli to form ribonucleoprotein fragments in which 32 S, 28 S and 18 S RNA could be found. Under the conditions of these cleavage reactions, extrinsic 32 S, 28 S and 18 S RNA underwent little or no degradation. The exact constitution of the RNP particles and the mechanism of their release into the cytoplasm remains to be defined (*26b′, 26b″, 40″*). These reactions have been summarized recently by Scherrer (*43b*) as follows:

$$
\begin{array}{ccccc}
 & & \nearrow \begin{matrix}55\ S\ RNP\\(32\ S\ RNA)\end{matrix} & \rightarrow & \begin{matrix}50\ S\ RNP\\(28\ S\ RNA)\end{matrix}\\
\begin{matrix}80\ S\ RNP\\(45\ S\ RNA)\end{matrix} & \rightarrow & \begin{matrix}70\ S\ RNP\\(41,\ 36\ RNA)\end{matrix} & & \\
 & & \searrow \begin{matrix}40\ S\ RNP\\(28\ S\ RNA)\end{matrix} & \rightarrow & \begin{matrix}30\ S\ RNP\\(18\ S\ RNA)\end{matrix}
\end{array}
$$

The 80 S RNP particle would be the nucleolar element that contains the 45 S RNA precursor of the ribosomal particles. These would undergo the cleavage reactions indicated in Chapter V for the RNA classes (shown in the RNP particle). This concept indicates that the newly formed RNA would be laid down on nucleolar proteins and that the cleavage reactions involved in the formation of the smaller products occur on such particles. Presumably such RNP particles would be released from nucleoli and nuclei as intact elements (*26b′, 40″*).

B. Relationship of Nucleolar Protein Synthesis and Nucleolar RNA Synthesis

In studies on regenerating liver, Higashi *et al.* (*26b*) found that treatment of the livers with cycloheximide in doses sufficient to block protein synthesis also produced marked decreases in the biosynthesis of RNA of the nucleolus. This experiment would suggest that at least in the nucleolus, RNA and protein synthesis are mutually obligatory reactions, but the further study of this problem should give more conclusive answers to this fascinating idea. If it is true, the underlying mechanisms would be of particular interest in their relationship to the synthesis of "informosomes" and other nuclear products.

When HeLa cells were deprived of methionine (*54b*), there was a failure of completion of the ribosomes. Whether this resulted from a failure of methylation of newly synthesized RNA (*46′*) or a failure of synthesis of critical proteins necessary for completion of the ribosomal particle is not clear at present. As indicated in Chapter V, the position of the methyl groups in nucleolar RNA is highly specific and may be critical for further synthetic reactions as well as cleavage reactions.

V. Nucleolar Enzymes

As might be expected from a structure so active in the biosynthesis of cytoplasmic elements, the nucleolus contains a number of biosynthetic enzymes for RNA including the RNA polymerase discussed in Chapter V. In addition, the nucleolus contains the important biosynthetic enzyme NAD pyrophosphorylase or NAD synthetase.* The nucleolus contains a number of other enzymes that may be important in the functional activity of the ribosomes or polysomes and others (such as nucleolar ribonucleases and nucleolar ATPases) that seem to be primarily destructive. Although the nucleolus would also be a logical site for protein synthesis it is only recently that evidence has been obtained for the synthesis of protein in nucleoli (*64b″*).

A. Purification of the Nucleolar RNA Polymerase of the Novikoff Hepatoma

The previous studies on the RNA polymerase of normal liver and of tumors in this laboratory (*41a*) demonstrated that this enzyme has a pH optimum of 8.2, requires the presence of DNA and all four types of nucleoside triphosphates, is dependent on divalent ions and is inhibited by actinomycin D. The kinetics of labeling or biosynthetic reactions are linear for 10–20 minutes and labeling is linear with respect to enzyme concentration within restricted limits. Recent studies in this laboratory (*32a*) have demonstrated that this nucleolar enzyme is stable for several months in buffers containing 3 m*M* 2-mercaptoethanol, 1 m*M* EDTA, 50% glycerol when buffered at pH 7–8 with Tris. The enzyme activity was enhanced by polyamines, including spermine, spermidine, and putrescine and by ammonium sulfate in concentrations up to 30 m*M*. Like the nuclear enzyme studied by Cunningham (*12d*) and by Goldberg (*20a*) the nucleolar RNA polymerase was solubilized at pH 10. Simply by isolation of nucleoli there was a purification of 32-fold over the nuclear activity. By treatment of the nucleoli with Tris buffer at pH 10, the enzyme was purified by a factor of 1.5. Saturation with ammonium sulfate to precipitate the enzyme (30–45% ammonium sulfate) resulted in a purification of approximately 4 to 4.5-fold over the isolated nucleoli and by 143-fold over the whole nuclear preparation. However, this enzyme is by no means pure at this point.

Much debate has existed about the possibility that there are two types of nucleolar RNA polymerase, one of which is activated by divalent ions, usually Mg^{2+}, and the other activated by ammonium sulfate and Mn^{2+} (*26c, 64a*). Considerable questions exist about this because up to this point no "pure" enzyme has been isolated that has a specific requirement

* Some of the high levels of RNA methylase activity found in tumors (*17a, 54a*) may reflect high activities of nucleolar RNA synthesis or synthesis of specific nucleolar RNAs of tumors (see Chapter V).

for one divalent ion. The possibility exists that there are competing reactions; i.e., RNase cleavages or other types of reactions that are either blocked or inhibited by one or more divalent ions or ammonium sulfate. Further studies on this point can be made (*26c, 64a*), now that Cunningham *et al.* (*12c'*) have shown that it is possible to separate out a Mn^{2+} requiring RNA polymerase from rat liver nuclei. There seems little doubt from the studies of Pogo (*40'*) that different salt concentrations markedly affect the activity of the RNA polymerases. Moulé (*32b*) has found that in the usual isotonic media employed for RNA polymerase studies, primarily G-C rich nucleolar type RNA is synthesized. However, in higher salt concentrations, A-U rich RNA is synthesized. The relationship of these RNA polymerases to the extranucleolar RNA polymerases (*28a, 42a*) is not clear at the present time. How either or both of these are stimulated by hormones (Chapters V and X), such as growth hormone (*26c'*), estrogen (*36a*) or cortisone (*64b'*) is unknown at present.

B. The Convertase Enzyme(s)

Although little has been directly studied on the activity of the enzyme that converts 45 S RNA to the products, 35 S and 28 S RNA, there is no question that such an enzyme must exist (*29, 55a*). It becomes of considerable significance to isolate the enzyme(s) and determine the mechanism of action because of the conflicting hypotheses shown in Chapter V. Such enzymes may function by cleavage of 45 S RNA or its oligomers in specific regions and finally the rRNA molecules would be produced.*

Although there has been much discussion about these reactions, there has been no satisfactory isolation of such enzymes. There are serious experimental difficulties in such studies. Even under optimal conditions, 45 S RNA undergoes a mild to severe breakdown in nucleolar preparations by random cleavages that result in a smear of heterodisperse elements over the gradient profiles. These random breaks indicate that both the "convertase" enzymes and RNases are active, and thus far, a system for studying convertases alone has not been developed (*12c*). Evidence has now been obtained that the cleavage enzymes remain in nucleolar preparations under conditions when nonspecific RNases are removed (*29*).

* In recent studies in this laboratory it has been found that low molecular weight RNAs including a 7 S, an 8 S and a U3 RNA are hydrogen-bonded to nucleolar 28 S RNA. Interestingly, neither the 35 S nor the 45 S nucleolar RNA are hydrogen-bonded to low molecular weight RNAs. Presumably the cleavage reactions producing such low molecular weight RNAs may occur on the 35 S nucleolar RNA (Prestayko, Tonato, and Busch, unpublished).

VI. Nucleolar Ribonuclease

The study of nucleolar RNases is complicated by the two basic problems that pertain to any study on intracellular distribution of enzymes; i.e., there is almost no way to insure that any enzyme found in an isolated organelle is not an adventitious contaminant, particularly when other organelles, such as the lysosomes, that contain high concentrations of a given enzyme may adhere or be broken; secondly, there may very well be serious losses of soluble enzymes from any organelle into the isolation media. Another precautionary note is that one cannot be certain that a given enzyme is in fact functioning *in vivo* in the manner defined in the *in vitro* system used for testing. In the case of ribonuclease, the

TABLE 7.9

FREE RIBONUCLEASE ACTIVITY IN THE NUCLEI AND NUCLEAR SUBFRACTIONS FROM NORMAL AND NEOPLASTIC TISSUES (*12a*)

Fraction[a]	Specific activity (ng RNase/mg protein)
Normal rat liver	
Nuclei	2.1
S_1	2.8
S_2	7.4
Nucleoli	18.0
Regenerating rat liver	
Nuclei	9.0
S_1	11.2
S_2	17.6
Nucleoli	30.8
Walker 256 carcinosarcoma	
Nuclei	0
S_1	0.6
S_2	9.1
Nucleoli	10.6
Novikoff hepatoma	
Nuclei	0
S_1	0
S_2	5.7
Nucleoli	8.6
Ehrlich ascites tumor	
Nuclei	0
S_1	0
S_2	12.0
Nucleoli	14.2

[a] S_1 and S_3 are supernatant layers of 0.25 and 0.88 *M* sucrose above the nucleolar pellets.

enzyme concentration varies in a curious way which would not seem teleologically to be in the best interests of the cell.

There are two examples of increased RNase activity that have been studied in the authors' laboratory, one in regenerating liver and the other in thioacetamide-treated livers. Table 7.9 presents the data on free RNase (alkaline) activity in nuclei and nuclear subfractions including the isolated nucleoli of a number of tissues such as the regenerating liver and some tumors. The RNase activity of the nuclei of the regenerating liver cells is almost four times greater than that of normal rat liver and in addition, the nucleolar RNase activity is almost twice that of the normal rat liver.

A. Free and Latent Ribonuclease

Much of the RNase activity in cells is either bound to ribonuclease inhibitors present in both the cytoplasm and the nucleus and much seems to be either membrane bound or localized within the substructures of the cell (*43*). As shown in Table 7.10, the total activity of nuclear RNase is almost four times greater than the nuclear free RNase activity in the normal liver since 60% is inhibitor-bound and the rest is latent (*12a, 12b*). In the nucleoli, more is free than inhibitor-bound. In regenerating liver nuclei, the total RNase is almost twice that of normal liver nuclei, and in the nucleoli, there is also a marked increase in the free and total activity. When these data are compared to the increased rates of synthesis of the nucleolar RNA (see Chapter V) in nucleoli of regenerating liver, it seems illogical for the RNase activity to rise. The increase in RNase activity is paralleled by the increase in the ribonuclease inhibitor activity which is 2.5 times that of the control. Presumably this increased activity suffices to decrease the destructive effects of the nucleolar ribonuclease.

On the other hand, transplantable tumors exhibit active cell growth and as noted earlier by Roth (*43*) their levels of ribonuclease activity are markedly diminished by comparison with the normal liver (Table 7.10). In these cells, which have enormous nucleoli by comparison to the normal or regenerating liver, and in which the production of RNA is 30 fg (10^{-15} gm) per nucleolus per minute; i.e., almost twice that of the regenerating liver and ten times that of the normal liver, the nucleolar RNase is decreased to approximately one-half that of the normal liver in the Novikoff hepatoma and the Walker tumor. In the Ehrlich ascites tumor of the mouse, the content of nucleolar RNase is intermediate between that of the normal liver and the other tumors (Table 7.10).

In addition, there is a markedly increased level of ribonuclease inhibitor in the supernate obtained from the tumors (Table 7.10). By com-

TABLE 7.10

RIBONUCLEASE AND RIBONUCLEASE INHIBITOR OF NUCLEI FROM NORMAL AND NEOPLASTIC TISSUES (*12a*)[a]

Fraction	Free activity	Inhibitor-bound activity	Latent activity	Total activity of nuclear subfractions	Free inhibitor (nuclear high speed supernate)
Normal rat liver					
Nuclei	168	382	142	692	
S_1	140	265	130	535	
S_2	50	10	18	78	
Isolated nucleoli	100	70	10	180	
Supernate		60			80
Regenerating rat liver (18 hr).					
Nuclei	720	510	140	1370	
S_1	670	510	140	1320	
S_2	77	55	25	157	
Isolated nucleoli	185	56	20	261	
Supernate		100			192
Walker 256 carcinosarcoma					
Nuclei	0	500	10	510	
S_1	36	252	0	324	
S_2	90	60	0	150	
Isolated nucleoli	63	70	7	140	
Supernate		78			450
Novikoff hepatoma					
Nuclei	0	500	20	520	
S_1	0	368	15	383	
S_2	68	20	2	90	
Isolated nucleoli	60	80	20	160	
Supernate		60			400
Ehrlich ascites tumor					
Nuclei	0	530	0	530	
S_1	0	400	80	480	
S_2	120	200	10	330	
Isolated nucleoli	85	70	10	165	
Supernate		100			415

[a] Inhibitor-bound activity was estimated after pretreatment with 0.2 m*M* *N*-ethylmaleimide for 10 minutes at 20°C. Latent activity was determined after freezing and thawing (once). Activity is expressed as nanograms per gram wet weight of nuclear preparations. Nuclei were suspended in 4 volumes of 0.44 *M* sucrose, disrupted by sonication, and centrifuged at 100,000*g* for 1 hour. The supernate thus obtained was used for inhibitor assay. The results are averages from 2–6 determinations.

TABLE 7.11

Ratios of Free Inhibitor to Free and Total RNase Activity (*12a*)[a]

Fraction	Free inhibitor: Total free activity	Free inhibitor: Total activity
Normal rat-liver nuclei	1.4	0.5
Regenerating rat-liver (18 hr.) nuclei	1.0	0.55
Walker 256 carcinosarcoma nuclei	12.0	3.6
Novikoff hepatoma nuclei	16.0	3.1
Ehrlich ascites tumor nuclei	10.1	2.1

[a] The values obtained for inhibitor and RNase concentrations (Table 7.10) were recalculated on the basis that 1 unit of inhibitor causes a 50% inhibition of 5 ng of pancreatic RNase. The values presented are ratios of the amount of inhibitor sufficient to cause a 50% inhibition of RNase activity to the total free RNase activity (ng RNase) and to the total RNase activity (ng RNase).

parison with the normal liver, the level of free inhibitor was increased by almost six fold and even by comparison with the regenerating liver, there was a marked increase in the level of free inhibitor. The values for the levels of free inhibitor were quite similar in each of the tumors analyzed. As shown in Table 7.11, these analyses may be calculated either on the basis of free inhibitor over total free activity or free inhibitor over total activity. This ratio is almost ten times greater in the tumors than in normal or regenerating liver.

From the point of view of the biosynthetic activity of the cells, then, the tumors seem to make the most teleological sense; i.e., their levels of ribonuclease are markedly reduced at the same time as the levels of ribonuclease inhibitor are markedly increased (*12a, 12b*).

These tumors are among the most rapidly growing tissues available; perhaps, in tumors with slower growth rates, the levels of nuclear and nucleolar RNases are higher. Whether there is a correlation between the levels of RNA synthesis, RNase activity, growth rates of the tumors, and lethality of neoplastic cells is not yet clear. However, suggestions have been made (*13a*) that the levels of RNase may be, in part, the cause rather than the result of the neoplastic process.

B. Relationship of Nucleolar RNase to Nucleolar Size

The interesting finding (*56*) that the level of nucleolar RNase increased rapidly after administration of thioacetamide to rats showed that the increases in rates of synthesis that were correlated with increased nucleolar size in normal liver were also related at least in part to an increased level of RNase. However, the increase in the level of activity of this enzyme was not entirely correlated with size since the level of RNase increased faster than the mass of the nucleolus. In addition, the level of RNase was independent of the nucleolar volume; i.e., this activity increased faster than the increment in nucleolar volume since the RNase activity increased almost tenfold at a time when the nucleolar volume was increasing by a factor of 2. This nucleolar increase was not associated with a marked increase in the activity of the cytoplasmic RNase although there were changes in the activity of the RNases of various cycloplasmic fractions.

To ascertain whether other substances might decrease the activity of the nucleolar RNase, the RNase activity of nucleoli was studied after treatment with thioacetamide or with actinomycin and thioacetamide (*45*). After treatment with actinomycin D (Table 7.12) the RNase activity of the nucleoli decreased to one-third to one-sixth the normal level. This change correlated with a similar decrease in nuclear RNase activity. In addition to these decreases, actinomycin D also induced a decrease in the nucleolar RNase in animals treated with thioacetamide (to about one-half the level of the thioacetamide-treated nucleoli). The low dose of actinomycin D employed should not have altered the RNase activity of other cellular elements. If actinomycin blocks the synthesis of RNA and either interferes with the synthesis of nucleolar proteins or blocks utilization of preformed precursors, it would appear that both synthesis and organization occur simultaneously or in very close sequence. In addition Higashi *et al.* (*26b*) showed in livers of regenerating rats treated with cycloheximide, administration of labeled precursors resulted in the uptake of much less isotope into both the nucleolar RNA and the nucleolar proteins than in control regenerating livers.

Strangely, there seems to be a direct relationship between the activity of the normal liver nucleolar RNA polymerase system and the nucleolar RNase activity. Apparently, this relationship is not very logical unless there is a built-in "suicide" mechanism in the ribosomes or the apparent RNase has another function in the nucleolus or the ribosomes.

VII. Relationship of RNase in the Cytoplasm to Nucleolar Activity

To investigate the possible relationship of nucleolar and nuclear RNase activity that occur in the cell as a whole, studies were made on the total ribonuclease of various cellular fractions of control and thioacetamide-treated cells (Table 7.13). Along with the increase in the nuclear RNase activity which is paralleled by an increased RNase in the nucleoli, there is a marked alteration in the enzyme levels in the various cytoplasmic

TABLE 7.12

INTRANUCLEAR DISTRIBUTION OF ENZYMES WHICH ARE KNOWN NUCLEAR CONSTITUENTS (*45*)

Enzyme	Fraction		Distribution			Change in optical density		Specific activity
		cpm/gm liver/20 min	*% of nuclei*	*cpm/mg RNA × 10^{-3}/20 min*	*cpm/mg protein/20 min*	*ΔA/30 min/gm liver*	*ΔA/30 min/mg protein*	*mμmole/ min/gm liver*
RNA polymerase	Nuclei	574	100	9.9	140			
	Nuclei disrupted[a]	503	88	8.4	157			
	Nu S_1	82	14	1.5	37			
	Nu S_2	316	55	53.9	263			
	Nucleoli	263	45	49.7	3300			
Ribonuclease (normal rats)	Nuclei		100			3.3	0.82	
	Nuclei disrupted		151			5.0	1.25	
	Nu S_1		106			3.5	1.3	
	Nu S_2		166			5.5	2.3	
	Nucleoli		37			1.3	5.0	
Ribonuclease (rats treated with thioacetamide)	Nuclei		100			6.5	1.5	
	Nuclei disrupted		242			15.7	3.1	
	Nu S_1		200			13.0	3.5	
	Nu S_2		165			10.7	3.6	
	Nucleoli		85			5.5	3.1	
Ribonuclease (rats treated with thioacetamide plus actinomycin D)	Nuclei		100			4.5	1.5	
	Nuclei disrupted		195			8.8	2.7	
	Nu S_1		140			6.3	2.6	
	Nu S_2		164			7.4	2.9	
	Nucleoli		77			3.5	4.4	

Ribonuclease (rats treated with actinomycin D)	Nuclei	100	0.94	0.90		
	Nuclei disrupted	310	2.9	2.3		
	Nu S_1	500	4.7	1.8		
	Nu S_2	75	0.70	0.40		
	Nucleoli	4.5	0.042	0.80		
NAD synthetase (DPN pyrophosphorylase)	Nuclei	100			27	7.3
	Nuclei disrupted	148			45	15
	Nu S_1	67			18	7.2
	Nu S_2	104			28	38
	Nucleoli	29			8	22
ATPase A	Nuclei	100			41	14
	Nuclei disrupted	142			58	24
	Nu S_1	100			41	23
	Nu S_2	22			9	8
	Nucleoli	15			6.5	31
Acid deoxyribonuclease	Nuclei	100	0.28	0.061		
	Nuclei disrupted	257	0.72	0.20		
	Nu S_1	160	0.45	0.18		
	Nu S_2	21	0.06	0.05		
	Nucleoli	11	0.03	0.12		
Alkaline deoxyribonuclease	Nuclei	100	0.24	0.078		
	Nuclei disrupted	310	0.75	0.29		
	Nu S_1	162	0.39	0.20		
	Nu S_2		0	0		
	Nucleoli	10	0.025	0.12		
ATPase B	Nuclei	100			61	21
	Nuclei disrupted	124			75	32
	Nu S_1	102			62	35
	Nu S_2	11			6.5	5
	Nucleoli	5			2.9	6

[a] Disrupted by sonic oscillation. S_1 and S_2 are supernatant layers of 0.25 and 0.88 *M* sucrose above the nucleolar pellets.

TABLE 7.13

EFFECT OF THIOACETAMIDE ON THE TOTAL RIBONUCLEASE ACTIVITY OF THE PARTICULATE FRACTIONS AND RIBONUCLEASE INHIBITOR OF THE CELL-FREE SUPERNATANT FRACTION (*12b*)[a]

Fraction	Free activity	Inhibitor-bound activity	Latent activity	Total (particulate fractions)	Free inhibitor (cell-free supernatant)	Free inhibitor / Free ribonuclease	Free inhibitor / Total ribonuclease (particulate fractions)
Control							
Nuclear fraction	40	60	80	180	0		
Mitochondrial fraction	145	30	190	365	0	19.6	6.9
Lysosomal fraction	370	180	480	1030	0		
Microsomal fraction	55	50	55	160	0		
Total	610	320	805	1735			
Supernatant	0	650	0	0	2400		
Thioacetamide (24 hr.)							
Nuclear fraction	460	240	93	793	0		
Mitochondrial fraction	345	35	120	500	0		
Lysosomal fraction	600	80	300	980	0	5.6	3.4
Microsomal fraction	262	128	42	432	0		
Total	1667	483	555	2705			
Supernatant	0	620	0	0	1870		

[a] Inhibitor-bound ribonuclease activity was estimated after preincubating the fractions with 0.3 m*M* N-ethylmaleimide for 10 minutes at 20°C. Total ribonuclease activity is the sum of free, inhibitor-bound, and latent (activity released by freezing and thawing) activities. In all cases, activity is expressed as ribonuclease units (or inhibitor units) per gram of liver (fresh weight). Inhibitor: ribonuclease ratios were calculated on the basis of 5 ng of ribonuclease being 50% inhibited by 1 unit of inhibitor. Free inhibitor is the quantity of inhibitor capable (theoretically) of causing a 50% inhibition of the free RNase; e.g., for control 2400/122 (610/5) = 19.6.

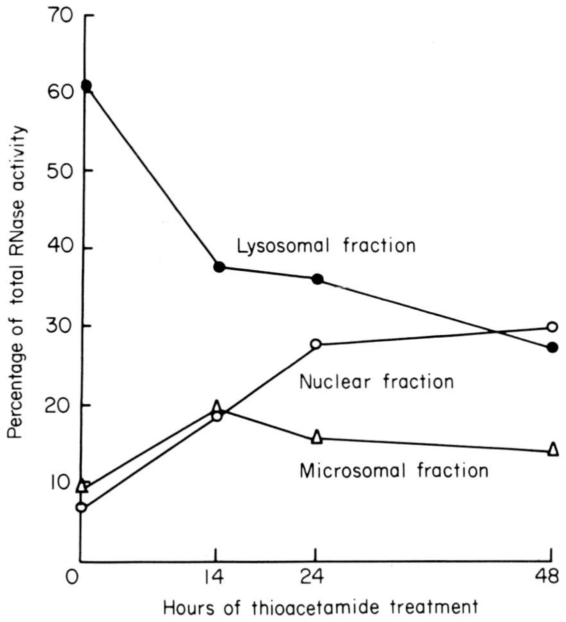

Fig. 7.8. Redistribution of RNase after thioacetamide treatment. (From *12b*).

fractions, particularly the "microsomal" fraction in which the ribonuclease activity increases more than three times and the mitochondrial fraction in which the activity more than doubles (*12a, 12b*). The specific activity of the enzyme in the lysosomal fraction (nanograms RNase per milligram protein) is not altered. However, as shown in Fig. 7.8, there is a marked redistribution of the total enzyme; i.e., the relative total amount of the enzyme in the lysosomal fraction diminished and the percentage in the nuclear and microsomal fraction increased.

In most of the cell fractions, thioacetamide diminished the amount of latent enzyme and, moreover, the total free inhibitor diminished sharply in the cell in relationship to the total ribonuclease. Either there is a change in the binding of ribonuclease in the cellular fractions of the thioacetamide-treated cells or there is a change in the character, amount, or function of the ribonuclease inhibitor. In any event, the total ribonuclease activity increases in the thioacetamide-treated cells and the intracellular ribonuclease inhibitor decreases.

These events in the cytoplasm are probably related to the altered composition of the cytoplasm; i.e., the total RNA in the cytoplasm diminishes sharply and there is increased turnover of the cytoplasmic ribosomal RNA; the half-life is 3.5 days vs. 5.5 days in the controls (*46a*). The picture emerges of an indirect effect of the ribonuclease on the increased nucleolar function; i.e., in response to reduction of "feedback repression," the nucleolus may be activated to maximal function with respect to RNA synthesis (see Chapter X).

One cautionary note must accompany any discussion of RNase or other enzymes of isolated cellular particles. Ever since Novikoff (*37a*) demonstrated that the alkaline phosphatase could adhere to the nuclear membrane, there has been concern among biochemists about the adherence of soluble enzymes to isolated particles. It is possible that much of the nuclear or nucleolar ribonuclease might originate from the cytoplasmic elements, since in the course of isolation of nuclei after homogenization there is a juxtaposition of the nuclear membrane to cytoplasmic elements with which the nucleus may not be associated during usual cell function. Moreover, during sonication, the nucleolus is exposed to both the nucleoplasm and nuclear membranes in an abnormal way. Accordingly, it is not certain that the nucleolar content of ribonuclease is unaffected by the medium in which it is isolated. The effects with actinomycin D suggest that this hazard is not serious, but further experiments are required to eliminate the possibility, particularly since the specific activity of the ribonuclease in the lysosomes is always greater than that of the nucleus. If the lysosomes should break during the process of homogenization of the cells or as a result of thioacetamide treatment, it would not be surprising if some of the free enzyme would enter the nuclear and, possibly, the nucleolar compartment.

VIII. Other Nucleolar Enzymes

A. Polyriboadenylic Acid Synthetase

In normal adult or fetal liver or livers treated with thioacetamide or actinomycin D, there is synthesis of RNA rich in adenylic acid both in the nucleus and nucleolus. Another type of product formed is the poly A (polyriboadenylic acid) recently characterized (*17, 26, 37, 40a*) as:

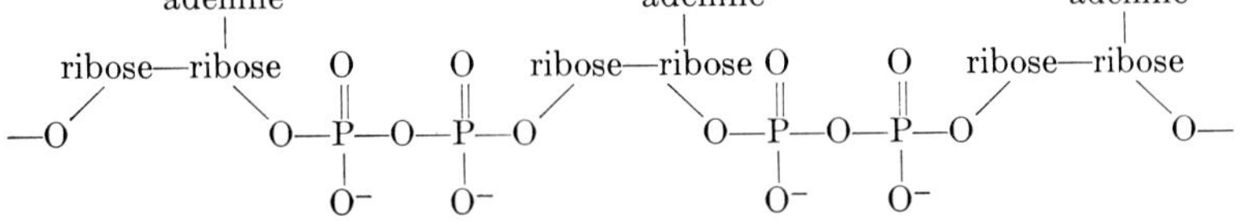

Synthesis of this interesting compound also takes place in the nucleolus (*64b*). Interestingly, Otake *et al.* (*37b*) reported that this ADP ribose polymer is bound to histones.

B. NAD Synthesis

Many years ago, Schneider and Hogeboom (*44*) reported that the only clearcut enzyme synthesis that occurred in the nucleus was that of DPN or NAD. In essence they found that there was no cellular source of this enzyme other than the nucleus. Later studies in this and other laboratories (*45, 52a*) showed that this biosynthetic activity was localized primarily to the nucleolus; i.e., the specific activity of the NAD synthetase of nucleoli was three times that of the remainder of the nuclei. Since the nuclear activity was increased after the nuclei were disrupted, it would appear that some of this enzyme may be bound (Table 7.12). Although the role of NAD is well defined in hydrogen and electron transport systems, the reasons for the synthesis of this product in the nucleolus are quite mysterious.

Although the significance of NAD pyrophosphorylase in the nucleolus and nucleus has not been understood, the recent studies of Little *et al.* (*29a*) and Zimmerman *et al.* (*64c*) have shown that NAD exerts an important role in repair of DNA. Apparently, either the ligase enzyme or the repair enzyme requires addition of AMP, which is derived specifically from NAD (*27a′*). It has been recently suggested (*44a*) that there may be more than one type of NAD synthetase and other forms may be in sites other than the nucleolus.

C. Nucleolar Enzymes of Unknown Function

Histochemical analysis has shown that acid phosphatase (*33, 60*), alkaline phosphatase (*20, 33, 65*), as well as ATPase (*13, 28*) and a 5′ nucleotidase (*60*) were localized to the nucleolus (Chapter II). Supportive evidence for the presence of ATPase in the nucleolus was obtained from the biochemical analyses made in this laboratory in connection with the overall study of the nucleolar enzymes by Siebert *et al.* (*45*) who showed that two ATPases; i.e., A and B were localized to the nucleus and the former was localized to the nucleolus (Table 7.14). At this moment, none of the functions of these apparently hydrolytic enzymes is understood though it is possible as has been found for other systems that the ATPase is related to ion transport in the nucleolus.

Among the other enzymes that have been reported in the nucleolus in mammalian cells are succinic dehydrogenase (*14*) and a polynucleotide phosphorylase which was found in isolated nucleoli. This enzyme catalyzes the cleavage of RNA to nucleoside diphosphates and its presence may be related in part to the theory of Harris (*23–25*) that there is intranuclear degradation of newly synthesized RNA. On the other hand, this enzyme may have biosynthetic activity (*45*).

TABLE 7.14

SUBSTRATE SPECIFICITES OF NUCLEAR AND NUCLEOLAR ATPases A AND B (*45*)

	ATPase A		ATPase B	
Substrate	Nuclei disrupted by sonic oscillation	Nucleoli	Nuclei disrupted by sonic oscillation	Nucleoli
ATP	100	100	100	100
GTP	33	24	14	70
CTP	29	52	17	20
UTP	200	56	107	160
dATP	138	92	86	170
dATP	100	100	100	100
dGTP	142	91	152	53
dCTP	52	78	48	41
TTP	41	83	24	6

In isolated nucleoli of starfish oocytes, Vincent and Baltus have found acid phosphatase (*57–59*), nucleoside phosphorylase (*2, 3*), DPN synthetase (*2, 3*), and RNase (*2, 3*). Most of these same enzymes are also present in the nucleoli of the mammalian cells.

D. NUCLEAR ENZYMES AND THE NUCLEOLUS

One important question about nucleolar enzymes is whether they are unique in any way or whether they may be simply activated forms of the nuclear enzymes. Since some nuclear enzymes, such as DNase I and II,

TABLE 7.15

COMPARISON OF NUCLEAR AND NUCLEOLAR RIBONUCLEASE ACTIVITY WITH DIFFERENT SUBSTRATES (*45*)

	Hydrolysis rates		
Substrate	Nuclei disrupted by sonic oscillation	Nucleoli	Nuclei disrupted by sonic oscillation: nucleoli
Ribosomal RNA	100	100	1.00
Nuclear 18 S RNA	79	41	1.93
Nuclear 28 S RNA	87	92	0.94
Nuclear 35 S RNA	108	88	1.23
Nuclear 45 S RNA	161	147	1.09
Polyadenylic acid	33	16	2.06
Polyuridylic acid	0	0	

and ATPase B are concentrated in the nucleolus, presumably completely free interchange of nuclear and nucleolar enzymes does not occur. Moreover, nucleolar enzymes may have different substrate specificities than the extranucleolar nuclear enzymes (Tables 7.14, 7.15). This fact, coupled with the concentration of RNA polymerase (24-fold in nucleoli), RNase (6-fold in nucleoli), DPN pyrophosphorylase (3-fold in nucleoli), and ATPase A (2-fold in nucleoli), suggests that some nucleolar enzymes may be unique. To prove the point, evidence is required for differences in protein composition, in sequences of amino acids in isozymes, or in cofactor requirements.

REFERENCES

1. Adams, H. R., and Busch, H., Effects of thioacetamide on incorporation of orotic acid-2-C^{14} into RNA fractions in liver. *Cancer Res.* **23**, 576–582 (1963).
2. Baltus, E., Observation sur le rôle biochimique du nucléole. *Biochim. Biophys. Acta* **15**, 263–267 (1954).
3. Baltus, E., Etude de la ribonucléase de nucléolus isolés a partie des oocytes d'astéries. *Biochim. Biophys. Acta* **55**, 82–87 (1962).
4. Beck, J. S., Anderson, J. R., McElhinney, A. J., and Rowell, N. R., Antinucleolar antibodies. *Lancet* **II**, 575–577 (1962).
5. Bielka, H., Welfle, H., and Schneiders, I., Studies on tissue and species specificity of ribosomal proteins. *Acta Biol. Med. Ger.* **19**, K13–K14 (1967).
6. Birnstiel, M. L., Chipchase, M., and Bonner, J., Incorporation of Leucine-H^3 into subnuclear components of isolated pea nuclei. *Biochem. Biophys. Res. Commun.* **6**, 161–166 (1961).
7. Birnstiel, M. L., Chipchase, M., and Flamm, W. G., On the chemistry and organisation of nucleolar proteins. *Biochim. Biophys. Acta* **87**, 111–122 (1964).
8. Birnstiel, M. L., Chipchase, M., and Hayes, R. J., Incorporation of L-(^{14}C) leucine by isolated nuclei. *Biochim. Biophys. Acta* **55**, 728–733 (1962).
9. Birnstiel, M. L., and Hyde, B. B., Protein synthesis by isolated pea nucleoli. *J. Cell Biol.* **18**, 41–50 (1963).
10. Birnstiel, M. L., Rho, J. H., and Chipchase, M., Fractionation of isolated pea nuclei. *Biochim. Biophys. Acta* **55**, 734–740 (1962).
11. Brecher, P. I., and Wotiz, H. H., Stereospecificity of the uterine nuclear hormone receptors. *Proc. Soc. Exptl. Biol. Med.* **128**, 470–473 (1968).
12. Busch, H., "Histones and Other Nuclear Proteins." Academic Press, New York, 1965.

12a. Chakravorty, A. K., and Busch, H., Alkaline ribonuclease and ribonuclease inhibitor in nuclear and nucleolar preparations from normal and neoplastic tissues. *Cancer Res.* **27**, 780–792 (1967).

12b. Chakravorty, A. K., and Busch, H., Effect of thioacetamide on the activity of alkaline ribonuclease and ribonuclease inhibitor in rat liver. *Biochem. Pharmacol.* **16**, 1711–1719 (1967).

12c. Choi, Y. C., Smetana, K., and Busch, H., Studies on RNA isolated from nucleolar fractions of Novikoff hepatoma cells by sequential extractions with dilute and concentrated salt solutions. *Exptl. Cell Res.* **53**, 582–602 (1968).

12c′. Cunningham, D. D., Cho, S., and Steiner, D. F., Soluble RNA polymerase from rat liver nuclei. *Biochim. Biophys. Acta* **171,** 67–74 (1969).
12d. Cunningham, D. D., and Steiner, D. F., Extraction of RNA polymerase from rat liver nuclei in a soluble form. *Biochim. Biophys. Acta* **145,** 834–836 (1967).
13. Dalcq, A. M., The cytochemical localization of adenosinetriphosphatase in the eggs of mammals and its relation to their morphogenetic organization. *Bull. Acad. Roy. Med. Belg.* [6] **24,** 825–898 (1959).
13a. Daoust, R., and Simard, A., RNA synthesis and neoplastic transformation in rat liver. *Proc. Am. Assoc. Cancer Res.* **7,** 17 (1966).
14. De, P., Chatterjee, R., and Mitra, S., Nucleolar localization of succinic dehydrogenase in normal rat hepatic cells. *Nature* **191,** 925 (1961).
15. Ermolaeva, L. P., and Zbarsky, I. B., Amino acid incorporation into nucleolar proteins of animal cells *in vivo.* Proc. 7th Int. Cong. Biochemistry, Tokyo, 1967. Sci. Council Japan, 1968.
16. Fogel, S., and Sypherd, P. S., Chemical basis for heterogeneity of ribosomal proteins. *Proc. Natl. Acad. Sci. U. S.* **59,** 1329–1336 (1968).
17. Fujimura, S., Hasegawa, S., Shimizu, Y., and Sugimura, T., Polymerization of the adenosine 5′-diphosphate ribose moiety of nicotinamide-adenine dinucleotide by nuclear enzyme. I. Enzymatic reactions. *Biochim. Biophys. Acta* **145,** 247–259 (1967).
17a. Gantt, R., and Evans, V. J., Comparison of soluble RNA methylase capacity in paired neoplastic and nonneoplastic cell lines *in vitro. Cancer Res.* **29,** 536–541 (1969).
18. Georgiev, G. P., Samarina, O. P., Lerman, M. I., Smirnov, M. N., and Seretzov, A. N., Biosynthesis of messenger and ribosomal ribonucleic acids in the nucleolochromosomal apparatus of animal cells. *Nature* **200,** 1291–1294 (1963).
19. Georgiev, G. P., Samarina, O. P., Mantieva, V. L., and Zbarsky, I. B., Further studies on the nuclear ribonucleoproteins. *Biochim. Biophys. Acta* **46,** 399–400 (1961).
20. Golarz, M. N., and Bourne, G. H., Induction and accentuation of phosphatase activity in the nucleoli of muscle nuclei by denervation and injected nucleotides. *Exptl. Cell Res.* **25,** 691–693 (1961).
20a. Goldberg, M. L., Moon, H. D., and Rosenau, W., Partial purification of RNA polymerase from rat liver nuclei. *Biochim. Biophys. Acta* **171,** 192–194 (1969).
21. Grogan, D. E., and Busch, H., Studies on fractionation of saline-soluble nucleolar proteins on diethylaminoethyl-cellulose. *Biochemistry* **6,** 573–578 (1967).
22. Grogan, D. E., Desjardins, R., and Busch, H., Nucleolar proteins of rat liver and Walker tumor. *Cancer Res.* **26,** 775–779 (1966).
23. Harris, H., Turnover of nuclear and cytoplasmic ribonucleic acid in two types of animal cell, with some further observations on the nucleolus. *Biochem. J.* **73,** 362–369 (1959).
24. Harris, H., Nuclear ribonucleic acids. *Progr. Nucleic Acid Res.* **2,** 19–59 (1963).
25. Harris, H., Function of the short-lived ribonucleic acid in the cell nucleus. *Nature* **201,** 863–867 (1964).
26. Hasegawa, S., Fujimura, S., Shimizu, Y., and Sugimura, T., The polymerization of adenosine 5′-diphosphate-ribose moiety of NAD by nuclear enzyme. II. Properties of the reaction product. *Biochim. Biophys. Acta* **149,** 369–376 (1967).
26a. Healy, G. M., and Parker, R. C., Selective nucleolar uptake of α_1-acid glycoprotein by mammalian cells in tissue culture. *Biochim. Biophys. Acta* **148,** 556–558 (1967).
26b. Higashi, K., Matsuhisa, T., Kitao, A., and Sakamoto, Y., Selective suppression

of nucleolar RNA metabolism in the absence of protein synthesis. *Biochim. Biophys. Acta* **166**, 388–393 (1968).

26b′. Ishikawa, K., Kuroda, C., and Ogata, K., Release of ribonucleoprotein particles containing rapidly labeled ribonucleic acid from rat liver nuclei. Effect of adenosine 5′-triphosphate and some properties of the particles. *Biochim. Biophys. Acta* **179**, 316–331 (1969).

26b″. Izawa, M., and Kawashima, K., Some properties of ribonucleoprotein particles in the isolated nucleoli of mouse ascites tumor cells. *Biochim. Biophys. Acta* **174**, 124–136 (1969).

26c. Jacob, S. T., Sajdel, E. M., and Munro, H. N., Presence of two RNA polymerase activities in liver nucleoli. *Biochim. Biophys. Acta* **157**, 421–423 (1968).

26c′. Janne, J., and Raina, A., On the stimulation of ornithine decarboxylase and RNA polymerase activity in rat liver after treatment with growth hormone. *Biochim. Biophys. Acta* **174**, 769–772 (1969).

26d. Kashnig, D. M., and Kasper, C. B., Isolation, morphology, and composition of the nuclear membrane from rat liver. *J. Biol. Chem.* **244**, 3786–3792 (1969).

27. Krichevskaya, A. A., Molnar, J., Bruskov, V. I., Samarina, O. P., and Georgiev, G. P., Some properties of nuclear ribonucleoprotein particles containing mRNA. *Proc. 7th Inter. Biochim. Congr. Tokyo, 1967.* Sci. Council Japan, Tokyo, 1968.

27a. Kuntzel, H., Proteins of mitochondrial and cytoplasmic ribosomes from *neurospora crassa. Nature* **222**, 142–146 (1969).

27a′. Laipis, P. J., Olivera, B. M., and Ganesan, A. T., Enzymatic cleavage and repair of transforming DNA. *Proc. Natl. Acad. Sci.* **62**, 389–396 (1969).

28. Landschütz, C. C., Zytochemischer nachweis der adenosintriphosphatase [ATP-Ase] in den zellen des asictestumors der maus. *Experientia* **6**, 232–234 (1950).

28a. Liao, S., Sagher, D., and Fang, S., Isolation of chromatin-free RNA polymerase from mammalian cell nuclei. *Nature* **220**, 1336–1337 (1968).

29. Liau, M. C., Craig, N. C., and Perry, R. P., The production of ribosomal RNA from high molecular weight precursors. I. Factors which influence the ability of isolated nucleoli to process 45 S RNA. *Biochim. Biophys. Acta* **169**, 196–205 (1968).

29a. Little, J. W., Zimmerman, S. B., Oshinsky, C. K., and Gellert, M., Enzymatic joining of DNA strands. II. An enzyme-adenylate intermediate in the DPN-dependent DNA ligase reaction. *Proc. Natl. Acad. Sci. U. S.* **58**, 2004–2011 (1967).

29a′. Love, R., and Walsh, R. J., The relation of nucleolini to nucleolar vacuoles in the living cell. *Exptl. Cell Res.* **53**, 432–446 (1968).

29a″. Maden, B. E. H., and Vaughan, M. H., Synthesis of ribosomal proteins in the absence of ribosome maturation in methionine-deficient HeLa cells. *J. Mol. Biol.* **38**, 431–435 (1968).

29b. Maggio, R., Progress report on the characterization of nucleoli from guinea pig liver. *Natl. Cancer Inst. Monograph* **23**, 213–222 (1966).

30. Maisel, J. C., Nucleolar reaction with normal serum shown by nonimmune fluorescent staining. *J. Lab. Clin. Med.* **60**, 357–364 (1962).

31. Maisel, J. C., and Lytle, R. I., Affinity of animal cell nucleoli for normal serum. Preliminary characterization of serum and cell components. *J. Cell Biol.* **29**, 461–473 (1966).

32. Moore, P. B., Traut, R. R., Noller, H., Pearson, P., and Delius, H., Ribosomal proteins of *Escherichia coli.* II. Proteins from the 30 S subunit. *J. Mol. Biol.* **31**, 441–461 (1968).

32a. Moskowitz, G., and Busch, H., Unpublished studies (1969).
32b. Moulé, Y., Personal communication. Nucleolus Workshop, Prague (1969).
33. Mulnard, J., Presence d'inclusions Feulgen positives dans les nucleoles larvaires de quelques Dipteres. *Arch. Biol.* **67**, 485–498 (1956).
33a. Mundell, R. D., Studies on nucleolar and ribosomal basic proteins and their relationship to nucleolar function. *Exptl. Cell Res.* **53**, 395–400 (1968).
34. Mundkur, B., and Brauer, B., Selective localization of nucleolar protein with amido black 10B. *J. Histochem. Cytochem.* **14**, 94–103 (1966).
35. Muramatsu, M., Hodnett, J. L., and Busch, H., Studies on the "Independence" of nucleolar ribonucleic acid synthesis. *Biochim. Biophys. Acta* **91**, 592–597 (1964).
36. Muramatsu, M., Smetana, K., and Busch, H., Quantitative aspects of isolation of nucleoli of the Walker carcinosarcoma and liver of the rat. *Cancer Res.* **23**, 510–518 (1963).
36a. Nicolette, J. A., Lemahieu, M. A., and Mueller, G. C., A role of estrogens in the regulation of RNA polymerase in surviving rat uteri. *Biochim. Biophys. Acta* **166**, 403–409 (1968).
37. Nishizuka, Y., Ueda, K., Nakazawa, K., and Hayaishi, O., Studies on the polymer of adenosine diphosphate ribose. I. Enzymic formation from nicotinamide adenine dinucleotide in mammalian nuclei. *J. Biol. Chem.* **242**, 3164–3171 (1967).
37a. Novikoff, A. B., The validity of histochemical phosphatase methods on the intracellular level. *Science* **113**, 320–325 (1951).
37b. Otake, H., Miwa, M., Funimura, S., and Sugimura, T., Binding of ADP-ribose polymer with histone. *J. Biochem.* **65**, 145–146 (1969).
38. Parson, J. T., and McCarty, K. S., Rapidly labeled messenger ribonucleic acid-protein complex of rat liver nuclei. *J. Biol. Chem.* **243**, 5377–5384 (1968).
39. Perry, R. P., On ribosome biogenesis. *Natl. Cancer Inst. Monograph* **23**, 527–546 (1966).
40. Perry, R. P., The nucleolus and the synthesis of ribosomes. *Progr. Nucleic Acid Res. Mol. Biol.* **6**, 219–257 (1967).
40′. Pogo, A. O., Modification of ribonucleic acid synthesis in isolated rat liver nuclei by low salt concentrations and specific divalent cations. *Biochim. Biophys. Acta* **182**, 57–65 (1969).
40″. Puvion, E. de, Defert-Kobus, W., Kerckaert, J.-P., Puvion-Dutilleul, F., Segard, E., and Montreuil, J., Etudes sur les acides nucleiques spontanement diffusibles des noyaux du foi de rat. Etude morphologique. Demonstration de l'origine nucleolaire des particles diffusibles. *C. R. Acad. Sc. Paris* **266**, 278–281 (1968).
40a. Reeder, R. H., Ueda, K., Honjo, T., Nishizuka, Y., and Hayaishi, O., Studies on the polymer of adenosine diphosphate ribose. II. Characterization of the polymer. *J. Biol. Chem.* **242**, 3172–3179 (1967).
41. Reid, B. R., and Cole, R. D., Biosynthesis of a lysine-rich histone in isolated calf thymus nuclei. *Proc. Natl. Acad. Sci. U. S.* **51**, 1044–1050 (1964).
41′. Ritchie, R. F., Nucleolar structure by immunologic technics. *Cancer Res.* **29**, 267–269 (1969).
41a. Ro, T. S., and Busch, H., *In vitro* labeling of RNA in isolated nucleoli of the Walker tumor and liver. *Cancer Res.* **24**, 1630–1633 (1964).
42. Robbins, E., and Borun, T. W., The cytoplasmic synthesis of histones in HeLa cells and its temporal relationship to DNA replication. *Proc. Natl. Acad. Sci. U. S.* **57**, 409–416 (1967).

42a. Roeder, R. G., and Rutter, W. J., Multiple forms of DNA-dependent RNA polymerase in eukaryotic organisms. *Nature* **224**, 234–237 (1969).
43. Roth, J. S., Some observations on the assay and properties of ribonucleases in normal and tumor tissues. *Methods Cancer Res.* 3, 153–242 (1968).
43a. Sadowski, P. D., and Howden, J. A., Isolation of two distinct classes of polysomes from a nuclear fraction of rat liver. *J. Cell Biol.* **37**, 163–181 (1968).
43b. Scherrer, K., Personal communication, Nucleolus Workshop, Prague, 1969.
44. Schneider, W. C., and Hogeboom, G. H., Biochemistry of cellular particles. *Ann. Rev. Biochem.* **25**, 201–224 (1956).
44a. Siebert, G., Personal communication, 1969.
45. Siebert, G., Villalobos, J., Ro, T. S., Steele, W. J., Lindenmayer, G., Adams, H., and Busch, H., Enzymatic studies on isolated nucleoli of rat liver. *J. Biol. Chem.* **241**, 71–78 (1966).
45a. Smetana, K., and Busch, H., Studies on staining and localization of acidic nuclear proteins in the Walker 256 carcinosarcoma. *Cancer Res.* **26**, 331–337 (1966).
45b. Smith, S. J., Adams, H. R., Smetana, K., and Busch, H., Isolation of the outer layer of the nuclear envelope. *Exptl. Cell Res.* **55**, 185–197 (1969).
46. Soeiro, R., Vaughan, M. H., and Darnell, J. E., The effect of puromycin on intranuclear steps in ribosome biosynthesis. *J. Cell Biol.* **36**, 91–101 (1968).
46′. Starr, J. L., and Sells, B. H., Methylated ribonucleic acids. *Physiol. Rev.* **49**, 623–669 (1969).
46a. Steele, W. J., Effects of thioacetamide on nuclear and nucleolar RNA. In, The Cell Nucleus—Metabolism and Radiosensitivity. Proc. Int. Symp., Rijswijk, The Netherlands, Taylor & Francis Ltd., London, 1966, pp. 203–220.
47. Sypherd, P. S., Site of synthesis of ribosomal proteins. *J. Mol. Biol.* **17**, 358–369 (1966).
47a. Taborsky, G., Personal communication, 1967.
48. Tamaoki, T., The particulate fraction containing 45 S RNA in L cell nuclei. *J. Mol. Biol.* **15**, 624–639 (1966).
49. Tamaoki, T., and Mueller, G. C., The effects of actinomycin D and puromycin on the formation of ribosomes in HeLa cells. *Biochim. Biophys. Acta* **108**, 73–80 (1965).
50. Tamaoki, T., and Mueller, G. C., A rapidly labeled ribonucleoprotein in HeLa cell nuclei. *Biochim. Biophys. Acta* **108**, 81–94 (1965).
51. Tissieres, A., Structure of *E. coli* ribosomes and the incorporation of amino acids. *In* "The Molecular Basis of Neoplasia," pp. 507–518. Univ. of Texas Press, Austin, Texas, 1962.
52. Tissieres, A., and Hopkins, J. W., Factors affecting amino acid incorporation into proteins by *Escherichia coli* ribosomes. *Proc. Natl. Acad. Sci. U. S.* **47**, 2015–2023 (1961).
52a. Traub, A., Kaufmann, E., and Ginzburg-Tietz, Y., Studies on nuclear ribosomes. I. Association of DPN-pyrophosphorylase with nuclear ribosomes in normal and neoplastic tissues. *Exptl. Cell Res.* **34**, 371–383 (1964).
53. Traub, P., and Nomura, M., Structure and function of *E. coli* ribosomes. VI. Reconstitution of functionally active 30 S ribosomal particles from RNA and proteins. *Proc. Natl. Acad. Sci. U. S.* **59**, 777–784 (1968).
54. Traut, R. R., Moore, P. B., Delius, H., Noller, H., and Tissieres, A., Ribosomal proteins of *Escherichia coli*, I. Demonstration of different primary structures. *Proc. Natl. Acad. Sci. U. S.* **57**, 1294–1301 (1967).

54a. Tsutsui, E., Srinivasan, P. R., and Borek, E., tRNA methylases in tumors of animal and human origin. *Proc. Natl. Acad. Sci. U. S.* **56,** 1003–1009 (1966).

54b. Vaughan, M. H., Soeiro, R., Warner, J. R., and Darnell, J. E., The effect of methionine deprivation on ribosome synthesis in HeLa cells. *Proc. Natl. Acad. Sci. U. S.* **58,** 1527–1534 (1967).

55. Vaughan, M., Warner, J., and Darnell, J., Ribosomal precursor particles in the HeLa cell nucleus. *J. Mol. Biol.* **25,** 235–251 (1967).

55a. Vesco, C., and Penman, S., The fractionation of nuclei and the integrity of purified nucleoli in HeLa cells. *Biochim. Biophys. Acta* **169,** 188–195 (1968).

56. Villalobos, J., Jr., Steele, W. J., and Busch, H., Increased ribonuclease activity of nuclei of liver cells of rats treated with thioacetamide. *Biochem. Biophys. Res. Commun.* **17,** 723–729 (1964).

57. Vincent, W. S., Isolation and chemical properties of nucleoli of Starfish oocytes. *Proc. Natl. Acad. Sci. U. S.* **38,** 139–145 (1952).

58. Vincent, W. S., Structure and chemistry of nucleoli. *Intern. Rev. Cytol.* **5,** 269–299 (1955).

59. Vincent, W. S., Proteins of starfish nucleoli. *Biol. Bull.* **113,** 334 (1957).

60. Vorbrodt, A., Histochemically demonstrable phosphatases and protein synthesis. *Exptl. Cell Res.* **15,** 1–20 (1958).

61. Warner, J. R., The assembly of ribosomes in HeLa cells. *J. Mol. Biol.* **19,** 383–398 (1966).

62. Warner, J. R., and Soeiro, R., Nascent ribosomes from HeLa cells. *Proc. Natl. Acad. Sci. U. S.* **58,** 1984–1990 (1967).

63. Warner, J. R., Girard, M., Latham, H., and Darnell, J. E., Ribosome formation in HeLa cells in the absence of protein synthesis. *J. Mol. Biol.* **19,** 373–382 (1966).

63a. Warner, J. R., Soeiro, R., Birnboim, H. C., Girard, M., and Darnell, J. E., Rapidly labeled HeLa cell nuclear RNA. I. Identification by zone sedimentation of a heterogeneous fraction separate from ribosomal precursor RNA. *J. Mol. Biol.* **19,** 349–361 (1966).

64. Whittle, E. D., Bushness, D. E., and Potter, V. R., RNA associated with the outer membrane of rat liver nuclei. *Biochim. Biophys. Acta* **161,** 41–50 (1968).

64a. Widnell, C. C., and Tata, J. R., Studies on the stimulation by ammonium sulphate of the DNA-dependent RNA polymerase of isolated rat liver nuclei. *Biochim. Biophys. Acta* **123,** 478–492 (1966).

64b. Wintzerith, M., Personal communication (1967).

64b′. Yu, Fu-Li, and Feigelson, P., The sequential stimulation of uracil-rich and guanine-rich RNA species during cortisone induction of hepatic enzymes. *Biochem. Biophys. Res. Comm.* **35,** 499–504 (1969).

64b″. Zimmerman, E. F., Hackney, J., Nelson, P., and Arias, I. M., Protein synthesis in isolated nuclei and nucleoli of HeLa cells. *Biochemistry* **8,** 2636–2644 (1969).

64c. Zimmerman, S. B., Little, J. W., Oshinsky, C. K., and Gellert, M., Enzymatic joining of DNA strands: A novel reaction of diphosphopyridine nucleotide. *Proc. Natl. Acad. Sci. U. S.* **57,** 1841–1848 (1967).

65. Zjadela, F., and Morin, G. A., Etude cytologique et histochimique du noyau hépatique de souris isolé. *Rev. Hématol.* **7,** 628–651 (1952).

CHAPTER VIII

The Nuclear Ribonucleoprotein Network and the Nuclear Residue

I. Introduction

The nucleolus is generally visualized within the nuclear matrix as a dense structure apparently separated from the remainder of the nucleus although it is clear that the nucleolus lacks a membrane (Chapter II). Since genetic, cytochemical, and biochemical studies show that the nucleolus is a most active site of synthesis of RNA, one persistent interesting question is how does the nucleolar product move to the cytoplasm. The present day concepts are that either whole ribosomes or larger complexes of polysomes or interchromatinic granules move through the nuclear pores into the cytoplasm (Chapter I). Some physical mechanisms are necessary since in phase microscopic studies, nucleoli of both normal and tumor cells appear almost always to be located centrally within the nuclear matrix (Chapter IX). Even in cells with very active ribosome synthesis, phase microscopy only rarely reveals physical approximation of the nucleolus to the nuclear membrane.

Early but now unacceptable ideas included (a) the nucleolus discharges its products directly into the cytoplasm and (b) the nucleolus is extruded into the cytoplasm. One early idea was that after physical migration of the nucleolus to the nuclear membrane, there is a "dumping" of the nucleolar products into the cytoplasm; this idea was based on time-lapse photography of tumor cells in tissue culture. Lettré *et al.* (*8–10*) observed occasionally that the nucleolus approached the nuclear membrane and "extruded" its products into the cytoplasm. Other studies (*1, 1a*) suggested that nucleoli themselves might pass from the nuclear membrane to the cytoplasm. In tissue culture, such extrusions may reflect agonal states of some cells.

A misleading finding is physical approximation of the nucleolus to the cytoplasm at the nuclear membrane which is observed commonly in electron micrographs. In such pictures, many tumor nucleoli (Fig. 5.9) appear to adjoin the inner layer of the nuclear membrane. However, the frequency with which this phenomenon is observed as contrasted with its rarity in tissue culture or in isolated cells and nuclei suggests that it represents a dehydration artifact of electron microscopic preparative techniques; i.e., in the course of the fixation procedures, shrinkage of cell constituents occurs during the interactions of the cells, the nucleoli, the fixatives, and the stains (Chapter II).

A. Light Microscopic Observations on the Nuclear Ribonucleoprotein Network

In the course of experiments designed to evaluate the effects of a variety of procedures on nuclear composition and nucleolar products, it was found that the extraction procedure of Sibatani *et al.* (*17*) which employs successive extraction with 0.15 *M* and 2 *M* NaCl was of value for the successive extraction of both the nuclear "sap" fractions and the "chromatin fraction." Figure 8.1a shows that extraction of the nuclei with 0.14 *M* NaCl produces a nuclear appearance markedly different from the rather amorphous appearance of the nuclei isolated by the Chauveau technique (Fig. 8.1b) or of nuclei *in situ* (Fig. 11.1A). In normal liver nuclei, there are large ribonucleoprotein strands that extend from the

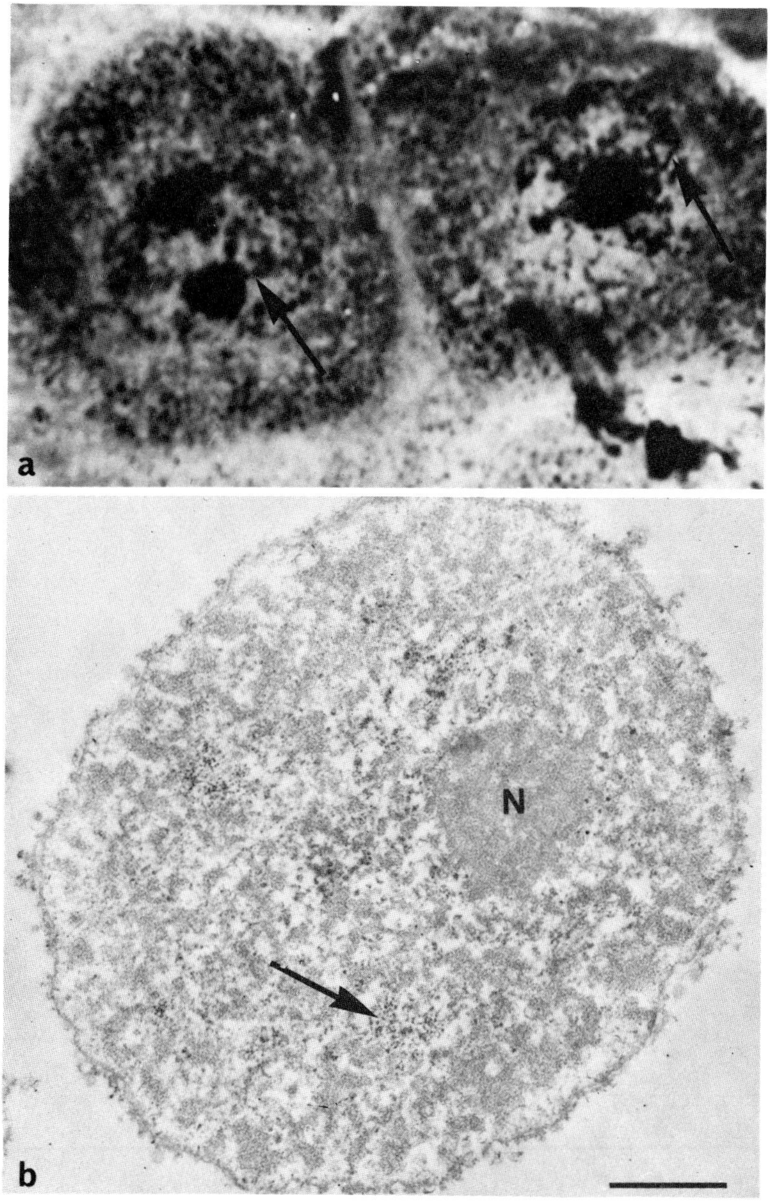

Fig. 8.1. a. Smear of liver cells extracted with 0.14 *M* NaCl stained with toluidine blue without previous fixation. The nuclear RNP network seems to radiate from the nucleolus. ×2200 (*18*). b. Liver cell nucleus isolated by the Chauveau technique (Chapter XI). Note the presence of clusters of interchromatinic dark particles (arrow). Some are localized around the nucleolus and may represent elements of the nuclear ribonucleoprotein network. N, nucleolus. ×16,000 (*16′*).

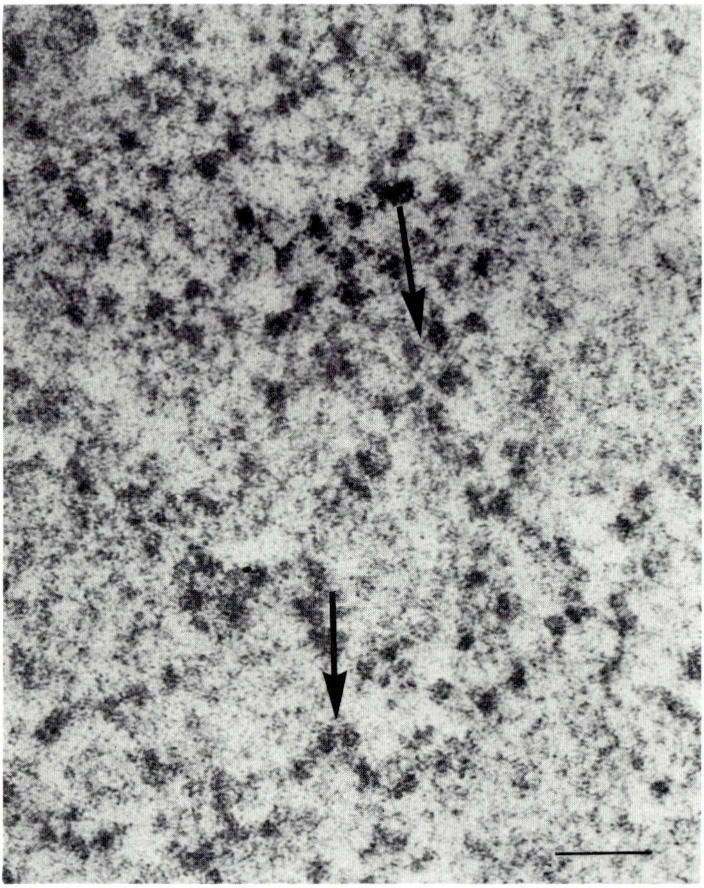

Fig. 8.2. Interchromatinic area of a human leukemic myeloblast containing numerous dark granules resembling ribosomes. Some of these granules are connected with a filament (arrows) which may be template RNA. ×123,000.

nucleolus to the nuclear membrane; similar ribonucleoprotein strands were found in tumor cells. In some places this "network of ribonucleoprotein elements" appears to be thickened so that there are masses of ribonucleoproteins as well as finer fibrils. Figure 8.1b shows that in the electron micrographs of normal liver there are nuclear elements apparently composed of dense granules that are not joined to the nucleolus; on closer inspection of the nucleolus and its periphery, there are many of these fine interchromatinic granules (*3, 16, 18, 22*). The reason why these elements of the nuclear reticulum or nuclear ribonucleoprotein network are not usually visible is that they are obscured by masses of salt-extractable chromatin and soluble ribonucleoproteins in normal nuclei.* In occasional cases, interchromatinic masses are visualized that appear to be nuclear polysomes as in the human myeloblast shown in Fig. 8.2. These masses may be portions of the interchromatinic dense masses seen in Fig. 8.1b.

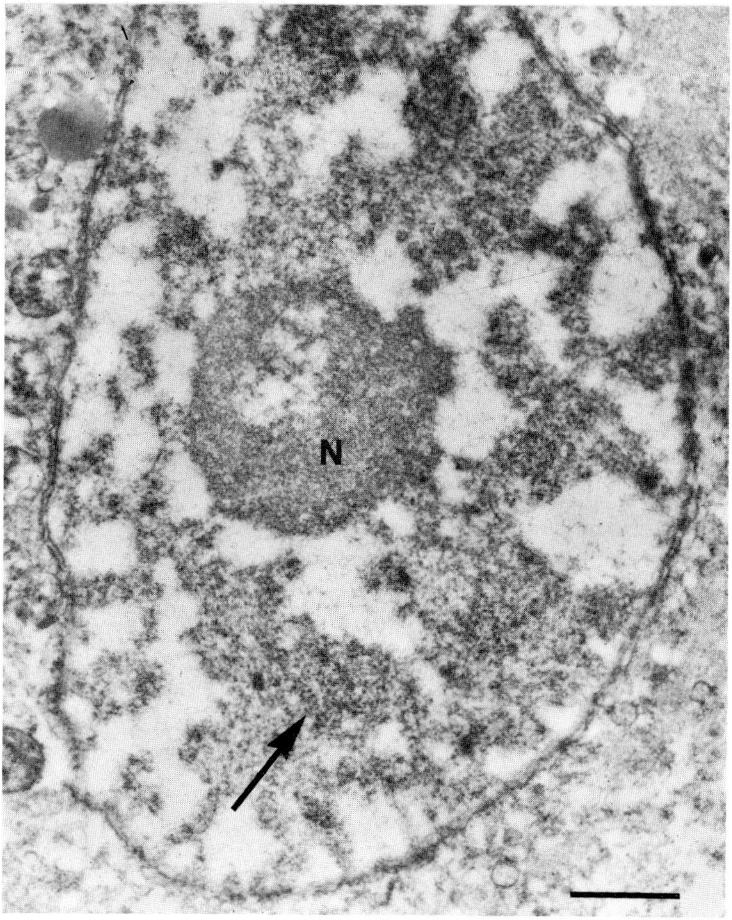

Fig. 8.3. Nuclear residue from Walker tumor after treatment with dilute (0.14 *M*) and concentrated (2 *M*) saline solutions. The nucleolus (N) is attached to a network (arrow) composed of particles and filaments. ×14,000 (*16'*).

* Three different classes of nuclear ribonucleoproteins have solubility differences; i.e., one group of ribonucleoproteins is readily soluble in 0.15 *M* NaCl, a second is insoluble in 0.15 *M* NaCl but is soluble in 2 *M* NaCl, and a third group is insoluble in either 0.15 *M* NaCl or 2 *M* NaCl (*17, 21*).

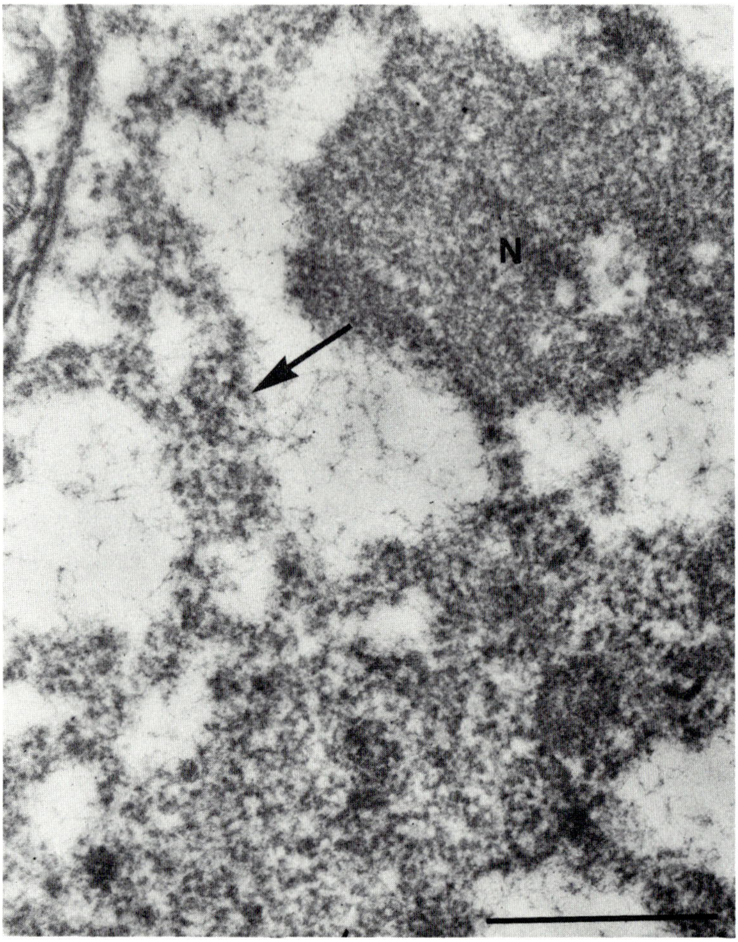

Fig. 8.4. Nuclear residue from the same specimen as that in Fig. 8.3. Ribonucleoprotein particles are present in the network (arrow). ×30,000 (*16′*).

B. Electron Microscopic Observations on the Nuclear Ribonucleoprotein Network

Figure 8.3 shows the presence of the nuclear ribonucleoprotein network in the residues of the nuclei of Walker tumor preparations after successive treatment with dilute and concentrated saline solutions; i.e., 0.14 *M* and 2 *M* NaCl, respectively (*16′*). The nucleolar residue (N) appears to be at the center of the masses of ribonucleoprotein elements. In several places, there are direct connections between the nucleolar substance and the more dispersed elements of the network. At higher magnifications employed in this study, the granular components of the extranucleolar masses are more dense than intranucleolar elements (Fig. 8.4). The density of the extranucleolar ribonucleoprotein elements approximates that of the interchromatinic dense granules. Interestingly, the nuclear membrane apparently separates these elements from the cytoplasm; i.e., neither the elements of the nuclear ribonucleoprotein network nor the interchromatinic granules are seen in the cytoplasm.

To ascertain whether the dense granules were produced in the nuclear ribonucleoprotein network or nuclear matrix by the treatment of the preparations with salt solutions, studies were made with DNase (Fig. 8.5). The nuclear ribonucleoprotein network and the dense granules in the extranucleolar fibrils of the matrix are readily apparent. These dense

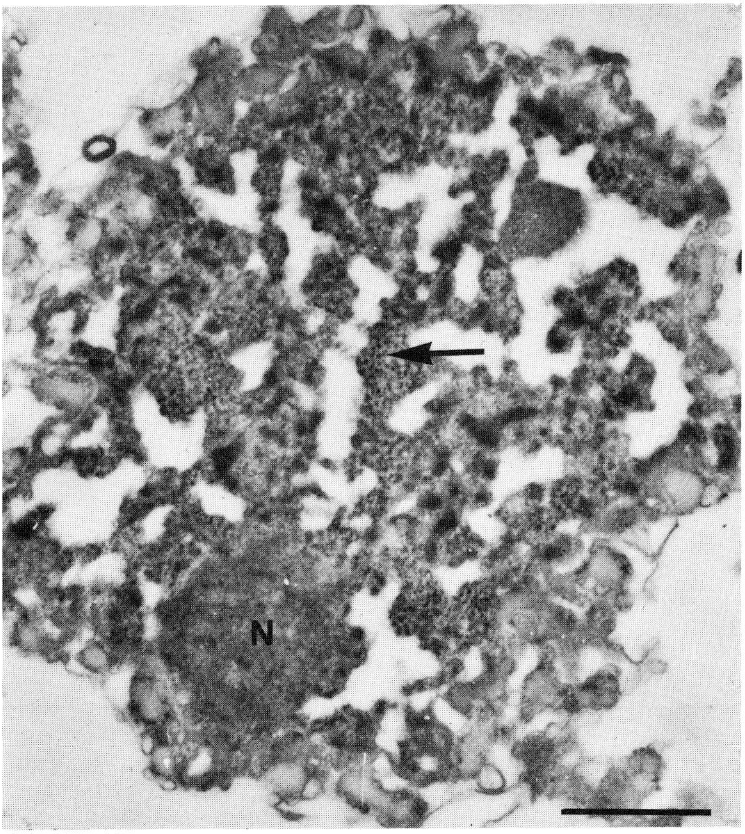

Fig. 8.5. Isolated liver nucleus treated with deoxyribonuclease. The interchromatinic particles in the network (arrow) are not destroyed by this treatment. ×19,000 (*16'*).

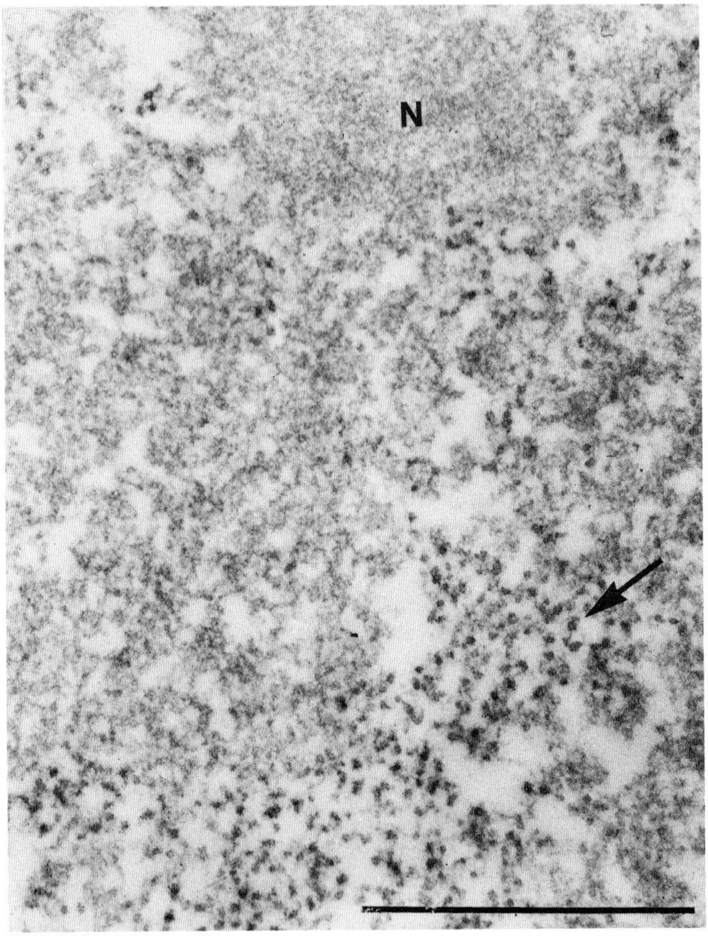

Fig. 8.6. Isolated liver nucleus digested with ribonuclease. The interchromatinic particles of the network are present (arrow) while those in nucleoli (N) are practically removed. ×45,000 (*16'*).

elements of the nuclear ribonucleoprotein network are quite resistant to RNase unlike the nucleolar ribonucleoproteins which are rapidly and readily destroyed by RNase. These ribonucleoprotein particles have a protective protein coat. When preparations of the type shown in Fig. 8.6 were treated first with pepsin and then with ribonuclease the pepsin produced a marked decrease in the density of these elements and the combined treatment with pepsin and RNase (Figs. 8.7–8.9) dissolved the particulate elements.

The precise nature of the coat surrounding the dense ribonucleoprotein

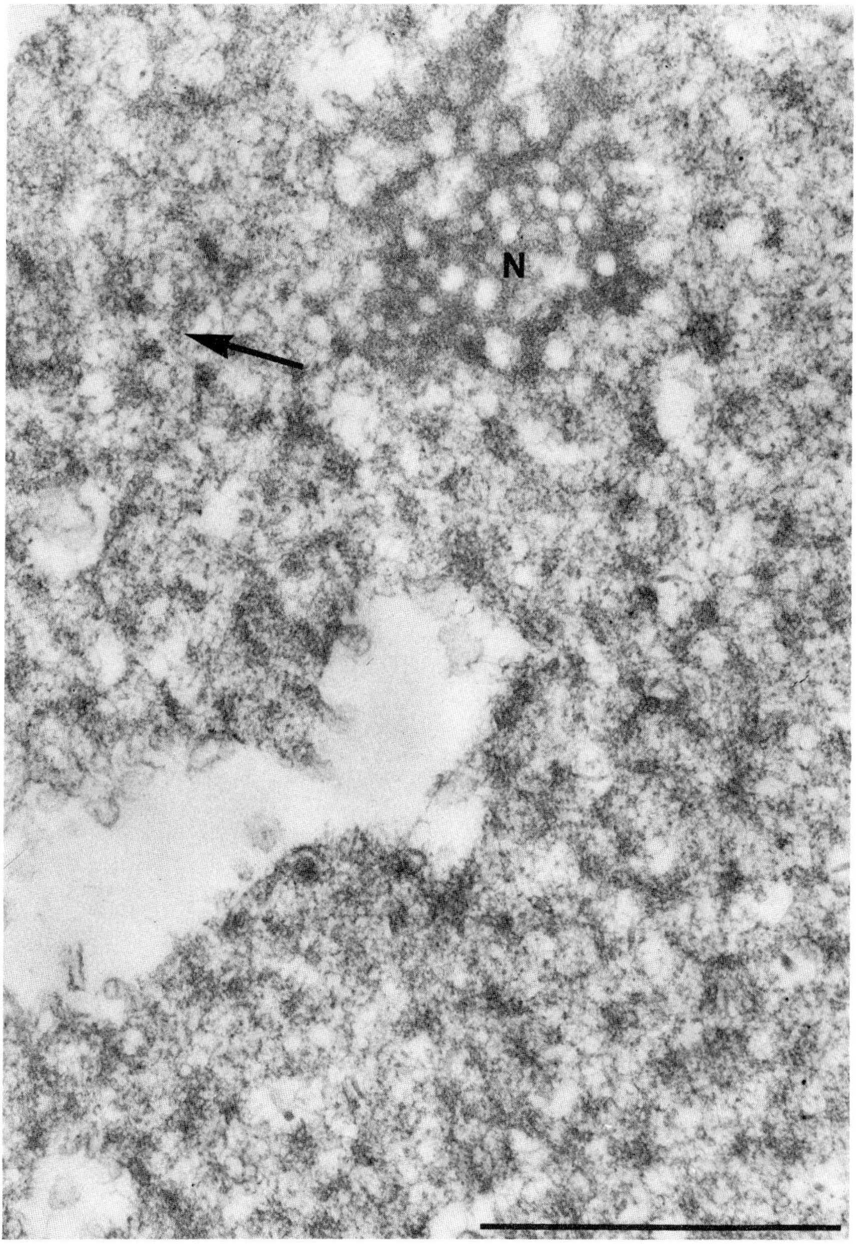

Fig. 8.7. Isolated liver nucleus digested with pepsin. The density of the interchromatinic granules (arrow) is reduced. N, nucleolus. ×47,000 (*16'*).

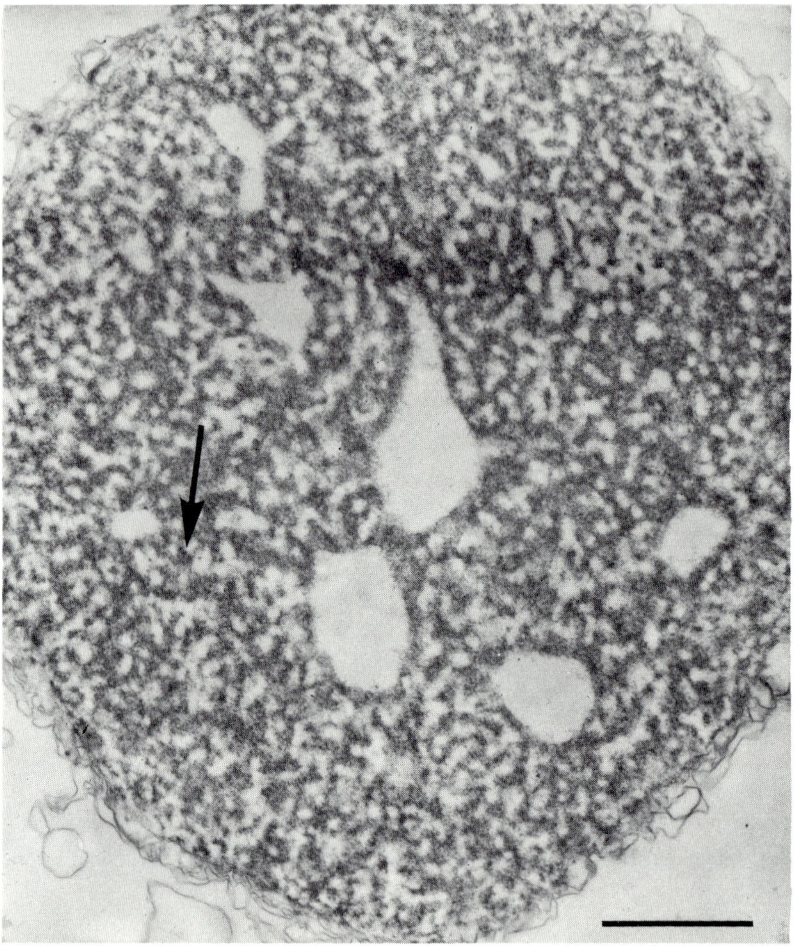

Fig. 8.8. Isolated liver cell nucleus digested with ribonuclease followed by pepsin. Residues of interchromatinic particles are still present (arrow). ×19,800 (*16'*).

granules or "interchromatinic" granules in the nuclear ribonucleoprotein network is not defined. It seems likely that some are acidic proteins (*4*) which differ in their solubility characteristics from those of the remainder of the proteins in the nucleolus (Chapter VII).

II. RNA in the Nuclear Ribonucleoprotein Network

One of the most serious problems in dealing with the nuclear ribonucleoprotein network has been the difficulty of isolation of the network elements as distinct morphological or cytological entities. Extraction of the nuclei with appropriately buffered 0.15 *M* NaCl and 2 *M* NaCl leaves a residue of nucleolar mass and network elements. Very mild sonication to shake the fibers of nuclear ribonucleoprotein network free from the nucleoli has been attempted but neither the proper energy for sonication nor the appropriate ionic environment has been established. Unfortunately, methods are not yet available for separation of the network elements in a fibrillar form from the nucleoli at the same time that the integrity of nucleoli is maintained. Accordingly, all of the results obtained are on the residue fraction that is insoluble in 0.15 *M* NaCl and 2 *M* NaCl.

Only partial data are available on the components of the nuclear ribonucleoprotein network. The ^{32}P nucleotide compositions of the nuclear

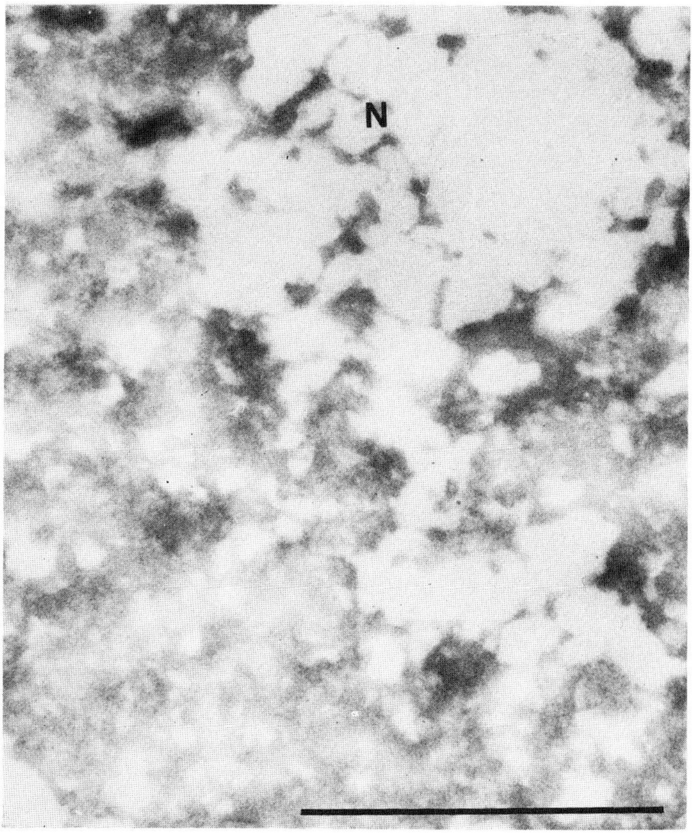

Fig. 8.9. Isolated liver nucleus after sequential treatment with pepsin and ribonuclease. The interchromatinic granules are destroyed. N, nucleolus. ×48,000.

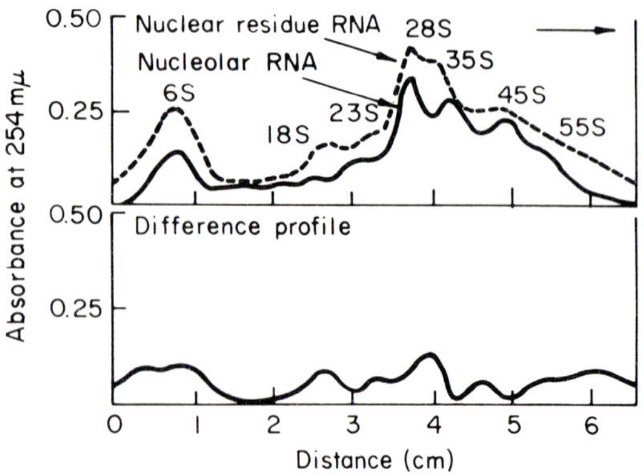

Fig. 8.10. Sedimentation profiles of RNA of the nuclear residue fraction and isolated nucleoli (top profiles). The difference profile (bottom) was obtained by means of subtraction of the profile of nucleolar RNA from the profile of RNA of the nuclear residue fraction (*21*).

TABLE 8.1

Nucleotide Composition (^{32}P) of RNA of the Nucleolar and Nuclear Residue Fractions of Rat Liver[a]

	AMP	UMP	GMP	CMP	AMP + UMP / GMP + CMP
Nucleoli	20.1	16.5	39.0	24.4	0.58
Nuclear residue fraction	26.1	21.9	26.7	25.4	0.92

[a] From Steele and Busch (*21*).

residue fraction reflect the labeling of the rapidly sedimenting RNA (Table 8.1).

In addition to the nucleolar RNA, the residue fraction clearly contains a type of RNA that is rich in adenylic and uridylic acids, despite the relatively minor differences in the profile obtained for the nucleolar RNA and the RNA of the nuclear residue fraction (Fig. 8.10). It is possible that some nuclear 45 and 55 S RNA of the DNA–RNA fraction which has a ^{32}P nucleotide composition of A, 30; U, 22.6; G, 21.7; and C, 25.3 may be contaminating the RNA of the nuclear residue fraction (Fig. 8.10). Such RNA may be bound by protein and may account for as much as 60–70% of the total RNA in the fraction.

A. On the Source of 18 S rRNA

As noted in Chapter V, in all probability neither 18 S nor 28 S rRNA is produced in final form in the nucleoli. However, the labeling of the 18 S and 28 S RNA in the ribonucleoprotein elements of the perinuclear ribosomal layer and the labeling of the RNA in the ribosomes of the cytoplasm is essentially the same for both the 18 S and 28 S RNA (*19*). With the finding of the nuclear ribonucleoprotein network, it was thought that this nuclear fraction rather than the nucleolus might be the site of final processing of the nucleolar 28 S and 23 S rRNA precursors to the 28 S and 18 S rRNA (*21*). Unequivocal evidence for this idea has not yet been obtained.

Recent studies in this laboratory (*19*) and in others (*12a, 13, 25*) have led to the conclusion that the concept presented from studies on HeLa cells (*12, 20*) that 18 S RNA-containing ribonucleoproteins are rapidly transported to the cytoplasm is not valid for all types of cells. These conclusions are based on kinetic studies of labeling of cytoplasmic 18 S and 28 S RNA which proceeds at virtually identical rates. Accordingly, as noted in Chapter V, it would appear that not only do 18 S and 28 S rRNA have common precursors but their release into the cytoplasm is simultaneous in a number of cell types.

Current studies are in progress to evaluate the number of types of 18 S RNA (*5*) in the nucleus and cytoplasm because the probability exists that there are a number of species of such RNA molecules, some of which are probably messenger RNA's (*7*).

III. RNA of the Nuclear Residue Fraction of Novikoff Hepatoma

A. Sedimentation Peaks and Coefficients

To obtain the least degraded RNA of the nuclear residue fraction, the extraction procedures were carried out on nuclear preparations of the Novikoff hepatoma ascites cells (*16a*). Zonal ultracentrifugation and repurification of the fractions provided the sucrose density gradient patterns shown in Fig. 8.11. Figure 8.11 shows that little 4–7 S RNA remained in the residue or chromatin fractions after the initial extraction with 0.15 *M* NaCl. However, both 18 S and 28 S RNA peaks were found and in each case a shoulder of 45 S RNA was noted as well as a considerable fraction that sedimented more rapidly. As noted in the portions of Fig. 8.11, designated as a, b, and c, rapidly sedimenting RNA in the cuts 1, 2, and 3 sedimented as rather discrete peaks with sedimentation coefficients of 60 S and 85 S. Figure 8.12 shows that treatment of these RNA fractions with DNase and pronase did not affect their sedimentation coefficients which means that neither protein nor DNA is part of these large RNA molecules (*2, 6, 12, 14, 15, 23, 24, 26*). The large size

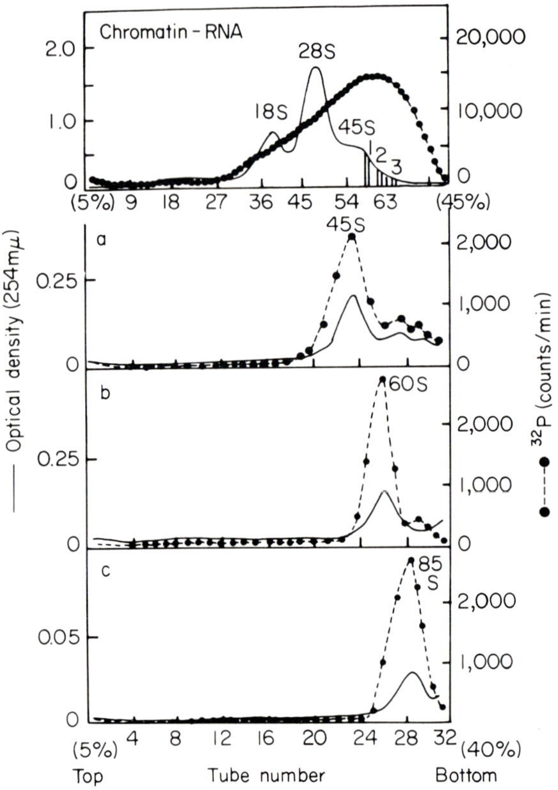

Fig. 8.11. Zonal ultracentrifugation patterns for RNA of the chromatin (A) and residue (B) fractions. Samples containing 100 mg of RNA were centrifuged through a 5.45% linear sucrose gradient at 35,000 rpm for 15 hours at 4°C. RNA from the rapidly sedimenting fractions 1, 2, and 3 thus obtained (top) was precipitated and rerun on 5–40% sucrose gradient in a SB 110 rotor of International B-35 centrifuge at 25,000 rpm for 15 hours at 4°C. Sections a, b, and c show the analytical patterns for fractions 1, 2, and 3, respectively. Optical density (———); radioactivity (- - - ● - - - ● - - -).

of some of these molecules is demonstrable by electron microscopy (Fig. 5.7.).

B. Labeling Rates of the RNA of the Nuclear Residue Fraction

The specific activities of the RNA of both the residue and the chromatin fraction of the Novikoff hepatoma are shown in Table 8.2. These data indicate that the rates of labeling are significantly greater in the 60 S and 85 S RNA than in the 45 S RNA, presumably because of the greater turnover rates of the more rapidly sedimenting RNA. Moreover, the labeling rates of the residue fraction are apparently considerably greater

TABLE 8.2

SPECIFIC ACTIVITY OF ^{32}P IN RAPIDLY SEDIMENTING RNA[a]

Fraction	(cpm/mg RNA)		Residue / Chromatin
	Chromatin	*Residue*	
	20-min. pulse		
45 S	5.7×10^4	1.7×10^5	2.98
60 S	1.5×10^5	4.0×10^5	2.66
85 S	2.2×10^5	5.5×10^5	2.50
	6-hour labeling		
45 S	1.7×10^6	2.7×10^6	1.58
60 S	1.5×10^6	3.1×10^6	2.06
85 S	1.8×10^5	3.2×10^6	1.77

[a] Novikoff hepatoma ascites cells (*16a*).

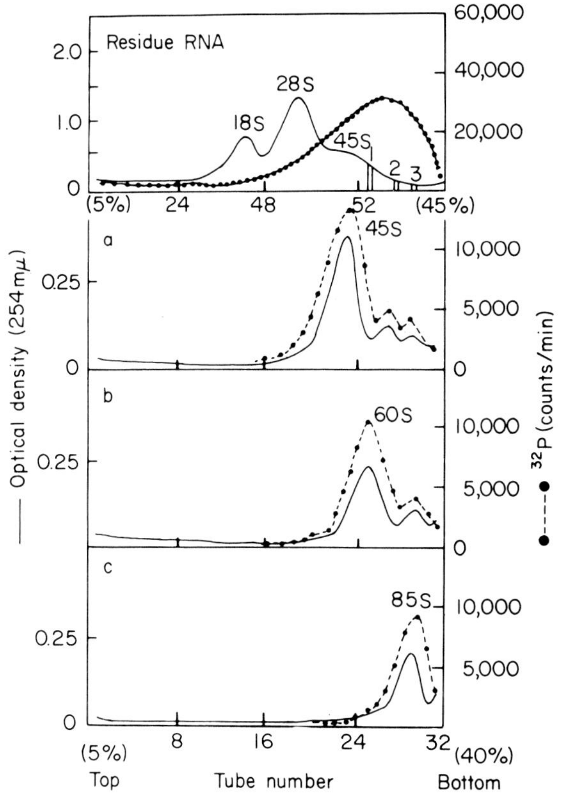

Fig. 8.11 (*continued*).

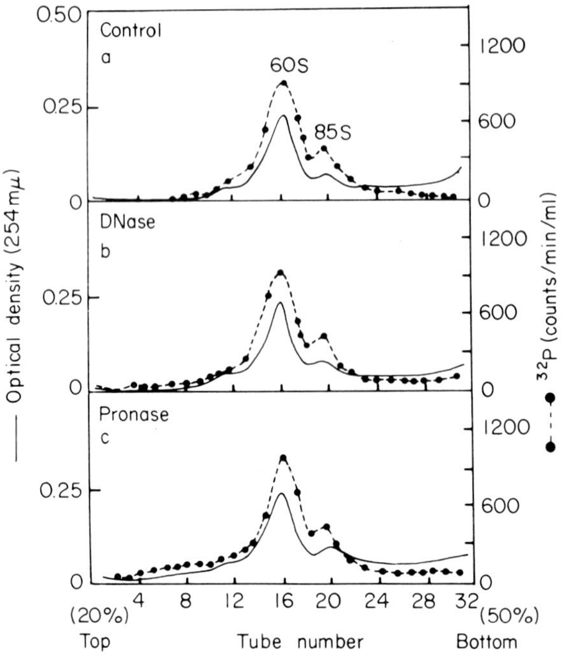

Fig. 8.12. Effect of DNase and pronase on the sedimentation of pooled 60 S and 85 S RNA. Analytical gradients were centrifuged as in Fig. 8.11. Optical density (———); radioactivity (- - - · - - - · - - -).

than those of the chromatin fraction reflecting a greater turnover rate of the elements in these less soluble pools (see Chapter VII, p. 336).

C. Nucleotide Compositions of the RNA in the Nuclear Residue Fraction

Table 8.3 shows that the ^{32}P nucleotide compositions of the RNA of the nuclear residue fraction are markedly different from that of the chromatin fraction. The high uridylic acid content of the chromatin fraction is similar to that reported for the "heterodisperse" nuclear RNA by a number of authors (*2, 11, 12, 14, 20*); in the chromatin fraction the percentage of isotope in the other nucleotides is not significantly different. The RNA of the nuclear residue fraction has considerably more isotope in guanylic and cytidylic acids and much less in adenylic and uridylic acids.

The UV nucleotide compositions are shown in Table 8.4. The data for the residue fraction resemble those obtained in the ^{32}P-nucleotide analysis in that the RNA is rich in guanylic and uridylic acids. However, the uridylic acid content is higher and the guanylic acid content is lower as

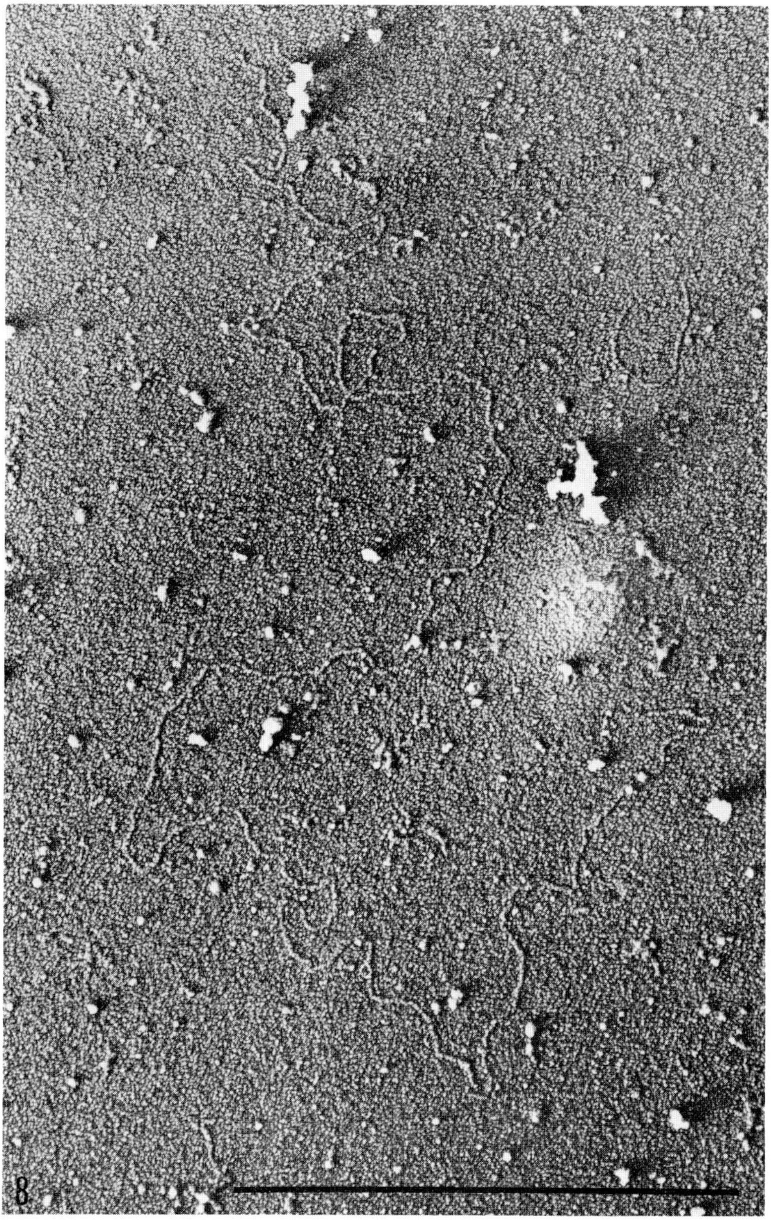

Fig. 8.13. RNA molecule from 85 S RNA with a length greater than 6 μ.

TABLE 8.3

^{32}P Distribution of Nucleotides in Rapidly Sedimenting RNA[a]

Fraction	AMP	UMP	GMP	CMP	$\frac{\text{AMP + UMP}}{\text{GMP + CMP}}$
			20-min. pulse		
Chromatin					
45 S	22.8	30.8	22.8	23.6	1.15
60 S	23.8	30.0	23.2	23.0	1.16
85 S	24.2	30.9	20.6	24.3	1.22
Residue					
45 S	16.2	20.5	36.0	27.3	0.58
85 S	17.1	20.3	34.6	28.9	0.60

[a] Novikoff hepatoma ascites cells (*16a*).

determined by the UV nucleotide analysis than by the ^{32}P nucleotide analysis. In the chromatin fraction only the 85 S RNA has a UV nucleotide composition that is similar to that of the ^{32}P nucleotide composition, although in this case, the uridylic acid content is lower and the guanylic acid content is higher than that obtained by ^{32}P nucleotide analysis.

D. Distribution of Oligonucleotides in the Newly Synthesized RNA of the Nucleolar and Nuclear Residue Fractions

Table 8.5 shows the percentage of isotope recovered in the various oligonucleotides of the fingerprints obtained by the method noted in

TABLE 8.4

UV Nucleotide Composition of Rapidly Sedimenting RNA[a]

Fraction	AMP	UMP	GMP	CMP	$\frac{\text{AMP + UMP}}{\text{GMP + CMP}}$
Chromatin					
45 S	19.3	19.1	33.1	28.5	0.73
60 S	21.8	23.5	28.6	26.1	0.82
85 S	22.7	27.3	26.6	23.4	1.0
Residue					
45 S	17.5	23.0	31.5	28.0	0.68
85 S	17.0	23.6	32.1	27.3	0.68
Ribosomal RNA (rat liver)					
18 S	20.7	21.4	32.2	25.7	0.72
28 S	17.2	19.2	34.9	28.7	0.57
Chromatin RNA					
18 S	21.3	18.6	33.7	26.4	0.66
28 S	18.5	16.9	35.3	29.3	0.55

[a] Novikoff hepatoma ascites cells (*16a*).

TABLE 8.5

PERCENTAGE DISTRIBUTION OF RADIOACTIVITY IN OLIGONUCLEOTIDES AFTER A 20-MINUTE LABELING PERIOD[a]

	Chromatin			Residue			Nucleolar
	45 S	60 S	85 S	45 S	60 S	85 S	45 S
C	19.00	19.40	19.20	24.70	23.50	23.60	27.70
U	22.70	22.90	23.10	18.30	17.50	18.50	13.70
ψ or Pseudo-U	0.41	0.27	0.32	0.37	0.46	0.41	0.80
AC	6.90	6.65	6.95	4.52	4.92	4.96	4.40
AU	10.30	9.93	10.10	5.46	5.61	5.48	2.90
GC	7.32	6.64	7.45	12.10	12.60	12.50	15.00
GU	10.30	9.09	9.19	10.40	11.30	11.30	9.80
A_2C	1.93	2.00	2.14	1.42	1.58	1.63	1.70
A_2U	2.38	2.30	2.70	1.59	1.61	1.49	1.30
(AG)C	3.92	4.23	4.44	4.07	4.80	3.98	4.90
(AG)U	4.30	4.69	4.58	3.44	3.75	3.29	3.60
G_2U	1.97	1.68	1.20	3.03	2.45	3.33	2.60
$A_3U + G_2C$	2.55	2.60	2.65	3.26	2.85	3.26	4.40
A_3C	1.02	1.14	1.13	0.81	0.85	1.01	0.90
$(A_2G)C$	1.65	2.08	2.03	1.98	1.64	1.62	1.70
$(AG_2)C$	1.50	1.27	1.27	1.65	1.25	1.30	1.50
$(A_2G)U$	1.97	1.74	1.39	1.74	1.13	1.20	1.60
$(AG_2)U$	0.52	0.46	0.41	1.26	0.80	1.00	1.00

[a] Novikoff hepatoma ascites cells (*16a*).

Chapter V (p. 256), following isolation of RNA after a 20-minute pulse labeling period. In the residue fraction, particularly in the 85 S RNA, there was more isotope in C, GC, GU, G_2U, and AG_2U as might be expected from the higher content of GMP and CMP in the RNA of the residue fraction. On the other hand, there was more isotope in U, AC, AU, A_2U, and AGU of the chromatin fraction as might be expected from the higher content of AMP and UMP in the RNA of this fraction.

When compared with the oligonucleotide composition of nucleolar 45 S RNA (Table 8.5), the isotope in C and U in the residue fraction is virtually a mean between that of the nucleolar and chromatin fractions. However, the values for AC and AU are much closer to those of the nucleolar fraction than those of the chromatin fraction and the value for GU in the residue fraction is higher than that of the nucleolar or chromatin RNA. Similar types of differences found for the tri- and tetranucleotides led to the conclusion that the RNA in the residue fraction is not just a mixture of RNA's of the chromatin and nucleolar fractions but instead contains specific large-size RNA which is either unique to this fraction or is concentrated in it. It remains for future studies to determine whether there are few or many types of RNA in the residue fraction and whether these RNA's are messengers on their way to the cytoplasm as parts of the polysome or informosome complexes.

IV. Summary

Although the functions of the nuclear ribonucleoprotein network are not defined, some possibilities are that it serves as:

1. a pathway for transportation of the nucleolar ribonucleoproteins to the cytoplasm;
2. a site of formation of some messenger RNA;
3. a site of coupling of messenger RNA to ribosomal precursors;
4. a site of synthesis of the nuclear and nucleolar proteins; or
5. a site of synthesis or assembly of whole polysomes or polysomes prior to their release to the cytoplasm (Fig. 8.2). This latter idea is supported by the presence in masses, insoluble in both dilute and concentrated saline solutions, that abound in both AU- and GC-rich RNA fraction and also in 18 S and 28 S rRNA.

REFERENCES

1. Albanese, M. P., Passagio di nucleoli attraverso la membrana nucleare negli ovociti in accrescimento di *Patella coerulea L. Experientia* **20**, 550–551 (1964).

1a. Anteunis, A., Fautrez-Firlefyn, N., and Fautrez, J., Ultrastructure du nucléole expulsé dans le cytoplasme de l'oocyte d'Artemia salina. *Compt. Rend.* **266**, 1862–1863 (1968).

2. Attardi, G., Parnas, H., Huang, M.-I. H., and Attardi, B., Giant size rapidly labeled nuclear ribonucleic acid and cytoplasmic messenger ribonucleic acid in immature duck erythrocytes. *J. Mol. Biol.* **20**, 145–182 (1966).

3. Bernhard, W., and Granboulan, N., The fine structure of the cancer cell nucleus. *Exptl. Cell Res.* Suppl. 9, 19–53 (1962).

4. Busch, H., "Histones and Other Nuclear Proteins." Academic Press, New York, 1965.

5. Dingman, C. W., and Peacock, A. C., Analytical studies on nuclear ribonucleic acid using polyacrylamide gel electrophoresis. *Biochemistry* **7**, 659–668 (1968).

6. Houssais, J.-F., and Attardi, G., High molecular weight nonribosomal-type nuclear RNA and cytoplasmic messenger RNA in HeLa cells. *Proc. Natl. Acad. Sci. U. S.* **56**, 616–623 (1966).

7. Jacob, S. T., and Busch, H., Template activity of nuclear RNA fractions of the Walker tumor. *Biochim. Biophys. Acta* **138**, 249–257 (1967).

8. Lettré, R., and Siebs, W., Beobachtungen am nucleolus *in vitro* gezuchteter zellen. *Z. Krebsforsch.* **60**, 19–20 (1954).

9. Lettré, R., and Siebs, W., Film sur des études du nucléole de cellules cultivées *in vitro. Chemotherapia* **2**, 223–225 (1961).

10. Lettré, R., and Siebs, W., Some studies of the nucleolus of cells cultivated *in vitro*. *Pathol. Biol., Semaine Hop.* [N.S.] **9,** 819–823 (1961).
11. Okamura, N., and Busch, H., Base composition of high molecular weight nuclear RNA of Walker tumor and liver of the rat. *Cancer Res.* **25,** 693–697 (1965).
12. Penman, S., Vesco, C., and Penman, M., Localization and kinetics of formation of nuclear heterodisperse RNA, cytoplasmic heterodisperse RNA and polyribosome-associated messenger RNA in HeLa cells. *J. Mol. Biol.* **34,** 49–69 (1968).
12a. Roberts, W. K., and D'Ari, L., Base sequence differences between the ribosomal and "ribosomal precursor" ribonucleic acids from Ehrlich ascites cells. *Biochemistry* **7,** 592–600 (1968).
13. Sadowski, P. D., and Howden, J. A., Isolation of two distinct classes of polysomes from nuclear fraction of rat liver. *J. Cell Biol.* **37,** 163–181 (1968).
14. Scherrer, K., Marcaud, L., Zajdela, F., London, I. M., and Gros, F., Patterns of RNA metabolism in a differentiated cell: A rapidly labeled, unstable 60 S RNA with messenger properties in duck erythroblasts. *Proc. Natl. Acad. Sci. U. S.* **56,** 1571–1578 (1966).
15. Schutz, G., Gallwitz, D., and Sekeris, C. E., Rapidly labeled high molecular RNA from rat liver. *European J. Biochem.* **4,** 149–156 (1968).
16. Shankar Narayan, K., Muramatsu, M., Smetana, K., and Busch, H., Ultrastructural studies on RNA and DNA components of isolated nucleoli of Walker 256 carcinosarcoma. *Exptl. Cell Res.* **41,** 81–98 (1966).
16′. Shankara narayan, K., Steele, W. J., Smetana, K., and Busch, H., Ultrastructural aspects of the ribonucleoprotein network in nuclei of Walker tumor and rat liver. *Exptl. Cell Res.* **46,** 65–77 (1967).
16a. Sharma, O. K., Arendell, J. P., Hidvegi, E. J., Marks, F., Prestayko, A., Smetana, K., and Busch, H., Isolation and characterization of 60 S and 85 S nuclear RNA of Novikoff hepatoma cells. *Physiol. Chem. Phys.* **1,** 185–209 (1969).
17. Sibatani, A., deKloet, S. R., Allfrey, V. G., and Mirsky, A. E., Isolation of a nuclear RNA fraction resembling DNA in its base composition. *Proc. Natl. Acad. Sci. U. S.* **48,** 471–477 (1962).
18. Smetana, K., Steele, W. J., and Busch, H., A nuclear ribonucleoprotein network. *Exptl. Cell Res.* **31,** 198–202 (1963).
19. Smith, S. J., Adams, H. R., Smetana, K., and Busch, H., Isolation of the outer layer of the nuclear envelope. *Exptl. Cell Res.* **55,** 185–197 (1969).
20. Soeiro, R., Birnboim, C., and Darnell, J., Rapidly labeled HeLa cell nuclear RNA. II. Base composition and cellular localization of a heterogeneous RNA fraction. *J. Mol. Biol.* **19,** 362–372 (1966).
21. Steele, W. J., and Busch, H., Studies on the ribonucleic acid components of the nuclear ribonucleoprotein network. *Biochim. Biophys. Acta* **129,** 54–67 (1966).
22. Swift, H., Cytochemical studies on nuclear fine structure. *Exptl. Cell Res.* Suppl. 9, 54–67 (1963).
23. Tamaoki, T., and Mueller, G. C., A rapidly labeled ribonucleoprotein in HeLa cell nuclei. *Biochim. Biophys. Acta* **108,** 81–94 (1965).
24. Vesco, C., and Penman, S., The fractionation of nuclei and the integrity of purified nucleoli in HeLa cells. *Biochim. Biophys. Acta* **169,** 188–195 (1968).
25. Whittle, E. D., Bushnell, D. E., and Potter, V. R., RNA associated with the outer membrane of rat liver nuclei. *Biochim. Biophys. Acta* **161,** 41–50 (1968).
26. Yoshikawa-Fukada, M., Fukada, T., and Kawade, Y., Characterization of rapidly labeled ribonucleic acid of animal cells in culture. *Biochim. Biophys. Acta* **103,** 383–398 (1965).

CHAPTER IX

*Nucleoli of Specialized Cells**

* In this chapter, the references are presented after each Part.

Although studies have been made on nucleoli of many kinds of cells, they have been extensively studied from a morphological and chemical point of view in only a small number of cell types (see also Chapter I). The aim of this chapter is to discuss in some detail the kinds of information available on nucleoli of blood cells (Part I), nerve cells (Part II), liver cells (Part III), and tumor cells (Part IV). Special studies have been made on such subjects as nucleolar abundance in myocardial cells (*7a*) which have shown a tendency toward an increased nucleolar number in larger hearts and on the role of the nucleolus in viral synthesis in cells of primary renal tumors in leopard frogs (*20a*) in which there is a redistribution of the fibrillar region followed by their transformation into electron dense clumps which appear to be the sites of formation of viral nucleoids. Extensive studies have also been made on nucleoli in plant cells, e.g., lemon fruit and *Zea mays* on the relationship of nucleolar changes to starch granule formation, (*23b,c;* see Chapters I and II). Recently, remarkable morphological appearances have been described of stacked granular lamellae of different lengths and intranuclear membranes in crayfish oocytes (23a). Such studies are an interesting evolution of the extensive review of nucleoli developed by Montgomery in 1898 (*28a*); however, the availability of more sophisticated methods has provided a means for more incisive studies into the ultrastructure and biochemistry. The following sections deal with areas in which much more detailed information is available.

Part I NUCLEOLI OF BLOOD CELLS

I. Introduction

Blood cells represent convenient models for cytological studies of structural changes during cell maturation and differentiation (Table 9.1). These studies are important not only for basic biological research but also for practical purposes in medical diagnosis. The changes related to maturation and differentiation of various blood cell lines have been studied intensively for a long time. However, the lack of satisfactory methods for visualization of nucleoli has led to controversies about the presence of nucleoli, particularly in more mature cell forms. In the beginning and first part of this century, nucleoli in blood cells were visualized by means of procedures used originally for supravital staining or by means of staining with basic dyes under special conditions. Later, nucleoli were differentiated from other nuclear structures by their different staining patterns in panoptic or polychromatic staining procedures or they appeared as empty light areas in the stained chromatin structure of the nucleus. The majority of such methods showed the presence of nucleoli in immature cells but in more mature cells nucleoli could not be distinguished from interchromatinic areas. This was probably the reason

TABLE 9.1

Terminology of Blood Cells and Intermediates in Maturation

Erythropoietic line		Granulocytic line	Thrombopoietic line	Lymphocytic line	Monocytic line	Plasmacytic line
Pronormoblast (proerythroblast)	Promegaloblast	Myeloblast	Megakaryoblast	Lymphoblast	Monoblast	Plasmablast
↓	↓	↓	↓	↓	↓	↓
(Macroblast)	Basophilic megaloblast	Promyelocyte	Promegakaryocyte	Prolymphocyte	Promonocyte	Proplasmacyte
↓	|	↓	↓	↓	↓	↓
Basophilic normoblast (basophilic erythroblast)	Polychromatic megaloblast	Myelocyte	Mature megakaryocyte	Mature lymphocyte	Mature monocyte	Mature plasmacyte
↓	↓	↓	↓			
Polychromatic normoblast	Orthochromatic megaloblast	Metamyelocyte	Thrombocyte			
|	↓	↓				
Orthochromatic normoblast (orthochromatic erythroblast)	Reticulocyte (promegalocyte)	Mature granulocyte				
↓	↓					
Reticulocyte (pronormocyte)	Mature megalocyte					
↓						
Mature normocyte (mature erythrocyte)						

TABLE 9.2

NUCLEOLI IN BLOOD CELLS[a]

Cell	Number of nucleoli	Size	Shape
Lymphoblast	1–2	Large	Regular
Prolymphocyte	1–2	Medium or small	Regular
Mature lymphocyte	1–2	Small	Regular
Monoblast	1–6	Large or small	Regular or irregular
Monocyte	1–6	Small	Regular
Myeloblast	2–6	Large	Regular
Promyelocyte	1–4	Large or small	Regular
Myelocyte	?	Small	Regular
Pronormoblast (proerythroblast)	2–5	Large	Irregular
Macronormoblast	3–6	Large or small	Irregular or regular
Basophilic normoblast	1–3	Medium or small	Regular
Polychromatophilic normoblast	?	Small	Regular
Orthochromatophilic normoblast	?	?	?

[a] From Asano (*3*), Davidson (*11*), Gonzalez Guzman (*14, 15*), Rind (*39*), Schneiberg (*43*), Smetana and Janovsky (*58, 59*), Smetana *et al.* (*52*), Smetana and Potměšil (*62*), Thorell (*67*), Undritz (*69, 70*), Vendrely and Vendrely (*72*), and Wintrobe (*73*).

why only a few studies on nucleoli in blood cells were published. Some time ago, the approximate number and shape of nucleoli were established in immature cells (Table 9.2) and provided a valuable tool for classification of these cells (*69, 70*). The presence of nucleoli in mature leukocytes, however, still remained a subject of controversy. According to many hematologists, the nucleoli disappear during maturation of these cells (*5*) and the presence of nucleoli in stained leukemic lymphocytes was regarded as one sign of malignant growth (*42*). Although phase contrast microscopy is a very useful method for demonstration of the number and shape of nucleoli in immature blood cells, particularly for pathological forms (*39*), it was difficult to distinguish nucleoli from chromocenters in mature lymphocytes. Even staining with acridine orange did not provide satisfactory results since the nucleoli in mature normal lymphocytes were usually not stained (*21*).

Progress in the visualization of nucleoli was achieved by the development of simple procedures which selectively stained nucleoli while other nuclear basophilic structures were less stained and chromatin structures were completely negative. The majority of the methods used Pischinger's procedure for the determination of isoelectric point (*31*); the methods now used mainly stain structures containing RNA by means of buffered or specially prepared basic dyes such as methylene blue or toluidine blue

(*32, 44, 47, 62, 65*), or other basic dyes (*61*). However, the suitability of these dyes for staining of nucleoli and structures containing RNA should be tested on control systems since the result of staining can be influenced substantially by the quality of the dye used (*48*). The microspectrophotometric methods provide basic information on the quantitative changes of the amount of ribonucleic acid related to the nuclcolar areas during the maturation of blood cells; they have been used to study granulopoiesis and erythropoiesis (*17, 67*).

Electron microscopic studies of blood cells have verified light microscopic findings on the presence of nucleoli in certain blood cells. They have helped to distinguish nucleoli from other nuclear structures, and partially from ribonucleoprotein structures in interchromatinic areas. Furthermore, electron microscopic studies have shown that the ultrastructural appearance of nucleoli in blood cells depends on their stage of maturation. Questions about the number and presence of nucleoli could not be readily answered by electron microscopy because of the limited number of cells that can be observed by electron microscopic procedures (*26*). Moreover, the sections of the cells do not always contain the structures which are being investigated.

II. The Lymphocytic Line

A. Lymphoblasts and Immature Lymphocytes

These are characterized by the presence of one or two large nucleoli (Fig. 9.1a and b) (*69*). In some leukemic lymphoblasts, however, even more nucleoli can be observed. The shape of these nucleoli is usually round or oval. The distribution of structures containing RNA is rather uniform, and fine filaments are present as well as small light areas between them (Fig. 9.1a) (*50*). In the electron micrographs, nucleoli with nucleolonemas (Fig. 9.2) or compact nucleoli with uniform distributions (Fig. 9.3) of ribonucleoprotein structures were noted in immature lymphocytes and lymphoblasts (*13, 16, 52, 53, 60*). In nucleoli of some leukemic lymphoblasts in patients with rapidly advancing leukemia, ring-shaped nucleoli have been noted beside these nucleoli (*53*). This seems to indicate that active and "resting" nucleoli could be present in the same nucleus since ring-shaped nucleoli have been observed in cells with inhibited RNA synthesis. In nucleoli of lymphoblasts in long-term tissue cultures, dense granules appeared after 6 weeks of cultivation (*71*). Their significance and composition is not known, but they are much like the dense granules observed in Walker tumor (*41, 45, 64*).

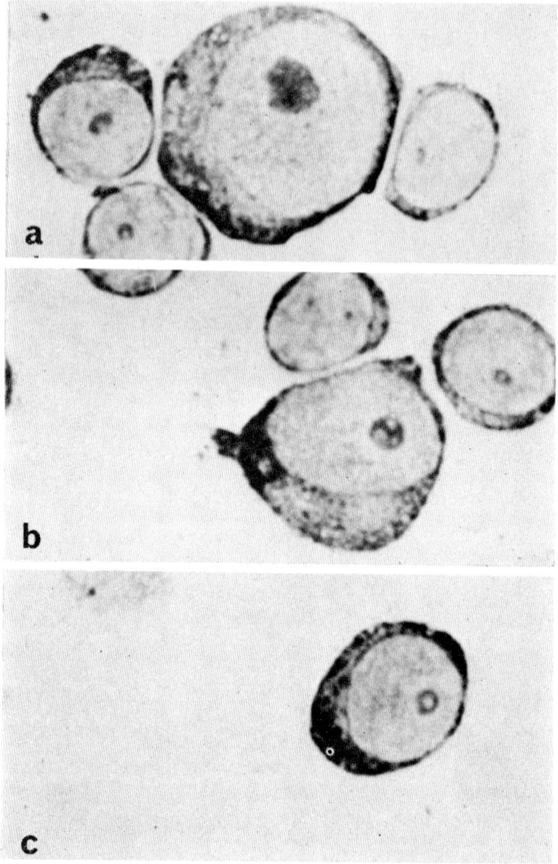

Fig. 9.1. a. Leukemic lymphoblast stained with toluidine blue. The distribution of nucleolar RNA-containing structures is rather uniform and fine trabecular structures can be observed. ×2000 (*60*). b. Leukemic prolymphocyte. The same specimen as in Fig. 9.1a. Note filaments and light areas in the nucleolus. ×2000 (*60*). c. Mature leukemic lymphocyte from the same specimen as in previous figures. Note the ring-shaped appearance of the nucleolus. ×2000 (*60*).

B. Mature Lymphocytes

Although the majority of hematologists described mature lymphocytes as cells without nucleoli, in the past 20 years studies of mature lymphocytes with special staining procedures have shown the presence of nucleoli in these cells (*44, 47, 66*). The nucleoli in mature lymphocytes usually have a ring-shaped appearance, due to the presence of ribonucleic acid in their periphery (Figs. 9.1c, 9.4, 9.5, 9.6) (*60*). The majority of lymphocytes contain one nucleolus; two or more nucleoli were noted in relatively few of these cells (*58*). According to Grundmann (*18, 19*) lymphocytes with one or two large nucleoli are from lymphatic nodules. However, the possibility exists that the different appearance of nucleoli in lymphocytes might be due to the maturation rather than a different origin of these cells (*4*). Such small nucleoli may reflect irreversible nucleolar changes related to the inhibition of RNA synthesis (*34*); the larger ring-shaped nucleoli may undergo reversible changes in size and structure under various types of stimulation including that of phytohemagglutinin.

Electron microscopic studies verified the existence of ring-shaped nucleoli in mature lymphocytes (*60*). The ribonucleoprotein particles and even the ribonucleoprotein fibrillar components were present only in the peripheral part of the nucleolus resulting in its characteristic appearance (Figs. 9.5, 9.6). According to ultrastructural cytochemical observations, the central light areas of ring-shaped nucleoli are composed of chromatin structures and protein filaments (Figs. 9.7, 9.8). The chromatin structures present in nucleolar central light areas are organized as

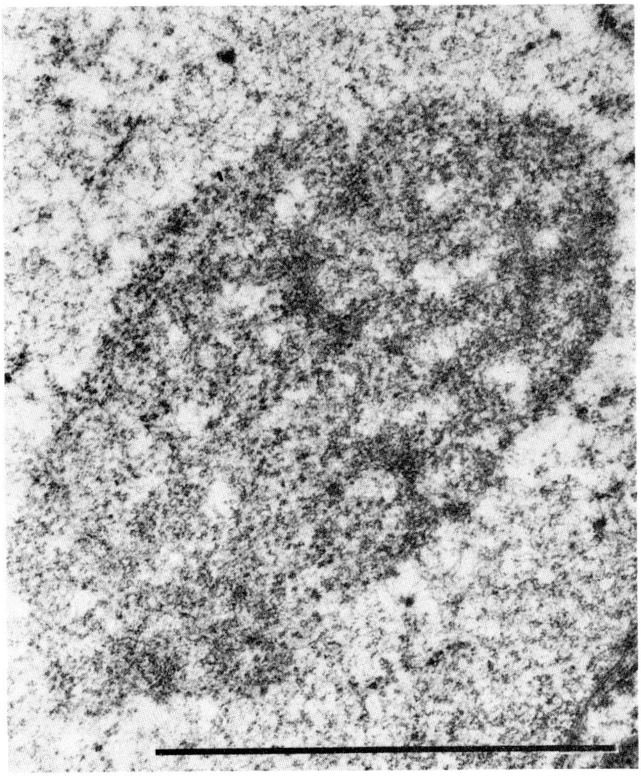

Fig. 9.2. Nucleolus with nucleolonemas in leukemic lymphoblast. ×61,500 (53). The marker line is 1 μ.

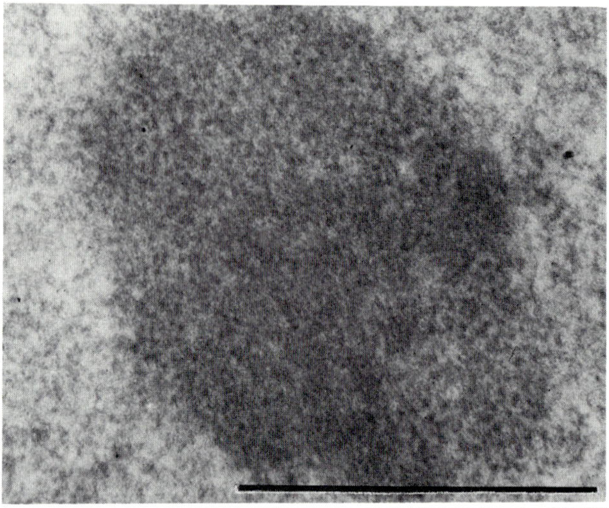

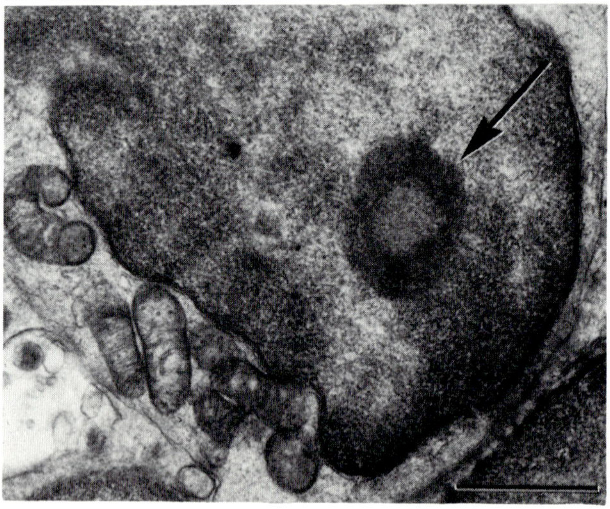

Fig. 9.3. Compact nucleolus of a leukemic lymphoblast. ×48,000 (*60*).

Fig. 9.4. Ring-shaped nucleolus (arrow) in a mature lymphocyte of a patient with rubeola. ×19,000 (*56*).

clusters of fibrils or solitary fine filaments probably representing uncoiled molecules of DNA (*49*). The connections of the intranucleolar chromatin structures, including those in central light areas of ring-shaped nucleoli (Fig. 9.8) with the nucleolus-associated chromatin, have also been noted. The ultrastructural studies were not helpful for studies of small nucleoli because of the confusion arising from the fact that the peripheral portion of a large nucleolus seen in tangential sections could appear to be a small nucleolus.

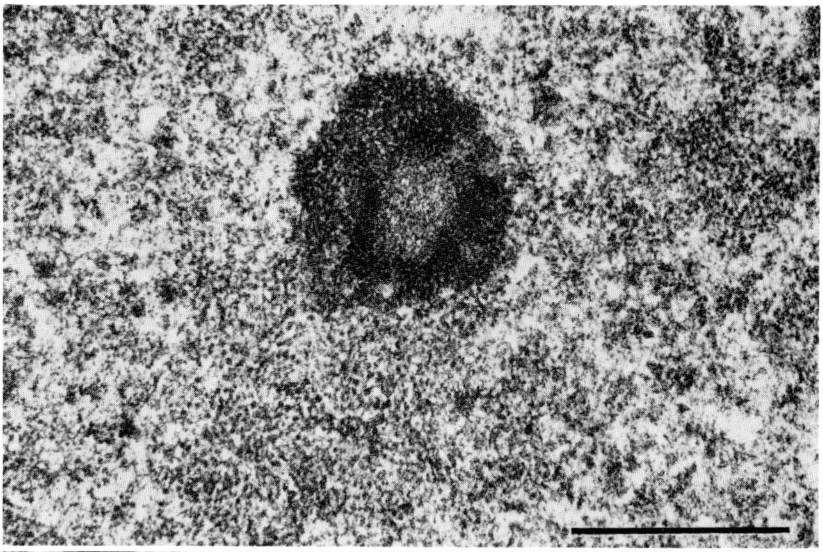

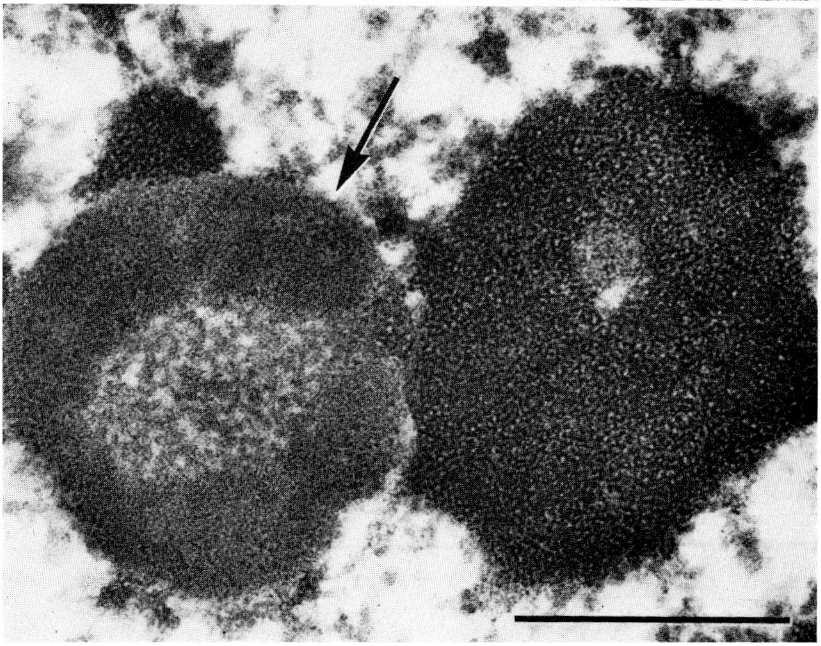

Fig. 9.5. Ring-shaped nucleolus of a leukemic lymphocyte. Note the presence of RNP structures only in the peripheral part of the nucleolus. ×28,500 (*49*).

Fig. 9.6. Ring-shaped nucleolus (arrow) containing a trabecular core and an adjacent chromatin clump (*49*). The sample was fixed in formaldehyde and treated with cold trichloracetic acid. ×36,000.

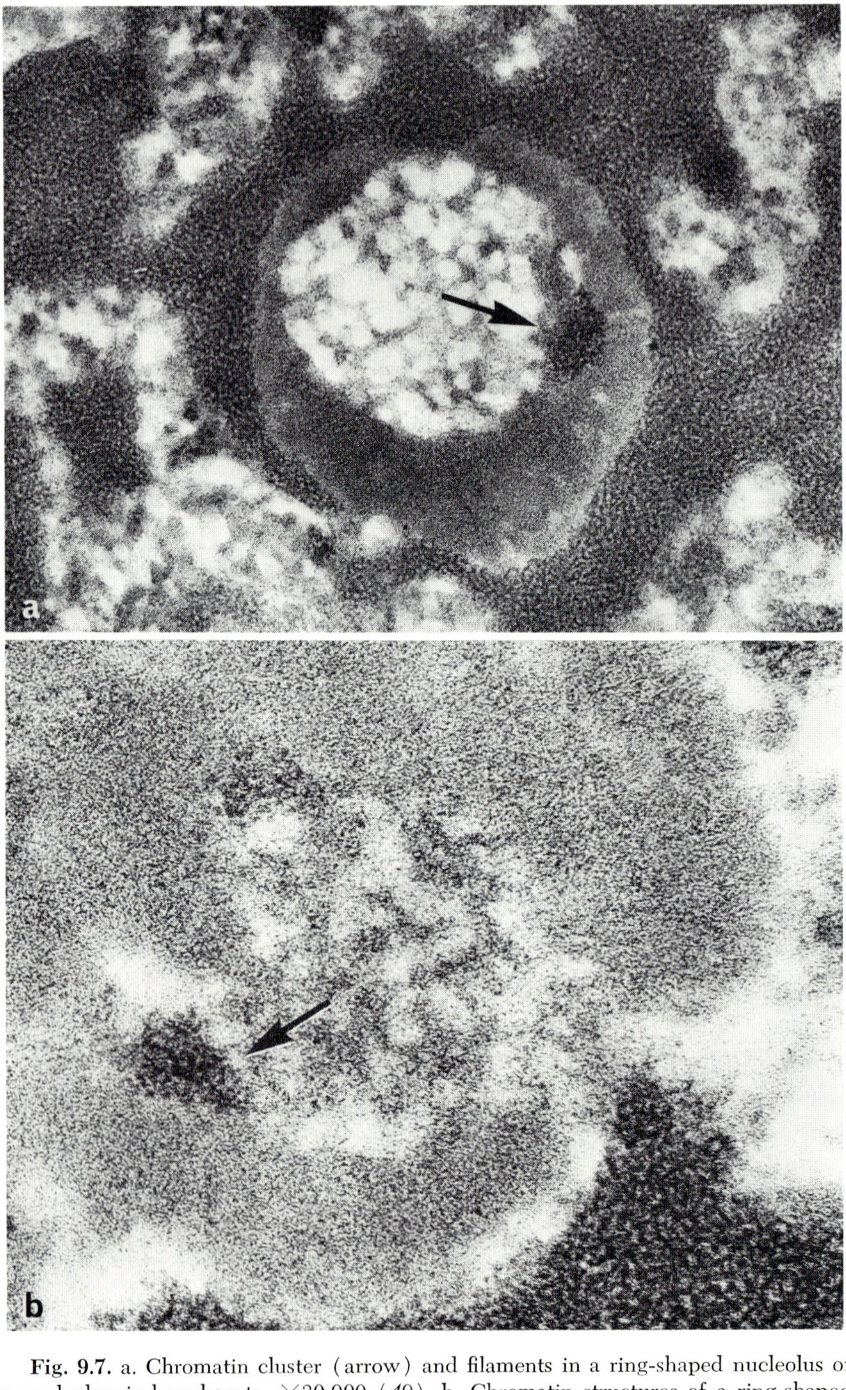

Fig. 9.7. a. Chromatin cluster (arrow) and filaments in a ring-shaped nucleolus of a nonleukemic lymphocyte. ×30,000 (*49*). b. Chromatin structures of a ring-shaped nucleolus in leukemic lymphocyte (*49*). Chromatin cluster (arrow). ×88,000.

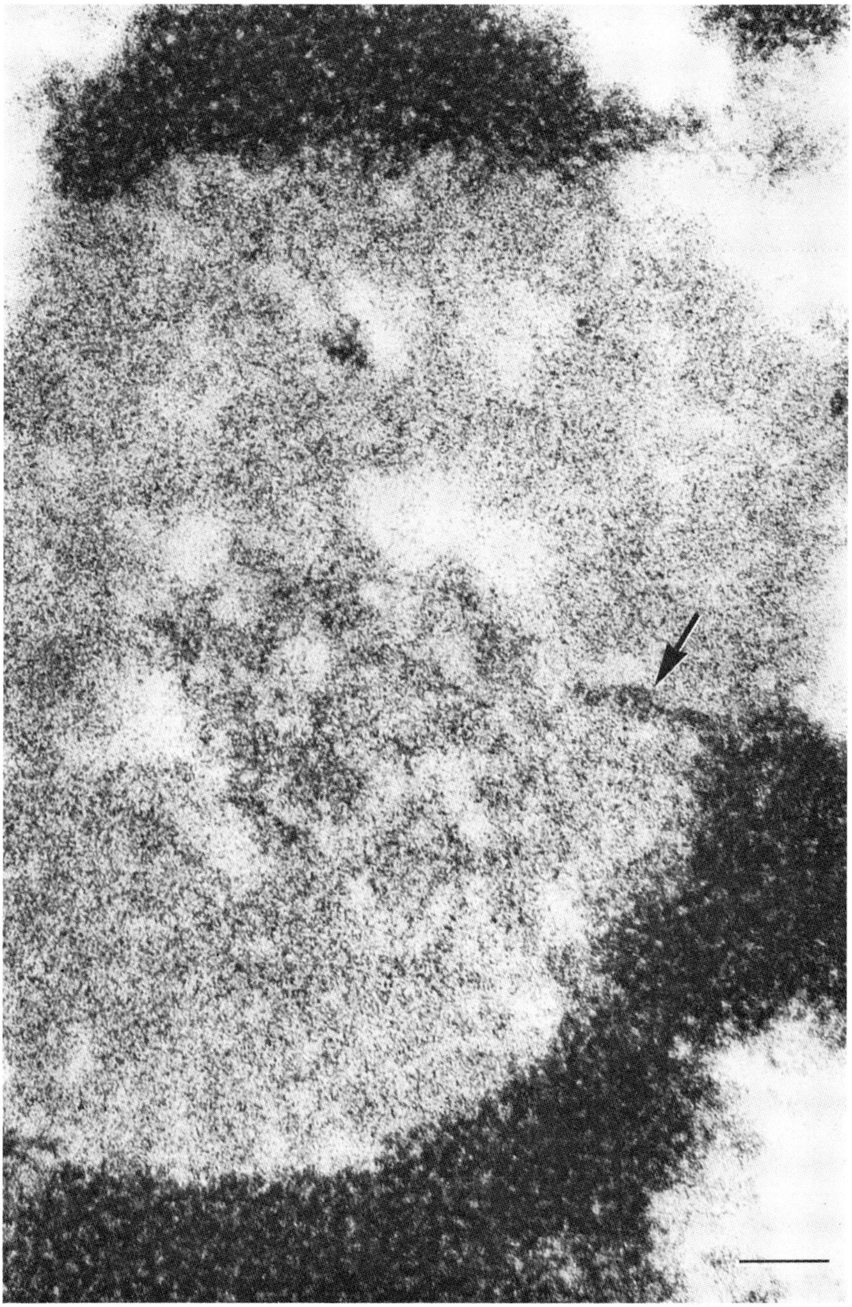

Fig. 9.8. A chromatin strip (arrow) penetrating from nucleolus-associated chromatin into the central light area in a ring-shaped nucleolus. ×123,000 (*49*).

The incidence of ring-shaped nucleoli in mature lymphocytes is probably related to the cessation of the formation of new RNA-containing structures during maturation, since cytoautoradiographic studies have shown that there is an inhibition of RNA synthesis in these cells (*9*, *12*, *29*). Moreover, the inhibition of RNA synthesis with actinomycin D or chromomycin A_3 or mithramycin *in vitro* and *in vivo* induces the formation of ring-shaped nucleoli from nucleoli of other types [(*23*, *24*, *30*) see Chapter II]. Similarly, the incorporation of ^{3}H-uridine was reduced or inhibited in ring-shaped nucleoli (*34*).

Transformation of ring-shaped nucleoli of mature lymphocytes to nucleoli similar to those in lymphoblasts was observed after stimulation with phytohemagglutinin (Table 9.3) (*33*). The change is related to

TABLE 9.3

THE EFFECT OF PHYTOHEMAGGLUTININ ON THE PRESENCE OF COMPACT AND RING-SHAPED NUCLEOLI IN CULTIVATED LYMPHOCYTES[a]

Nucleolar type	No. of Expt.	Controls	Phytohemagglutinin-treated
Ring-shaped nucleoli	6	88.4	6.9
Compact nucleoli		11.6	93.1

[a] From Pössnerová and Smetana (*33*).

stimulation of RNA synthesis in these cells and their differentiation. These ring-shaped nucleoli seem to reflect the reversible inhibition of RNA synthesis during the morphological maturation of nucleoli. The transformation of ring-shaped nucleoli of mature lymphocytes in nucleoli similar to those in lymphoblasts was also observed after antigenic stimulation in mixed leukocyte cultures (*54*). The majority of leukemic lymphocytes cannot be restimulated with phytohemagglutinin (*1*) despite the fact that they contain ring-shaped nucleoli (*60*). This effect may be related to losses of phenotypically important receptors related to RNA synthesis in these tumor cells (*8*). DNA templates for RNA synthesis are apparently present in ring-shaped nucleoli of both leukemic and normal lymphocytes (*49*) and RNA synthesis in stimulated leukemic lymphocytes has been noted (*40*).

C. THE NUCLEOLAR COEFFICIENT OF LYMPHOCYTES IN HUMAN PERIPHERAL BLOOD

The nucleolar coefficient was introduced by Gonzalez Guzman (*14*, *15*) to indicate the number of nucleoli per lymphocyte; i.e., nucleolar coefficient = No/C (no = number of nucleoli, C = number of cells in which the nucleoli were counted). The determination of the nucleolar coefficient has been useful in clinical studies to provide basic information on the number of nucleoli of lymphocytes in the peripheral blood in health and disease. Recent studies, however, have shown the necessity of determining the absolute number of uninucleolar lymphocytes since the nucleolar coefficient can reflect completely different situations depending on the amount of uninucleolar or multinucleolar lymphocytes in the peripheral blood. For example, high values of the nucleolar coefficient can reflect either the higher incidence of multinucleolar lymphocytes (newborns) or a decrease of the number of uninucleolar lymphocytes (patients with breast cancer). Therefore, information on the values of the nucleolar coefficient should be accompanied by data about the number of uninucleolar and multinucleolar lymphocytes in the peripheral blood of study subjects.

In adults with normal laboratory and clinical findings the values of the nucleolar coefficient range from 0.9 to 1.4 with the mean values about 1 (*14, 15, 37, 58*). The number of uninucleolar lymphocytes is in the range of 900–2500 while the number of multinucleolar lymphocytes ranges from 30–90 in 1 mm of peripheral blood. Higher values of the nucleolar coefficient of lymphocytes have been found in infants, particularly in the newborn (*58, 59*). These higher values of the nucleolar coefficient are due to the higher incidence of multinucleolar lymphocytes.

Higher values of the nucleolar coefficient are found in patients with infectious diseases (*36*) as well as in patients with certain neoplastic diseases (*35, 52, 55, 57*). The higher values of the nucleolar coefficient of lymphocytes in the peripheral blood of female patients with breast cancer (*57*) or genital cancer (*55*) are due to a decrease in absolute uninucleolar lymphocyte counts; a similar result was found in patients with malignant lymphogranuloma (*35*). The absolute number of uninucleolar lymphocytes appears to be the most sensitive indicator as to the stage of progression of these neoplastic diseases (*57*). Since lymphocytes participate in the defense against the tumor growth and even in mature lymphocytes, ring-shaped nucleoli are activated after adequate stimulation, defects in these cells in patients with tumors might reflect the breakdown of the host defense systems in the peripheral blood (*55, 57*). However, the decrease of uninucleolar lymphocytes in peripheral blood is not specific for neoplastic diseases, because it has been observed also in patients with infections, hepatitis, varicella, and scarlet fever (*27*). High values of the nucleolar coefficient have been found also in patients with lymphocytic leukemia (*14, 52*). However, the higher values of the nucleolar coefficient were due to the increase of multinucleolar lymphocytes. In patients with a more malignant course of this disease, relatively low values of the coefficient were noted (*52*).

III. The Lymphoid Cells

Although there are many hematological and histological studies of the lymphoid cells, the descriptions of nucleoli are less numerous and sometimes controversial. The morphology of nucleoli may be important for the identification and classification of lymphoid cells (*22*). The shape of nucleoli in lymphoid cells is usually irregular when these cells are stained with special procedures that are selective for nucleoli and RNA-containing structures. The number of nucleoli ranges from 2 to 6 and sometimes cannot be determined exactly because of their irregular shape and the connections between them. The appearance of these nucleoli is useful for distinguishing lymphoid cells from other mononucleolar cells, particularly in the peripheral blood of patients with infectious disease. The shape of nucleoli of monoblasts also may be irregular (*58*). Nuclcoli of lymphoid cells stained with panoptic procedures may appear as round or oval bodies in the chromatin structure of the nucleus. These shapes are due to the overlapping of some portions of irregular nucleoli by chromatin structures. In electron micrographs, nucleoli in inactive forms of lymphoid cells seem to be small and an increased size of these structures has been noted in activated cells (*4*).

The formation of irregular nucleoli from nucleoli of regular shape has been observed in the transitional forms between cells of lymphoblastic character and less differentiated lymphoid cells (Fig. 9.9a) in mixed leukocyte culture reaction (*54*). A similar phenomenon has been found in leukocyte cultures stimulated with phytohemagglutinin (*33*). Some nucleoli resulting from such transformation were linked by RNA-containing filaments (Figs. 9.9a and 9.9b).

IV. The Plasmacytic Cell Line

A. Plasmablasts and Immature Plasmacytes

The number of nucleoli in plasmablasts ranges from 1 to 6 (*43, 69, 70*). The nucleoli in plasmablasts or immature plasmacytes are usually dense and stain uniformly with toluidine blue; they may contain fine trabecular structures with small light areas between them (*60*) (Figs. 9.10a and b). In electron micrographs, nucleoli in plasmablasts and immature plasmacytes are composed of nucleolonemas with very small light areas between them or they are compact without nucleolonemas and have uniform distributions of ribonucleoprotein structures (*6, 25, 60*). In plasmablasts from plasmacytomas, large dense granules on the periphery of

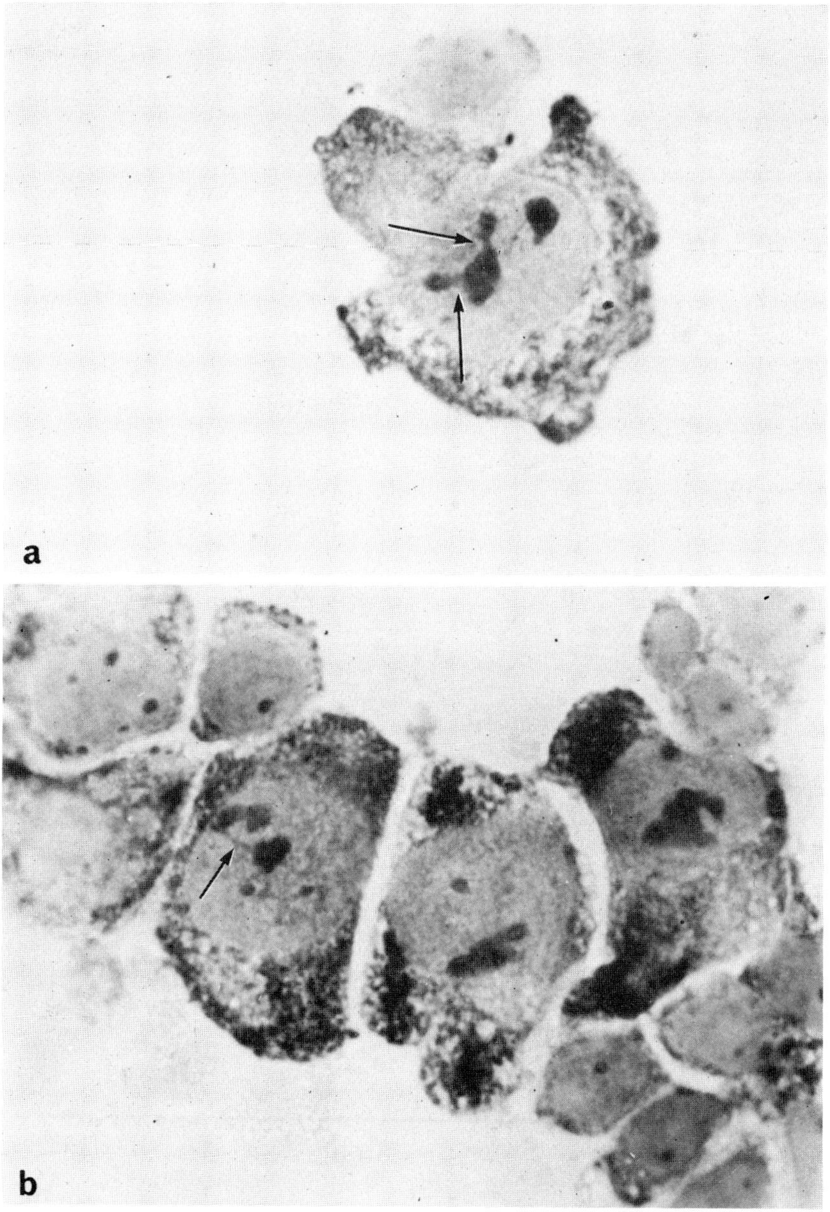

Fig. 9.9. a. Lymphoid cell. There is constriction (arrows) and separation of small nucleoli and a connection between large and small nucleolus. ×2000 (*54*). b. Lymphoid cells. There is an irregular shape of the nucleoli and a connection of some nucleoli by a fine RNA-containing filament (arrow). The same specimen as in Fig. 9.9a. ×1800.

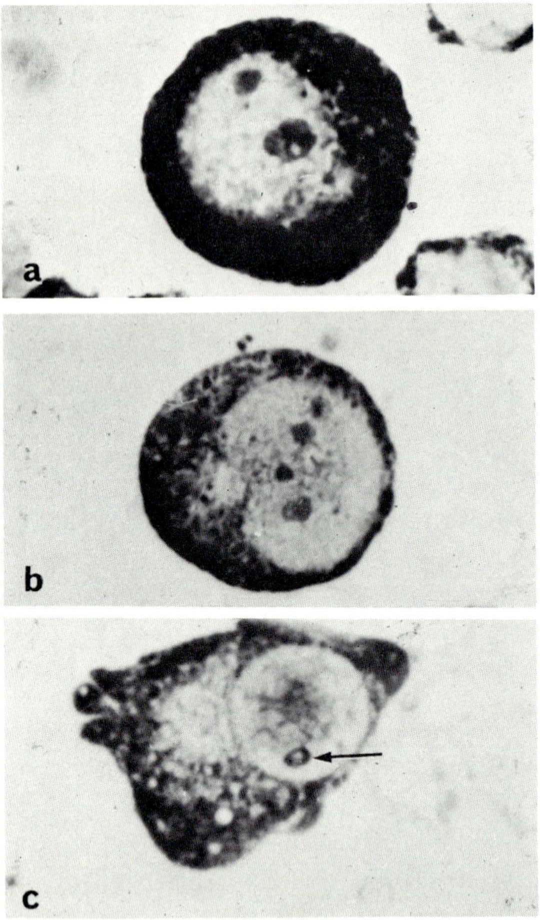

Fig. 9.10. Nucleolar changes during maturation of the plasmacytic cell line (*60*) a = plasmablast; b = proplasmacyte; c = mature plasmacyte with a ring-shaped nucleolus (arrow). ×2500.

nucleoli have been observed (*6*). Similar nucleolar structures have been noted also in plasmablasts of patients with rubeola (*56*). The significance and composition of these granules, however, is not clear.

B. Mature Plasmacytes

The special staining procedures based on the staining of structures containing ribonucleic acid have shown the presence of nucleoli in mature plasmacytes (*44, 47, 60*). The number of nucleoli in mature

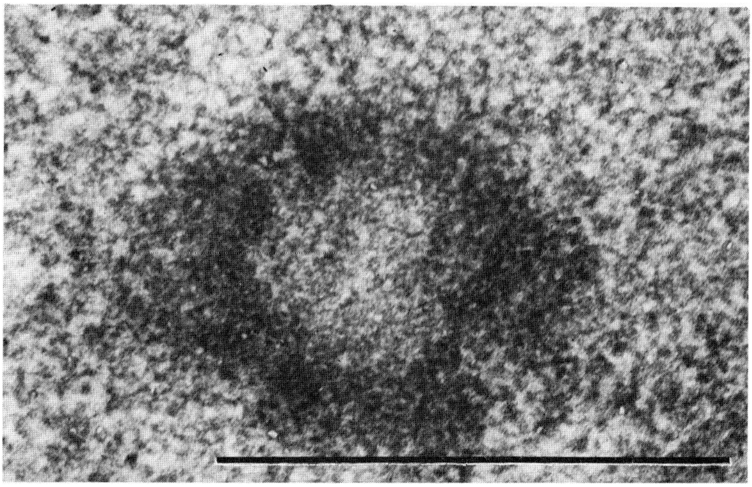

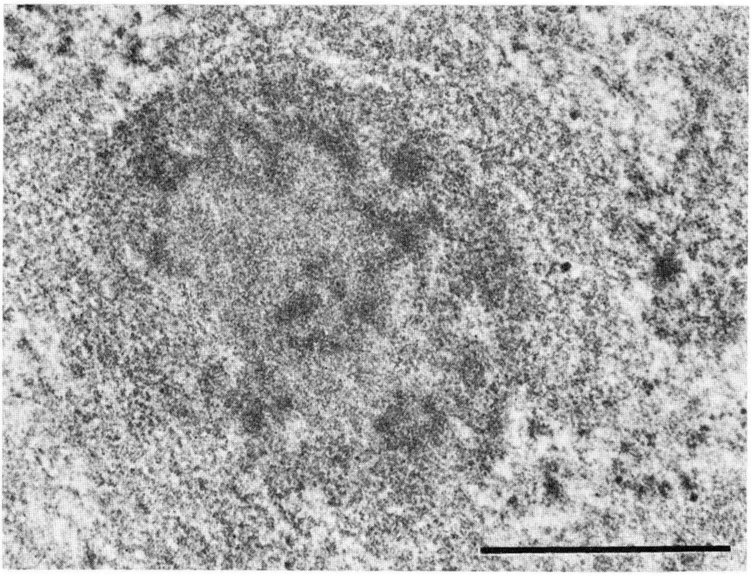

Fig. 9.11. Ring-shaped nucleolus of mature plasmacyte. ×68,000 (*56*).

Fig. 9.12. Ring-shaped nucleolus of a mature plasmacyte with residual nucleolonemas and a fibrillar core. ×33,000 (*60*).

plasmacytes has not been established. The ring-shaped appearance of nucleoli (Fig. 9.10c) seems to be very similar to that in mature lymphocytes (*47, 60*). Electron microscopic studies (Figs. 9.11 and 9.12) have also shown the presence of ribonucleoprotein structures in the peripheral part of nucleoli in mature plasmacytes which determine their ringlike shape (*60*). In some ring-shaped nucleoli of mature plasmocytes, fragments of nucleolonemas were still preserved in their periphery (Fig. 9.12). Although extensive ultrastructural cytochemical studies have not been carried out on ring-shaped nucleoli of mature plasmacytes, the central light area of these nucleoli contains chromatin-like structures. As in mature lymphocytes, the ring-shaped nucleoli in mature plasmacytes can be related to the inhibition of RNA synthesis during the maturation of these cells. This conclusion is supported by autoradiographic studies which showed that RNA synthesis in mature plasmacytes is inhibited during maturation while the synthesis of protein continues (*68*). However, in high proportions of the myelomatous plasma cells, autoradiography has shown a persistence of varying degrees of RNA synthesis (*38*).

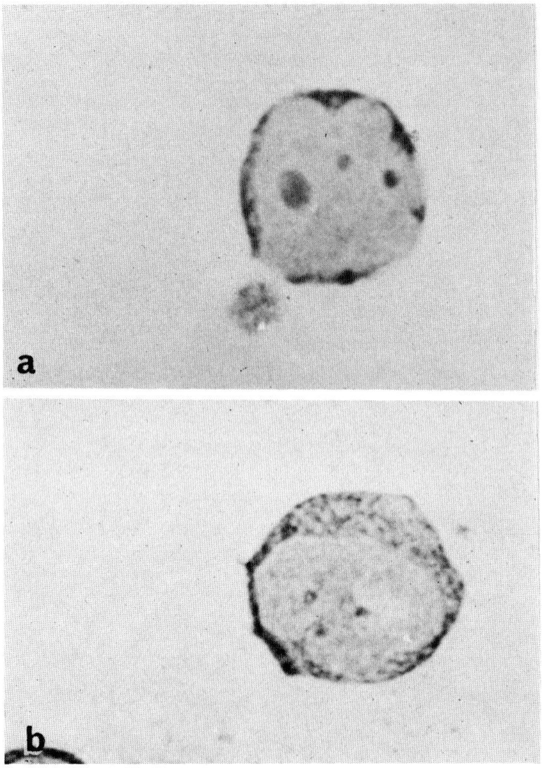

Fig. 9.13. Nucleolar changes during maturation of monocytes. a = leukemic monoblast; b = leukemic monocyte. ×2200 (*60*).

V. The Monocytic Cell Line

A. Monoblasts and Immature Monocytes

In these cells (Fig. 9.13) the number of nucleoli ranges from three to six (*69*). Their shape is usually described as round or oval according to the morphology of nucleolus-associated chromatin. Nucleoli of irregular shape, however, have been also observed in immature monocytes or monoblasts in the peripheral blood of newborns (*58*). In electron micrographs, these nucleoli were shown to be composed of nucleolonemas and light areas between them (Fig. 9.14a) or they appeared to be compact (Fig. 9.14b) with a rather uniform distribution of ribonucleoprotein structures (*26, 51, 60*). In some immature leukemic monocytes as in immature leukemic lymphocytes, ring-shaped nucleoli and compact nucleoli were noted in the same nucleus (*51*).

B. Mature Monocytes

The nucleoli in mature monocytes are very small in comparison with those in mature lymphocytes (*47, 62*). Therefore, they usually appear in the light microscope after special staining as structureless small bodies. Some nucleoli seem to be ring-shaped, particularly when they are large enough to permit studies on their distribution of RNA (Fig. 9.13b). The number of nucleoli ranges from one to seven and the nucleolar coefficient in monocytes in the peripheral blood of healthy adult persons is 2.3 (*62*). Some nucleoli in leukemic monocytes are ring-shaped and contain ribonucleoprotein particles in their periphery (Fig. 9.15) (*51, 60*).

VI. The Granulocytic Line

A. Myeloblasts

The number of nucleoli in these cells ranges from two to six. Their shape is usually regular, round, or oval (Fig. 9.16). In leukemic myeloblasts, particularly in acute leukemia, the number of nucleoli may be reduced (*70*), and their size may be increased (*7, 28, 43*), possibly by fusion of nucleoli of normal size (*70*).

The size and the number of nucleoli may be influenced by antileukemic therapy. For example, nucleoli of leukemic myeloblasts may disappear on treatment with aminopterin (*20*). An increased number of nucleoli has been noted in leukemic paramyeloblasts after previous therapy with 6-mercaptopurine (*39*).

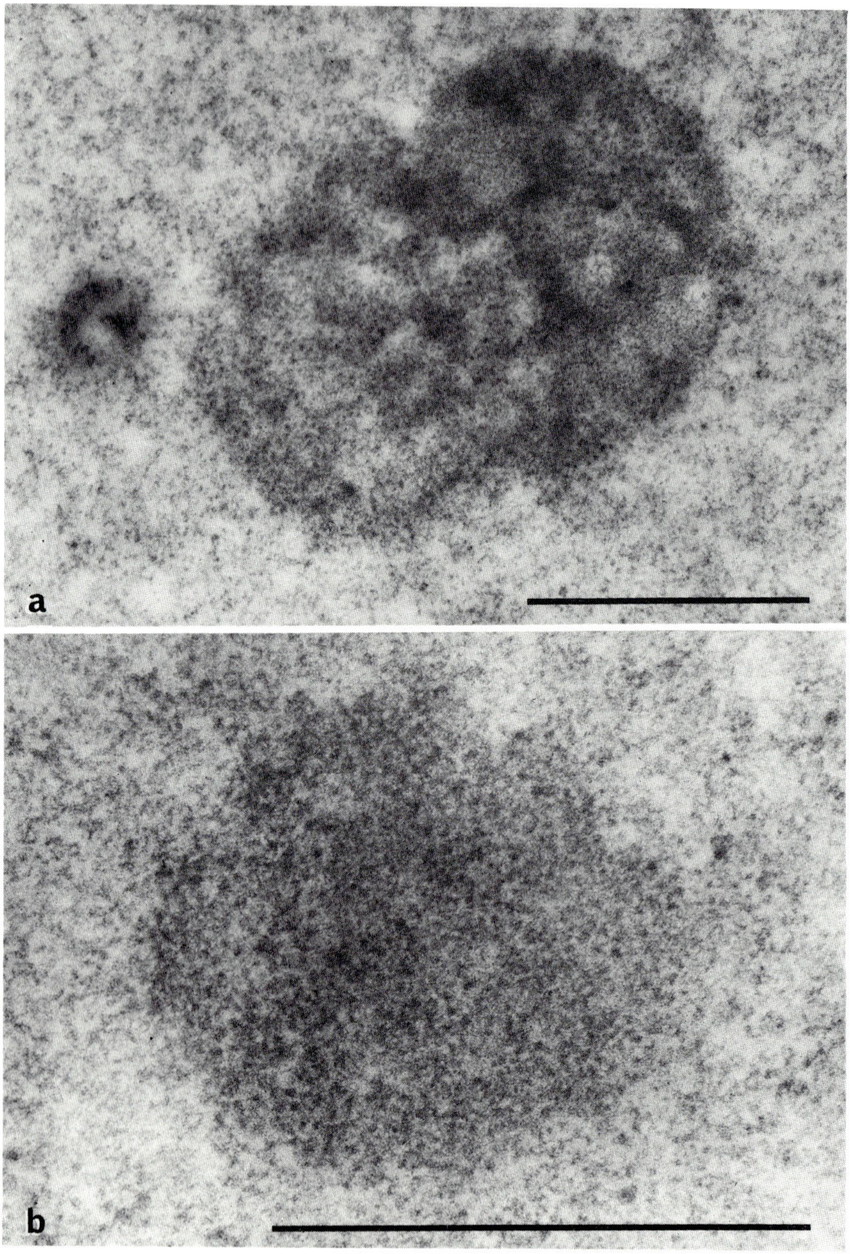

Fig. 9.14. Nucleoli in leukemic monoblasts. a = nucleolus with nucleolonemas ×36,800; b = nucleolus with a rather uniform distribution of RNP structures. ×70,000 (*51*).

The distribution of ribonucleoprotein structures in nucleoli of myeloblasts is usually rather uniform when they are stained with basic dyes (Figs. 9.16a and b). The nucleolus-associated chromatin appears to be less developed in basophilic myeloblasts than in eosinophilic or neutrophilic myeloblasts (*70*). In electron micrographs, nucleoli of myeloblasts are either composed of nucleolonemas and light areas between them or they are rather compact and contain several vacuoles (Fig. 9.17). In nucleoli of myeloblasts in acute leukemias, ring-shaped nucleoli can also be found (Fig. 9.18).

B. Promyelocytes

The nucleoli of these cells are usually smaller than those of myeloblasts (Fig. 9.16c) (*67*). In eosinophilic promyelocytes, however, giant nucleoli are occasionally visible (*70*). The number of nucleoli ranges from two to four (*39*). Promyeloblasts as well as myeloblasts with more nucleoli, however, have also been described (*43*). In the electron

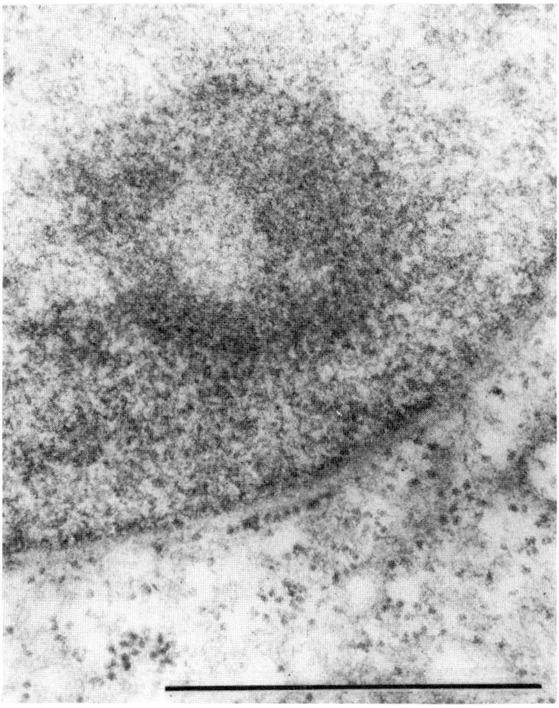

Fig. 9.15. The ultrastructure of a ring-shaped nucleolus of a leukemic monocyte. ×50,000.

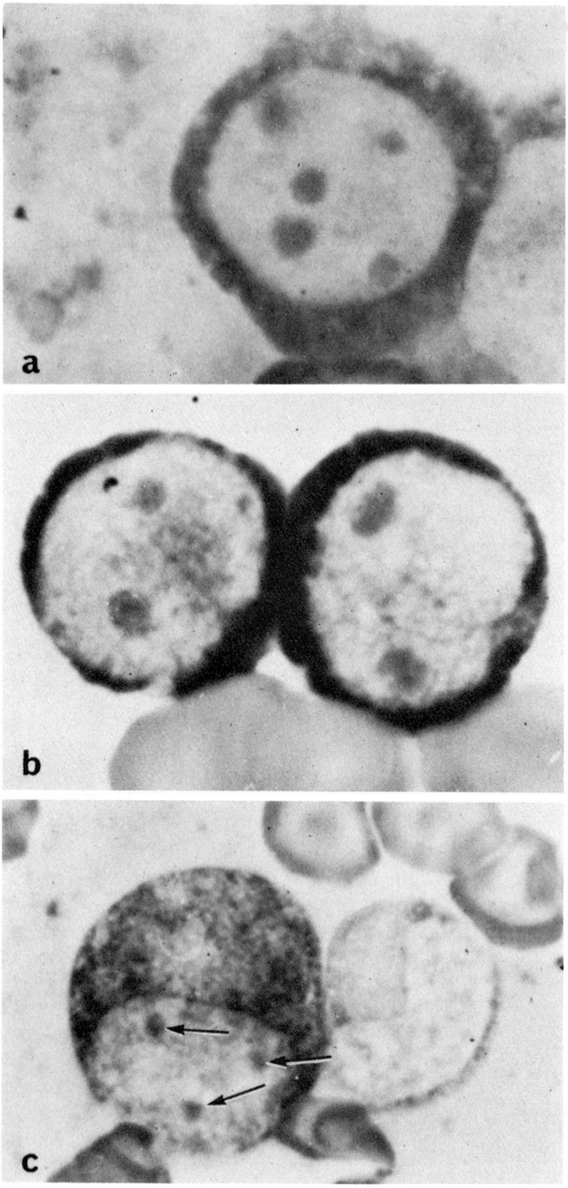

Fig. 9.16. a. A myeloblast with five nucleoli. ×2500. b. Myeloblasts from acute myeloblastic leukemia. ×2800. c. Leukemic promyelocyte and metamyelocyte. There are distinct nucleoli in the promyelocyte (arrows). ×2300.

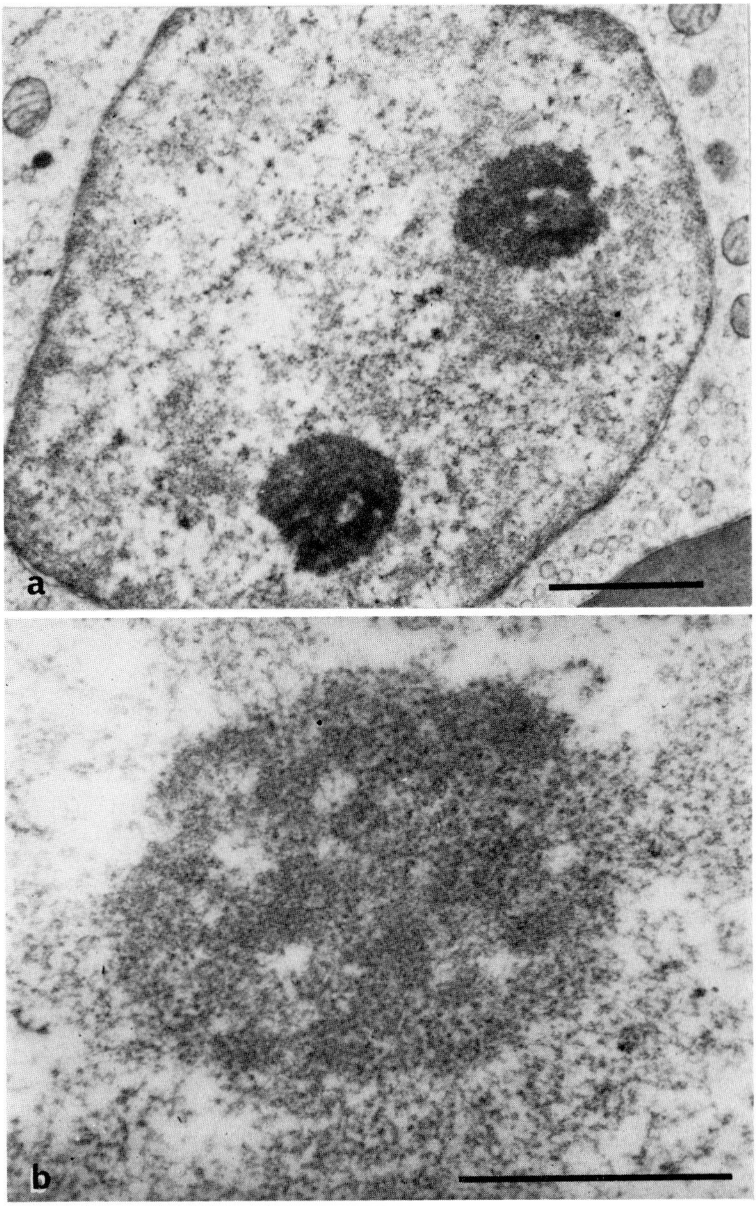

Fig. 9.17. Nucleolar ultrastructure in myeloblasts. a = ×20,000; b = ×35,000.

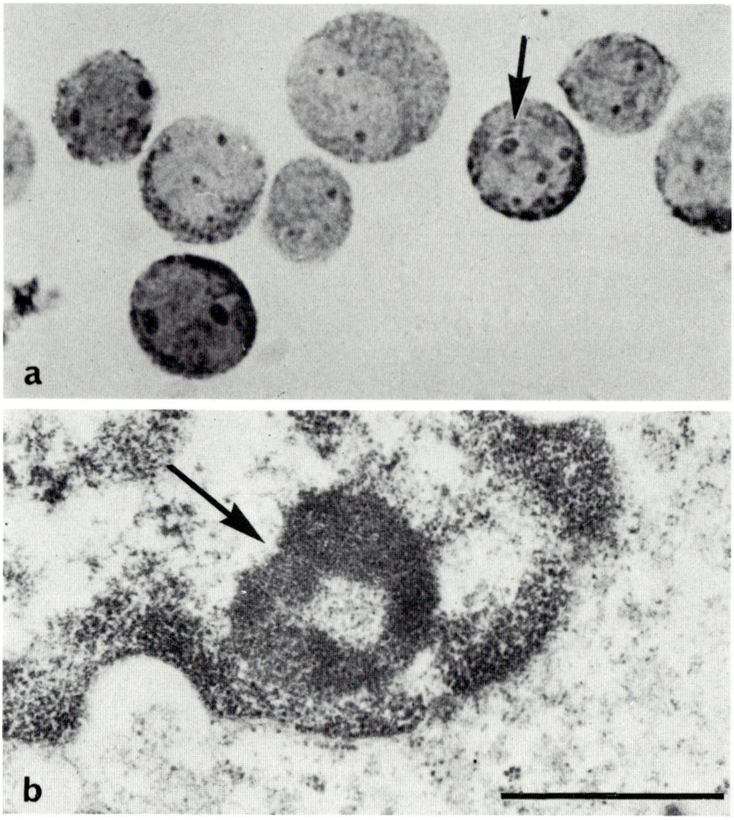

Fig. 9.18. Ring-shaped nucleoli (arrow) in myeloblasts of acute myeloblastic leukemia. a = ×1000; b = ×30,000.

microscope, the nucleolar ultrastructure is similar to that of myeloblasts (Fig. 9.19).

C. Myelocytes, Metamyelocytes, and Granulocytes

Nucleoli disappear during maturation of the granulocytic line in the myelocyte stage (*67*). At this stage, nucleoli appear to be structureless small bodies or they transform into "postnucleolar chromatin masses" (*11, 65, 70*). In electron micrographs, these small nucleoli are composed of typical RNP structures (Fig. 9.20). In the more mature cell forms (metamylocytes and granulocytes), nucleoli are not present although the ribonucleoprotein structures in the interchromatinic areas resemble them (Fig. 9.16c). According to some (*10*), the ribonucleoprotein structures in interchromatinic areas are products of the nucleolar apparatus and possibly the nuclear ribonucleoprotein network (*46, 54;* see Chapter VIII). Thc "postnukleolare chromatin Schollen" (postnucleolar chromatin masses) in nuclei of metamyelocytes or granulocytes may also be residual structures of previously active nucleoli. In any event, the possibility cannot be excluded that a small number of pathological meta-

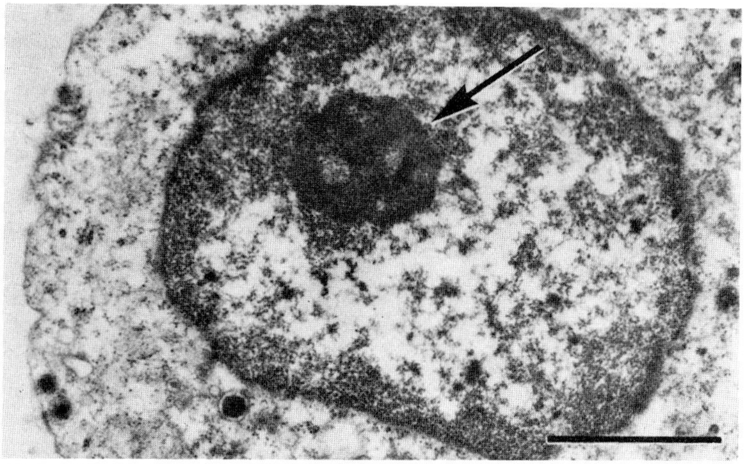

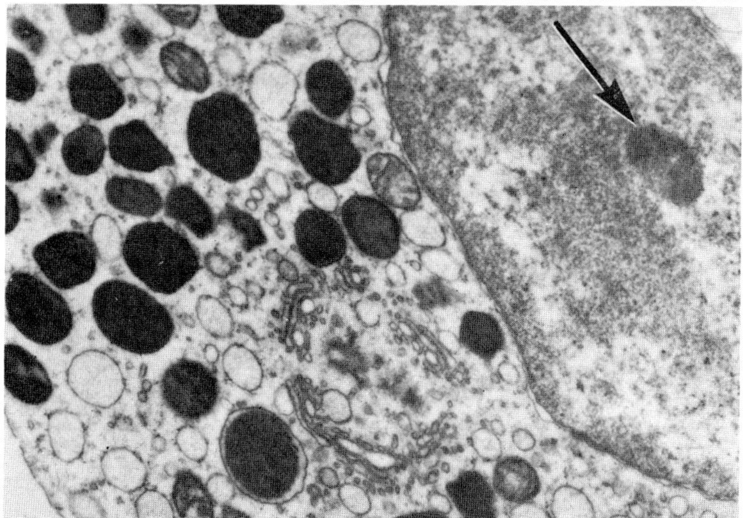

Fig. 9.19. Nucleolus (arrow) in a transitional cell form between myeloblast and promyelocyte. ×23,000.

Fig. 9.20. Nucleolus (arrow) in a myelocyte. ×18,000.

myelocytes or even granulocytes have small nucleoli or nucleolar fragments in their nuclei.

VII. Erythropoietic Line

A. The Normoblastic Series

The nucleoli of proerythroblasts are large, usually polymorphous and irregular in regard to the shape either as stained for RNA (Fig. 9.21) or in phase contrast. The morphology of these nucleoli resembles that of nucleoli in rapidly growing tumor cells and might be related to the functional activity of pronormoblasts which is mainly cell division. These cells provide new cell generations of erythroblasts and erythrocytes necessary for the whole organism. The irregular shape of nucleoli in pronormoblasts is probably the reason why these structures are less distinct after staining with panoptic methods (May–Grünwald–Giemsa or Wright procedures) than those in myeloblasts and why their periphery seems to be covered by chromatin (*65, 70*).

The number of nucleoli in proerythroblasts (pronormoblasts) ranges from three to seven (*43, 69*). In electron micrographs, large nucleoli in proerythroblasts of patients with erythroleukemia are composed of nucleolonemas and their shape seems to be very irregular (*2*) as noted in light microscopic observations of proerythroblastic nucleoli. The nucleoli of macroblasts contain several large or small nucleoli, the outlines of which glisten through the chromatin if the cells are stained with panoptic procedures (*70*). In phase contrast, the nucleoli of macroblasts seem to have an irregular polymorphous shape similar to that of proerythroblast nucleoli. The nucleoli in basophilic normoblasts are usually smaller and less visible; some of them seem to be ring-shaped (Fig. 9.22). The nucleoli disappear in this stage and seem to be destroyed (*2*). The incorporation of tritiated cytidine in nucleoli, however, can still be observed in some basophilic erythroblasts (*17*).

Beyond the stage of polychromatophilic normoblasts, the nucleoli are not visible in smears stained with panoptic procedures although small basophilic bodies are present in nuclei of these cells when stained for RNA. In electron micrographs, some of these structures appear to be nucleolar fragments. Ultraviolet spectrophotometric studies of erythroblasts have shown that the regression of nucleoli during maturation of these cells is accompanied by decrease in cell content of RNA (*17, 67*). Furthermore, the loss of nucleolar RNA synthesis preceded the gradual loss of cytoplasmic RNA (*17*). The different staining properties of nucleoli in pathologic pronormoblasts are due to a decreased content of

RNA and are related to the stage of anemia and the number of erythrocytes in patients studied (*20*). In contrast, an increase of RNA content in nucleoli of megaloblasts demonstrated by means of staining with methylene blue has been observed in patients with pernicious anemia treated with liver extracts (*20*).

B. The Megaloblastic Series

The number of nucleoli in promegaloblasts ranges from three to six (*69*). Their shape is similar to that of those in proerythroblasts (pronormoblasts) (Fig. 9.23). They are less basophilic as compared to nucleoli of proerythroblasts. At the stage of the basophilic megaloblasts, the nucleoli disappear. However, if special methods for staining nucleoli are used, small basophilic bodies containing RNA are still present in nucleoli even at the stage of polychromatophilic megaloblasts (Fig. 9.24). Typical nucleoli have been noted by electron microscopy in more mature megalo-

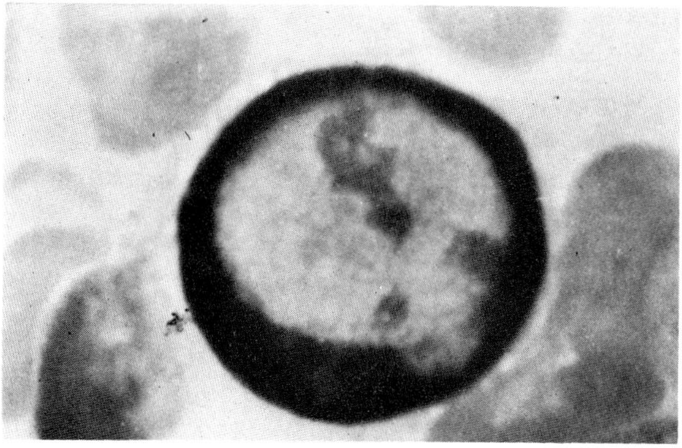

Fig. 9.21. Nucleoli in a proerythroblast. ×2200.

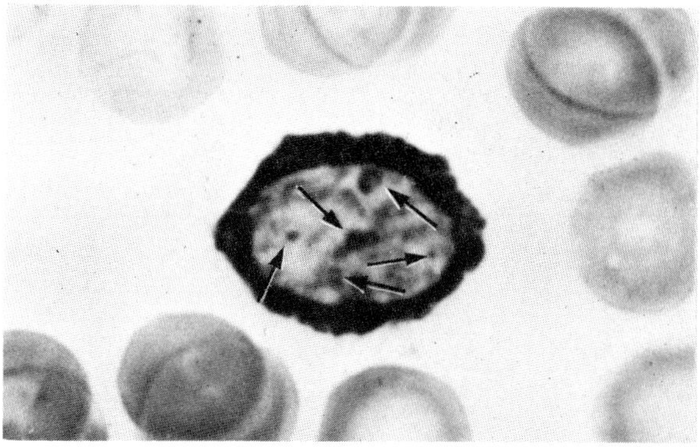

Fig. 9.22. Nucleoli in a basophilic erythroblast (arrows). ×2200.

blasts of a guinea pig with experimental megaloblastosis (*2*). Therefore, some possibility exists that the small basophilic bodies beyond the stage of basophilic megaloblasts, as well as the basophilic normoblasts, are micronucleoli or are nucleolar fragments that represent final irreversible stages of the nucleolar regression during the maturation of nucleated mammalian red cells. This suggestion is supported by the fact that the incorporation of RNA precursors has not been observed beyond the stage of basophilic erythroblasts (*17*). Furthermore, the number of small nucleoli in other cells can be increased by inhibition of RNA synthesis with RNA inhibitors (*34, 63*).

VIII. The Thrombopoietic Megakaryocytic Line

Nucleoli are present in all stages of megakaryocytic maturation and development (*20*). In megakaryoblasts, several small nucleoli are usually

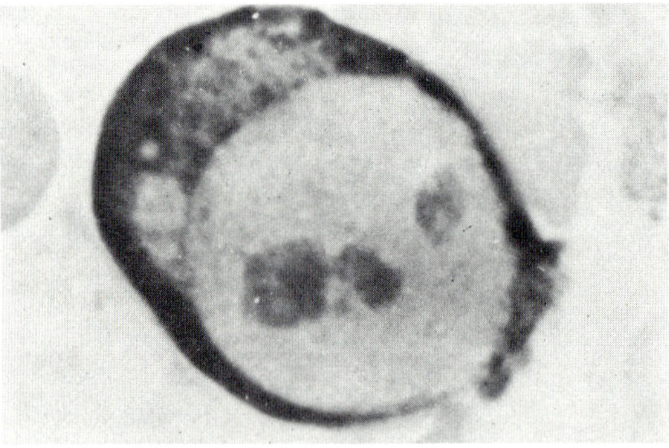

Fig. 9.23. Nucleoli in a promegaloblast. ×2300.

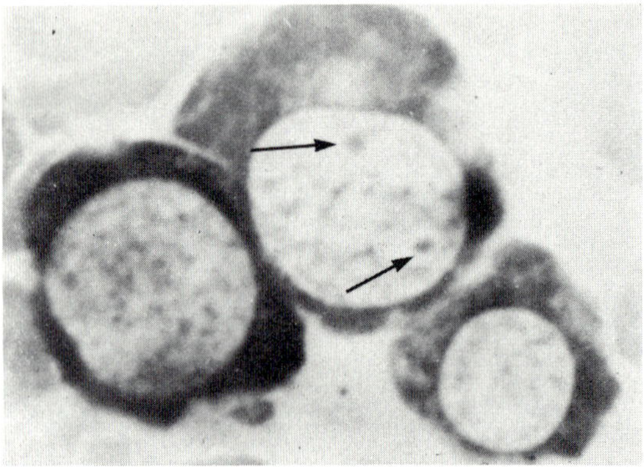

Fig. 9.24. RNP structures in a polychromatophilic megaloblast (arrows). ×2300.

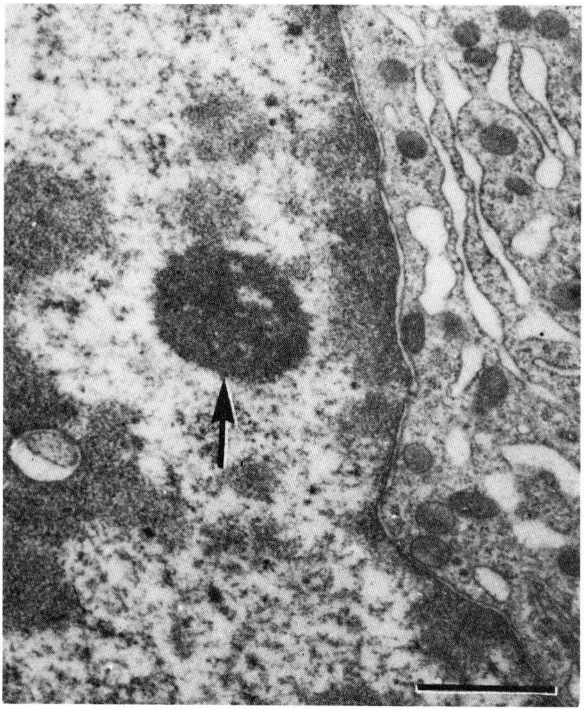

Fig. 9.25. Nucleolus in a megakaryocyte (arrow). ×18,000.

present. They can be covered by the chromatin and, therefore, it is sometimes very difficult to see them in cells stained with panoptic methods. In immature as well as in mature megakaryocytes, numerous small nucleoli can be observed, particularly when special procedures are used that stain only RNA structures (*20, 46*). In the electron micrographs, these small nucleoli seem to have a rather compact appearance (Fig. 9.25).

IX. Summary

The studies of blood cells have shown that there are gradual changes of nucleolar morphology during the differentiation and maturation of these cells. The changes of the nucleolar morphology are characterized by a decrease in the size and number of nucleoli as well as by a decrease in the amount of ribonucleic acid. The nucleoli apparently disappear during the development and maturation of the erythrocytic and granulocytic cell lines and they persist even in the mature stages of lymphocytic, plasmacytic, and monocytic development. However, the morphology of nucleoli in mature lymphocytes, plasmacytes, and monocytes differs from that in immature cells. They are smaller in size and they have a ring-shaped appearance in the majority of the mature lymphocytes and plasmacytes in which inhibition of RNA synthesis has been noted. The ring-shaped appearance of nucleoli probably reflects the reversible inhibition of RNA synthesis since their transformation into nucleoli similar to those in immature cells have been observed in lymphocytes cultivated after antigenic stimulation or stimulation with phytohemagglutinin.

The investigation of nucleoli in blood cells is important not only from the theoretical point of view but also because it can contribute to studies of pathogenesis of certain diseases and, moreover, it can provide useful information in medical diagnosis.

REFERENCES

1. Airo, R., and Astaldi, G., Recenti acquiszioni sulla reattivita del linfocito. *Haematol. Latina* **10**, 209–270 (1967).
2. Akasaka, K., An electron microscopic study of blood cells. Ultrastructure of nucleated red cells observed mainly in erythroleukemic cells. *Acta Haematol. Japon.* **24**, 519–535 (1961).
3. Asano, M., Kawahara, I., and Akabane, T., Classification of acute leukemia cells in children based on electron microscopic observations and relevancy to their involved findings in organs. *Med. J. Shinshu Univ.* **9**, 107–133 (1964).
4. Bernhard, W., and Leplus, R., "Fine Structure of the Normal and Malignant Human Lymph Node." Pergamon Press, Oxford, 1964.
5. Bessis, M., "Cytology of the Blood and Blood Forming Organs." Grune & Stratton, New York, 1956.
6. Bessis, M., and Thiery, J. P., Etude au microscope électronique des hémosarcomes humaines. *Nouvelle Rev. Franc. Hematol.* **2**, 577–601 (1962).
7. Bessis, M., and Thiery, J. P., Etudes au microscope electronique sur les leucemies humaines. *Nouvelle Rev. Franc. Hematol.* **1**, 703–728 (1961).
7a. Bloom, S., and Egli, D., Variation of myocardial nucleolar abundance with heart weight. *Proc. Soc. Exptl. Biol. Med.* **130**, 1019–1021 (1969).
8. Busch, H., "An Introduction to the Biochemistry of the Cancer Cell." Academic Press, New York, 1962.
9. Cooper, H. L., and Rubin, A. D., RNA metabolism in lymphocytes stimulated by phytohemagglutinin: Initial responses to phytohemagglutinin. *Blood* **25**, 1014–1027 (1965).
10. Costachel, O., and Micronescu, S., Observations on adult polynuclear leukocyte nucleus stained with basic dyes. *Folia Haematol.* **86**, 21–34 (1966).
11. Davidson, J. N., Leslie, J., and White, J. C., The cytoplasmatic basophilia of marrow cells; the distribution of nucleic acids. *J. Pathol. Bacteriol.* **60**, 1–20 (1948).
12. Epstein, L. B., and Stohlman, F., RNA synthesis in cultures of normal human peripheral blood. *Blood* **24**, 69–75 (1964).
13. Fukuhara, F., Electron Microscopic study on several characteristic cells in children's diseases. *Ann. Paediat. Japon.* **9**, 43–65 (1963).

14. Gonzalez Guzman, J., Generalities on the nuclear content of some blood cells. *Blood* **3**, Suppl. 2, 57–65 (1948).
15. Gonzalez Guzman, J., Contribution à la reconaissance de l'appareil nucléolaire des cellules leucémiques. *Sang* **20**, 225–238 (1949).
16. Granboulan, N., Etudes au microscope électronique des cellules de la lignée lymphocytaire normale. *Rev. Hematol.* **15**, 52–71 (1960).
17. Grasso, J., Woodard, J., and Swift, H., Cytochemical studies of nucleic acid and proteins in erythrocytic development. *Proc. Natl. Acad. Sci., U. S.* **50**, 134–140 (1963).
18. Grundmann, E., Über die Unterscheidung von zwei Lymphozytentypen im Phasenkontrastmikroskop. *Arch. Pathol. Anat. Physiol.* **332**, 17–24 (1959).
19. Grundmann, E., Weitere Untersuchungen über die Lymphozytenbildung. *Verhandl. Deut. Ges. Pathol.* **42**, 211–215 (1959).
20. Haenel, U., Veränderungen der nukleolar Substanz am Knochenmark bei perniziöser Anämie und ihre pathologische Bedeutung. *Helv. Med. Acta* **17**, 627–650 (1951).
20a. Jacob, J., Involvement of the nucleolus in viral synthesis in the cells of primary renal tumors of leopard frogs. *Cancer Res.* **28**, 2126–2136 (1968).
21. Johnson, D. M., Pratt, P. T., and Rigby, P. G., An evaluation of human lymphocyte nuclear RNA with acridine orange. *Blood* **29**, 800–807 (1967).
22. Jörke, D., "Die Lymphoidzellen des Blutes." Akademie Verlag, Berlin, 1963.
23. Journey, L. J., and Goldstein, M. N., Electron microscope studies on HeLa cell lines sensitive and resistant to actinomycin D. *Cancer Res.* **21**, 929–932 (1961).
23a. Kessel, R. G., and Beams, H. W., Intranucleolar membranes and nuclear-cytoplasmic exchange in young crayfish oocytes. *J. Cell Biol.* **39**, 735–741 (1968).
23b. Kordan, H. A., Nucleolar activity and starch synthesis in lemon fruit tissue *in vitro*. *Bull. Torrey Botanical Club* **92**, 21–37 (1965).
23c. Kordan, H. A., Nucleolar changes in nongrowing and growing lemon fruit explants (*Citrus limon* L.). *Experientia* **25**, 517–518 (1969).
24. Kume, F., Maruyama, S., D'Agostino, A. N., and Chiga, M., Nucleolar change produced by mithramycin in rat hepatic cell. *Exptl. Mol. Pathol.* **6**, 254–260 (1967).
25. Lievre, J. A., Camus, J. P., Levy, K., Badin, J., Bessis, M., and Tessier, F., Myelome a globuline cristallisable. *Nouvelle Rev. Franc. Hematol.* **1**, 23–35 (1961).
26. Low, F. N., and Freeman, J. A., "Electron Microscopic Atlas of Normal and Leukemic Human Blood." McGraw-Hill, New York, 1958.
27. Michalowski, H., and Plenert, W., Studien zur Differenzierung der Lymphozyten des peripheren Blutes im Kindesalter. II. *Folia Haematol.* **85**, 227–235 (1965).
28. Micu, D., Grozea, P., Maxmilian, S., and Mihalescu, E., Punctia ganglionaria in clinica. *Ed. Acad. Rep. Populare Romine, Bucuresti* (1964).
28a. Montgomery, T. H., Comparative cytological studies with special regard to the nucleolus. *J. Morphol.* **15**, 265–564 (1898).
29. Mueller, G. C., and Le Mahieu, M., Induction of ribonucleic acid synthesis in human leukocytes. *Biochim. Biophys. Acta* **114**, 100–107 (1966).
30. Panteleakis, P. N., and Mateyko, G. M., Effects of actinomycin D on the nucleoli of different cells *in vitro*. *J. Cell Biol.* **31**, 154A (1966).
31. Pischinger, A., Die Lage des isoelektrischen Punktes histologischer Elemente als Ursache ihrer verschiedenen Färbbarkeit. *Z. Zellforsch. Mikroskop. Anat.* **3**, 169–197 (1926).

32. Plenert, W., Michalowski, H., and Kehnscherper, M., Studien zur Differenzierung der Lymphozyten des peripheren Blutes im Kindesalter. I. *Folia Haematol.* **85**, 218–226 (1965).
33. Pössnerová, V., and Smetana, K., The morphology of nucleoli of lymphocytes in peripheral human blood after short-term cultivation *in vitro*. *Folia Morphol.* **14**, 240–244 (1966).
34. Potměšil, M., and Smetana, K., The effect of Chromomycin A_3 on the formation of ring shaped nucleoli and incorporation of H^3 uridine in cell cultures. *Folia Biol.* (*Prague*) **14**, 132–139 (1968).
35. Potměšil, M., Smetana, K., and Wienerová, E., Nucleolar coefficient of the lymphocytes in the peripheral blood of patients with Hodgkin's disease. *Neoplasma* **12**, 205–214 (1965).
36. Potměšil, M., Smetana, K., and Wienerová, E., Nucleolar coefficient of lymphocytes in the peripheral blood of patients with acute tonsillitis and infectious hepatitis. *Folia Haematol.* **88**, 275–286 (1967).
37. Potměšil, M., and Wienerová, E., The nucleolar coefficient of lymphocytes in the peripheral blood of healthy subjects. *Folia Morphol.* **14**, 352–361 (1965).
38. Quaglino, D., Torelli, U., Sauli, S., and Mauri, C., Cytochemical and autoradiographic investigations on normal and myelomatous plasma cells. *Acta Haematol.* **38**, 79–94 (1967).
39. Rind, H., "Atlas der Phasenkontrasthämatologie." Akademie Verlag, Berlin, 1958.
40. Rubin, A. D., Havemann, K., Kochwa, S., and Damaschek, W., Differences in nucleic acid metabolism of PHA stimulated lymphocytes from patients with infectious mononucleosis and chronic lymphocytic leukemia. *Blood* **28**, 998 (1966).
41. Sankanarayanan, K., and Busch, H., Dense granules of Walker 256 carcinosarcoma cells of the rat. *Exptl. Cell Res.* **38**, 434–437 (1965).
42. Scherer, E., and Fiebelkorn, H. J., Über die Darstellbarkeit der Nukleinsäuren der Zellkornes und ihre Bedeutung für Diagnose und Prognose maligner Bluterkrankungen. *Folia Haematol.* **72**, 143–148 (1954).
43. Schneiberg, K., Badania nad jaderkami interfazy komorok szpikowych w przypadkach prawidlowych i w przypadkach ostrych i przewleklych bialaczek szpikowych. *Patol. Polska* **6**, 19–23 (1955).
44. Seman, G., La mise en évidence des nucléoles sur frottis par le bleu de méthylène boraté. *Rev. Franc. Etudes Clin. Biol.* **5**, 196–198 (1960).
45. Shankar, K. N., Steele, W. J., Smetana, K., and Busch, H., Ultrastructural aspects of the ribonucleoprotein network in nuclei of Walker tumor and rat liver. *Exptl. Cell Res.* **46**, 65–77 (1967).
46. Shea, J. R., and Leblond, C. P., Number of nucleoli in various cell types of the mouse. *J. Morphol.* **119**, 425–434 (1966).
47. Smetana, K., A further contribution to the question of the incidence of nucleoli in the lymphocytes in men. *Folia Biol.* (*Prague*) **7**, 268–274 (1961).
48. Smetana, K., Basic histochemical and cytochemical methods. *Methods Cancer Res.* **2**, 361–443 (1967).
49. Smetana, K., Freireich, E. J., and Busch, H., Chromatin structures in ring shaped nucleoli of human lymphocytes. *Exptl. Cell Res.* (1969) **52**, 112–128 (1968).
50. Smetana, K., and Heřmanský, F., A contribution to the knowledge of the ultrastructure of leukemic lymphocytes. *Neoplasma* **10**, 405–411 (1963).
51. Smetana, K., and Heřmanský, F., On the ultrastructure of leukemic monocytes. *Folia Haematol.* **86**, 35–46 (1966).

52. Smetana, K., Heřmanský, F., and Pössnerová, V., Nuclear coefficient and glycogen content of lymphocytes in chronic lymphadenoses. *Neoplasma* **9**, 549–554 (1962).
53. Smetana, K., Heřmanský, F., Janele, J., and Busch, H., A further note on the ultrastructure of leukemic lymphocytes. *Folia Haematol.* **89**, 1–14 (1968).
54. Smetana, K., and Iványi, D., Cytological analysis of mixed leukocyte cultures. *Folia Biol. (Prague)* **13**, 150–154 (1967).
55. Smetana, K., Janele, J., and Kafka, V., The nucleolar coefficient of the lymphocytes in the peripheral blood in women with carcinoma of genitalia. *Folia Biol. (Prague)* **8**, 390–395 (1962).
56. Smetana, K., Janele, J., and Busch, H., A contribution to the problem of lymphocytes with plasmatic reaction. *Folia Morphol. (Prague)* **14**, 86–89 (1965).
57. Smetana, K., Janele, J., and Malínský, L., Nucleolar coefficient of lymphocytes in the peripheral blood of female patients with breast carcinoma. *Neoplasma* **13**, 646–648 (1966).
58. Smetana, K., and Janovský, M., Nukleolární koeficient lymfocytů periferní krve novorozenců v prvních dvou dnech života (The nucleolar coefficient of lymphocytes in the peripheral blood of newborns in the first two days of life). *Cesk. Morfol.* **9**, 266–273 (1961).
59. Smetana, K., and Janovský, M., Nukleolární koeficient lymfocytů periferní krve kojenců (The nucleolar coefficient of the lymphocytes in the peripheral blood of sucklings). *Cesk. Morfol.* **11**, 131–134 (1962).
60. Smetana, K., Lane, M., and Busch, H., Studies on nucleoli of leukemic agranulocytes and plasmacytes in multiple myeloma. *Exptl. Mol. Pathol.* **5**, 236–244 (1966).
61. Smetana, K., Lejnar, J., and Potměšil, M., A further contribution to the demonstration of RNA and nucleoli of blood cells in smear preparations. *Folia Haematol.* **91**, 381–394 (1969).
62. Smetana, K., and Potměšil, M., A contribution to the demonstration of nucleoli and their nucleolar coefficient in the mature monocytes of human peripheral blood. *Folia Morphol. (Prague)* **15**, 98–103 (1967).
63. Smetana, K., and Potměšil, M., Further note to the question of the incidence of ring shaped nucleoli in liver cells of rats after treatment with actinomycin D. *Z. Zellforsch. Mikroskop. Anat.* **92**, 62–69 (1968).
64. Smetana, K., Steele, W. J., and Busch, H., A nuclear ribonucleoprotein network. *Exptl. Cell Res.* **31**, 198–201 (1963).
65. Stobbe, H., "Hämatologischer Atlas." Akademie Verlag, Berlin, 1959.
66. Stockinger, L., and Kellner, G., Der Lymphozytennukleolus. *Wien. Z. Inn. Med. Grenzg.* **33**, 135–141 (1952).
67. Thorell, B., Studies on the formation of cellular substances during blood cell production. *Acta Med. Scand.* **129**, Suppl. 200, 1–120 (1947).
68. Torelli, U., Quaglino, D., Artusi, T., Emilia, G., Ferrari, G., and Mauri, C., An autoradiographic study of the RNA and protein metabolism of normal plasma cells and phytohaemagglutinin-stimulated lymphocytes. *Exptl. Cell Res.* **42**, 1–9 (1960).
69. Undritz, E., Die Stamzellen der Blutkörperchen, ihre Unterscheide und Nomenklatur. *Schweiz. Med. Wochschr.* **78**, 993–1001 (1948).
70. Undritz, E., "Sandoz Atlas of Haematology." Sandoz, Basel, 1952.
71. Uzman, B. G., Foley, G. E., Farber, S., and Lazarus, H., Morphologic variations in human leukemic lymphoblasts (CCRF-CEM cells) after long-term culture and after exposure to chemotherapeutic agents. A study with the electron microscope. *Cancer* **19**, 1725–1742 (1966).

72. Vendrely, C., and Vendrely, R., Localisation de l'acide ribonucleique dans les différent tissus et organes de vertèbres. *In* "Handbuch der Histochemie" (W. Graumann and R. Neumann, eds.), Vol. 3, Part 2, pp. 84–243. Fischer, Stuttgart, 1959.
73. Wintrobe, M. M., "Clinical Hematology." Lea & Febiger, Philadelphia, Pennsylvania, 1961.

Part II NUCLEOLI OF NERVE CELLS

I. Introduction

Nucleoli in nerve cells are very prominent regardless of the procedures used for their visualization. It is not surprising that these nucleoli have been a subject of great interest for numerous investigators.

In "supravital" preparations as well as in fixed and stained specimens, nucleoli in nerve cells appear to be relatively large bodies, usually with a smooth surface (*100*). They are distinct even when stained with procedures commonly used for the demonstration of nuclei which appear pale and less stained in comparison with them. However, in some nerve cells, the chromatin clusters (chromocenters) or Barr bodies (sex chromatin bodies) may look like nucleoli or nucleolar fragments. Such chromatin structures can be distinguished easily from nucleoli by light microscopic cytochemical procedures or electron microscopy. The chromatin structures are Feulgen positive and do not contain RNA. In contrast, nucleoli in nerve cells are not stained by the Feulgen reaction, but they contain RNA and are surrounded by the Feulgen-positive nucleolus-associated (perinucleolar) chromatin. The characteristic nucleolar ultrastructure is also different from that of chromatin structures (see Chapter II).

Nucleoli in nerve cells can be observed first during differentiation of neuroblasts into neurocytes before Nissl bodies can be detected in the cytoplasm of these cells (*71, 98, 99*). According to Hydén (*56*), nucleoli cannot be detected in the multiplying neuroblasts; however, they can be seen in young embryonic neurocytes which are a later stage of nerve cell development. The nucleoli in these young cells seem to originate from chromatin structures and a positive Feulgen reaction is noted in the early stages of their development (*99, 109*). In grasshopper neuroblasts, a secondary constriction near the distal ends of one of the longer pairs of chromosomes is a nucleolar organizer (*22*). The "young" nucleoli stain intensely with basic dyes because of their high RNA content. In aging cells the nucleolar staining decreases and vacuoles are present (*7, 8, 71*).

Nucleoli seem to be present in all mature nerve cells except the smallest nerve cells of the human cerebellar and cerebral cortex (*100*). The number of nucleoli in nerve cells can vary, particularly with age. With maturation, the number of nucleoli decreases and only one nucleolus is usually present in most mature nerve cells (*14, 49, 100, 102*). In some nerve cells, however, more than two nucleoli were found (*67, 102*). Three possible mechanisms may lead to the reduction of the number of nucleoli in nerve cells: (a) the size of one may be reduced, possibly by lysis, while the size of the main nucleolus increases (*65, 91, 122*); (b) fusion of nucleoli may occur to form a solitary nucleolus (*111*); and (c) the nucleoli may migrate out of the nucleus (*66*).

The nucleolar size and the nucleolar volume are also not constant during the life of the nerve cell (*19, 39–41, 43, 49, 62, 63, 68, 100, 102, 126*). During cell growth, the nucleolar size and volume increase. This enlargement seems to be independent of nuclear growth; therefore, the nucleus–nucleolus size ratio is not constant in nerve cells and changes can be observed during normal cell life and maturation (*49, 100, 102*). According to Hydén (*58*), there is no definite relation between the volume of the nucleus and that of the nucleolus.

II. Nucleolar Structure and Cytochemistry

At the end of the last century several investigators noted that the nucleolus and Nissl bodies seemed to have similar staining properties (*102*) and could be stained with basic as well as acidic dyes. Timofeew (*118*) noted that the nucleoli of ganglia cells are composed of acidophilic granules embedded in a basophilic ground substance. According to Smirnow (*113*), the fine structures in the ground substance could be stained with fuchsin and large nucleolar granules were characterized by their affinity to the toluidine blue, methylene blue, or Victoria blue. The staining properties of some of these granules were similar to chromatin. The peripheral part of the nucleolus was basophilic (*73, 74, 109*) and the central part stained with acidic dyes. The composition of the nucleolar components seemed to change during the development of nerve cells (*85*). In the first stage of nucleolar development these structures were stained uniformly. In later stages, the nucleolus seemed to be composed of the peripheral portion and central "nucleolar body." The peripheral part could be stained with methyl green and the nucleolar body seemed to be oxyphilic as well as basophilic. "Lipoidosomes" (or lipid-containing bodies) appeared in the nucleolar bodies during the late stages of the embryonal development of the nerve cells and seemed to be replaced by vacuoles. The peripheral part of the nucleolus was said to represent active chromatin (*109*). The central part of the nucleolus, according to Havet (*45*), appeared to have a complicated structure, being composed of trabecular and reticular structures. Mjassojedoff's studies (*82*) indicated that the nucleolus was composed of three components: basichromatin, oxychromatin, and ground substance (matrix). The basichromatin was stained with safranin O, the oxychromatin with "wasserblue" and the ground substance with picric acid. The morphology and amount of basichromatin or oxychromatin were not constant depending on the state of the nucleolar cycle* or the animal species investigated. In mammalian nerve cells, the amount of basichromatin was very low or absent and only acidophilic nucleoli were present in some nerve cells (*73, 118*).

A. Nucleolar Chromatin

The introduction of modern cytochemical methods facilitated the differentiation between the main nucleoprotein structures in the nucleoli of nerve cells and also provided basic information on the presence of other substances in the nucleolus. The studies of nucleoli in nerve cells by means of the Feulgen reaction showed that the large basophilic granules or clusters at the periphery of the nucleolus described as the peripheral part of the "basichromatin" were the only structures which gave real Feulgen-positive reactions. However, in some nerve cells (*64*), the chromatin filaments seen by the light microscope might be present even within the nucleolus. The "oxychromatin" and the "basichromatin" of the nucleolar body apparently did not contain DNA and were Feulgen negative. Therefore, nucleoli in nerve cells seem to be Feulgen-negative bodies surrounded by perinucleolar nucleolus-associated chromatin (*46*). Hamberger and Hydén (*36*) observed different organizations and amounts of nucleolus-associated chromatin around nucleoli of ganglion cells. They found that the nucleolar bodies were surrounded by complete or incomplete rings of nucleolus-associated chromatin.

* According to Mjassojedoff (*82*), the peripheral nucleolar basophilic structures migrated into the nucleoplasm in which their growth could be observed as well as the formation of a basophilic membrane on their surface. The lifetime of such bodies (basophilic nucleoli) was limited and they developed into acidophilic ones. This cycle, however, was less pronounced in mammalian nerve cells.

B. Nucleolar RNA

Methods demonstrating RNA-containing structures (*56*) showed that these structures are the main nucleolar elements in nerve cells as in other cells (see Chapter II). The amount of RNA, however, has been reported to be small in nucleoli of nerve cells although it is higher than in the rest of the nucleus. The RNA concentration in nucleoli was found to be about 1% in mammalian spinal ganglion cells, 2% in anterior horn cells, and 0.5% in fish ganglion cells (*58, 90*). The nucleoli of nerve cells in the supraoptic nucleus were estimated to contain 1.7 pg of RNA (*25a, 58*). In the nucleoli of "heavy" Purkinje cells 0.16 pg RNA/μ^3 were found. The values for intermediate Purkinje cells were about 0.19 pg/μ^3 and for light cells, 0.02 pg/μ^3 (*16*).

C. Proteins

The presence of basic as well as acidic proteins in nucleoli of nerve cells (*75, 76*) might explain why these structures could be stained under certain conditions with both basic and acidic dyes (*49, 100, 102*). The protein content of the nucleolus in the nerve cell is higher than that of the rest of the nucleus or cytoplasm (*60, 90*). In nucleoli of spinal ganglion cells about 0.40 pg/μ^3 of protein was found on the basis of X-radiographs (*60*). This seems to represent a 40% concentration of organic material assuming a density close to 1 (*58*). Higher values (50%) were reported in nucleoli of nerve cells from Deiters' nucleus or (70%) in anterior horn cells (*90*). The high concentration of proteins could explain the migration of nucleoli in nerve cells to the centripetal pole of the nucleus on ultracentrifugation at 400,000 *g* which shows that these structures have the highest densities (*13*).

D. Enzymes

There are few cytochemical reports on the presence of enzymes in the nucleoli of nerve cells. Alkaline phosphatase has been demonstrated in nucleoli of peripheral ganglion cells or in nerve cells of hypothalamic nuclei (*9, 92, 116*). Acid phosphatase has been found in nucleoli of anterior horn cells as well as in nucleoli of nerve cells in the stellate ganglion (*15, 114*) or hypothalamic nuclei (*28*). The presence of cholinesterase, oxalic acid dehydrogenase, and glucose-6-phosphatase has been observed in the spinal ganglia cells (*117*). 5′-Nucleotidase, adenosine triphosphatase, succinic dehydrogenese, lactic dehydrogenese, amylophosphorylase, and aldolase have been detected in nucleoli of hypothalamic nerve cells (*61*). Negligible activity of monoamino oxidase and glucose 6-phosphate dehydrogenase has been also observed in these nucleoli (*61*). Some of the findings of enzymes in nerve cell nucleoli, however, should be verified by further studies since the nuclear localization of enzymes detected by the Gomori-type procedures has been questioned. For example, a high activity of alkaline phosphatase has been observed in nucleoli of hypothalamic nerve cells by means of Gomori's method but no activity has been noted in those nucleoli by means of Burstone's naphtol AS phosphate method (*61*).

Only a few inorganic substances have been detected in nucleoli of nerve cells. Nucleoli of nerve cells contain potassium, calcium, silicates, and iron (*50, 100*). Nucleolar affinity to silver has been demonstrated but its basis has not been satisfactorily clarified in nerve cells. This question, however, was studied in other cell types (*96*). Two structural components can be seen usually in nucleoli of nerve cells after silver impregnation (*49, 100–102*): granules and a less dense ground substance. Estable and Sotelo (*29*) described in nerve cell nucleoli, the presence of heavily impregnated trabecular structures (nucleolonemas).

III. Nucleolar Vacuoles and "Inner Bodies"

The nucleolar inner bodies and vacuoles in nerve cells have been studied by numerous investigators in the past because these structures are very prominent and their significance and composition are not satisfactorily clarified. Furthermore, the incidence of nucleolar inner bodies and vacuoles in nerve seemed to be related (*53*) and they were often confused by the procedures used (*100*). Vacuoles or vacuolelike structures are normal components of the nucleolus, although they can also represent pathological structures. Their formation can be induced by experimental stimulation of nerve cells (*11, 49, 53, 65, 100, 123*). The size of nucleolar vacuoles or vacuolelike structures may be very different in various nerve cells and one solitary large vacuole, as well as multiple small vacuoles, may be present in nerve cell nucleoli. The large vacuoles might originate by the fusion of smaller ones (*69*). The presence of multiple small vacuoles in these nucleoli may produce a mulberrylike appearance (*53, 100*) which is related to the high synthetic and secretory activities of this structure (*53, 123*). The vacuoles can migrate from the nucleolus into the nucleoplasm and cytoplasm or their content is secreted into the nucleus (*53, 65, 123*).

Nucleololus and Nucleolar Satellites

A solitary intranucleolar body (*104*) in nerve cells described often as "nucleololus" (*102*) has been observed in the form of a single large vacuole or in the form of a single large solid body about 1 μ in diameter. The differences in the appearance of nucleololi are probably due to the different procedures used for their visualization (*102, 104, 108*). According to cytochemical studies (*25*), "nucleololi" seem to contain bound lipids and proteins and their incidence can be related to the accumulation of proteins.

Multiple intranucleolar bodies studied by Höpker (*53*) appear to be related to the formation of nucleolar vacuoles. Unlike nucleolini (*100*), these bodies do not contain DNA. These Feulgen-positive nucleolini, however, seem to be different from the nucleolini described by Love and Bharadwaj (*79*).

The nucleolar satellites (Fig. 9.26) described in nerve cells under different terms (paranuclear body, corpus paranucleolare, accessory body, zolle basophile, clot of Levi) (*18, 49, 100, 102*) might represent the "sex chromatin." However, according to Lindsday and Barr (*77*), the sex chromatin bodies are not identical to the basophile clots of Levi nor the accessory bodies because of the different behavior of these structures. The accessory body decreases in size during chromatolysis and does not change its position. Moreover, it cannot be stained with basic dyes or the Feulgen method. The basophilic clots of Levi are always at the periphery of the nucleolus and do not change position, although they can be stained with basophilic dyes or the Feulgen reaction. The sex chromatin bodies (nucleolar satellites), 1 μ in diameter, are present mainly in female cells. In males, the percentage of nerve cells possessing the nucleolar satellites is significantly lower (*17, 83*). Unlike the nucleolus, nucleolar satellites have positive Feulgen reactions and do not contain RNA. The position of the nucleolar satellites in the nucleus can vary widely. They are often associated with the nucleolus, although they can also be found at the nuclear membrane or in the nucleoplasm (*12, 33*). This difference in position of the nucleolar satellites seems to depend on the functional state of the nerve cell. Barr and Bertram (*10, 11*) have shown that the nucleolar satellites associated originally with the nucleolus migrated to the nuclear membrane after electric stimulation of nerve cells and returned during the repair phase.

Electron microscopic studies demonstrated that nucleoli in nerve cells are composed of trabecular structures (nucleolonemas) surrounded by small light areas (Fig. 9.26) (*2, 4, 24, 47, 54, 55, 86, 89, 93, 119*). In cerebropleural ganglion cells of several nudibranchs, an unusual zonal organization of nucleolar components was observed (*103*). In these nucleoli, nucleolonema-like structures were present in the nucleolar periphery. In some nerve cells, the nucleoli are compact without distinct nucleolonemas (*42, 119*). The pronucleoli in neuroblasts appeared to be

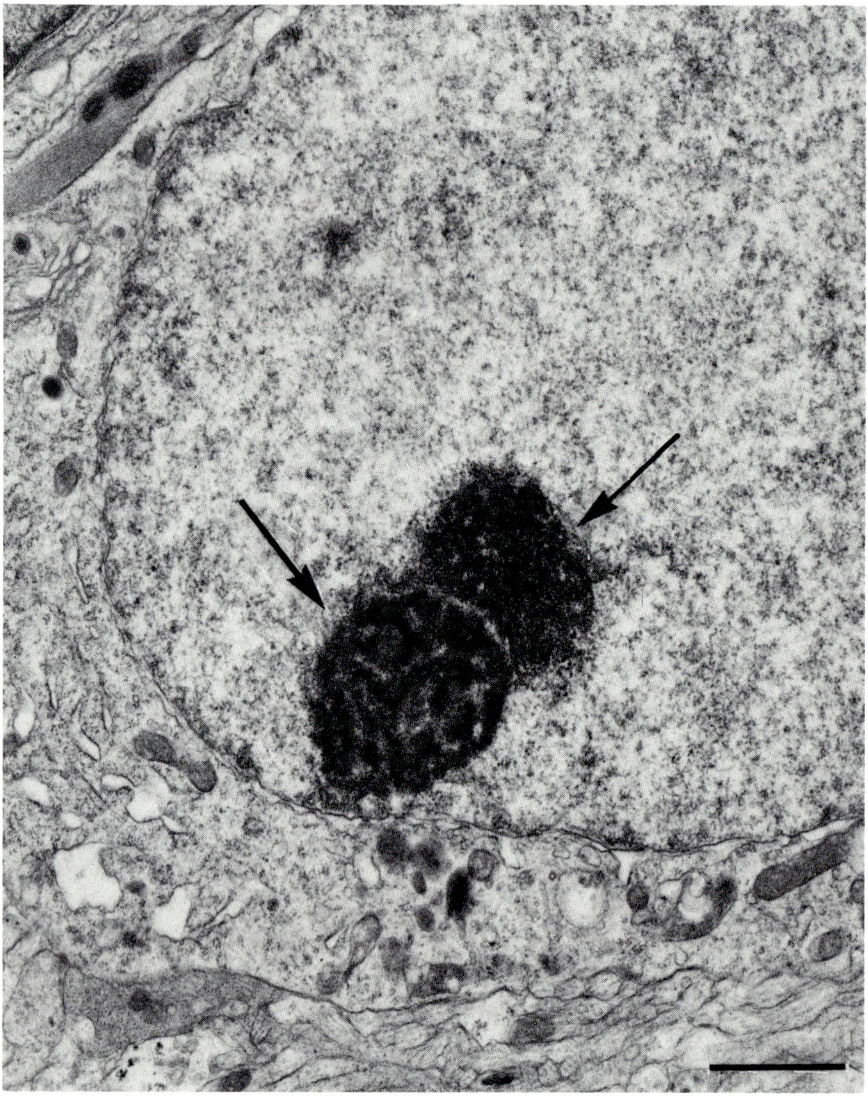

Fig. 9.26. Nucleolus (left arrow) with satellite body (right arrow) in a pyramidal neuron in the cerebral cortex of a rat. The specimens were postfixed in osmium tetroxide after perfusion of the rat with 2% formaldehyde and glutaraldehyde. ×17,700 (courtesy of Dr. W. Caley, Department of Anatomy, Baylor College of Medicine).

dense small bodies composed mainly of RNP granules (*115*). However, a nucleolar membrane was not found (*42, 93*). Electron microscopic studies also have shown that chromatin structures are attached to the nucleolar body and may represent such chromatin structures as chromosomal residues on nucleolar satellites (Fig. 9.26) (*2, 47, 55*). Such studies also suggested the presence of intranucleolar chromatin structures within nucleoli of nerve cells (*2, 24*). The nucleololus of nucleoli in nerve cells appears as a large light area when studied with the electron microscope (*25*). Such light areas are composed of fine filaments and granules corresponding to the proteins determined by light microscopic cytochemistry (see above).

IV. The Nucleolar Role in Synthesis of Nissl Substance

The nucleolar morphology has been studied intensively in different states of the nerve cell. These studies, however, were usually based on the relationships of fine morphological and cytochemical studies since corresponding biochemical studies on nucleoli isolated from nerve cells have not been reported.

The similar staining properties of the nucleolus and cytoplasmic Nissl substance discovered at the end of the last century indicated that both structures could have a similar composition. Modern cytochemical methods demonstrated that both nucleoli and cytoplasmic Nissl substances contain RNA. Since the nucleoli can be detected in young nerve cells before the Nissl substance appears in the cytoplasm (*85, 98, 99*), it has been suggested that the nucleolus might play a principal role in its formation.

Numerous observations suggest that the nucleolar products migrate to the cytoplasm; the morphology of the product is influenced by the procedures used for its visualization. In silver-impregnated or stained specimens, the nucleolar material migrating to the cytoplasm appears to be fine granules. However, large round bodies, vacuolelike structures, and even whole nucleoli have been observed migrating to the cytoplasm (*1, 44, 46, 49, 52, 69, 100–102, 110, 111, 121;* see Chapter VIII).

The possible pathways of the migration of the nucleolar material to the neural cytoplasm have been described. The composition of the product, however, was usually not indicated. According to some observations, the nucleoli are connected to strands that lead to the nuclear membrane surface (*26*). Roncoronischen fibrils which are like the nuclear ribonucleoprotein network (Chapter VIII) probably represent such pathways; these fibrils connect the nucleolus with the region of the nuclear membrane (*81*). The Feulgen-positive channels connecting the nucleolar surface with the nuclear membrane (*1*) apparently represent chromatin structures surrounding the interchromatinic areas in which the ribonucleoprotein structures are present, forming a network-like structure (see Chapter VIII). In any event, autoradiographic studies of nerve cells (*72*) showed clearly that RNA is synthesized in the nucleolus and migrates to the cytoplasm. These dynamic studies also verified quantitative cytochemical studies on RNA in different states of nerve cells, indicating that the RNA is synthesized in the nucleolus.

Some of the observations indicating the migration of the nucleolar material to the cytoplasm suggested that proteins as well as RNA could be of nucleolar origin (*58, 100*). Furthermore, autoradiographic studies on nerve cells have suggested that proteins can be synthesized in the region of the nucleolus-associated chromatin (*105*). However, better biochemical as well as cytochemical and autoradiographic studies are required in this field.

All these observations suggest that the RNA of cytoplasmic Nissl bodies is synthesized in the nucleolus. The Nissl substance consists of membranes, endoplasmic reticulum, and ribosomes responsible for its basophilic properties (*35, 93*); biochemical as well as morphological studies on nucleoli demonstrated that the nucleolus is the site of the synthesis of ribosomal RNA (see Chapters I, II, V, and VI).

V. Nucleolar Changes under Different Experimental and Pathological Conditions

Morphological as well as cytochemical studies on nucleoli in nerve cells under different experimental and pathological conditions have shown that the nucleolus is highly responsive to cellular reactions and hence, is an important indicator of a cell's function (*23, 49, 56, 58*). Similarly, Nissl substance and the region of the nuclear membrane are reactive to both normal and pathological stimuli. However, the changes in the amount and morphological organization of Nissl substance are most striking during the cell reaction to different injuries, i.e., chemical, electrical, acoustical, or mechanical injury, and to reactions to infection, radiation, and vitamin deficiencies (*49, 125*). The depletion of the Nissl substance characterized by the decrease of its staining properties, which is referred to as chromatolysis or chromatolytic reaction of the nerve cell, seems to stimulate the nucleolus to elaborate more material for replacement of the depleted RNP of this substance (*11*). Usually the nucleolus reacts by structural changes and enlargement as well as by the increase of its RNA and protein content. However, a decrease was also noted of the nucleolar basophilic staining properties, presumably a decrease of the RNA-containing structures during the marked chromatolysis (*27, 31*).

Mild chromatolysis of nerve cells corresponds to the moderately chromophobic cells of Einarson and Krogh's evaluation of basophilia in nerve cells of human brain (*27*). Moderately chromophobic cells, according to these authors, seems to be characterized by a slightly decreased stainability of the Nissl substance and by nucleolar enlargement. The extremely chromophobic cells would represent extensive chromatolytic reactions characterized by the depletion of the Nissl substance and the decrease of nucleolar stainability. Such chromophobic cells were described in the human central nervous system of patients with acute delirium, traumatic encephalopathies, symptomatic epilepsy, and after shock therapy (*27*).

A. Chemical Irritation

The nucleolar response induced by chemical irritation or intoxication is minor except for the nucleolar reaction of living Purkinje cells to treatment with ribonuclease. In these cells, the nucleolus was much more sensitive than the cytoplasmic Nissl substance (*48*). The nucleolar changes were characterized by reversible losses of the basophilic nucleolar material containing RNA. The RNA already incorporated into the Nissl substance was unaffected. Streptomycin produced a slight decrease of absorption of nerve cells of the vestibular ganglion or Deiters' nucleus (*30*). The "stimulation" of nucleoprotein production in nerve cells by malonanitrite apparently did not produce marked nucleolar changes in nerve cells of rabbits or of patients with mental disorders. In all of these cells a strong absorption by nucleoli at 2600 and 2800 Å was noted (*59*). Acute or subacute intoxication with methoxypyridoxine induced an increase of nucleolar granular components as well as cytoplasmic changes in the ultrastructural appearance of Nissl substance (*88*). The stimulation of ganglion cells of the supraoptic nucleus with sodium chloride (*25a*, *58*) increased the amount of RNA in the nucleolus as well as in the cytoplasm. However, the RNA concentration in the nucleolus decreased since its volume increased like that of the nucleus and cytoplasm.

B. Ischemic Changes of Nerve Cells

These changes, produced by arterial occlusion, are characterized by dramatic chromatolysis, the intensity of which depends on the length of the occlusion. The intensity of the chromatolysis of nerve cells is also related to the degree of the spastic paralysis of corresponding muscles (*51*). The reaction of the nucleolus is less pronounced than that of the Nissl bodies. No changes in the stainability of nucleoli with basic dyes can be observed during mild chromatolysis (*32*). When slight spastic paralysis was induced, the nucleoprotein content in the cytoplasm decreased (chromatolysis) and the nucleoprotein content increased along with the size of the nucleoli (*51, 120*). After longer ischemia, however, the nucleoprotein content of the nucleolus decreased and the RNA-containing structures in the cytoplasm disappeared (*51*). Thus RNA remains in nucleoli for longer periods than in the cytoplasm. Anoxia induced by placing experimental animals in an air-tight chamber produced chromatolysis. The nucleoli, however, were less altered and a gearlike irregularity of their shape was noted (*84*). These experiments also showed that the chromatolysis was reversible as long as the nucleoprotein content of nucleoli was maintained and the nucleolus could provide the cytoplasm with RNA.

C. Electrical Stimulation

Electrical stimulation of nerve axons produces a marked chromatolysis. After electrical stimulation, the nucleolar satellite migrates first toward the nuclear membrane and then returns during the recovery of the nerve cell (*10, 11, 18, 56, 77*). The nucleolar changes were characterized by increases of nucleolar size and vacuolization. The maximal changes of the nucleolus were noted during the period of most active restoration of depleted Nissl substance in the cytoplasm.

Experimental motor stimulation up to exhaustion of running animals results in chromatolysis in motor nerve cells. The nucleoli in these cells are enlarged and contain increased amounts of the ribonucleoproteins (*56*). In other experiments, the motor activity of swimming animals produced a slight decrease of the nucleolar stainability in motor nerve cells as well as a marked decrease of the stainability of the cytoplasmic Nissl substance (*70*). The decreased stainability of these structures was apparently due to the decrease of RNA. These changes seemed to be reversible since the number of cells with lower stainability of the Nissl substance and nucleoli decreased during the recovery of experimental animals (*70*).

Sensory stimulation by rotating experimental animals did not produce chromatolysis or nucleolar changes of nerve cells in the ganglion vestibulare or Deiters' nucleus. Repeated rotation, including fatigue, however, produced chromatolysis; the nucleolar borderline seemed blurred, but its strong absorption at 2570 Å was preserved (*37*).

The nerve cells of the cochlear ganglion reacted to acoustic stimulation by reversible chromatolysis. The nucleoli, however, apparently were not changed. On the other hand, acoustic trauma produced by a repeating series of revolver shots induced not only chromatolysis but also a decrease of nucleolar RNA. Some disintegration of nucleolus-associated chromatin was also observed in these experiments since the number of nucleoli surrounded by an incomplete ring of nucleolus-associated chromatin increased in comparison with controls (*36, 38*).

D. Mechanical Injury

The regeneration of the axon after different injuries (amputation, transsection, crush) produces striking changes in the corresponding "cell bodies" including reversible chromatolysis. During axon regeneration the nucleoli react by a reversible enlargement and vacuolization as well as by changes of their ultrastructural appearance; these nucleolar changes seem to precede those in the cell bodies (*123*). The initial ultrastructural changes of nucleoli are characterized by the decrease of light areas and an increase of granular structures producing compact nucleoli (*3, 41*). In further stages of axon regeneration a light area appears in the central part of the nucleoli (Fig. 9.27) which persists to the stage in which the depleted Nissl substance is reconstituted. In the last stages of axon regeneration, characterized by the hyperbasophilic staining properties of the cytoplasm due to the increase of cytoplasmic ribonucleoproteins, compact nucleoli reappeared (*94, 95, 95a*). These nucleoli, as well as nucleoli with a large central area, were also observed in spinal ganglion cells undergoing hypertrophy after axon regeneration (*95a*). The central light areas, however, seemed to possess characteristic RNP structures as compared with similar nucleoli during axon regeneration (Fig. 9.27c). Neither the significance of these nucleolar ultrastructural changes in nerve cells during axon regeneration and hypertrophy nor the behavior of the nucleolar satellites or accessory bodies has been clarified. The size of nucleolar satellites increased and the size of accessory bodies decreased during axon regeneration. The changes in the position of nucleolar satellites during axon regeneration were similar to those in nerve cells after electric stimulation. The nucleolar satellites which had migrated toward the nuclear membrane returned to their original position during the recovery of nerve cell (*34*).

E. Radiation

The number of reports on the nucleolar changes produced in nerve cells by radiation is limited, although numerous studies have dealt with radiation effects on the nervous system (*106, 107*). The majority of these studies are histological (*106, 107*). UV radiation produced striking changes of the nucleolar morphology in grasshopper neuroblasts (*22*);

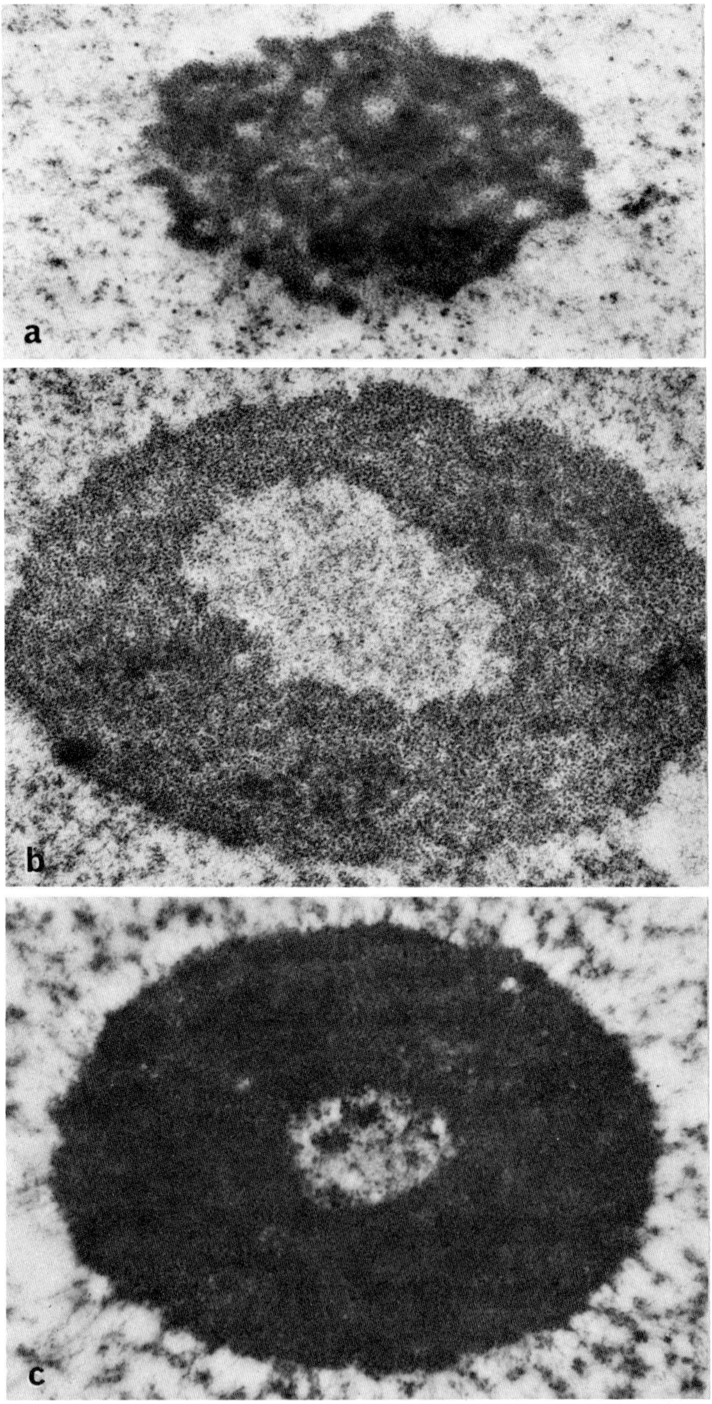
a
b
c

these were characterized by the fragmentation which could be followed either by fusion and restitution of the usual shape or by formation of small spherical bodies (*22*). Wavelengths ranging from 2650 to 2804 Å seemed to be most effective to produce such changes. Similar changes have been observed after radiation with X-rays although they were less frequent (*22*). Direct microirradiation of nucleoli in grasshopper neuroblasts resulted in the inhibition of the mitotic capacity of these cells (*97*). Single high doses of X-rays either induced chromatolysis in nerve cells accompanied by nucleolar enlargement or no marked nucleolar changes were observed (*21, 107*). The β rays apparently produced nucleolar hypertrophy and increased basophilic properties of the cytoplasm in nerve cells indicating an increase of RNA-containing structures in these cells (*20*). The irradiation of experimental animals with 185 MeV protons resulted in a chromatolysis of ganglion cells. Nucleoli in these cells initially became more compact, later lost their characteristic appearance, and finally fragmented (*5, 6*). The fine morphological changes of nucleoli were represented by the formation of protrusions on their surface and budding toward the nucleoplasm. The ringlike bodies composed of granular and/or fibrillar structures seemed to be of nucleolar origin. The ringlike bodies composed of fibrillar structures were much like those observed after treatment with actinomycin D. Some of the ringlike bodies which were composed of granular nucleolar components resembled small ring-shaped nucleoli of mature monocytes.

F. Viruses

The effect of viruses on the nucleolus of nerve cells has not been studied extensively. Poliomylitis virus produces an increase of the nucleolar size in anterior horn cells followed by chromatolysis and further degenerative changes (*57*). Rabies and neurovaccinia viruses apparently do not induce marked changes of the nucleolar body in nerve cells. However, new substances have been observed at the site of the nucleolus-associated chromatin in infected cells (*57*).

G. Expulsion of Nucleoli

Expulsion of nucleoli out of nuclei has been noted in ganglion cells of patients with poliomyelitis and in pyramidal cells of patients with

Fig. 9.27. Nucleoli of spinal ganglia cells of lizard (*Lacerta muralis*) fixed in osmium tetroxide. ×22,500. A = nucleolus of a normal cell; B = nucleolus of a nerve cell, 4 days after axon section; C = nucleolus of a nerve cell undergoing hypertrophy (courtesy of Prof. E. Pannese, II Cattedra di Anatomia Umana dell' Universita di Milano).

mental diseases (*78, 124*). The percentage of ejected nucleoli seemed to be very small in nerve cells of schizophrenic patients since less than 1% of nucleoli were outside the nucleus or in the region of the nuclear membrane. The nucleolus was ejected from nuclei either by the formation of a nuclear bud or simply by the rupture of the nuclear membrane. However, nucleolar expulsion has also been found in apparently "normal" cells (*49, 71, 80, 112*). "Nucleolar expulsion" from the nuclei of nerve cells may also be an artifact of the preparative techniques used (*49*). Nemiloff observed (*87*) that nucleolar ejection could be a result of the fixation procedure or pressure produced by the forceps during the preparation or sectioning with microtome. Electron microscopic studies on nucleolar bodies in the cytoplasm of nerve cells showed that extruded nucleoli differ from those in nuclei; i.e., they lack nucleolonemas and appear to be more compact. No relation of ejected nucleoli to the cytoplasmic organelles could be observed. However, some electron micrographs indicated a possible release of ribonucleoprotein particles from the nucleolar periphery to the adjacent cytoplasm (*112*).

VI. Conclusion

The nucleoli in nerve cells are very prominent regardless of the methods used for their visualization. The morphology, number, and size of nucleoli depend on the age and functional state of the nerve cell.

Cytochemical studies of nucleoli in nerve cells have shown that the nucleoprotein structures determine the nucleolar appearance in light microscopic as well as electron microscopic preparations. Beside RNA and DNA, which are associated with the nucleolus, acidic and basic proteins as well as several enzymes and inorganic substances have been found in the nucleoli of nerve cells.

Light microscopic structural studies of nerve cells indicated that the nucleolus can appear to be a homogeneous structure or it may possess vacuoles, granules, and trabecular structures. The morphological appearance of the nucleololus seems to depend on the method used for its demonstration.

Electron microscopic studies demonstrated that the nucleolus in nerve cells is usually composed of nucleolonemas and light areas, although compact nucleoli without distinct nucleolonemas were also found. The nucleololus seen in electron micrographs appears to be a vacuolelike structure containing fine filaments and granules of lower density.

The studies of nucleoli in nerve cells under different experimental and pathological conditions suggested that the nucleolus may be highly responsive to cell reactions to different stimulations or injuries. The Nissl substance in the cytoplasm, however, seems to be more sensitive. These studies also indicated that the depletion of the Nissl substance in the cytoplasm (chromatolysis of the nerve cell) was reversible so long as the nucleolar nucleoprotein content was adequate and the nucleolus could provide the cytoplasm with RNA.

The numerous fine morphological as well as cytochemical studies suggest that one principal role of the nucleolus in nerve cells is the synthesis of the basophilic component of the cytoplasmic Nissl substance.

REFERENCES

1. Altmann, H. W., Zur morphologie der wechselwirkung von kern and cytoplasma. *Klin. Wochschr.* **33**, 306–314 (1955).
2. Andres, K. H., Untersuchungen über den Feinbau von spinal ganglien. *Z. Zellforsch. Mikroskop. Anat.* **55**, 1–48 (1961).
3. Andres, K. H., Untersuchungen über morphologische veränderungen in spinalganglien während der retrograden degeneration. *Z. Zellforsch. Mikroskop. Anat.* **55**, 49–79 (1961).
4. Andres, K. H., Elektronenmikroskopische untersuchungen über präparatorisch bedingte and postmortale strukturveränderungen in spinalganglien zellen. *Z. Zellforsch. Mikroskop. Anat.* **59**, 78–115 (1963).
5. Andres, K. H., Elektronenmikroskopische untersuchungen uber strukturveränderungen in der kernen von spinalganglien zellen der rattle nach bestrahlung with 185 MEV-protonen. *Z. Zellforsch. Mikroskop. Anat.* **60**, 560–581 (1963).
6. Andres, K. H., Larsson, B., and Rexed, B., Zur morphogenese der akuten strahlenschädigungen in der kernen von spinal ganglien zellen der ratte nach bestrahlung mit 185 MeV-protonen. *Z. Zellforsch. Mikroskop. Anat.* **60**, 532–559 (1963).
7. Andrew, W., The Purkinje cell in man from birth to senility. *Z. Zellforsch. Mikroskop. Anat.* **28**, 292–304 (1938).
8. Andres, W., Cytological changes in senility in the trigeminal ganglion, spinal cord and brain of the mouse. *J. Anat.* **75**, 406–418 (1940–1941).
9. Bargmann, W., Hild, W., Ortman, R., and Schiebler, T. H., Morphologische und experimentalle untersuchungen über das hypothalamisch-hypophysäre system. *Acta Neuroveget.* (*Vienna*) **1**, 233–275 (1950).
10. Barr, M. L., and Bertram, E. G., A morphological distinction between neurones of the male and female and the behaviour of the nucleolar satellite during accelerated nucleoprotein synthesis. *Nature* **163**, 676–677 (1949).
11. Barr, M. L., and Bertram, E. G., The behaviour of nuclear structures during depletion and restoration of Nissl material in motor neurons. *J. Anat.* **85**, 171–181 (1951).
12. Barr, M. L., Bertram, E. G., and Lindsay, H. A., The morphology of the nerve cell nucleus according to sex. *Anat. Record* **107**, 283–297 (1950).
13. Beams, H. W., and King, R. S., The effects of ultracentrifuging the spinal ganglion cells of the rat with special reference to Nissl bodies. *J. Comp. Neurol.* **61**, 175–189 (1935).

14. Beheim-Schwarzbach, D., Morphologische beobachtungen an nervenzellen. *J. Hirnforsch.* **2**, 1–35 (1955–1956).
15. Bodian, D., and Mellors, R. C., The regenerative cycle of microneurons with special reference to phosphatase activity. *J. Exptl. Med.* **81**, 469–489 (1945).
16. Brattgard, S. O., and Hydén, H., Mass, lipids, pentose nucleoproteins and proteins determined in nerve cells by X-ray microradiography. *Acta Radiol.* Suppl. 94, 1–48 (1952).
17. Brüning, E. J., Zur orthologie und pathologie des plexus coeliacus beim menschen. *Acta Neuroveget.* (*Vienna*) **17**, 40–62 (1958).
18. Brusar, A., A propos de la structure du noyau de la cellule nerveuse. *Anat. Anz.* **98**, 343–352 (1952).
19. Buttlar-Brentano, V. K., Zur lebensgeschichte des nucleus basalis tuberomammilaris, supraopticus und paraventricularis unter normalen und pathogenen bedingungen. *J. Hirnforsch.* **1**, 337–419 (1954–1955).
20. Cambell, B., and Novick, R., Effects of beta rays on central nervous tissues. *Proc. Soc. Exptl. Biol. Med.* **72**, 34–38 (1949).
21. Cambell, B., Peterson, S. C., and Novick, R., Early changes induced in Purkinje cells of rabbit by single massive doses of roentgen rays. *Proc. Soc. Exptl. Biol. Med.* **61**, 353–355 (1946).
22. Carlson, J. G., and McMaster, R. D., Nucleolar changes in the grasshopper neuroblast by different wavelength of ultraviolet radiation and their capacity for photorecovery. *Exptl. Cell Res.* **2**, 434–444 (1951).
23. Cook, W. H., Walker, J. H., and Barr, M. L., A cytological study of transneuronal atrophy in the cat and rabbit. *J. Comp. Neurol.* **94**, 267–291 (1917).
24. Cravioto, H., Elecktronenmikrokopische untersuchungen an sympatischen nervensystem des menschen. I. Nervenzellen. *Z. Zellforsch. Mikroskop. Anat.* **58**, 312–330 (1962).
25. Dutta, C., Siegesmund, K. A., and Fox, C. A., Light and electron microscope observation on an intranucleolar body in nerve cells. *J. Ultrastruct. Res.* **8**, 542–551 (1963).
25a. Edström, J. E., and Eichner, D., Quantitative ribonukleinsäure-Untersuchungen an den Ganglienzellen des Nucleus supraopticus der Albino-Ratte unter experimentellen Bedingungen [kochsalz-Belastung]. *Z. Zellforsch. Mikroskop. Anat.* **48**, 187–200 (1958).
26. Einarson, L., Notes on the morphology of the chromaphil material of nerve cells and its relation to nuclear substances. *Am. J. Anat.* **53**, 141–175 (1933).
27. Einarson, L., and Krogh, E., Variations in the basophilia of nerve cells associated with increased cell activity and functional stress. *J. Neurol., Neurosurg., Psychiat.* [N.S.] **18**, 1–12 (1955).
28. Eränkö, O., Histochemical evidence of intense phosphatase activity in the hypothalamic magnocellular nuclei of the rat. *Acta Physiol. Scand.* **24**, 1–6 (1951).
29. Estable, C., and Sotelo, J. R., Una nueva estrutura cellulare: El nucleolonema. *Publ. Inst. Invest. Sci. Biol.* **1**, 105–126 (1951).
30. Floberg, L. E., Hamberger, C. A., and Hydén, H., Inhibition of nucleic acid production in vestibular nerve cells by streptomycin. *Acta Oto-Laryngol.* Suppl. 75, 36–52 (1949).
31. Gersch, J., and Bodian, D., Some chemical mechanisms in chromatolysis. *J. Cellular Comp. Physiol.* **21**, 253–279 (1943).
32. Gomer, L., and Pike, F. H., The histological changes in nerve cells due to

total temporary anaemia of the central nervous system. *J. Exptl. Med.* **11,** 257 (1909).

33. Graham, M. A., Sex chromatin in cell nuclei of the cat from the embryo to maturity. *Anat. Record* **119,** 469–491 (1954).
34. Haggar, R. A., Behavior of the accessory body of Cajal during axon reaction. *J. Comp. Neurol.* **108,** 269–283 (1957).
35. Haguenau, F., and Bernhard, W., Aspect de la substance de Nissl au microscope electronique. *Exptl. Cell Res.* **4,** 496–498 (1953).
36. Hamberger, C. A., and Hydén, H., Changes in the cochlear ganglion caused by acoustic stimulation and trauma. *Acta Oto-Laryngol.* Suppl. 61, 1–89 (1945).
37. Hamberger, C. A., and Hydén, H., Transneuronal chemical changes in ascites nucleus. *Acta Oto-Laryngol.* Suppl. 75, 82–114 (1949).
38. Hamberger, C. A., Hydén, H., and Nillsson, G., The correlation between cytochemical changes in the cochlear ganglion and functional tests after acoustic stimulation and trauma. *Acta Oto-Laryngol.* Suppl. 75, 124–133 (1949).
39. Harting, K., Uber das grossenverhaltnis von kernkörperchen zu kernen in sympatischen nervenzellen des menschen. *Z. Zellforsch. Mikroskop. Anat.* **36,** 361–470 (1951).
40. Harting, K., Zum bau und grossenbeziehung des nucleolus vegetativer ganglien zellen. *Anat. Anz.* **100,** 375–379 (1953–1954).
41. Hartmann, J. F., Electron microscopy of motor nerve cells following section of axones. *Anat. Record* **118,** 19–34 (1954).
42. Hartmann, J. F., An electron optical study of sections of central nervous system. *J. Comp. Neurol.* **99,** 201–249 (1953).
43. Hartmann, O., Über das verhalten der zell, kern und nukleolengrösse. *Arch. Exptl. Zellforsch. Gewebezuecht.* **15,** 1–94 (1919).
44. Hatai, S., A note on the significance of the form and contents of the nucleus in the spinal ganglion cells of the foetal rat. *J. Comp. Neurol.* **14,** 27–48 (1904).
45. Havet, J., L'origine des nucléoles vrais ou plasmosomes des cellules nerveuses. *Anat. Anz.* **29,** 258–266 (1906).
46. Hertl, M., Zum nukleolus-problem. *Z. Zellforsch. Mikroskop. Anat.* **46,** 18–51 (1957).
47. Hess, A., The fine structure of young and old spinal ganglia. *Anat. Record* **123,** 399–423 (1955).
48. Hild, W., The effect of ribonuclease on living neurons. *Anat. Record* **127,** Suppl., 308–309 (1957).
49. Hild, W., Das neuron. *In* "Handbuch der mikroskopischen Anatomie des Menschen" (W. von Möllendorf and W. Bergmann, eds.), Vol. 4, Sect. 1, Part 4, pp. 1–184. Springer, Berlin, 1959.
50. Hintzsche, E., Die verteilung anorganischer stoffe im nervengewebe. *Z. Mikroskop.-Anat. Forsch.* **46,** 203–222 (1939).
51. Hochberg, J., and Hydén, H., The cytochemical correlate of motor nerve cells in spastic paralysis. *Acta Physiol. Scand.* **17,** Suppl. 60, 1–63 (1949).
52. Holmgreen, E., Zur kenntnis der spinal ganglien zellen des kaninchen und des frosches. *Anat. Anz.* **16,** 161–171 (1899).
53. Höpker, W., Uber den nucleolus der nervenzelle. *Z. Zellforsch. Mikroskop. Anat.* **38,** 218–229 (1953).

54. Horstmann, E., and Knoop, A., Zur struktur des nucleolus und des kerns. *Z. Zellforsch. Mikroskop. Anat.* **46,** 100–107 (1957).
55. Hossack, J., and Wyburn, G. M., Electron microscopic studies of spinal ganglion cells. *Proc. Roy. Soc. Edinburgh* **B65,** 239–250 (1954).
56. Hydén, H., Protein metabolism in nerve cell during growth and function. *Acta Physiol. Scand.* **6,** Suppl. 17, 1–136 (1943).
57. Hydén, H., The nucleoproteins in virus reproduction. *Cold Spring Harbor Symp. Quant. Biol.* **12,** 104–114 (1947).
58. Hydén, H., The neuron. *In* "The Cell" (J. Brachet and A. E. Mirsky, eds.), Vol. 4, pp. 215–324. Academic Press, New York, 1960.
59. Hydén, H., and Hartelius, H., Stimulation of the nucleoprotein production in nerve cells by malononitrite and its effect on psychic functions in mental disorders. *Acta Psychiat. Neurol.* Suppl. 48, 1–117 (1948).
60. Hydén, H., and Larsson, S., The application of a scanning and computing cell analyser to neurocytological problems. *J. Neurochem.* **1,** 134–144 (1956).
61. Iijima, K., Shantha, T. R., and Bourne, G. H., Enzymehistochemical studies on the hypothalamus with special reference to the supraoptic and paraventricular nuclei of squirrel monkey (Saimiri sciureus). *Z. Zellforsch. Mikroskop. Anat.* **79,** 76–91 (1967).
62. Jerusalem, C., Wachstum und funktionelle grossenänderunden der nukleolen. *Arch. Entwicklungsmech. Organ.* **149,** 504–527 (1957).
63. Jonas, F. J., Über die nucleus-nucleolus relation. Biometrische untersuchungen an zellkern und nucleolus von nervenzellen des menschlichen sympatischen grenzstranges. *Z. Zellforsch. Mikroskop. Anat.* **35,** 333–356 (1951).
64. Kaufmann, B. P., Somatic mitosis of *Drosophila melanogaster. Z. Morphol.* **56,** 125–154 (1934).
65. Klatzo, J., Uber das verhalten des nucleolar apparates in den menschlichen pallidum zellen. *J. Hirnforsch.* **1,** 47–60 (1954–1955).
66. Köhler, H., Histologie und histopathologie der wichtigsten vegeativen ganglien unserer haussaügertiere. *Arch. Exptl. Veterinaer. med.* **6,** 373–378 (1952).
67. Köhler, H., Mehrkernigkeit in sympatischen ganglien zellen bei sehr alten tieren. *Z. Alternsforsch.* **7,** 18–24 (1953).
68. Körner, F., Variations statistische untersuchungen über die grösse der kerne und kernkörperchen menschlicher nervenzellen. *Z. Mikroskop.-Anat. Forsch.* **42,** 81–115 (1937).
69. Körner, F., Beobachtungen über den austrit geformter substanzen aus der kernkörperchen bei menschlichen nerven zellen. *Z. Mikroskop.-Anat. Forsch.* **42,** 362–378 (1937).
70. Kulenkampff, H., Das verhalten der vorder wurzenzellen der weissen maus unter dem reiz physiologischer tätigheit. *Z. Anat. Entwicklungs Geschichte* **116,** 143–156 (1951).
71. La Velle, A., Nucleolar changes and development of Nissl substance in the cerebral cortex of fetal guinea pigs. *J. Comp. Neurol.* **94,** 453–473 (1951).
72. Leblond, C. P., and Amano, M., Synthetic processes in the cell nucleus. IV. Synthetic activity in the nucleolus as compared to that in the rest of the cell. *J. Cytochem. Histochem.* **10,** 162–174 (1962).
73. Levi, G., Su alcune particolarita di struttura der nucleo delle cellule nervose. *Riv. Patol. Nervosa Mentale* **1,** 141–149 (1896).
74. Levi, G., Ricerce sperimentali sopra elementi nervosi. *Arch. Exptl. Zellforsch. Gewebezuecht.* **2,** 244–272 (1926).
75. Liisberg, M. F., Residual protein. *Acta Anat.* **53,** 240–258 (1963).

76. Limarenko, I. M., Uber die chemische zusammensetzung der tygroide im kern von nerven zellen. *Dokl. Akad. Nauk SSSR* **107**, 859–861 (1956).
77. Lindsday, H. A., and Barr, M. L., Further observations on the behaviour of nuclear structures during depletion and restoration of Nissl material. *J. Anat.* **89**, 47–63 (1955).
78. Loerbroks, E., Beobachtungen an grenzstrang ganglien poliomyelitis kranker. *Z. Mikroskop.-Anat. Forsch.* **58**, 1–36 (1952).
79. Love, R., and Bharadwaj, L. P., Two types of ribonucleoprotein in the nucleolus of mammalian cells. *Nature* **183**, 1453–1454 (1959).
80. Marx, V., and Höpker, W., Zum problem der nebennukleolen in nervenzellen. *Z. Zellforsch. Mikroskop. Anat.* **47**, 43–52 (1957).
81. Mencl, E., Einige beobachtungen über die roncoronischen fibrillen der nervenzellkerne. *Arch. Mikroskop. Anat. Entwicklungsmech.* **68**, 527–539 (1906).
82. Mjassejedoff, S. W., Das kernkörperchen und seine beziehung zu den chromatinelementen des kernes. *Z. Mikroskop.-Anat. Forsch.* **9**, 404–467 (1927).
83. Moore, K. L., and Barr, M. L., Morphology of the nerve cell nucleus in mammals, with special reference to the sex chromatin. *J. Comp. Neurol.* **98**, 213–231 (1953).
84. Morrison, L. R., Histopathologic effect of anoxia on the central nervous system. *Arch. Neurol. Psychiat.* **55**, 1–34 (1946).
85. Muhlmann, M., Mikrochemische untersuchungen au der wachsenden nervenzellen. *Arch. Mikroskop. Anat. Entwicklungsmech.* **79**, 175–205 (1912).
86. Murakami, M., Elektronenmikroskopiche untersuchung der neurosekretorische zellen im hypothalamus der maus. *Z. Zellforsch. Mikroskop. Anat.* **56**, 277–299 (1962).
87. Nemiloff, A., Beobachtungen uber die nervenelemente bei ganoiden und knochenfischen. Teil I. Bau der nervernzellen. *Arch. Mikroskop. Anat. Entwicklungsmech.* **72**, 1–46 (1908).
88. Niklowitz, W., Elektronen mikroskopische untersuchungen am ammonshorn. II. Über die substruktur der pyramiden zellen nach methoxypyridoxinvergiftung. *Z. Zellforsch. Mikroskop. Anat.* **70**, 220–239 (1966).
89. Niklowitz, W., and Bak, T. J., Elektronenmikroskopische untersuchungen an ammonshorn. I. Die normale substruktur der pyramiden zellen. *Z. Zellforsch. Mikroskop. Anat.* **66**, 529–547 (1965).
90. Nürnberger, J., Engström, A., and Lindström, B., A study of the anterior horn cells of the adult cat by two independent cytochemical micro absorption techniques. *J. Cellular Comp. Physiol.* **39**, 215–254 (1952).
91. Olszewski, J., Über die gestaltung der randkörperchen der nucleoli einiger menschlicher nervenzellarten. *J. Hirnforsch.* **1**, 206–207 (1954–1955).
92. Ortmann, R., Histochemische untersuchungen an den kerneinschlusskörpern der ganglien zellen im nucleus praeopticus des karpfens (cyprinus carpio, Linne). *Z. Anat. Entwicklungsgeschichte* **119**, 485–499 (1956).
93. Palay, S. L., and Palade, G. E., The fine structure of neurons. *J. Biophys. Biochem. Cytol.* **1**, 69–88 (1955).
94. Pannese, E., Electron microscopic study of ganglion nerve cells undergoing hypertrophy. *Proc. 5th Intern. Congr. Electron Microscopy, Philadelphia, 1962,* Vol. 2, ss-5. Academic Press, New York, 1962.
95. Pannese, E., Investigations on the ultrastructural changes of the spinal ganglion neurons in the course of axon regeneration and cell hypertrophy. I. Changes during axon regeneration. *Z. Zellforsch. Mikroskop. Anat.* **60**, 711–740 (1963).

95a. Pannese, E., Investigations on the ultrastructural changes of the spinal ganglion neurons in the course of axon regeneration and cell hypertrophy. *Z. Zellforsch. Mikroskop. Anat.* **61**, 561–586 (1963).
96. Paweletz, N., Siebs, W., and Lettre, R., Untersuchungen zur argentataffin reaction des nukleolus. *Z. Zellforsch. Mikroskop. Anat.* **76**, 577–605 (1967).
97. Peters, K., Nucleolus. *In* "Strahlenpathologie der Zelle" (E. Scherer and H. S. Stender, eds.), pp. 32–46. Thieme, Stuttgart, 1963.
98. Preto-Parvis, V., Modificazioni del nucleolo e del contenuto in acids nucleinici durante lo sviluppo di diversi tipi di cellule nervose. *Atti Soc. Ital. Anat.* **60**, 258–262 (1952).
99. Preto-Parvis, V., Fenomeni nucleari e modificazioni citochimiche nello sviluppo di grandi neuroni. *Z. Zellforsch. Mikroskop. Anat.* **39**, 550–587 (1954).
100. Reiser, K., Die nervenzelle. *In* "Handbuch der mikroskopischen Anatomie des Menschen" (W. von Möllendorf and W. Bargmann, eds.), Vol. 4, Part 4, pp. 185–514. Springer, Berlin, 1959.
101. Saguchi, S., "Zytologische studien," Nos. 1–2. Tokyo, Kanazawa Med. Coll. 1927–1928.
102. Scharf, J. H., Sensiblen ganglien. *In* "Handbuch der mikroskopischen Anatomie des Menschen" (W. von Möllendorf and W. Bargmann, eds.), Vol. 4, Part 3, pp. 1–485. Springer, Berlin, 1958.
103. Schmekel, L., and Wechsler, W., Elektronenmikroskopische untersuchungen an cerebropleural ganglien von nudibranchiern. *Z. Zellforsch. Mikroskop. Anat.* **89**, 112–132 (1968).
104. Schrön, O., Über das korn in keimfleck and im den kernkörperchen der ganglienzellen bei der säugethieren. *Untersuch. Naturlehre Menschen Tiere* **9**, 209–216 (1865).
105. Schultze, B., Oehlert, W., and Maurer, W., Autoradiographische untersuchung zum mechanismu der eiweissneubildung in ganglien zellen. *Beitr. Pathol. Anat. Allgem. Pathol.* **120**, 58–84 (1959).
106. Schummelfeder, N., Zentralnervensystem. *In* "Strahlenpathologie der Zell" (E. Scherer and H. S. Stender, eds.), pp. 292–316. Thieme, Stuttgart, 1963.
107. Schummelfeder, N., Ebschner, K. T., Bergedar, H. D., and Krogh, E., Strahlennekrose am kleinhirn nach einzeitiger, hochdosierter röntgenbestrahlung. *Arch. Pathol. Anat. Physiol.* **334**, 32–55 (1961).
108. Schwalbe, G., Uber den bau der spinalganglien nebst bemerkungen uber die sympatischen ganglienzellen. *Arch. Mikroskop. Anat. Entwicklungsmech.* **4**, 45–72 (1868).
109. Scott, G. H., On the structure, microchemistry and development of nerve cells, with special reference to their nuclein compounds. *Trans. Roy. Can. Inst.* **6**, 405–438 (1899).
110. Seite, R., Elaborations figurées dans la cellule ganglionaire vegetative. *Arch. Anat. Microscop. Morphol. Exptl.* **44**, 89–139 (1955).
111. Seite, R., Chambot, G., and Picard, D., Images d'une neurosecretion d'origine nucléaire dans les ganglions du systeme nerveux vegetatif chez le chat. *Compt. Rend. Soc. Biol.* **148**, 558–560 (1954).
112. Shimizu, N., and Ishii, S., Electron-microscopic observations on nucleolar extrusion in nerve cells of the rat hypothalamus. *Z. Zellforsch. Mikroskop. Anat.* **67**, 367–372 (1965).
113. Smirnow, A. E., Einige beobachtungen über den bau der spinal ganglien zellen bei einem viermonatlichen menschlichen embryo. *Arch. Mikroskop. Anat. Entwicklungsmech.* **59**, 459–470 (1902).

114. Smith, W. K., and Luttrell, C., The acid phosphatase content of normal autonomic neurons and its increase following section of the cell processes. *Anat. Record* **97**, 426–427 (1947).
115. Stevens, B. J., The fine structure of the nucleolus during mitosis in the grasshopper neuroblast cells. *J. Cell Biol.* **24**, 349–368 (1965).
116. Sulkin, N. M., and Kunz, A., A histochemical study of the autonomic ganglia of the cat following prolonged preganglionic stimulation. *Anat. Record* **108**, 255–277 (1950).
117. Tewari, H. B., and Bourne, G. H., The histochemistry of the nucleus with reference to nucleo-cytoplasmic relations in the spinal ganglion neuron of the rat. *Acta Histochem.* **13**, 323 (1962).
118. Timofeew, D., Beobachtungen über der bau den nervenzellen der spinal ganglien und des sympaticus beim vogel. *Intern. Monatsschr. Anat. Physiol.* **15**, 259–268 and 273–281 (1898).
119. Trujillo-Cenóz, O., Some aspects of the structural organization of the arthropod ganglia. *Z. Zellforsch. Mikroskop. Anat.* **56**, 649–682 (1962).
120. Tureen, L. L., Effect of experimental temporary vascular inclusion on the spinal cord, correlation between structural and functional changes. *Arch. Neurol. Psychiat.* **35**, 789–807 (1936).
121. Vogt, C. U. O., Eine neurohistologische beleuchtung der nucleolus function. *Biol. Zentr.* **65**, 61–69 (1946).
122. Vogt, C. U. O., Lebensgeschichte, Funktion und Tätigkeitsregulierung des nucleolus. *Arzneimittel-Forsch.* **1**, 8–14 and 43–50 (1947).
123. Watson, W. E., The nucleic acid content of neuronal nucleoli and of neuronal cell bodies after axolony. *J. Physiol.* (*London*) **194**, 66–67 (1968).
124. Weber, A., Expulsions des nucléoles hors des noyaux dans les cellules pyramidales du cortex prefrontal chez malades attients de troubles mentaux. *Arch. Neuroveget.* **13**, 4–17 (1956).
125. Wehling, H., Morphologische veränderungen an motorischen vorderhonzellen von frosch und kröte nach application von trypaflavin. *Z. Zellforsch. Mikroskop. Anat.* **36**, 171–197 (1951).
126. Zambrano, D., and Mordoh, J., Neurosecretory activity in supraoptic nucleus of normal rats. *Z. Zellforsch. Mikroskop. Anat.* **73**, 405–413 (1966).

Part III NUCLEOLI OF LIVER CELLS

I. Introduction

Nucleoli of liver cells have been studied extensively both by morphological and biochemical techniques. Since livers can be obtained in satisfactory quantities from small laboratory animals and are susceptible to both physiological and pathological changes, they serve as very convenient models (*17*). The introduction of methods for isolation of nucleoli (Chapter XI) has also facilitated a correlation of morphological and biochemical efforts to explore the role of nucleoli in liver cells. Electron and light microscopic methods have been used to control the purity of nucleolar preparations as well as to study nucleolar fine morphology and cytochemistry (*12, 20, 33, 38, 48, 65*).

Nucleoli in liver cells are very prominent and can be readily visualized by the usual cytological procedures, although there may be difficulties in distinguishing these structures from large chromatin clusters present in some cells. Accordingly, the best results for the reliable demonstration of nucleoli have been achieved by procedures (Chapter XII) which preferentially stain RNA-containing structures in the cell (*8a, 38a, 47*) (Fig. 2.1). In phase-contrast microscopy, nucleoli in liver cells appear as large dense bodies which can be easily identified in most nuclei.

The number of nucleoli in adult liver cells usually ranges from one to seven in most nuclei although as many as 11 or more nucleoli have been observed per nucleus (*47*). Nucleoli of human embryonic hepatoblasts contain one to three nucleoli and in rat embryonic liver cells the mean number of nucleoli per nucleus was 1.3 (*21, 66*). In adult human and rat liver two to three nucleoli were present in most hepatocytes (*53, 66*). The nucleoli in adult animals, however, were smaller as compared with those in embryonal liver cells (*66*). In adult mice, a mode of three nucleoli per nucleus has been observed in diploid nuclei; in tetraploid nuclei, the mode of nucleoli per nucleus was six (*47*). Nucleoli of senile liver cells are enlarged and characterized by increases in nucleolar number. In human senile hepatocytes, more than 12 nucleoli could be found in some nuclei (*4*). In senile mice, the enlarged nuclei contain more than six nucleoli and the nucleolar number occasionally increased up to 30 nucleoli in such nuclei (*4*).

The nucleolar diameter in liver cells ranges from 0.8 to 1.8 μ and the nucleolar volume ranges from 1 to 8 μ^3 (*14, 27, 28, 47, 66*). The variation of the values of the nucleolar volume is partly related to the different methods under which the measurement was made including the preparation of specimens for microscopy (*39*).

II. Cytochemistry

Light microscopic studies of liver cells by means of Feulgen reaction and UV-absorption methods indicated that both DNA and RNA are present in the nucleolus. Feulgen staining is more pronounced in the peripheral part of the nucleolus as is the staining with basic dyes or UV absorption after digestion with ribonuclease (*17, 17a*). The peripheral staining is related to the presence of the perinucleolar nucleolus-associated chromatin which forms a shell around the nucleolar body (*20, 29, 31, 32, 38a, 65*). Examination of nucleoli in liver cells at lower magnification might suggest that the nucleoli are Feulgen positive throughout (*20, 29, 43, 68*). However, the relatively large area of the nucleolus-associated chromatin is probably responsible for this effect; the more intense staining of the nucleolar periphery indicates that the DNA-containing structures are largely perinucleolar (*68*). Cytochemical studies also indicated that the RNA-containing structures are responsible for the basophilic staining properties of nucleoli in liver cells as in other cell types (*68*); the cytochemical methods used for the demonstration of proteins showed high concentrations of protein in liver cell nucleoli (*71*).

The presence of alkaline phosphatases has been demonstrated in nucleoli of rat liver cells, including cirrhotic livers and hepatomas, by means of Gomori's methods (*44, 59*). Increased activity of ATPase in nucleoli of liver cells has been noted after treatment of experimental animals with 2,4 dinitrophenol or Janus green (*69*). An interesting observation has been made by Vorbrodt (*70*) that GTPase in nucleoli of liver cells in adult rats is bound to both nucleolar fibrillar and granular components. In Morris hepatomas, the activity of this enzyme was lower and it was absent in liver cells of embryos. Succinic dehydrogenase has been detected in normal rat hepatocytes (*19*). Glycogen and lipids were rarely observed in nucleoli of liver cells (*63, 74*).

III. Ultrastructure

The ultrastructural appearance of nucleoli in "normal" liver cells is characterized by the presence of distinct nucleolonemas well separated by light areas (*2, 7–9, 16, 18, 23, 24, 45, 50*) (Figs. 2.5, 2.18b). The appearance of nucleolar ultrastructure depends on the fixation used. The nucleolonemas are more prominent after fixation with osmium tetroxide. After fixation with formaldehyde (*2*) or glutaraldehyde (*64*), nucleoli in liver cells appear to be more compact although nucleolonemas and light areas can be seen. Nucleolonemas seem to be branching and anastamosing fiber-like structures about 0.09–0.2 μ in width; they contain fine ribonucleoprotein structures embedded in the protein matrix. These ribonucleoprotein structures have been described in liver cell nucleoli as granules or filaments containing structures of variable density. Ultrastructural cytochemical studies have shown that both granular and fibrillar RNA-containing structures are present in nucleoli of liver cells as in nucleoli of other cell types (see Chapter II) (*34*). The areas containing fibrillar components represented about 19% of the whole nucleolar area and the areas containing granular components were about 77% of the whole nucleolar area (Fig. 2.18a) (*50*). Compact nucleoli without nucleolonemas have been observed in salamander liver cells (*24*). Such nucleoli are composed of a homogeneous central region surrounded by a peripheral zone containing numerous ribonucleoprotein particles (*24*).

Nucleoli of embryonic rat liver cells are larger and may be more compact than those of adult rats (*22*). Nevertheless, the nucleolonemas are readily identified and the ratio of areas containing fibrillar components to the total nucleolar area was almost the same as in adult liver nucleoli (*66*).

Nucleoli in liver cells of senile mice also contain normal nucleolonemas or they resemble rings with a large central area (*3*). Although the significance of the ringlike nucleoli in liver cells of senile animals has not been investigated, similar nucleoli are produced in liver cells experimentally by inhibition of RNA synthesis with a low dose of actinomycin D injected into the experimental rats (150 μg/kg body weight) (*49*). Both nucleoli with nucleolonemas and ring-shaped nucleoli have been observed in a single nucleus of senile animals (*3*). Since ring-shaped nucleoli have low activity in formation of new nucleolar RNA-containing structures, the presence of a ring-shaped nucleolus and nucleoli with nucleolonemas in one nucleus can indicate that a resting and an active nucleolus may be present in the same nucleus of liver cells in senile animals (see also p. 387).

The chromatin structures in liver cells visualized by electron microscopic procedures, are present not only at the nucleolar periphery, but also within the nucleolus (*1, 2, 48, 67*). Electron microscopic studies of isolated liver cell nucleoli demonstrated that only a part of the nucleolus-associated chromatin was removed during the isolation procedure (*48, 67*). The proximal portion of this chromatin (Fig. 2.12) could not be removed with either sonication or compression-decompression methods (Chapter XI). This effect might be explained by the observation that in liver nucleoli, the proximal portion of the nucleolus-associated chromatin is connected with the intranucleolar chromatin structures. In these nucleoli the nucleolus-associated chromatin penetrates into the nucleolus in the form of septalike structures (Fig. 2.12) (see Chapter II). All these observations provide morphological evidence that the nucleolus-associated chromatin or at least its proximal portion is an integral part of the nucleolar apparatus as Caspersson and his associates suggested on the basis of light microscopic cytochemical studies (*10*).

The intranucleolar chromatin structures (Fig. 2.24) are present in nucleoli of liver cells either in the form of clumped fibrils about 100 Å in width or in the form of dispersed fine filaments 20 Å in diameter. At very high magnifications, chromatin fibrils seem to be composed of similar filaments about 20 Å in width except that these filaments were twisted or coiled (*67*).

IV. Nucleolar Changes Under Different Experimental and Pathological Conditions

Nucleoli are small in starving animals (*29, 53*). Refeeding of starved animals produces nucleolar enlargement and an increase of the nucleolar mass (*29, 53*). When the starved animals were refed with a high protein diet, the nucleolar mass increased up to normal values; feeding with low protein diets did not restore normal liver size (*29*). The nucleolar size did not change after 6 hours of refeeding starved animals with a nonprotein diet as compared to animals refed on high protein diets. Later, however, the nucleolar size in liver cells of animals refed with nonprotein diets increased above values found in refed animals on high protein diets (*53*). The ultrastructure of enlarged liver cell nucleoli in refed animals was normal (*23*). However, nucleoli of liver cells of one refed mouse were surrounded by a "membrane" (*15*). This membrane appeared to be composed of several sublayers and its width ranged from 350 to 550 Å.

The nonprotein or low protein diet produced an increase of liver nucleolar volume compared with results on a high protein diet (*51–56, 60*). Interpherometric determinations of nucleolar dry matter indicated that in liver cells of animals on nonprotein diets the enlargement was due to increases of substances other than water (*54*). When the concentration of RNA in rat liver nucleoli was estimated by the interpherometric determination of dry matter before and after digestion with ribonuclease, the values in enlarged nucleoli of animals on nonprotein diets did not differ from those found in animals on high protein diets. These results indicate an absolute increase of RNA in the large nucleoli of liver cells in protein-deprived animals (*54*). The ultrastructural morphology of enlarged nucleoli in liver cells of animals on a low protein diet was similar to that in animals on nonprotein diets. In some enlarged nucleoli, the nucleolonemas seemed to be more prominent and separated from other nucleolar components (*60*).

A choline-deficient diet produced nucleolar changes similar to those of the low protein or nonprotein diet. However, the increase in the nucleolar volume seemed to be more pronounced (*25*).

The effect of hormones on the morphology of liver cell nucleoli has not been studied extensively although there are numerous reports on hormonal effects on RNA and proteins (*62*). In pigeons the injection of estradiol, like ovoposition, produced a threefold increase of the nucleolar volume (*13*). This nucleolar hypertrophy was accompanied by an increase of RNase-sensitive basophilic staining properties of the cytoplasm reflecting an increase of cytoplasmic RNA. Similar nucleolar hypertrophy and increases of cytoplasmic RNA have been produced in ovoviviparous male fishes by estradiol (*75*). Hypophysectomy reduces nucleolar volume and the frequency of large nucleoli in liver cells (*40*). However, the nucleolar size in hepatocytes of regenerating liver increased despite hypophysectomy. The decrease of nucleolar size and number produced by hypophysectomy was not reversed by protein deficiency which induces some nucleolar enlargement in normal animals. On the other hand, nucleolar size and number in hepatocytes of hepatectomized rats were quite close to normal values when somatotropic hormone was administered to those animals (*40*).

The Effect of Various Substances on Nucleoli in Liver Cells

There are numerous studies on the effects of various substances on liver cells. Some of them have striking effects on nucleolar composition, function, and morphology (see Chapter X). Nucleolar vacuolization and the formation of inclusion bodies by transformation of nucleoli were reported in mice treated with antithyroid drugs (*40a, 40b*). Similar inclusions, however, have been observed in hepatocytes and hepatoma cells of mice on methionine-rich diets. However, histochemical studies showed that such inclusions were of cytoplasmic origin and the Feulgen-positive layer around the inclusions did not represent the nucleolus-associated chromatin but rather resembled chromatin associated with the nuclear membrane (*31*). Although inclusions produced by treatment of experimental animals with colchicine resembled nucleoli observed with the light microscope, electron microscopy showed that these inclusions were surrounded by double-layered membranes like those of the nucleus; hence, such inclusions are of cytoplasmic origin (*72*). Similarly, the nucleolar lipid inclusions in animals treated with thioacetamide are of cytoplasmic origin (*63*); they are formed by invagination of the cytoplasm into the nucleus followed by the breaking of the nuclear membrane and migration of lipid droplets from nuclei into the nucleolus.

In regenerating liver there are marked increases of nucleolar activities as demonstrated by morphological and biochemical methods (see Chapters II and V). The nucleoli enlarge, decrease in number, and increase in mean mass and RNA (*11, 35–37, 58*). The shape of these enlarged nucleoli is less regular but the inner nucleolar structure and the ratios of fibrillar components to the whole nucleolar areas are apparently unchanged (Fig. 2.19) (*27, 36, 50*). The nucleolar changes which are associated with increased nucleolar function may be enhanced or repressed. Actinomycin D or aflatoxin represses the increases in nucleolar size induced by the partial hepatectomy or thioacetamide. The resulting nucleoli are small and the nucleolar labeling with cytidine and lysine decreased (*41, 57*). Moreover, the nucleoli are more compact and the nucleolonemas decrease or are absent (*57*). On the other hand, thioacetamide enhanced the nucleolar enlargement produced by partial hepatectomy and a great increase of RNA was noted in these nucleoli (*27, 37*). The administration of aflatoxin, inhibiting compensatory changes in regenerating liver, produced segregation of nucleolar components (*6*). Thioacetamide, 4-dimethylaminoazobenzene, and 3-methyl-4-dimethylaminoazobenzene produced marked anisonucleolinosis in nucleoli of liver cells stimulated by partial hepatectomy (*37*). The anisonucleolinosis in nucleoli of liver cells stimulated by partial hepatectomy was also observed after total body irradiation with X-rays (*36*).

Radiation produced either nucleolar enlargement or fragmentation in liver cells depending on the dose. Total body irradiation with X-rays (400 r) induced nucleolar enlargement (*46*). On the other hand 800–9600 r produced nucleolar fragmentation (*16*). Local irradiation of the liver produced necrosis of liver cells at the periphery of lobules or portal areas; nucleoli in the other cells were enlarged (*73*). The long term irradiation of experimental animals with strontium (^{90}Sr) increased the nucleolar number, nucleolar vacuoles, and nucleolar size in liver cells (*26*).

Studies on liver cells in various forms of infectious hepatitis have shown that nucleoli in these cells usually increase in size (*5, 14, 42, 45, 61*). However, in liver cells of mice infected by *Ectromelia* virus, fragmentation of nucleoli was observed (*30*). In addition, very dense bodies of various sizes appeared within nucleoli and occasionally nucleolar fragments were scattered in the nucleoplasm (*30*). Infection of experimental animals with contagious canine hepatitis produced inclusions associated with the nucleolus (*42*). These inclusions were formed around nucleoli which were initially enlarged. Later, the nucleoli decreased in size and were surrounded by inclusion bodies which also decreased in size during their further development. The inclusions surrounding nucleoli were Feulgen positive and probably represented viruses or products utilized for virus reproduction (*42*). In human liver cells of patients with virus hepatitis, changes in the nucleolar size and number have been reported (*14*), but the nucleolar ultrastructure was unaltered. The nucleolar diameter in normal persons ranged from 0.8 to 1.8 μ and in patients with hepatitis ranged from 1.5 to 2.3 μ. The maximal nucleolar diameter observed in normal persons was 2.7 μ and in patients with hepatitis, 3.3 μ. In sections of diseased liver, more nucleoli (two nucleoli per nucleus) were found as compared to normal liver (one nucleolus per 1.5 nucleus) (*14*).

V. Conclusion

Since nucleoli in liver cells are prominent, they can be visualized by commonly employed cytological procedures. Methods for demonstration of RNA-containing structures provide optimal results. Chromatin structures are present not only around nucleoli, but also within these organelles as shown by light microscopic and electron microscopic procedures. In addition to proteins, several enzymes have been found in nucleoli of liver cells. Occasionally glycogen and lipids are observed in liver cell nucleoli. Electron microscopic studies have shown that nucleoli in liver cells are usually composed of nucleolonemas and small light areas. The ultrastructure of chromatin structures and ribonucleoprotein fibrillar or granular components is similar to that of other types of cells.

Studies on liver cells under different experimental conditions or pathological states have shown that marked changes of the nucleolar morphology occur when the synthesis of nucleic acids or proteins is either stimulated or inhibited. The mechanisms of these nucleolar changes are discussed in Chapters V and X.

REFERENCES

1. Aldridge, W. C., and Watson, M. L., Perchloric acid extraction as a histochemical technique. *J. Histochem. Cytochem.* **11,** 773–781 (1963).
2. Altmann, H., Stöcker, E., and Thoenes, W., Über chromatin and DNS-synthese in nukleolus. Elektronenmikroskopische, autoradiographische und lichtmikroskopische untersuchungen an leberzellen von ratten. *Z. Zellforsch. Mikroskop. Anat.* **59,** 116–133 (1963).
3. Andrew, W., An electron microscope study of age changes in the liver of mouse. *Am. J. Anat.* **110,** 1–18 (1962).
4. Andrew, W., Brown, H. M., and Johnson, J. B., Senile changes in the liver of mouse and man, with special reference to the similarity of the nuclear alterations. *Am. J. Anat.* **72,** 199–221 (1943).

5. Bearcroft, W. G. C., The liver in infective hepatitis. *Nature* **190,** 549–550 (1961).
6. Bernhard, W., Frayssinet, C., Lafarge, C., and Le Breton, E., Lesions nucléolaires précoces provoquées par l'aflatoxine dans les cellules hépatoques du rat. *Compt. Rend.* **261,** 1785–1788 (1965).
7. Bernhard, W., Bauer, A., Cropp, A., Haguenau, F., and Oberling, C., L'ultrastructure du nucléole de cellules normales et cancéreuses. Etude au microscope électronique. *Exptl. Cell Res.* **9,** 88–100 (1955).
8. Borysko, E., and Bang, F. B., Structure of the nucleolus as revealed by the electron microscope. *Bull. Johns Hopkins Hosp.* **89,** 468–473 (1951).
8a. Busch, H., Muramatsu, M., Adams, H. R., Smetana, K., Steele, W. C., and Liau, M. C., Isolation of nucleoli. *Exptl. Cell Res.* Suppl. 9, 150–163 (1963).
9. Cachera, R., and Darnis, F., Examen du foie humain (prélevé par ponction biopsie) au microscope électronique. *Rev. Franc. Etudes Clin. Biol.* **1,** 296–317 (1956).
10. Caspersson, T. O., "Cell Growth and Cell Function: A Cytochemical Study." Norton, New York, 1950.
11. Castro, S., and Foraker, A., Nucleolar mass in rat livers. *Arch. Pathol.* **74,** 495–498 (1962).
12. Clarke, J. A., Salsbury, A. J., and Rowland, G. F., Surface ultrastructure of human leucocytes, mouse macrophages and rat liver cells and of isolated nuclei and nucleoli. *Brit. J. Haematol.* **14,** 533–542 (1968).
13. Clavert, J., and Randavel, C., Action de la folliculine sur le foie de Pigeon. Modifications histologiques du noyau de la cellule hépatique et de la basophilie du cytoplasm. *Compt. Rend. Soc. Biol.* **141,** 1258–1259 (1947).
14. Cossel, L., Elektronenmikroskopische befundean den leberepithelzellen bei virus hepatitis. *Acta Hepato-Splenol.* **8,** 333–355 (1961).
15. David, H., Uber einen nukleolus mit membran. *Z. Zellforsch. Mikroskop. Anat.* **53,** 50–54 (1960).
16. David, H., "Submikroskopische ortho-und pathomorphologie der leber." Akademie Verlag, Berlin, 1964.
17. Davidson, J. N., Some factors influencing the nucleic acid content of cells and tissues. *Cold Spring Harbor Symp. Quant. Biol.* **12,** 50–59 (1947).
17a. Davidson, J. N., and Waymouth, C., The nucleoproteins of the liver cells demonstrated by ultraviolet microscopy. *J. Physiol. (London)* **105,** 191–196 (1946).
18. Davis, J. M., The ultrastructure of the mammalian nucleolus. *In* "The Cell Nucleus" (J. S. Mitchell, ed.), pp. 3–13. Academic Press, New York, 1960.
19. De, P., Chatterjee, R., and Mitra, S., Nucleolar localization of succinic dehydrogenase in normal rat hepatic cells. *Nature* **191,** 925 (1961).
20. Desjardins, R., Smetana, K., and Busch, H., A large scale procedure for isolation of nucleoli from rat liver. *Exptl. Cell Res.* **40,** 127–137 (1965).
21. Du Bois, A. M., The embryonic liver. *In* "The Liver" (C. Rouiller, ed.), Vol. 1, pp. 1–39. Academic Press, New York, 1963.
22. Dvorak, M., Elektronmikroskopische untersuchungen an embryonalen leberzellen. *Z. Zellforsch. Mikroskop. Anat.* **62,** 655–666 (1964).
23. Fawcett, D. W., Observations on the cytology and electron microscopy of hepatic cells. *J. Natl. Cancer Inst.* **15,** Suppl., 1475–1502 (1955).
24. Fawcett, D. W., "The Cell. An Atlas of Fine Structure." Saunders, Philadelphia, Pennsylvania, 1966.

25. Grisham, J. W., Banson, B. B., and Hartroff, W. S., Nuclear and nucleolar alterations in hepatic parenchymal cells of the choline deficient rat. *Arch. Pathol.* **70**, 50–58 (1960).
26. Hunstein, W., Veranderungen der rattenleber nach bestrahlung mit radioactivem strontium (Sr^{90}). *Beitr. Pathol. Anat. Allgem. Pathol.* **122**, 345–362 (1960).
27. Kleinfeld, R. G., and Haam, E., Effect of thioacetamide on rat liver regeneration. I. Cytological studies. *Cancer Res.* **19**, 769–778 (1959).
28. Koulish, S., and Kleinfeld, R. G., The role of the nucleolus. I. Tritiated cytidine activity in liver parenchymal cells of thioacetamide treated rats. *J. Cell Biol.* **23**, 39–51 (1964).
29. Lagerstedt, S., Cytological studies on the protein metabolism on the liver in rat. *Acta Anat.* **7**, Suppl. 9, 1–116 (1949).
30. Leduc, E. H., and Bernhard, W., Electron microscope study of mouse liver infected by etromelia virus. *J. Ultrastruct. Res.* **6**, 466–488 (1962).
31. Leduc, E. H., and Wilson, J. W., A histochemical study of intranuclear inclusions in mouse liver and hepatoma. *J. Histochem. Cytochem.* **7**, 8–16 (1959).
32. Macary, J., Les nucléoles vacuolaires de la cellule hépatique humaine. *Compt. Rend. Soc. Biol.* **148**, 39–41 (1954).
33. Maggio, R., Siekevitz, P., and Palade, G. E., Studies on isolated nuclei. II. Isolation and chemical characterization of nucleolar and nucleoplasmic subfractions. *J. Cell Biol.* **18**, 293–312 (1963).
34. Marinozzi, V., Cytochimie ultrastructurale du nucléole-RNA et proteins intranucléolaire. *J. Ultrastruct. Res.* **10**, 433–456 (1964).
35. Mironescu, S., and Dragomir, C., Nucleolar number variations in rat liver cells after partial hepatectomy. *Exptl. Cell Res.* **43**, 217–220 (1966).
36. Mironescu, S., and Dragomir, C., Nucleolar behavior in regenerating liver of normal and whole body irradiated rats. *Cancer Res.* **27**, 1819–1830 (1967).
37. Mironescu, S., Encut, J., Mironescu, K., and Liciu, F., Nucleolar behavior in regenerating liver of rats receiving intraabdominal injections of azo dyes and thioacetamide. *J. Natl. Cancer Inst.* **40**, 917–933 (1968).
38. Monti, K. J., Litt, M., Kay, E. R., and Dounce, A. L., Isolation and properties of liver cell nucleoli. *J. Biophys. Biochem. Cytol.* **2**, 127–145 (1956).
38a. Muramatsu, M., Smetana, K., and Busch, H., Quantitative aspects of isolation of nucleoli of Walker carcinosarcoma and liver of the rat. *Cancer Res.* **23**, 510–518 (1963).
39. Nadal, C., Polyploidie hépatique du rat. III. Etude des nucleoles hépatiques du rat normal. *Exptl. Cell Res.* **408**, 507–517 (1967).
40. Nadal, C., and Zajdela, F., Polyploidie hépatique du rat. IV. Variation expérimentale du volume nucléaire des hépatocytes leur méchanismes de regulation. *Exptl. Cell Res.* **48**, 518–528 (1967).
40a. Noel, R., and Drouin, M., Modifications nucléaires reversibles apparues dans la cellule hépatique après l'action de certains antithyroidiens de synthèse. *Compt. Rend. Assoc. Anatomistes* **39**, 1–5 (1952).
40b. Noel, R., and Drouin, M., La vacuolisation nucléolaire des cellules hépatiques provoquée chez la souris blanche par l'ingestion de thiourée. *Arch. Anat.* (*Strasbourg*) **34**, 329–338 (1951–1952).
41. Oda, A., and Chiga, M., Effect of actinomycin D on the hepatic cell of partially hepatectomized rats. *Lab. Invest.* **14**, 1419–1427 (1965).
42. Ohno, S., and Kinosita, R., Morphology of intranuclear inclusions in liver cells infected with contagious canine hepatitis. *Exptl. Cell Res.* **7**, 578–580 (1954).

43. Randavel, C., Contribution à l'étude des acides nucléiques de la cellule hépatique chez le pigeon folliculinisé. *Compt. Rend. Soc. Biol.* **142**, 235–237 (1948).
44. Ros, M., and Ely, J., Alkaline phosphatases in fixed plant cells. *Exptl. Cell Res.* **2**, 339–348 (1951).
45. Rouiller, C., and Jezequel, A. M., Electron microscopy of the liver. *In* "The Liver" (C. Rouiller, ed.), Vol. 1, pp. 195–265. Academic Press, New York, 1963.
46. Scherer, E., Cytologische und karyometrische untersuchungen zur strahlenwirkung anf leber und milz bei anwendung von total und teil bestrahlung. I. Mitteilung: Untersuchungen an der leber der weissen maus. *Strahlentherapie* **100**, 86–103 (1956).
47. Shea, J. R., and Leblond, C. P., Number of nucleoli in various cell types of the mouse. *J. Morphol.* **119**, 425–434 (1966).
48. Smetana, K., and Busch, H., Studies on the ultrastructure of the nucleoli of the Walker tumor and rat liver. *Cancer Res.* **24**, 537–557 (1964).
49. Smetana, K., and Potmesil, M., Ring-shaped nucleoli in liver cells of rats after treatment with actinomycin D. *Z. Zellforsch. Mikroskop. Anat.* **92**, 62–69 (1968).
50. Smetana, K., Shankar Narayan, K., and Busch, H., Quantitative analysis of ultrastructural components of nucleoli of the Walker tumor and liver. *Cancer Res.* **26**, 786–796 (1966).
51. Stenram, U., The nucleolar size in the liver cell of rats fed high and non protein diets. *Exptl. Cell Res.* **5**, 539–541 (1953).
52. Stenram, U., Interferometric dry matter determinations on liver nucleoli of protein-fed, protein-deprived and throid-fed rats. *Exptl. Cell Res.* **12**, 626–632 (1957).
53. Stenram, U., Nucleolar size in the liver of rats fed on high and nonprotein diets after starvation. *Acta Anat.* **26**, 352–361 (1956).
54. Stenram, U., Interpherometric determinations of the ribose nucleic acid concentration in liver nucleoli of protein fed and protein deprived animals. *Exptl. Cell Res.* **15**, 174–183 (1958).
55. Stenram, U., The volume and ribose nucleic acid concentration of nucleoli in liver and hepatoma cells of rats fed on high and nonprotein diets. *Acta Pathol. Microbiol. Scand.* **44**, 239–246 (1958).
56. Stenram, U., On protein deficiency in mice with special reference to liver nucleolar size. *Z. Zellforsch. Mikroskop. Anat.* **56**, 125–129 (1962).
57. Stenram, U., and Willen, R., The effect of actinomycin D on ultrastructure and radiographic ribonucleic acid and protein labeling in rat liver after partial hepatectomy. *Cancer Res.* **26**, 765–772 (1966).
58. Stowell, R. E., Nucleic acids and cytologic changes in regenerating rat liver. *A.M.A. Arch. Pathol.* **46**, 164–178 (1948).
59. Stowell, R. E., Lee, C. S., Tsuboi, K. K., and Villasana, A., Histochemical and macrochemical changes in experimental cirrhosis and hepatoma formation in mice by carbon tetrachloride. *Cancer Res.* **11**, 345–354 (1951).
60. Svoboda, D., and Higginson, J., Ultrastructural changes produced by protein and related deficiencies in the rat liver. *Am. J. Pathol.* **45**, 353–379 (1964).
61. Svoboda, D., Nielson, A., Werder, A., and Higginson, J., An electron microscopic study of viral hepatitis in mice. *Am. J. Pathol.* **41**, 205–224 (1962).
62. Tata, J. R., Hormones and the synthesis and utilization of ribonucleic acids. *Progr. Nucleic Acid Res. Mol. Biol.* **5**, 191–251 (1966).

63. Thoenes, W., Fett in nucleolus. *J. Ultrastruct. Res.* **10,** 194–206 (1964).
64. Trump, B. F., Goldblatt, P. J., and Stowell, R. E., Studies of mouse liver necrosis *in vitro. Lab. Invest.* **14,** 1969–1999 (1965).
65. Unuma, T., Floyd, L., and Busch, H., Selective removal of the perinucleolar nucleolus associated chromatin from nucleoli of livers of thioacetamide-treated rats. *Exptl. Cell Res.* **52,** 101–111 (1968).
66. Unuma, T., Morris, H. P., and Busch, H., Comparative studies of the nucleoli of Morris hepatomas, embryonic liver and aflatoxin B_1 treated liver of rats. *Cancer Res.* **27,** 2121–2233 (1967).
67. Unuma, T., Smetana, K., and Busch, H., A morphologic study on the chromatin areas associated with the nucleolus. *Exptl. Cell Res.* **48,** 665–671 (1967).
68. Vendrely, C., and Vendrely, R., Localisation de l'acide ribonucléique dans les différents tissus et organes de vertèbres. *In* "Handbuch der Histochemie" (W. Graumann and K. Neumann, eds.), Vol. 3, Part 2, pp. 84–244. Fischer, Stuttgart, 1959.
69. Vorbrodt, A., The effect of some inhibitors of oxidative phosphorylation on the histochemically demonstrable phosphatases. *Exptl. Cell Res.* **12,** 154–162 (1957).
70. Vorbrodt, A., Electron microscopic studies on the localization of guanosine-5-triphosphatase. *Acta Histochem.* **28,** 215–226 (1967).
71. Wachstein, M., Cyto- and histochemistry of the liver. *In* "The Liver" (C. Rouiller, ed.), Vol. 1, pp. 137–194. Academic Press, New York, 1963.
72. Wessel, W., Elektronen mikroskopische untersuchungen von intranukleären einschlusskörper. *Arch. Pathol. Anat. Physiol.* **331,** 314–328 (1958).
73. Wilson, M. E., and Stowell, R. E., Cytological changes following roentgen irradiation of the liver in mice. *J. Natl. Cancer Inst.* **13,** 1123–1137 (1953).
74. Yokoyama, H. O., Tsubai, K. K., Wilson, M. E., and Stowell, R. E., Histochemical studies on regenerating mouse liver. *Lab. Invest.* **2,** 91–108 (1953).
75. Zahnd, D., Contribution à l'étude des modifications hépatiques au cours du cycle ovarien chez deux poissons ovovivipares Xiphophorus bellini et Lebistes reticulatus. Thesis, Strasbourg. Quoted in Vendrely and Vendrely (68).

Part IV NUCLEOLI OF TUMOR CELLS

I. Introduction

Abnormalities of the nucleolar morphology were used by cytologists as criteria for the diagnosis of cancer in the nineteenth century (73). Subsequently, numerous reports on the nucleolar morphology of cancer cells have been published. The introduction of new techniques for the isolation of nucleoli from neoplastic cells has greatly enhanced the use of biochemical methods for the exploration of nucleoli in both experimental and human neoplasms (see Chapters I, II, IV–VII, XI). The most remarkable differences found between nucleoli of tumor cells and other cells are the differences in the ^{32}P nucleotide compositions and frequency distribution of oligonucleotides as shown in Chapter V.

II. Nucleolar Size

Nucleoli are usually enlarged in most neoplastic cells. The diameter of nucleoli in various cancer cells ranges from 1.5–3.0 μ, the mean nucleolar area ranges from 1.4–8.7 μ^2 and the nucleolar volume is between 10–25 μ^3 (*57, 59, 65, 94, 108, 114, 126, 134, 143*). The nucleolar hypertrophy in cancer cells usually is not accompanied by a proportional increase of the nuclear size. Therefore, the values of the nucleolus:nucleus size ratios in cancer cells differ from those found in normal cells or cells of benign lesions (*59, 64, 87, 89, 90*). A nucleolus:nucleus ratio greater than 0.25 and the presence of nucleoli with diameters greater than one third of the nuclear diameter can be helpful in the identification of malignant (neoplastic) cells (*47, 95*). The enlargement of nucleoli, however, is not pathognomonic for cancer since large nucleoli have been found in growing or protein-secreting non-neoplastic cells as well as in cells infected with viruses (*19, 22, 24, 26, 55, 112*).

Nucleolar enlargement has also been observed during the cell cycle of normal cells (*50*); the nucleolus:nucleus volume ratios may not be significantly different between cells of normal or hyperplastic tissues and those of benign or malignant tumors (*57*); the nucleolar size in cells of adjacent nontumor tissues may be increased by the growing tumor (*60, 100*). The effects of therapy should also be considered, since radiation or some drugs may significantly influence the nucleolar size in neoplastic cells. Despite these considerations, studies of malignant tumor and "corresponding" normal cells indicated in general that nucleolar size, mean nucleolar area, and nucleus:nucleolus area ratios show increases in many malignant tumor cells.

The nucleolar size in neoplastic cells may be related to their proliferative activity (Figs. 9.28 and 9.29). Caspersson (*24, 25*) reported that the nucleolar volume is relatively small in neoplastic cells of the A-type (suggested to be extremely active cells at the periphery of the growing tumor) as compared with less active cells of B-type (suggested to be in the deeper portions of the tumor). Similarly, Siering and Ader-

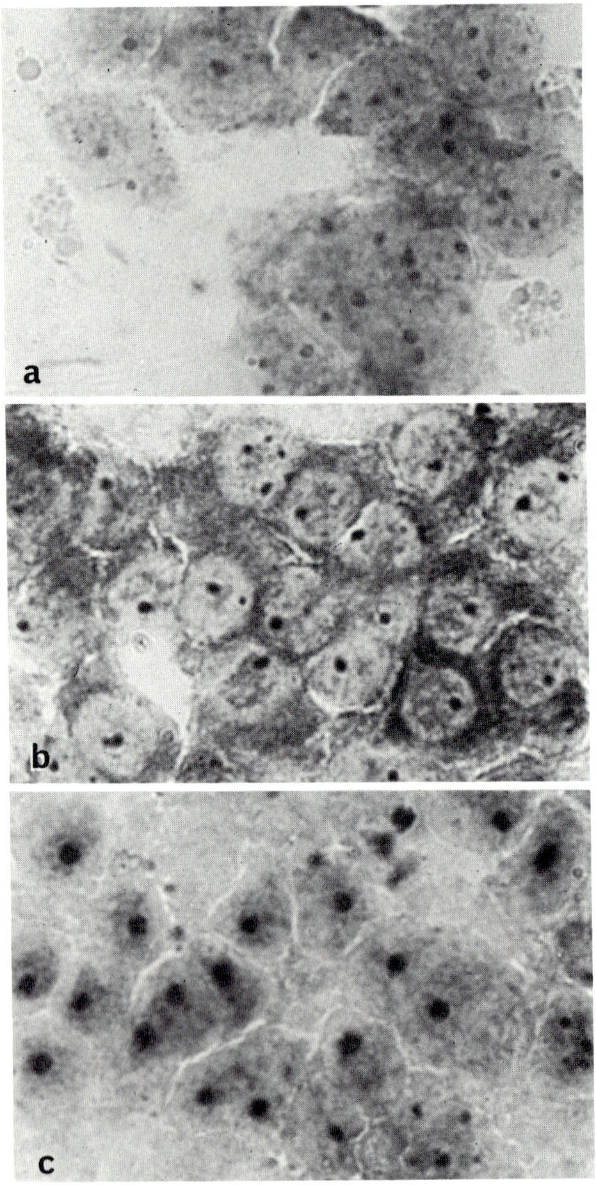

Fig. 9.28. Thyroid gland. (a) Cells of diffuse simple goiter; (b) Follicular adenoma cells; (c) Follicular carcinoma cells. ×1000. (Courtesy of Dr. V. Smejkal, Institut of Endocrinology, Prague.)

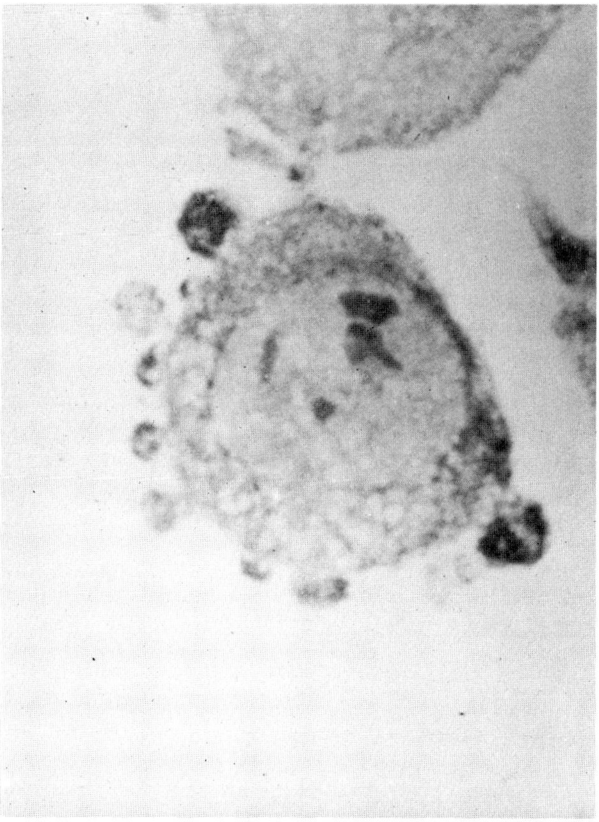

Fig. 9.29. Irregularly shaped nucleoli of different sizes in a human bronchial carcinoma cell. ×1000.

hold (*120*) suggested that young tumor cells (tumorkeimzellen) are not characterized by large nucleoli and by striking changes in the nucleolar-nuclear ratios. On the other hand, nucleolar size has been found to be larger in progressing malignant tumor cells as compared with those in less rapidly progressing or regressing neoplastic cells (*27, 84, 94, 99, 117*) and nucleolar diameter seems to correspond to the grade of malignancy (*27, 44, 103, 115*) (Figs. 9.28 and 9.29). For example, the mean nucleolar diameter was about 1.49 μ in ovarian cystadenocarcinoma of the less malignant grade I and 2.0 and 2.97 μ in the same tumor type of the more malignant grades II and III as compared with 1.3 μ found for benign ovarian cystadenoma (*84*). The values of the mean nucleolar area have also been related to the histological grading of the tumor (*44*) (Table 9.4).

Increased nucleolar size has been described during experimental

TABLE 9.4

MEAN NUCLEOLAR AREA IN VARIOUS HUMAN MALIGNANT TUMORS RELATED TO THE HISTOLOGICAL GRADING OF THE MALIGNANCY[a]

Grade of malignancy	1	2	3	4
Carcinoma of the thyroid	2.4	3.7	5.0	5.9
Carcinoma of the breast	2.6	3.3	5.6	6.6
Carcinoma of the liver	—	3.2	6.7	8.5
Carcinoma of the stomach	—	3.2	3.8	4.9
Carcinoma of the rectum	3.8	3.7	7.9	14.6
Carcinoma of the ovary	2.1	4.7	4.6	—
Carcinoma of the uterine cervix	—	5.3	9.7	6.0
Carcinoma of the endometrium	2.6	3.4	4.5	8.3

[a] Values in μ^2; (from Ferreira, *44*).

carcinogenesis in epidermal cells or hepatocytes. Application of carcinogens to mouse skin induced a marked increase of the nucleolar size in hypertrophic epidermal cells (*31, 32, 101, 102, 107a*). However, nucleolar enlargement can also be induced by noncarcinogenic substances. It is less than that produced by potent carcinogens (*33*). In squamous cell carcinoma induced by cholanthrene or methylcholanthrene, the nucleolar size was larger than that of cells which failed to undergo malignant change (*102*). However, nucleoli in such malignant tumors can also be small when compared with those in normal or hypertrophic spinous and basal cells of the epidermis (*33*).

Several observations have shown that nucleolar size and morphology can change during chemical carcinogenesis. Treatment of experimental animals with azo dyes, acetylaminofluorene, or ethionine resulted in the enlargement of the nucleoli in liver cells and induced hepatomas (*18, 28, 46, 54, 60, 71, 101, 128*). In hepatomas produced by acetylaminofluorene, nucleoli were variable in size, ranging from normal size to markedly enlarged; in invasive rapidly growing tumors, the nucleolar size was quite variable (*46*). However, a marked increase of the nucleolar size was also produced by protein-free diet or partial hepatectomy (also see Chapters I, V, and IX—Part III). By contrast, small nucleoli were observed in hepatoma cells and large nucleoli were seen in nontumor hepatocytes of rats fed with *p*-dimethylaminobenzene (*128*).

Nucleolar enlargement has been observed not only in preneoplastic or neoplastic cells produced by chemical carcinogens but also in virus-infected cells and in transformed or neoplastic cells produced by some oncogenic viruses (*7, 35, 35a, 38–40, 42, 53, 70, 91, 96, 107, 110, 111*). However, the changes of the nucleolar size in transformed or neoplastic cells during viral carcinogenesis seemed to be less pronounced as compared with the changes in the nucleolar shape, number, and structure (*7, 11, 12, 38–42, 51–53, 63, 70, 76, 91, 92, 96, 107, 109, 119*).

III. The Nucleolar Number

The increase of nucleolar number in some neoplastic cells may be the result of polyploidy of NOR's (*75;* see also Chapter III). A particularly large number of nucleoli has been noted in sarcoma cells (*23, 120*). Defined nucleoli also may be absent in some sarcomas (*115*); in such sarcomas, there is dispersion of RNA-containing nuclear structures (*115*). An increased number of nucleoli may be observed in "more malignant" cells. An increase in the nucleolar number is not typical for malignant tumor growth, since single nucleoli or decreased numbers of nucleoli are also found in cancer cells and multinucleolar nuclei occur in many nontumor cells (*1, 8, 22, 23, 45, 69, 108, 118, 120, 121, 134*).

In cystadenomas of the ovary, the number of nucleoli ranged from one to four per nucleus and 86% of the nuclei had two nucleoli. In ovarian cystadenocarcinomas, 1–24 nucleoli could be found and the percentage of nuclei with increased numbers of nucleoli increased with the degree of malignancy (*84*). On the other hand, the nucleolar number may be reduced in spontaneous as well as experimental tumors. For example, one nucleolus has been found in the majority of cancer cells in human hepatomas; two to four nucleoli per nucleus were observed only occasionally as noted in Part III of this chapter; two to three nucleoli are normally present in hepatocytes (*1, 3, 8*). Similarly, the number of nucleoli per nucleus averages 1.3 in different types of Morris hepatomas in rats as compared with 2.7 nucleoli per nucleus in hepatocytes of adult rats (*134*). The number of nucleoli in thyroid tumors was also reduced as compared with thyreocytes of benign goiter (Fig. 9.28) (*121*). The absence of visible nucleoli in nuclei with a "young chromatin pattern" may be one evidence of malignancy (*23*). For example, in very undifferentiated prostatic cancers, nucleoli may not be visualized with the usual stains. Small nucleoli, however, can be obscured by the chromatin structure when the usual cytological procedures are used for staining (*115*) both nucleoli and chromatin structures. On the other hand, the number of nucleoli may be increased in moderately undifferentiated cancer of the prostate (*23*).

In diagnosis, changes in the nucleolar number in cells may provide evidence for malignancy (*5*). The number of nucleoli, however, must be increased or decreased beyond normal values of corresponding normal cells (*95*). In drawing diagnostic conclusions, pathologists recognize that radiation or antitumor chemotherapy may alter the nucleolar number and size of the tumor cells (*48, 56, 61, 105, 106, 115, 116;* see Chapter X).

IV. Nucleolar Shape

An irregular shape or budding of the nucleolus is one criterion of malignancy (Fig. 9.29, 9.30) (*8, 20, 45, 69, 75, 82, 84, 88, 108, 112, 125, 143*). Most descriptions of malignant tumor cells indicate that the shape of nucleoli in these cells is irregular more often than in normal cells or cells of benign tumors (*45, 59, 69, 75, 112*). Variability of the nucleolar shape exists in a tumor from cell to cell. However, young nontumor cells may also have irregularly shaped nucleoli; e.g., regenerating liver cells, proliferating fibroblasts of young granulation tissue, proerythroblasts, and other cells in early prophase (*5, 35a, 74, 83*). Some cultured non-tumor cells have irregularly shaped nucleoli. Many of these cells in well established tissue cultures may have similar appearances and properties to those of tumor cells after numerous passages (*29, 30, 79, 81, 82, 93, 97*). A marked similarity in the nucleolar shape has been observed between fibroblasts and sarcoma cells when cultured under the same conditions (*87*). In some instances, it has been found that such cultured cells have undergone malignant transformation.

On the other hand, regular or spherical nucleoli have also been noted in numerous malignant tumor cells (*23, 66, 69*). Such nucleoli are usually small and multiple (*143*), although large spherical nucleoli may be present in malignant tumor cells; e.g., in human cancerous effusions (*59, 88, 108*). However, in human effusions, nucleoli of cancer cells also may be irregular in shape as in ascites tumor cells of rodents (*59, 88, 108*). In such tumors, the irregular nucleolar shape seems to be charac-

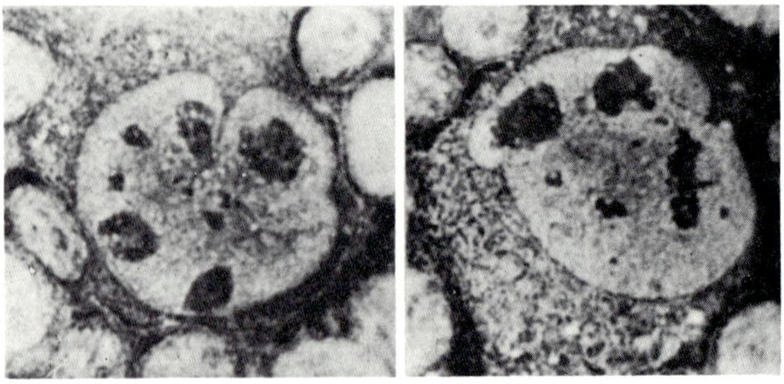

Fig. 9.30. Nucleoli of Reed-Sternberg cells. ×600. (Courtesy of Dr. G. Seman, Institut for Cancer Research, Villejuif.)

teristic and differentiates them from the spherical or oval nucleoli of proliferating mesothelial cells (*112*). Occasionally both a regularly (spherical or oval) and irregularly shaped nucleolus is present in the same nucleus of a cancer cell.

V. Light Microscopic Structure and Cytochemistry

Few light microscopic studies have been reported on the structural differences of nucleoli of nontumor and tumor cells. The reported increase of refractive or argentaphilic bodies in tumor nucleoli (*104*) is difficult to explain since the significance of these structures has not been satisfactorily clarified. The increase of the number of argentaphilic bodies may be related to malignancy since the majority of nucleoli in normal tissues possessed two, benign tumor cells three to five, and cancer cells 6–12 such bodies (*104*).

By cytochemical studies, no substantial differences have been found between the general composition of nucleoli in malignant and nonmalignant cells with the exceptions of changes in the contents of nucleic acids and proteins (*24, 49, 127, 128;* and Chapters V, VII). In some experimental tumors, some ribonucleoprotein-containing bodies within nucleoli referred to as "nucleolini" seemed to be more prominent (*85, 86*). In cultured tumor cells, these structures vary considerably in size within the nucleus (p. 39). This variation called "anisonucleolinosis" appears to be related to the aneuploidy of the tumor cells (*85*). Anisonucleolinosis also has been observed in aneuploid nontumor cells and has been induced by the carcinogen 4-nitroquinoline N-oxide, by actinomycin D, or 2-deoxy-5-fluorouridine, and by virus infection (herpes, polyoma, SV 40) (*85*).

Some light microscopic studies on DNA associated with the nucleolus indicated that intranucleolar chromatin structures were more prominent in cancer cells (*75, 87, 122, 123*). This finding agrees with electron microscopic observations which show that intranucleolar areas containing chromatin structures are increased in neoplastic cells as compared with corresponding normal cells (*135*). Areas containing chromatin structures also are increased in cultured cells infected with oncogenic viruses (*53*). Unusual Feulgen-positive bodies 0.5–1.0 μ in diameter have been observed in precancerous and cancer cells of the Lucké renal adenocarcinoma which is of viral origin (*38, 40*). Feulgen-positive bodies have also been found in nucleoli of regenerating liver cells treated with carcinogens (*96a*). Light microscopic differences in the organization of the nucleolus-associated chromatin have been frequently found in malignant tumor cells (*75, 87*). Large areas or clusters of nucleolus-associated chromatin appear to be increased in cancer cells. However, the nucleolus-associated chromatin was also observed as a fine thin layer around tumor nucleoli (*75, 87*) and its relative decrease was noted by electron microscopy (*135*). A similar organization of chromatin has also been observed around nuclear inclusions found in some tumor cells. The cytochemistry of such inclusions in hepatomas of mice demonstrated that they originated in the cytoplasm although their nucleolar origin in various cells was also considered (*2, 49, 78*).

The results of staining of nucleolar proteins with amido black 10B suggested that these proteins in some neoplastic cells may possess certain unique properties (*98*). Positive staining was noted in nucleoli of Ehrlich ascites tumor and some avian tumors. Nucleoli in nontumor cells did not provide positive results. Although cytochemical studies were made on enzymes in various tumor cells (*2, 21, 129, 139*), the number of reports on differences between the localization of enzymes in nucleoli of malignant and nonmalignant cells is extremely limited. The ATPase activity seemed highest in nucleoli of Ehrlich ascites tumor in mice as compared with the whole nucleus and cytoplasm (*77*). In aging tumor cells, the ATPase activity decreased first in the cytoplasm and nuclei and then in the nucleoli. The activity of GTPase was lower in nucleoli of Morris hepatoma cells than in normal hepatocytes. In liver cells of rat embryos, however, GTPase was not demonstrated (*138*). Succinic dehydrogenase activity was detected in nucleoli of mouse mammary (*36*) and human uterine cervical carcinomas (*37*). The activity of 5′-nucleotidase seemed to be higher in nucleoli of some ascites cells of mouse sarcoma as compared with Ehrlich ascites tumor (*137*).

Various metal ions have been found in nucleoli of malignant tumor and nontumor cells. In prostatic carcinoma cells a low concentration of zinc was found localized to nucleoli (Fig. 9.31). In hyperplastic or normal prostatic cells higher concentrations of zinc were found as compared with cancer cells, but no distinct nucleolar localization for this metal was observed (*58*).

VI. Ultrastructural Morphology

Electron microscopic studies have not shown any constant or specific differences between the general nucleolar ultrastructure of tumor and nontumor cells (*12, 34, 140*). Nucleoli with nucleolonemas, compact nucleoli without distinct nucleolonemas, and ring-shaped nucleoli have been described in tumor cells (Figs. 9.32–9.35) as well as in nontumor cells. However, in some tumor cells, i.e., in uterine and mammary cancer,

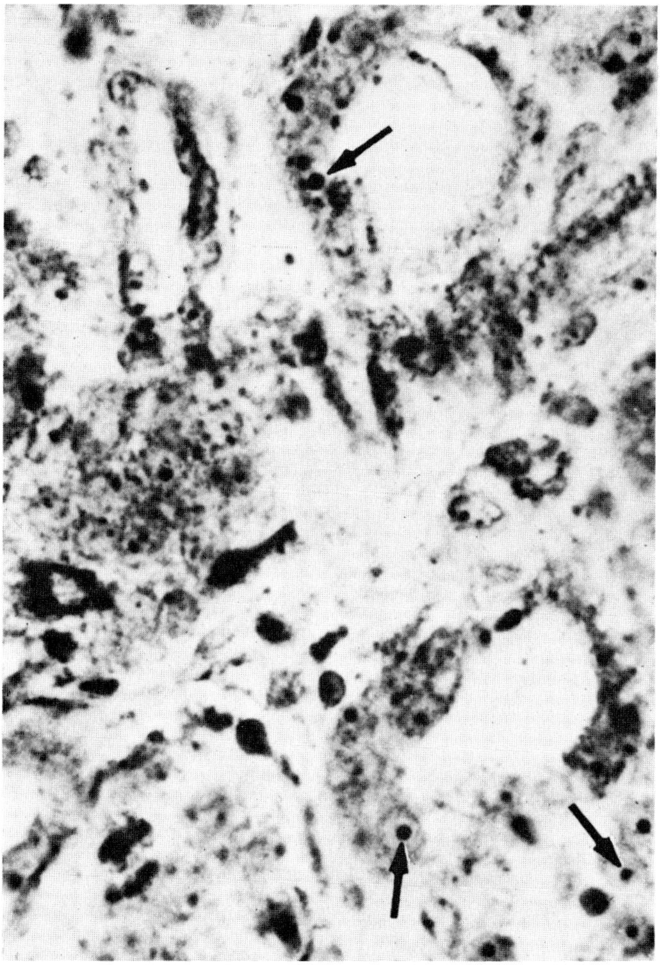

Fig. 9.31. Human prostatic cancer stained with dithiozone for zinc. The nucleoli are densely stained (arrows). ×650 (*58*). Courtesy of Dr. F. Gyorkey, Veterans Administration Hospital, Houston, Texas.

malignant schwannoma, leukemic myeloblasts, S_2 carcinomas, and cultured KB cells (human carcinoma of the oral cavity) some nucleoli were characterized by the filamentous appearance of the nucleolonemas (Fig. 9.35a,b) (*3, 14, 50a, 62, 67, 131*). Compact nucleoli (Fig. 9.33) without distinct nucleolonemas were found more often in tumor cells than in nontumor cells. Such nucleoli were observed in cells of human cancerous peritonitis, mammary cancer, cancer of the uterine cervix, pheochromocytoma, and different experimental tumors (*6, 10, 13, 19, 34, 38, 43, 62, 67, 68, 141, 142*).

Several studies suggested that the size of nucleolar areas containing

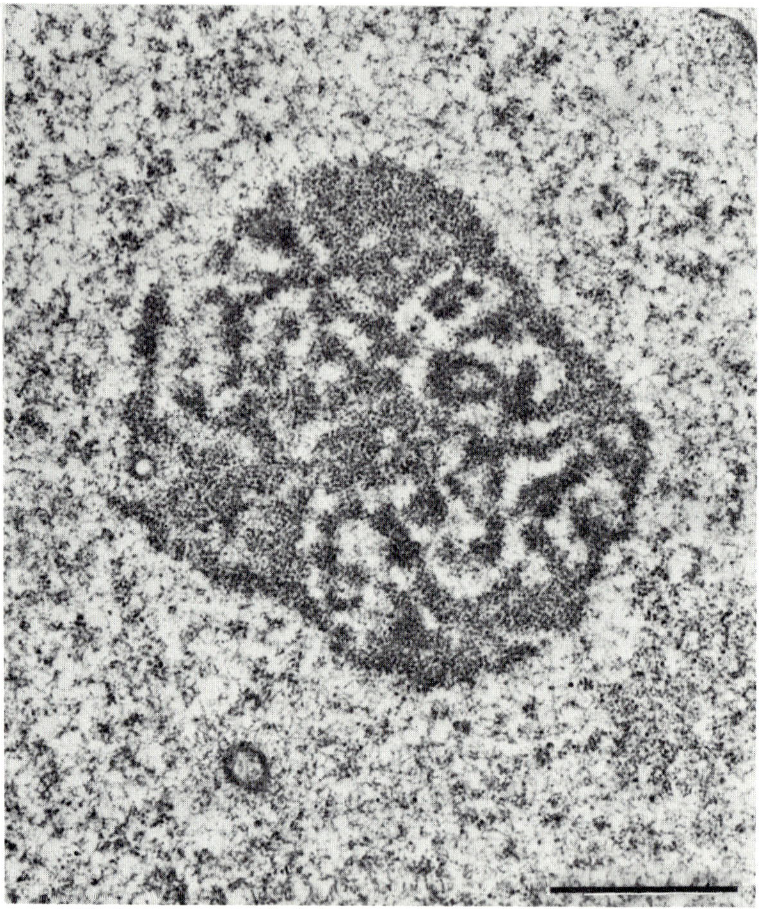

Fig. 9.32. Nucleolus with prominent nucleolonemas in a Walker tumor cell. ×25,000.

chromatin structures and the distribution of both granular and fibrillar nucleolar components may be different in tumor cells as compared to corresponding nontumor cells. The relative percentages of the intranucleolar chromatin areas was twice as large in Novikoff hepatomas as in normal liver cells, although no differences were found in the ratio of the areas occupied by nucleolar body to the area of the total nucleolar chromatin (*135*). In some human malignant tumors, as well as in experimental animal tumors, the nucleolar ribonucleoprotein granular components seemed to be more prominent and fibrillar components appeared to be reduced (*4, 14, 34, 124, 134*). Quantitative planimetric studies on areas containing nucleolar fibrillar components indicated that

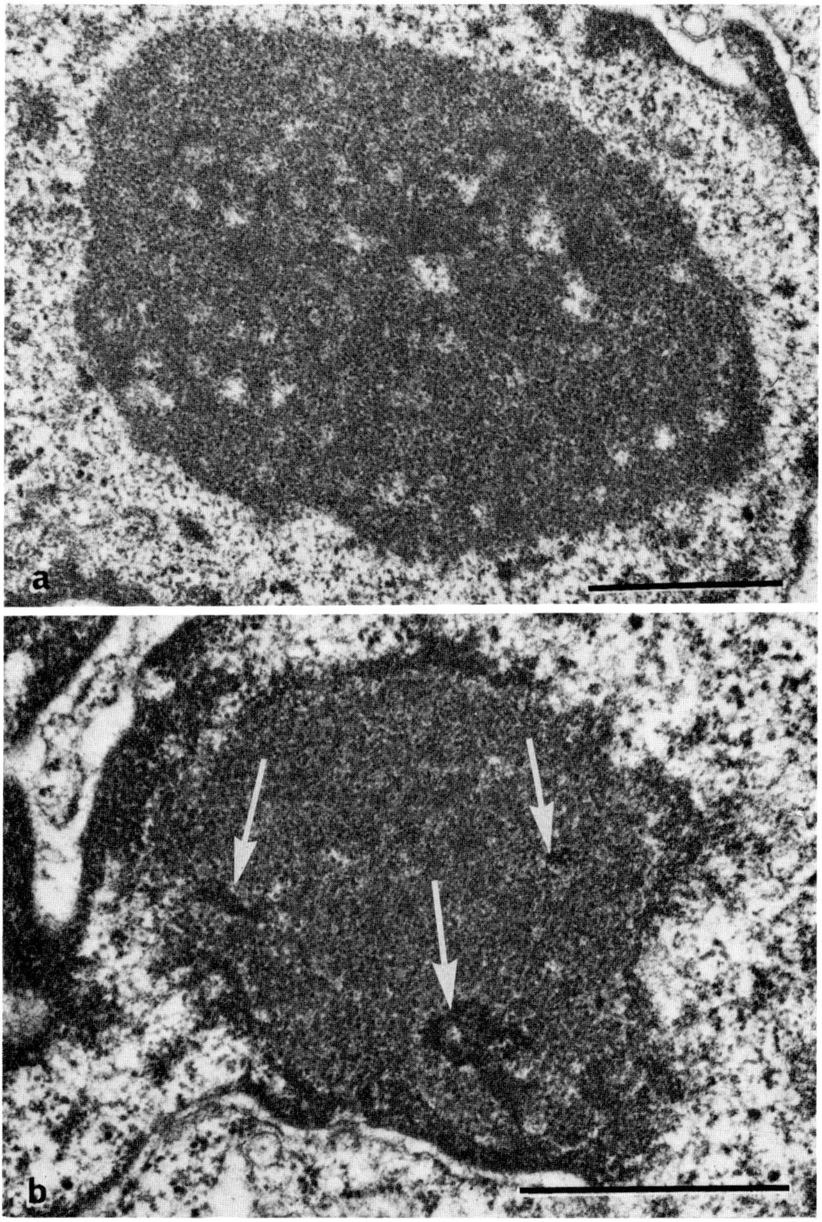

Fig. 9.33. Human lymphosarcoma cells. (a) Large nucleolus of compact appearance with less distinct nucleolonemas and small light areas. ×26,000. (b) Compact nucleolus without nucleolonemas. Intranucleolar chromatin (arrows). ×36,000.

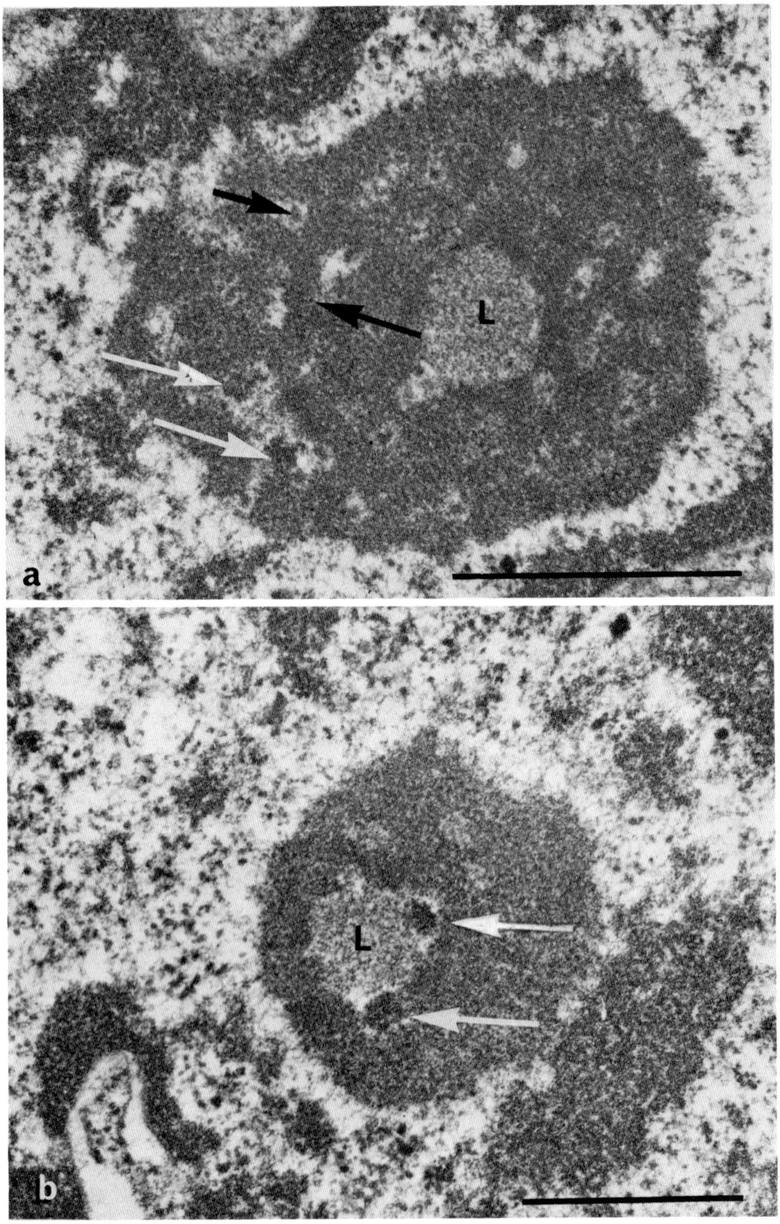

Fig. 9.34. Ring-shaped nucleoli in human lymphosarcoma cells. (a) Nucleolus with a large central light area (L). Nucleolonemas can be still distinguished (black arrow). Intranucleolar chromatin (white arrows); microspherules in light areas (black point). ×38,000. (b) Nucleolus with a large light central area without nucleolonemas in its peripheral part. Intranucleolar chromatin (white arrows). ×29,000.

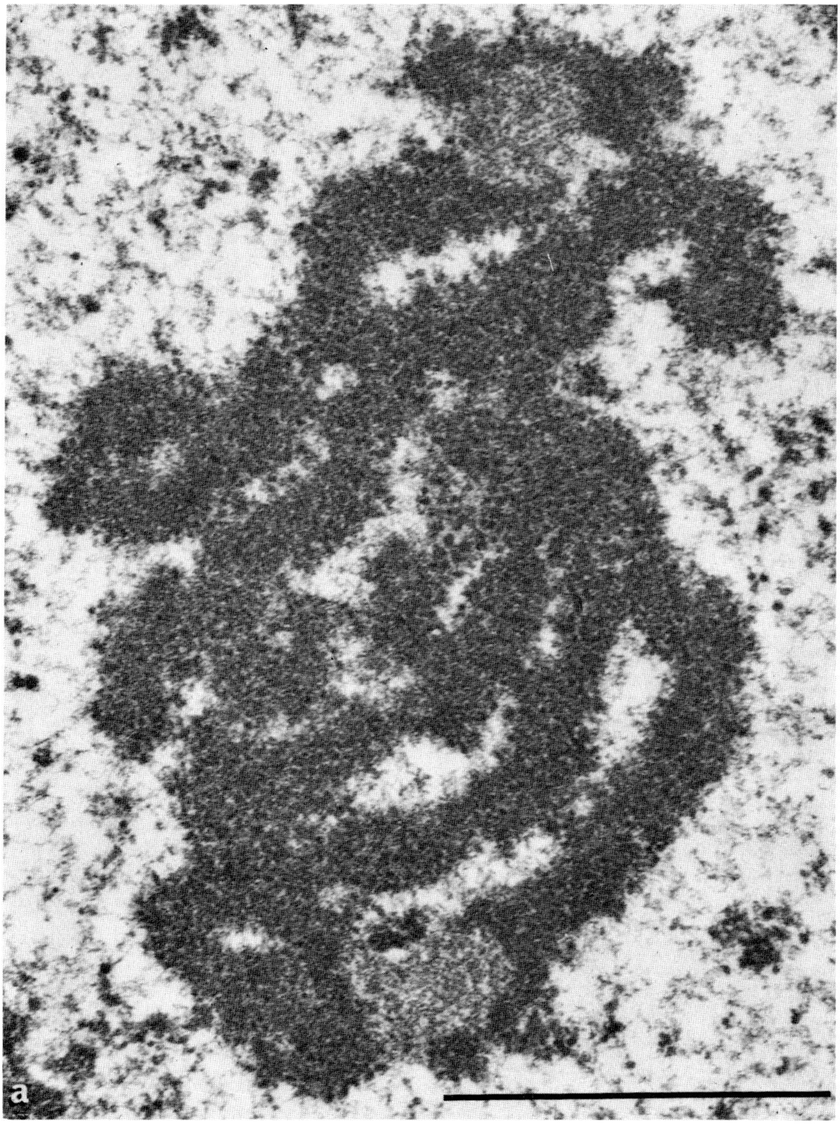

Fig. 9.35. a. Nucleolus with filamentous appearance of nucleolonemas in a human lymphosarcoma cell. ×51,500. b. Nucleolus with filamentous appearance of one nucleolonema in a leukemic cell. ×12,000. (Courtesy of Dr. John Lazlo, Duke University School of Medicine, Durham, North Carolina.)

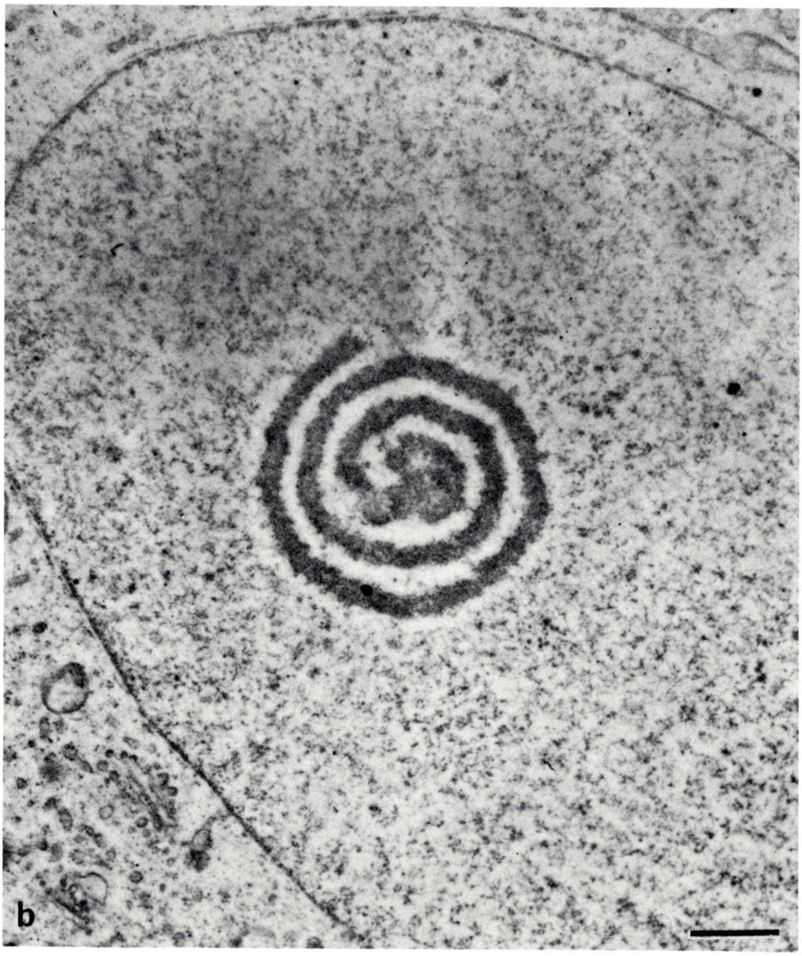

Fig. 9.35. (*Continued*).

the relative size of areas containing fibrillar components was markedly reduced in the Novikoff hepatoma but not in a variety of differentiated Morris hepatomas (*134*) (Table 9.5). In some tumor nucleoli the demarcation of areas containing fibrillar or granular components was less distinct and fibrillogranular areas containing both ribonucleoprotein components were noted (*14, 15, 124, 134*).

Dense structures in nucleoli of some neoplastic cells may appear as granules or small bodies of different size. Dense atypical granules of various sizes ranging between 200 and 400 Å in diameter were observed in nucleoli of Walker tumors in rats (Fig. 2.10a) and cultured human lymphoblasts (*113, 136*). Such granules, however, were also found in nucleoli of nontumor cells such as in plant cells, amoebas, and thio-

TABLE 9.5

RATIO OF THE AREA CONTAINING FIBRILLAR COMPONENTS TO THE TOTAL NUCLEOLAR AREA[a]

Cell type	Ratio[b]
Adult liver cells	19.0 ± 4.0
Embryonic liver cells	19.7 ± 3.3
Morris 9121 hepatoma cells	20.5 ± 4.6
Morris 8995 hepatoma cells	15.5 ± 4.4
Morris 7787 hepatoma cells	17.1 ± 4.9
Morris R-7 hepatoma cells	18.3 ± 2.4
Novikoff hepatoma cells	13.4 ± 3.0

[a] (*134*).
[b] Plus minus standard deviation.

acetamide-treated hepatocytes (*14, 130;* Chapter II). The composition of these granules is not known. Multiple large dense granules or bodies of high density (0.1–0.3 μ in diameter) were noted in nucleoli or at their periphery in Reed-Sternberg cells of Hodgkin's disease, human lymphosarcoma or reticulosarcoma cells, human myeloma, spinous cells of viral benign human skin tumors of molluscum contagiosum, and the cells of some experimental tumors (*11, 12, 14, 16, 17, 39, 52, 53, 63, 72, 80*). Such "spotted" nucleoli (Fig. 9.34a) were also found in various cells infected with some RNA as well as DNA viruses (*11, 12, 14–16, 53, 63, 132*). Some of these viruses, such as Rous sarcoma virus, SV 40, and polyoma virus, are oncogenic (*11, 14, 110a*). In virus-infected cells, the dense granules usually appeared in the early phases of the infection before virus particles could be detected (*11, 14*).

"Spotted" nucleoli were also found in cells of experimental tumors, such as rat hepatoma or mouse sarcoma, produced by chemical carcinogens (*12, 52*). These dense structures appeared to originate from dense parts of nucleolonemas (*53*). Similar atypical nucleolar structures were produced by UV radiation (*96b*) or actinomycin D and resemble the dense microspherules of actinomycin D-treated nucleoli of Ehrlich ascites tumor and Novikoff hepatoma (*133;* Chapter X). Increased density of some nucleolar components has also been observed as a result of necrobiosis and autolysis (*9*). In interpretation of such structures, both therapy and the status of the cells under investigation must be considered.

VII. Conclusions

No constant general differences in nucleolar morphology have been found between tumor and nontumor cells by means of light microscopy.

Moreover, the general ultrastructure of nucleoli in tumor cells does not differ substantially from that in nontumor cells. On the other hand, nonspecific variations, as well as aberrations, of the distribution and organization of various nucleolar components have been noted in tumor cells as compared with corresponding nontumor cells. In comparison with nontumor cells, nucleoli in tumor cells seem to be characterized by nonspecific changes of the nucleolar morphology. Nucleoli in malignant tumor cells vary widely in regard to size, number, shape, and structure. Generally, in experimental tumors, the nucleoli are enlarged, compact, and irregular. However, in the wide variety of human tumors, nucleolar morphology is quite variable. Characterization of nucleoli in a given tumor may provide useful data for studies of nucleolar function as well as for diagnosis.

REFERENCES

1. Abdine, F. H., and Mokhtar, N. A., Primary tumors of the liver in infancy and childhood. *Arch. Pathol.* **72**, 98–103 (1961).
2. Altmann, H. W., Zur morphologie der wechselwirkung von kern und cytoplasma. *Klin. Wochschr.* **33**, 306–314 (1955).
3. Andrew, W., Brown, H. M., and Johnson, J. B., Senile changes in the liver of mouse and man, with special reference to the similarity of the nuclear alterations. *Am. J. Anat.* **72**, 199–221 (1943).
4. Ashworth, C. T., Luibel, F. J., and Sanders, E., The fine structure of nuclei in certain human malignant neoplasms. *Am. J. Clin. Pathol.* **34**, 9–20 (1960).
5. Ayres, W. W., A method of staining of nucleoli of cells in fresh benign and malignant tissues. *Cancer Res.* **8**, 352–356 (1948).
6. Benedeczky, J., and Lapis, K., Vergleichende elektronenmikroskopische untersuchungen am nebennieren mark und phäochromotcytom des menschen. *Beitr. Pathol. Anat. Allgem. Pathol.* **137**, 403–438 (1968).
7. Bereczsky, E., Dmochowski, L., and Gray, C. E., Study of host virus relationship. I. Light phase and fluorescence microscopy of mouse embryo cells infected with polyoma virus. *J. Natl. Cancer Inst.* **27**, 99–129 (1961).
8. Berman, C., "Primary Carcinoma of the Liver." Lewis, London, 1951.
9. Bernhard, W., Ultrastructural aspects of the normal and pathological nucleolus in mammalian cells. *Natl. Cancer Inst. Monograph* **23**, 13–38 (1966).
10. Bernhard, W., Ultrastructural aspects of nucleocytoplasmic relationship. *Exptl. Cell Res.* Suppl. 6, 17–50 (1958).
11. Bernhard, W., Electronmicroscopic studies of tumor viruses. *Cancer Res.* **20**, 712–727 (1960).
12. Bernhard, W., Some problems of fine structure in tumor cells. *Progr. Exptl. Tumor Res.* **3**, 1–34 (1963).
13. Bernhard, W., Bauer, A., Cropp, A., Haguenau, F., and Oberling, C., L'ultrastructure du nucléole des cellules normales et cancéreuses. Etude au microscope électronique. *Exptl. Cell Res.* **9**, 88–100 (1955).
14. Bernhard, W., and Granboulan, N., Electron microscopy of the nucleolus in

vertebrate cells. *In* "The Nucleus" (A. J. Dalton and F. Haguenau, eds.), pp. 81–149. Academic Press, New York, 1968.
15. Bernhard, W., and Granboulan, N., The fine structure of the cancer cell nucleus. *Exptl. Cell Res.* Suppl. 9, 19–53 (1963).
16. Bernhard, W., and Leplus, R., "Fine Structure of the Normal and Malignant Lymph Node." Pergamon Press, Oxford, 1964.
17. Bessis, M., and Thiéry, P., Etude au microscope électronique des hemosarcomes humains. *Nouvelle Rev. Franc. Hematol.* **2**, 577–601 (1962).
18. Brux, de, J., and Marchand, C., Etude sur le fonctionnement. La régéneration et la cancérisation de la cellule hépatique. *Presse Med.* **49**, 704–705 (1949).
19. Bucher, O., and Gattiker, R., Karyometrische untersuchungen an gewebekulturen. Uber die grossenbeziehungen zwischen kern und kernkörperchen. *Z. Mikroskop.-Anat. Forsch.* **60**, 467–502 (1954).
20. Burke, M. D., and Melamed, M. R., Exfoliative cytology of metastatic cancer in lung. *Acta Cytol.* **12**, 61–74 (1968).
21. Burstone, M. S., "Enzyme Histochemistry and its Application in the Study of Neoplasms." Academic Press, New York, 1962.
22. Busch, H., Byvoet, P., and Smetana, K., The nucleolus of the cancer cell: A review. *Cancer Res.* **23**, 313–339 (1963).
23. Cardozo Lopes, P., "Clinical Cytology." L. Staflen, Leyden, 1954.
24. Caspersson, T., "Cell Growth and Cell Function." Norton, New York, 1950.
25. Caspersson, T., and Santesson, L., Studies on protein metabolism in the cells of epthelial tumours. *Acta Radiol.* **46**, 1–105 (1942).
26. Caspersson, T. O., and Thorsson, K. G., Virus and zellstoffwechsel. *Klin. Wochschr.* **31**, 205–212 (1953).
27. Castren, H., Uber die struktur der zellen der bindegewebegeschwülste beim menschen. *Arb. Pathol. Inst. Univ. Helsingfors* **4**, 240–318 (1926).
28. Catchpole, H. R., and Gersch, I., Components of normal and abnormal liver cells studied by ultraviolet microscopy and by differential centrifugation. *Discussions Faraday Soc.* **9**, 471–480 (1950).
29. Chlopin, N. G., Uber *in vitro* kulturen von geweben der säugetiere mit besonderer berücksichtigung des epithels. I. Kulturen von submaxillaris. *Arch. Pathol. Anat. Physiol.* **243**, 373–387 (1923).
30. Chlopin, N. G., Uber *in vitro* kulturen von geweben der säugetiere mit besonderer berücksichtigung des epithels. II. Kulturen der harnblasenschleimhaut. *Arch. Pathol. Anat. Physiol.* **252**, 748–773 (1924).
31. Cooper, N. C., Early cytological changes produced by carcinogens. *Bull. N. Y. Acad. Med.* [2] **32**, 79–82 (1956).
32. Cowdry, E. V., Malignant properties of cancer cells. *Ann. N. Y. Acad. Sci.* **63**, 1043–1053 (1956).
33. Cowdry, E. V., and Paletta, F. X., Changes in cellular, nuclear and nucleolar sizes during methylcholanthrene epidermal carcinogenesis. *J. Natl. Cancer Inst.* **1**, 745–759 (1941).
34. Dalton, A. J., and Felix, M. D., The electron microscopy of normal and malignant cells. *Ann. N. Y. Acad. Sci.* **63**, 1117–1140 (1956).
35. Dave, C. J., and Law, L. W., Morphologic changes in salivary gland tissue of the newborn mouse exposed to parotid tumor agent. *J. Natl. Cancer Inst.* **23**, 1157–1165 (1959).
35a. Davis, J. M. G., The ultrastructure of nucleoli and chromosomes during the early stages of liver regeneration and the changes produced in these struc-

tures by X-radiation. *Phil. Trans. Roy. Microscop. Soc.* [B] **246**, 291–303 (1963).

36. De, P., Chatterjee, R., and Mitra, S., Nucleolar succinic dehydrogenase in mouse mammary carcinoma cells. *Nature* **192**, 286–287 (1961).
37. De, P., Chatterjee, R., and Mitra, S., Nucleolar localization of succinic dehydrogenase. *J. Histochem. Cytochem.* **10**, 6–7 (1962).
38. Dmochowski, L., Viruses and tumors in the light of electron microscope studies: A review. *Cancer Res.* **20**, 977–1015 (1960).
39. Dourmashkin, R., and Bernhard, W., A study with the electron microscope of the skin tumour of molluscum contagiosum. *J. Ultrastruct. Res.* **3**, 11–38 (1959).
40. Duryee, W. R., Precancer cell in amphibian adenocarcinoma. *Ann. N. Y. Acad. Sci.* **63**, 1280–1302 (1956).
41. Edwards, G. A., Comparative micromorphology of normal, transforming and malignant cells. *Natl. Cancer Inst. Monograph* **4**, 313–321 (1960).
42. Edwards, G. A., Ruska, C., Ruska, H., and Skiff, J. V., The micromorphology of a human bronchogenic carcinoma. *Cancer* **12**, 982–1002 (1959).
43. Fasske, E., and Themann, H., Die elektronenmikroskopische struktur menschlicher carcinome. *Beitr. Pathol. Anat. Allgem. Pathol.* **122**, 313–344 (1960).
44. Ferreira, A. E. M., Relation of the macronucleolus and nucleonucleolar ratio of cancer to histological grading. *J. Lab. Clin. Med.* **26**, 1612–1628 (1941).
45. Fidler, H., A comparative cytological study of benign and malignant tissues. *Am. J. Cancer* **25**, 772–779 (1935).
46. Flaks, B., Fine structure of primary rat hepatoma induced by 2-acetylaminofluorene feeding. *European J. Cancer* **4**, 513–521 (1968).
47. Foot, N. C., The identification of tumor cells in sediments of serous effusions. *Am. J. Pathol.* **13**, 1–11 (1937).
48. Gati, E., Inke, G., Bajtai, A., and Gyarfas, J., Cytological changes in ascites carcinoma cells on the effect of nitrogen mustard, TEM or BCM, with special reference to changes in nuclear and nucleolar volume. *Acta Morphol. Acad. Sci. Hung.* **7**, 343–350 (1967).
49. Giordano, A., Pecchiai, L., and Rilke, F., Le nucleoproteine delle cellule neoplastiche. *Riv. Istochim. Normale Pathol.* **4**, 275–317 (1958).
50. Gonzalez, P., and Nardone, R. M., Cyclic nucleolar changes during the cell cycle. *Exptl. Cell Res.* **50**, 599–615 (1968).

50a. Gonzalez Angulo, A., Smetana, K., and Martin, S. A., Estudio con microscopio electronico de un caso de sarcoma neurogeno. *Bol. Inst. Estud. Med. Biol. (Mex.)* **22**, 429–441 (1964).

51. Granboulan, N., and Riviére, M. R., Etude au microscope électronique des particules virales présentes dans les lymphomatoses spontanées de la souris. *J. Microscopie* **1**, 23–37 (1962).
52. Granboulan, N., Riviére, M. R., and Bernhard, W., Présence de particules d'aspect viral dans un sarcoma greffable de la souris provaqué par le methylcholanthrene. *Bull. Cancer* **47**, 291–307 (1960).
53. Granboulan, N., Tournier, P., Wicker, R., and Bernhard, W., An electron microscope study of the development of SV 40 virus. *J. Cell Biol.* **17**, 423–440 (1963).
54. Grishem, J. W., Electron microscope study of carcinoma of the liver in ethionine-fed rats. *Gastroenterology* **38**, 792–793 (1960).
55. Grundmann, E., "Allgemeine Cytologie." Thieme, Stuttgart, 1964.

56. Gusberg, S. B., Tovel, H. M. M., Long, M., and Hill, J. C., Studies on nucleoprotein patterns in radiosensitivity testing. *Ann. N. Y. Acad. Sci.* **63**, 1447–1457 (1956).
57. Guttman, P. H., Nuclear-nucleolar volume ratio in cancer. *Am. J. Cancer* **25**, 802–806 (1935).
58. Györkey, F., Min, K. W., Huff, J. A., and Györkey, P., Zinc and magnesium in human prostate gland, normal, hyperplastic and neoplastic. *Cancer Res.* **27**, 1348–1353 (1967).
59. Haam, E. von, and Alexander, H. G., Cytological studies of malignant tumors. *Am. J. Clin. Pathol.* **6**, 394–414 (1936).
60. Hadjiolov, A. A., Uber die veränderungen der nucleinsäuren in der leber von ratten bei cancerogenese durch 4-dimethylaminoazobenzol. *Z. Krebsforsch.* **62**, 361–369 (1958).
61. Haenel, U., Veränderungen der nucleolarsubstanz an knochenmarkzellen bei perniciöser anämie und ihre pathogenetische bedeutung. *Helv. Med. Acta* **17**, 627–650 (1950).
62. Haguenau, F., Le cancer mammaire de la souris et de la femme. Etude comparative au microscopeélectronique. *Pathol. Biol. Semaine Hop.* [N.S.] **7**, 989–1015 (1959).
63. Haguenau, F., Significance of ultrastructure in virus induced tumors. *Natl. Cancer Inst. Monograph* **4**, 211–249 (1960).
64. Haumeder, E., Vergleichende kern und nucleolenmessungen an verschiedenen organgeweben mit besonderen berücksichtigung der malignen tumorzellen. *Z. Krebsforsch.* **40**, 105–116 (1934).
65. Heilmayer, L., and Begemann, H., "Atlas der klinischen hämatologie und cytologie." Springer, Berlin, 1955.
66. Herbut, P. A., and Clerf, L. H., Bronchogenic carcinoma. Diagnosis by cytologic study of bronchoscopically removed secretions. *J. Am. Med. Assoc.* **130**, 1006–1012 (1946).
67. Hinglais-Guillaud, N., Moricard, R., and Bernhard, W., Ultrastructure des cancers pavimenteux invasifs du col utérin ches la femme. *Bull. Cancer* **48**, 293–316 (1961).
68. Howatson, A. F., and Ham, A. W., Electron microscopic studies of sections of two rat liver tumors. *Cancer Res.* **15**, 62–69 (1955).
69. Hughes, H. E., and Dodds, T. S., "Handbook of Diagnostic Cytology." Williams & Wilkins, Baltimore, Maryland, 1968.
70. Jensen, F., Koprowski, H., Pagano, J. S., Ponten, J., and Ravdin, R. G., Antalogous and homologous implantation of human cells transformed *in vitro* by simian virus 40. *J. Natl. Cancer Inst.* **32**, 917–937 (1964).
71. Karasaki, S., The fine structure of proliferating cells in preneoplastic rat livers during azo-dye carcinogenesis. *J. Cell Biol.* **40**, 322–335 (1969).
72. Klug, H., Licht und elektronen mikroskopische untersuchungen an ascites tumor zellen. *Z. Krebsforch.* **64**, 313–322 (1961).
73. Koller, P. C., The nucleus of the cancer cell. *Exptl. Cell Res.* Suppl. 9, 3–14 (1963).
74. Kopac, M. J., and Mateyko, G. M., Nucleolar chromosomes. *Advan. Cancer Res.* **8**, 122–190 (1964).
75. Kopac, M. J., and Mateyko, G. M., Malignant nucleoli. Cytological studies and perspectives. *Ann. N. Y. Acad. Sci.* **73**, 237–282 (1958).
76. Koprowski, H., Ponten, J. A., Jensen, F., Ravdin, R. G., Morhead, P., and

Saksela, E., Transformation of cultures of human tissue infected with simian virus SV_{40}. *J. Cellular Comp. Physiol.* **59**, 281–292 (1962).

77. Landschütz, C., Zytochemischer nachweis der adenosintriphosphatase (ATP-ase) in den zellen des ascites tumors der maus. *Experientia* **6**, 232–233 (1950).
78. Leduc, E. H., and Wilson, J. W., A histochemical study of intranuclear inclusions in mouse liver and hepatoma. *J. Histochem. Cytochem.* **7**, 8–16 (1959).
79. Leighton, J., Kline, J., and Orr, H. C., Transformation of normal human fibroblasts into histologically malignant tissue *in vitro*. *Science* **123**, 502–503 (1956).
80. Leplus, R., Debray, J., Pinet, J., and Bernhard, W., Lésions nucleaires décelées au microscope électronique dans les cellules de lymphomes malins ches l'homme. *Compt. Rend.* **253**, 2788–2790 (1961).
81. Lettré, R., and Siebs, W., Zur struktur des nucleolus. *Naturwissenschaften* **41**, 458 (1954).
82. Lettré, R., and Siebs, W., Beobachtungen zur struktur des nucleolus in normalen zellen sowie in tumorzellen. *Z. Krebsforsch.* **60**, 564–580 (1955).
83. Loele, W., Beziehungen zwischen oxydasen, vitalfärbung, postmortalfärbung und morphologie der zelle. *Arch. Pathol. Anat. Physiol.* **265**, 827–844 (1927).
84. Long, M. E., and Taylor, H. C., Nucleolar variability in human neoplastic cells. *Ann. N. Y. Acad. Sci.* **63**, 1093–1106 (1955–1956).
85. Love, R., Anisonucleolinosis in mammalian cell cultures. *Natl. Cancer Inst. Monograph* **23**, 167–180 (1966).
86. Love, R., and Bharadvaj, T. P., Two types of ribonucleoprotein in the nucleolus of mammalian cells. *Nature* **183**, 1453–1454 (1959).
87. Ludford, R. J., Nuclear structure and its modification in tumours. *Brit. J. Cancer* **8**, 112–131 (1954).
88. Luse, S. H., and Reagan, J. W., A histocytological study of effusions. II. Effusions associated with malignant tumors. *Cancer* **7**, 1167–1181 (1954).
89. MacCarty, W. C., A cytological key to the diagnosis and prognosis of neoplasms. *J. Lab. Clin. Med.* **13**, 354–365 (1928).
90. MacCarty, W. C., The value of the macronucleolus in the cancer problem. *Am. J. Cancer* **26**, 529–532 (1936).
91. Mayor, H. D., A labile intranuclear RNA associated with the development of adenoviruses. *J. Exptl. Med.* **119**, 433–441 (1964).
92. Mayor, H. D., Stinebaugh, S. E., Jamison, R. M., Jordan, L. E., and Melnick, J., Immunofluorescent, cytochemical and microcytological studies in the growth of the simian vacuolating virus (SV-40) in tissue culture. *Exptl. Mol. Pathol.* **1**, 397–416 (1962).
93. Maximow, A., Uber krebs ähnliche verwandlung der milchdrüse in gewebskulturen. *Arch. Pathol. Anat. Physiol.* **256**, 813–845 (1925).
94. McCormach, C. J., Biometric studies of endometrial cells. *Proc. Staff Meetings Mayo Clinic* **10**, 24–29 (1935).
95. McGrew, E., Criteria for the recognition of malignant cells in circulating blood. *Acta Cytol.* **9**, 58–60 (1965).
96. Melnick, J. L., Papova virus group. *Science* **135**, 1128–1130 (1962).

96a. Mironescu, S., Encut, J., Mironescu, K., and Liciu, F., Nucleolar behavior in regenerating liver of rats receiving intraabdominal injections of azo dyes and thioacetamide. *J. Natl. Cancer Inst.* **40**, 917–933 (1968).

96b. Montgomery, P. O'B., and Reynolds, R. C., Cellular and subcellular responses to ultraviolet radiation. *Lab. Invest.* **13**, 1243–1253 (1964).

97. Moore, A. E., Southam, C. M., and Sternberg, S. S., Neoplastic changes developing in epithelial cell lives derived from normal persons. *Science* **124**, 127–129 (1956).

98. Mundkur, B., and Brauer, B., Selective localization of nucleolar protein with amido black 10B. *J. Histochem. Cytochem.* **14**, 94–103 (1966).

99. Naidu, V. R., The value of enlarged nucleoli in diagnosis of malignancy. *Proc. Staff Meetings Mayo Clinic* **10**, 356–362 (1935).

100. Nieburgs, H. E., Parets, A. D., Perez, V., and Boudreau, C., Cellular changes in liver tissue adjacent and distant to malignant tumors. *Arch. Pathol.* **80**, 262–272 (1965).

101. Opie, E., Mobilization of basophile substance (ribonucleic acid) in the cytoplasm of liver cells with the production of tumors by butter yellow. *J. Exptl. Med.* **84**, 91–106 (1946).

102. Page, R. C., Cytologic changes in the skin of mice during application of carcinogenic agents. *A.M.A. Arch. Pathol.* **26**, 800–813 (1938).

103. Page, R. C., Carpenter, W., and MacCarty, W. C., Cytology of ovarian tumors. *A.M.A. Arch. Pathol.* **24**, 1–7 (1937).

104. Page, R. C., Regan, J. E., and MacCarty, W. C., Intranucleolar bodies in normal and neoplastic human tissue. *Am. J. Cancer* **32**, 383–394 (1938).

105. Parchwitz, H. K., Strahlungwirkung auf Tumorzellen. *In* "Strahlenpathologie der Zelle" (E. Scherer and H. S. Stender, eds.), pp. 334–363. Thieme, Stuttgart, 1963.

106. Peters, K., Nukleolus. *In* "Strahlenpathologie der Zelle" (E. Scherer and H. S. Stender, eds.), pp. 32–46. Thieme, Stuttgart, 1963.

107. Pontén, J., Jensen, F., and Koprowski, H., Morphological and virological investigation of human tissue cultures transformed with SV_{40}. *J. Cellular Comp. Physiol.* **61**, 145–154 (1963).

107a. Pullinger, B. D., The first effects on mouse skin of some polycyclic hydrocarbons. *J. Pathol. Bacteriol.* **50**, 463–471 (1940).

108. Quensel, U., Zytologische untersuchungen von ergüssen der brust und bauchhöhlen mit besonderen berücksichtigung der karzinomatösen exsudate. *Acta Med. Scand.* **68**, 458–501 (1928).

109. Rabson, A. S., Kirschstein, R. L., and Legallais, F. Y., Autologous implantation of rhesus monkey cells transformed *in vitro* by semian virus 40. *J. Natl. Cancer Inst.* **35**, 981–991 (1965).

110. Rake, G., and Blank, H., The relationship of host and virus in molluscum contagiosum. *J. Invest. Dermatol.* **15**, 81 (1950).

110a. Rapp, F., Viral oncogenesis. *Methods Cancer Res.* **1**, 451–544 (1967).

111. Reissig, M., and Melnick, J. L., The cellular changes produced in tissue cultures by herpes B virus correlated with the concurrent multiplication of the virus. *J. Exptl. Med.* **101**, 341–352 (1955).

112. Reitalu, J., The appearance of nucleoli and heterochromatin in mesothelial cells and cancer cells of ascites tumours of the mouse. *Acta Pathol. Microbiol. Scand.* **41**, 257–266 (1957).

113. Sankaranarayan, K., and Busch, H., Dense granules in nucleoli of Walker 256 carcinosarcoma cells of the rat. *Exptl. Cell Res.* **38**, 434–437 (1965).

114. Saxen, A., Pathologisch-anatomischen und klinische studien über die pri-

mären von der nasen-kavität und den angrenzenden nebenhöhlen ausgehenden papillome und karzinome. *Arb. Pathol. Inst. Univ. Helsingfors* **4,** 1–130 (1926).
115. Scherer, E., and Fiebelkorn, H. J., Uber die darstellbarkeit der nukleinsäuren des zellkernes und ihre bedeutung für diagnose und prognose maligner bluterkrankungen. *Folia Haematol.* **72,** 143–148 (1954).
116. Scherer, E., Ringleb, D., and Ventzke, L. E., Uber den einfluss der röntgenstrhlen auf den nucleolar-apparat der zellen. *Strahlentherapie* **90,** 41–52 (1953).
117. Seybolt, J. F., Papanicolaou, G. N., and Cooper, W. A., Cytology in the diagnosis of gastric cancer. *Cancer* **4,** 286–295 (1951).
118. Shea, J. R., and Leblond, C. P., Number of nucleoli in various cell types of the mouse. *J. Morphol.* **119,** 425–434 (1966).
119. Shein, H. M., and Enders, J. F., Transformation induced by simian virus 40 in human cell cultures. I. Morphology and growth characteristics. *Proc. Natl. Acad. Sci. U. S.* **48,** 1164–1172 (1962).
120. Siering, H., and Aderhold, K., "Leitfaden der nativcytomorphologie maligner tumoren." Fischer, Jena, 1956.
121. Smejkal, V., and Frahm, H., Zytochemische darstellung von RNS and DNS im aspirationspunktat der schilddrüse mit dünner nadel. *Zeit. Inn. Med. (Wien)* **23,** 399–403 (1968).
122. Smetana, K., and Busch, H., Studies on the ultrastructure of the nucleoli of the Walker tumor and rat liver. *Cancer Res.* **24,** 537–557 (1964).
123. Smetana, K., and Busch, H., Studies on staining and localization of acidic nuclear proteins in the Walker 256 carcinosarcoma. *Cancer Res.* **26,** 331–337 (1966).
124. Smetana, K., Shankar Narayan, K., and Busch, H., Quantitative analysis of ultrastructural components of nucleoli of the Walker tumor and liver. *Cancer Res.* **26,** 786–798 (1966).
125. Stenius, F., Studien über die pathologie und klinik der papillome und karzinome der harnblase. *Arb. Pathol. Inst. Univ. Helsingfors* **3,** 27–190 (1923).
126. Stenram, U., The volume and ribose nucleic acid concentration of nucleoli in liver and hepatoma cells of rat fed on high and non-protein diets. *Acta Pathol. Microbiol. Scand.* **44,** 239–246 (1958).
127. Stowell, R. E., Nucleic acid in human tumors. *Cancer Res.* **6,** 426–435 (1946).
128. Stowell, R. E., Alterations in nucleic acids during hepatoma formation in rats fed p-dimethylaminoazobenzene. *Cancer* **2,** 121–131 (1949).
129. Stowell, R. E., Lee, C. S., Tsuboi, K. K., and Villasana, H., Histochemical and microchemical changes in experimental cirrhosis and hepatoma formation in mice by carbon tetrachloride. *Cancer Res.* **11,** 345–354 (1951).
130. Sun, C. N., The occurrence of dense granules of unknown function in the nucleoli of certain plant cells. *Exptl. Cell Res.* **25,** 213–215 (1966).
131. The de, G., Rivière, M., and Bernhard, W., Examen au microscope électronique de la tumeur VX2 du lapin domestique derivée du papillome de shope. *Bull. Cancer* **47,** 569–584 (1960).
132. Tournier, P., Cathala, F., and Bernhard, W., Ultrastructure et développement intracellulaire du virus de la varicelle observé au microscope électronique. *Presse Med.* **65,** 1229–1234 (1957).
133. Unuma, T., and Busch, H., Formation of microspherules in nucleoli of tumor cells treated with high doses of actinomycin D. *Cancer Res.* **27,** 1232–1242 (1967).

134. Unuma, T., Morris, H. P., and Busch, H., Comparative studies of the nucleoli of Morris hepatomas, embryonic liver and aflatoxin B treated liver of rats. *Cancer Res.* **27**, 2221–2233 (1967).
135. Unuma, T., Smetana, K., and Busch, H., A morphologic study on the chromatin areas associated with the nucleolus. *Exptl. Cell Res.* **48**, 665–671 (1967).
136. Uzman, B. G., Foley, G. E., Farber, S., and Lazarus, H., Morphologic variations in human leukemic lymphoblasts after long term culture and exposure to chemotherapeutic agents. *Cancer* **19**, 1725–1742 (1966).
137. Vorbrodt, A., Histochemically demonstrable phosphatases and protein synthesis. *Exptl. Cell Res.* **15**, 1–20 (1958).
138. Vorbrodt, A., Electron microscopic studies on the localization of guanosine-5-triphosphatase activity in cell nuclei. *Acta Histochem.* **28**, 215–226 (1967).
139. Wachstein, M., Histochemistry of enzymes in tumors. *In* "Handbuch der Histochemie" (W. Graumann and K. Newmann, eds.), Vol. 7, Part 2, pp. 73–153. Fischer, Stuttgart, 1962.
140. Wessel, W., and Bernhard, W., Vergleichende elektronen mikroskopische untersuchung von Ehrlich und Yoshida-ascites tumor zellen. *Z. Krebsforsch.* **62**, 140–162 (1957).
141. Yasuzumi, G., and Suhigara, R., A comparative electron microscopic study on Ehrlich ascites tumor cells, Yoshida sarcoma cells and human cancerous peritonitis ascites cells. *Cancer Res.* **18**, 1167–1170 (1958).
142. Yasuzumi, Y., and Suhigara, R., The fine structure of nuclei as revealed by electron microscopy. *Exptl. Cell Res.* **37**, 207–229 (1965).
143. Zadek, I., Die cytodiagnostischen kernzeichen der krebs zellen. *Acta Med. Scand.* **80**, 78–92 (1933).

CHAPTER X

Effects of Drugs and Other Agents on the Nucleolus

I. Introduction

Like virtually all other cell structures, the nucleolus may either be functionally stimulated or inhibited by a variety of types of drugs. Although no new functions are introduced into nucleoli by drugs, the types of reactions that drugs induce are of great interest from both morphological and biochemical points of view. As noted in Chapters I and II, the metabolic activities of the nucleolus are normally very sensitive to the demands of the cell to the point that even the morphological composition of the nucleolus varies from cell type to cell type.

In the most extreme cases noted in Chapter IX, the composition of nucleoli of normal liver cells and normal lymphocytes are vastly different. In normal liver cells, the major nucleolar elements include the granular and fibrillar elements (nucleolonemas), discrete light areas (nucleolar vacuoles), and the perinucleolar as well as intranucleolar chromatin masses. In most cell types, these various structural elements are present in rather definite proportions. In the mature lymphocyte, there is a central core of light and apparently hypofunctional nucleolar mass which is surrounded by a peripheral ring fibrillar mass containing few granular elements. In the presence of inhibitory and stimulatory agents, these two cell types undergo remarkably different changes which exemplify the greatest variations in the cellular responses to drugs and other agents. At present, many studies on RNA inhibitors are being made on other cells including cells in tissue culture, pancreatic cells, and other cell types.

In liver nucleoli, the most common response to inhibitory stimuli includes a series of morphological changes which are referred to as (*a*) segregation—the masses containing the granular and fibrillar elements separate; (*b*) atrophy—the granular elements decrease in amount and the fibrillar residue undergoes shrinkage; and (*c*) dissolution or fragmentation—the nucleolar mass disintegrates into a variety of masses of smaller size and less defined cytoarchitecture. In the presence of stimulatory agents, liver nucleoli undergo marked granular hypertrophy; i.e., the relative amount of granular elements increases markedly over that of the fibrillar portions of the nucleolus and there may be disappearance of the nucleolar vacuoles and the light spaces so that a "compact" nucleolus develops.

In lymphocyte nucleoli, inhibitory agents may produce effects (*b*) and (*c*) above, but the relatively low activity of these nucleoli is reflected in their already limited rim of active surface which surrounds the central light area. In the presence of stimulatory agents, however, the lymphocyte nucleoli undergo truly remarkable transformation from the "hypofunctional" state to a state of great activity as indicated by the profusion of granular elements, the development of a "compact" nucleolus, and the enlargement of the nucleolus as a whole.

Although these types of responses are in a sense the prototypes of the nucleolar changes seen after stimulatory or inhibitory activity, they are only part of a wide spectrum of changes, which is one reason why the effects of these agents has been so interesting to students of the nucleolus.

II. The Inhibitory Agents

Important contributions to the field of nucleolar ultrastructure and nucleolar function have been made by morphological and biochemical studies on the inhibitory effects of various physical and chemical agents. It is now clear that there is a remarkable interdependence of normal nucleolar biosynthetic functions and morphological appearance on both protein and RNA synthesis. Interference with these synthetic reactions results in both inhibition of function and morphological distortion of the nucleolus. The uses of actinomycin D in obtaining biochemical information on synthetic rates in nucleoli by inhibition of nucleolar RNA synthesis was presented in Chapter V (see also p. 18).

A. Actinomycin D

By far the most commonly used agent for inhibitory studies on nucleoli is actinomycin D (Table 10.1). In their most interesting early report on

TABLE 10.1

References to Effects of Antibiotics Causing Nucleolar Segregation and Inhibition of RNA Synthesis

Antibiotic	Segregation	Inhibition of RNA synthesis	Direct inhibition of the RNA polymerase system
Actinomycin D	*4, 14, 34, 48, 58, 65, 95, 98, 112, 117, 118, 123, 136*	*4, 8, 10, 17, 22, 23, 30, 35, 36, 36a, 36b, 41, 46, 47, 58, 64b, 83–85, 99*	*10, 97, 97a, 144*
Aflatoxin B_1	*60, 89, 131–133, 143*	*5, 9, 15, 31, 55, 143*	*33, 56, 113*
Anthramycin	*37*		
Chromomycin A_3		*47*	*50*
Cordecypin		*13, 104b*	
Cycloheximide		*29, 47, 141a*	
Daunomycin	*19, 20, 105*		
Echinomycin		*32*	
Mithramycin	*54a*		
Mitomycin C	*57*		
Nogalomycin	*4, 47*		
Olivomycin	*47*	*32*	
Puromycin	*63*	*30, 40a, 130, 134, 135*	
Quinomycin A	*36c*		
Rifamycin		*32*	
Streptovaricin		*32*	
Triostin C	*36c*		

the effects of actinomycin D (*11*) on the nucleolus, Journey and Goldstein (*48*) noted a significant change in the appearance of the nucleoli. Their study was followed by a number of reports which showed that striking changes in nucleoli occurred as seen by light and electron microscopy. For example, Reynolds and Montgomery (*94–96*) reported that the actinomycin D-treated nucleolus seemed to be producing light and dark regions on its surfaces (Fig. 10.1); they called these regions the "nucleolar caps." Electron microscopic correlates of the formation of these caps revealed corresponding structures that were readily visible and the appearance of the caps could be correlated with structures that were apparently segregating. For example, the light caps appeared to be composed of the "fibrillar nucleolar elements" and the dark caps appeared to be composed of the "granular nucleolar elements." Some of these correlations are illustrated in Figs. 10.1 and 10.2 for actinomycin D and other agents.

After these investigations with actinomycin D, a number of other investigations on different cell types were made with a large number of other agents, including many other antibiotics (Table 10.1), toxic products such as aflatoxin (*18, 40*), other toxic substances (Table 10.2), and various physical agents and viruses (Table 10.3).

B. Mechanisms of Nucleolar Segregation

These agents all produce similar types of lesions as has been pointed out by Simard and Bernhard (*107*) and by Svoboda and Higginson (*132*). The granular elements of the nucleolus seem to separate from the fibrillar elements and the latter remain behind and ultimately disappear. Evidence that these events are morphologically similar as a result of treatment of cells with carcinogens, plant extracts, and general toxins is presented in Figs. 10.3–10.6. These morphological events are noted even in isolated nucleoli (Fig. 10.7).

To provide some clearer picture of the morphological events that were occurring in the nucleoli in which segregation of the morphological elements was occurring, Simard and Bernhard (*107*) provided the schematic pattern shown in Fig. 10.3. In Fig. 10.3a the granular and fibrillar masses are intermixed. However, inhibitors initially produce segregation of these elements (Fig. 10.3b) with new regions being visible, the pars amorpha (P) and a dense mass (SC). Later, the granular elements decrease and the pars amorpha enlarges (Fig. 10.3c). When the granular elements disappear, only the fibrillar regions remain as shown in Fig. 10.3d. Svoboda *et al.* (*131–133*) coined the term "macrosegregation" for the separation of the granular and fibrillar elements in large and distinct zones (Fig. 10.6). On the other hand, "micro-

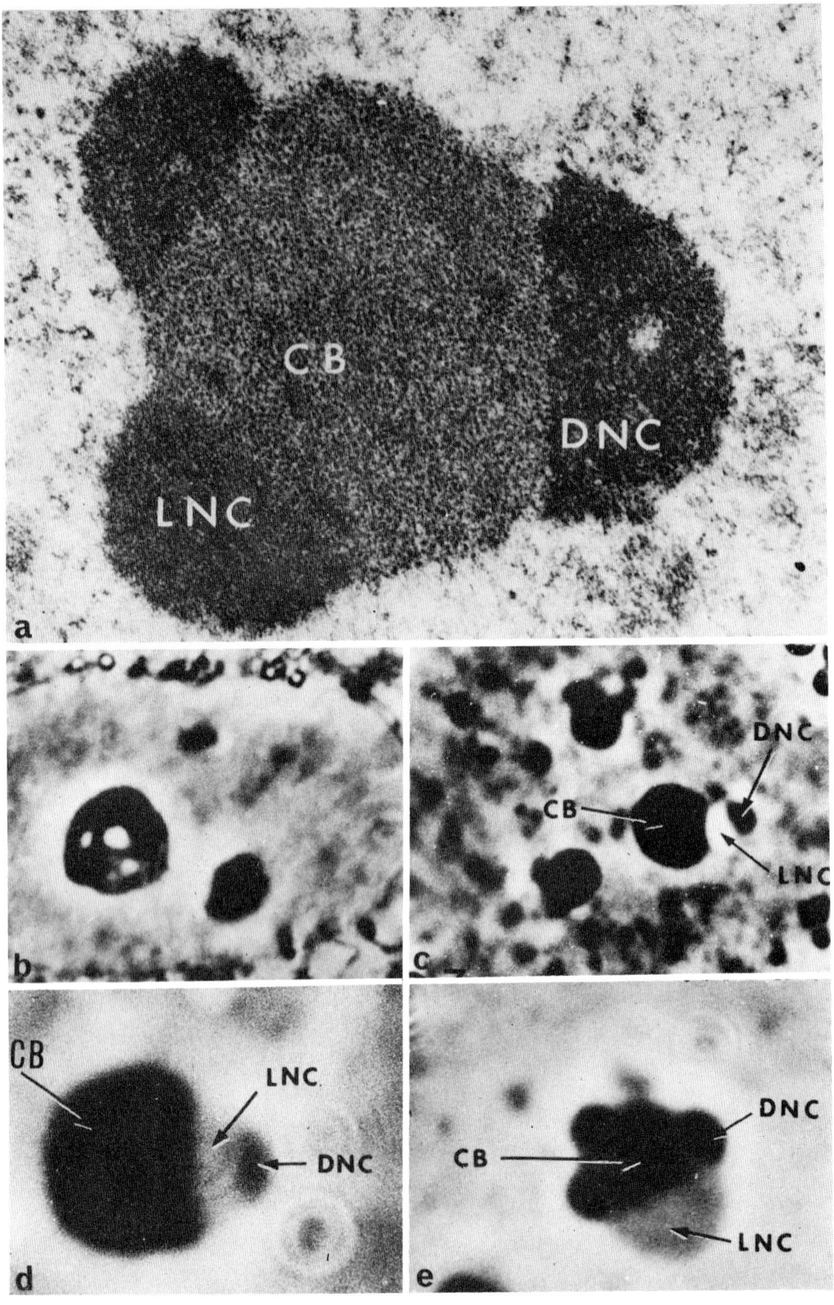

Fig. 10.1. a. Nucleolus of Chang liver cell treated with actinomycin D. Note nucleolar caps composed of granules of uniform density. There tends to be a very

TABLE 10.2

References to Various Chemical Substances Causing Nucleolar Segregation and Inhibition of RNA Synthesis

Compound	Segregation	Inhibition of RNA synthesis	Direct inhibition of the RNA polymerase system
Acridine orange	*94*		
Adenosine	*61, 62*		
Azaserine	*107*		
Benzimidazoles	*6, 109, 110*		
Cycloheximide	*141*		
Dimethylnitrosamine		*54*	
Ethidium bromide	*105*	*47*	
Fluorouracil	*119, 120*	*3, 42, 43*	
Lasiocarpine	*89, 91, 133*		*89, 90*
3′-Methyl-4-dimethylamino-azobenzene	*81a, 91*		
Nickel carbonyl	*130a*		*130a*
Nicotinamide		*93*	
Nitrogen mustard		*8, 88*	
4-Nitroquinoline-1-oxide	*2, 24–28, 36c, 49, 59, 96*	*66, 82a*	*82, 82a*
Proflavin	*94, 105, 121, 122*	*32, 47*	
Pyrrolizidine alkaloids	*133a*		
Tannic acid	*89, 133*		

segregation" is characterized by condensation of fibrillar components forming an interrupted or continuous rim around the granular components or by formation of plaque-like condensed fibrillar components in granular areas; the "caps" are not present (*131–133*) (Fig. 10.6).

distinct line of demarcation between dark nucleolar caps and the central body of nucleolus. ×30,000. b. Phase photomicrograph of normal Chang liver-cell nucleolus. Note clear vacuoles which may represent either fibrillar zones or amorphous areas seen in electron micrographs. ×3500. c and d. Nucleoli of Chang liver cells after treatment with proflavin and acridine orange, respectively. Light nucleolar caps rest on central body of nucleolus. Small dark nucleolar caps are seen resting on light nucleolar caps. Dark caps produced by acridine derivatives are generally smaller and more difficult to see with phase microscopy than those produced by actinomycin D and 4-nitroquinoline N-oxide. Figure 10.1c, ×3500; Fig. 10.1d, ×7500. e. Nucleolus of Chang liver cell after exposure to actinomycin D. Note prominent dark nucleolar caps resting on central body of nucleolus. ×6500 (*94*). CB = central body, LNC = light nucleolar caps, DNC = dark nucleolar cap (courtesy of Drs. Rollin Reynolds and Philip O'B. Montgomery, Veterans Administration Hospital and Southwestern Medical School, Dallas, Texas).

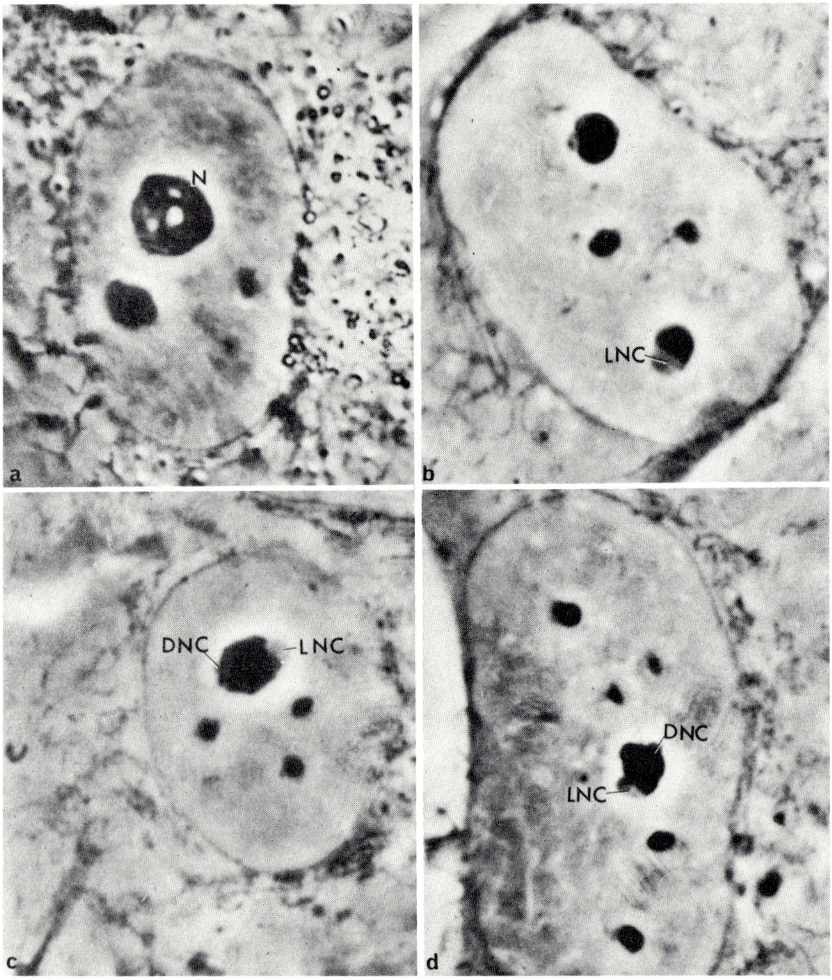

Fig. 10.2. a. Phase-contrast photomicrograph of untreated Chang liver cell seen with the oil-immersion phase microscope. The nucleolus (N) contains intertwining strands of dark contrast material. ×3500. b–d. Phase-contrast photomicrographs of Chang liver cells after treatment with 4-nitroquinoline N-oxide. Light nucleolar "caps" (LNC) have formed, and the margination of dark contrast material forms dark nucleolar "caps" (DNC). ×3500 (*96*) (courtesy of Drs. Rollin Reynolds and Philip O'B. Montgomery, Veterans Administration Hospital and Southwestern Medical School, Dallas, Texas).

TABLE 10.3

REFERENCES TO RADIATION AND INFECTIOUS AGENTS PRODUCING NUCLEOLAR SEGREGATION

	Segregation
UV-radiation	*7, 71–76, 85–87*
X-radiation	*39, 44, 69, 88*
Heat	*108*
Herpes simplex virus	*79, 111*
Mycoplasma	*46a*

C. Nucleolar Aggregates

Not infrequently, there are observed masses of segregating nucleolar elements that are referred to as "nucleolar aggregates." These are part of the process of "microsegregation" according to Svoboda *et al.* (*131–*

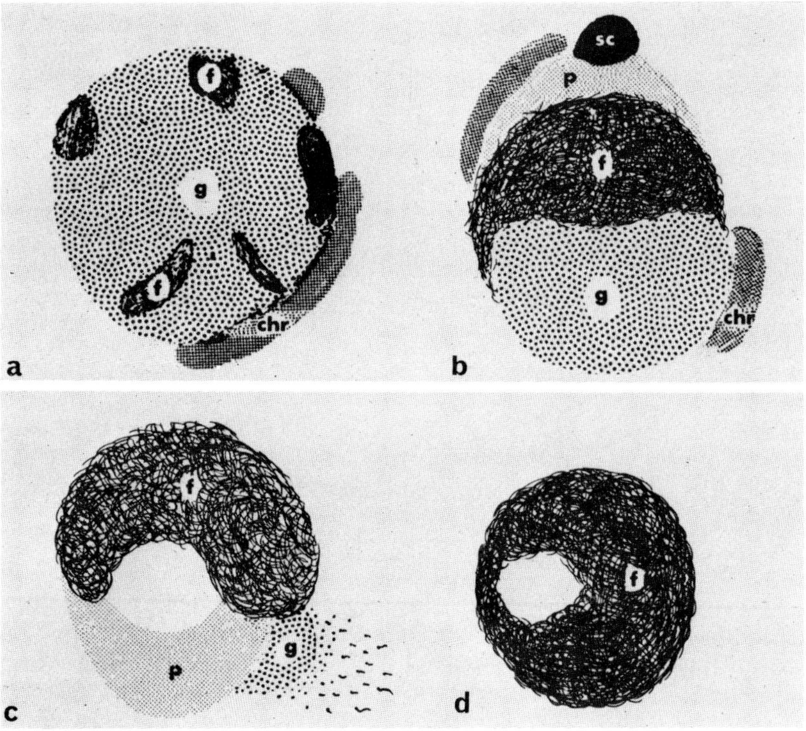

Fig. 10.3. Morphological sequence of nucleolar segregation. g, granular elements; f, fibrillar elements; p, pars amorpha; sc, dense substance; chr, nucleolus-associated chromatin (*107*).

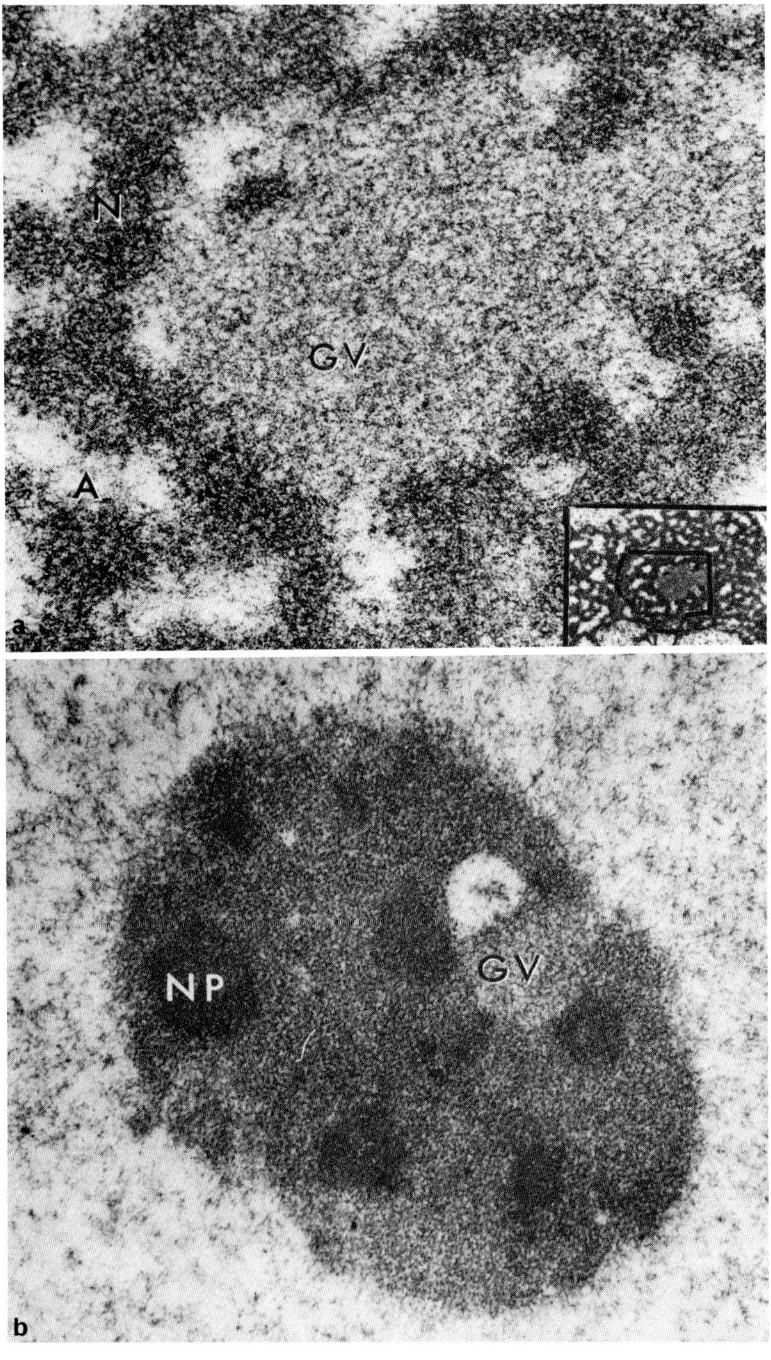
N
GV
A
a
NP
GV
b

133). These masses are usually dense structures which may contain relatively small amounts of fibrillar and granular elements of varying degrees of condensation (Fig. 10.9).

D. Nucleolar Dark Substance

Although the segregation of the granular and fibrillar elements of the nucleoli are the most readily recognized effect of nucleolar inhibitory agents, other changes are frequently noted. For example, in Fig. 10.8, there is the formation of a dark substance adjacent to the nucleolus (Fig. 10.3 SC). This may be the substance IV or the dark substance described by Bernhard and his associates (*5*). Although its composition is not defined, it may resemble the components of the nucleolar microspherules in that it contains dense ribonucleoproteins.

E. Formation of Nucleolar "Microspherules"

The macrosegregation effects of drugs that inhibit nucleolar RNA synthesis are generally seen in liver nucleoli or other nucleoli that have well defined nucleolonemas and light spaces. In very active, more compact nucleoli, such as those of cancer cells or thioacetamide-treated liver cells, the segregations are characterized by formation of multiple small bodies which are referred to as "microspherules" or "nucleolar plaques" (*94–96, 102, 136*). These are apparently composed of dense fibrillar elements. They are usually distributed around the nucleolar periphery (Figs. 10.10–10.12), but may be found within the nucleolar mass.

Although formation of "microspherules" may represent a special case different from the general segregation effects induced by inhibitors of nucleolar RNA synthesis, biochemical studies do not differentiate between these two effects (*102, 136;* see also Chapter V). Another possi-

Fig. 10.4. a. Normal Chang liver cell nucleolus. Inset at lower right shows entire nucleolus at lower magnification. Interlacing strands of nucleolonema are composed of granules measuring up to 200 Å in diameter. Between strands are spaces resembling surrounding nucleoplasm and designated amorphous areas (A). Other areas composed of granules of intermediate density are designated fibrillar zones. Fibrillar zones give rise to light nucleolar caps. ×87,000 (*94*). b. Nucleolus of Chang liver cell 30 minutes after exposure to proflavine. Note segregation of darker granules of nucleolus and disappearance of nucleolonema as an organized structure. Segregation of darker granules results in formation of extremely dense bodies (previously designated nucleolar plaques—NP) scattered throughout nucleolus. Amorphous areas have largely disappeared. GV—nucleolar vacuole ×35,000 (*94*). (Courtesy of Drs. Rollin Reynolds and Philip O'B. Montgomery, Veterans Administration Hospital and Southwestern Medical School, Dallas, Texas.)

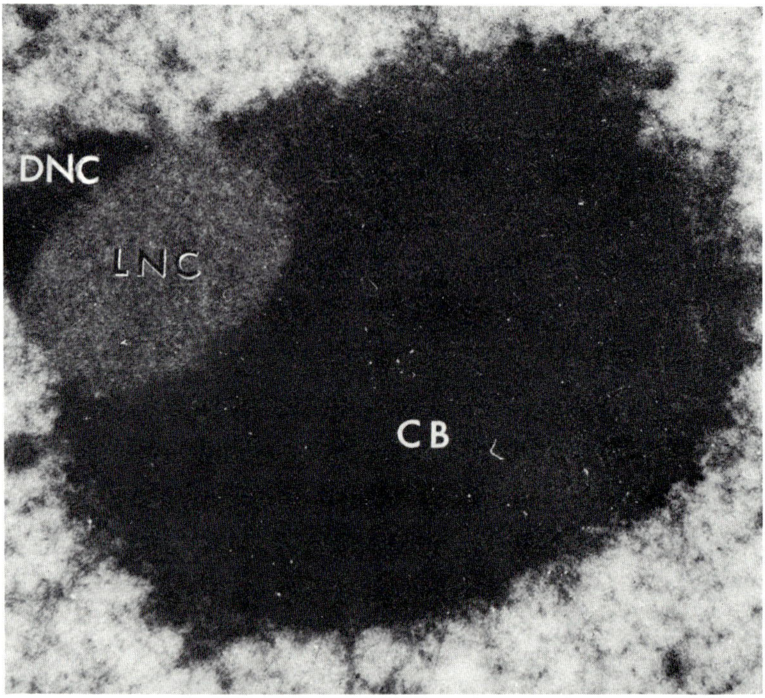

Fig. 10.5. Segregation of nucleolus produced by UV-radiation (*94*). (Courtesy of Drs. Rollin Reynolds and Philip O'B. Montgomery, Veterans Administration Hospital and Southwestern Medical School, Dallas, Texas.)

bility is that the nucleoli which show this phenomenon arise from fusion of a number of nucleolus-forming sites and the multiple segregations result from segregational events that occur at each of the NOR's involved (*100a;* see also Chapter III).

F. The Nature of the Segregation Products and Microspherules

Studies that have been made on the variety of nucleolar segregation products have utilized two major types of methods; i.e., enzymatic digestion of the nucleolar elements with DNase, proteinases, and RNase, and analysis of the nucleolar RNA after treatment with the drugs. Generally, treatment of the segregated nucleolar masses with either DNase or pepsin has little effect on their electron microscopic appearance or density of the structures (Figs. 10.12 and 10.13).

On the other hand, treatment of the sections with RNase resulted in markedly decreased density in the sections (Fig. 10.12) in both "macrosegregation" or microspherules. Simultaneous treatment of the micro-

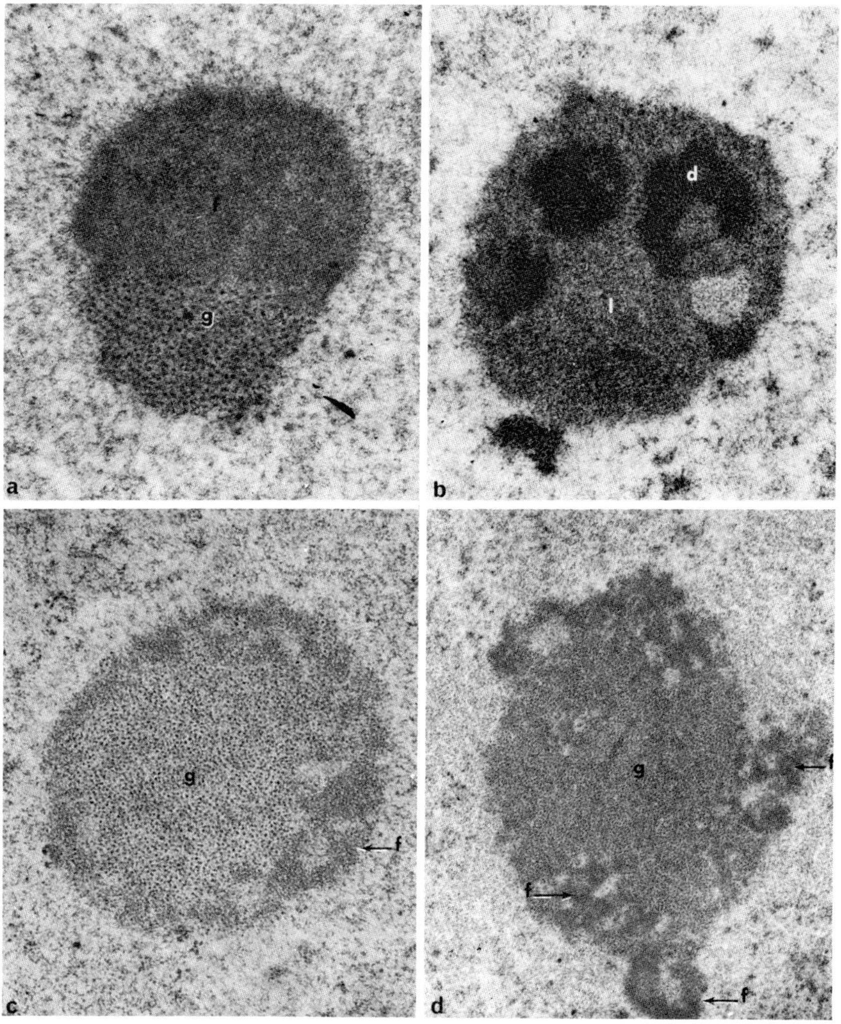

Fig. 10.6. a. Segregation of nucleolar components in liver cells. Macrosegregation produced by lasiocarpine is characterized by distinct demarcation of fibrillar (f) and granular (g) zones. ×41,500. b. Macrosegregation produced by tannic acid. In this form of segregation dense aggregates (white d) remain in the interior of the nucleolus, and light zones (white I) contain both fibrillar and granular components. ×41,500. c. Microsegregation produced by 3′ methyl-dimethylaminoazobenzene. The granular area (g) is surrounded by fibrillar components (f). ×21,000. d. Microsegregation produced by dimethylnitrosamine. The granular area (g) is associated with interrupted aggregates composed predominantly of fibrillar components (f). ×21,000 (*131–133*). (Courtesy of Drs. D. Svoboda, A. Racela, and J. Higginson, Department of Pathology and Oncology, University of Kansas School of Medicine, Kansas City, Kansas.)

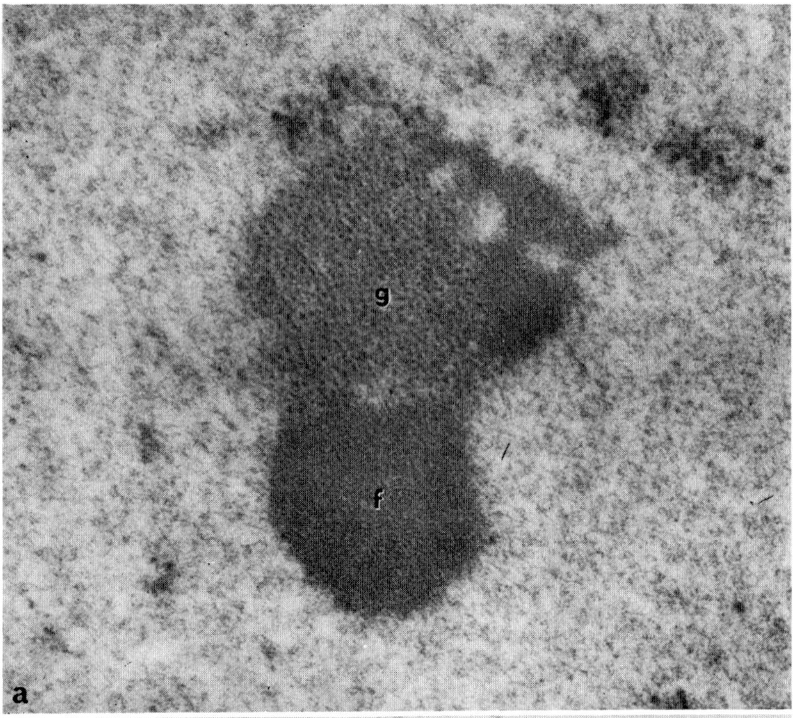
g
f
a

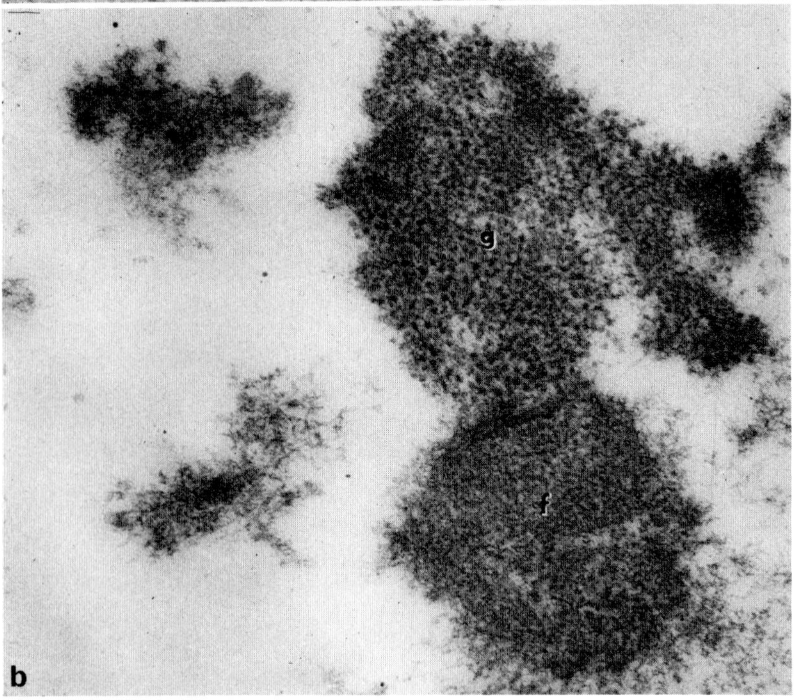
g
f
b

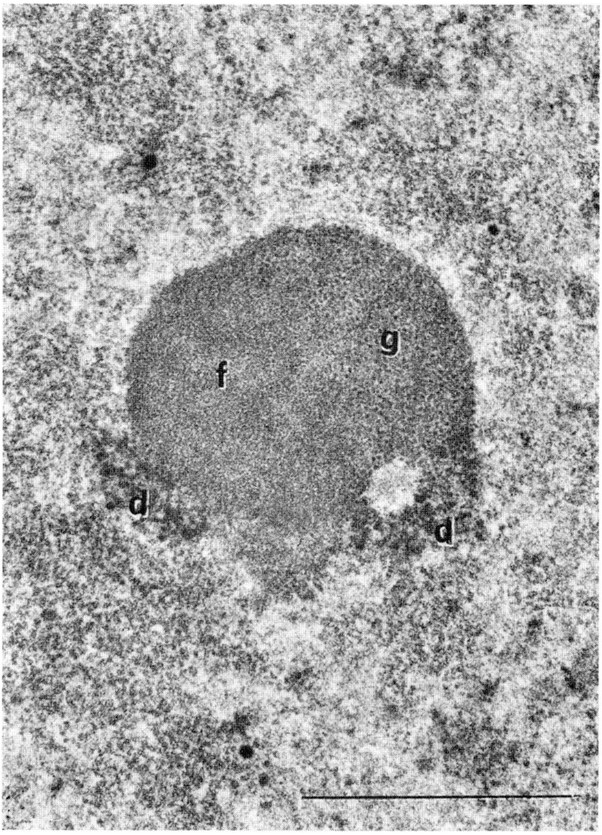

Fig. 10.8. Nucleolus of rat liver showing effects of 500 μg/kg of aflatoxin B_1 at 1 hour after administration of the drug. The nucleolus was compact and rounded with segregation of granular (g) from fibrillar (f) components; the presence of dark material (d) is noted. Particles (arrows) were observed inside the dark material. The specimens were fixed with osmium tetroxide, followed by 10% formalin, embedded in Epon Araldite, and stained with uranyl acetate and lead citrate. ×37,000 (*30a*).

Fig. 10.7. a. Nucleolus *in situ* 24 hours after injection of lasiocarpine. The fibrillar (f) and granular (g) constituents of the *in situ* nucleolus are segregated. ×60,000 (*37a*). b. Isolated nucleolus 24 hours after injection of lasiocarpine. The segregated state of the nucleolar granules (g) and fibrils (f) is preserved in this isolated nucleolus. ×60,000 (*37a*). The method for isolation is the sonication procedure (Chapter XI). (Courtesy of Drs. D. Svoboda, A. Racela, and J. Higginson, Department of Pathology and Oncology, University of Kansas School of Medicine, Kansas City, Kansas.)

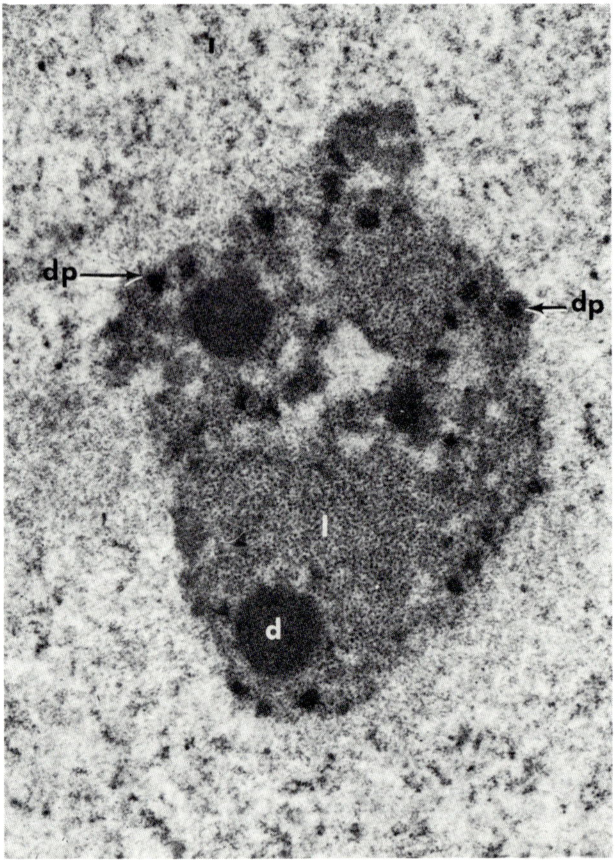

Fig. 10.9. Nucleolus of a rat liver cell 18 hours after the administration of tannic acid. Dense areas (white a), lighter areas composed of fibrillar as well as granular components (white 1) and dense plaques (black dp) are apparent. ×41,500 (*133*). (Courtesy of Drs. D. Svoboda, A. Racela, and J. Higginson, Department of Pathology and Oncology, University of Kansas School of Medicine, Kansas City, Kansas.)

spherules with pepsin and ribonuclease did not produce any greater decrease in density than that observed with ribonuclease alone (Fig. 10.13).

It is of interest that even after drastic treatment with both pepsin and RNase, the microspherules in the nucleoli still retain some of their ultrastructural features; i.e., they still appear to be faint fibrillar masses. This retention of structure has given rise to the speculation that other substances, such as lipids, may be present in these abnormal nucleolar bodies (*136*).

Analytical studies on the nucleolar RNA after treatment with actino-

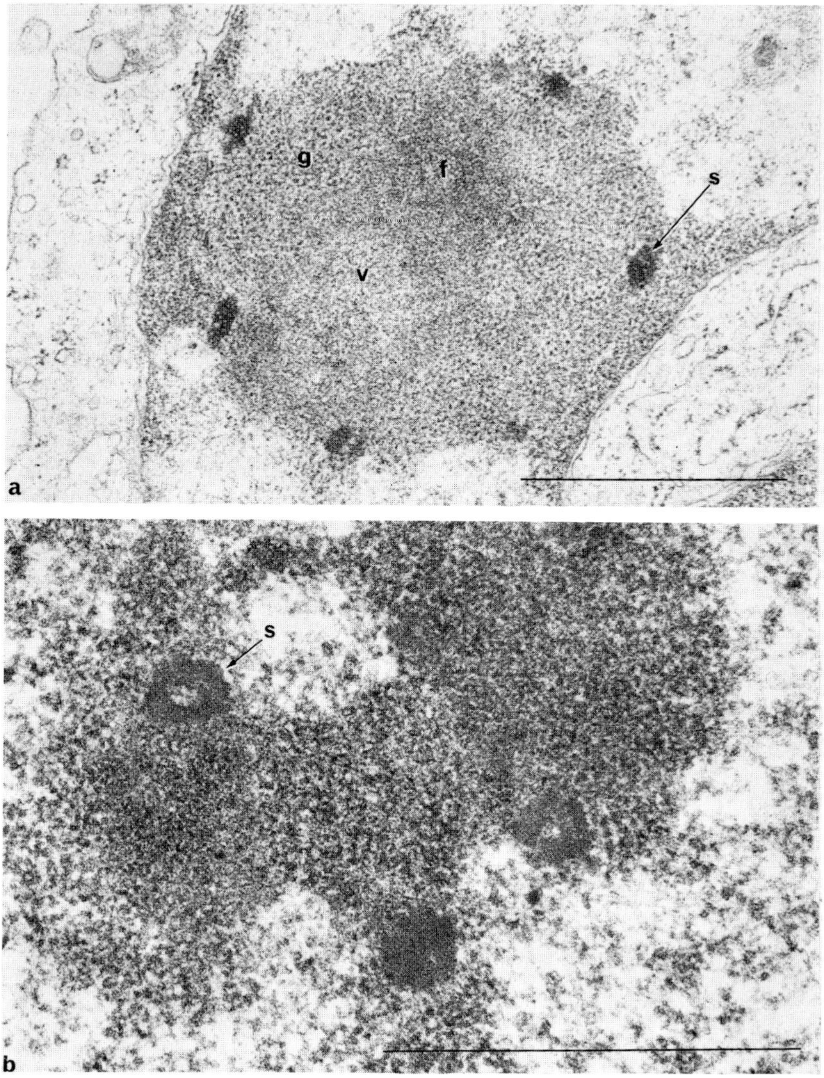

Fig. 10.10. a. A nucleolus of an Ehrlich ascites cell 60 minutes after administration of 1.6 mg/kg of actinomycin D. Fixation was carried out in osmium tetroxide and formalin. Several dense microspherules (s) are apparent. ×35,000. b. High magnification of the microspherules seen in the nucleoli of Ehrlich ascites cells treated with actinomycin D. The microspherules (s) contain dense fibrils. Some of them have a centrally located light area. The specimen was fixed in osmium tetroxide and postfixed in formalin. ×55,000 (*136*).

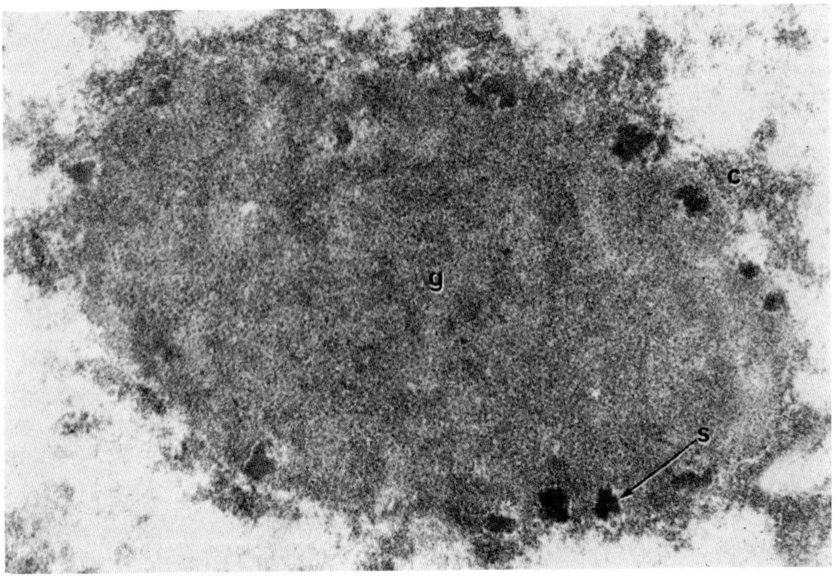

Fig. 10.11. A nucleolus of a Novikoff hepatoma cell, 60 minutes after treatment with 10 μg/ml of actinomycin D *in vitro*. Microspherules (s) are evident as are the halos around them. The microspherules are located almost exclusively at the periphery of this nucleolus. The specimen was fixed in glutaraldehyde and osmium tetroxide. ×34,000 (*136*).

mycin D and aflatoxin B_1 have shown that the main effect is the disappearance of the bulk of the rapidly sedimenting nucleolar RNA, including the 35 S, 45 S, and the 55–60 S RNA (Fig. 10.14, also Chapter V). The 28 S and 4–8 S nucleolar RNA persist longer. Because of the increasing proportion of the 4–8 S RNA in the analytical gradients (Fig. 10.15), as the proportion of fibrillar mass increases, some of the elements in the fibrillar mass probably consist of the nucleolar 4–8 S RNA. As noted in Chapter VI, this portion of the nucleolar RNA has a slower turnover rate than the rapidly sedimenting RNA and presumably represents a more stable nucleolar component than the export products which are the precursors of rRNA (see Chapter V) .

III. Mechanisms of Nucleolar Segregation

A. Mechanism of Inhibition of Synthesis of RNA

Although a wide variety of agents of dissimilar chemical structures and physical characteristics cause the nucleolar elements to segregate, in practically all cases the only common feature of these events is the inhibition of nucleolar RNA polymerase. Changes in the primer, alterations in the subunits or total structure of the RNA polymerase, or changes in the availability of nucleotides may alter the activity of the RNA polymerase system.

Most inhibitory drugs apparently either inactivate or destroy the DNA templates involved in the synthesis or ribosomal precursors (*21, 38, 66, 77, 92, 100, 101, 106*).* As summarized by Waring (*140*), mitomycin C, actinomycin D (*96a*), proflavine (*64a*) and ethidium appear to intercalate into the DNA structure. Proflavin is apparently intercalated in a planar relationship between the bases of DNA and the strains induced result in the inactivity of the DNA as a template. In a similar manner, mitomycin, acridine, and other agents may serve to bridge the DNA strands and thereby function to alter the template activity of the molecule. Actinomycin D, which binds to deoxyguanosine residues, also serves to alter the secondary and tertiary structure of the DNA double helix and thereby alters its priming ability. Inhibition of nucleolar activity by heat, X-radiation, or UV radiation results from modification of nucleolar DNA by scission of DNA strands, base changes, or strand separation.

Secondary effects ensue after the DNA is altered including such physical changes as the condensation of the DNA-protein complex. The loss of activity of this complex is associated with such marked changes as macrosegregation in which the fibrillar mass apparently condenses as a single mass or as microspherules consisting of many small masses.

Although the synthesis of 35 S and 45 S nucleolar RNA either slows or ceases and finally these precursors of rRNA disappear from the nucleolus (Fig. 10.14), the various inhibitory agents differ in their chemical effects. For example, as shown in Table 10.4, aflatoxin B_1, which produces a marked shutoff of synthesis of 35 S and 45 S nucleolar RNA, does not alter the ^{32}P nucleotide composition of the newly synthesized RNA. Such data suggest that this inhibitor produces a general block of biosynthetic reactions of the nucleolus. On the other hand, actinomycin D produces a block that has a higher degree of specificity for the biosynthesis of the nucleolar GC-rich 45 S RNA and, as a result, the RNA produced in the presence of this inhibitor is higher in its content of AMP and UMP than the whole nucleolar RNA. Whether this difference is related to the carcinogenicity of aflatoxin B_1 is not clear at present.

* Some antitumor agents, such as mustards, not only bind to DNA but also inhibit the replication of nucleolar DNA. Vinblastine sulfate induced a greater inhibition of nucleolar DNA synthesis than extranucleolar DNA synthesis (*16*).

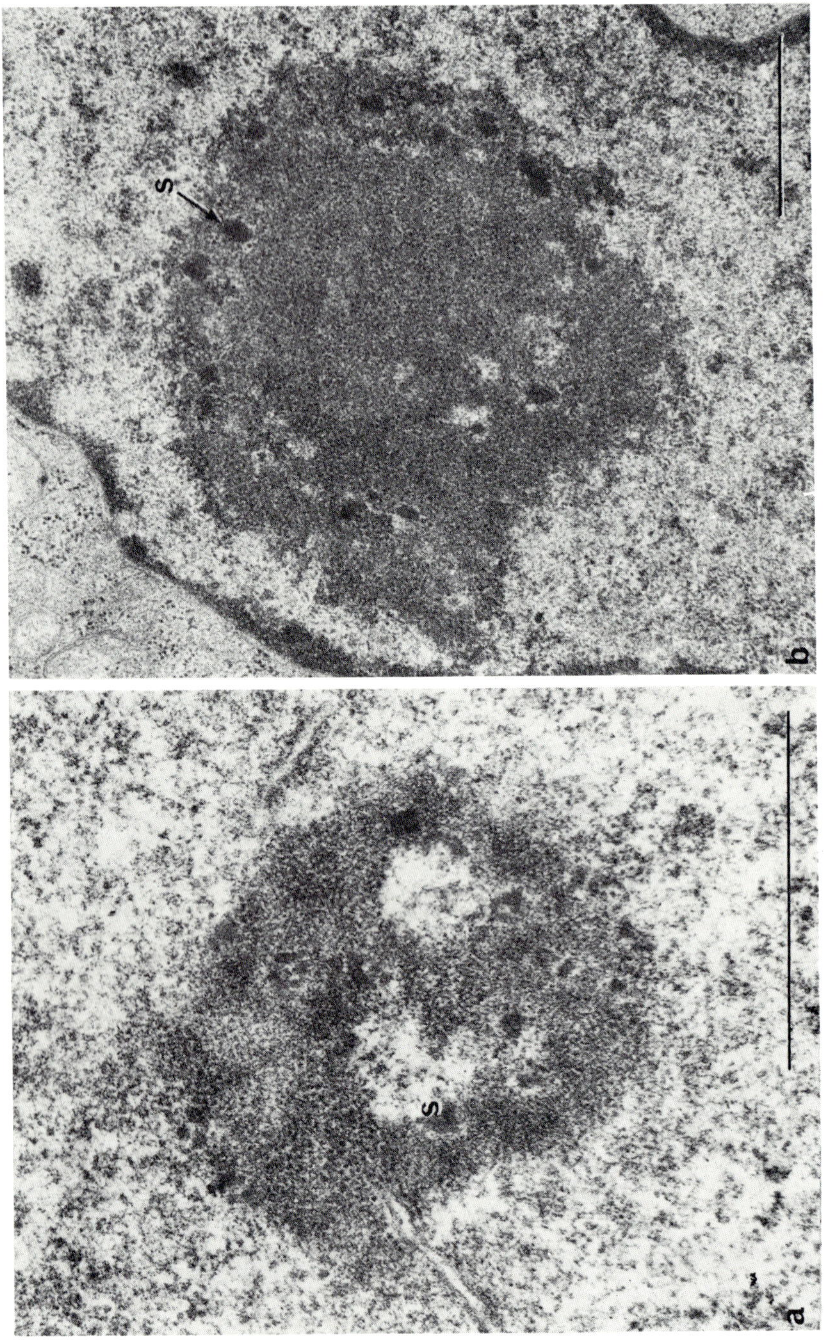
s
b
s
a

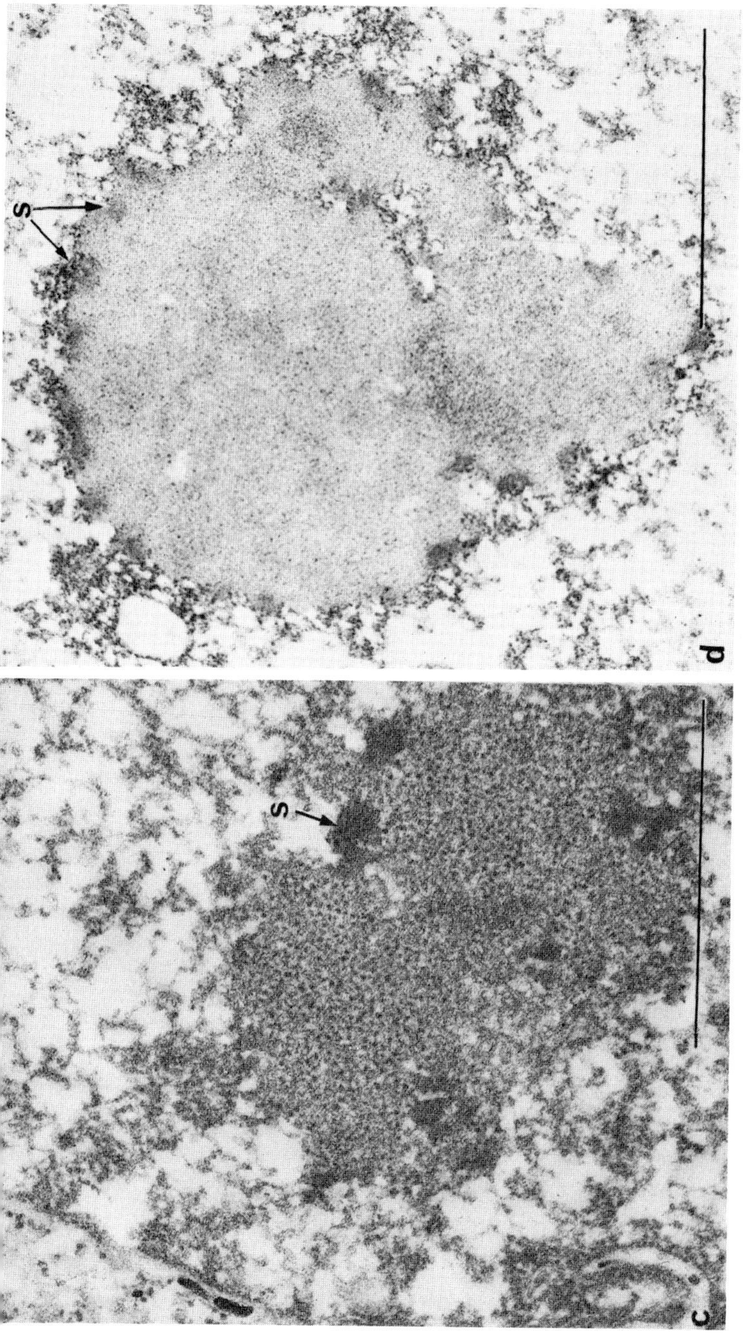

Fig. 10.12. a. A nucleolus of a Novikoff hepatoma cell, 60 minutes after treatment with a dose of actinomycin D of 800 μg/kg of body weight *in vivo*. The specimen was fixed in osmium tetroxide and formalin. $\times$48,000. b. The same specimen as in Fig. 10.12a, but fixation was carried out in formalin. $\times$24,000. c. Pepsin treatment. The density of the microspherules (s) was reduced slightly. The granular and fibrillar components of the nucleolus appeared more sharply after treatment. The specimen was the same as in Fig. 10.12b. d. Ribonuclease treatment. The density of the microspherules was decreased remarkably. All the granular elements inside the nucleolus disappeared. The perinucleolar chromatin pattern was preserved. The specimen was the same as in Fig. 10.12c. $\times$40,000 (*136*).

TABLE 10.4

^{32}P-NUCLEOTIDE COMPOSITION OF HIGH MOLECULAR WEIGHT NUCLEOLAR RNA AFTER ADMINISTRATION OF AFLATOXIN B_1 AND ACTINOMYCIN D (*30a*)

	45 S	55 S
Control		
AMP	18.7	20.2
UMP	16.6	17.3
GMP	38.1	36.4
CMP	26.5	26.0
AMP + UMP/GMP + CMP	0.57	0.60
Aflatoxin B_1 (50 μg/kg body weight)		
AMP	18.2	18.4
UMP	17.3	17.4
GMP	37.4	37.7
CMP	27.0	26.5
AMP + UMP/GMP + CMP	0.55	0.56
Actinomycin D (20 μg/kg body weight)		
AMP	22.4	22.3
UMP	21.9	21.8
GMP	30.0	32.6
CMP	25.8	23.2
AMP + UMP/GMP + CMP	0.79	0.79

B. Continuing Migration of Nucleolar Elements

Although the mechanisms of movement of nucleolar products to the cytoplasm are not defined (Chapter VIII), the effects of agents producing segregation of the nucleolar elements support the idea that dynamic rather than passive processes are involved. It appears that the movement of the nucleolar products is centrifugal and that some sort of "push-pull" mechanism, possibly involving protein binding and contraction, is active. Whether these involve fine fibrillar elements that are "guy-wires" from the cytoplasm that pull the nucleolar products outward or a continuing internal pressure that extrudes nucleolar elements to the cytoplasm is not yet clear. In any event, these mechanisms operate to produce a slow loss of dense nucleolar granular elements and the conversion of the nucleolus to the fibrillar mass as noted previously (Fig. 10.3).

IV. Stimulation of the Nucleolus

None of the mechanisms involved in the stimulation of the nucleolar synthesis of rapidly sedimenting RNA precursors of rRNA has been

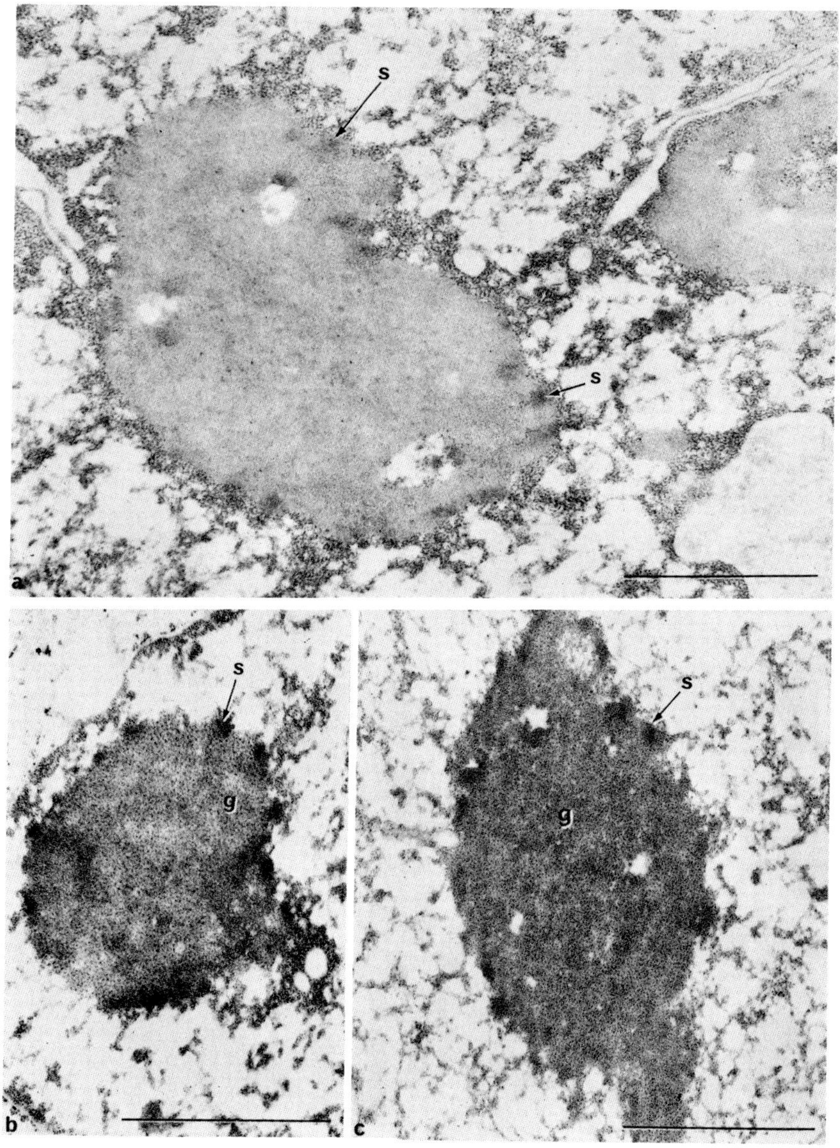

Fig. 10.13. a. Pepsin and ribonuclease treatment of same specimen as Fig. 10.12. The density of the microspherules (s) was decreased markedly compared with Fig. 10.12c. All the granular elements of the nucleolus and all the cytoplasmic ribosomes disappeared. The pepsin-resistant matrix of the nucleolus is visible. ×26,000 (*136*). b. Deoxyribonuclease treatment. The density of the microspherules (s) was not significantly decreased. The granular elements of the nucleolus were intact but the nucleolus-associated chromatin was removed. The specimen was the same as in Fig. 10.12b. ×28,000 (*136*). c. Pepsin and deoxyribonuclease treatment. The density of the microspherules (s) was decreased slightly, compared with Fig. 10.13b. The granular elements in the nucleolus appeared more sharply. No nucleolus-associated chromatin was observed but the nuclear ribonucleoprotein network in the nucleus was preserved. The specimen was the same as in Fig. 10.12b. ×26,000 (*136*).

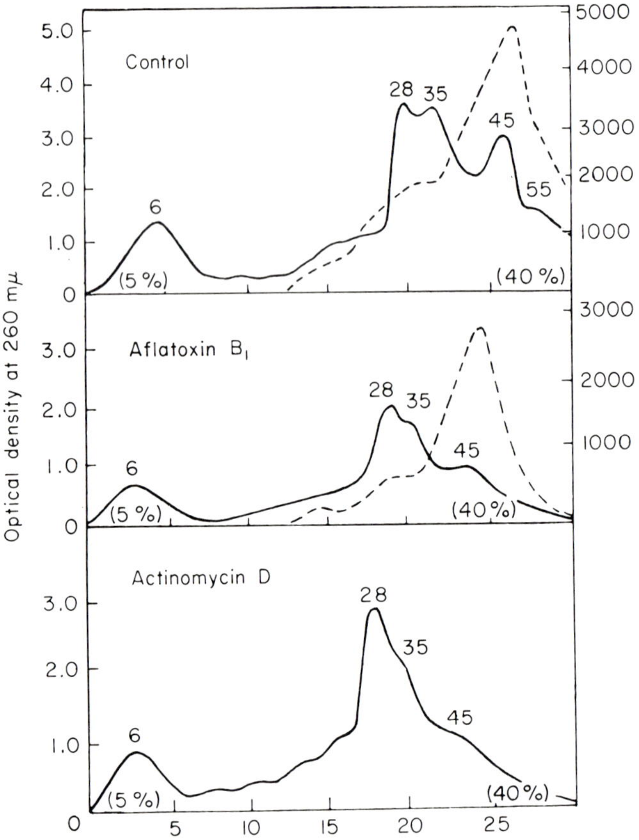

Fig. 10.14. Sucrose density gradient sedimentation patterns for nucleolar RNA of normal liver. The approximate sedimentation coefficients are shown above the peaks and the shoulders. The distributions of radioactivity after a 10-minute pulse of ^{14}C-2-orotic acid are shown as dashed lines. The dose for aflatoxin B_1 was 50 μg/kg, i.p. while that for actinomycin D was 20 μg/kg, i.v. Abscissa: Fraction number; Ordinate: (left) OD at 260 mμ; (right) dpm/fraction (*30a*).

adequately clarified but almost any condition that requires either increased protein synthesis or new cell formation produces increased nucleolar activity. A few drugs and some hormones have been found to produce increased nucleolar activity either by feedback mechanisms or by direct influences on the nucleolus. One example of the former will be discussed in this chapter; i.e., the effects of thioacetamide on the normal rat liver. The effects of phytohemagglutinin on the lymphocytes are similar but the mechanisms are not clear.

A. Thioacetamide

Thioacetamide, which produces huge increases in nucleolar size, causes a variety of effects on the liver and kidney (*125–129*) cells, including cytoplasmic alterations of the lysosomes (Chapter VII) and the release of proteinases and RNase, which presumably destroy the RNA and protein of the ribosomes (*51, 52, 78, 124*). As pointed out by Steele (*114*), the half-life of ribosomal RNA in the thioacetamide-treated liver cell is only 60% that of the normal liver cell; i.e., about 3.5 days as compared to the normal half-life 5.5 days. This effect is associated with a

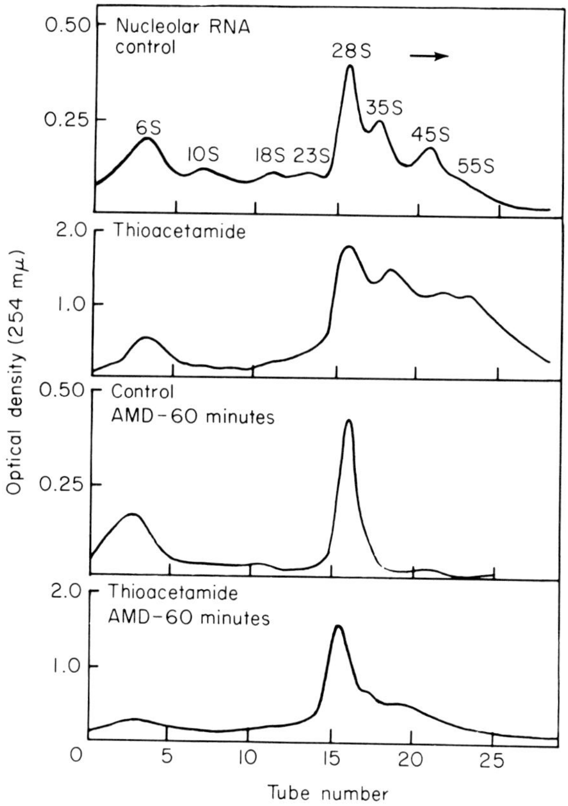

Fig. 10.15. Sedimentation profiles in sucrose density gradients (5–40%) of RNA of isolated nucleoli of control and thioacetamide-treated rats. In samples labeled AMD, animals were treated 60 minutes previously with actinomycin D (150 μg/kg body weight, i.v.). Thioacetamide (50 μg/kg body weight) was administered daily for 9 days. The numbers above the peaks represent approximate sedimentation coefficients. The arrow indicates the direction of sedimentation (*102*).

TABLE 10.5

WEIGHTS OF PROTEINS OF CELLULAR FRACTIONS[a] (*78*)

Group	Whole homog-enate	Mito-chondria	Micro-somes	Cyto-plasmic sap	Nuclear: Acid-in-soluble	Nuclear: NaCl soluble	Nuclear: His-tones
Control	185	14.9	25.6	48.5	8.3	1.6	5.7
Thioacetamide[b]							
1 day	173	12.5	20.8	44.0	7.8	1.6	5.2
9 days	178	14.3	15.3	45.5	6.6	1.5	4.6
Regenerating liver	159	12.1	20.3	40.5	7.9	1.7	5.0
Thioacetamide-treated (9 day) regenerating liver[c]	148	9.5	12.7	36.3	5.9	1.0	2.8

[a] Expressed as milligrams per gram wet weight of liver.

[b] Thioacetamide was administered in a daily dose of 50 mg/kg body weight.

[c] For the thioacetamide-treated regenerating liver experiments, the animals were treated with thioacetamide for 9 days prior to hepatectomy.

remarkably reduced level of ribosomal and other cytoplasmic proteins in these cells (Table 10.5). It would appear that feedback mechanisms which are yet undefined serve to inform the nucleus and the nucleolus that there is a need for new synthesis of rRNA and ribosomes and their response is associated with remarkable changes in morphology and increased functional activity of the nucleolus (*1, 51, 52, 102, 115, 125–129, 137–139*).

By contrast with the normal liver nucleolus with its readily distinguishable nucleolonemas, defined amounts of fibrillar and granular elements, and many light spaces, the thioacetamide-treated nucleolus is characterized by its compactness (Fig. 2.21) high concentrations of granular elements, its loss of light spaces, and the apparent disappearance of fibrillar elements. Along with the enlargement of the nucleoli and changes in its ultrastructural morphology in the thioacetamide-treated rat livers, there are marked changes in the nucleolar RNA (*1, 115*). The concentrations of the precursors of rRNA markedly increase, as shown in Fig. 10.15. In addition, the relative proportions of RNA ranging in sedimentation coefficient from 28 to 85 S are much greater than those of RNA with sedimentation coefficients of 4–8 S. Such results indicate that the granular elements of the nucleolus are the elements that contain the rapidly sedimenting RNA (*53, 102*).

The fibrillar elements decrease in their relative proportions in the thioacetamide-treated nucleolus. This result correlates with the concept that these elements contain the 4–8 S RNA. These results are consistent with the idea developed from similar studies on actinomycin D-treated nucleoli (*102;* Chapters I, II, V, VI; see Figs. 2.21, 2.22).

B. The Fate of the Nucleolar Product in Thioacetamide-Treated Cells

What happens to the nucleolar product in thioacetamide-treated cells is not clear. The nuclear ribonucleoprotein network which was readily demonstrated in tumor cells and liver cells appears to be attenuated in these cells. This finding may reflect the destructive effects of the released lysosomal enzymes on these preparations. It would appear that the newly synthesized RNA is protected from destructive enzymes in the environment as it is formed and while it remains in the nucleolus. However, when the RNA migrates toward the nuclear membrane of thioacetamide-treated cells, it would seem to be more exposed to RNases than in either normal liver or neoplastic cells. This may reflect a deficiency of protective systems for RNA, possibly because of lack of formation of protective proteins. On the other hand, as noted in Chapter VII, the increased concentrations of nuclear and nucleolar RNase may overwhelm the usual protective mechanisms.

The increased activity of nucleoli of thioacetamide-treated cells seems to reflect feedback stimulation which results in markedly increased synthesis of nucleolar products. These biosynthetic reactions apparently progress unmolested in the nucleolus but when the products leave the nucleolus, they are apparently subjected to destruction in the nucleoplasm and in the cytoplasm by enzymes released from lysosomes.

C. Phytohemagglutinin and the Lymphocytes

Although the total functions of lymphocytes have not been fully defined (*1a, 12a*), both the lymphocytes in lymph nodes and those in the peripheral blood may have several functions. These cells have been assigned the very important role of synthesis of the gamma globulins and/or cellular immunity and hence they play a vital role in host defense mechanisms. The lymphocyte may also serve as a pericellular and intracellular scavenger of debris and infectious agents (Chapter IX).

Normally, lymphocytopoiesis is limited to specific regions, including defined lymph nodes, intestinal Peyer's patches, tonsils, the appendix, and small lymphoid masses that are generally distributed throughout the body. Lymphocytes are probably not formed in the peripheral blood. Although the mature lymphocytes in the tissues and peripheral blood were thought of as "end-stage" cells, incapable of further division, the

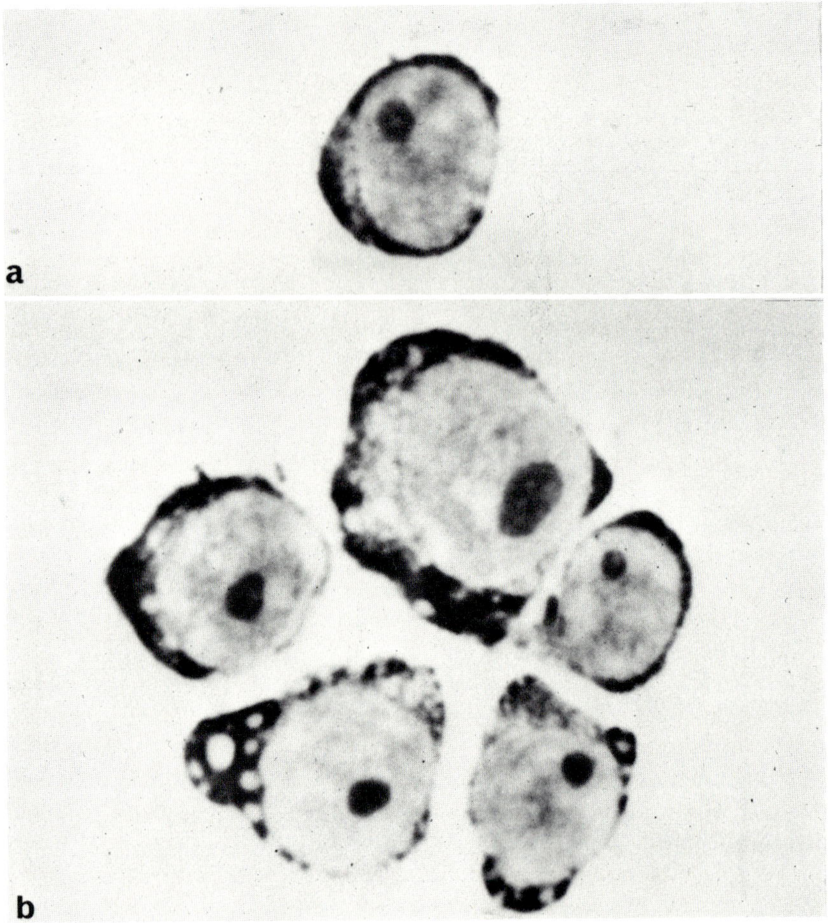

Fig. 10.16. a. Ring-shaped nucleolus of a mature lymphocyte. b. Nucleolar enlargement in lymphocytes cultured 6 hours in the presence of phytohemagglutinin (from Potmesil and Smetana, *88a*).

recent findings of the remarkable stimulation of the nucleolus by antigenic stimuli or phytohemagglutinin have served to dispel this concept. When exposed to this drug, the lymphocyte not only undergoes remarkable "activation" with respect to its internal synthetic reactions but, in addition, reproduction occurs to form new lymphocytes (*1a, 12a, 89a, 143a*). Figure 10.16 shows the remarkable increase in nuclear and nucleolar size of phytohemagglutin-stimulated lymphocytes.

The mechanisms by which phytohemagglutinin activates the lymphocytes are still under investigation (*99*). Virtually all types of effects have been postulated from direct effects on the genome to a variety of nucleolar, nucleoplasmic, cytoplasmic, and membrane stimuli (*12, 99*). Which, if any of these, constitutes the essential stimulus remains to be determined.

D. Electron Microscopic Analysis—Lymphocytes

Electron microscopic analysis of the changes in nucleolar structure after incubation of lymphocytes with phytohemagglutinin have been made by Tokuyasu (*135a*). Figure 10.17 shows that in the normal lymphocyte, the chromatin is clumped and the nucleolus is the small ring-shaped structure discussed in Chapter IX. Under the influence of phytohemagglutinin, there are minor changes in the lymphocyte at 4 hours. At 24 hours, there are marked nuclear changes; i.e., the clumped chromatin now has become largely dispersed and there is a huge increase in nucleolar activity associated with dispersion of the chromatin masses within the nucleolus. A marked increase in the granular and fibrillar elements of the nucleolus occurs (Fig. 10.17). These changes persist for at least 48 hours as shown in Fig. 10.17. The marked irregularity of these nucleoli and their increased granular elements are similar to the changes noted in the nucleolus of thioacetamide-treated cells and may reflect the preparation of the cell for markedly increased ribosome synthesis, protein synthesis, and cell division. These treated cells resemble lymphoblasts in their ultrastructure and functional state.

Biochemical studies on the nuclear products of the lymphocytes stimulated by phytohemagglutinin have supported the concept that there is markedly enhanced synthesis of ribosomal precursors (*12′, 38a*). These studies lend support to the idea that the nucleoli of these cells have markedly increased activity. Thus far, however, studies have not been reported on isolated nucleoli of these cells.

V. Stimulatory Effects of Hormones on the Nucleolus

As noted in Chapter V, a number of hormones increase the biosynthesis of ribosomal RNA. Relatively few specific studies on nucleolar changes are reported (p. 46). It is well established that ecdysone, a growth hormone of *Chironomus,* produces puffing (*64*), and hydrocortisone and adrenalectomy enhance RNA synthesis and probably increase nucleolar activity (*45, 81, 142*). Like protein deficiency (*103, 104, 116*), these hormones produce increased stresses on ribosomal syn-

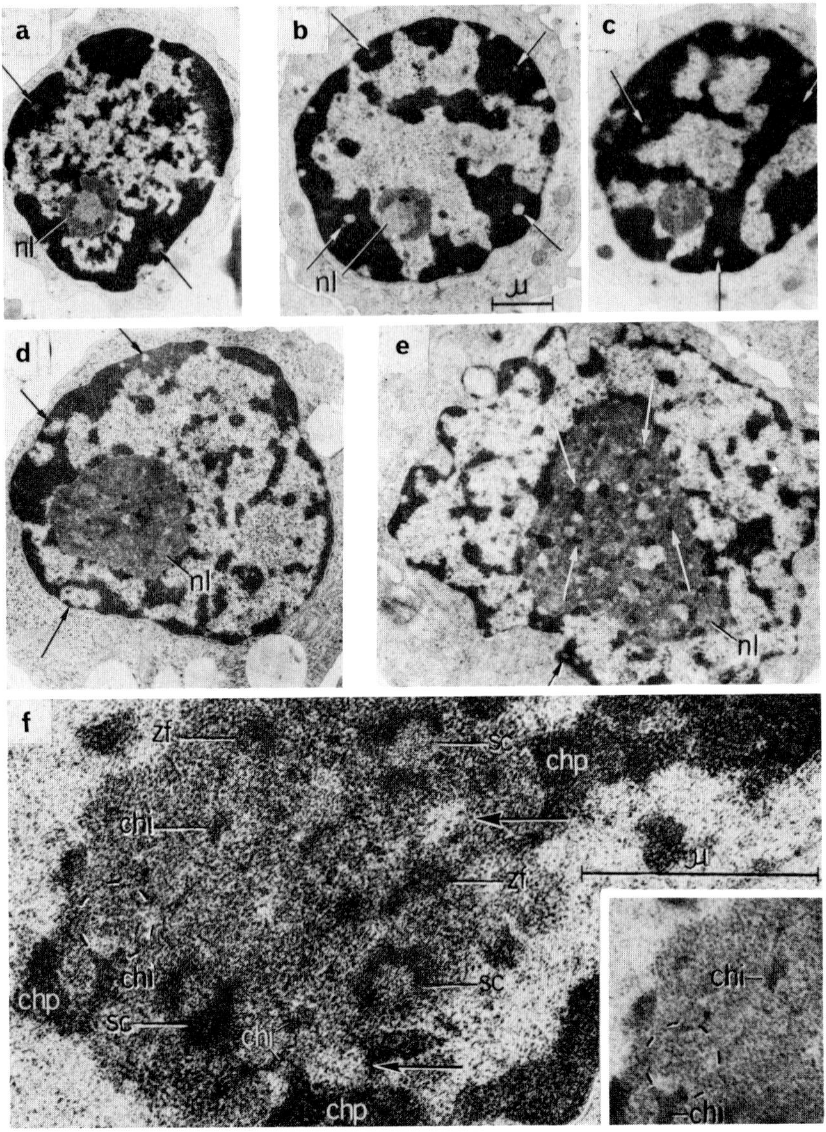

Fig. 10.17. Electron micrographs of cultured lymphocytes stimulated with phytohemagglutinin. a. Lymphocyte at 0 hour. nl = nucleolus, arrows = round spaces in the peripheral chromatin. ×8,000. b and c. Cross profiles of a lymphocyte after 4 hours of cultivation. nl = nucleolus, arrows = light areas in the chromatin structure. d. Lymphocyte after 24 hours of cultivation characterized by a marked enlargement of the nucleolus (nl). Arrows = light areas in the peripheral chromatin. ×8,000. e. Differentiated lymphocyte after 48 hours of cultivation. Areas (white arrows) of intranucleolar chromatin in the enlarged nucleolus are increased. ×8,000. f. En-

thesis of proteins and hence create a demand for more ribosomes.* Whether they have any direct nuclear effects such as those reported for the estrogens (*104a*) and androgens (*5a*) is not known. Even in the case of the estrogens, in which direct nuclear binding seems established, no definitive specific effect on nucleolus organizers or specific nucleolar genes has been found. Thymidine (4 m*M*) inhibits nucleolar as well as chromatin RNA synthesis (*29a*). Other substances have been reported to stimulate the nucleolus. Among these are several drugs, including pilocarpine (*70*), which has been reported to have a stimulatory effect on the nucleoli of the pancreas. Growth hormone and hydrocortisone enhance the synthesis of high molecular weight nuclear RNA; some of the stimulation by growth hormone is nucleolar (*5′*). Radiation has also been reported to increase nucleolar size (*80*) as does infection with some viruses such as encephalomyocarditis and adenovirus 12 (*67, 68*).

The effects of heat on nucleolar components were studied by Simard and Bernhard (*108*) and more recently by Simard (*106a*). Apparently heat produces an unraveling of the nucleonemas and the nucleonemal fine structures. However, segregation of the nucleolar elements does not occur. This effect is similar to the effects produced by some antibiotics.

REFERENCES

1. Adams, H. R., and Busch, H., Labeling of nucleolar RNA with orotic acid-2-C^{14} in livers of normal and thioacetamide-treated rats. *Biochem. Biophys. Res. Commun.* **9**, 578–583 (1962).
1a. Airo, R., and Ustaldi, G., Recenti acquisizioni sulla reattivita del linfocit. *Haematolo. Latina* (*Milan*) **10**, 209–270 (1967).
2. Amsterdam, D., Berek, J., Zahalsky, A. C., and Lazarus, S. S., Differential inhibition by 4-nitroquinoline-N oxide in serially propagated mammalian cell lines. *Exptl. Cell Res.* **48**, 499–504 (1967).
3. Andoh, T., and Chargaff, E., Formation and fate of abnormal ribosomes of

* A very interesting observation has been made on nucleoli in liver cells in ethionine-intoxicated animals treated with adenine or methionine (*103, 104*). This treatment induced a new formation of nucleolonemas from short, rod-like structures which elongated and twisted during the restitution of the usual nucleolar ultra-structural appearance.

larged nucleolus of a lymphocyte cultured for 24 hours at higher magnification (×32,000). The areas containing fibrillar components (zf, sc) as well as granular components are increased. Fibrillar component surrounding light intranucleolar areas = sc; intranucleolar chromatin = sc, perinucleolar chromatin = chp, spaces of low density = arrows. The electron micrographs are moderately underfocussed. The circled area demonstrating fibrillar components is shown at sharp focus in the insert (*135a*). (Courtesy of Dr. K. Tokuyasu, Department of Pathology, University of California, Los Angeles, California.)

E. coli cells treated with 5-fluorouracil. *Proc. Natl. Acad. Sci. U. S.* **54,** 1181–1189 (1965).
4. Arrighi, F. E., Nucleolar RNA synthetic activity in Chinese hamster cells *in vitro* and the effects of actinomycin D and nogalamycin. *J. Cellular Physiol.* **69,** 45–52 (1967).
5. Bernhard, W., Frayssinet, C., LaFarge, C., and LeBreton, E., Lésions Nucléolaires précoces provoquées par L'aflatoxine dans les cellules hépatiques du rat. *Compt. Rend.* **261,** 1785–1788 (1965).
5'. Brossard, M., and Nicole, L., Effect of growth hormone and hydrocortisone on the synthesis of rapidly labeled, high molecular weight, nuclear RNA in rat liver. *Canadian J. Biochem.* **47,** 226–229 (1969).
5a. Bruchovsky, N., and Wilson, J. D., The intranuclear binding of testosterone and 5α-androstan-17β-ol-3-one by rat prostate. *J. Biol. Chem.* **243,** 5953–5960 (1968).
6. Bucknall, R. A., and Carter, S. B., A reversible inhibitor of nucleic acid synthesis. *Nature* **213,** 1099–1101 (1967).
7. Budnitskaya, E. V., Brunfaut, M., and Errera, M., Effect of radiation on RNA and RNA metabolism in human cancer cells. *Federation Proc.* **24,** Transl. Suppl. 218–220 (1965).
8. Caspersson, T. O., Farber, S., Foley, G. E., and Killander, D., Cytochemical observations on the nucleolus-ribosome system. Effects of actinomycin D and nitrogen mustard. *Exptl. Cell Res.* **32,** 529–552 (1963).
9. Chang, S. B., Kader, M. M. A., Wick, E. L., and Wogan, G. N., Aflatoxin B_2: Chemical identity and biological activity. *Science* **142,** 1191–1192 (1963).
10. Chiga, M., Kume, F., and Millar, R. C., Nucleolar alteration produced by actinomycin D and the delayed onset of hepatic regeneration in rats. *Lab. Invest.* **15,** 1403–1408 (1966).
11. Ciferri, O., Albertini, A., and Cassani, G., Origin of the sarcosine molecules of actinomycins. *Biochem. J.* **96,** 853–861 (1965).
12. Cooper, H. L., Ribonucleic acid metabolism in lymphocytes stimulated by phytohemagglutinin. II. Rapidly synthesized ribonucleic acid and the production of ribosomal ribonucleic acid. *J. Biol. Chem.* **243,** 34–43 (1968).
12'. Cooper, H. L., Ribosomal ribonucleic acid production and growth regulation in human lymphocytes. *J. Biol. Chem.* **244,** 1946–1952 (1969).
12a. Daniels, J. C., Ritzmann, S. E., and Levin, W. C., Lymphocytes: Morphological, developmental and functional characteristics in health, disease and experimental study—analytical review. *Texas Rept. Biol. Med.* **26,** 5–92 (1968).
13. Darlix, J. L., Sentenac, A., and Fromageot, P., Etude du RNA synthetise *in vitro. Biochim. Biophys. Acta* **166,** 438–458 (1968).
14. DeMan, J. C. H., and Noorduyn, N. J. A., Light and electron microscopic radioautography of hepatic cell nucleoli in mice treated with actinomycin D. *J. Cell Biol.* **33,** 489–596 (1967).
15. DeRecondo, A. M., Frayssinet, C., LaFarge, C., and LeBreton, E., Action de l'aflatoxine sur le métabolisme du DNA au cours de l'hypertrophie compensatrice du foie après hepatectomie partielle. *Biochim. Biophys. Acta* **119,** 322–330 (1966).
16. Desjardins, R., Grogan, D. E., Arendell, J. P., and Busch, H., Effects of antitumor agents on the synthesis of nucleolar DNA. *Cancer Res.* **27,** 159–164 (1967).

17. De Zoeten, G. A., and Schlegel, D. E., Nucleolar and cytoplasmic uridine-^{3}H incorporation in virus-infected plants. *Virology* **32**, 416–427 (1967).
18. Diener, U. L., and Davis, N. D., Aflatoxin production by isolates of Aspergillus flavus. *Phytopathology* **56**, 1390–1393 (1966).
19. DiMarco, A., Silvestrini, R., Dasdia, T., and DiMarco, S., Nucleolar and chromosomal origin of nuclear RNA and its passage to cytoplasm. *Riv. Istochim.* **11**, 211–220 (1965).
20. DiMarco, A., Silverstrini, R., DiMarco, S., and Dasdia, T., Inhibiting effect of the new cytotoxic antibiotic daunomycin on nucleic acids and mitotic activity of Hela cells. *J. Cell Biol.* **27**, 545–550 (1965).
21. Ebstein, B. S., Tritiated actinomycin D as a cytochemical label for small amounts of DNA. *J. Cell Biol.* **35**, 709–713 (1967).
22. Edström, J. E., and Baneholt, B., Sedimentation properties of the newly synthesized RNA from isolated nuclear components of Chironomus tentans salivary gland cells. *J. Mol. Biol.* **28**, 331–343 (1967).
23. Elkind, M. M., Whitmore, G. F., and Alescio, T., Actinomycin D: Suppression of recovery in X-irradiated mammalian cells. *Science* **143**, 1454–1457 (1964).
24. Endo, H., On the relation between carcinogenic potency of 4-nitroquinoline N-oxides and the reactivity of their nitro-groups with SH-compounds. *Gann* **49**, 151–156 (1958).
25. Endo, H., Aoki, M., and Aoyama, Y., Formation of nuclear inclusion bodies in tissue culture cells by 4-nitroquinoline N-oxide. *Gann* **50**, 209–217 (1959).
26. Endo, H., and Kume, F., Comparative studies on the biological actions of 4-nitroquinoline 1-oxide and its reduced compound, 4-hydroxyaminoquinoline 1-oxide. *Gann* **54**, 443–453 (1963).
27. Endo, H., and Kume, F., Induction of sarcoma in rats by a single injection of 4-hydroxyaminoquinoline 1-oxide. *Gann* **56**, 261–265 (1965).
28. Endo, H., Takayama, S., Kasuga, T., and Ohashi, M., Histochemical studies on nuclear inclusion in tissue culture cells induced by 4-nitroquinoline N-oxide. *Gann* **52**, 173–177 (1961).
29. Ennis, H. L., Synthesis of ribonucleic acid in L cells during inhibition of protein synthesis by cycloheximide. *Mol. Pharmacol.* **2**, 543–557 (1966).
29a. Fakan, F., Inhibition of nucleolar and chromatin RNA synthesis in HeLa cells by excess thymidine. *Histochemie* **17**, 64–72 (1969).
30. Ficq, A., Effects de l'actinomycine D et de la puromycine sur le métabolisme de l'oocyte en croissance. Etude autoradiographique. *Exptl. Cell Res.* **34**, 581–594 (1964).
30a. Floyd, L., Unuma, T., and Busch, H., Effects of aflatoxin B_1 and other carcinogens on nucleolar RNA of various tissues of the rat. *Exptl. Cell Res.* **51**, 423–438 (1968).
31. Frayssinet, C., LaFarge, C., DeRecondo, A. M., and LeBreton, E., Inhibition de l'hypertrophie compensatrice du foie chez le Rat par les toxines d'Aspergillus flavus. *Compt. Rend.* **259**, 2143–2146 (1964).
32. Gause, Jr., G. G., Loshkareva, N. P., and Zbarsky, I. B., Effect of olivomycin and echinomycin on initiation and growth of RNA chains catalyzed by RNA polymerase. *Biochim. Biophys. Acta* **166**, 752–754 (1968).
33. Gelboin, H. V., Wortham, J. S., Wilson, R. G., Friedman, M., and Wogan, G. N., Rapid and marked inhibition of rat-liver RNA polymerase by Aflatoxin B. *Science* **154**, 1205–1206 (1966).

34. Geuskens, M., Etude autoradiographique et ultrastructurale de l'action de l'actinomycine D sur les oocytes d'astérie. *Exptl. Cell Res.* **39**, 400–412 (1965).
35. Girard, M., Penman, S., and Darnell, J. E., The effect of actinomycin on Ribosome formation in HeLa cells. *Proc. Natl. Acad. Sci. U. S.* **51**, 205–211 (1964).
36. Goldberg, I. H., and Reich, E., Actinomycin inhibition of RNA synthesis directed by DNA. *Federation Proc.* **23**, 958–964 (1964).
36a. Goldblatt, P. J., Sullivan, R. J., and Farber, E., Morphologic and metabolic alterations in hepatic cell nucleoli induced by varying doses of actinomycin D. *Cancer Res.* **29**, 124–135 (1969).
36b. Goldblatt, P. J., Sullivan, R. J., and Farber, E., Induction of partial nucleolar segregation in hepatic parenchymal cells by actinomycin D following inhibition of ribonucleic acid synthesis by ethionine. *Lab. Invest.* **20**, 283–291 (1969).
36c. Harada, Y., Sunagawa, N., and Katagiri, K., Segregation of the nucleolar materials produced by quinoxaline antibiotics in JTC-13 cells—Comparison of effects among 4-nitroquinoline 1-oxide, actinomycin-B, and quinoxaline antibiotics. *Gann* **59**, 513–522 (1968).
37. Harris, C., Grady, H., and Svoboda, D., Segregation of the nucleolus produced by anthramycin. *Cancer Res.* **28**, 81–90 (1968).
37a. Harris, C., Reddy, J., and Svoboda, D., Isolation and ultrastructure of nucleoli altered *in vivo*. *Exptl. Cell Res.* **51**, 268–274 (1968).
38. Haselkorn, R., Actinomycin D as a probe for nucleic acid secondary structure. *Science* **143**, 682–684 (1964).
38a. Hausen, P., and Stein, H., On the synthesis of RNA in lymphocytes stimulated by phytohemagglutinin. I. Induction of uridine-kinase and the conversion of uridine to UTP. *European J. Biochem.* **4**, 401–406 (1968).
39. Herich, R., The action of X-rays on the differentiation and development of nucleolar structures in relation to cell division. *Flora (Jena)* **155**, 505–510 (1965).
40. Hesseltine, C. W., Shotwell, O. L., Ellis, J. J., and Stubblefield, R. D., Aflatoxin formation by Aspergillus flavus. *Bacteriol. Rev.* **30**, 795–805 (1966).
40a. Higashi, K., Matsuhisa, T., Kitao, A., and Sakamoto, Y., Selective suppression of nucleolar RNA metabolism in the absence of protein synthesis. *Biochim. Biophys. Acta* **166**, 388–393 (1968).
41. Holtzer, R. L., Oda, A., and Chiga, M., Effect of actinomycin D on deoxycytidylate deaminase activity of rat liver after partial hepatectomy. *Lab. Invest.* **13**, 1514–1519 (1964).
42. Hsu, T. C., Arrighi, F. E., Klevecz, R. R., and Brinkley, B. R., The nucleoli in mitotic divisions of mammalian cells *in vitro*. *J. Cell Biol.* **26**, 539–553 (1965).
43. Hsu, T. C., Humphrey, R. M., and Somers, C., Persistent nucleoli in animal cells following treatments with fluorodeoxyuridine and thymidine. *Exptl. Cell Res.* **33**, 74–77 (1964).
44. Hugon, J., and Borgers, M., Fine structure of the nuclei of the duodenal crypt cells after X-irradiation. *Am. J. Pathol.* **52**, 701–723 (1968).
45. Ittel, M. E., Wintzerith, M., and Mandel, P., Localisation des synthèses des RNA nucléaires du foie de Rat. *Bull. Soc. Chim. Biol.* **49**, 13–23 (1967).
46. Jezequel, A. M., and Bernhard, W., Modifications ultrastructurales du pancréas exocrine de rat sous l'effet de l'actinomycine D. *J. Microscopie* **3**, 279–296 (1964).

46a. Jezequel, A. M., Seeve, M. M., and Steiner, J. W., Segregation of nucleolar components in mycoplasma infected cells. *Lab. Invest.* **16**, 287–304 (1967).

47. Jones, K. W., The induction of paracrystalline thread-complexes in the nuclei of amphibian cells by actinomycin D and other DNA-binding antibiotics. *J. Ultrastruct. Res.* **18**, 71–84 (1967).

48. Journey, L. J., and Goldstein, M. N., Electron microscope studies on HeLa cell lines sensitive and resistant to actinomycin D. *Cancer Res.* **21**, 929–932 (1961).

49. Kamahora, J., and Kakunaga, T., *In vitro* carcinogenesis of 4-nitroquinoline-1-oxide with golden hamster embryonic cells. *Proc. Japan Acad.* **42**, 1079–1081 (1966).

50. Kaziro, Y., and Kamiyama, M., Mechanism of action of Chromomycin A_3. II. Inhibition of RNA polymerase reaction. *J. Biochem.* (*Tokyo*) **62**, 424–429 (1967).

51. Kleinfeld, R. G., and von Haam, E., Effect of thioacetamide on rat liver regeneration. II. Nuclear RNA in mitosis. *J. Biophys. Biochem. Cytol.* **6**, 393–398 (1959).

52. Koulish, S., and Kleinfeld, R. G., The role of the nucleolus. I. Tritiated cytidine activity in liver parenchymal cells of thioacetamide-treated rats. *J. Cell Biol.* **23**, 39–51 (1964).

53. Kume, F., and Chiga, M., Inhibition of the actinomycin-induced nucleolar change by thioacetamide in mouse hepatic parenchymal cell. *Lab. Invest.* **17**, 767–771 (1967).

54. Kume, F., and Chiga, M., Acute nucleolar morphological change produced by dimethylnitrosamine in mouse hepatic parenchymal cells. *Gann* **59**, 151–153 (1968).

54a. Kume, F., Maruyama, S., D'Agastino, A. N., and Chiga, M., Nucleolar changes produced by mithramycin in rat hepatic cells. *Exptl. Mol. Pathol.* **6**, 254–260 (1967).

55. LaFarge, C., Frayssinet, C., and DeRecondo, A. M., Inhibition par l'aflatoxine de la synthèse de RNA hépatique chez le Rat. *Bull. Soc. Chim. Biol.* **47**, 1724–1725 (1965).

56. LaFarge, C., Frayssinet, C., and Simard, R., Inhibition préférentielle des synthèses de RNA nucléolaire provoquée par l'aflatoxine dans les cellules hépatiques du Rat. *Compt. Rend.* **263**, 1011–1014 (1966).

57. Lapis, K., and Bernhard, W., The effect of mitomycin C on the nucleolar fine structure of KB cells in cell culture. *Cancer Res.* **25**, 628–645 (1965).

58. Laufer, H., Nakase, Y., and Vanderberg, J., Developmental studies of the dipteran salivary gland. 1. The effects of actinomycin D on larval development, enzyme activity, and chromosomal differentiation in *Chironomous thummi*. *Develop. Biol.* **9**, 367–384 (1964).

59. Lazarus, S. S., Vethamany, V. G., Shapiro, S. H., and Amsterdam, D., Histochemistry and fine structure of 4-nitroquinoline-N-oxide-induced nuclear inclusions. *Cancer Res.* **26**, 2229–2238 (1966).

60. Legator, M., Biological effects of aflatoxin in cell culture. *Bacteriol. Rev.* **30**, 471–477 (1966).

61. Lettré, R., Observations on the behavior of the nucleolus of cells *in vitro*. *8th Congr. Cell Biol. Leiden, 1954* pp. 141–150. Interscience, New York.

62. Lettré, R., and Siebs, W., Beobachtungen zur Struktur des nucleolus in normalen zellen sowie in Tumorzellen. *Z. Krebsforsch.* **60**, 564–580 (1955).

63. Lewin, P. K., and Moscarello, M. A., Nucleolar changes in HeLa cells grown in the presence of aminonucleoside. *Lab. Invest.* **19**, 265–272 (1968).

64. Lezzi, M., Induktion eines Ecdyson-Aktivierbaren Puff in isolierten Zellkernen von *Chironomus* durch KCl. *Exptl. Cell Res.* **43**, 571–577 (1966).
64a. Li, H. J., and Crothers, D. M., Relaxation studies of the proflavine-DNA complex: The kinetics of an intercalation reaction. *J. Mol. Biol.* **39**, 461–477 (1969).
64b. Lorch, I. J., and Jeon, K. W., Reversible effect of actinomycin D on nucleoli of *Amoeba discoides:* Nuclear transplantation study. *Nature* **221**, 1073–1074 (1969).
65. Love, R., Clark, A. M., and Studzinski, G. P., Cytochemical investigations of deoxyribonucleic acid-mediated control of ribonucleoprotein metabolism. *Nature* **203**, 1384 (1964).
66. Malkin, M. F., and Zahalsky, A. C., Interaction of the water-soluble carcinogen 4-nitroquinoline N-oxide with DNA. *Science* **154**, 1665–1667 (1966).
67. Martin, E. M., The effect of encephalomyocarditis virus infection on the ribonucleic acid metabolism of Krebs II mouse ascites cells. *Biochem. J.* **71**, P12–P13 (1959).
68. Martinez-Palomo, A., and Granboulan, N., Electron microscopy of Adenovirus 12 replication. II. High-resolution autoradiography of infected KB cells labeled with tritiated thymidine. *J. Virol.* **1**, 1010–1018 (1967).
69. Matsudaira, H., Sekiguchi, T., and Nakamura, C., Effects of X-irradiation on the synthesis of rapidly labelled RNA in an experimental tumour of the rat. *Int. J. Radiat. Biol.* **13**, 21–34 (1967).
70. Misuraca, G., Limatola, E., and D'Uva, V., Rilievi sperimentali sul ciclo secretorio della cellula pancreatica esocrina di anfibio anuro con particolare riguardo alle interazioni nucleo-citoplasmatiche. I. Effetti della pilocarpina. *Boll. Soc. Ital. Biol. Sper.* **62**, 2034–2038 (1966).
71. Montgomery, P. O'B., and Bonner, W. A., A new technique for ultraviolet microbeam irradiation of living cells. *A.M.A. Arch. Pathol.* **66**, 418–421 (1958).
72. Montgomery, P. O'B., and Hundley, L. L., Ultraviolet microbeam irradiation of the nucleoli of living cells. *Exptl. Cell Res.* **24**, 1–5 (1961).
73. Montgomery, P. O'B., Karney, B. S., Reynolds, R. C., and McClendon, D., Cellular and subcellular effects of ionizing radiation. *Am. J. Pathol.* **44**, 727–746 (1964).
74. Montgomery, P. O'B., and Reynolds, R. C., Cellular and subcellular responses to ultraviolet radiation. *Lab. Invest.* **13**, 1243–1253 (1964).
75. Montgomery, P. O'B., Reynolds, R. C., and Cook, J. E., Nucleolar "caps" induced by flying spot ultraviolet nuclear irradiation. *Am. J. Pathol.* **49**, 555–567 (1966).
76. Moreno, G., Etude au microscope électronique de l'irradiation ultraviolette localisée au noyau, au nucléole et au cytoplasme. *Compt. Rend. Soc. Biol.* **161**, 1866–1868 (1967).
77. Muller, W., and Crothers, D. M., Studies of the binding of actinomycin and related compounds to DNA. *J. Mol. Biol.* **35**, 251–290 (1968).
78. Muramatsu, M., and Busch, H., Effects of thioacetamide on metabolism of proteins of normal and regenerating liver. *Cancer Res.* **22**, 1100–1104 (1962).
79. Nii, S., and Kamahora, J., Intranuclear changes in FL cells induced by herpes simplex virus and Actinomycin S. *Biken's J.* **5**, 301–303 (1962).
80. Nix, T. E., Nordquist, R. E., Scott, J. R., and Everett, M. A., An ultrastructural study of nucleolar enlargement following ultraviolet irradiation of human epidermis. *J. Invest. Dermatol.* **45**, 114–118 (1965).
81. Noorduyn, N. J. A., and DeMan, J. C. H., RNA synthesis in rat and mouse

hepatic cells as studied with light and electron microscope radioautography. *J. Cell Biol.* **30**, 655–660 (1966).

81a. O'Hegarty, M. T., and Harman, J. W., The inhibitory effect of 3-methylcholanthrene on nucleolar alterations induced in rat liver cells by 3′-methyl-4-dimethylaminoazobenzene. *Cancer Res.* **29**, 521–528 (1969).

82. Paul, J. S., Reynolds, R. C., and Montgomery, P. O'B., Inhibition of DNA-dependent RNA polymerase by 4-nitroquinoline-N-oxide in isolated nuclei. *Nature* **215**, 749–750 (1967).

82a. Paul, J. S., Reynolds, R. C., and Montgomery, P. O'B., Effects of 4-nitroquinoline-N-oxide on RNA synthesis. *Cancer Res.* **29**, 558–570 (1969).

83. Perry, R. P., The cellular sites of synthesis of ribosomal and 4 S RNA. *Proc. Natl. Acad. Sci. U. S.* **48**, 2179–2186 (1962).

84. Perry, R. P., Selective effects of actinomycin D on the intracellular distribution of RNA synthesis in tissue culture cells. *Exptl. Cell Res.* **29**, 400–406 (1963).

85. Perry, R. P., Role of the nucleolus in ribonucleic acid metabolism and other cellular processes. *Natl. Cancer Inst. Monograph* **14**, 73–89 (1964).

86. Perry, R. P., and Errera, M., The influence of nucleolar ribonucleic acid metabolism on that of the nucleus and cytoplasm. *In* "The Cell Nucleus" (J. S. Mitchell, ed.), pp. 24–29. Butterworth, London and Washington, D. C., 1960.

87. Perry, R. P., Hell, A., and Errera, M., The role of the nucleolus in ribonucleic acid and protein synthesis. I. Incorporation of cytidine into normal and nucleolar inactivated HeLa cells. *Biochim. Biophys. Acta* **49**, 47–57 (1961).

88. Peters, K., and Stein, G., Experimentelle Untersuchungen uber die Wirkung von Co^{60}-Gammastrahlen und Trenimon—einzeln und in Kombination—auf den Kern und den Nukleolus von Krebszellen (Stamm HeLa) in der Interphase *in vitro. Strahlentherapie* **125**, 258–272 (1964).

88a. Potmesil, M., and Smetana, K., Significance of ring-shaped nucleoli and micronucleoli in human lymphocytes. *Folia Biol.* (*Prague*) **15**[4], 300–305 (1969).

89. Racela, A., Grady, H., and Svoboda, D., Ultrastructural nuclear changes due to tannic acid. *Cancer Res.* **27**, 1658–1671 (1967).

89a. Rebuck, J. W., Monto, R. W., Monaghan, E. A., and Riddle, J. M., Potentialities of the lymphocyte, with an additional reference to its disfunction in Hodgkins disease. *Ann. N. Y. Acad. Sci.* **73**, 8–38 (1958).

90. Reddy, J., Harris, C., and Svoboda, D., Inhibition of lasiocarpine of RNA synthesis, RNA polymerase and induction of tryptophan pyrrolase activity. *Nature* **217**, 659–661 (1968).

91. Reddy, J., and Svoboda, D., The relationship of nucleolar segregation to ribonucleic acid synthesis following the administration of selected hepatocarcinogens. *Lab. Invest.* **19**, 132–145 (1968).

92. Reich, E., Actinomycin: Correlation of structure and function of its complexes with purines and DNA. *Science* **143**, 684–689 (1964).

93. Revel, M., Mandel, L., Wintzerith, M., Klein, N., and Mandel, P., Regulation de la biosynthèse des acides ribonucléiques. Action de la nicotinamide. *Bull. Soc. Chim. Biol.* **43**, 91–101 (1961).

94. Reynolds, R. C., and Montgomery, P. O'B., Nucleolar pathology produced by acridine orange and proflavine. *Am. J. Pathol.* **51**, 323–339 (1967).

95. Reynolds, R. C., Montgomery, P. O'B., and Hughes, B., Nucleolar "caps" produced by actinomycin D. *Cancer Res.* **24**, 1269–1277 (1964).

96. Reynolds, R. C., Montgomery, P. O'B., and Karney, D. H., Nucleolar "caps"—

a morphologic entity produced by the carcinogen 4-nitroquinoline N-oxide. *Cancer Res.* **23**, 535–538 (1963).
96a. Ringertz, N. R., and Bolund, L., Actinomycin binding capacity of deoxyribonucleoprotein. *Biochim. Biophys. Acta* **174**, 147–154 (1969).
97. Ro, T. S., and Busch, H., *In vitro* labeling of RNA in isolated nucleoli of the Walker tumor and liver. *Cancer Res.* **24**, 1630–1633 (1964).
97a. Ro, T. S., and Busch, H., Studies on RNA polymerase of nuclear fractions of regenerating liver. *Biochim. Biophys. Acta* **134**, 184–187 (1967).
98. Rodriguez, T. G., Ultrastructural changes in the mouse exocrine pancreas induced by prolonged treatment with actinomycin D. *J. Ultrastruct. Res.* **19**, 116–129 (1967).
99. Rubin, A. D., and Cooper, H. L., Evolving patterns of RNA metabolism during transition from resting state to active growth in lymphocytes stimulated by phytohemagglutinin. *Proc. Natl. Acad. Sci. U. S.* **54**, 569–576 (1965).
100. Schwartz, H. S., Sondergren, J. E., Garofalo, M., and Sternberg, S. S., Actinomycin D effects on nucleic acid and protein metabolism in intact and regenerating liver of rats. *Cancer Res.* **25**, 307–317 (1965).
100a. Scott, D., and Evans, H. J., Influence of the nucleolus on DNA synthesis and mitosis in *vicia faba. Exptl. Cell Res.* **36**, 145–159 (1964).
101. Semmel, M., and Huppert, J., Interaction *in vitro* entre l'actinomycine D et le RNA ribosomal. *Biochim. Biophys. Acta* **103**, 702–704 (1965).
102. Shankar Narayan, K. S., Steele, W. J., and Busch, H., Evidence that the granular and fibrillar elements of nucleoli contain 28 S and 6 S RNA, respectively. *Exptl. Cell Res.* **43**, 483–492 (1966).
103. Shinozuka, H., Goldblatt, P. J., and Farber, E., The disorganization of hepatic cell nucleoli induced by ethionine and its reversal by adenine. *J. Cell Biol.* **36**, 313–328 (1968).
104. Shinozuka, H., Verney, E., and Sidransky, H., Alterations in the liver due to acute amino acid deficiency. An electron microscopic study of young rats force-fed a threonine-devoid diet. *Lab. Invest.* **18**, 72–85 (1968).
104a. Shyamala, G., and Gorski, J., Estrogen receptors in the rat uterus. Studies on the interaction of cytosol and nuclear binding sites. *J. Biol. Chem.* **244**, 1097–1103 (1969).
104b. Siev, M., Weinberg, R., and Penman, S., The selective interruption of nucleolar RNA synthesis in HeLa cells by cordycepin. *J. Cell Biol.* **41**, 510–520 (1969).
105. Simard, R., Specific nuclear and nucleolar ultrastructural lesions induced by proflavin and similarly acting antimetabolites in tissue culture. *Cancer Res.* **26**, 2316–2328 (1966).
106. Simard, R., The binding of actinomycin D-H^3 to heterochromatin as studied by quantitative high resolution radioautography. *J. Cell Biol.* **35**, 716–722 (1967).
106a. Simard, R., Personal communication, Nucleolus Workshop, Prague, 1969.
107. Simard, R., and Bernhard, W., The phenomenon of nucleolar segregation: Specific action of certain antimetabolites. *Intern. J. Cancer* **1**, 463–479 (1966).
108. Simard, R., and Bernhard, W., A heat-sensitive-cellular function located in the nucleolus. *J. Cell Biol.* **34**, 61–76 (1967).
109. Sirlin, J. L., and Jacob, J., Function, development and evolution of the nucleolus. *Nature* **195**, 114–117 (1962).
110. Sirlin, J. L., and Jacob, J., Sequential and reversible inhibition of synthesis of ribonucleic acid in the nucleolus and chromosomes: Effect of benzamide and substituted benzimidazoles on dipteran salivary glands. *Nature* **204**, 545–547 (1964).

111. Sirtori, C., and Bosisio-Bestetti, M., Nucleolar changes in KB tumor cells infected with *Herpes simplex* virus. *Cancer Res.* **27**, 367–376 (1967).

112. Smuckler, E. A., and Benditt, E. P., The early effects of actinomycin on rat liver. *Lab. Invest.* **14**, 1699–1709 (1965).

113. Sporn, M. B., Dingman, C. W., Phelps, H. L., and Wogan, G. N., Aflatoxin B_1: Binding to DNA *in vitro* and alteration of RNA metabolism *in vivo*. *Science* **151**, 1539–1541 (1966).

114. Steele, W. J., Effects of thioacetamide on nuclear and nucleolar RNA. *In* "The Cell Nucleus" (H. M. Klouwen, ed.), pp. 203–220. Taylor & Francis, London, 1966.

115. Steele, W. J., and Busch, H., Increased content of high molecular weight RNA fraction in nuclei and nucleoli of livers of thioacetamide-treated rats. *Biochim. Biophys. Acta* **119**, 501–509 (1966).

116. Stenram, U., On protein deficiency in mice with special reference to liver nucleolar size. *Z. Zellforsch. Mikroskop. Anat.* **56**, 125–129 (1962).

117. Stenram, U., Radioautographic RNA and protein labeling and the nucleolar volume in rats following administration of moderate doses of actinomycin D. *Exptl. Cell Res.* **36**, 242–255 (1964).

118. Stenram, U., Electron-microscopic study of liver cells of rats treated with actinomycin D. *Z. Zellforsch. Mikroskop. Anat.* **65**, 211–219 (1965).

119. Stenram, U., Cytological, radioautographic and ultrastructural studies on the effect of 5-fluorouracil on rat liver. *Z. Zellforsch. Mikroskop. Anat.* **71**, 207–216 (1966).

120. Stenram, U., and Willen, R., Radioautographic, ultrastructural and biochemical studies on the effect of fluorouracil on the RNA synthesis in the liver of rats. *Z. Zellforsch. Mikroskop. Anat.* **82**, 270–281 (1967).

121. Stenram, U., and Willen, R., Effects of proflavin on ultrastructure and RNA and protein synthesis in the liver of rats. *Exptl. Cell Res.* **50**, 505–514 (1968).

122. Stenram, U., and Willen, R., The effect of actinomycin D on RNA synthesis and nucleolar ultrastructure in the liver of rats treated with fluorouracil. *Z. Zellforsch. Mikroskop. Anat.* **90**, 403–413 (1968).

123. Stewart, G. A., and Farber, E., The rapid acceleration of hepatic nuclear ribonucleic acid breakdown by actinomycin but not by ethionine. *J. Biol. Chem.* **243**, 4479–4485 (1968).

124. St. Mironescu, I. E., Mironescu, K., and Liciu, F., Nucleolar behavior in regenerating liver of rats receiving intra-abdominal injections of azo dyes and thioacetamide. *J. Natl. Cancer Inst.* **40**, 917–933 (1968).

125. Stöcker, E., Autoradiographische Untersuchungen zur Aminosäure-Inkorporation im Nukleolus. *Z. Zellforsch. Mikroskop. Anat.* **58**, 790–797 (1963).

126. Stöcker, E., Zur Frage der funktionellen und pathologischen Kernschwellung (Autoradiographische Untersuchungen mit H^3-Phenylalanin und H^3-Cytidin an der Rattenleber nach Thioacetamid). *Naturwissenschaften* **50**, 130 (1963).

127. Stöcker, E., and Altmann, H.-W., Die DNA-Synthese in Leberparenchymzellen und Gallengangsepithelien von normalen und mit Thioacetamid behandelten Ratten. *Naturwissenschaften* **51**, 15 (1964).

128. Stöcker, E., Altmann, H.-W., and Bödefeld, P., Zelluläre RNS- und Protein-Synthese in atropinisierten exokrinen Pankreas der Maus. *Naturwissenschaften* **54**, 371–372 (1967).

129. Stöcker, E., and Klärner, P., Autoradiographische Untersuchungen über die zelluläre Nukleinsaure- und Eiweissynthese in der Niere unter Thioacetamid. *Naturwissenschaften* **52**, 498–499 (1965).

130. Studzinski, G. P., and Jackson, L. G., Inhibition of puromycin of incorporation of tritiated uridine into nucleolar and cytoplasmic ribonucleic acids. *Nature* **212**, 194–196 (1966).
130a. Sunderman, F. W., Jr., and Esfahani, M., Nickel carbonyl inhibition of RNA polymerase activity in hepatic nuclei. *Cancer Res.* **28**, 2565–2567 (1968).
131. Svoboda, D., Grady, H., and Higginson, J., Aflatoxin B_1 injury in rat and monkey liver. *Am. J. Pathol.* **49**, 1023–1051 (1966).
132. Svoboda, D., and Higginson, J., A comparison of ultrastructural changes in rat liver due to chemical carcinogens. *Cancer Res.* **28**, 1703–1733 (1968).
133. Svoboda, D., Racela, A., and Higginson, J., Variations in ultrastructural nuclear changes in hepatocarcinogenesis. *Biochem. Pharmacol.* **16**, 651–657 (1967).
133a. Svoboda, D., and Soga, J., Early effects of myrolizidine on the fine structures of rat liver cells. *Am. J. Pathol.* **48**, 347–374 (1966).
134. Tamaoki, T., and Mueller, G. C., Effect of puromycin on RNA synthesis in HeLa cells. *Biochem. Biophys. Res. Commun.* **11**, 404–410 (1963).
135. Tamaoki, T., and Mueller, G. C., The effects of actinomycin D and puromycin on the formation of ribosomes in HeLa cells. *Biochim. Biophys. Acta* **108**, 73–80 (1965).
135a. Tokuyasu, K., Fine structural alterations of interphase nuclei of lymphocytes stimulated to growth activity *in vitro*. Personal communication (1968).
136. Unuma, T., and Busch, H., Formation of microspherules in nucleoli of tumor cells treated with high doses of actinomycin D. *Cancer Res.* **27**, 1232–1243 (1967).
137. Vorbrodt, K., Microinterferometric and cytophotometric studies of the effect of thioacetamide on the nuclei and nucleoli of parenchymal cells of the rat liver. *Folia Histochem. Cytochem.* **4**, 179–194 (1966).
138. Vorbrodt, A., Steplewski, Z., and Krzyozowska-Gruca, S., Electron histochemical investigations of the effect of thioacetamide on the regenerating rat liver cell. *Folia Histochem. Cytochem.* (*Cracow*) **4**, 157–162 (1966).
139. Vobrodt, A., Vorbrodt, K., Krzyzowska-Gruca, S., Steplewski, Z., and Chorazy, M., Cytophotometric and cytochemical studies on the effect of actinomycin D and thioacetamide on rat liver cells. *Arch. Immunol. Therap. Exp.* **14**, 818–824 (1966).
140. Waring, M. J., Drugs which affect the structure and function of DNA. *Nature* **219**, 1320–1325 (1968).
141. Warner, J. R., The assembly of ribosomes in HeLa cells. *J. Mol. Biol.* **19**, 383–398 (1966).
141a. Willems, M., Penman, M., and Penman, S., The regulation of RNA synthesis and processing in the nucleolus during inhibition of protein synthesis. *J. Cell Biol.* **41**, 177–187 (1969).
142. Wintzerith, M., Ittel, M. E., and Mandel, P., Effect of adrenalectomy on RNA biosynthesis in rat liver nuclear fractions. *Life Sci.* **6**, 867–871 (1967).
143. Wogan, G. N., Chemical nature and biological effects of the aflatoxins. *Bacteriol. Rev.* **30**, 460–470 (1966).
143a. Yoffey, J. M., Hanks, G. A., and Kelly, L., Some problems of lymphocyte production. *Ann. N. Y. Acad. Sci.* **73**, 67–78 (1958).
144. Zbarsky, S. H., The effect of actinomycin D on the incorporation of C^{14}-formate into nucleic acids of Ehrlich ascites carcinoma and Novikoff hepatoma *in vitro. Can. J. Biochem.* **42**, 563–566 (1964).

CHAPTER XI

Isolation and Purification of Nuclei and Nucleoli

I. Isolation of Nuclei

Procedures for the isolation of nuclei have been important in studies on histones and other nuclear proteins (*2, 6*), nuclear enzymes (*6, 47, 50, 54, 60*), and nuclear RNA and DNA (*7, 43*). Since nuclei are a starting material for isolation of nucleoli, procedures for isolation of nuclei must provide products that are satisfactory with respect to yield, morphology, and chemical composition if isolation of nucleoli is to be satisfactory.

The objectives of procedures for isolation of nuclei are to secure a product in which (*a*) the nuclei are anatomically identical to those of the whole cell, (*b*) the contents of the nuclei as they exist in the cell are all present in the isolated product, and (*c*) the isolated nuclei do not contain cytoplasmic constituents. At the present time, the procedures commonly used for isolation of nuclei are (1) modifications of the procedure of Chauveau *et al.* (*15*) in which the tissue is homogenized in sucrose solutions containing Ca^{2+} followed by centrifugation of the homogenate; (2) treatment of the tissue with citric acid following the technique of Dounce and his associates (*20–23*); and (3) techniques employing "hypotonic shock" followed by treatment with detergents and homogenization (*31a, 43a*). Another technique used is the nonaqueous technique in which the tissue is subjected to rapid freezing and lyophilization followed by a milling procedure in nonaqueous solvents (*3, 53*). Each of the procedures has advantages from the point of view of morphology or retention of enzymatic or other macromolecular constituents in the product, but each has such significant disadvantages that no ideal procedure exists for isolation of the nuclei from any kind of cell.

A. Appearance and Composition of the Nuclei

The only satisfactory method for routine assessment of the appearance of the nuclei *in vivo* is the use of the light and phase microscope. By phase microscopy of either living cells in tissue culture or in smears of tissue preparations, nuclei are seen as structures that are homogeneous and less dense than the cytoplasm. The nuclear membrane is not visible as a dense, thick, defined structure although the nucleoli are usually different in density from the remainder of the nucleoplasm.

The most valuable morphological evaluation of nuclear structures is made by analysis of the electron microscopic appearance of the nuclear substructures, including the nuclear membrane, the nucleolus, and the chromatin (*7, 25, 56, 62*). The nuclear membrane appears to be a double-layered structure, as suggested by light microscopists, and its appearance by electron microscopy is characteristic for both the layers of the membrane (Fig. 11.1) and the appearance of the "nuclear pores." The nucleoli have characteristic appearances due to the granular and fibrillar components of the nucleolonemas and the light spaces between these structures. Clumped masses of chromatin are also visualized. These may be present either on the periphery of the nucleus apparently matted around the inner layer of the nuclear membrane (Fig. 11.1a), or randomly distributed within the nucleus. Ribonucleoproteins of varying density are seen. Some have diameters of 500 Å or greater and these are referred to as the interchromatinic dense particles or Swift and Watson granules (*62*); other ribonucleoprotein particles of 200–400 Å diameter and of equal density are also found in interchromatinic regions (Fig. 11.1). These smaller particles are apparently cross sections of the nuclear ribonucleoprotein network (*57*) (Chapter VIII).

The appearance of the nucleoli constitutes one measure of the nativeness of the nuclei. The nucleoli are distinguished by some variability from tissue to tissue (*4, 5, 12–14, 32–34, 38*). The location of the nucleoli may be central, attached to the nuclear membrane, or randomly in the nucleus. In addition, the nucleoli vary in size and in number. They may be so minute as to be submicroscopic or they may be so large as to practically fill the nuclear volume and leave only a small rim of nucleoplasm and chromatin within the nuclear membrane in cells of thioacetamide-treated animals (10, Chapter X). Their number generally varies from one to ten (Chapter III).

From the biochemical point of view, certain criteria for nuclear contents now seem established. The nuclear content of DNA, RNA, and protein have now been defined within limits; isolated nuclei of normal rat liver cells contain approximately 8–11 pg of DNA, 2.2–3.0 pg of RNA, and 35–45 pg of protein per nucleus. Nuclei of tumor cells contain amounts of DNA ranging from below the diploid value for liver nuclei to as high as octaploid values. Values for RNA of Walker tumor nuclei are 1.5–2 times those of liver nuclei, but the amounts of protein present

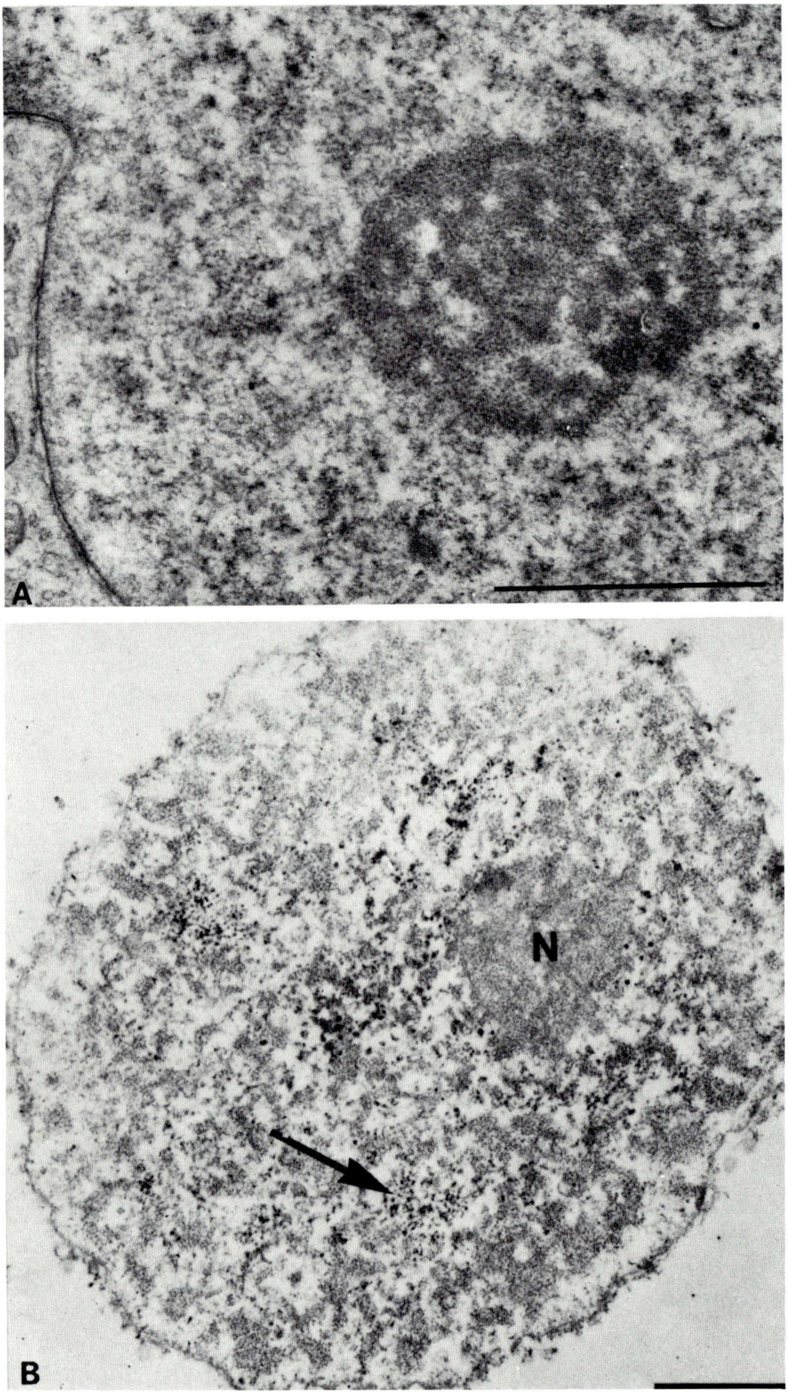
A
B
N

are not markedly different. The amounts of histones are approximately equivalent to the amounts of DNA (*6*).

B. Nuclear Enzymes

Certain enzymes are known to be localized to the nucleus and nucleolus (Chapters I, VII). These include NAD pyrophosphorylase, RNA polymerase, and enzymes involved in conversion of 45 S and 55 S RNA to 28 S RNA (*6, 54, 55*). Since a variety of marker enzymes have been reported to be present solely in certain cytoplasmic particles, their presence in nuclear preparations is evidence of contamination of the preparations (*54, 55*).

II. The Sucrose–Calcium Procedures for Isolation of Nuclei

Although there are many questions about the nativeness of the nuclei isolated by procedures employing "sucrose-calcium" solutions, this method has become the most widely used one for the preparation of nuclei. Initially, this procedure was developed by Hogeboom *et al.* (*29, 30*); it was later modified to its present form by Chauveau *et al.* (*15*). The nuclei obtained are free of cytoplasmic tags and adherent clumps of cytoplasm. By phase microscopy, the nuclear membrane is doubly refractile and sharp and the nucleoli are highly refractile and readily visualized (Fig. 11.2). With proper purification, no cytoplasmic contamination is visible by light or electron microscopy (Fig. 11.3).

A. Preparation of the Tissue for Homogenization

The liver normally contains large amounts of blood which can be removed by perfusion through the portal vein. Initially, 20 ml of ice-cold 0.15 *M* NaCl is perfused through the liver followed by 20 ml of 0.25 *M* sucrose. The liver is then rapidly removed, placed in ice-cold 0.25 *M* sucrose, and transferred to a cold room (2–4°) for further procedures. Inasmuch as liver tissue contains considerable arborization of the portal system as well as the hepatic blood vessels, it is necessary to separate the parenchymal cells from these more fibrous structures by passing the tissue through a mincer. In this laboratory, the Harvard

---

Fig. 11.1. A. Nucleolus within the nucleus of a liver cell *in situ*. ×37,000. Chromatin masses are composed of the faint fibrils. The marker line represents 1 μ. B. Isolated nucleus of rat liver stained with uranyl acetate showing the nucleolus (N). ×18,000. The arrow points to the interchromatinic dense granules. The marker line is 1 μ.

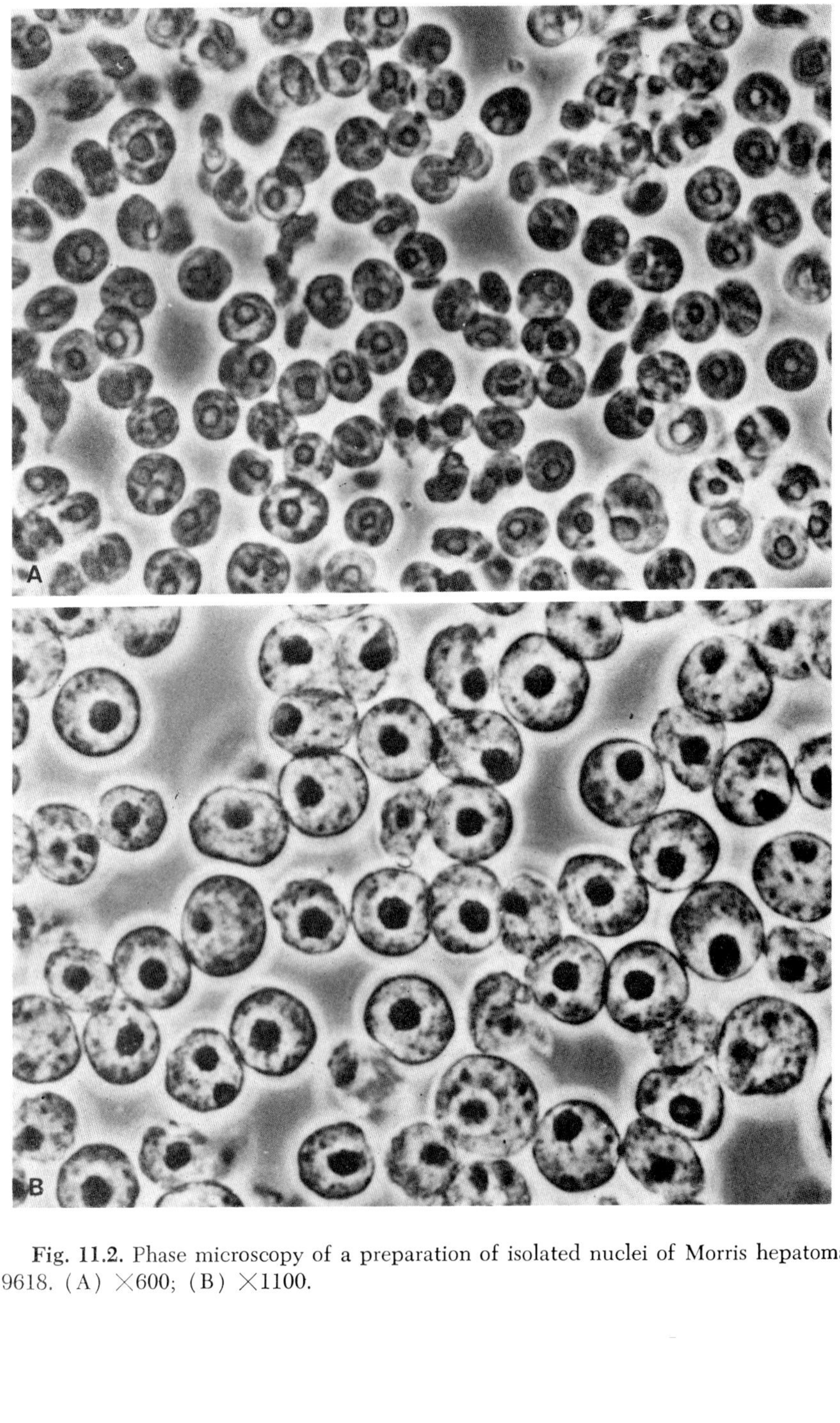

Fig. 11.2. Phase microscopy of a preparation of isolated nuclei of Morris hepatoma 9618. (A) ×600; (B) ×1100.

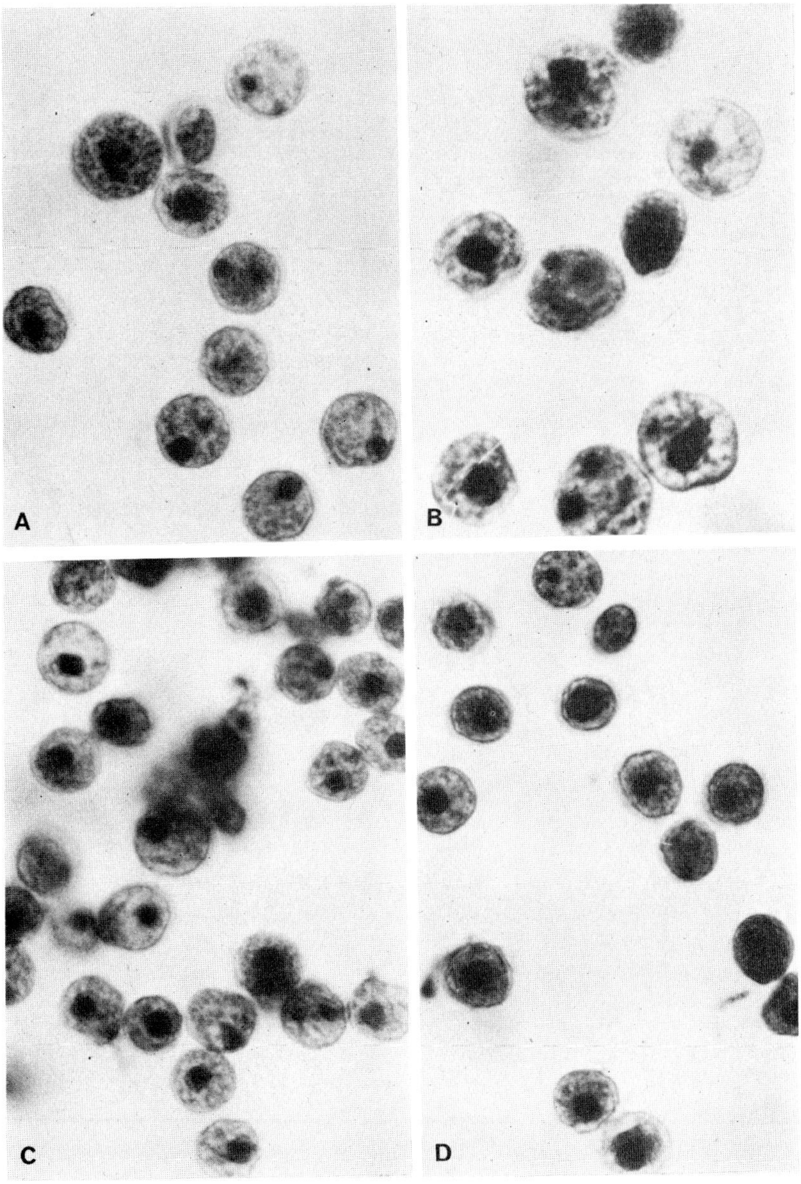

Fig. 11.3. Isolated nuclei of Morris hepatomas stained with Azure C. (A) Hepatoma 9618A, ×1500; (B) Hepatoma 9121, ×2000; (C) Hepatoma 7787, ×1500; (D) 9108A, ×1500.

tissue press is employed for this purpose. In essence, this instrument is a stainless-steel cylinder with a perforated plate at the base and a cylinder driven by a screw at the top. Although this is a useful instrument, one modification employed includes a handle on the screw plate so as to provide a greater leverage on the handle. A more practical model is the large-size press (*16–18*) which employs the same principle except that force is applied with a rack and pinion device which works much more rapidly and is much less tiring.

The medium for homogenization varies a great deal and in this laboratory, the original system of Chauveau *et al.* (*15*) has been used in which the minced or fragmented tissue is suspended in 10 vol of 2.4 *M* sucrose containing 3.3 m*M* calcium acetate or $CaCl_2$. The concentration of the sucrose is quite critical, and care is taken that the percentage of sucrose is 60% by weight as determined either with a saccharimeter or a direct reading refractometer (Abbé). In other laboratories, 0.25 *M* sucrose, 0.88 *M* sucrose, and 2.0 *M* sucrose have been employed and amounts of Ca^{2+} ranging from 0.5 m*M* to 10 m*M* (*29, 35, 36*) have been employed. In an endeavor to isolate nuclei with optimal enzymatic activity, Widnell and Tata (*65*) used 0.32 *M* sucrose containing 3 m*M* $MgCl_2$ as the homogenizing medium. When large amounts of sucrose are employed, as in the mass isolation of nuclei and nucleoli, it is a considerable saving to obtain sucrose from commercial sources. Preparations used for mixers with soft drinks have been obtained from the Imperial Sugar Company (Sugarland, Texas) and in addition, granular preparations of high purity (better than is usually available from some chemical supply companies) can be obtained in 100-lb bags at low cost.

The amount of homogenizing medium has also varied, ranging from 3 to 15 ml of sucrose solutions per gram of tissue. In this laboratory, optimal preparations were obtained with 10–15 vol of 2.2 *M* or 2.4 *M* sucrose per gram of tissue.

B. Homogenization

The all-glass homogenizer used by Potter and Elvehjem (*46*) was a widely used type of homogenizing device, but with the development of polymers, such as Teflon, which are not detrimental to tissues, the use of glass homogenizers has largely been discontinued. Large test tubes made of glass pipe (1.6 inch) expanded at the top are fitted with Teflon pestles containing a stainless-steel shaft for insertion into the chuck of a drill motor (Montgomery Ward & Co.). These motors are inexpensive and yet have substantial torque which is essential for homogenization of the liver cells, which normally is carried out at approximately 1200 rpm. As indicated earlier (*11*), the pestle clearance is very critical for some purposes; for the homogenization of the liver, clearances of 0.006 to 0.010 inches are satisfactory. If the pestle clearance is less, considerable losses of nuclei occur inasmuch as they are fragile, particularly when "hardened" by Ca^{2+}. If the pestle clearance is too great, many whole cells remain in the homogenate.

For tissues other than the liver, such as tumors, spleen, and kidney, the conditions vary considerably inasmuch as the degree of cytoplasmic adherence is much greater, especially in transplanted tumors and in some lymphocytes. For spleen, the pestle clearance may be somewhat greater and for the Walker tumor it should be considerably less; i.e., 0.003 inch. Even with these pestle clearances, good nuclear preparations of the Walker tumor have not been attained in this laboratory as long as there is divalent ion in the medium (*11*). Nuclei obtained with media in which divalent ions were omitted were not contaminated by cytoplasmic tags, but the nuclei were considerably swollen and, in addition, the nucleoli were not clearly visible. With the Morris hepatomas 9618A, 9121, and 9108A, nuclear preparations have been obtained of equal quality to those obtained with rat liver (Fig. 11.3).

Other types of homogenizers have been developed that are of considerable value although they have been used to a lesser extent in this laboratory than the Teflon-glass homogenizer. For small-scale studies, the Dounce homogenizer (*23*) has been utilized and the process of "Douncing" consists of several up-and-down strokes of the hand homogenizer (either of the Ten Broeck type or the Dounce ball type) in which both the pestle and the tube are ground glass with a clearance of approximately 0.006–0.010 inch. The pestle of the Ten Broeck homogenizer has a crossbar at the top which is manually twisted to provide a homogenizing effect.

For large-scale homogenization, the continuous tissue homogenizer (*8*) is useful. This instrument (Glenco, Houston) consists of a cylinder made of either stainless steel or glass pipe which has a screw-feed at one end and a tapered pestle at the other with very little dead space between them. The instrument is powered by a relatively low speed motor (800 rpm) on the end containing the feeding screw and on the other end, the motor for the homogenizing pestle operates at 1700 rpm. Both the feeding screw and the homogenizing pestle are made of PVC (polyvinyl chloride) or Plexiglas and the sample flows through the instrument at a rate of 100–150 ml per minute either by gravity or under pressure of a pump.

C. Number of Strokes or Passes for Homogenization

For homogenization with the Teflon-glass homogenizer, three strokes of the homogenizer are required to produce fragmentation of most of the cells of minced liver tissue. With the standard Potter-Elvehjem homogenizer, 6 to 10 up-and-down strokes of the homogenizer are required, and the same is true for the Dounce procedure. With the continuous tissue homogenizer, one or two passes of the suspension suffice to produce a preparation that is equivalent to that produced by the usual homogenization in the Teflon-glass homogenizer in which approximately 90% of the nuclei are released from the surrounding cytoplasm. In its present form, the continuous homogenizer is a practical device for homogenizing 250 or more grams of tissue.

D. Temperature Control

The shearing forces produced by homogenization may vary from those that are minor when homogenizers of the Ten Broeck or Dounce type (23) are used to marked when motors with high speeds and torque are employed. With hand homogenization, there is relatively little damage to the nuclei, although the products have been less satisfactory in this laboratory. With high speed homogenization and excessive homogenization, there has been frequently very serious damage to the nuclear membranes and, in addition, the sedimentation profiles of nuclear RNA are markedly abnormal. With the high speed, tight-fitting, motor-driven homogenizers there is also considerable friction which results in heating of the homogenate such that the temperature rises from 0–3° to 10–15°. To protect the tissue against the damaging effects of heating from internal friction as well as the external heating from the operator's hand, jacketed homogenizers have been developed. These consist of the usual glass pipe with a glass jacket through which a cooling solution of ethylene glycol–water (1:2) is pumped from a bath that is kept at —5°. Such baths can be obtained from Wilkens-Anderson or the Forma Companies and are particularly useful for experiments with small amounts of homogenates. Cracked ice is a poor medium for removing the heat from the homogenizer.

In the continuous tissue homogenizer, the sample is cooled with the aid of copper tubing wound around the stationary cylinder, by a jacket of circulating medium kept at —3 to —5° (or by a layer of ice slush kept around the cylinder). After the samples are homogenized, they are passed through eight layers of gauze and a 90-mesh stainless-steel wire screen prior to centrifugation.

E. Centrifugation

The initial homogenates are subjected to a variety of centrifugation procedures depending upon the initial suspending solutions and on the quality of the product that is desired. In the procedure of Chauveau *et al.* (*15*), the homogenate is poured into 30-ml nitrocellulose tubes and centrifuged at 40,000 *g* for 60 minutes. The density of the 60% sucrose solution is such that only the nuclei are sedimented under these conditions and the debris, whole cells, and cytoplasmic constituents are floated to the top of the solution or else are left on the side of the tube. Nuclear pellets prepared from normal liver using these conditions are pale amber to white in color, approaching a whitish color. Contamination is apparent when the surface of the pellet is either coated with a brownish layer or has brown streaks. In the procedure generally used in this laboratory, the pellet is resuspended in 1.0 *M* sucrose containing 0.001 *M* calcium acetate in a volume of 1 ml per gram of original tissue. The sample is rehomogenized with a loosely fitting Teflon pestle (0.010–0.012-inch clearance) attached to a motor operating at 500 rpm. This homogenate is centrifuged at 3000 *g* for 5 minutes, and the precipitate consists of highly purified nuclei (Fig. 11.1–11.3).

Although at one time, it was a common practice to refer to the "600 *g* precipitate" of the homogenates prepared in 0.25 *M* sucrose or 0.88 *M* sucrose as "nuclear preparations," this terminology has now been largely abandoned. The precipitates obtained by centrifugation of homogenates for 10 minutes at 600 *g* contain whole cells, much debris, and considerable cytoplasmic contamination. This centrifugation procedure is of value only if the mitochondrial or other more slowly sedimenting cellular components are desired and the supernatant fraction is to be saved.

To improve the yield of nuclei in the Chauveau procedure, Blobel and Potter (*3b, 3c*) employed a discontinuous sucrose gradient. They indicated that with this type of gradient the yield of nucleoli was increased from approximately 35 to 90%.

F. Large Scale Preparation of Liver Nuclei (*28a*)

The weighed (ca. 1 kg) liver tissue is disrupted in a Hobart model 4612 meat chopper (grinder) fitted with a specially made stainless-steel chopper plate (1-mm diameter holes). A tissue–sucrose slurry (8–12 liters) is prepared and maintained at 2–4° by continuous stirring of the macerated tissue into 1.6 *M* sucrose (1:9 w/v) containing 3.3 m*M* $CaCl_2$. Homogenization is effected by pumping this slurry (150 ml per minute) through a continuous tissue homogenizer equipped with a lucite pestle having a total clearance of 0.006 inch. A low temperature (8–12°) is maintained during homogenization by circulating a water–antifreeze mixture at −2° through a copper cooling coil wrapped around the stainless-steel tube of the homogenizer. At this temperature virtually no degradation of nuclear RNA occurs. Fibrous tissue and incompletely homogenized liver tissue are removed from the homogenate by filtration through two layers of 20-mesh stainless-steel wire cloth. The filtered homogenate is cooled and maintained at 2° prior to isolation of nuclei.

One homogenization with 1.6 *M* sucrose releases 90% of the liver nuclei from cytoplasm in 12 liters of tissue–sucrose slurry in approximately 90 minutes. Less concentrated sucrose is not so effective in stripping the cytoplasm from the nuclei; more concentrated sucrose produces excess heat.

G. Isolation of Nuclei

Nuclei are isolated by means of two air turbine-drive Sharples T-1 open-type laboratory supercentrifuges equipped with clarifier bowl assemblies and cooling coils in the frame assembly. Heating of the bowls is prevented by circulation of water–antifreeze (−2°) through the cooling coils and free passage of air into the centrifuge chamber through holes (⅜ inch each) drilled at the top and bottom of the frame assembly. Sucrose solutions and homogenate are passed into the two centrifuge bowls at equal rates with the use of a pair of 45-drill feed nozzles connected with ½-inch Tygon tubing and a Y tube to a single variable speed pump. The clarifier bowls are filled with 2.4 *M* sucrose during acceleration of the centrifuges to 50,000 rpm (60,000 *g*). The combination of the high density of 2.4 *M* sucrose and high gravitational force produces a large layer of 2.4 *M* sucrose between the walls of the bowl and the flow-through core as the less dense homogenate passes through the bowl. This standing discontinuous gradient is effective in producing morphologically satisfactory nuclei. At 180 ml per minute, the homogenate passes through the bowl as a thin stream confined to the core of the bowl with little if any mixing of the two sucrose layers. Nuclei without cytoplasmic tags are sufficiently dense to penetrate the 2.4 *M* sucrose and sediment against the walls while cytoplasmic materials and whole cells pass out with the 1.6 *M* sucrose homogenate as a supernatant fraction. Nuclei are collected from 12 liters of homogenate with the use of two instruments in 35 to 40 minutes. At the end of each run, fresh 1.6 *M* sucrose is used to flush the remaining homogenate from the tubing, pump, and clarifier bowls before deceleration and collection of the nuclei from the walls of the bowls. Under these conditions, approximately 40% of the total nuclei is collected during one centrifugation run. A second batch of nuclei is collected by centrifugation of the first supernatant fraction under identical conditions.

Approximately 80% of the free nuclei are collected from 12 liters of homogenate by two such centrifugations in two instruments in a total time of 90 to 100 minutes. Centrifugation is usually started when the homogenization is two thirds complete. In experiments with 120 animals, it is possible to kill the animals, perfuse and dissect the livers, disrupt and homogenize the tissue, and collect the nuclei in good yields and high quality in a total time of 4 to 4½ hours.

H. Criteria of Purity

The most satisfactory initial criterion of morphological purity of the preparations is direct examination of the product by phase microscopy. Under the conditions of the phase examination in which the pellet is suspended in 0.25 *M* sucrose alone or containing 3.3 m*M* $CaCl_2$, the optimal preparations contain only rounded nuclei with no visible mitochondrial particles and clear, sharp nucleoli. For counting the nuclei in a hemacytometer, they may be stained with either Azure C (*10, 42*) or Giemsa stain (*19*). Direct determinations of the recovery of nuclei can be made by light microscopic analysis.

When the morphological criteria of purity are satisfactory (Figs. 11.1–11.3); i.e., less than one whole cell per 4000 nuclei and no visible cytoplasmic particles, enzyme analyses for contaminating cytoplasm may be useful. Widnell and Tata (*65*) utilized glucose-6-phosphatase, cytochrome oxidase, glucose 6-phosphate dehydrogenase, and NADH–cytochrome *c* reductase as cytoplasmic enzyme markers. In nuclear preparations obtained in this laboratory, Siebert *et al.* (*55*) found that the activity of the cytoplasmic marker enzymes, glutamate dehydrogenase and adenylate kinase, were 0.4% and 0.5% of the whole homogenate, respectively.

Electron microscopic analysis of nuclear preparations is difficult to evaluate on a quantitative basis inasmuch as the procedures for preparation of the samples on the grids and the techniques for fixation and staining may produce some damage to the samples (Figs. 11.4 and 11.5). Nonetheless, the purity can be ascertained on the basis of either total or recognizable particle counts. With the optimal preparations, contamination of the nuclear preparations with either mitochondria or lysosomal particles is not found, or these particles are present in amounts of less than 0.1% of the total particles in the preparation. However, small

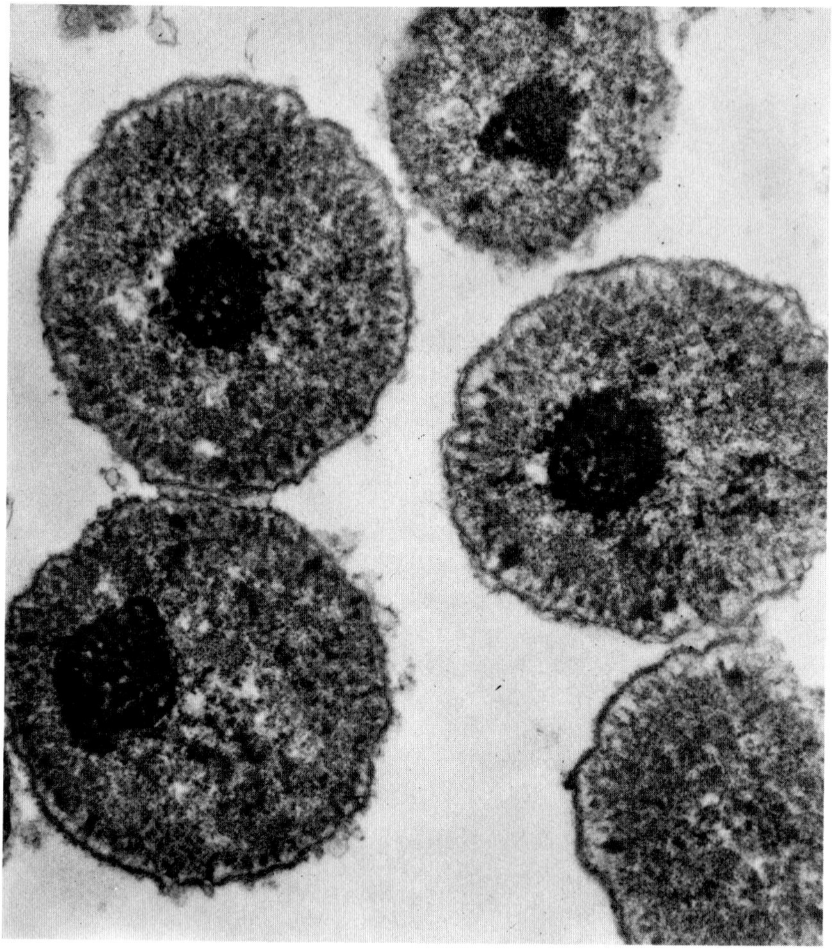

Fig. 11.4. Isolated nuclei of rat liver prepared by the sucrose-calcium procedure and stained with uranyl acetate. ×9000.

amorphous components are present, generally as a very small percentage of the total nuclear area in amounts ranging from 1 to 8% of the total particles in the preparation.

The yields of nuclei with the Chauveau procedure (*15*) are approximately 25–35% of the total in the homogenate, and with the continuous homogenization procedure they approximate 50–60% of the total in the homogenate. The former, however, have a higher degree of purity. Potter (*3b, 3c*) which employs discontinuous sucrose gradients.

Analysis of the constituents of isolated liver nuclei provides the follow-

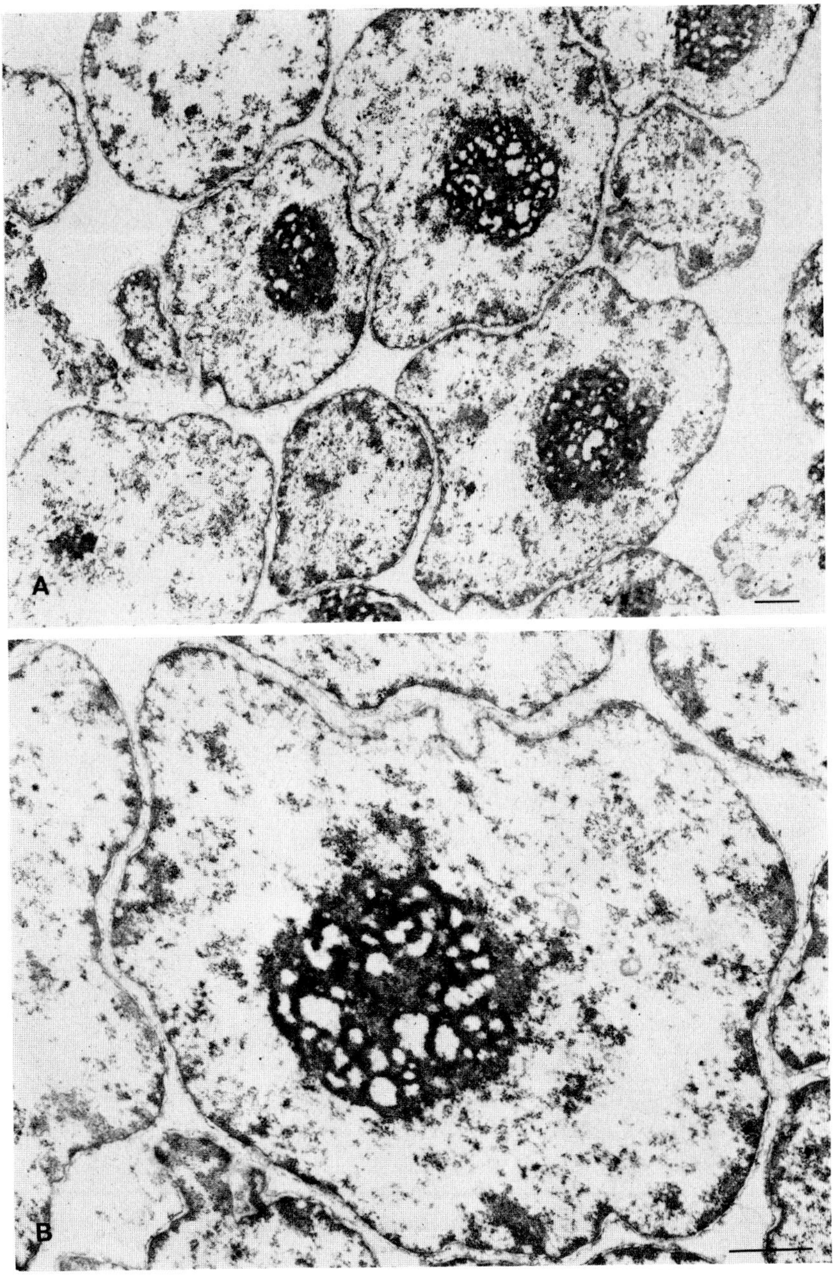

Fig. 11.5. Electron micrographs of isolated nuclei of Morris hepatoma 9618A. (A) Little or no cytoplasm is apparent, but outer nuclear membranes are still attached to the nuclei. The large nucleoli of these cells are visible in several of the nuclei. ×6000. (B) Focusing on the largest nucleolus. This is seen to consist mainly of granular elements and one ribonucleoprotein branch of the nuclear ribonucleoprotein network is seen emerging from the nucleolus toward the nuclear membrane. ×11,000.

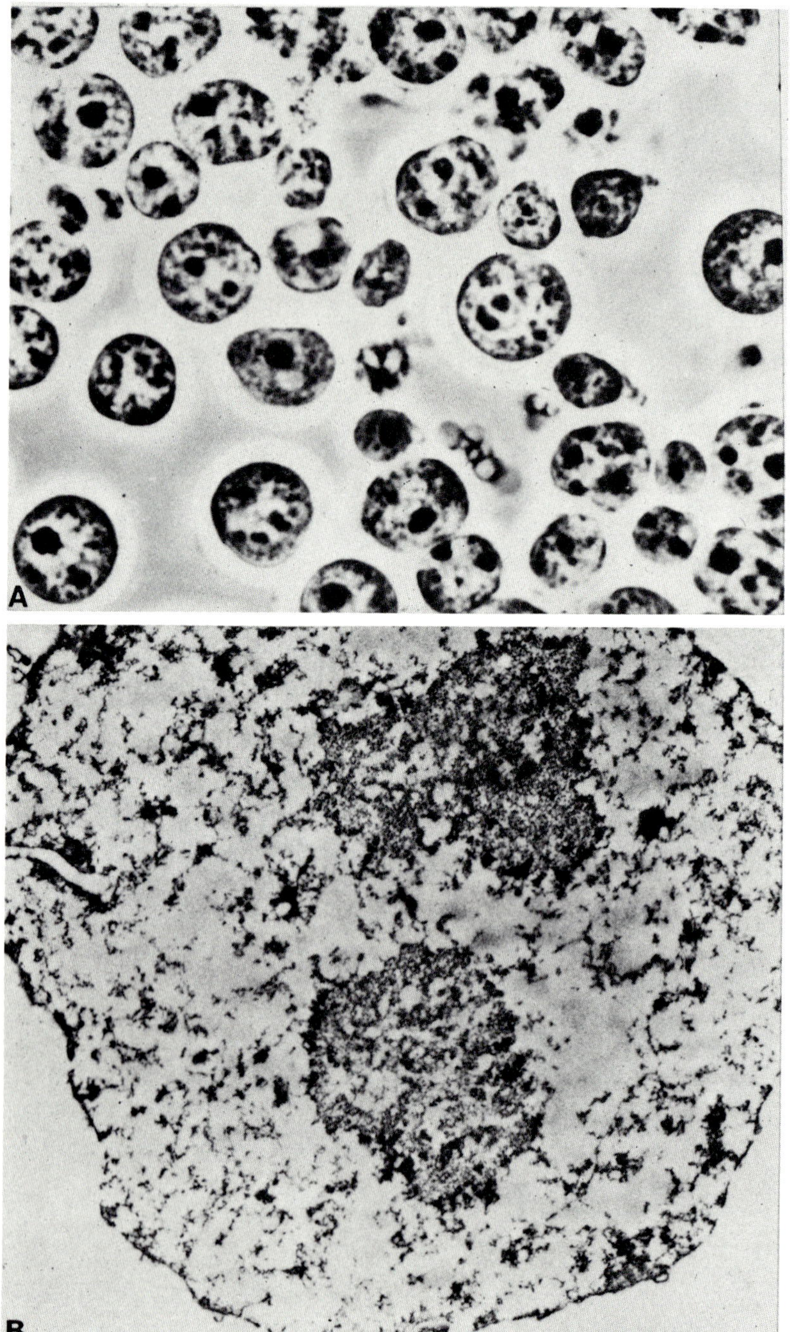
A
B

ing values for DNA, RNA, and proteins as picograms per nucleus: 11, 3, and 44. The ratio of RNA to DNA and protein to DNA are approximately 0.25 and 4, respectively (see p. 530).

III. The Citric Acid Procedure for Isolation of Nuclei

The use of citric acid as a suspending medium for preparation of nuclei from either whole homogenates or in the initial extraction of cells was popularized by Dounce (*20–23*) after Marshak (*37*) had reported that homogenization in citric acid removed the cytoplasm from the nuclear membrane. This method has been used particularly in preparation of nuclei from cells that represented difficult problems with the sucrose procedure, such as tumor cells, thymocytes, and cells of skin. Although the precise properties are not established that make citric acid valuable in removal of cytoplasm, reduction of pH, chelation of calcium ion, and extraction of acid-soluble proteins and lipid may be involved as factors reducing the adherence of the cytoplasm to the nuclear membrane.

In this laboratory, the citric acid procedure has been used recently for preparation of the nuclei of the Walker tumor cells in both small scale and mass preparations (*28*). With this procedure, RNA of high molecular weight, particularly the 45 S RNA can readily be isolated from tumor nuclei and, by comparison with the sucrose–calcium procedure, the amount of associated ribosomal RNA is reduced to a minimum. The preservation of the 45 S RNA may occur because ribonuclease, a basic protein, is removed by citric acid or is inactivated at the low pH or because Ca^{2+} is chelated by the citric acid. In any event, this method is of particular value for preservation of nuclear RNA.

A. Morphology of Nuclei Isolated with Citric Acid Procedure

The morphology of these nuclei is quite different from that which is seen by phase microscopy of the nuclei in whole cells in that the nuclear membrane is pale, the nucleolus is enlarged, less refractile, and apparently flattened; the chromatin appears to be clumped (Fig. 11.6A). The cytoplasm is largely removed, but many clumps of apparently denatured cytoplasm are found in the preparations; these clumps can

Fig. 11.6. (A) Nuclei of Walker tumor cells isolated by the citric acid procedure. The nucleoli are large and appear to be intact. Little cytoplasmic contamination is seen. The nuclei are pale and appear to be extracted. (B) Electron micrograph of a nucleus of a Walker tumor cell isolated by the citric acid procedure. ×12,000.

be removed by centrifugation of the preparation after it is layered over 0.88 *M* sucrose containing 1.5% citric acid.

Electron microscopy of these preparations reveals that the organization of the chromatin and the subcomponents of the nucleoli are different from those seen with the sucrose–calcium procedure (Fig. 11.6B). The borders of the nucleoli are considerably more hazy than those observed after isolation with the sucrose–calcium procedure and the nucleoli appear to be larger. The chromatin is clumped. These changes may be due to the considerable extraction of protein with citric acid in the concentrations used and to the low pH necessary to effect removal of the cytoplasm. Since protein exerts an important effect on the stainability of the citric acid preparations for electron microscopy, it is not surprising that there are changes in the electron micrographs of citric acid-treated preparations (Fig. 11.6B).

B. Solutions for Homogenization

The homogenizing medium employed has varied only in the concentration of citric acid employed and in other components added. Marshak (*37*) who first used the method for liver and tumor tissue utilized 5% citric acid solutions, although Stoneburg and Haven (*61*) used the method in conjunction with pepsin digestion (in 1940). The pH of the homogenizing solution is dependent upon the concentration of the citric acid employed so that 1.5% citric acid solutions have a pH of 2.3 and 4% citric acid solutions have a pH of 2.2. Since there is variation of the tenacity of the cytoplasmic adherence to the nuclear membrane in various tissues, for some tissues lower concentrations of citric acid may be employed than for others; e.g., for the Walker tumor a concentration of 2.5% was found to be optimal (*23, 28*).

C. Temperature

As in the case of isolation of nuclei with either the sucrose–calcium procedure or the nonaqueous media, it is generally believed to be optimal to keep the temperature as close to 0° as possible throughout the homogenization and centrifugation. When high molecular weight RNA is isolated, it is of much importance to minimize RNase activity, and cold conditions are an absolute requirement.

D. Homogenization

In this laboratory, tumor tissue is homogenized in 2.5% citric acid solutions in a weight-to-volume ratio of 1:10. Liver preparations are made in 1.5% citric acid in the same ratio of weight to volume. There are no special problems involved in the citric acid procedure that have not already been dealt with in connection with the sucrose–calcium method. The same Teflon–glass system is used for small-scale preparations in this laboratory. For larger preparations, the continuous homogenizers are used and satisfactory preparations are obtained with either stainless-steel or glass cylinders.

Although years ago Schneider and Hogeboom (*51*) pointed out the unsuitability of the Waring Blendor and similar instruments for most biochemical preparations, such instruments as well as some oscillators are still being used. These instruments are unsuited to the isolation of nuclei by any method because they not only destroy nuclei, but also destroy high molecular weight RNA.

E. Centrifugation

Initially, the homogenates of tumor preparations are centrifuged at 600 *g* for 10 minutes and then are rehomogenized for short periods in 0.25 *M* sucrose containing 1.5% citric acid; these suspensions are then layered over 0.88 *M* sucrose containing 1.5% citric acid. The samples are centrifuged at 900 *g* for 10 minutes after which the supernatant solutions are discarded and the sediment which contains the nuclear preparation is saved. For liver, the procedure may be simpler, depending upon the starting preparation; i.e., the layering procedure is not necessary if a fairly good preparation is obtained initially.

F. Yield and Purity

The yields of nuclei of high purity obtained with the citric acid procedure are somewhat higher than those obtained with the sucrose–calcium procedure; i.e., they are approximately 43%, which is slightly greater than the 35% usually obtained with the Chauveau technique (*15*). Whether this difference is significant is not yet defined. Although the purity of these preparations is reasonable, it has not yet been possible to obtain a preparation which is completely free of cytoplasmic contamination as detected by light microscopy. The separation of the cytoplasm from the nucleus occurs by a process which resembles the shelling of a seed or hard-boiled egg yolk from the surface shell of denatured protein. Since this shelling process is not always complete, an occasional nucleus is found with some of the cytoplasm partially adherent to it. These cytoplasmic tags differ from those found in the Chauveau procedure (*15*) in that they are wedge-shaped with the larger end of the wedge farther from the nuclear membranes; in the Chauveau procedure, the opposite is generally true; i.e., the larger part of the wedge is closer to the nuclear membrane.

Electron microscopic analysis shows that there are generally a number of small segments of the denatured cytoplasm associated with these preparations, so that the purity index by particle count (nuclear particles/total particles) is not so satisfactory as would be desirable. These particles are generally bits of cytoplasm that apparently become as dense as the nuclei by dehydration or denaturation. With nuclear preparations obtained by the Chauveau procedure (*15*), the purity of the preparation may be as high as 99% by particle count in the phase microscope and as high as 92% by particle count in the electron micrographs. The best purity of the nuclear preparations obtained with the citric acid procedure in this laboratory was about 97% by particle count with the phase microscope and about 85% by particle count of electron micrographs.

For further assays of purity, chemical methods have been employed and assays of DNA, RNA, and protein have been reported. However, no assays have been presented of enzymes in the nuclei obtained by this procedure, probably because of the assumption that they are denatured.

In this laboratory, values for DNA, RNA, and protein have been obtained in comparison with the sucrose–calcium procedure (see Table 11.1). Dounce (*23*) has reported that in nuclei isolated by the citric acid procedure, DNA comprised 20–23% of the dry weight, a value that agrees well with the results obtained in studies in this laboratory. Citric acid–sucrose solutions at pH 5.8 were then used and the DNA was markedly reduced in amount. Dounce (*23*) has indicated that the procedure employing buffers at pH 5.8 results in a more satisfactory prepara-

TABLE 11.1

NUCLEAR CONTENTS

Sample	Sucrose–calcium (*42*) (pg/nucleus)	Citric acid (*28*) (pg/nucleus)
Liver		
DNA	10.7	8.8
RNA	2.9	2.2
Protein	43.9	35.1
Tumor (Walker 256)		
DNA	15.4	11.9
RNA	23.5	5.7
Protein	171.5	34.8

tion from the point of view of the protein content and minimal proteolytic activity in the preparation.*

IV. Isolation and Purification of Nucleoli

Morphologically satisfactory nucleoli have been isolated by microdissection under phase microscopy (*24, 31*). Such techniques are both laborious and demanding of special skills; they are unsuitable for biochemical analysis because only a few nucleoli can be isolated in any given time. For studies on nucleoli of *Drosophila*, however, microdissection has been of use in conjunction with ultramicromethods (*24*) for analysis of RNA.

The first satisfactory method for isolation of nucleoli from nonmammalian species came from the studies of Vincent (*63, 64*) on nucleoli of starfish eggs. Monty *et al.* (*39*) in Dounce's laboratory announced the first isolation of nucleoli from mammalian cells using sonication to break up the cells and cell nuclei.

Maggio *et al.* (*35, 36*) reinvestigated the procedure of Monty *et al.* (*39*) and found that the sonication procedure could be markedly improved provided the container was cooled and Ca^{2+} was added to the medium.† Details of their procedure were provided to us at a time when efforts were under way for isolation of nucleoli from tumors and precancerous livers in the authors' laboratory. A reinvestigation of the variables of the procedure employing sonication led to the method (*9,*

* Recently, Berkowitz *et al.* (*3a*) reported that they were able to isolate highly purified nuclei from HeLa cells by the following procedure: The cell pellet was washed with a solution containing 0.32 *M* sucrose, 0.002 *M* $MgCl_2$, 0.001 *M* potassium phosphate, at pH 6.8. The cells were then suspended in a solution containing 0.01 *M* NaCl and 0.001 *M* potassium phosphate, at pH 6.8. After centrifugation at 800 *g* for 10 minutes, the pellet was suspended in a solution containing 0.32 *M* sucrose, 0.001 *M* $MgCl_2$, and 0.3% Triton N-101, adjusted to pH 6.2–6.4 with 0.001 *M* potassium phosphate. The cell suspension was homogenized in a Dounce homogenizer and then centrifuged at 800 *g* for 10 minutes to provide the nuclear preparation. This preparation was reported to have a well defined ultrastructure and high RNA polymerase activities (*3a*).

† Among the early and less successful procedures tested for isolation of nucleoli were such methods as treatment of nuclei with citric acid and with 50% acetic acid (*10*), a method useful for preparation of chromosomes (*58*). A "panker" was also developed which consists of an upper rotating head which is kept at a defined distance from a lower stationary plate by means of a micrometer screw (*44, 45*). For isolation of nuclei, this head was set at a greater distance from the plate. The reported isolation of nucleoli by this method has not been verified.

10, 42, 48, 49) now successfully employed in many laboratories for isolation of nucleoli of normal and tumor tissues.*

V. The Sonication Procedure for Isolation of Nucleoli

A. Isolation and Sonication of Nuclei

For the purposes of isolation of nucleoli, it is essential that a divalent ion be present in the medium for isolation of the nuclei and hence the sucrose–calcium procedure (v.s.), for isolation of nuclei is employed.

For sonication of the nuclei, the Branson sonifier is now employed; it has a power of 0–150 W and operates at 20 kc/second. The Raytheon Sonic Oscillator, Model DF-101, which has a power of 200 W and operates at 10 kc/sec may also be used. The containers for these two instruments differ as well as the size of the plate from which the sonic vibrations emanate. The Raytheon instrument contains a vibrating reed which oscillates in an alternating electromagnetic field. The reed must be screwed tightly into the bottom of a stainless-steel diaphragm approximately 2.5 cm in diameter which forms the base of a stainless-steel cup with a volume of approximately 50 ml. This container is surrounded by a stainless-steel jacket which can be connected conveniently to a pump carrying cooling fluid which forms part of a recirculating cooling system. The cup of a Branson Sonifier is an all-glass tapered vessel which has several glass semicircular tubes emerging from its base in the form of a rosette. This rosette must be placed into a circulating ice or a cooling bath which operates at a temperature of −5 to −10°. For the Raytheon sonicator, the temperature of the cooling bath (Wilkens-Anderson Low Temp) is maintained at 2°; each of these baths contains a mixture of ethylene glycol and water in a 1:3 ratio. With the Branson Sonifier, the tip must be examined periodically for micro to macro pitting. It should be burnished frequently to provide maximum output.

The Raytheon Sonic Oscillator is adjusted in this laboratory to provide 1.0–1.1 amp of output current for periods of 15–30 seconds. At each time interval, a drop of the solution is used to determine the extent of destruction of the nuclei and the sonic oscillation is discontinued when the nuclei are virtually destroyed. The destruction of the nuclei does not ordinarily require longer than 45–120 seconds in the Raytheon Sonic Oscillator, and with the Branson Sonifier, it generally requires from 25 to

* Although methods have been reported for isolation of nuclei and nucleoli from plant cells, at present such preparations are in relatively little use. Apparently the methods either lacked adequate specificities or do not readily provide as satisfactory a product as is obtainable from mammalian systems.

120 seconds. Although the Branson Sonifier operates at higher amperages; i.e., 7.5–8.0, and at more kilocycles per second, the sonic oscillations emanate from what is virtually a point source. The sonic oscillations in the Branson Sonifier are of sufficient strength to set up waves throughout the rosette which circulate the contained suspension. This circulation permits cooling of the entire contents of the sample.

B. Staining of the Nuclei and Nucleoli

It is necessary to follow the stages of the destruction of the nuclei quite closely inasmuch as excessive sonic oscillations result in destruction of the nucleoli. The stain most commonly used is azure C which is dissolved in 0.25 *M* sucrose to a final concentration of 0.1% of the dye. This stain permits rapid visualization of the nucleoli as well as ribonucleoprotein components of the cytoplasm. The pH of the stain is 3.4 after direct addition of azure C to the sucrose solution. Other basic dyes, including toluidine blue, Giemsa stain, and methylene blue, may also be used.

C. Effect of Increased Temperature

The rate of destruction of the nucleoli and nuclei is dependent upon the temperature of the sonication medium. At 2°, the rate of destruction of the nuclei is markedly reduced by comparison with that at 15°; i.e., 50% nuclear destruction was found approximately 10 seconds earlier and 90% nuclear destruction occurred 20 seconds earlier. However, the number of nucleoli recovered was reduced and the rate of nucleolar destruction was markedly accelerated at 15° so that the higher temperature is not utilized.

D. Volume of the Solutions for Sonication

The optimal volume is 20–40 ml. In systems employing the Branson Sonifier, the rosette supplied is used for sonication of 35 ml of homogenate or a beaker can be substituted in which the sample is stirred with the aid of a magnetic stirrer; the temperature is maintained below 4°. Under the conditions of the early experiments, a greater time was required for destruction of the nuclei with a greater volume of suspending medium. However, within 3–4 minutes, the nuclei in a total volume of 150–200 ml could be destroyed with this instrument. With the Raytheon Sonic Oscillator, it was found that the time for destruction of liver nuclei was 30 seconds in a volume of 10 ml, 1 minute in a volume of 20 ml, and 5 minutes in a volume of 40 ml.

E. PURIFICATION OF THE NUCLEOLI FROM THE SONICATE

The whole sonicate in a volume of 20 ml is layered over 20–25 ml of 0.88 *M* sucrose and the mixture is centrifuged for 20 minutes at 2000 *g* in a swinging bucket rotor (*48*, *49*). The sedimented nucleoli may be

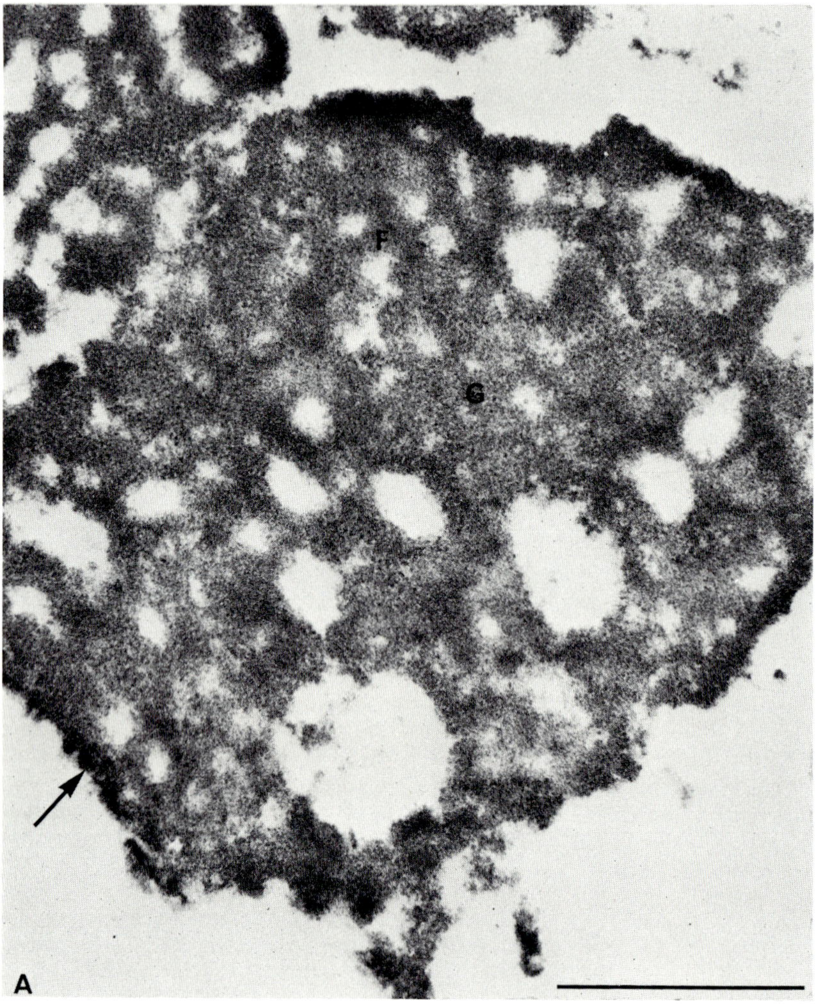

Fig. 11.7. A. Isolated nucleoli of normal rat liver stained with toluidine blue. ×1200. B. Electron micrograph of isolated nucleoli showing high level of purity by the lack of cytoplasmic contamination. The marker line is 1 μ.

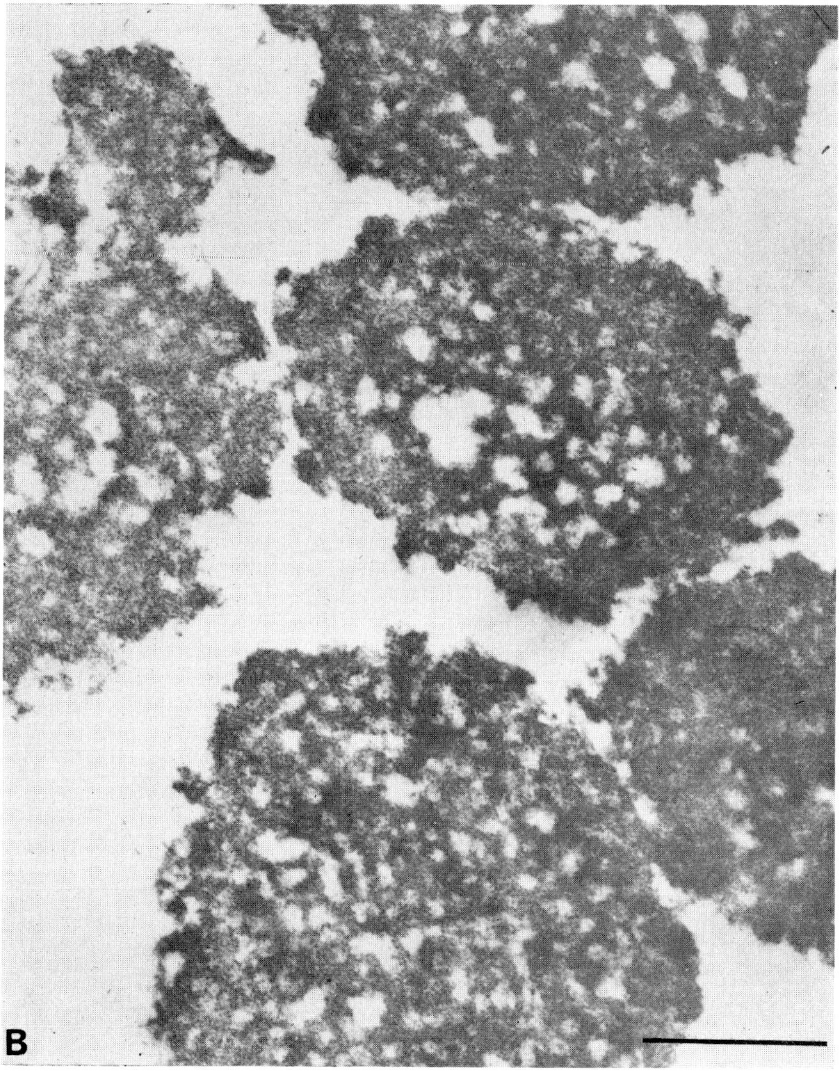

Fig. 11.7 (*Continued*).

resuspended in 0.25 *M* or 0.88 *M* sucrose and centrifuged once more to remove coprecipitated nucleoprotein. This preparation constitutes the type of preparation shown in Fig. 11.7, which is a satisfactory nucleolar preparation.

F. Criteria of Purity

Criteria for isolated nucleoli of liver and Walker tumor include morphological, chemical, and enzymatic analyses. In view of the substructure of the nucleoli, which includes the well defined nucleolonemas as well as the light spaces which lend to the whole structure a sponge-like appearance, it is possible to identify the nucleoli more exactly by electron microscopy than by light microscopy. In addition, the nucleolonemas contain ribonucleoprotein aggregates that appear granular in electron micrographs (Figs. 11.7B and 11.8).

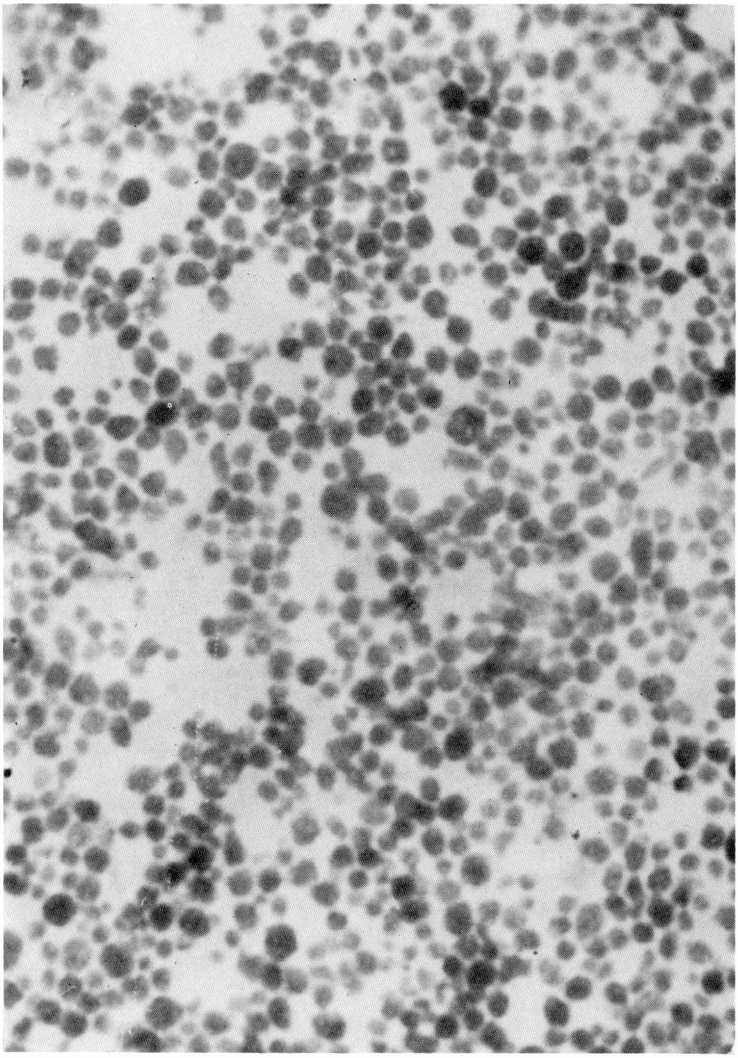

Fig. 11.8. Electron micrograph of an isolated nucleolus of a Novikoff hepatoma cell showing preservation of the outer layer of the nuclear chromatin (arrow), the fibrillar elements (F), and the granular elements (G). ×25,500.

By light microscopy, nucleoli have a reasonably characteristic appearance as seen by their spherical or irregular shapes when stained with basic dyes. By phase microscopy, however, it is possible that ribonucleoprotein aggregates from the cytoplasm may be aggregated and produce a similar appearance. For this reason, electron microscopy is a more satisfactory index of the type of product isolated. From the standpoint of purity as indicated by the presence of whole nuclei, the minimal standard of preparation in this laboratory is a maximum of 1 nucleus per 4000 nucleoli. However, in many preparations, no nuclei are found and electron microscopic analysis is used for particle counts of contaminants. Since both large and small contaminants are counted, purities of 85% are considered satisfactory; i.e., of all particles present, 85% are nucleoli. In some instances, purities of 92% have been obtained.

Chemical analyses of nucleoli have shown that the major nucleolar components are RNA, DNA, and protein in the approximate ratio of 1:1:10 (Chapter V). The high DNA content may reflect the presence in the nucleolar preparations of either perinucleolar or intranucleolar chromatin. Evidence for the presence of intranucleolar DNA strands has been provided by electron microscopic studies (Chapters II and IV). Perinucleolar DNA and associated proteins appear to be essential for the maintenance of the nucleolar structure during the course of isolation of the nucleoli. Differences have not yet been found between the intra- and extranucleolar DNA or nuclear proteins on the basis of elementary base or amino acid analysis.

Nucleolar RNA differs in sedimentation characteristics from whole nuclear RNA (*27, 40, 41*) as indicated in the sucrose density sedimentation diagrams (Figs. 5.1, 5.2). RNA of isolated nucleoli has much larger amounts of both 35 S and 45 S RNA than whole nuclear RNA, but the relative amount of 18 S RNA is much less. In fact, in the isolated liver nucleoli, there is little, if any, RNA in the 18 S region. In nucleoli isolated from the Walker tumor a similar type of pattern is observed. The larger amounts of higher molecular weight RNA and limited amount of 18 S RNA provide a distinguishing feature of nucleolar RNA. Differences between nucleolar and extranucleolar proteins are now being studied (*26;* see also Chapter VII).

G. Enzymatic Assays for Nucleolar Purity

There are no specific nucleolar enzymes that are readily distinguishable from the nuclear enzymes as a whole. However, by comparison with nuclei, purified nucleoli from liver cells contain lower specific activities and disproportionately smaller amounts of the enzymes glutamate dehydrogenase, adenylate kinase, catalase, and glucose-6-phosphatase (*55*). In addition, they contain larger amounts of RNA polymerase, ribonuclease, DPN pyrophosphorylase, and, to a lesser extent, ATPase A than do whole nuclei (Chapter VII). The activity of nucleoli in conversion of GC-rich 45 S to 28 S RNA is remarkable, but the specific activities of the responsible enzymes have not yet been evaluated (*31b, 62a*).

VI. Mass Isolation of Nucleoli Using the Sonication Procedure

There are no special requirements for the mass isolation of nucleoli from the nuclei obtained by large scale techniques since batch processes are used in this laboratory for this purpose (*8, 17, 18, 28a*). However, the sonic oscillation is carried out in a Branson Sonifier because it can hold a greater volume than the Raytheon Sonic Oscillator. Moreover, centrifugation of the sonicate which is layered over 0.88 *M* sucrose is carried out in the rotor of the RC-3 centrifuge which contains four swinging buckets, each of which has a volume of 650 ml. The total amount of nucleoli that has been sedimented in the course of any one centrifugation is 7 gm. If the degree of nucleolar purification is not quite satisfactory, the sediment is resuspended in 0.25 *M* sucrose and either the sonic oscillation, the centrifugation, or both may be repeated.

VII. Compression–Decompression Procedure for Isolation of Nucleoli (*48, 49*)

The basis for the compression–decompression procedure for isolation of nucleoli was the concept of Somers *et al.* (*58*). They showed that such fragile structures as chromosomes could be obtained from nuclei without serious damage to their morphology by alternate aspiration and ejection of nuclear suspensions from a syringe. It was then found (*19*) that when either nuclei or whole cells were incubated with Ca^{2+} in concentrations of 0.005–0.010 *M* for periods of approximately 60 minutes, the nuclei or whole cells became selectively "hardened," which permits them to rupture on release from a French pressure cell. Under these conditions, the nucleoli were not destroyed and could then be isolated from the "pressate" by differential centrifugation.

A. Homogenization

The tissues are homogenized with Teflon–glass homogenizers as described for isolation of nuclei. The homogenization medium consists of 0.25 *M* sucrose containing 0.005–0.010 *M* $CaCl_2$. The pestle clearance is 0.005–0.006 inch regardless of whether Walker tumor or liver is studied. The homogenate is filtered through eight layers of gauze over a 90-mesh stainless-steel screen. The degree of purification varies considerably, but highly purified nuclear preparations are not essential for continuation of the procedure.

B. Incubation

All homogenates are kept for periods of 40–60 minutes at 0–2° in the cold laboratory or in the refrigerator. If the samples are kept for longer or shorter times, the ratio of nuclei to nucleoli in the final product is increased considerably.

C. Disruption of the Nuclei

Forty milliliters of the homogenate are placed in a French pressure cell which is then set into a Carver hydraulic press. The pressure is increased to 7000–9000 psi. The needle valve of the pressure cell is opened a crack, and pressure is then adjusted to 5000–7000 psi as the sample is permitted to emerge from the pressure cell. A variety of pressures, ranging from 3000 to 9000 psi, were utilized in this step. At the lower end of the pressure scale, the destruction of the nuclei was not satisfactory, and large numbers of whole or partially broken nuclei were found. At the higher pressures, there was an alteration in the color of the preparations which developed a grayish hue. The duration of the release of the sample from the French pressure cell varies from 5 to 15 minutes.

D. Differential Centrifugation

A discontinuous gradient of sucrose utilized for the preparation of the nucleoli from the pressed product consisted of an upper layer of 5 ml of 1.5 *M* sucrose and a lower layer of 3 ml of 2.2 *M* sucrose in a 30-ml centrifuge tube of the 25.1 swinging bucket rotor; 20 ml of pressed product were layered over the gradient. After centrifugation of the discontinuous gradient at 25,000 rpm for 10–20 minutes, the supernatant solutions were removed by aspiration with a water pump and the pellet that remained at the bottom of the tube constituted the nucleolar preparation (Fig. 11.9).

The optimal conditions for centrifugation of the pressed product were studied for the Walker tumor and the liver. Inasmuch as the nucleoli of the Walker tumor are larger and probably more dense than those of the

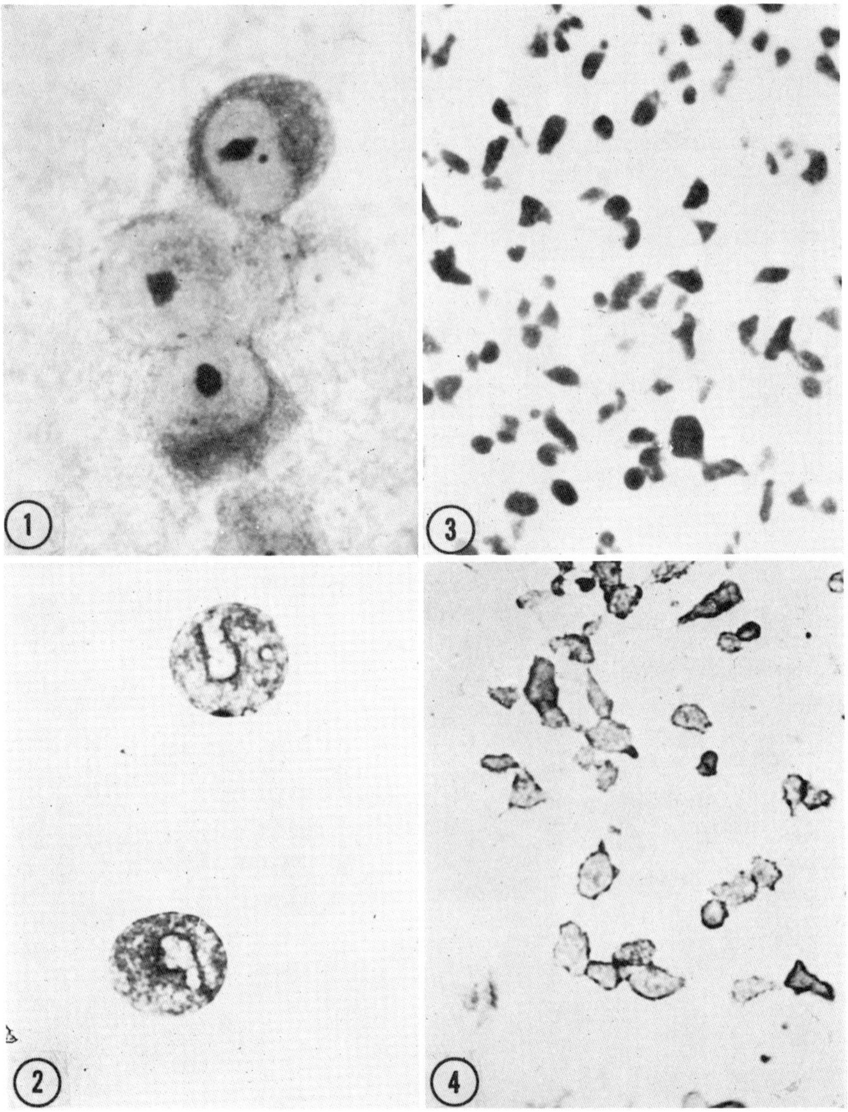

Fig. 11.9. 1. Smear of isolated nuclei from Walker tumor stained with toluidine blue. The nucleoli and the cytoplasmic basophilic structures are stained intensely. ×1600. 2. Smear of isolated nuclei from Walker tumor stained wtih toluidine blue after hydrolysis with 1 *N* HCl. Chromatin-containing structures are stained. The nucleolus is not stained, but the perinucleolar nucleolus-associated chromatin is stained. ×1600. 3. Smear of isolated nucleoli from Walker tumor, stained with toluidine blue at pH 5.0. Note the similar morphology and staining properties of isolated nucleoli and nucleoli *in situ*. ×1600. 4. Isolated nucleoli from the Walker

liver, 50% of the total nucleoli are sedimented in approximately 7½ minutes in a force field of 64,000 *g*. For the nucleoli of the liver, the time for sedimentation of 50% of the nucleoli is approximately 20 minutes. The type of product varies inasmuch as the longer the period of sedimentation, the more chromatin tends to sediment with the nucleoli. At the optimal time points indicated above, the percentage of the total nucleic acid and proteins in nucleoli obtained from the Walker tumor are 72, 16, and 12 for protein, RNA, and DNA, respectively. In the nucleoli obtained from the liver, these values are 89, 7, and 3%, respectively. At longer times more contamination is visible by direct particle counts of the nucleolar preparations in the light microscope and in electron micrographs.

E. Composition of Nucleoli Isolated by the Press Procedure

As indicated in Table 5.1, some differences were found in the composition of the individual nucleoli isolated by the sonication or press (compression–decompression) procedure. Less DNA was found in the nucleoli of the liver and less protein was found in the nucleoli of the Walker tumor with the compression–decompression procedure than with the sonication technique. However, it is likely that different size populations of the nucleoli may be studied with these methods and, in addition, the long duration of the incubation in the course of the compression–decompression procedure may have produced some changes due to enzymatic activity. The similarities of the results for the different preparations are striking.

F. Recovery

The overall recovery of the nucleoli under the centrifugation conditions indicated above was approximately 50% that of the whole homogenate in studies on both the Walker tumor and the liver.

VIII. "Nucleolar" Preparations Obtained by Chemical Extractions

After nuclei are extracted with 0.15 *M* NaCl and 1 or 2 *M* NaCl, a residue fraction is left in which the labeling of RNA is significantly greater than that of the extract (*1*). A similar result is obtained when

---

tumor, stained with toluidine blue after hydrolysis with 1 *N* HCl. The nucleolus associated chromatin is stained, but the rest of the nucleolus is not stained. Septa penetrating from the nucleolus-associated chromatin into the nucleolus are visible in both the isolated nucleoli and the nucleoli *in situ*. × 1800.

the nuclei are initially extracted with 0.1–0.2 *M* phosphate buffer. The residue fraction has been referred to as the "nucleolar fraction" because of the presence of structures that stain with the density of nucleoli when the samples are stained with pyronine. The basis of the chemical technique for preparation of the "nucleolar fraction" is, then, successive extraction of the nuclei with dilute and concentrated saline solutions (*52*, *59*).

A. Extraction of Liver Nuclei

All operations are carried out in the cold laboratory at 2–4°. The Tris–NaCl extract is obtained by treatment of the isolated nuclei twice with 0.14 *M* NaCl containing 0.05 *M* Tris-HCl buffer at pH 7.6 and 0.001 *M* Mg acetate; 0.5 ml of the solution is used per gram of original liver (*59*). The nuclei are homogenized in this solution for a period of 1 minute with a loosely fitting Teflon–glass homogenizer (pestle clearance 0.009–0.012 inch). Addition of sodium polyvinyl sulfate (PVS) at this step produces a much more satisfactory yield of high molecular weight RNA in the product (*59*) than is obtained when no RNase inhibitor is added; the final concentration of PVS is 0.2 mg/ml of solution. The extracted nuclei are centrifuged at 4000 *g* for 10 minutes.

The nuclear residues are then suspended in a small volume of the Tris-saline solution, 0.2 ml per gram of original liver, after which 2 *M* NaCl is added in the amount of 4.0 ml per gram of liver. This mixture is homogenized in a Teflon–glass homogenizer. This suspension is then shaken for 30 minutes in an equipoise shaker. To prevent foaming, a few drops of octanol are added. The suspension is then centrifuged for 30 minutes at 40,000 *g*. The supernatant solution contains the deoxyribonucleoproteins and ribonucleoproteins, and the pellet is the "nucleolar" or residue fraction.

B. Extraction of Calf Thymus Nuclei

In the studies of Sibatani *et al.* (*52*), the nuclei were suspended in 10 vol of 0.1 *M* Tris–HCl at pH 7.6 containing 3 m*M* $CaCl_2$. This mixture was blended at 8000 rpm for 1 minute and centrifuged at 1000 *g* for 15 minutes.

The extraction of DNA was effected by stirring with 18–20 vol of 1 *M* NaCl. The gel was stirred at high speed, and extraction was carried out for 16 hours at 2°, which was the temperature of the entire procedure. Sibatani *et al.* (*52*) noted that the Waring Blendor should not be utilized at this point because the organization of the nucleoli is destroyed. The sample was then centrifuged at 40,000 *g* in order to sediment the "nucleolar" fraction. In the case of calf thymus, this fraction accounted for about 2% of the dry weight of the nucleus, but the specific activity of the RNA was high by comparison with that extracted with the dilute or concentrated saline solutions.*

C. Nature of the Product

Morphological studies on the product obtained after saline extraction revealed that a number of ribonucleoproteins remained in the residue; most of those remaining are components of the nuclear ribonucleoprotein network (*57*). This network apparently consists of the nucleolus and ribonucleoprotein strands that emanate from it (Chapter VIII).

The sucrose density sedimentation diagrams of the RNA extracted from isolated nucleoli are substantially different from those of the "nucleolar residues" obtained in the chemical extraction procedures. However, some similarities are apparent, such as the small amount of 18 S RNA and the predominant amounts of higher molecular weight RNA. Although Sibatani *et al.* (*52*) were able to isolate RNA from the fraction obtained from calf thymus nuclei that had virtually the same base composition as DNA, the "nucleolar preparations" from rat liver have not been found to contain such a fraction. Evidence obtained in this laboratory has indicated that some AU-rich RNA is present in the "nucleolar residues" (*15a;* see also p. 372, 376).

D. Extraction Methods Employing Detergents

Penman *et al.* (*43a*) developed a procedure for isolation of "nucleolar preparations" which has been used in a number of their studies. Following swelling of cells in a hypotonic medium containing 0.01 *M* NaCl, 0.003 *M* $MgCl_2$, and 0.02 *M* Tris (pH 7.4), the cells were broken in a stainless-steel ball homogenizer. After centrifugation, the nuclei were resuspended in this buffer which was made 0.5% with respect to Tween 40. After centrifugation, the nuclei were treated with Tween 40 and sodium deoxycholate. After recentrifugation, the detergent treated nuclei were suspended in 0.5 *M* NaCl, 0.05 *M* $MgCl_2$ and 0.01 *M* Tris (pH 7.4). These centrifuged preparations were then treated with DNase. The digested preparations were centrifuged at 10,000 *g* for 5 minutes and the resulting pellet was the nucleolar preparation.

In attempting to account for the marked differences in the RNA classes

* In recent studies in this laboratory it has been found that a number of structures that appear to be nucleoli in calf thymus nuclei are really regions of condensed chromatin, i.e., "chromo-centers."

recovered in the preparations obtained by their method and those of the sonication procedure presented above, Vesco and Penman (*62a*) have indicated that the losses of all the nucleolar 28 S RNA and some of the 32 S RNA resulted from the high salt concentration employed earlier. Presumably, the low molecular weight nucleolar RNA would also be removed. When they isolated nucleolar preparations in a medium of 0.1 *M* NaCl, 0.01 *M* $MgCl_2$, and 0.01 *M* Tris (pH 7.4), their nucleoli contained some of the missing RNA sedimentation classes (*62a*).

REFERENCES

1. Allfrey, V. G., and Mirsky, A. E., Some aspects of ribonucleic acid synthesis in isolated cell nuclei. *Proc. Natl. Acad. Sci. U. S.* **43**, 821–826 (1957).
2. Allfrey, V. G., and Mirsky, A. E., Role of histone in nuclear function. *In* "The Nucleohistones" (J. Bonner and P. Ts'o, eds.), pp. 267–288. Holden-Day, San Francisco, California, 1964.
3. Allfrey, V. G., Stern, H., Mirsky, A. E., and Saetren, H., The isolation of cell nuclei in non-aqueous media. *J. Gen. Physiol.* **35**, 529–554 (1951).

3a. Berkowitz, D. M., Kakefuda, T., and Sporn, M. B., A simple and rapid method for the isolation of enzymatically active HeLa cell nuclei. *J. Cell Biol.* **42**, 851–854 (1969).

3b. Blobel, G., and Potter, V. R., Nuclei from rat liver: Isolation method that combines purity with high yield. *Science* **154**, 1662–1665 (1966).

3c. Blobel, G., and Potter, V. R., Studies on free and membrane-bound ribosomes in rat liver. I and II. *J. Mol. Biol.* **26**, 279–301 (1967).

4. Brachet, J., La détection histochimique des acides pentonucléiques. *Compt. Rend. Soc. Biol.* **133**, 88–90 (1940).
5. Brachet, J., La localisation des acides pentosenucléiques dans les tissus animaux et les oeufs d'amphibiens en voie de developpement. *Arch. Biol. (Liege)* **53**, 207–257 (1942).
6. Busch, H., "Histones and Other Nuclear Proteins." Academic Press, New York, 1965.
7. Busch, H., Byvoet, P., and Smetana, K., The nucleolus of the cancer cell: A review. *Cancer Res.* **23**, 313–339 (1963).
8. Busch, H., and Desjardins, R., A continuous tissue homogenizer. *Exptl. Cell Res.* **40**, 353–359 (1965).
9. Busch, H., Lane, M., Adams, H. R., DeBakey, M. E., and Muramatsu, M., Isolation of nucleoli from human tumors. *Cancer Res.* **25**, 225–233 (1965).
10. Busch, H., Muramatsu, M., Adams, H. R., Smetana, K., Steele, W. J., and Liau, M. C., Isolation of nucleoli. *Exptl. Cell Res.* Suppl. 9, 150–163 (1963).
11. Busch, H., Starbuck, W. C., and Davis, J. R., A method for isolation of nuclei from cells of the Walker 256 carcinosarcoma. *Cancer Res.* **19**, 684–687 (1959).
12. Caspersson, T. O., "Cell Growth and Cell Function. A Cytochemical Study." Norton, New York, 1950.
13. Caspersson, T. O., and Santesson, L., Studies on protein metabolism in the cells of epithelial tumors. *Acta Radiol.* Suppl. 46, 1–105 (1942).
14. Caspersson, T. O., and Schultz, J., Ribonucleic acids in both nucleus and

cytoplasm and the function of the nucleolus. *Proc. Natl. Acad. Sci. U. S.* **26**, 507–515 (1940).

15. Chauveau, J., Moule, Y., and Rouiller, C., Isolation of pure and unaltered liver nuclei, morphology and biochemical composition. *Exptl. Cell Res.* **11**, 317–321 (1956).

15a. Choi, Y. C., and Busch, H., Effects of actinomycin D on the oligonucleotide composition of nucleolar 45 S RNA of Novikoff hepatoma ascites cells. *Biochim. Biophys. Acta* **174**, 766–769 (1969).

16. Desjardins, R., Grogan, D. E., and Busch, H., Effects of antitumor agents on synthesis of nucleolar DNA. *Cancer Res.* **27**, 159–164 (1967).

17. Desjardins, R., Smetana, K., and Busch, H., A large scale procedure for isolation of nucleoli from rat liver. *Exptl. Cell Res.* **40**, 127–137 (1965).

18. Desjardins, R., Smetana, K., Grogan, D., Higashi, K., and Busch, H., A large scale procedure for isolation of nucleoli from Walker tumor. *Cancer Res.* **26**, 97–104 (1966).

19. Desjardins, R., Smetana, K., Steele, W. J., and Busch, H., Isolation of nucleoli of the Walker carcinosarcoma and liver of the rat following nuclear disruption in a French pressure cell. *Cancer Res.* **23**, 1819–1823 (1963).

20. Dounce, A. L., Enzyme systems of isolated cell nuclei. *Ann. N. Y. Acad. Sci.* **50**, 982–999 (1950).

21. Dounce, A. L., The significance of enzyme studies on isolated cell nuclei. *Intern. Rev. Cytol.* **3**, 199–223 (1954).

22. Dounce, A. L., The isolation and composition of cell nuclei and nucleoli. *In* "The Nucleic Acids" (E. Chargaff and J. N. Davidson, eds.), Vol. 2, pp. 93–154. Academic Press, New York, 1955.

23. Dounce, A. L., The isolation of nuclei from tumor cells. *Exptl. Cell Res.* Suppl. 9, 126–143 (1963).

24. Edström, J.-E., Chromosomal RNA and other nuclear RNA fractions. *In* "The Role of Chromosomes in Development" (M. Locke, ed.), pp. 137–152. Academic Press, New York, 1964.

25. Granboulan, N., and Granboulan, N. P., Cytochimie ultrastructurale du nucléoli. I. Mise en évidence de chromatine a l'intérieur du nucléole. *Exptl. Cell Res.* **34**, 71–87 (1964).

26. Grogan, D. E., Desjardins, R., and Busch, H., Nucleolar proteins of Walker tumor and rat liver. *Cancer Res.* **26**, 775–779 (1966).

27. Hiatt, H. H., A rapidly labeled RNA in rat liver nuclei. *J. Mol. Biol.* **5**, 217–229 (1962).

28. Higashi, K., Shankar Narayanan, K., Adams, H. R., and Busch, H., Utilization of the citric acid procedure and zonal ultracentrifugation for mass isolation of nuclear RNA from Walker 256 carcinosarcoma. *Cancer Res.* **26**, 1582–1590 (1966).

28a. Hodnett, J. L., and Busch, H., Isolation and characterization of uridylic acid-rich 7 S ribonucleic acid of rat liver nuclei. *J. Biol. Chem.* **243**, 6336–6344 (1968).

29. Hogeboom, G. H., Schneider, W. C., and Striebich, M. J., Cytochemical studies on isolation and biochemical properties of liver cell nuclei. *J. Biol. Chem.* **196**, 111–120 (1952).

30. Hogeboom, G. H., Schneider, W. C., and Striebich, M. J., Localization and integration of cellular function. *Cancer Res.* **13**, 617–632 (1953).

31. Kopac, M. J., and Mateyko, G. M., Nucleolar chromosomes: Structures, interactions, and perspectives. *Advan. Cancer Res.* **8**, 121–190 (1964).
31a. Lazarus, H. M., and Sporn, M. B., Purification and properties of a nuclear exoribonuclease from Ehrlich ascites tumor cells. *Proc. Natl. Acad. Sci. U. S.* **57**, 1386–1393 (1967).
31b. Liau, M. C., Craig, N. C., and Perry, R. P., The production of ribosomal RNA from high molecular weight precursors. I. Factors which influence the ability of isolated nucleoli to process 45 S RNA. *Biochim. Biophys. Acta* **169**, 196–205 (1968).
32. MacCarty, W. C., The value of the macronucleolus in the cancer problem. *Am. J. Cancer* **26**, 529–532 (1936).
33. MacCarty, W. C., Further observations on the macronucleolus of cancer. *Am. J. Cancer* **31**, 104–106 (1937).
34. MacCarty, W. C., and Haumeder, E., Has the cancer cell any differential characteristics? *Am. J. Cancer* **20**, 403–407 (1934).
35. Maggio, R., Siekevitz, P., and Palade, G. E., Studies on isolated nuclei. I. Isolation and chemical characterization of a nuclear fraction from guinea pig liver. *J. Cell Biol.* **18**, 267–293 (1963).
36. Maggio, R., Siekevitz, P., and Palade, G. E., Studies on isolated nuclei. II. Isolation and chemical characterization of nucleolar and nucleoplasmic subfractions. *J. Cell Biol.* **18**, 293–313 (1963).
37. Marshak, A., P^{32} uptake by nuclei. *J. Gen. Physiol.* **25**, 275–291 (1941).
38. Montgomery, T. H., Comparative cytological studies with special regard to the nucleolus. *J. Morphol.* **15**, 265–564 (1898).
39. Monty, K. J., Litt, M., Kay, E. R. M., and Dounce, A. L., Isolation and properties of liver cell nuclei. *J. Biophys. Biochem. Cytol.* **2**, 127–145 (1956).
40. Muramatsu, M., and Busch, H., Studies on nuclear and nucleolar RNA of regenerating rat liver. *J. Biol. Chem.* **240**, 3960–3966 (1965).
41. Muramatsu, M., Hodnett, J. L., Steele, W. J., and Busch, H., Molecular properties of RNA of isolated nucleoli of rat liver. I. Synthesis of 28 S RNA in the nucleolus. *Biochim. Biophys. Acta* **123**, 116–125 (1966).
42. Muramatsu, M., Smetana, K., and Busch, H., Quantitative aspects of isolation of nucleoli of the Walker carcinosarcoma and liver of the rat. *Cancer Res.* **23**, 510–518 (1963).
43. Okamura, N., and Busch, H., Base composition of high molecular weight nuclear RNA of Walker tumor and liver of the rat. *Cancer Res.* **25**, 693–697 (1965).
43a. Penman, S., Smith, I., and Holtzman, E., Ribosomal RNA synthesis and processing in a particulate site in the HeLa cell nucleus. *Science* **154**, 786–789 (1966).
44. Poort, C., Electrophoretic comparison of nuclear and nucleolar proteins. I. Beef pancreas. *Biochim. Biophys. Acta* **46**, 373–380 (1961).
45. Poort, C., Electrophoretic comparison of nuclear and nucleolar proteins. II. Rat pancreas. *Biochim. Biophys. Acta* **51**, 236–240 (1961).
46. Potter, V. R., and Elvehjem, C. A., Modified method for study of tissue oxidations. *J. Biol. Chem.* **114**, 495–504 (1936).
47. Reid, B. R., and Cole, R. D., Biosynthesis of a lysine-rich histone in isolated calf thymus nuclei. *Proc. Natl. Acad. Sci. U. S.* **51**, 1044–1050 (1964).
48. Ro, T. S., and Busch, H., *In vitro* labeling of RNA in isolated nucleoli of the Walker tumor and liver. *Cancer Res.* **24**, 1630–1633 (1964).

49. Ro, T. S., Muramatsu, M., and Busch, H., Labeling of RNA of isolated nucleoli with UTP-^{14}C. *Biochem. Biophys. Res. Commun.* **14**, 149–155 (1964).
50. Schneider, W. C., Enzymes of cancer nuclei. *Exptl. Cell Res.* Suppl. 9, 430–434 (1963).
51. Schneider, W. C., and Hogeboom, G. H., Biochemistry of cellular particles. *Ann. Rev. Biochem.* **25**, 201–224 (1956).
52. Sibatani, A., deKloet, S. R., Allfrey, V. G., and Mirsky, A. E., Isolation of a nuclear RNA fraction resembling DNA in its base composition. *Proc. Natl. Acad. Sci. U. S.* **48**, 471–477 (1962).
53. Siebert, G., Enzymes of cancer nuclei. *Exptl. Cell Res.* Suppl. 9, 389–418 (1963).
54. Siebert, G., and Humphrey, G. B., Enzymology of the nucleus. *Advan. Enzymol.* **27**, 239 (1965).
55. Siebert, G., Villalobos, J., Jr., Ro, T. S., Steele, W. J., Lindenmayer, G., Adams, H. R., and Busch, H., Enzymatic studies on isolated nucleoli of rat liver. *J. Biol. Chem.* **241**, 71–78 (1966).
56. Smetana, K., and Busch, H., Studies on the ultrastructure of the nucleolus of the Walker tumor and the rat liver. *Cancer Res.* **24**, 537–557 (1964).
57. Smetana, K., Steele, W. J., and Busch, H., A nuclear ribonucleoprotein network. *Exptl. Cell Res.* **31**, 198–202 (1963).
58. Somers, C. E., Cole, A., and Hsu, T. C., Isolation of chromosomes. *Exptl. Cell Res.* Suppl. 9, 220–234 (1963).
59. Steele, W. J., Okamura, N., and Busch, H., Effects of thioacetamide on the composition and biosynthesis of nuclear and nucleolar RNA. *J. Biol. Chem.* **240**, 1742–1749 (1965).
60. Stern, H., Allfrey, V. G., Mirsky, A. E., and Saetren, H., Some enzymes of isolated nuclei. *J. Gen. Physiol.* **35**, 559–578 (1951).
61. Stoneburg, C. A., and Haven, F. L., Tumor inhibition by anti-tumor nuclei serum. *Am. J. Cancer* **38**, 377–379 (1940).
62. Swift, H., Cytochemical studies on nuclear fine structure. *Exptl. Cell Res.* Suppl. 9, 54–67 (1963).
62a. Vesco, C., and Penman, S., The fractionation of nuclei and the integrity of purified nucleoli in HeLa cells. *Biochim. Biophys. Acta* **169**, 188–195 (1968).
63. Vincent, W. S., Isolation and chemical properties of nucleoli of starfish oocytes. *Proc. Natl. Acad. Sci. U. S.* **38**, 139–145 (1952).
64. Vincent, W. S., Structure and chemistry of nucleoli. *Intern. Rev. Cytol.* **5**, 269–299 (1955).
65. Widnell, C. C., and Tata, J. R., A procedure for the isolation of enzymically active rat-liver nuclei. *Biochem. J.* **92**, 313–317 (1964).

CHAPTER XII

Cytological and Cytochemical Methods

I. Introduction

Although numerous studies have been made during the last 60 years, no staining methods have been found that are absolutely specific for nucleoli. Nevertheless, some cytological and cytochemical methods stain selectively for nucleoli because of differences in their chemical properties or composition as compared with the whole nucleus. These methods can be carried out on cell suspensions or cell fractions, in smears, touch and squash preparations, and tissue sections including ultrathin sections.

The identification of nucleoli in ultrathin sections by means of the electron microscope is easier than with the light microscope because the nucleolar ultrastructural appearance is very characteristic even in sections prepared by the commonly used electron microscopic procedures. However, the number of nucleoli and their presence or absence in cells is best determined by light microscopic methods. Evaluation of nucleolar size in ultrathin sections is difficult and depends on the position of the nucleolus in the cell preparation.

The light microscopic methods used for the visualization of nucleoli in cell suspensions or cell fractions are now very simple. However, they are limited to the investigation of cells in fluids, tissue homogenates, and cell fractions. Furthermore, they do not represent permanent preparations and the time at which they are observed after staining can influence the result of staining and the nucleolar morphology. Nevertheless, these methods are very useful for the rapid control of biochemical isolation procedures and for clinical purposes. Moreover, some of these methods facilitate the investigation of nucleoli in living cells.

Smears, touch and squash preparations, or sections represent relatively stable preparations that eliminate the influence of the time of observation. Light microscopic observation of the cell organelles, including nucleoli, is much easier in smears than in touch preparations or sections. On the other hand, the disadvantage of studying smears lies in the impossibility of observing the relationship between cells. Therefore, morphological and functional classification of the cells may be difficult. The determination of the presence or absence of nucleoli in cells is more exact in smears or touch preparations than in various types of sections in which the part of the nucleus that contains a nucleolus may not be present in that plane of the section. In contrast, the overlapping of other cell structures over or under the nucleoli in thick smears or touch preparations can make it difficult to identify micronucleoli. With all these light microscopic methods, nucleoli appear as positively stained bodies with or without special structure or as empty areas surrounded by nucleolus-associated chromatin. The best results in visualization of the nucleoli have been achieved by methods for staining the nucleolar body or RNA without simultaneously staining chromatin structures or DNA. Such methods, however, do not demonstrate a relationship between the nucleolus and nucleolus-associated chromatin as do methods for the simultaneous staining of nucleoli or RNA and chromatin or DNA. Some disadvantage of these methods lies in the possible overlapping of small nucleoli by surrounding chromatin structures (*64*). Methods providing "a negative picture" of nucleoli by demonstration of the chromatin structures or DNA around the nucleolus are not satisfactory for nucleolar studies since the differentiation of nucleoli from light areas in the nuclear chromatin structures is very difficult (*24, 70*).

II. Identification of Nucleoli in Cell Suspensions or Cell Fractions

The simplest method to study nucleoli is by phase-contrast microscopy which permits investigations on nucleoli in living cells. In some cells, nucleoli cannot be distinguished from chromatin clusters by phase microscopy but they can be visualized by special methods in smears or in sections; i.e., nucleoli in some mature lymphocytes (*19, 58, 66, 69*). The number, size, and shape of nucleoli can be determined in most cells or isolated nuclei by phase-contrast microscopy. The inner nucleolar structure has been also observed in living cells as in fixed cells by this method (*21, 34, 87*).

The procedures for "supravital" staining are divided into two groups. In the first group of methods, the dyes are dissolved in suitable media and mixed with cell suspensions or cell fractions. In the second group, the dyes are dissolved in absolute ethanol or methanol and smears of such solutions are prepared on the microscopic slides. A small drop of the specimen is spread on the dry film of the dye which is redissolved in the medium of the specimens. This method is simpler than the first one and the influence of the dye-containing medium is eliminated (*2*). The staining, however, can be variable because the concentration of the dye can also vary in different locations on the smears. The first method usually provides more constant results. The final result of the staining depends on the dye, its concentration, pH, temperature, and time of staining. The dye solutions must be free of microbial contamination, contamination by precipitated dyes, or adventitious cells as shown by control slides.

The dyes used for the supravital staining of nucleoli are usually basic or neutral. Methylene blue, toluidine blue, cresyl blue, neutral red, and the fluorochrome acridine orange have been commonly used. Some procedures for staining of nucleoli by supravital techniques utilize dye mixtures (*20, 52*). The viability of cells with nucleoli stained by most supravital staining procedures is at least questionable (*23, 52*), with the possible exception of staining with acridine orange (*29*). In some procedures (*20*), the cells can be fixed also during staining. At present, some methods used for visualization of nucleoli by means of supravital staining are used for controls and/or identification of nucleoli during their isolation (*50*) or in clinical cytology (*2, 20, 68*).

A. The Methyl Alcohol-Leishman or Giemsa Procedure (*20*)

Procedure

1. Dilute the examined cell suspension or blood (e.g., in white cell-count pipettes 1:20) with the staining solution containing 1 ml of methyl alcohol, 2 ml of distilled water, 1–4 drops of Giemsa stain or 5–10 drops of Leishman stain.
2. Observe in a counting chamber or put a small drop on the microscopic slide and cover with a thin glass cover slip.

Result

Nucleoli are stained a deep blue and nucleoli pale blue. When nuclei are stained a deep blue, the concentration of the dye in the staining solution must be decreased.

B. The Supravital Methylene Blue Technique for Nucleoli in Blood Cells (*20*)

Procedure

1. Dilute blood to 2% with 0.5% methylene blue in isotonic saline solution.
2. Mix and stain for 10–15 minutes.
3. Centrifuge and observe the sediment.

Result

Nucleoli are deep blue; nuclei are pale blue.

C. The Azure C Methods (*2, 9, 51*)

Procedure I

1. Place a small drop of 0.25 or 0.5% azure C dissolved in absolute methanol on a cover slip and cover with another cover slip. The stain spreads between cover slips and the excess of the stain can be removed by the filter paper.

2. Pull apart the glass cover slips with a sliding movement and dry. Mark the prepared cover slip with a grease pencil on the reverse side.
3. Place a small drop of the suspension containing investigated cells on the microscopic slide and cover with the prepared cover slip (the marked side up).
4. Observe after 1–15 minutes.

Result

The nucleoli are intensely stained a bluish purple; the rest of the nucleus is less stained or unstained. The method can be modified by preparation of a thin film of the stain. Then a small drop of the investigated suspension is placed directly on the film of the stain and covered with a cover slip. Both procedures are simple and are useful in diagnostic cytology.

Procedure II

1. Mix a small drop of suspension containing the investigated specimens with a small drop of 0.1–0.6% azure C in 0.25 *M* sucrose or distilled water.
2. Examine in a counting chamber or place a small drop on a microscopic slide, cover with a thin cover slip and examine.

Result

The result of staining depends on the pH and concentration of the staining solution and the number of cells or cell structure examined. For example, at a concentration of 0.06% of the dye the staining is limited almost completely to the nucleoli. At a concentration of 0.08%, the nuclear membrane and basophilic cytoplasm are also stained. The differences between nucleoli are more prominent when the preparations are observed with a phase-contrast microscope. The original blue-violet or deep-purple color, however, is orange in the phase-contrast microscope. This procedure is a very convenient aid in biochemical preparations and is the one commonly employed in the authors' laboratory.

D. The Acridine Orange Method

This procedure was used particularly in hematological or diagnostic cytology (*29, 30, 68*). The staining of nuclei is influenced by the pH of the staining solution, the duration of staining, and the number and type of cells observed.

Procedure

1. Mix the suspension of cells with acridine orange dissolved in physiological solutions (e.g., saline, Ringer's, Tyrode's) in a ratio of 1:1 or 1:2. The concentration of the dye in the solution should be in a range of 1:1,000–1:10,000 depending on the number and type of cells present. The pH should be 7–8.
2. Place a small drop on a microscope slide; the fluorescence microscope should be used along with UV or blue part of the spectrum.

Result

Nucleoli are red; chromatin structures are green. In damaged cells, the chromatin structures may be stained yellow or red.

III. General Cytological Methods Used to Demonstrate Nucleoli in Sections, Smears, and Touch and Squash Preparations

Numerous methods are used for the staining of nuclear structures. Only a few procedures distinguish nucleoli from other nuclear structures and the mechanism is usually poorly understood. Some manuals of microscopic techniques utilize cytochemical procedures for demonstrating nucleoli (*45*). According to Bělař (*3*) nucleoli can be distinguished to some extent from the chromatin structures by means of triacid, Giemsa's, and Mann's procedures. Of these procedures, the Giemsa or May-Grünwald-Giemsa procedures are commonly used. Siering and Anderhold (*68*) noted that the newer procedures used in the diagnostic cytology did not provide better results. Giemsa's method can be carried out on smears, touch preparations, or sections, including "thick sections" of specimen embedded in plastics used for electron microscopy. For the staining of nucleoli in smears, the methylene blue procedure described by Seman (*66*) seems to be very convenient as well as the acetobasic fuchsin stain (*81*) or chlorazol black E (*63*) for squash preparations. The routine histologic procedure using hemalum–eosin and carried out in smears or paraffin sections is also recommended as a nucleolar stain (*46*), although differentiation of atypical or small nucleoli may be difficult.

In "thick sections" of specimens embedded in water-insoluble media used for the electron microscopy, the nucleoli can be easily distinguished after staining with most methods (see D-iii, E, F, G of this chapter). When the specimens are fixed in osmium tetroxide, the nucleoli are less preserved and therefore less stained.

The visualization of nucleoli by means of classical impregnation methods is based on the argentaffinic property of nucleolar structures.

The basis of this property is not clear; the result of impregnation procedures depends on the particular fixative used (*53*). The argentaffinic property of nucleolar structures is also demonstrated by their ability to reduce photographic emulsions (*14*, *82*). After impregnation with classic procedures, the nucleoli appear to be granular or filamentous structures after staining with "photographiclike" procedures they appear as dense black bodies.

In ultrathin sections, the usual impregnation ("staining") procedures (e.g., uranyl acetate, lead hydroxide, or lead citrate procedures) are satisfactory for contrast pictures of the main nucleolar structures. The result of these procedures depends on the previous fixation used, particularly in regard to the chromatin structures (*4*, *41*, *85*). The chromatin structures, particularly nucleolus-associated chromatin and intranucleolar chromatin are better preserved by aldehydic fixatives (formaldehyde or glutaraldehyde) than by osmium tetroxide.

A. The Methylene Blue Method for Nucleoli in Smears (*66*)

Procedure

1. Fix smears (thin cell films) in absolute methanol and air dry.
2. Stain for 5–8 seconds with a solution containing 5 parts of 0.2% methylene blue in distilled water and 2 parts of saturated sodium borate in distilled water. Both components of the staining solution must be mixed just before staining.
3. Wash rapidly in water and air dry.

Result

Nucleoli are blue or blue-violet.

B. Aceto-Basic Fuchsin Stain for Nucleoli and Chromosomes of Plants (*81*)

Procedure

1. Fix specimens (root tips) in 45% acetic acid for 10–30 minutes.
2. Hydrolyze in 1 *N* HCl for 15–60 seconds at 60°.
3. Transfer to aceto-basic fuchsin and stain for 1–3 hours. Prepare to stain by dissolving 1 gm of basic fuchsin in 50 ml of 40% acetic acid at 50°, cool to 25–30°, and filter. Since the quality of the dye from different sources varies, its concentration in the staining solution may be modified.
4. Make a squash preparation and decolorize the cytoplasm with 30% acetic acid to desired intensity instead of 45% solution which is usually used with the squash procedure. Seal the cover slip to preserve the preparation.

Result

Nucleoli are violet and chromosomes are dark violet.

C. Chlorazol Black E Method for Plant Cell Nucleoli (*63*)

Procedure

1. Fix specimens in formalin-aceto-alcohol (*37*) for 24 hours.
2. Wash in 70% ethanol and rehydrate.
3. Treat in 2% pectinase in the dark at room temperature until cell separation is obtained on squashing. The usual time is about 5–6 hours. The time of treatment in pectinase varies with the specimens investigated.
4. Make a squash preparation in 45% acetic acid and remove the cover slip by the dry-ice method (*12*).
5. Transfer to absolute ethanol, then to 90% and 70% ethanol.
6. Stain in 1% chlorazol black E in 70% ethanol for 10 minutes.
7. Wash in 70% ethanol and dehydrate.
8. Clear in xylene and mount in a synthetic resin.

Result

Nucleoli are stained intensely black; other structures are less stained or unstained.

D. The Giemsa Stain

Procedure for smears

1. Fix in methanol for 5–10 minutes and dry.
2. Stain with dilute Giemsa stain for 10–30 minutes. Prepare the staining solution by diluting the stock solution of the stain with distilled water or buffer solution (pH 6.4–7). Usually the proportion is 1–2 drops of the stain to 1 ml of distilled water of buffer (pH 6.4–7).
3. Wash in distilled water and dry.

Procedure for paraffin sections

1. Fix in alcohol-formalin (2 parts of 80% ethanol to 1 part of formalin or 9 parts of 96% ethanol to 1 part of formalin), Ragaud's fluid (*37*), or Maximow's fluid (*59*), and embed in paraffin.
2. Rehydrate deparaffinized sections.
3. Stain in diluted Giemsa stain (1 ml of the stock solution for 50 ml distilled water or buffer solution, pH 6.4–7.5).
4. Wash in distilled water or buffer solution, pH 6.4–7.5.
5. Treat successively in the differentiation and dehydration solutions:
   a. acetone 95 ml, acetic acid 5 ml

b. acetone 70 ml, xylene 30 ml
c. acetone 30 ml, xylene 70 ml
6. Clear in xylene and mount.

Result
Nuclei are blue-violet; nucleoli are blue.

Procedure for Vestopal sections (65):
1. Treat the section for 12 hours in twice-changed buffer solution at pH 6.6 (640 ml of 0.91% KH_2PO_4 in distilled water and 360 ml of 1.19% $Na_2HPO_4 \cdot 2H_2O$).
2. Stain in freshly prepared staining solution for 12–24 hours. Change once. Prepare the staining solution by mixing solution A and B at the same rate. Solution A is prepared by dissolving 1 gm of azure I in 125 ml of boiling distilled water and adding 0.5 gm of methylene blue after 5 minutes. After filtration, hot distilled water is added to reach 250 ml. Solution B is prepared by dissolving 1.5 gm of yellowish eosin in 250 ml of hot distilled water and filtration of this solution.
3. Rinse twice in distilled water, 30 seconds each time.
4. Differentiate and clear in following solutions (3 seconds each):
   a. methyl benzoate 25 ml, dioxane 75 ml
   b. methyl benzoate 35 ml, acetone 65 ml
   c. acetone 30 ml, methyl benzoate 70 ml
   d. acetone 5 ml, methyl benzoate 95 ml
5. Clear in three changes of methyl benzoate (10 minutes each) and mount.

Result
Chromatin is reddish-violet; nucleoli are reddish-blue to reddish-violet.

E. A Rapid Method for Staining "Thick" Sections of Vestopal-Embedded Tissues (*60*)

Procedure
1. Cover sections on slides with 5–7 drops of 99% ethanol and heat above a microflame until boiling.
2. Decant excess ethanol, and cover the sections with 3–5 drops of the stain (0.1% methylene blue, 0.1% thionin in 0.1 M Na_2HPO_4). Keep the slide over the microflame about 10 seconds (do not boil). The stain can be diluted in 0.1 M Na_2HPO_4 for some specimens, such as human peripheral blood (*85a*).
3. Wash rapidly in distilled water, dry, and mount. The overstained sections can be easily differentiated in 70–95% ethanol.

Result

Chromatin is violet; nucleoli are blue-violet to violet.

F. Azure II–Methylene Blue Method for Epoxy Sections (*57*)

Procedure

1. Flood the sections with 1% periodic acid for 5 minutes.
2. Rinse in distilled water and wipe the slides dry to within the margin of sections.
3. Cover the sections with equal volumes of 1% azure II in distilled water and 1% methylene blue in 1% borax solution. Heat the slide carefully on a hot plate and stain for 15 minutes.
4. Rinse and dry.
5. Rinse in xylene and mount.

Result

Nucleoli are stained blue-violet and are prominent when the specimens are fixed in OsO_4. After fixation with glutaraldehyde, the nucleoli may be less prominent since the chromatin structure is better preserved.

G. Toluidine Blue Method for Staining Epoxy Sections (*83*)

Procedure

1. Flood the sections on the slide with a freshly prepared solution of 0.1% toluidine blue in 2.5% Na_2CO_3 (pH 11.3) and stain for 30–120 minutes.
2. Remove the excess of the stain with filter paper and wash gently with tap water.
3. Differentiate and dehydrate in 90% ethanol and then in absolute ethanol under microscopic control.
4. Clear in xylene and mount.

Result

When osmium tetroxide is used for fixation, nucleoli are dark purple and prominent; chromatin is stained less.

H. Silver-Impregnation Procedure for Nucleolonemas (*18*)

Procedure

1. Fix the specimens (1–5 mm in diameter) in the solution containing 60 ml of 96% ethanol, 30 ml of distilled water, 5 ml of formalin, 5 ml of diethylether and 1 gm of uranium nitrate for 4–24 hours.
2. Impregnate in 1.5–2% silver nitrate in water for 36–48 hours at 37° in the dark.

3. Rinse in 50% alcohol for 3–5 minutes.
4. Reduce with a solution containing 1 gm of pyrogallic acid, 10 ml of formalin, 20 ml of 96% ethanol, and 70 ml of distilled water for 5–10 hours in the dark.
5. Wash in distilled water for 10 minutes to remove the reducing solution. The distilled water should be changed several times.
6. Cut frozen sections or embed in paraffin.
7. Treat the sections in 0.2% solution of gold chloride in distilled water.
8. Rinse in pure water and fix in 3–5% sodium thiosulphate solution for 10 minutes.
9. Dehydrate, clear in xylene–creosote and xylene.
10. Blot and mount in balsam.

This method (slightly modified) has been also used for electron microscopic studies (*26*). The "fixation" with sodium thiosulphate for 15–30 minutes follows step 5 and the specimens are then embedded in Epon with or without postfixation in osmium tetroxide.

Result

In the light microscope, nucleolonemas are strongly impregnated. In the electron microscope, the silver grains are deposited usually in great numbers in the peripheral part of the nucleolus. Although the specificity of the silver impregnation is obscure, the grains seem to localize peripherally in ribonucleoproteins.

I. Modified Gonzalez-Ramirez Method for Tissue Culture (*53*)

Procedure

1. Fix in 4% formaldehyde in veronal-acetate buffer (pH 7.4) for 50 minutes. For electron microscopic observations fix in 1% OsO_4 in veronal-acetate buffer (pH 7.2–7.4) at 4° for 30 minutes or in 2.5% glutaraldehyde in the same buffer at room temperature for 60 minutes. Wash the osmium- or glutaraldehyde-fixed specimens in running tap water for 15–60 minutes.
2. Rinse for 10 seconds in the solution containing 5 drops of 25% NH_4OH in 100 ml of distilled water and rinse briefly in distilled water.
3. Treat in 2% silver nitrate for 24 hours in the dark.
4. Rinse in pyridine solution (0.5 ml of pyridine in 100 ml of distilled water) for 5–10 seconds.
5. Treat in the dark for 20–24 hours in the solution containing 8–12

drops of pyridine in 20 ml of delRio-Hortega's silver carbonate. Prepare del Rio-Hortega's solution of silver carbonate by mixing 5 ml of 10% silver nitrate with 20 ml of 5% sodium carbonate; then add ammonium hydroxide drop by drop to dissolve the precipitate and add distilled water up to 45 ml (*59*).

6. Wash briefly in distilled water and fix in 2% sodium thiosulphate for 5 seconds.
7. Dehydrate and mount.
   For electron microscopic observations dehydrate and embed in water-insoluble plastic (methacrylates, Epon, Araldite).

Result

In the light microscope, nucleolonemas are prominent dark-brown or black, impregnated structures (trabecular and granular); the other cell structures are indistinct. In the electron microscope, the nucleolar structures seem to have a higher affinity for the silver than do the structures in the rest of the nucleus.

IV. Cytochemical Methods

Numerous cytochemical procedures for the demonstration of nucleic acids are particularly convenient for the visualization of nuclei since they stain the nucleolar body or nucleolus-associated chromatin, or both these structures. The methods for demonstration of acidic or basic nuclear proteins are used less although the nucleolar body or nucleolus-associated chromatin can also be stained by means of such procedures. All these methods can be carried out on smears, touch, or squash preparations, and tissue sections.

Nucleic acids can be demonstrated by reactions of phosphoric acid, sugar components, or bases (*61*). For visualization of bases, UV absorption is usually employed (*11*). The tetrazolium reaction after benzoylation or acetylation is still the subject of controversy (*54*). The sugar components of nucleic acids can be detected by reactions of aldehydes after acid hydrolysis. The negatively charged phosphate groups of nucleic acids are responsible for their basophilic staining properties. However, basophilic staining properties are not specific for nucleic acids. Nevertheless, specific staining of nucleic acids with basic dyes can be achieved under defined conditions, particularly in combination with extraction (enzymatic and nonenzymatic) procedures.

In ultrastructural chemistry, nucleic acids can be demonstrated by means of selective impregnation procedures (*25, 80, 86*) or extraction methods (*5*).

A. The Feulgen Reaction

The Feulgen reaction is based on mild hydrolysis of nucleic acid by 1 *N* HCl which removes purine bases from the pentose–phosphoric acid complexes and leaves a free aldehyde group that reacts with Schiff's reagent. The optimal time of hydrolysis depends on the fixation used. Five minutes are recommended for hydrolysis after fixation in Zenker's solution or absolute ethanol. Hydrolysis for 8 minutes is used after fixation in formalin, formol sublimate, and Carnoy's and Helly's solutions. For preparation of Schiff's reagent and hydrolysis, there are several methods available (*56*).

The method can be carried out on smears, touch or squash preparations, colloidin or paraffin sections, and also in sections of specimens embedded in Araldite or Epon (*13, 25*). The time of hydrolysis, however, should be 20 minutes.

Preparation of the stain (Schiff's reagent)

1. Dissolve 1 gm of basic fuchsin in 20 ml of boiling distilled water and stir.
2. Cool to 50° and filter.
3. Add 20 ml 1 *N* HCl to the filtrate.
4. Cool to 25° and add 1 gm sodium bisulfite. Sodium or potassium metabisulfite may be used instead of sodium bisulfite.
5. Store in the dark for 14 to 24 hours. The solution will turn orange or straw-colored.
6. Add 1 to 2 gm of activated charcoal, and shake for 1 minute.
7. Filter and use.

The solution may be stored at 4° in tightly stoppered bottles for several weeks or even months. If it turns pink or reddish, it should not be used.

Procedure

1. Hydrate sections or smears.
2. Rinse briefly in cold 1 *N* HCl.
3. Place slides with sections or smears in 1 *N* HCl at 60° for a period depending on the fixation of the specimen (see above).
4. Rinse briefly in cold 1 *N* HCl and then rinse in distilled water.
5. Stain slides with specimens in Schiff's reagent for 1 to 2 hours.
6. Wash 2 minutes.
7. Rinse in 3 changes of freshly prepared bisulfite solution consisting of 5 ml of 10% sodium bisulfite (or sodium or potassium metabisulfite), 5 ml 1 *N* HCl, and 100 ml distilled water.
8. Wash at least 5 minutes in running tap water.

9. Dehydrate in alcohols, clear in xylene, and mount in balsam.
10. Smears may be dried, and then mounted.

Result

DNA-containing structures (chromatin) are red-purple. Nucleoli are empty areas surrounded by nucleolus-associated chromatin. In some large nucleoli, there is penetration of the chromatin into the nucleolar body.

B. Staining of DNA-Containing Structures with Basic Dyes

For selective staining of DNA, methyl green is commonly used. The mechanism of the staining reaction is still uncertain (*56*), although some studies indicate that the selective staining of DNA-containing structures with methyl green is due to the affinity of this dye for highly polymerized DNA (*10, 27, 31*). The specificity of DNA staining with methyl green is limited. The quality of the dye may also influence the result of staining because of contamination with methyl violet which must be extracted by means of suitable procedures (*27, 54*).

The staining of DNA-containing structures with basic dyes at acidic pH is specific when RNA is removed previously from the investigated specimen (*15, 51, 62, 73, 74, 77*). In such procedures, the RNA is removed either by enzymatic digestion (*62, 77*) or by hydrolysis and extraction with HCl by means of the Feulgen technique (*15, 28, 51, 70, 73, 74*). When the treatment in HCl is long enough, the histones are also extracted (*73, 74*). Of the basic dyes, toluidine blue, methylene blue, azure I, azure II, azure C, and galloganin chromalum with adjusted pH have been used.

1. *A Simple Stain for DNA-Containing Structures in Smears*

This procedure is very similar to the Deitch method (*15*) in which acetylation is employed to block functional protein groups. Some studies indicated that the acetylation can be omitted since the interference by proteins may not be significant (*56*). The control experiments with deoxyribonuclease and methylation have shown that phosphate groups of DNA are responsible for the staining (*74*).

Procedure

1. Fix smears for 10 minutes to 24 hours in absolute ethanol or 10 to 15 minutes in absolute methanol. The fixation can be omitted for staining nuclei during nonaqueous isolation procedures (*67*).
2. Dry smears and rinse in distilled water.
3. Treat with 1 *N* HCl at 60° for 10 minutes.

4. Wash gently in distilled water and dry.
5. Stain for 10 minutes in 0.1–0.05% toluidine blue or 0.05% methylene blue at pH 5. The pH is adjusted with MacIlvaine's buffer. Azure I, azure II, or azure C can also be used (*74*).
6. Wash briefly in distilled water and dry.

Result

Chromatin structures are blue or blue-violet. Nucleoli appear as empty spaces surrounded by nucleolus-associated chromatin. In some nucleoli, the intranucleolar chromatin structures can also be defined.

2. *The Staining of DNA-Containing Structures with Some Basic Dyes after Ribonuclease* (*8, 76, 77*)

Procedure

1. Deparaffinize sections of specimens fixed in formol or Carnoy's solution.
2. Treat in 10% formol for 4 hours. This step can be omitted.
3. Incubate in ribonuclease for 1–3 hours at 37°. Ribonuclease can be dissolved in McIlvain's buffer pH 7 in the ratio 5 mg/100 ml (*77*) or in redistilled water (1 mg/1 ml).
4. Rinse briefly in buffer or distilled water.
5. Stain in 0.5% toluidine blue or 5% methylene blue for 12 hours. The pH of these stains was adjusted with HCl and NaOH to pH 3. When gallocyanin–chromalum (*16*) is used, the time of staining is about 24–48 hours. The sections are differentiated after staining with toluidine blue in 96% ethanol. After staining with methylene blue, the sections are washed briefly in distilled water and the stain is fixed in 5% ammonium acetate for 3 hours. The staining in gallocyanin–chromalum is followed by a short wash in distilled water.
6. Dehydrate and mount in a neutral medium.

Result

The chromatin structures are stained. Nucleoli seem to be empty areas surrounded by nucleolus-associated chromatin. Occasionally intranucleolar chromatin is visualized.

C. The Selective Staining of RNA-Containing Structures with Buffered Basic Dyes

These methods are specific for the demonstration of structures containing RNA only under certain conditions. The result and the specificity of staining depend not only on the pH of the staining solutions, the concentration of the dye, previous fixation, and treatment of the study specimens before staining, but also on the quality of the dye (*39, 40, 69, 75*). Therefore, testing of the suitability of the dye for staining of RNA is recommended for some procedures. A simple morphological test can be carried out on the smears of the peripheral blood (*69, 70*). A satisfactory dye for the staining RNA is one that, at a specific pH and concentration, stains the basophilic cytoplasmic structures and nucleolus in lymphocytes and does not stain nuclear chromatin.

1. *Staining with Buffered Solution of Methylene Blue* (*22*)

Procedure

1. Fix dry smears in absolute ethanol for 10 minutes.
2. Dry.
3. Stain with methylene blue at pH 4.9 for 10 minutes. Staining solution: 0.13% methylene blue is mixed with an equal volume of Michaelis' (0.1 *M* sodium–veronal) buffer at pH 4.9.
4. Rinse shortly in Michaelis' buffer at pH 4.9.
5. Dry or dehydrate rapidly after previous fixation of methylene blue with 5% ammonium molybdate for 4 minutes.

Result

The structures containing RNA-including nucleoli are stained blue.

2. *Staining of Smears with Buffered Solution of Toluidine Blue* (*69, 75*)

Procedure

1. Stain dry smears within 24 hours of their preparation with 0.05% toluidine blue solution in 1% alcohol at pH 4.9–5.5 for 5–15 minutes. Adjust the pH with MacIlvaine's buffer.
2. Pour off the stain.
3. Dry slides in vertical position.

Result

Structures containing RNA, including the nucleoli, stain blue or violet-blue. Acid mucopolysaccharides can be also stained. The metachromatic reaction or the anatomic position of these substances usually differ from that of RNA in the nucleus and nucleolus. Moreover, the RNA can be distinguished in questionable cases by the ribonuclease test.

3. *The Modified Stockinger and Kellner Procedure* (*55*)

Procedure

1. Fix smears in methanol–formol (methanol 85 ml, 5% formol 15 ml) for 5 minutes.
2. Stain in a hot solution of toluidine blue (70°) for 5 minutes. The

toluidine blue solution contains 300 mg of toluidine blue, 50 ml of distilled water, and 50 ml of citric acid buffer pH 4.9 (citric acid 20 gm, 1 *N* NaOH 200 ml, distilled water to 1000 ml).

3. Wash twice in citric acid buffer (see above) for 30–90 seconds and dry.

Result

Nucleoli and basophilic structures containing RNA are blue.

4. *Aniline–Azure B Method for Small Nucleoli in Paraffin Sections* (*48*)

In this procedure, the staining of nucleoli is due to the affinity of RNA or RNA and proteins to azure B; the DNA remains unstained. The RNA is stained with azure B at pH 2–3 and the higher pH produces nonspecific staining of RNA and proteins. The affinity of chromatin structures to azure B is suppressed by blocking liberated aldehyde groups of DNA with aniline.

Procedure

1. Fix the tissues in 10% formol–saline neutralized with $CaCO_3$ (*37*) and embed in paraffin. The paraffin–beeswax–stearic acid (*47*) embedding procedure is recommended.
2. Float sections on flattening bath containing 1% of a solution composed of 1% alcohol and 10% gelatin as an adhesive.
3. Mount sections on slides, dry at 37° for 18 hours, and deparaffinize.
4. Hydrolyze at 40° in 10% (w/v) HCl in tetrahydrofuran for 4–5 minutes.
5. Rinse gently in distilled water.
6. Treat in 10% (v/v) aniline in glacial acetic acid for 10 minutes.
7. Rinse carefully in distilled water.
8. Stain in 0.1% azure B diluted 1:10 in Tris buffer at pH 3 or 7.
9. Rinse, blot the excess water around the sections.
10. Dehydrate with acetone, clear in xylene, and mount in a polystyrene mounting medium.

Result

The maximum staining of nucleoli is achieved at pH 7. RNA and nucleoli stain weakly at pH 3.

5. *The Toluidine Blue–Molybdate Method* (*39, 40*)

Staining of ribonucleoprotein by this procedure involves three main stages:

1. Release of phosphate groups of RNA by inactivation of protein-bound amino groups.

2. Binding of toluidine blue by phosphate groups of RNA at pH 3.
3. Production of a metachromatic complex of toluidine blue with polymolybdate at some sites of dye binding.

Procedure

1. Fix wet smears in formol sublimate (10% formol saturated with mercuric chloride) at 37° for periods of 5, 10, 16, 24 hours, and 14 days. For paraffin sections, fixation for 1 hour in formol sublimate at 20° is recommended.
2. Treat successively for 5 minutes in Lugol's iodine (0.5% iodine in 1% aqueous potassium iodide), 5% sodium thiosulphate, and wash in running tap water.
3. Stain with 0.01% solution of toluidine blue in McIlvaine's buffer at pH 3 for 30 minutes.
4. Wash in tap water for 10 seconds.
5. Treat in 5% solution of ammonium molybdate for 40 minutes.
6. Wash in tap water for 10 seconds.
7. Dehydrate in tertiary butyl alcohol, clear in xylene, and mount in Permount or Canada balsam.

Result

The result of staining depends on the time of fixation. In nucleoli, the "pars amorpha" is green or blue-green and "nucleolini" (or stained vacuoles) are purple. After 24 hours of fixation nucleoli are blue-purple or blue.

D. Methods for the Simultaneous Demonstration of DNA and RNA

The advantage of these methods is their usefulness for studies on the morphological relationship between structures containing both nucleic acids, in the nucleolar body, nucleolus-associated chromatin, and intranucleolar chromatin structures. Some methods, however, do not differentiate between DNA and RNA. Therefore, methods in which both types of nucleic acids are stained different colors are more convenient and permit the differentiation of the nucleolar body from the nucleolus-associated chromatin and intranucleolar chromatin structures. The micronucleoli may be overlapped by surrounding chromatin structures and therefore, their visualization is usually not possible. Of the methods used for simultaneous demonstration of DNA and RNA, the methyl green–pyronin or acridine orange methods provide the most satisfactory results.

1. *The Methyl Green–Pyronin Method* (*33, 37, 38, 54*)

The specificity of this method depends on the degree of polymerization of nucleic acids, the quality of the dyes used, and staining conditions (*32*). Methyl green stains highly polymerized DNA, and pyronin stains depolymerized DNA and RNA. The specimens can be stained either with a mixture of methyl green and pyronin or first with methyl green and then with pyronin (*32, 61*).

Procedure

1. Hydrate fixed smears or deparaffinized sections.
2. Stain in the mixture of methyl green–pyronin for 20 minutes to 24 hours. The methyl green–pyronin mixture is prepared for use by addition of equal volumes of 0.2 *M* acetate buffer (pH 4.8) and the dye solution consisting of 17.5 ml of 5% aqueous pyronin Y, 10 ml 2% aqueous purified methyl green, and 250 ml distilled water. Kurnick (*32*) recommends staining for 6 minutes in the dye mixture containing 2% pyronin Y, 12.5 ml; 2% methyl green, 7.5 ml, and distilled water, 30 ml.
3. Wash briefly in cold distilled water or 0.2 *M* acetate buffer (pH 4.8). This step is omitted in the Kurnick modification (*33*).
4. Blot with filter paper.
5. Rinse in absolute acetone and transfer to a mixture of acetone and xylene, clear in two changes of xylene, and mount in Canada balsam or neutral mounting medium. In Kurnick's modification (*33*), the slides are immersed in two changes of *n*-butyl alcohol (5 minutes each), then cleared in xylene for 5 minutes, immersed in cedar oil for 5 minutes, and covered in Permount.

Result

RNA and nucleoli are red, and chromatin structures are green to bluish-green.

2. *Demonstration of DNA and RNA with Acridine Orange* (*6, 7, 27*)

The demonstration of nucleic acids with acridine orange is based on their secondary fluorescence after staining with this fluorochrome. The differential binding of acridine orange to nucleic acids seems to be due to the structure and molecular weight differences of DNA and RNA (*27*). In most methods, the green fluorescence is characteristic for DNA. Under unusual conditions, however, the fluorescence of DNA may also be red. Furthermore, under special conditions, proteins and acid mucopolysaccharides can also bind acridine orange (*27*). The staining with acridine orange can be carried out in smears, touch and squash preparations, frozen or paraffin sections, and in "thick" sections of specimens embedded in Epon after previous fixation in glutaraldehyde and removal of the embedding medium (*42, 43, 78*). The concentration of the acridine orange should be in the range of 0.005–0.01% at pH 4.

Procedure for smears

1. Fix smears in alcohol–ether (1:1), absolute ethanol, or methanol for more than 15 minutes.
2. Hydrate smears in graded alcohols.
3. Rinse in 1% acetic acid for 30 seconds.
4. Stain in 0.01% solution of acridine orange (prepared by diluting 0.1% aqueous solution of acridine orange with Michaelis' buffer, 0.1 *M* sodium veronal, at pH 6) (*52*).
5. Wash in Michaelis' buffer (pH 6) for 2 minutes.
6. Rinse in 0.1 *M* aqueous solution of $CaCl_2$ for 1–2 minutes.
7. Wash in Michaelis' buffer at pH 6.
8. Observe in the fluorescence microscope.

Procedure for sections

1. Treat deparaffinized and rehydrated sections in 1% acetic acid and wash in water (omit in frozen sections).
2. Stain in 0.1% solution of acridine orange buffered with Michaelis' buffer at pH 6 for 2–15 seconds.
3. Rinse in buffer (pH 6) for 1 minute.
4. Treat in 1 *M* $CaCl_2$ for 9–30 seconds.
5. Wash in buffer and observe in the fluorescence microscope.

Result

DNA and chromatin appear green; RNA and nucleolar bodies are red or orange.

E. Extraction Procedures

Extraction procedures are either enzymatic or nonenzymatic. In nonenzymatic procedures other substances, as well as nucleic acids, can also be extracted (*54*). Therefore, it is useful to follow and evaluate the results of extraction very carefully with regard to the structures in which the investigated nucleic acid is localized. The enzymatic procedures are usually more exact in regard to the specificity of the extraction. Their results, however, depend on the quality of the enzyme used and as for nonenzymatic procedures, on the conditions of extraction, fixation, and, in some instances, on the embedding medium. After extraction of DNA, nucleoli appear as positively stained structures. When RNA is removed, nucleoli seem to be empty spaces surrounded by nucleolus-associated chromatin and contain occasionally fine structures of the intranucleolar chromatin.

1. *Extraction of RNA with Perchloric Acid* (*17, 88*)

For the extraction of paraffin sections of tissue specimens fixed in formalin or formalin sublimate, the procedure of Erickson *et al.* (*17*) is commonly employed. For the extraction of smears, Wenderoth's procedure (*88*) is recommended.

Procedure for paraffin sections

1. Hydrate deparaffinized sections.
2. Treat in 10% $HClO_4$ at 4° for 12–18 hours.
3. Neutralize in 1% Na_2CO_3 for 1 to 5 minutes.
4. Wash in water.
5. Stain in 0.5% aqueous toluidine blue.
6. Dehydrate, clear, and mount.

Procedure for smears

1. Stain smears by Giemsa's method or toluidine blue, and mark the position of investigated cells.
2. Treat smears with $HClO_4$ at 40° for 3 minutes.
3. Wash 10 minutes in water and stain again by Giemsa's method or toluidine blue.

Result

Basophilic structures containing RNA and nucleoli are not stained.

2. *Digestion with Ribonuclease According to Brachet* (*8*)

In this procedure, common fixation methods can be used. However, a short fixation time is recommended (*38*).

Procedure

1. Hydrate deparaffinized sections or smears.
2. Incubate sections or smears in a 0.01–0.1% solution of crystalline ribonuclease (*44*), in distilled water at 60–65° for 1–3 hours. Incubate control sections or smears in distilled water at the same temperature and for the same time.
3. Wash briefly in distilled water and stain with methyl green–pyronin, toluidine blue, or gallocyanin.
4. Dehydrate, clear, and mount.

Result

RNA-containing basophilic structures, including nucleolar bodies, are not stained. The lack of staining of these structures in controls may be due to the extraction of RNA with distilled water (*37*).

3. *Digestion with Ribonuclease According to Amano (1)*

In this procedure, fixation in Carnoy's fluid for 24 hours is recommended. Shorter fixation may cause nonspecific loss of RNA-containing material. By contrast, fixation in formalin for 5 hours can inhibit the complete removal of RNA from investigated specimens.

Procedure

1. Incubate hydrated sections or smears in RNase solution containing 1 mg of the enzyme per 1 ml of distilled water for 4 hours at 40°.
2. Stain with 0.1% aqueous toluidine blue.
3. Wash in three changes of distilled water.
4. Dehydrate, clear, and mount.

Result

RNA-containing basophilic structures, including nucleoli, are not stained.

4. *Digestion with Deoxyribonuclease (1)*

Since the common fixatives inhibit the complete removal of DNA from nuclear structures, fixation in Carnoy's fluid for 3–24 hours is recommended.

Procedure

1. Incubate hydrated sections or smears for 24 hours at 37° in a deoxyribonuclease solution containing 0.05 mg crystallized enzyme per ml of Gomori's 0.2 *M* Tris–maleate buffer with 0.2 *M* $MgSO_4$ at pH 5.7.
2. Wash in three changes of distilled water.
3. Stain with 0.1% aqueous toluidine blue.
4. Dehydrate and mount.

Result

Structures containing DNA are not stained. Nucleoli are more prominent and stain intensely.

F. Extraction Procedures with Ribonuclease and Deoxyribonuclease in Electron Microscopic Cytochemistry

Two types of methods can be used for digestion of RNA or DNA for electron microscopic cytochemistry. The first method permits digestion in ultrathin sections. In the second method, the digestion of RNA or DNA either after fixation or without fixation is carried out in tissue blocks, frozen sections, cell suspensions, or isolated cell fractions before embedding in water-insoluble media.

In the procedures performed in ultrathin sections, the specimens are usually embedded in water-soluble media and glycol methacrylate is recommended for such purposes. In Durcupan, the nucleases do not function in contrast to pepsin or trypsin (*35*). The extraction methods can be also carried out in ultrathin sections of specimens embedded in water-insoluble media such as Epon and Vestopal (*49*). The results, however, are less satisfactory as compared with methods carried out in sections of specimens embedded in glycol methacrylate.

1. *Digestion with Nuclease in Ultrathin Sections* (*35, 36*)

Procedure

1. Fix blocks of tissues (smaller than 1 mm^3) or cell suspensions with 10% neutral formalin or in 6.5% glutaraldehyde at 4° for 15 minutes.
2. Embed in glycol methacrylate and cut ultrathin sections.
3. Incubate ultrathin sections by floating on the incubation fluid for 15 minutes to 3 hours at 37°. The incubation fluid contains 0.1–1% ribonuclease or 0.1–0.2 deoxyribonuclease in distilled water adjusted with 0.01 *M* NaOH to pH 6.8 for ribonuclease or to pH 6.2 to 6.6 for deoxyribonuclease. The deoxyribonuclease can be also dissolved in 0.003–0.0003 *M* $MgSO_4$ to the concentration 0.1–0.2% of the enzyme. The pH of this incubation solution can be also adjusted with 0.01 *N* NaOH to pH 6.2–6.6.
4. Float sections on distilled water.
5. Pick up sections on grids with the support membrane.
6. Dry and stain with uranyl acetate and lead citrate.

Result

The nucleolar components containing RNA are removed after digestion with ribonuclease. The action of deoxyribonuclease is not constant. The best results are obtained after short fixation with formaldehyde. The chromatin structures including nucleolus-associated chromatin and the intranucleolar chromatin are removed or their density is reduced.

2. *Digestion of Fixed Specimens with Nuclease before Embedding* (*79, 80*)

Procedure

1. Fix small tissue blocks (smaller than 1 mm^3) or small clumps of cells or cell suspensions in 10% formalin in 0.2 *M* phosphate buffer pH 7.2 to 7.6 at 4° for 15 to 30 minutes.

2. Wash in cold 0.2 *M* phosphate buffer at pH 7.2 to 7.6.
3. Digest with ribonuclease (1 mg/1 ml) or deoxyribonuclease (0.5–1 mg/1 ml) at pH 6.5 to 7 for 1 to 3 hours at 20–37° after adjusting pH with 0.01 *N* NaOH. The deoxyribonuclease can also be dissolved in 0.003 *M* magnesium acetate (*84*).
4. Treat with cold (3–5°) 5% trichloracetic acid for 15 minutes. This step can be omitted, particularly when ethanol used for dehydration contains uranyl acetate (*71, 72*).
5. Dehydrate and embed in a water-insoluble medium (Epon, Epon-Araldite, Vestopal).
6. Cut ultrathin sections and stain with uranyl acetate and lead citrate.

Result

The results are similar to those obtained by digestion procedures carried out in ultrathin sections.

REFERENCES

1. Amano, M., Improved techniques for the enzymatic digestion of nucleic acids from tissue sections. *J. Histochem. Cytochem.* **10,** 204–212 (1962).
2. Ayres, W. W., A method of staining nucleoli of cells in fresh benign and malignant tissues. *Cancer Res.* **8,** 352–356 (1948).
3. Bělař, K., Die Technik der deskriptiven Cytologie. *In* "Methodik der wissenschaftlichen Biologie" (T. Peterfi, ed.) Vol. I, pp. 638–734. Springer, Berlin, 1928.
4. Bernhard, W., Ultrastructural aspects of the normal and pathological nucleolus in mammalian cells. *Natl. Cancer Inst. Monograph* **23,** 13–38 (1966).
5. Bernhard, W., and Granboulan, N., The structure of the cancer cell nucleus. *Exptl. Cell Res.* Suppl. 9, 19–53 (1963).
6. Bertalanffy, F. D., Evaluation of the acridine-orange fluorescence microscope method for cytodiagnostic of cancer. *Ann. N. Y. Acad. Sci.* **93,** 717–749 (1962).
7. von Bertalanffy, L., Masin, M., and Masin, F., A new and rapid method for diagnosis of vaginal and cervical cancer by fluorescence microscopy. *Cancer* **11,** 873–887 (1958).
8. Brachet, J., The use of basic dyes and ribonuclease for the cytochemical detection of ribonucleic acid. *Quart. J. Microscop. Sci.* **94,** 1–10 (1953).
9. Busch, H., Muramatsu, M., Adams, H., Steele, W. J., Liau, M. C., and Smetana, K., Isolation of nucleoli. *Exptl. Cell Res.* Suppl. 9, 150–163 (1963).
10. de Carvalho, S., Données cytocymiques sur la liaison entre les acides nucléiques au cours de la maturation cellulaire et de la mito obtenues par l'analyse chromatique qualitative. *Compt. Rend. Soc. Biol.* **146,** 1461–1462 (1952).
11. Caspersson, I., "Cell Growth and Cell Function." Norton, New York, 1950.
12. Conger, A. D., and Fairchild, L. M., A quick freeze method for making smear slides permanent. *Stain Technol.* **28,** 281–283 (1953).
13. Chouinard, L. A., Nukleolonema and Pars Amorpha in Root Meristematic cells of Vicia Faba. *Can. J. Botany* **44,** 403–411 (1966).

14. Das, N. K., Demonstration of non-RNA nucleolar fraction by silver staining. *Exptl. Cell Res.* **26**, 428–431 (1962).
15. Deitch, A., A method for the cytophotometric estimation of nucleic acids using methylene blue. *J. Histochem. Cytochem.* **12**, 451–461 (1964).
16. Einarson, L., On the theory of gallocyanin-chromalum staining and its application for quantitative estimation of basophilia. *Acta Pathol. Microbiol. Scand.* **28**, 82–102 (1951).
17. Erickson, R. O., Sax, K. B., and Ogur, M., Perchloric acid in the cytochemistry of pentose nucleic acid. *Science* **110**, 472–473 (1949).
18. Estable, C., and Sotelo, J. R., Technical procedures for the study of the nucleolonema. *Stain Technol.* **27**, 307–312 (1952).
19. Fröhlich, K., "Phasenkontrastmikroskopie in der Medizin." Fischer, Jena, 1955.
20. Gonzalez-Guzman, J., Generalities on the nucleolar content of some blood cells. *Blood* **3**, Suppl. 2, 57–65 (1948).
21. Gonzalez-Ramirez, J., Zuniga, M., and Nunez, B. A., Nucleolar physiology. I. Passage of two different nucleolar components towards the cytoplasm in living cells. *Bol. Inst. Estud. Med. Biol.* (*Mex.*) **17**, 51–57 (1959).
22. Haenel, U., Veränderung der Nukleolarsubstanz an Knochenmarkzellen bei perniziöser Anämie und ihre pathogenetische Bedeutung. *Helv. Med. Acta* [A]**17**, 627–650 (1950).
23. Hammar, J. A., Lipoidbildung in den weissen Blutkörperchen. Mikroskopische Studie zur Autolyse des Blutes. *Kgl. Svenska Vetenskapsakad. Handl.* **49**, 3–44 (1912).
24. Hertl, M., Zum Nukleolus-System in hämatologischen Zellen. *Folia Haematol.* [N.S.] **8**, 1–4 (1963).
25. Huxley, H. E., and Zubay, G., Preferential staining of nucleic acid-containing structures for electron microscopy. *J. Biophys. Biochem. Cytol.* **11**, 273–296 (1961).
26. Izard, J., and Bernhard, W., Analyse ultrastructurale de l'argentophilie du nucléole. *J. Microscopie* **1**, 421–434 (1962).
27. Kasten, F. H., Cytochemical studies with acridine orange and the influence of dye contaminants in the staining of nucleic acids. *Intern. Rev. Cytol.* **21**, 141–202 (1967).
28. Kellner, G., and Stockinger, L., Schnellmethoden zum Nachweis raschwachsenden Zellen. *Z. Wiss. Mikroskopie* **62**, 265–272 (1959).
29. Kossenow, W., "Lebende Blutzellen im Fluorescenz und Phasenkontrast Mikroskop." Karger, Basel, 1956.
30. Kossenow, W., Die Fluorochromierung mit Acridinorange, eine Methode zur Lebend-beobachtung gefärbter Blutzellen. *Acta Haematol.* **7**, 217–221 (1952).
31. Kurnick, N. B., Methyl green-pyronin. I. Basis of selective staining of nucleic acids. *J. Gen. Physiol.* **33**, 243–274 (1950).
32. Kurnick, N. B., Histological staining with methyl green-pyronin. *Stain Technol.* **27**, 233–242 (1952).
33. Kurnick, N. B., Pyronin Y in the methyl green-pyronin histological stain. *Stain Technol.* **30**, 213–230 (1955).
34. La Cour, L. F., The internal structure of nucleoli. *Chromosomes Today, Proc. 1st Oxford Chromosome Conf., 1964* Vol. 1, pp. 150–160. Plenum Press, New York, 1966.
35. Leduc, E., and Bernhard, W., Water-soluble embedding media for ultrastructural cytochemistry. *Symp. Intern. Soc. Cell Biol.* **1**, 21–44 (1962).

36. Leduc, E. H., Marinozzi, V., and Bernhard, W., The use of water soluble glycol methacrylate in ultrastructural cytochemistry. *J. Roy. Microscop. Soc.* [3] **81**, 119–130 (1963).
37. Lillie, R. D., "Histopathologic Technic and Practical Histochemistry." McGraw-Hill (Blakiston), New York, 1954.
38. Lison, L., "Histochimie et Cytologie Animales." Gauthier-Villars, Paris, 1960.
39. Love, R., and Liles, R. H., Differentiation of nucleoproteins by inactivation of protein bound amino groups and staining with toluidine blue and ammonium molybdate. *J. Histochem. Cytochem.* **7**, 164–181 (1959).
40. Love, R., and Suskind, R. G., Further observation of ribonucleoproteins of mitotically dividing mammalian cells. *Exptl. Cell Res.* **22**, 193–207 (1961).
41. Marinozzi, V., The role of fixation in electron staining. *J. Roy. Microscop. Soc.* [3] **81**, 141–154 (1963).
42. Mayor, H. D., Cytochemical and fluorescent antibody studies on the growth of polio virus in tissue cultures. *Texas Rept. Biol. Med.* **19**, 106–122 (1961).
43. Mayor, H. D., Hampton, J. C., and Rosario, B., A simple method for removing the resin from epoxy-embedded tissue. *J. Biophys. Biochem. Cytol.* **9**, 909–910 (1961).
44. McDonald, M. R., A method for the preparation of "protease free" crystalline ribonuclease. *J. Gen. Physiol.* **32**, 39–42 (1949).
45. McManus, J. F. A., and Mowry, R. W., "Staining Methods Histologic and Histochemical." Harper (Hoeber), New York, 1960.
46. Menzies, D. W., Nucleolar definition in hemalum and eosin staining. *Stain Technol.* **37**, 41–42 (1962).
47. Menzies, D. W., Paraffin beeswax-stearic acid: An embedding mass for thin sections. *Stain Technol.* **37**, 235–238 (1962).
48. Menzies, D. W., Staining of small lymphoid nucleoli in paraffin sections by aniline-azur B. *Stain Technol.* **41**, 165–168 (1966).
49. Monneron, A., and Bernhard, W., Action de certaines enzymes sur des tissus inclus in epon. *J. Microscopie* **5**, 697–714 (1967).
50. Muramatsu, M., and Busch, H., Isolation, composition and function of nucleoli of tumors and other tissues. *Methods Cancer Res.* **2**, 303–361 (1967).
51. Muramatsu, M., Smetana, K., and Busch, H., Quantitative aspects of isolation of nucleoli of the Walker carcinosarcoma and liver of the rat. *Cancer Res.* **23**, 510–518 (1963).
52. Naegeli, O., "Blutkrankheiten und Blutdiagnostik. Sweite Auflage," pp. 23–26. Berlin, J. Springer, 1923.
53. Paweletz, N., Siebs, W., and Lettré, R., Untersuchungen zur Argentaffinreaktion des Nukleolus. *Z. Zellforsch. Mikroskop. Anat.* **76**, 577–605 (1967).
54. Pearse, A. G., "Histochemistry Theoretical and Applied." Churchill, London, 1961.
55. Plenert, W., Michalowski, H., and Kehnscherper, M., Studien zur Differenzierung der Lymphozyten des peripheren Blutes im Kindesalter. *Folia Haematol.* **85**, 218–226 (1966).
56. Richards, B. M., Cytochemistry of the nucleic acids. *In* "Protoplasmatologia. Handbuch der Protoplasmaforschung" (M. Alfert, H. Bauer, C. V. Harding, P. Sitte, eds.), 5/3b, 1–38 (1966).
57. Richardson, K. C., Jarett, L., and Finke, E. H., Embedding in epoxy resins for ultrathin sectioning in electron microscopy. *Stain Technol.* **35**, 312–323 (1960).

58. Rind, H., "Atlas der Phasenkontrashämatologie." Akademie Verlag, Berlin, 1958.
59. Romeis, B., "Mikroskopische Technik." Leibnitz, Munich, 1948.
60. Rüdeberg, C., A rapid method for staining thin sections of vestopal W-embedded tissues for light microscopy. *Experientia* **23**, 792 (1967).
61. Sandritter, W., Die Nachweismethoden der Nukleinsäuren. *Z. Wiss. Mikroskopie* **62**, 284–304 (1959).
62. Sandritter, W., Kiefer, G., and Rick, W., Stochiometrie von Gallocyanin-Chromalum mit Desoxyribonukleinsäure. *Histochemie* **3**, 315–340 (1963).
63. Savage, J. R. K., and Scott, D., Chlorazol black E for plant cell nucleoli. *Stain Technol.* **36**, 85–86 (1961).
64. Scherer, E., and Fiebelkorn, H. J., Über die Darstellbarkeit der Nukleinsäuren des Zellkernes und ihre Bedeutung für Diagnose und Prognose maligner Bluterkrankungen. *Folia Haematol.* **72**, 143–148 (1954).
65. Schwalbach, G., Lickfeld, G., and Hoffmeister, H., Differentiated staining of osmium tetroxide fixed, vestopal-W-embedded tissue in thin sections. *Stain Technol.* **38**, 15–22 (1963).
66. Seman, G., La mise en évidence des nucléoles sur frottis par le bleu de méthylène boraté. *Rev. Franc. Etudes Clin. Biol.* **5**, 196–198 (1960).
67. Siebert, G., Isolation of nuclei. *Methods Cancer Res.* **2**, 287–302 (1967).
68. Siering, H., and Anderhold, K., "Leitfaden der Nativcytomorfologie maligner Tumoren." Fischer, Jena, 1956.
69. Smetana, K., A further contribution to the question of the incidence of nucleoli of mature lymphocytes in man. *Folia Biol.* (*Prague*) **7**, 268–274 (1961).
70. Smetana, K., Basic histochemical and cytochemical methods. *Methods Cancer Res.* **2**, 362–445 (1967).
71. Smetana, K., and Busch, H., On the ultrastructure of the Walker 256 carcinoma. *Cancer Res.* **23**, 1600–1603 (1963).
72. Smetana, K., and Busch, H., Studies on the ultrastructure of the Walker tumor and rat liver. *Cancer Res.* **24**, 537–557 (1964).
73. Smetana, K., and Busch, H., Studies on staining and localization of acidic nuclear proteins in the Walker 256 carcinosarcoma. *Cancer Res.* **26**, 331–337 (1966).
74. Smetana, K., Lejnar, J., and Potměsil, M., A note to the demonstration of DNA in nuclei of blood cells in smear preparations. *Folia Haematol.* **88**, 305–317 (1967).
75. Smetana, K., and Potměsil, M., A contribution to the demonstration of nucleoli and their nucleolar coefficient in mature monocytes of human peripheral blood. *Folia Morphol.* **15**, 98–103 (1967).
76. Spannhof, L., "Einführung in die Praxis der Histochemie." Fischer, Jena, 1964.
77. Stenram, U., The specificity of the gallocyanin-chromalum stain for nucleic acids as studied by the ribonuclease technique. *Exptl. Cell Res.* **4**, 383–389 (1953).
78. Stevens, B., The effect of actinomycin D on nucleolar and nuclear fine structure in the salivery gland of chironomus thumni. *J. Ultrastruct. Res.* **11**, 329–352 (1964).
79. Swift, H., Studies on nuclear fine structure. *Brookhaven Symp. Biol.* **12**, 134–151 (1959).
80. Swift, H., Cytochemical studies on nuclear fine structure. *Exptl. Cell Res.* Suppl. 9, 54–67 (1963).

81. Tanaka, R., Aceto-basic fuchsin as a stain for nucleoli and chromosomes of plants. *Stain Technol.* **36**, 325–327 (1961).
82. Tandler, C. J., The silver-reducing property of the nucleolus and the formation of the prenucleolar material during mitosis. *Exptl. Cell Res.* **17**, 560–564 (1959).
83. Trump, B. F., Smuckler, E. A., and Benditt, E. P., A method for staining epoxy sections for light microscopy. *J. Ultrastruct. Res.* **5**, 343–348 (1961).
84. Unuma, T., and Busch, H., Formation of microspherules in nucleoli of tumor cells treated with high doses of actinomycin D. *Cancer Res.* **27**, 1232–1242 (1967).
85. Venable, J. H., and Coggeshall, R., A simplified lead citrate stain for use in electron microscopy. *J. Cell Biol.* **25**, 407–408 (1965).
85a. Vlastiborova, A., Personal communication (1967).
86. Watson, M. L., and Aldridge, W. G., Methods for the use of indium as an electron stain for nucleic acids. *J. Biophys. Biochem. Cytol.* **11**, 257–272 (1961).
87. Weissenfels, N., Struktur und Varhalten der Nukleolen von Hühnerherzmyoblasten in Gewebekultur während des Interphasenwachstums und der Mitose. *Z. Zellforsch. Mikroskop. Anat.* **62**, 667–700 (1964).
88. Wenderoth, H., Cytochemische Untersuchungen von Blutzellen mittels Perchlorsäure. *Acta Haematol.* **9**, 47–53 (1953).

AUTHOR INDEX

Numbers in parentheses are reference numbers and indicate that an author's work is referred to although his name is not cited in the text. Numbers in italic show the page on which the complete reference is listed.

A

C

D

E

F

G

L

N

O

P

Q

R

S

Y

SUBJECT INDEX

D

M

N

Q

R

S

T